Better Ceramics Through Chemistry VII: Organic/Inorganic Hybrid Materials

MATERIALS RESEARCH SOCIETY
SYMPOSIUM PROCEEDINGS VOLUME 435

Better Ceramics Through Chemistry VII: Organic/Inorganic Hybrid Materials

Symposium held April 8-12, 1996, San Francisco, California, U.S.A.

EDITORS:

Bradley K. Coltrain
Eastman Kodak Company
Rochester, New York, U.S.A.

Clément Sanchez
Universite Pierre et Marie Curie
Paris, France

Dale W. Schaefer
Sandia National Laboratories
Albuquerque, New Mexico, U.S.A.

Garth L. Wilkes
Virginia Tech
Blacksburg, Virginia, U.S.A.

MRS
MATERIALS RESEARCH SOCIETY
PITTSBURGH, PENNSYLVANIA

Single article reprints from this publication are available through
University Microfilms Inc., 300 North Zeeb Road, Ann Arbor, Michigan 48106

CODEN: MRSPDH

Published by:

Materials Research Society
9800 McKnight Road
Pittsburgh, Pennsylvania 15237
Telephone (412) 367-3003
Fax (412) 367-4373
Website: http://www.mrs.org/

Library of Congress Cataloging in Publication Data

Better ceramics through chemistry VII–organic/inorganic hybrid materials : symposium held April 8–12, 1996, San Francisco, California, U.S.A. / editors, Bradley K. Coltrain, Clément Sanchez, Dale W. Schaefer, Garth L. Wilkes

p. cm—(Materials Research Society symposium proceedings ; v. 435)
Includes bibliographical references and index.
ISBN 1-55899-338-X

I. Coltrain, Bradley K. II. Sanchez, Clément III. Schaefer, Dale W. IV. Wilkes, Garth L. V. Series: Materials Research Society symposium proceedings ; v. 435.

Manufactured in the United States of America

CONTENTS

Preface xv

Materials Research Society Symposium Proceedings xvi

PART I: ORGANIC/INORGANIC HYBRIDS BY CHEMICAL SYNTHESIS OR INTERCALATION

***Thermoplastic Hybrid Materials: Polyhedral Oligomeric Silsesquioxane (POSS) Reagents, Linear Polymers, and Blends** 3
Joseph D. Lichtenhan, Charles J. Noel, Alan G. Bolf, and Patrick N. Ruth

***Hybrid Sols as Intermediates to Inorganic-Organic Nanocomposites** 13
H.K. Schmidt, P.W. Oliveira, and H. Krug

***Hybrid Styryl-Based Polyhedral Oligomeric Silsesquioxane (POSS) Polymers** 25
T.S. Haddad, E. Choe, and J.D. Lichtenhan

***Intramolecular Condensation Reactions of α, ω-Bis(Triethoxy-Silyl)Alkanes. Formation of Cyclic Disilsesquioxanes** 33
Douglas A. Loy, Joseph P. Carpenter, Sharon A. Myers, Roger A. Assink, James H. Small, John Greaves, and Kenneth J. Shea

***Hybrid Organic-Inorganic Systems Derived from Organotin Nanobuilding Blocks** 43
F.O. Ribot, C. Eychenne-Baron, F. Banse, and C. Sanchez

***Studies on the Structure and Properties of Ceramic/Polymer Nanocomposites** 55
Kenneth E. Gonsalves and Xiaohe Chen

Novel Carboxy Functionalized Sol-Gel Precursors 67
H. Wolter, W. Storch, and C. Gellermann

Thermal Characteristics of Layered Double Hydroxide Intercalates-Comparison of Experiment and Computer Simulation 73
I.S. Bell, F. Kooli, W. Jones, and P.V. Coveney

Nanolayer Ordering in an Epoxy-Exfoliated Clay Hybrid Composite 79
Tie Lan and Thomas J. Pinnavaia

*Invited Paper

Silicate Clay Platelet Dispersion in a Polymer Matrix 85
C.D. Muzny, B.D. Butler, H.J.M. Hanley, F. Tsvetkov, and D.G. Peiffer

PART II: SYNTHESIS, CHARACTERIZATION, AND PROCESSING OF ORGANIC/INORGANIC HYBRID MATERIALS

***The Use of Functionalized Polybenzoxazoles and Polybenzobisthiazoles in Polymer-Silica Hybrid Materials** 93
J.E. Mark, J. Premachandra, C. Kumudinie, W. Zhao, T.D. Dang, J.P. Chen, and F.E. Arnold

***Star Gels. New Hybrid Network Materials from Polyfunctional Single Component Precursors** 105
Kenneth G. Sharp and Michael J. Michalczyk

Sol-Gel Kinetics for the Preparation of Inorganic/Organic Siloxane Copolymers 113
Stephen E. Rankin, Christopher W. Macosko, and Alon V. McCormick

***NMR Characterization of the Chemical Homogeneity in Sol-Gel Derived Siloxane-Silica Materials** 119
Florence Babonneau, Virginie Gualandris, and Monique Pauthe

Synthesis of Mesoporous Zirconia Using an Amphoteric Surfactant 131
A.Y. Kim, P.J. Bruinsma, Y.L. Chen, and J. Liu

Molecules with Polymerizable Ligands as Precursors to Porous Doped Materials 137
L.G. Hubert-Pfalzgraf, N. Pajot, R. Papiernik, and S. Parraud

***Organic-Rich Hybrid O/I Systems Based on Isocyanate Chemistry** 143
S. Cuney, J.F. Gerard, J.P. Pascault, and G. Vigier

***Structure/Property Behavior of Organic-Inorganic Semi-IPNS** 155
A.B. Brennan and T.M. Miller

***Preparation of Polyimide-Silica Hybrid Materials by High Pressure-Thermal Polymerization** 165
Kevin Gaw, Hironori Suzuki, Mitsutoshi Jikei, Masa-aki Kakimoto, and Yoshio Imai

Poly(Methyl Methacrylate)-Titania Hybrid Materials by Sol-Gel Processing 173
Jun Zhang, Shengcheng Luo, Linlin Gui, and Youqi Tang

*Invited Paper

A Study of the Mobility of Poly(Methyl Methacrylate)-Silicate Interpenetrating Networks with ^{2}H NMR 179
Clare P. Grey and Kenneth G. Sharp

*Molecular Probes of Physical and Chemical Properties of Sol-Gel Films 187
J.I. Zink, B. Dunn, B. Dave, and F. Akbarian

Tailoring the Microstructure of Polyimide-Silica Materials Using the Sol-Gel Process 199
J.C. Schrotter, M. Smaihi, and C. Guizard

PART III: ORGANIC/INORGANIC HYBRID COATINGS

Hybrid Organic/Inorganic Coatings for Abrasion Resistance on Plastic and Metal Substrates 207
J. Wen, K. Jordens, and G.L. Wilkes

Use of an Organosilane Coupling Agent in Colloidal Silica Coatings 215
M.W. Daniels, L. Chu, and L.F. Francis

Glycidoxypropyltrimethoxysilane Modified Colloidal Silica Coatings 221
L. Chu, M.W. Daniels, and L.F. Francis

PART IV: MECHANICAL PROPERTIES OF SELECTED ORGANIC/INORGANIC MATERIALS

*Effects of Temperature on Properties of Ormosils 229
J.D. MacKenzie, Q. Huang, F. Rubio-Alonso, and S.J. Kramer

Tailoring of Thermomechanical Properties of Thermoplastic Nanocomposites by Surface Modification of Nanoscale Silica Particles 237
C. Becker, H. Krug, and H. Schmidt

Improvement of Cryogenic Fracture Toughness of Epoxy by Hybridization 243
S. Nishijima, M. Hussain, A. Nakahira, T. Okada, and K. Niihara

Degree of Dispersion of Latex Particles in Cement Paste, as Assessed by Electrical Resistivity Measurement 249
Xuli Fu and D.D.L. Chung

Organic-Inorganic Hybrid Dental Restorative Material 255
Partha P. Paul, Scott F. Timmons, and Walter J. Machowski

*Invited Paper

PART V: ORGANIC/INORGANIC HYBRIDS–A ROUTE TO CONTROLLED POROSITY MATERIALS

*Synthesis of Ordered Silicates by the Use of Organic Structure-Directing Agents 263
Yoshihiro Kubota and Mark E. Davis

Controlling the Porosity of Microporous Silica by Sol-Gel Processing Using an Organic Template Approach 271
Yunfeng Lu, G.Z. Cao, Rahul P. Kale, L. Delattre, C. Jeffrey Brinker, and Gabriel P. Lopez

Controlling Porosity in Bridged Polysilsesquioxanes Through Elimination Reactions 277
Mark D. McClain, Douglas A. Loy, and Sheshasayana Prabakar

A New Category of Membranes Exploiting Potentiality of Organic/Inorganic Hybrid Materials 283
C. Guizard, P. Heckenbenner, J.C. Schrotter, N. Hovnanian, and M. Smaihi

Organically Modified Silicate Aerogels, "Aeromosils" 295
S.J. Kramer, F. Rubio-Alonso, and J.D. MacKenzie

Origin of Porosity in Arylene-Bridged Polysilsesquioxanes 301
Dale W. Schaefer, Greg B. Beaucage, Douglas A. Loy, Tamara A. Ulibarri, Eric Black, Kenneth J. Shea, and Richard J. Buss

General Routes to Microporous Thin Films: Formation of Organic/Inorganic Network 307
Y. Yan, Y. Hoshino, Z. Duan, S. Ray Chaudhuri, and A. Sarkar

PART VI: POSTER SESSION

Synthesis of Titanates $MTiO_3$ (M=Mn,Co,Ni) by the Sol-Gel Method 315
J. Moreno, G. Tavizon, L. Vicente, and T. Viveros

Synthesis of Inorganic-Organic Hybrids from Metal Alkoxides and Silanol-Terminated Polydimethylsiloxane 321
Shingo Katayama, Ikuko Yoshinaga, and Noriko Yamada

Electrical Properties of Ternary Si-C-N Ceramics 327
Christoph Haluschka, Christine Engel, and Ralf Riedel

New Phosphanyl-Substituted Titanium and Zirconium Alkoxide Precursors for Sol-Gel Processing 333
Anne Lorenz and Ulrich Schubert

*Invited Paper

Chemical Functionalization of Silica Aerogels 339
Nicola Hüsing, Ulrich Schubert, Bernhard Riegel, and Wolfgang Kiefer

Cross-Condensation Kinetics of Organically Modified Silica Sols 345
S. Prabakar and R.A. Assink

Processing Methyl Modified Silicate Materials for Optical Applications: A Structural Stability Study 351
A. Sarkar, Y. Yan, Z. Duan, Y. Hoshino, and S. Ray Chaudhuri

Shrinkage and Microstructural Development During Drying of Organically Modified Silica Xerogels 357
N.K. Raman, S. Wallace, and C.J. Brinker

Preparation, Characterization and Properties of New Ion-Conducting Ormolytes 363
K. Dahmouche, M. Atik, N.C. Mello, T.J. Bonagamba, H. Panepucci, M. Aegerter, and P. Judeinstein

Molecular State and Mechanical Properties of Epoxy Hybrid Composite 369
M. Hussain, S. Nishijima, A. Nakahira, T. Okada, and K. Niihara

On-Line Spectroscopic Studies of Group IV Alkoxides and Their Interactions with Organic Additives During the Sol-Gel Process 375
D. Wettling, S. Truchet, J. Guilment, and O. Poncelet

A TG/GC/MS Study of the Structural Transformation of Hybrid Gels Containing Si-H and Si-CH_3 Groups into Oxycarbide Glasses 381
G.D. Sorarù, R. Campostrini, G. D'Andrea, and S. Maurina

Hydrolysis-Condensation Behavior of Acetylacetone Modified Tin(IV) Tetra Tert-Amyloxide 387
L. Armelao, F.O. Ribot, and C. Sanchez

Effect of Processing Parameters on Second Order Nonlinearities of Azo Dye Grafted Hybrid Sol-Gel Coatings 395
B. Lebeau, C. Sanchez, S. Brasselet, and J. Zyss

Synthesis of Bioactive Ormosils by the Sol-Gel Method 403
K. Tsuru, C. Ohtsuki, A. Osaka, T. Iwamoto, and J.D. MacKenzie

Direct Deposition of Silica Films Containing Organic Groups and Dyes from Silicon Alkoxide Solutions 409
Junrok Oh, Hiroaki Imai, Hiroshi Hirashima, and Koji Tsukuma

Thermal Stability of Silicon-Based Hybrid Materials Containing Aluminum Studied by ^{27}Al and ^{29}Si Solid State MAS NMR 415
M.P.J. Peeters, A.P.M. Kentgens, and I.J.M. Snijkers-Hendrickx

Investigation of Hydrolysis and Condensation in Organically Modified Sol-Gel Systems: ^{29}SI NMR and the INEPT Sequence 421
T.M. Alam, R.A. Assink, and D.A. Loy

Schiff Base Mediated Sol-Gel Polymerization 427
D.A. Lindquist, C.M. Harrison, B. Williams, and R.D. Morris

Preceramic Polymer Applications - Processing and Modifications by Chemical Means 431
H.J. Wu, Y.D. Blum, S.M. Johnson, C. Kanazawa, J.R. Porter, and D.M. Wilson

NMR Characterization of Hybrid Systems Based on Functionalized Silsesquioxanes 437
C. Bonhomme, F. Babonneau, J. Maquet, C. Zhang, R. Baranwal, and R.M. Laine

Organically-Modified Eu^{3+}-Doped Silica Gels 443
V.C. Costa, B.T. Stone, and K.L. Bray

Detection of *Cryptosporidium Parvum* in Antibody-Doped Gels 449
E. Hong, E. Bescher, L. Garcia, and J.D. MacKenzie

Inorganic Organic Composite Materials as Absorbers for Organic Solvents 455
V. Gerhard, H. Schmidt, and U. Dreier

NMR and IR Spectroscopic Examination of the Hydrolytic Stability of Organic Ligands in Metal Alkoxide Complexes and of Oxygen Bridged Heterometal Bonds 461
D. Hoebbel, T. Reinert, and H. Schmidt

The *In-Situ* Generation of Silica Reinforcement in Modified Polydimethylsiloxane Elastomers 469
S. Prabakar, S.E. Bates, E.P. Black, T.A. Ulibarri, D.W. Schaefer, G..Beaucage, and R.A. Assink

Anisotropy in Hybrid Materials: An Alternative Tool for Characterization 475
V. Dessolle, E. Lafontaine, J.P. Bayle, and P. Judeinstein

Synthesis of Inorganic-Organic Hybrids from Metal Alkoxides and Ethyl Cellulose 481
Ikuko Yoshinaga, Noriko Yamada, and Shingo Katayama

Synthesis and Characterization of Titanium Oxo-Alkoxides Through Solvatothermal Process 487
N. Steunou, Y. Dromzee, F. Robert, and C. Sanchez

Displacement of Poly(Ethylene Oxide) from Layered Nanocomposites 495
C.O. Oriakhi and M.M. Lerner

Functional Phosphate Alkoxysilanes for Facilitated Transport Membrane Materials 501
N. Hovnanian, M. Smaihi, A. Cardenas, and C. Guizard

Polystyrene-Poly(Vinylphenol) Copolymers and Compatibilizers for Organic-Inorganic Composites 507
Christine J.T. Landry, Bradley K. Coltrain, and David M. Teegarden

Synthesis of an Inorganic/Organic Network Polymer by the Hydrolysis/Condensation of Poly(Diethoxysilylenemethylene) and Its Pyrolytic Conversion to Silicon Oxycarbide 513
Q. Liu, T. Apple, Z. Zheng, and L.V. Interrante

Micron Scale Patterning of Solution-Derived Ceramic Thin Films Directed by Self-Assembled Monolayers 521
P.G. Clem, N.L. Jeon, R.G. Nuzzo, and D.A. Payne

Vanadium-Oxo Based Hybrid Organic-Inorganic Copolymers 527
A. Campero, A.M. Soto, J. Maquet, and C. Sanchez

PART VII: ELECTRICAL AND OPTICAL PROPERTIES OF ORGANIC/INORGANIC HYBRID MATERIALS

***Nanostructured Materials for Photonics** 535
N.D. Kumar, G. Ruland, M. Yoshida, M. Lal, J. Bhawalkar, G.S. He and P.N. Prasad

Perfluoroaryl Substituted Inorganic-Organic Hybrid Materials 547
C. Roscher and M. Popall

Fabrication of GRIN-Materials by Photopolymerization of Diffusion-Controlled Organic-Inorganic Nanocomposite Materials 553
P.W. Oliveira, H. Krug, P. Müller, and H. Schmidt

Dielectric Properties of Organic-Inorganic Hybrids: PDMS-Based Systems 559
G. Teowee, K.C. McCarthy, C.D. Baertlein, J.M. Boulton, S. Motakef, T.J. Bukowski, T.P. Alexander, and D.R. Uhlmann

***Photochemical Studies Using Organic-Inorganic Sol-Gel Materials** 565
B.C. Dave, F. Akbarian, B. Dunn, and J.I. Zink

***The Formation of Laser Active Composite Films from Silicate Ceramics** 575
L.L. Beecroft, R.T. Leidner, C.K. Ober, D.B. Barber, and C.R. Pollock

*Invited Paper

Photorefractive Sol-Gel Materials 583
F. Chaput, B. Darracq, J.P. Boilot, D. Riehl, T. Gacoin, M. Canva, Y. Levy, and A. Brun

Silicone-Polyoxometalate (SiPOM) Hybrid Compounds 589
Dimitris E. Katsoulis and John R. Keryk

***Novel Sol-Gel Processed Photorefractive Materials** 595
Ryszard Burzynski, Saswati Ghosal, Martin K. Casstevens, and Yue Zhang

Dipolar Organic/Ferroelectric Oxide Hybrids 605
Eric Pascal Bescher, Edward Hong, Yu-Huan Xu, and John D. MacKenzie

Vanadium Oxide/Polypyrrole Aerogel Nanocomposites 611
B.C. Dave, B.S. Dunn, F. Leroux, L.F. Nazar, and H.P. Wong

Er^{3+}-Doped Silica and Hyrbrid Organic/Inorganic Silica Gels 617
B.T. Stone and K.L. Bray

PART VIII: PARTICULATES AND LAYERED FILMS

***Metal/Ceramic Nanocomposites by Sol-Gel Processing of Tethered Metal Ions: Optimization of the Particle-Forming Step** 625
Claus Görsmann, Ulrich Schubert, Jürgen Leyrer, and Egbert Lox

TEM-Characterization of Metallic Nanoparticles Embedded in Sol-Gel Produced Glass-Like Layers 637
U. Werner, M. Schmitt, and H. Schmidt

New Systems Related to CdS Nanoparticles in Sol-Gel Matrices 643
T. Gacoin, L. Malier, G. Counio, S. Esnouf, J.P. Boilot, L. Audinet, C. Ricolleau, and M. Gandais

Sol-Gel Process of Fluoride and Fluorobromide Materials 649
O. Poncelet, J. Guilment, and G. Paz-Pujalt

Investigation of Second Harmonic Generation in Glutamic Acid-Metal Complexes 655
Thomas M. Cooper, Steven M. Cline, David E. Zelmon, Rama Vuppuladhadium, Samhita Das Gupta, and Uma B. Ramabadran

*Invited Paper

Second Harmonic Generation from Multilayers of Oriented Metal Bisphosphonates 661
Grace Ann Neff, Timothy M. Mahon, Travis A. Abshere, and Catherine J. Page

Fiber Optic Sensing of Cyanides in Solutions 667
S.S. Park, J.D. MacKenzie, C.Y. Li, P. Guerreiro, and N. Peyghambarian

Author Index 673

Subject Index 677

PREFACE

The papers contained in this volume were presented at the seventh MRS symposium on "Better Ceramics Through Chemistry," held April 8–12, 1996 in San Francisco, California. This symposium has been held biennially since 1984.

The intent of this symposium series has been to foster discussion between ceramists, physicists, engineers, and chemists on the design, synthesis, and properties of new materials with molecular-level structural control. Over the years the symposium has been a forum for the presentation of multidisciplinary research.

This year the symposium focused on organic-inorganic nanocomposites. This rapidly expanding field explores unique materials arising from the coupling of organic polymers or monomers with inorganic materials. The inorganic components are frequently produced *in situ* via sol-gel chemistry beginning with inorganic monomers.

The symposium began with a tutorial on the synthesis and structure of hybrid organic-inorganic materials. Sessions were devoted to chemical synthesis or intercalation, characterization and processing of hybrids, coatings, mechanical properties, controlled porosity materials, electrical and optical properties, and particulates and layered films. Approximately 60 posters were presented. It is evident that this area of research is truly interdisciplinary and of interest to a wide range of scientists.

Approximately 125 papers were presented either orally or as posters. Ninety-two of the papers are included in this volume. All papers were subject to peer review, after which the authors were given an opportunity to respond to the reviewer's comments. Due to rapid and timely publication some of these papers were accepted without further revision. Thus, the papers comprising this volume represent the views and standards of the authors and not necessarily those of the editors.

Bradley K. Coltrain
Clément Sanchez
Dale W. Schaefer
Garth L. Wilkes

June, 1996

MATERIALS RESEARCH SOCIETY SYMPOSIUM PROCEEDINGS

Volume 395— Gallium Nitride and Related Materials, F.A. Ponce, R.D. Dupuis, S.J. Nakamura, J.A. Edmond, 1996, ISBN: 1-55899-298-7

Volume 396— Ion-Solid Interactions for Materials Modification and Processing, D.B. Poker, D. Ila, Y-T. Cheng, L.R. Harriott, T.W. Sigmon, 1996, ISBN: 1-55899-299-5

Volume 397— Advanced Laser Processing of Materials—Fundamentals and Applications, R. Singh, D. Norton, L. Laude, J. Narayan, J. Cheung, 1996, ISBN: 1-55899-300-2

Volume 398— Thermodynamics and Kinetics of Phase Transformations, J.S. Im, B. Park, A.L. Greer, G.B. Stephenson, 1996, ISBN: 1-55899-301-0

Volume 399— Evolution of Epitaxial Structure and Morphology, A. Zangwill, D. Jesson, D. Chambliss, R. Clarke, 1996, ISBN: 1-55899-302-9

Volume 400— Metastable Phases and Microstructures, R. Bormann, G. Mazzone, R.D. Shull, R.S. Averback, R.F. Ziolo, 1996 ISBN: 1-55899-303-7

Volume 401— Epitaxial Oxide Thin Films II, J.S. Speck, D.K. Fork, R.M. Wolf, T. Shiosaki, 1996, ISBN: 1-55899-304-5

Volume 402— Silicide Thin Films—Fabrication, Properties, and Applications, R. Tung, K. Maex, P.W. Pellegrini, L.H. Allen, 1996, ISBN: 1-55899-305-3

Volume 403— Polycrystalline Thin Films: Structure, Texture, Properties, and Applications II, H.J. Frost, M.A. Parker, C.A. Ross, E.A. Holm, 1996, ISBN: 1-55899-306-1

Volume 404— *In Situ* Electron and Tunneling Microscopy of Dynamic Processes, R. Sharma, P.L. Gai, M. Gajdardziska-Josifovska, R. Sinclair, L.J. Whitman, 1996, ISBN: 1-55899-307-X

Volume 405— Surface/Interface and Stress Effects in Electronic Materials Nanostructures, S.M. Prokes, R.C. Cammarata, K.L. Wang, A. Christou, 1996, ISBN: 1-55899-308-8

Volume 406— Diagnostic Techniques for Semiconductor Materials Processing II, S.W. Pang, O.J. Glembocki, F.H. Pollack, F.G. Celii, C.M. Sotomayor Torres, 1996, ISBN 1-55899-309-6

Volume 407— Disordered Materials and Interfaces, H.Z. Cummins, D.J. Durian, D.L. Johnson, H.E. Stanley, 1996, ISBN: 1-55899-310-X

Volume 408— Materials Theory, Simulations, and Parallel Algorithms, E. Kaxiras, J. Joannopoulos, P. Vashishta, R.K. Kalia, 1996, ISBN: 1-55899-311-8

Volume 409— Fracture—Instability Dynamics, Scaling, and Ductile/Brittle Behavior, R.L. Blumberg Selinger, J.J. Mecholsky, A.E. Carlsson, E.R. Fuller, Jr., 1996, ISBN: 1-55899-312-6

Volume 410— Covalent Ceramics III—Science and Technology of Non-Oxides, A.F. Hepp, P.N. Kumta, J.J. Sullivan, G.S. Fischman, A.E. Kaloyeros, 1996, ISBN: 1-55899-313-4

Volume 411— Electrically Based Microstructural Characterization, R.A. Gerhardt, S.R. Taylor, E.J. Garboczi, 1996, ISBN: 155899-314-2

Volume 412— Scientific Basis for Nuclear Waste Management XIX, W.M. Murphy, D.A. Knecht, 1996, ISBN: 1-55899-315-0

Volume 413— Electrical, Optical, and Magnetic Properties of Organic Solid State Materials III, A.K-Y. Jen, C.Y-C. Lee, L.R. Dalton, M.F. Rubner, G.E. Wnek, L.Y. Chiang, 1996, ISBN: 1-55899-316-9

Volume 414— Thin Films and Surfaces for Bioactivity and Biomedical Applications, C.M. Cotell, A.E. Meyer, S.M. Gorbatkin, G.L. Grobe, III, 1996, ISBN: 1-55899-317-7

Volume 415— Metal-Organic Chemical Vapor Deposition of Electronic Ceramics II, S.B. Desu, D.B. Beach, P.C. Van Buskirk, 1996, ISBN: 1-55899-318-5

MATERIALS RESEARCH SOCIETY SYMPOSIUM PROCEEDINGS

Volume 416— Diamond for Electronic Applications, D. Dreifus, A. Collins, T. Humphreys, K. Das, P. Pehrsson, 1996, ISBN: 1-55899-319-3

Volume 417— Optoelectronic Materials: Ordering, Composition Modulation, and Self-Assembled Structures, E.D. Jones, A. Mascarenhas, P. Petroff, R. Bhat, 1996, ISBN: 1-55899-320-7

Volume 418— Decomposition, Combustion, and Detonation Chemistry of Energetic Materials, T.B. Brill, T.P. Russell, W.C. Tao, R.B. Wardle, 1996 ISBN: 1-55899-321-5

Volume 420— Amorphous Silicon Technology—1996, M. Hack, E.A. Schiff, S. Wagner, R. Schropp, M. Matsuda, 1996, ISBN: 1-55899-323-1

Volume 421— Compound Semiconductor Electronics and Photonics, R.J. Shul, S.J. Pearton, F. Ren, C.-S. Wu, 1996, ISBN: 1-55899-324-X

Volume 422— Rare Earth Doped Semiconductors II, S. Coffa, A. Polman, R.N. Schwartz, 1996, ISBN: 1-55899-325-8

Volume 423— III-Nitride, SiC, and Diamond Materials for Electronic Devices, D.K. Gaskill, C. Brandt, R.J. Nemanich, 1996, ISBN: 1-55899-326-6

Volume 424— Flat Panel Display Materials II, M. Hatalis, J. Kanicki, C.J. Summers, F. Funada, 1996, ISBN: 1-55899-327-4

Volume 425— Liquid Crystals for Advanced Technologies, T.J. Bunning, S.H. Chen, W. Hawthorne, N. Koide, T. Kajiyama, 1996, ISBN: 1-55899-328-2

Volume 426— Thin Films for Photovoltaic and Related Device Applications, D. Ginley, A. Catalano, H.W. Schock, C. Eberspacher, T.M. Peterson, T. Wada, 1996, ISBN: 1-55899-329-0

Volume 427— Advanced Metallization for Future ULSI, K.N. Tu, J.W. Mayer, J.M. Poate, L.J. Chen, 1996, ISBN: 1-5899-330-4

Volume 428— Materials Reliability in Microelectronics VI, W.F. Filter, J.J. Clement, A.S. Oates, R. Rosenberg, P.M. Lenahan, 1996, ISBN: 1-55899-331-2

Volume 429— Rapid Thermal and Integrated Processing V, J.C. Gelpey, M. Öztürk, R.P.S. Thakur, A.T. Fiory, F. Roozeboom, 1996, ISBN: 1-55899-332-0

Volume 430— Microwave Processing of Materials V, M.F. Iskander, J.O. Kiggans, E.R. Peterson, J.Ch. Bolomey, 1996, ISBN: 1-55899-333-9

Volume 431— Microporous and Macroporous Materials, R.F. Lobo, J.S. Beck, S. Suib, D.R. Corbin, M.E. Davis, L.E. Iton, S.I. Zones, 1996, ISBN: 1-55899-334-7

Volume 432— Aqueous Chemistry and Geochemistry of Oxides, Oxyhydroxides, and Related Materials J.A. Voight, B.C. Bunker, W.H. Casey, T.E. Wood, L.J. Crossey, 1996, ISBN: 1-55899-335-5

Volume 433— Ferroelectric Thin Films V, S.B. Desu, R. Ramesh, B.A. Tuttle, R.E. Jones, I.K. Yoo, 1996, ISBN: 1-55899-336-3

Volume 434— Layered Materials for Structural Applications, J.J. Lewandowski, C.H. Ward, W.H. Hunt, Jr., M.R. Jackson, 1996, ISBN: 1-55899-337-1

Volume 435— Better Ceramics Through Chemistry VII—Organic/Inorganic Hybrid Materials, B. Coltrain, C. Sanchez, D.W. Schaefer, G.L. Wilkes, 1996, ISBN: 1-55899-338-X

Volume 436— Thin Films: Stresses and Mechanical Properties VI, W.W. Gerberich, H. Gao, J–E. Sundgren, S.P. Baker 1996, ISBN: 1-55899-339-8

Volume 437— Applications of Synchrotron Radiation to Materials Science III, L. Terminello, S. Mini, D.L. Perry, H. Ade, 1996, ISBN: 1-55899-340-1

Prior Materials Research Society Symposium Proceedings available by contacting Materials Research Society

Part I

Organic/Inorganic Hybrids by Chemical Synthesis or Intercalation

THERMOPLASTIC HYBRID MATERIALS: POLYHEDRAL OLIGOMERIC SILSESQUIOXANE (POSS) REAGENTS, LINEAR POLYMERS, AND BLENDS

JOSEPH D. LICHTENHAN+, CHARLES J. NOEL†, ALAN G. BOLF‡, PATRICK N. RUTH‡
+Phillips Laboratory, Propulsion Directorate, Edwards AFB, CA 93524
†Ohio State University, Department of Consumer and Textile Sciences, Columbus, OH 43210
‡Hughes STX, Phillips Laboratory, Edwards AFB, CA 93524

ABSTRACT

Polyhedral Oligomeric SilSesquioxanes (POSS) are structurally well defined cage-like molecules represented by the generic formula $(RSiO_{1.5})_n$. POSS compounds possess a unique hybrid composition with an oxygen to silicon ratio of 1.5, intermediate between that for silica and silicone. An entire monomer catalogue (chemical tree) of hybrid POSS-based reagents suitable for polymerization and grafting reactions has been developed from $R_7Si_7O_9(OH)_3$ and related precursors. POSS reagents containing no more than one or two reactive groups enable the preparation of hybrid materials with desirable physical properties such as thermoplasticity, and elasticity. An overview of the synthesis of POSS monomers and thermoplastic POSS-acrylic polymers is given. The thermal and physical properties for POSS-acrylic monomers, homopolymers, copolymers, and blends with poly(methyl methacrylate) are described.

INTRODUCTION

Thermoplastic polymers are highly desirable in manufacturing and engineering operations because of their wide range of tailorable properties and their ability to be rapidly thermoformed into finished articles via solvent free processing. Thermoplastics are also highly recyclable even after molding because the finished article and process scrap can be remelted, mixed with virgin resin (if so desired) and remolded. As applications for thermoplastics continue to grow, so does the need for new additives and resin systems. The development of a catalogue of hybrid monomers and thermoplastic hybrid polymers based on POSS-reagents offers considerable potential for the establishment of several new or improved properties in traditional thermoplastic resins.

EXPERIMENT

The POSS-methacrylate monomer and homopolymer were prepared according to literature methods.[1] PMMA was provided by Rohm and Haas. The POSS-methacrylate copolymer was prepared in a dry-box by dissolving POSSMA monomer (16.56g, 14.7mmol) and methyl methacrylate monomer (3.44g, 34.3mmol) in 80ml of toluene followed by the addition of 21mg (0.25mol %) of AIBN. The reaction was removed from the dry-box and allowed to react at 60 °C for 3 days. THF (50ml) was then added to the reaction solution and the polymer was precipitated through the addition of methanol (50ml). The precipitated polymer was collected through filtration and dried in a vacuum oven at 80 °C for 12 h (17.54 g, 88%). Copolymer $(POSSMA)_{18}(MMA)_{82}$: ^{1}H-NMR (300.13 MHz, $CDCl_3$), ppm: 3.89

Mat. Res. Soc. Symp. Proc. Vol. 435 © 1996 Materials Research Society

(b, $-CH_2-O-$), 3.58 (b, $-O-CH_3$), 1.71 (b), 1.23 (b), 1.01 (b), 0.85 (b), 0.75 (b), 0.60 (b). ^{29}Si-NMR (59.62 MHz, $CDCl_3$), ppm: -66.94 (1 Si), -68.58 (7 Si). Elemental anal. (calcd.) %C, 53.99 (54.55), %H, 8.06 (7.93).

Blends were prepared by dissolving the materials to be blended in a common solvent (such as chloroform), thoroughly mixing the solution, and recovering the solid blend by evaporating the solvent. Removal of virtually all traces of solvent from all of the blend compositions was accomplished by heating the materials to 196°C, at 20 torr, for 20 minutes. This was followed by compression molding the material at 500 psi and 196 °C for another 20 minutes during which time the mold cavity was evacuated to 20 torr. The thick disk samples were then cut and molded a second time into thinner specimens for testing. Solvent removal was confirmed by TGA analysis for each blend composition.

Characterization of the blends was accomplished using conventional thermal analytical techniques. DSC experiments were run at 10°C/min under nitrogen on a TA Instruments module 910 and TMA experiments were run at 3°C/min under nitrogen on a TA Instruments module 2940. TGA experiments were run on a TA Instruments (model TGA-951). Samples (10-20 mg) were first flushed with nitrogen at 100 cc/min for 30 minutes then heated at 10°C/min from room temperature to 1000°C under nitrogen. Molecular weight determinations were made by size exclusion chromatography with DAWN® light scattering and refractive index detectors.

Oxygen permeability measurements were performed on a Delta/Vertex Scientific Model 2110 dissolved oxygen analyzer. Disks (13mm x 0.25mm) of the polymer blends were compression molded (176°C at 500psi). Samples were equilibrated in phosphate buffered saline for 24 h at ambient temperature. Measurements were taken over a period of 2 hours or until oxygen readings stabilized. The dissolved oxygen permeability was calculated based on current flow during electrolytic reaction and was corrected for sample thickness.

DISCUSSION

Development of POSS Reagents

In an effort to develop polymeric hybrid materials with enhanced properties that can be understood and tailored at the molecular level, several classes of monomeric reagents based on polyhedral oligomeric silsesquioxanes (POSS) have been prepared. A number of initial design criteria have been considered in the development of these reagents. For example, the compatibility and solubility of the monomers with "conventional" organic comonomers is important for reasons of practicality and for property tailoring. Similarly, restricting the number of reactive sites on each POSS cage to no more than one or two is necessary in order to avoid the formation of intractable networks and for the incorporation of these reagents into linear polymers and thermoplastic systems. It was concluded that these preconsiderations could be met through the incorporation of nonreactive organic substituents on the cage (Figure 1).

Figure 1. Octomeric POSS monomer tree derived from POSS trisilanol precursors.

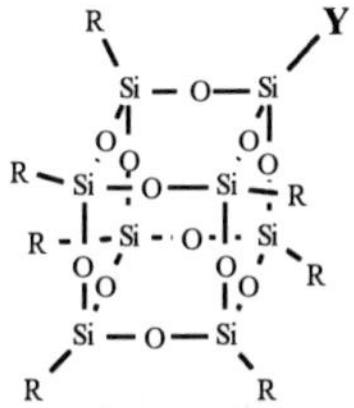

R = *c* - C_7H_{13}
c - C_6H_{11}
c - C_5H_9

Incorporated Y Functionalities

mono & di-halides	nitriles	silanes	styryls	epoxides
mono & di-alcohols	amines	silanols	α-olefins	cyclopropyls
mono & di-esters	isocyanates	silylchlorides & dichlorides	acrylics	norbornyls
mono & di-acids	arylbisamines			
acid chlorides	aryldiisocyanates			
bisphenols				
aryldiacids				

The preparation of POSS-based monomers is carried out through the corner capping of POSS-trisilanols (e.g. $R_7Si_7O_9(OH)_3$) with a variety of silane coupling agents (Figure 2).[2] Isolated yields from these corner capping reactions average around 90 % or more when care is taken to exclude moisture from the system. To date, the corner capping of incompletely condensed POSS trisilanols appears to be the most general and synthetically useful manipulation of POSS silicon-oxygen frameworks.

Through variation of the Y group on the silane a variety of functionalities can be placed off the corner of POSS frameworks (Figure 1). Subsequent transformations of the Y group are possible through the use of standard organic manipulations thereby affording a rich derivative chemistry to $R_7Si_8O_{12}Y$ reagents. Through a combination of corner capping and derivatization chemistry a monomer tree of polymerizable and graftable POSS reagents has been established from POSS trisilanol precursors.

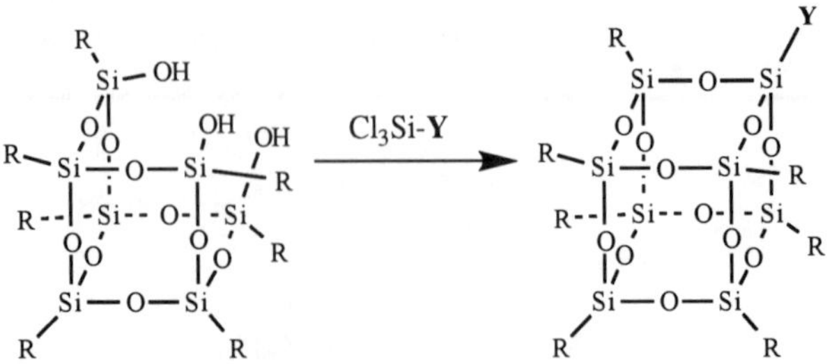

Figure 2. Corner capping of a POSS-trisilanol with a silane coupling agent to form a fully condensed POSS macromer ($R_7Si_8O_{12}Y$) containing seven non-reactive R substituents and with only one corner functionalized for polymerization / grafting reactions.

Synthetic methods which afford the preparation and isolation of incompletely condensed POSS molecules (e.g. trisilanols) and other useful POSS frameworks are under development. Improvement of the synthetic routes currently used for the preparation of POSS systems[2] is still required before their commercial development can be realized.

POSS reagents show a number of desirable physical properties. For example, they are highly soluble in common solvents like THF, toluene, chloroform, hexane[1], have high thermal stabilities[3], and can be polymerized using standard "organic" polymerization methodologies. POSS-reagents and their resins improve properties such as thermal stability, glass transition, heat distortion temperature, and oxygen permeability for many classes of common plastics. These property enhancements are believed to result from the ability of POSS segments to dominate polymer chain motions. This assumption is partly supported by the fact that octomeric POSS reagents containing cyclohexyl substituents are approximately 15 Å in diameter and have formula weights of 1000 amu or greater. In comparison, the crystal lamellae dimensions in semicrystalline polymers are on the order of 10-25nm x 0.1-1μm with amorphous layer thicknesses ranging from 5-10nm.[4] POSS reagents should therefore be considered macromeric with respect to polymer chain dimensions.

Of nearly equal importance to the development of new monomeric reagents and hybrid plastics is their compatibility with conventional resin systems. The blending of polymeric materials to achieve a balance of properties not available in individual components has become increasingly important in the polymer and materials science fields. Currently there is relatively little literature concerning the blending of organic polymers with polymers having inorganic and/or hybrid character.[5]

POSS-Methacrylate Macromers, Polymers, and Blends

POSS reagents have been incorporated through copolymerzation into a wide variety of linear polymeric resins.[6] In this work, blending studies were conducted with the extensively used thermoplastic poly(methyl methacrylate) (PMMA) and three different POSS macromers and two different POSS-methacrylate polymers (Figure 3 and Table 1). The objectives were to determine the extent of blending possible and to determine the effects of blending on the properties of PMMA. The rational for selection of this series of POSS-reagents was that the compatibility / miscibility would be observed to improve as the chemical similarity of the POSS system approached that of PMMA.

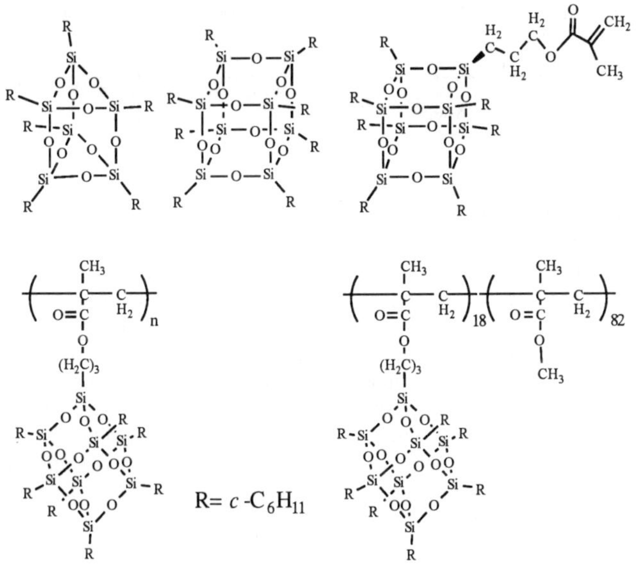

Figure 3. POSS macromers and polymer systems blended with PMMA include: $(c\text{-}C_6H_{11})_6Si_6O_9$, $(c\text{-}C_6H_{11})_8Si_8O_{12}$, $(c\text{-}C_6H_{11})_7Si_8O_{12}$propyl methacrylate (POSSMA), POSS-methacrylate homopolymer (poly-POSSMA), POSS-methacrylate/methylmethacrylate copolymer $(POSSMA)_{18}(MMA)_{82}$.

Blends containing the macromers $Cy_6Si_6O_9$, $Cy_8Si_8O_{12}$, and POSSMA with PMMA were prepared to determine if enhancement of the thermal stability or glass transition could be realized in the resin by simply using the POSS macromers as fillers. Furthermore, it was reasoned that the methacrylate functionalized macromer (POSSMA) may show enhanced compatibility with PMMA because of the presence of the one methacrylate group per POSS cage. Individual blends of these three macromers with PMMA were prepared at the 3, 6, 9, 12, 15 and 30% (w/w) levels.

Table 1. Partial listing of properties for POSS macromers and polymers.

Compound	M_w/M_n (x10^3)	Transition (°C)	T_{dec} (°C)[a]
$(c\text{-}C_6H_{11})_6Si_6O_9$	0.811 / 0.811[b]	sublimation	359[h]
$(c\text{-}C_6H_{11})_8Si_8O_{12}$	1.082 / 1.082[b]	sublimation	463[h]
POSSMA	1.126 / 1.126[b]	186-257, 387-416[f]	410
poly-POSSMA	68 / 42[c]	none	388
$(POSSMA)_{18}(MMA)_{82}$	247 / 143[d]	251[g]	364
PMMA	82 / 60[e]	139[g]	351

(a) Represents a 10% mass loss under nitrogen. (b) Formula weight. (c) Dp = 37. (d) Dp = 503 corresponding to 90 (POSSMA) units and 413 MMA units. (e) Dp = 600. (f) Melt/polymerization. (g) Glass transition by TMA. (h) Sublimation temperature.

Clear films were obtained only at the 3 and 6% (w/w) loading levels for the $Cy_6Si_6O_9$ and $Cy_8Si_8O_{12}$ macromers. Blends containing higher loadings were opaque and contained a white powdery material (indicative of the POSS macromer) on their surface. In contrast, blends with the POSSMA and PMMA were clear throughout the 3-30 % loading range but also showed a tendency for strain-induced phase separation at the higher loading levels. Preparation of such blends from the melt may result in higher levels of macromer incorporation. Attempts to incorporate these macromers into other thermoplastics such as PEBAX™ (a polyether/nylon block copolymer) have allowed for loadings to approximately 50 % but these compositions were also susceptible to strain-induced phase separation.

Analysis of the thermal properties for the range of PMMA/POSS-macromer blends was carried out using TGA, DSC and TMA. At the 3-6 % loading levels the TGA and DSC traces were observed to be similar to those of unblended PMMA in terms of the onset temperature of decomposition, char yield, and glass transition. This suggests that despite their transparency and apparent homogeneity there was no significant effect on thermal properties and little if any interaction between the PMMA and the POSS-macromers. The visibly phase-separated 15% and higher loaded blends (with $Cy_6Si_6O_9$ and $Cy_8Si_8O_{12}$) did show a modest 5-10 °C increase in glass transition by DSC over unblended PMMA. This increase was presumably a result associated with the POSS rich surface layer on these materials rather than a true property enhancement. The TGA trace for the 15 and 30 % macromer loadings also showed a second mass loss region associated with the POSS macromer. The intensity of mass loss for this second region was observed to correlate with increased POSS macromer content in the blends and char yields were observed to be higher in blends containing the POSSMA macromer.

Blends of PMMA with the poly-POSSMA homopolymer and the $(POSSMA)_{18}/(MMA)_{82}$ copolymer were anticipated to show more compatibility with PMMA since unlike the crystalline macromers they are of comparable molecular weights and amorphous. However blends of the POSS-methacrylate homopolymer with PMMA were observed to form visibly phase separated (opaque) blends at levels above 9%. A similar result was observed when blending the POSS macromers (except POSSMA) with PMMA. However, it appears that the homopolymer (poly-POSSMA) does show a 3 % enhanced miscibility with PMMA relative to that of the macromers. Analysis of the 3, 6, and 9 %

(w/w) blends of the poly-POSSMA homopolymer with PMMA by TGA showed similar onset of decomposition temperatures but revealed increases in the char yield as compared to the same percentage blends of the POSS-macromers with PMMA. As occurs for PMMA and related methacrylates, the POSSMA homopolymer, being a substituted poly(n-propylmethacrylate), would be expected to depolymerize to monomer upon pyrolysis. For the case of poly-POSSMA, this depolymerzation / volatilization process may be altered by the presence of the $(c\text{-}C_6H_{11})_7Si_8O_{12}$ group. If depolymerization does occur, the POSSMA monomer being unsymmetrical and of high mass, does not sublime or volatilize as readily as do the $(c\text{-}C_6H_{11})_6Si_6O_{12}$ and $(c\text{-}C_6H_{11})_8Si_8O_{12}$ macromers or the organic methacrylates and thereby contributes to the formation of residual char through decomposition (Table 2).

Table 2. Char yields from blends of PMMA with P(POSS)MA

Compound	Char Yield %
PMMA	0.0
POSSMA	7.7
poly-(POSS)MA	43
PMMA/P(POSS)MA 97/3	0.6
PMMA/P(POSS)MA 94/6	1.5
PMMA/P(POSS)MA 91/9	1.8
PMMA/P(POSS)MA 85/15	3.3
PMMA/P(POSS)MA 25/75	9.7

(a) TGA pyrolysis to 1000 °C under nitrogen.

Blends of PMMA with the $(POSSMA)_{18}/(MMA)_{82}$ copolymer were anticipated to show more compatibility since the copolymer itself contains PMMA segments. The $(POSSMA)_{18}/(MMA)_{82}$ copolymer is a visually transparent material when compression molded or cast as a film from solution. The glass transition temperature of the $(POSSMA)_{18}/(MMA)_{82}$ copolymer used in this study is approximately 100°C higher than that for PMMA and its decomposition temperature is also higher (Table 1).

Incorporating the POSS entity into a copolymer with methyl methacrylate greatly enhanced the amount of POSS which can be blended with PMMA. Clear films were obtained over a broad range including the 50/50 w/w blend level; resulting in a 35 % POSS content by weight for this blend composition. When the homopolymer (poly-POSSMA) was blended with PMMA, clear films could only be obtained with a POSS loading of 9 % or less. A series of $(POSSMA)_{18}/(MMA)_{82}$-copolymer blends with PMMA ranging from 14 to 35 % by weight of POSS were prepared and characterized using TGA, DSC, and TMA.

When two polymers are blended a popular criterion for determining that a single-phase blend is present is usually established by measuring the glass transition by DSC. If one phase is present, a single T_g is observed at a temperature between the T_g's of the two polymers, and depending on the composition of the blend. If the two polymers are phase separated, they behave independently and two T_g's are observed. In the case of POSS-polymers containing greater than 50 wt % POSS they typically do not exhibit a T_g by DSC. Similarly DSC experiments on the $(PMMA)_{82}(POSSMA)_{18}$ copolymer which contains 71 % by weight POSS did not show a T_g. However, TMA can be used to determine a softening temperature for POSS-polymers and their blends although at high loadings of POSS the materials are often to viscous to completely flow during the time scale of the experiment. Furthermore, the identification of multiple glass transitions by TMA is difficult unless the material undergoes two distinctive flow regions at each temperature. While it is believed that the blends reported in Table 3 behave as a single material, additional characterization is underway.

Table 3. Partial list of thermal properties for blends of PMMA with the $(POSSMA)_{18}(PMMA)_{82}$ copolymer.

PMMA/Copolymer Blend	Transition (°C)[a]	T_{dec} (°C)[b]	Char Yield (%)[c]
PMMA	139	351	0
80:20	148	364	4
70:30	144	358	5
60:40	163	364	8
50:50	186	360	14

(a) Determined by TMA in penetration mode with a 0.1 N force and compression molded samples. (b) Represents a 10% mass loss under nitrogen. (c) Determined by TGA after pyrolysis to 800 °C under nitrogen.

In addition to examining the compatibility and thermal properties of POSS-acrylic resins with PMMA, the oxygen permeability of the various blends was also examined. It was expected that the large volume of the POSS cage would enhance the permeability of resins like PMMA. Therefore, oxygen permeabilities were measured on the 3, 6, and 9 % (w/w) blends of the poly-POSSMA/PMMA polymers relative to PMMA.

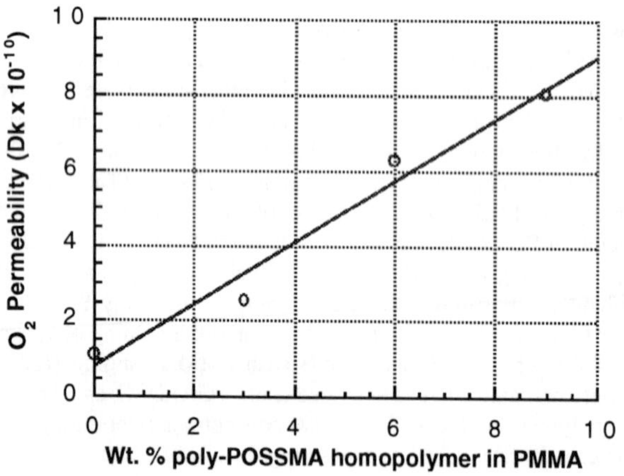

Figure 4. Plot of oxygen permeability relative to weight % of POSS incorporation in PMMA/poly-POSSMA homopolymer blends.

Figure 4 shows a plot of increasing oxygen permeability with increased POSS content. Permeability measurements were also conducted on the series of $(POSSMA)_{18}(MMA)_{82}$ copolymer/PMMA blends. The permeability of these blends was also enhanced relative to that of PMMA. The mechanism responsible for the increased permeability in these materials is not yet known, however it is suspected that the bulky POSS groups serve to increase the interchain spacings in the blend and thereby enhance the diffusion of gases.

CONCLUSIONS

The development of POSS-based reagents offers a wide variety of new hybrid building blocks for material chemists, scientists, and engineers. These systems are amenable to conventional usage with common "organic" monomers and polymers. POSS reagents also enable the preparation of a wide variety of "hybrid" thermoplastics. While the scope and range of property enhancement afforded by these systems has not been fully investigated, glass transitions, decomposition temperatures, char yields and permeabilities of methacrylics can be enhanced via incorporation of POSS entities either through blending or copolymerization. The miscibility of POSS macromers with poly(methyl methacrylate) was greatly enhanced through their polymerization. In particular, copolymerization of the POSS-entity with methyl methacrylate monomer resulted in a resin that showed a greatly enhanced miscibility with PMMA relative to the corresponding POSS-methacrylate homopolymer.

ACKNOWLEDGMENTS

This research has been supported by the Phillips Laboratory (Propulsion Directorate), and the Air Force Office of Scientific Research, Directorate of Chemistry and Life Sciences. We also thank Dr. Phil Corkhill, Mr. Albert Yul, Dr. Khushroo Gandhi of Pilkington Barnes Hind for permeability measurements.

REFERENCES

1. J. D. Lichtenhan, Y. A. Otonari, and M. J. Carr, Macromolecules **28**, 8435, (1995).

2. (a) The synthesis of $R_7Si_7O_9(OH)_3$ and cornercapping of POSS trisilanols was reported by Brown and Vogt in 1965 see: (a) J. F. Brown, Jr. and L. H. Vogt, Jr, J. Am. Chem. Soc. **87**, 4313 (1965). (b) F. J. Feher, D. A. Newman and J. F. Walzer, J. Am. Chem. Soc. **111**, 1741 (1989). (c) Also refer to J. D. Lichtenhan, Comments on Inorg. Chem., **17**, 115, (1995) and references therein.

3. J. D. Lichtenhan, R. A. Mantz, P. F. Jones, K. P. Chaffee, J. W. Gilman, I. M. K. Ismail, M. J. Burmeister, Chemistry of Materials accepted. J. D. Lichtenhan, R. A. Mantz, P. F. Jones, J. W. Gilman, K. P. Chaffee, I. M. K. Ismail, M. J. Burmeister. Polymer Preprints (Am. Chem. Soc., Div. Polym. Chem.) **36**, 334 (1995).

4. F. J. Baltá Calleja, Trends in Polym. Sci. **2**, 419 (1994).

5. For example see: H. R. Allcock, and K. B. Visscher, Chem. Mater. **4**, 1182 (1992). H. R. Allcock, K. B. Visscher and I. Manners, Chem. Mater. **4**, 1188 (1992). C. J. T. Landry, W. T. Ferrar, D. M. Teegarden and B. K. Coltrain, Macromolecules **26**, 35 (1993). T. Asuke, Y. Chien-Hua and R. West, Macromolecules **27**, 3023 (1994). P. B. Messersmith and E. P. Giannelis, J. Polym. Sci. Part A, Polym. Chem. **33**, 1047 (1995) and references therein.

(6). J. D. Lichtenhan, N. Q. Vu, J. A. Carter, J. W. Gilman and F. J. Feher, Macromolecules **26**, 2141 (1993). T. S. Haddad and J. D. Lichtenhan, J. Inorg. Organomet. Polym. **5**, 237 (1995). T. S. Haddad and J. D. Lichtenhan, Macromolecules submitted.

HYBRID SOLS AS INTERMEDIATES TO INORGANIC-ORGANIC NANOCOMPOSITES

H. K. SCHMIDT, P. W. OLIVEIRA, H. KRUG
Institut für Neue Materialien gem. GmbH, Im Stadtwald, Geb. 43 A, D-66123 Saarbrücken, Germany

ABSTRACT

For the preparation of inorganic-organic hybrid materials, synthesis processes have been developed to fabricate so-called hybrid sols, which contain an inorganic core (ceramic or glass) with nano-scale dimensions surface modified by organic groupings. These groupings have been reacted to the particle surface either by amino functional silanes (e.g. in case of iron oxide nanoparticles for amino group containing silanes) reacting with aliphatic acids to make surfaces unreactive and reduce the particle-to-particle interaction by acids, or silanes with polymerizable groupings reacted with organic monomers. In the paper a summary of the recent work and the development of structural models is given. Moreover, it could be shown that surface charges can be generated in the particles within already gelled systems and thus, particles can be transported by electric fields to form gradient index materials.

INTRODUCTION

Inorganic-organic hybrid materials fabricated by the sol-gel process are looking back now on an almost 20 years old history. However, the idea of a synthesis of organically modified inorganics or inorganically modified silicones (heteropolysiloxanes) is much older and goes back to the synthesis of silicones [1, 2], but never was exploited systematically for the synthesis of new materials for industrial applications. In [1], synthesis of heteropolysiloxanes is compared to silicon preparation, and silicones have been judged to be more advantageous. In [2], titanium dioxide was incorporated into silicones to improve the temperature resistance, but not much success was obtained. In the 80's, pioneered by D. Uhlmann and G. Wilkes, polycerams and ceramers were developed in order to improve mechanical properties of polymers. Schmidt developed heteropolysiloxanes, ormosils and ormocers for very special uses, such as contact lenses, functionalized surfaces for immunoassays or special adsorbents [3 - 5]. Slowly, the field was gaining increasing interest from the scientific community, and more and more types of materials and applications were proposed [6 - 8], and in various conferences so-called hybrid materials started to play an increasing role [9 - 13]. More and more materials have been developed and are already used in industry or are being developed.

One of the most interesting questions remaining is the structure of these materials, especially related to the phase dimension. Most of the materials still are considered to be molecular composites where an inorganic component exists in a molecular dispersion within a polymer type of network [14]. In this type of structure (e.g. SiO_4^{4-} tetrahedron), the function of the inorganic unit is to act as a three-dimensional crosslinker, leading to more brittleness. Due to the hydrolytic stability of SiC bonds, silicon has an exceptional position within these materials. Other elements have to be incorporated by different mechanisms, such as by forming complex compounds, e.g. ß-diketones or amines in order to disperse them molecularly. Inorganic properties resulting from the electronic properties of the inorganic component, of course, can be

Mat. Res. Soc. Symp. Proc. Vol. 435 © 1996 Materials Research Society

incorporated into polymeric networks, too (e.g. spectroscopic properties in transition metals). But all properties to be attributed to solid state inorganic materials properties are excluded in the molecular type of composite. That means, the range of property tailoring of molecular inorganic-organic hybrid materials is rather limited. Despite restrictions, very interesting applications have been developed based on the molecular approach, e.g. porous materials [15] or scratch-resistant coatings on plastics [16]. Due to the small phase dimensions, all these materials can be prepared with highly transparent properties, and they have been proposed for optical applications [17, 18].

The question, however, arises, by using sol-gel techniques can composite materials be obtained with larger phase dimensions but still showing high optical transparency. For this reason, in order to avoid disturbing Rayleigh scattering, as a general rule the phase dimension of the dispersed phase should be below 1/20 of the applied light wavelength, meaning particle sizes below 5 - 8 nm in the visible range. From these considerations the concept of the so-called nanocomposites have been developed. These composites, if the phase dimension of the inorganic part is below the Rayleigh scattering limit, should be able to be prepared in the form of transparent materials and, due to the extended inorganic phase their solid state properties (e.g. refractive index), should be able to be incorporated. Additional quantum size effects of small particles (e.g. semiconductor or metal colloids) become of interest as well as properties of the interfaces, which can become significant due to the interfacial volume.

The aim of this paper is to give some basic ideas of nanocomposite synthesis, properties and materials tailoring and to summarize some interesting results.

GENERAL ASPECTS

In order to exploit the possibilities of nanocomposite properties, it is necessary to develop methods for a cost-effective synthesis and for a perfect dispersion of the particles in a polymeric or inorganic-organic network. The composite can be considered as a nanoparticle-reinforced, nanoparticle-modified or a simple nanoparticle-containing polymeric network and shall be called "Nanomers" (nanoparticle and organic component containing polymer type materials).

Synthesis of Nano-Scale Particulate Phases

As described elsewhere, the sol-gel process can be used advantageously for the preparation of ultrafine particulate systems in the lower nanometer range [19, 20]. This route seems to be quite simple since if one starts from a molecular solution in a system able to precipitate, for example, after nucleation a growth reaction takes place involving all size ranges. If the thermodynamical parameters are suitable, large crystals can be formed. One of the major objectives of the sol-gel process, however, is to keep small particles in solution in order to carry out various moulding processes, such as film or coating formation. Due to the surface area, small particle size and the high number of particle interactions, the integral particle forces per volume unit becomes very large (e.g. van der Waal's forces). Therefore, a system has to be generated in which the agglomeration by attraction of the particles can be controlled. In general, in sol-gel ractions this is carried out by pH control which produces surface charges (Stern's potentials) [21], ending up with repulsing forces between the charged surfaces. In order to form compact materials from sols, the system has to be destabilized, either by change of pH, loss of solvent (overcoming the critical radius for repulsion from the Stern's potential), or by generating chemical reactions between the particles. In all cases where these reactions do not run under controlled conditions, random arrangements of the particles are formed, leading to gels with low density microstructures. The stabilization condition for sols is that the surface free energy generated by any type of surface active medium (e.g. electric charges) generates a lower Gibbs free energy minimum than the aggregation. This is demonstrated in fig. 1.

Fig. 1: Schematics of the stabilization of sol particles by electric charges and destabilization followed by aggregation.

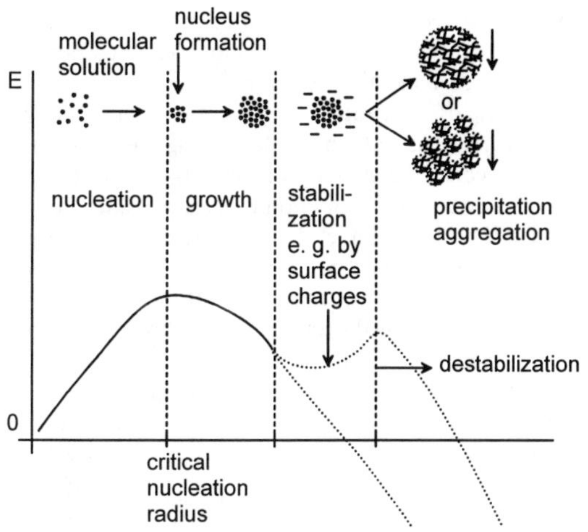

This type of sol-gel process (colloidal type) is typical for most of the elements being able to form insoluble compounds starting from a solution. The only difference between a precipitation process and a sol-gel process is that the sol-gel process, in general, is carried out far away from the point of zero charge (pzc). If well-crystallized precipitates shall be obtained, the reaction has to be carried out close to the point of zero charge. One of the disadvantages of charge-stabilized systems is that after the removal of the charges the forces between the particles in solution are not only van der Waal's forces, but in most cases (especially with sol-gel materials with reactive OH groups) also are based on hydrogen bridges or even the formation of chemical bonds. This leads to so-called hard agglomerates and gels if the whole volume is agglomerated "monolithically", which cannot be redispersed. For this reason, we have introduced [22, 23] the concept of chemical surface modification with the possibility to establish controlled interaction between the particle surfaces. The schematics are shown in fig. 2.

Fig. 2: Schematics of the chemical surface modification to stabilize sols and redispersion.

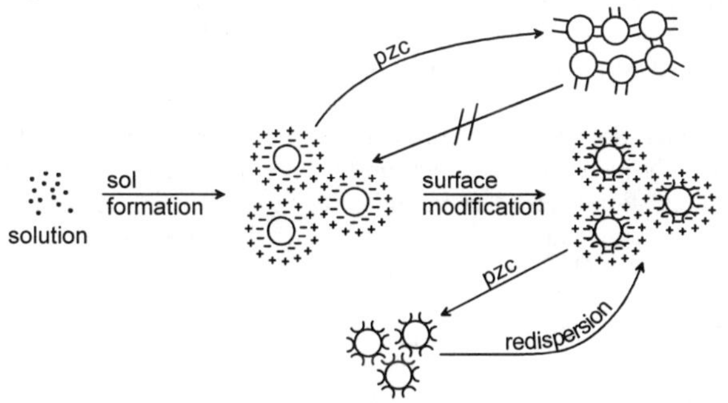

The chemical stabilization in addition to the charge stabilization also allows establishment of specific surface reactivities to control the polarity of the particles and by this, to control the interaction of the particles to other components (such as organic monomers) as well as to protect the surface from undesired reactions like absorption of components from the solution.

This concept leads to new type type of a sol-gel precursor, the so-called hybrid sols, which can be considered as a specific type of sol-gel precursor suitable for various reactions.

The preparation of this type of sols can be carried out by different routes. Some of them will be described.

Modification of Alkoxides

This route describes a more or less stoichiometric formation of alkoxide complexes, as described elsewhere [24]. Interesting complex formers with an interesting potential to act as surface modifiers for nanoparticles are ß-diketones or ether alcoholates, alkoxides or carboxylic acids for oxidic, and amines or aminoacids for transition metallic systems.

In order to obtain appropriately modified particles, this route requires very special conditions. First, a complex bond to the alkoxide precursor has to be possible. The above mentioned compounds are stable in many alkoxides [24]. Second, the stability of the bond has to be low enough to allow hydrolysis, condensation and growth. Third, the bond has maintained to the particle surface during the growth process and still has to be intact if small particles are formed. If all of the three conditions are fulfilled, this method can lead to chemically surface-modified colloidal particles. The schematics of this assumption are shown in figure 3. If the bonds to the surface are stable enough, they should be able to be established even after the colloid formation, thus replacing electric charges.

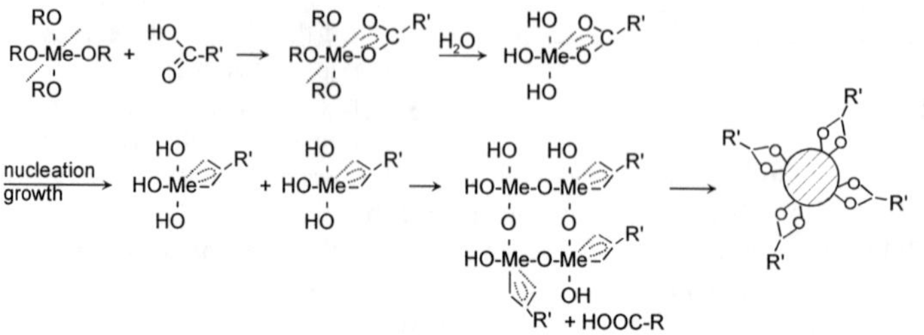

Fig. 3: General scheme of modification of alkoxides with subsequent hydrolysis and condensation with carboxylic acids as example.

It has been shown elsewhere [25] that even already weakly agglomerated nano-scale systems can be redispersed using this approach, for example agglomerated commercially available nano-scale boehmite refluxing with carboxylic acid groups such as propionic or acetic acid. The resulting system is a boehmite powder covered with carboxylic acid groups, which are stable enough not to be washed away by water. Sols prepared from these surface-modified systems do not show a gel point but only a slowly increasing viscosity up to more than 50 vol. % solid content (fig. 4). Without acid, gelation takes place in any cases [26]. Identical results are obtained for propionic acid. These systems can be moulded by plastic deformation and have been used for extrusion of tubings [27]. The surface modification can be followed by IR spectroscopy.

The >C=O frequency is shifted from the 1700 cm^{-1} regime of the free acid to the carboxylate form below 1600 cm^{-1}, typical for salts.

This example demonstrates the usefulness of chemical surface modification of nano-scale particles in order to enable appropriate processing.

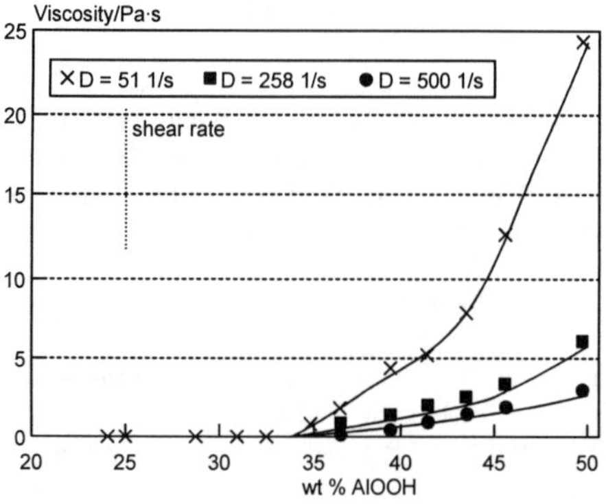

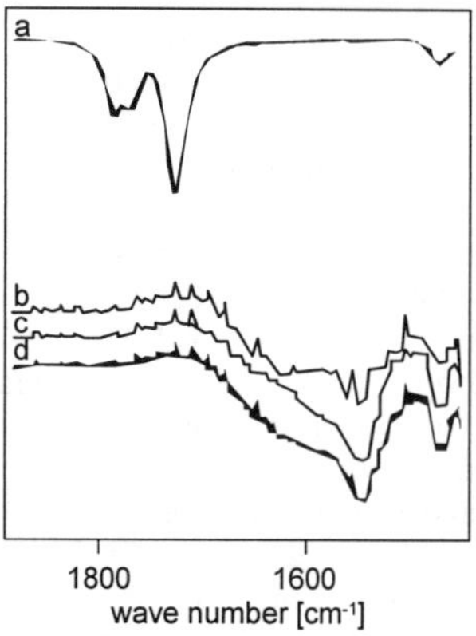

Fig. 4: Rheological behavior of aqueous boehmite sols with 15 nm particle diameter, surface modified with 1 wt% acetic acid.

Fig. 5: IR spectra of a: propionic acid, b: 1 wt%, c: 2 wt%, e: 5 wt% boehmite + propionic acid; Al propionate.

For the surface modification of small particles, different prerequirements have to be fulfilled. One of the most important is the use of small molecules in the range of below 10 or 15 carbon atoms; otherwise the volume required by the surface modifiers (e.g. if polymers are used) leads to a drastic decrease of the nanophase content of the whole system, and ceramic processing, for example, becomes very difficult, due to the volume required by the absorbed surface modifiers. If small molecules are used, the binding force to the surface has to be high enough to avoid high concentrations of the surface modifiers in the surrounding liquid. For this reason, the above mentioned systems have proven suitable for a variety of modifications.

Another type of surface modification is the reactive type, for example, the reaction with silanes which directly form bonds to the surface or which can be reacted in a way to surround particles with a coating. An interesting example for this type of surface modification has been developed by Lesniak et al. [28] who were able to develop a surprisingly hydrolytically (between pH 1 and 11) stable aminosilane coating by reacting it on the surface of 10 nm magnetite particles by using ultrasonic energy in the presence of γ-aminopropyl triethoxysilane to cover the magnetite surface completely. In fig. 6 the ξ potential of modified and unmodified magnetite particles are given, clearly showing the effect of modification.

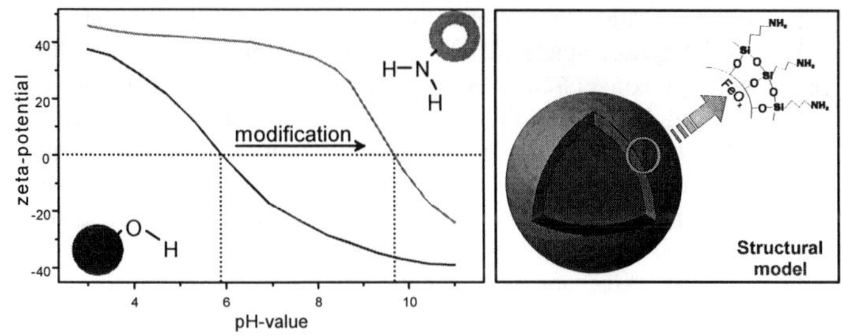

Fig. 6: ξ potential curves of surface modified and unmodified magnetite particles.

As known from the literature [29], polycondensed

aminosilanes rapidly depolymerize in aqueous solutions and never form stable systems. The high stability of the coated magnetite is attributed to a very high degree of polymerization of the aminosilane shell, since the Fe-O-Si bond is very sensitive to hydrolysis.

This material represents a variable precursor for different reactions. It can be used as a sol-gel precursor easily incorporated into sol-gel glasses to obtain superparamagnetic glasses. It also can be used as a precursor for polymer fabrication, for example, together with epoxides, to produce superparamagnetic polymers. Another way of attaching reactive bonds to the surface of iron oxide particles again utilizes carboxylic acids. The first successful sample has been obtained from reacting the iron oxide sols under ultrasonic treatment with citric acid [30]. These samples are hydrolytically stable at room temperature as well as in aqueous solutions, but now show -COOH groupings at the surface.

The modification of oxidic nanoparticles by reaction of carboxylic acids with zirconia particles has been described elsewhere [31 - 33]. The use of carboxylic acid with functional groups such as methacrylate leads to hybrid sols with a potential to be copolymerized with a variety of organic monomers and, of course, hybrid precursors such as silanes containing polymerizable groups, e.g. methacryloxysilanes. This route is started by reacting alkoxides with carboxylic acids and methacrylic acid. The question arises, however, how far during the subsequent hydrolysis and condensation does the bond between the particle surface and the carboxylic acid remain intact, and how much of an oxidic network is formed. As shown elsewhere [34], under conditions of hydrolysis and condensation and subsequent polymerisation when methacrylic acid is used as carboxylic acid, monoclinic nanoparticles ($\approx$ 2 nm) are formed. Various optical materials have been prepared using this approach [35 - 36]. The evidence for the existence of Zr carboxylic acid bonds throughout the process could be shown by IR spectroscopy.

In figure 7, the IR-spectra of ZrO_x/methacrylic (ma) acid system is shown in various states. It shows that the >C=O frequency remains unchanged throughout all processing steps (complexation, hydrolysis, polymerization) and that the reaction of ZrO_2 surfaces with carboxylic acids leads to the same C=O frequency.

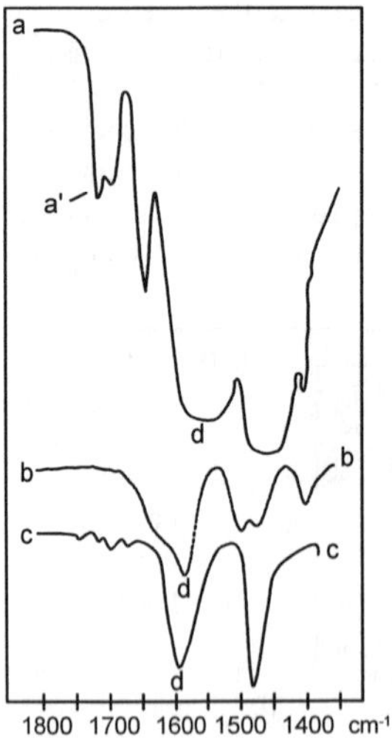

Fig. 7: IR spectra of a: $Zr(OR)_4$ complexed with excess of ma (> 2:1); b: sub µm ZrO_2 reacted with aliphatic carboxylic acid; c: after hydrolysis and condensation; a': >C=O frequency of the free acid; d: of the carboxylic and reacted to the particle surface.

One of the interesting questions in this connection is related to the polymerization mechanism since the "molecular weight" of „monomers" consisting of nanoparticles linked to double bonds is high compared to organic monomers. As shown by Krug [35], the polymerization kinetic of a mixture of ma-reacted ZrO_2 nanos and methacryloxy silane (MPTS) is slower than that of pure MPTS, but high conversion rates are obtained (up to 95 %). This leads to the conclusion that the nanoparticles show rather high diffusion rates almost comparable to those determined in polymerizing systems. From these findings, the question was asked how nanoscale particles in soft organic matrices can be used to provide selective diffusion in order to enforce phase separation in such systems. Mechanistic examinations during the fabrication of optical gratings or microlense arrays by holographic methods proved that a selective diffusion of nano-scale particles can be initiated by a photopolymerization process with an intensity gradient of the light beam [37]. The schematics of the process are shown in fig. 8, indicating the enrichment of areas with increased nanoparticle contents schematically.

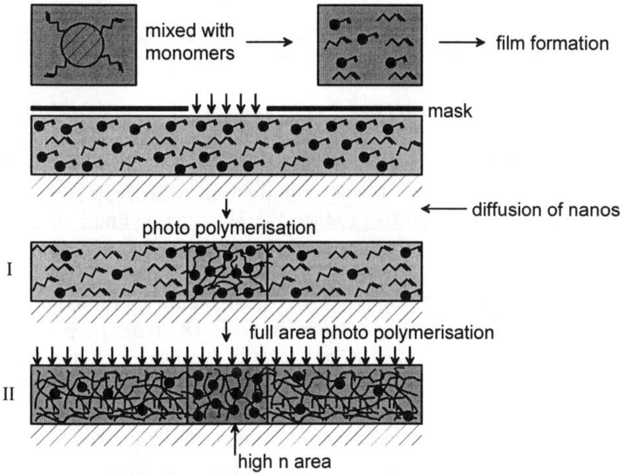

Fig. 8: Schematics of the preparation of gradients in photopolymerized nanocomposite films.

If this assumption is true, after full area polymerization, a difference in the refractive index in preliminary radiated and non-irradiated areas should remain. Moreover, the Δ n should depend on particle size (due to differences in diffusivity) and particle content. Oliveira et al. [37] have shown that the diffusion of ZrO_2 nanoparticles can be followed by on-line determination of the diffraction efficiency during an irradiation process using the interference pattern of a two-wave mixing process to generate intensity fluctuations. He found that the diffusivity increases by decreasing particle size. Moreover, in these experiments the ma coated ZrO_2 particles have been copolymerized with methacryloxy silane, and the diffractive efficiency also depends on the ZrO_2 content. With $C_{ZrO2} = 0$ (pure methacryloxy silane) a diffractive pattern could be obtained in the first step due to the Colburn-Haines effect. After full area polymerization, however, the diffractive efficiency returns to zero. These findings actually prove that ZrO_2 nanoparticles are "transported" by the attached double bonds due to a concentration gradient established by the "consumption" of double bonds through the photopolymerization step. Moreover, the concentration gradient of ZrO_2 could be directly measured by micro-Raman spectroscopy as well as by scanning EDX. This increase of nanoparticle content (in this case ZrO_2) is accompanied by an increase of the refractive index up to $1.5 \cdot 10^{-2}$ depending on composition [37].

Additional experiments have been carried out by the use of artificially generated gradients of the chemical potential, or even another potential within a material, to initiate diffusion processes and to fabricate gradient materials. Using this approach seems to be attractive, as shown by the example of an holographic process using the diffusion of high refractive index nano-scale particles. Another approach would be to use electric fields and for this case, the system described in [37] was used to investigate how electrophoresis can be used for building gradients. In fig. 9 the experimental set-up for this investigation is shown.

The question arises how nanoparticles in non-densified organic or composite matrices can be "moved" by other means. If charges are present on the nanoparticles, they should be able to move in an electric field. Various possibilities may exist with respect to the changed particles (fig. 9). Depending on the degree of surface coverage, residual ≡ZrOH groupings may exist, leading to negative surface charges. In this case, the particles should move to the anode. If the salt form is dominant, the particles should move to the cathode. In fig. 10, the schematics of the electrophoretic system are shown. The set-up was used to produce GRIN lenses.

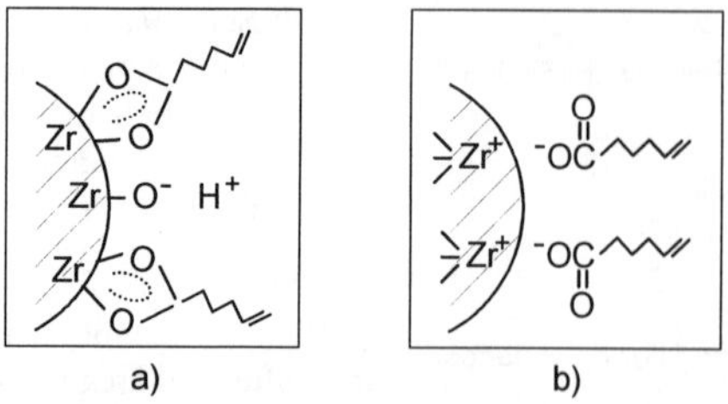

Fig. 9: Different models for surface charges on ZrO_2 nanoparticles.

To produce the lenses, a cylindrical receptacle has been filled with the photopolymerizable composite containing nano-scale ZrO_2 particles. The composite material used in the experiment is based on methacrylpropyl trimethoxysilane (MPTS), zirconium n-propoxide (ZR) and methacrylic acid (MA) in 10/2/2 (MPTS/ZR/MA) mole-%. Irgacure 184 (Ciba) was used as a photoinitiator in concentration of 0.4 mol%/C=C. The synthesis of the basic composition is described in [37].

The convergent lens has the radial concentration gradient of the ZrO_2 particles pointing to the geometrical center of the lens. To induce this concentration gradient of the positive ZrO_2 by electrophoresis, the window of the receptacle were defined as the cathode and the centered gold wire the anode. The voltage was employed for 36 hours.

The photopolymerization process using a Hg-Xe lamp was divided in two steps. First, the nanocomposite was gelated by a two minute treatment with a UV, and the cathode, consisting of a 50 μm gold wire, was removed. The second step includes the complete polymerization of the sol with subsequent 8 minutes of UV irradiation. After photopolymerization, the rod was cut into disks, and the refractive index was measured by Spectroscopic Ellipsometer (ES 4G-Sopra). Δ n was detected to be 0.07 (fig. 10). Since the particles move to the anode, the model a) in fig. 9 is more likely.

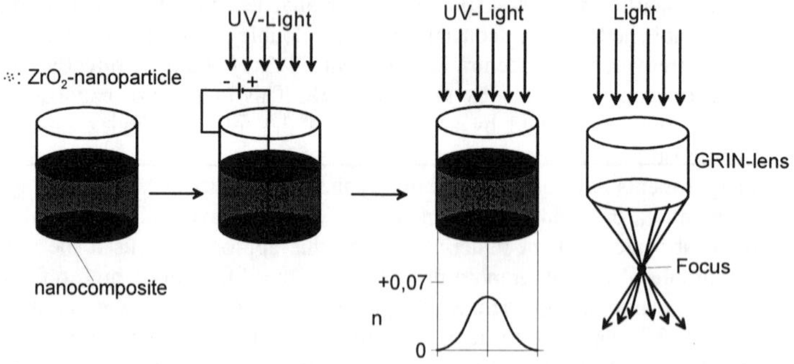

Fig. 10: Experimental set-up for electrophoretic gradient material fabrication (direct current voltage).

One of the most important issues of materials synthesis using surface-tailored nanoparticles is to prevent aggregation, especially if high optical quality (transparency) is required. Another reason for looking for high dispersion is to implement high interfacial volumes into materials properties. Whereas interfacial phases, in general, do not play an important role in systems with micrometer-size particles, the interfacial phase can become a remarkable component if particle size decreases to the nm range. In fig. 11 this is demonstrated on a simple example.

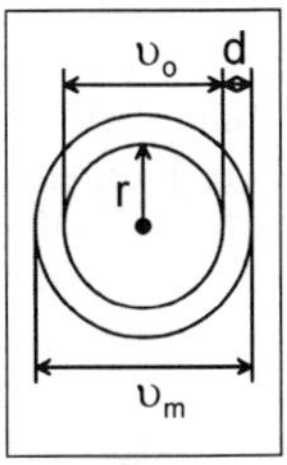

$$\upsilon = \frac{F}{(\upsilon_o \cdot (1-F) \cdot (\upsilon_m - \upsilon_o) \cdot 100}$$

υ = interfacial volume fraction
υ_o = particle volume without layer
υ_m = particle volume with layer
F = filler volume fraction

Fig. 11: Estimation of the volume of interfacial phases in nano-composites. Assumed data: d = 2 nm, F = 0.15 nm, r = 7.5 nm. Calculated value for v = 30 % [after 38].

As one can see, for the assumed filler volume fraction F = 0.15, the volume of the interfacial phase exceeds the degree of filling by the factor of 2 if the thickness of this phase is estimated to 2 nm. These data are estimated according to investigations of Kendall and coworkers [39]. This means that in nano-scale particle filled polymers effects have to be expected caused by an interfacial phase. In fig. 12 the storage modulus of three types of composites are compared. The composite is a copolymer of MMA (metal methacrylate) and HEMA (hydroxyethylmethoxylate) [40].

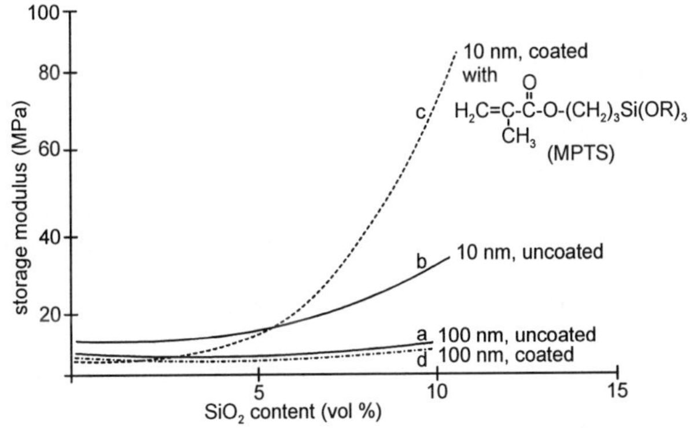

Fig. 12: Storage modulus as a function of the coating of 10 and 100 nm particles of SiO_2 with methacryloxy silane and subsequent polymerization.

A one-to-one composition of MMA to HEMA was used. Azubis-isobutyronitril (AIBN) was used as a thermal initiator. Fig. 12 shows clearly that 10 % filler that 100 nm, uncoated as well as the coated species, curve a and d, do not effect the mechanical properties. The use of 10 nm particles gives a slight increase of the storage modulus, but the particles coated with MPTS results in a strong increase of the storage modulus. Especially the difference in the coated and the uncoated case leads to the conclusion that these effects cannot only result from the pure filling effect since the degree of filling is very low at 10 %. The formation of a interfacial structure is postulated (curves b and c). It has to be mentioned that the composite systems b and c both are completely transparent, and no light scattering can be measured.

In the uncoated case the good dispersion can be attributed to the presence of HEMA which generates SiO_2 interactions with the surface OH groups by the alcoholic OH groups of the hydroxyethyl group. If one compares the Tg values (obtained by thermomecanical analysis from the tan δ maximum) for 100 nm coated and uncoated, for the 10 nm MPTS and APTS coated and 10 nm uncoated particles, corresponding results are obtained (figure 13).

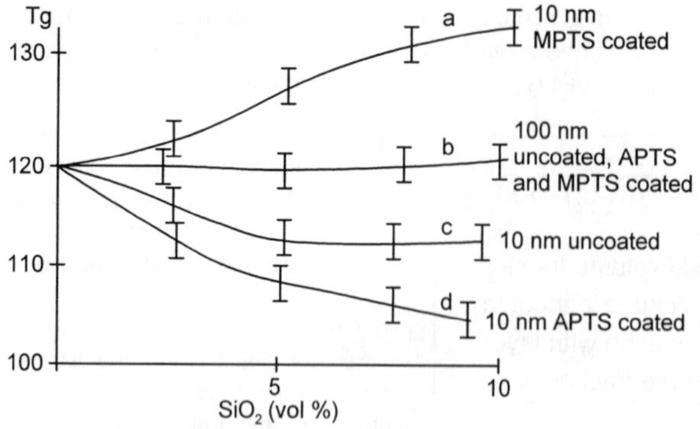

Fig. 13: Tg values of different SiO_2 MMA/HEMA copolymer composites.

As expected from the storage data, the 10 nm MPTS composite leads to increased Tg values, whereas the 100 nm systems are almost unaffected by the composite formation. For the 10 nm uncoated system, a slight decrease in the Tg value is observed, which does not match exactly with the storage modulus observation. However, in the 10 nm case, if another surface modifier is used, in this case acetoxypropyl trimethoxysilane (APTS), the Tg value decreases. This is attributed to the formation of a "softer interface", due to the fact that no bonds are formed between the matrix and the particles, which seems to have less order than the polymer by itself. A model is given in fig. 14.

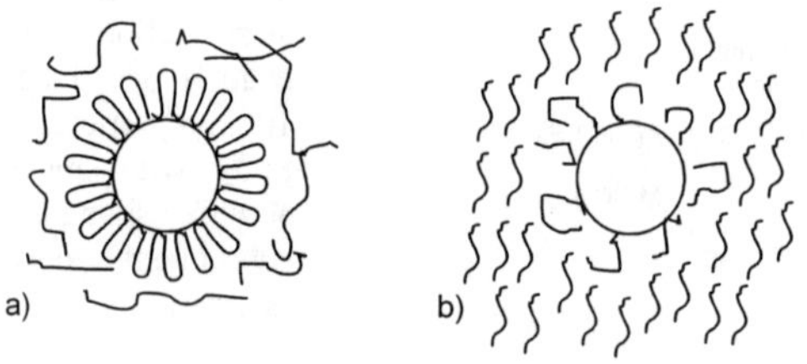

Fig. 14: Model for the formation of "soft" and "rigid" interfaces around nanoparticles by different surface modifiers. a): polymerizable, b): unpolymerizable.

These results can only be interpreted by the presume of a remarkable volume fraction attributed to an interfacial phase, too, according to figure 11. These first experiments show that by using nanoparticles as fillers in a perfectly dispersed state, it seems to be possible to use interfacial structures for tailoring polymer properties, which can be used as a new approach for manipulating polymer properties. But these investigations still are at their infancy.

As shown by work of Mennig and Nass, hybrid sols, that means nano-scale particles with tailored surfaces dispersed in solutions, not only can be used for ceramic polymer nanocomposites, but also for metal polymer composites [41] for the fabrication of high χ^3 value composites, or for colloidal processing of ceramics with extremely low sintering temperatures [42].

CONCLUSION

It can be concluded that the sol-gel process is a suitable means for the fabrication of nano-scale inorganic particles. If chemical surface modification with organic molecules is used for steric or electrosteric stabilization instead of charge stabilization, the hybrid sols are obtained.

These hybrid sols are characterized by functionalized surfaces which can be used for a variety of materials tailoring, for example, superparamagnetic composites, optical gradient materials or new types of interfacial phase-determined ceramic polymer composites. This class of materials can be considered nanoparticle-containing organic polymers (nanomers).

REFERENCES

1. W. Noll, Chemie und Technologie der Silicone, 2nd edition, Verlag Chemie, Weinheim, 1968.
2. K. A. Andrianov in Organic Silicon Compounds, State Scientific Publishing House for Chemical Literature, Moscow, 1955.
3. H. Scholze, H. Schmidt and H. Böttner, Poröse Membranen und Adsorbentien, Verfahren zu ihrer Herstellung und ihre Verwendung zur Stofftrennung. DOS 29 25 969 (1981).
4. G. Philipp and H. Schmidt, J. Non-Cryst. Solids **63**, p. 283 (1984).
5. H. Schmidt, O. von Stetten, G. Kellermann, H. Patzelt and W. Naegele, IAEA-SM-259/67, Wien 1982, p. 111.
6. J. D. Mackenzie, Y. J. Chung and Y. hu, J. Non-Cryst. Solids **147&148**, p. 271 (1992).
7. J. McKiernan, J. I. Zink, B. S. Dunn, SPIE Vol. **1758**, Sol-Gel Optics II, p. 381 (1992).
8. H. Schmidt, R. Kasemann, T. Burkhart, G. Wagner, E. Arpac and E. Geiter in ACS Symposium Series No. 585: Hybrid Organic-Inorganic Composites, edited by J. E. Mark, C. Y.-C. Lee and P. A. Bianconi (American Chemical Society, Washington, 1995), p. 331.
9. SPIE Proc. Sol-Gel Optics II, edited by J. D. Mackenzie, Vol. **1758**, 1992.
10. SPIE Proc. Sol-Gel Optics II, edited by J. D. Mackenzie, Vol. **2288**, 1994.
11. Interdisciplinary Workshop „Organic/Inorganic Polymer Systems", Feb. 12 - 14, 1995, Napa Valley, California/USA.
12. Proc. "Workshop on Organic-Inorganic Composite/Hybrid Materials", 07.12.95, Kyoto, Japan, edited by M. Toki, Kansai Research Institute, Kyoto, Japan.
13. Proceedings "First European Workshop on Hybrid Organic-inorganic Materials", Bierville France, edited by C. Sanchez und F. Ribot, Paris, Frankreich, 1993.
14. H. Schmidt in: Ultrastructure Processing of Advanced Materials, edited by D. R. Uhlmann und D. R. Ulrich (John Wiley and Sons, New York, 1992), p. 409.
15. H. Schmidt and H. Böttner in The Colloid Chemistry of Silica. Advances in Chemistry Series 234. Edited by H. E. Bergna (American Chemical Society, Washington, 1994), p. 419.
16. W. F. Maier, F. M. Bohnen, J. Heilmann, S. Klein, H.-C. Ko, M. F. Mark, S. Thorimbert, I.-C. Tilgner and M. Wiedorn, in NATO ASI Series E: Applied Sciences - Vol. 297: Applications of Organometallic Chemistry in the Preparation and Processing of Advanced Materials, edited by J. F. Harrod and R. Laine (Kluwer Academic Publishers, Dordrecht/Netherlands, 1995), p. 27.
17. R. Reisfeld in Sol-Gel Science and Technology, edited by M. A. Aegerter, M. Jafelici Jr., D. F. Souza and E. D. Zanotto (World Scientific Publishing Co. PTE Ltd., Singapore, 1989), p. 322.
18. B. S. Dunn and J. D. Mackenzie, SPIE Proc. **1328**, p. 174 (1990).
19. G.W. Scherer, C. J. Brinker, The Physics and Chemistry of Sol-Gel Processing. Academic Press, New York - Boston, 1990.

20. B. E. Yoldas, U. S. Patent 4 346 131, Aug. 24, 1982.
21. O. Z. Stern: Electrochem. 30 (1924) 508.
22. H. Schmidt, in: Proc. 8th Intl. Workshop on Glasses and Ceramics from Gels, Faro, Portugal, 18.-22.09.95. Edited by R. M. Almeida, J. Sol-Gel Science and Technology (in print).
23. H. Schmidt, H. Krug and P. W. Oliveira in Applied Organometallic Chemistry (in print).
24. D. C. Bradley, R. G. Mehrotra and D. P. Gaur, Metal Alkoxides, Academic Press, London, 1978.
25. H. Schmidt, R. Naß, M. Aslan, K.-P. Schmitt, T. Benthien und S. Albayrak, J. de Physique IV, Coll. C7, Vol. 3 (1993) 1251.
26. K. Färber, Master's Thesis, University of Saarland, Saarbrücken, 1992.
27. K.-P. Schmitt, Master's Thesis, University of Saarland, Saarbrücken, 1994.
28. C. Lesniak, T. Schiestel, R. Naß and H. Schmidt, Proceedings of the 1996 MRS Spring Meeting, San Francisco, April 08 - 12, 1996 (Mat. Res. Soc., Symp. Proc., in print).
29. Y. Charbouillon, Ph. D. Thesis, Institut National Polytechnique, Grenoble, 1987.
30. C. Lesniak, T. Schiestel, private communication.
31. P. W. Oliveira, H. Krug, H. Künstle and H. Schmidt in SPIE Vol. 2288 "Sol-Gel Optics III", edited by J. D. Mackenzie (SPIE, Bellingham/Washington, 1994), p. 554.
32. H. Schmidt in NATO ASI Series E: Applied Sciences - Vol. 297: Applications of Organometallic Chemistry in the Preparation and Processing of Advanced Materials, edited by J. F. Harrod and R. Laine (Kluwer Academic Publishers, Dordrecht/Netherlands, 1995), p.47.
33. H. Schmidt and H. Krug in ACS Symposium Series 572 "Inorganic and Organometallic Polymers II: Advanced Materials and Intermedioates", edited by P. Wisian-Neilson, H. R. Allcock, K. J. Wynne (American Chemical Society, Washington, 1994), p. 183.
34. H. Krug and H. Schmidt in Proceedings First European Workshop on Hybrid Organic-Inorganic Materials, Chateau de Bierville, France, edited by C. Sanchez and F. Robit (Université Pierre et Marie Curie, Paris, 1993) p. 127.
35. C. Becker, M. Zahnhausen, H. Krug and H. Schmidt in Ceramic Transactions Volume 55: Sol-Gel Science and Technology, edited by E. Pope, S. Sakka and L. Klein (American Ceramic Society, 1995), p. 299.
36. C. Becker, H. Krug and H. Schmidt in: Proc. 8th Intl. Workshop on Glasses and Ceramics from Gels, Faro, Portugal, 18.-22.09.95. Edited by R. M. Almeida, J. Sol-Gel Science and Technology (in print).
37. P. W. Oliveira, H. Krug and H. Schmidt, Proceedings of the 1996 MRS Spring Meeting, San Francisco, April 08 - 12, 1996 (in print).
38. C. Becker, private communication.
39. K. Kendall, F. R. Sherliker, Brit. Polym. J. **12**, 85 (1980).
40. C. Becker, private communication.
41. M. Mennig, M. Schmitt, U. Becker, G. Jung und H. Schmidt, in: SPIE Vol. 2288 „Sol-Gel Optics III", ed.: J. D. Mackenzie. SPIE, Bellingham/ Washington, 1994, 130.
42. R. Naß, S. Albayrak, M. Aslan and H. Schmidt in Advances in Science and Technology 11: Advanced Materials in Optics, Electro-Optics and Communication Technologies, edited by P. Vincenzini and G. C. Righini (Techna Srl, Faenza, Italy, 1995), p. 47.

HYBRID STYRYL-BASED POLYHEDRAL OLIGOMERIC SILSESQUIOXANE (POSS) POLYMERS

T.S. HADDAD*, E. CHOE†, J.D. LICHTENHAN†
*Hughes STX, Phillips Laboratory, Edwards Air Force Base, CA 93524
†Phillips Laboratory, Propulsion Directorate, Edwards Air Force Base, CA 93524

ABSTRACT

We have taken a unique approach to the synthesis and study of hybrid organic/inorganic materials. Our method involves synthesizing nano-size inorganic $P_1R_7Si_8O_{12}$ clusters which contain seven inert "R" groups for solubility and only one functional "P" group for polymerization. This strategy permits the synthesis of melt processable, linear hybrid polymers containing pendent inorganic clusters and allows us to study the effect these clusters have on chain motions and polymer properties. The synthesis of styrene-based polyhedral oligomeric silsesquioxane (POSS) macromers, their free radical homopolymerization and copolymerizations with varying amounts of 4-methylstyrene, and analysis of the effect of the pendent POSS group is presented. All of these polymers decompose under nitrogen between 365 and 400 °C, and the glass transitions for these materials vary from around 110 °C up to the decomposition point. Both T_{dec} and T_g increase with increasing POSS content. The shorter the spacer unit between the POSS group and the polymer chain the higher the T_g. Interestingly, a slight change in the inert "R" groups on the POSS cluster has a large effect on the glass transition indicating that POSS-POSS interactions have an effect on chain mobility.

INTRODUCTION

The design of new materials with enhanced properties continues to be a driver for the investigation of hybrid materials. As hybrid materials are copolymers based on inorganic and organic comonomers, they display enhanced properties by bridging the property space between two dissimilar types of materials.[1] A typical hybrid material will contain a crosslinked inorganic phase bound (often covalently) with an organic phase. Depending on the relative amounts of the two components, the properties of the resulting hybrid are intermediate between that of an inorganic and an organic polymer. Such methodology can be used to create either plastic inorganics or toughened plastics, and is superior to traditional blending methods.[1b] However, as most hybrid materials are obtained through a sol-gel type process, they tend to be highly crosslinked and difficult to process.

Our approach to the synthesis of easily processed hybrid materials is to design well-defined inorganic *oligomers* with only a single polymerization site per cluster. Each oligomeric cluster has an exactly defined degree of polymerization of eight, $(RSiO_{1.5})_8$, or more precisely, $P_1R_7Si_8O_{12}$. These polyhedral oligomeric silsesquioxane (POSS) macromers have an inorganic silica-like core and are surrounded by eight organic groups, of which seven are inert and just one is reactive. Polymerization at the single reactive "P" site, results in a linear polymer containing monodisperse, nano-size inorganic clusters[2] pendent to an organic polymer backbone[3].

This paper presents our research on the polymerization and copolymerization (with 4-methylstyrene) of four styryl-based POSS macromers, $P_1R_7Si_8O_{12}$ with R = cyclohexyl or cyclopentyl and P = styryl or ethylstyryl. These slight variations in both the inert "R" groups and in the "P" functionality result in significant property differences for the polymers.

Mat. Res. Soc. Symp. Proc. Vol. 435 © 1996 Materials Research Society

EXPERIMENT

POSS Macromers. The four POSS macromers **2a-2d**, $P_1R_7Si_8O_{12}$ with P = $C_6H_4CH{=}CH_2$ or $CH_2CH_2C_6H_4CH{=}CH_2$ and R = *c*-C_6H_{11} or *c*-C_5H_9, were all synthesized in the same manner from the known trisilanols[4], $R_7Si_7O_9(OH)_3$, and the appropriate trichlorosilane, $PSiCl_3$. The synthesis and characterization of $(C_6H_4CH{=}CH_2)(c\text{-}C_6H_{11})_7Si_8O_{12}$, **2c**, is given as a representative example. To a 150 mL dry THF solution of $(C_6H_{11})_7Si_7O_9(OH)_3$ **1a** (9.77 g, 10.0 mmol) and triethylamine (3.14 g, 31.0 mmol) a slight excess of styryltrichlorosilane (2.51 g, 10.6 mmol) was slowly added. The reaction flask was stirred under nitrogen for 16 hours, followed by filtration to remove the $HNEt_3Cl$ byproduct. The product was isolated by concentrating the THF filtrate to ≈25 mL, filtering through celite to remove a small amount of insolubles and then precipitating the product into rapidly stirred methanol (100 mL). The product was collected by filtration and dried in vacuo to yield 8.6 g (78 % yield). ^{1}H NMR ($CDCl_3$, 300.13 MHz) δ = 7.65 (d, $^3J_{H\text{-}H}$ = 8.0 Hz, aromatic C-H, 2H), 7.44 (d, $^3J_{H\text{-}H}$ = 8.0 Hz, aromatic C-H, 2H), 6.75 (dd, $^3J_{H\text{-}H}$ = 17.7 Hz, 10.9 Hz, vinyl C-H, 1H), 5.82 (d, $^3J_{H\text{-}H}$ = 17.7 Hz, vinyl C-H, 1H), 5.30 (d, $^3J_{H\text{-}H}$ = 10.9 Hz, vinyl C-H, 1H), 1.76 (m, Cy-CH_2, 35H), 1.25 (m, Cy-CH_2, 35H), 0.80 (m, Cy-CH, 7H). ^{13}C NMR ($CDCl_3$, 75.47 MHz) δ = 139.23 (aromatic), 136.88 (vinyl), 134.34 (aromatic), 131.54 (aromatic), 125.48 (aromatic), 114.63 (vinyl), 27.52 (CH_2), 27.47 (CH_2), 26.93 (CH_2), 26.86 (CH_2), 26.69 (CH_2), 26.62 (CH_2), 23.22 (CH), 23.13 (CH). ^{29}Si NMR ($CDCl_3$, 59.6 MHz) δ = -68.2 (3 Si-Cy), -68.5 (4 Si-Cy), -79.6 (1 Si-Styryl). Elemental analysis (calculated.): % C, 54.50 (54.13); % H, 7.68 (7.92).

Polymerizations. The four homopolymers **3a-3d**, the copolymers **4a-7d** and poly(4-methylstyrene) **8**, were all synthesized under nitrogen using a total of 3.00 mmol of monomers in 4 - 8 mL of dry oxygen-free toluene with 0.012 mmol AIBN initiator, and a reaction time of 64 hours at 60 °C. The polymers were isolated (in air) by transferring the toluene solution into methanol (100 mL) and collecting the precipitate. Separation from unreacted **2** was achieved by precipitating the pure polymer from a dilute THF solution (approximately 1 gram of crude product per mL of THF) by addition of an equivalent volume of methanol; purity was checked using ^{1}H NMR spectroscopy. Typical yields of purified polymer were between 30 and 50 %. All polymers were characterized by ^{1}H, and ^{29}Si NMR. NMR characterization data for polymer **6c** is given as a representative example. ^{1}H NMR ($CDCl_3$, 300.13 MHz) δ = 6.87 (br, aromatic C-H), 6.48 (br, aromatic C-H), 2.28 (br, Ph-CH_3), 1.77 (br, Cy-CH_2, backbone CH), 1.28 (br, Cy-CH_2, backbone CH_2), 0.83 (br, Cy-CH). ^{29}Si NMR ($CDCl_3$, 59.6 MHz) δ = -68.2 (3 Si-Cy), -68.5 (4 Si-Cy), -78.9 (1 Si-Styryl). Molecular weights (see Table 1) were determined using gel permeation chromatography and a combination of refractive index and multi-angle laser-light scattering measurements with a Wyatt Technologies DAWN spectrometer.

Thermal Analysis. All thermal analysis were conducted using a Dupont 2000 Thermal Analyzer. Thermogravimetric Analysis (TGA) was carried out on a DuPont TGA 951 under nitrogen with a heating rate of 10 °C per minute. Thermomechanical Analysis (TMA) was carried out on a DuPont TMA 2940 under nitrogen with a heating rate of 3 °C per minute and a mechanical force of 0.1 N. Differential Scanning Calorimetry (DSC) was carried out on a DuPont DSC 912 under nitrogen with a heating rate of 10 °C per minute.

RESULTS

Synthesis and Characterization of Polymerizable POSS Macromers. The synthesis of POSS macromers containing a single styrene group has been achieved by derivatizing the known trisilanols $R_7Si_7O_9(OH)_3$, **1a** (R = cyclohexyl) and **1b** (R = cyclopentyl).[4] These trisilanols are quantitatively "corner capped" with either styryltrichlorosilane or styrylethyltrichlorosilane (a 3:1 mixture of *para*- and *meta*- isomers), in high isolated yields (Figure 1).[5] The resulting macromers **2a-2d** have a spherical (Si_8O_{12}) inorganic core, surrounded by seven inert alkyl groups for solubility and one reactive styryl group for polymerization. These four similar macromers allows us to synthesize and compare polymers which differ only by the distance between the POSS cage and the styrene backbone, or by the inert "R" groups. An immediately obvious difference between the macromers is that the cyclohexyl derivatives **2a** and **2c** are about twice as soluble the cyclopentyl counterparts **2b** and **2d** and these solubility differences are magnified for the homopolymers of **2a-2d**. Solubility differences among various $R_8Si_8O_{12}$ molecules has been noted before and is believed to be related to the lattice energy of the crystal. A subtle change in R-group can result in significant solubility differences.[2b,3d]

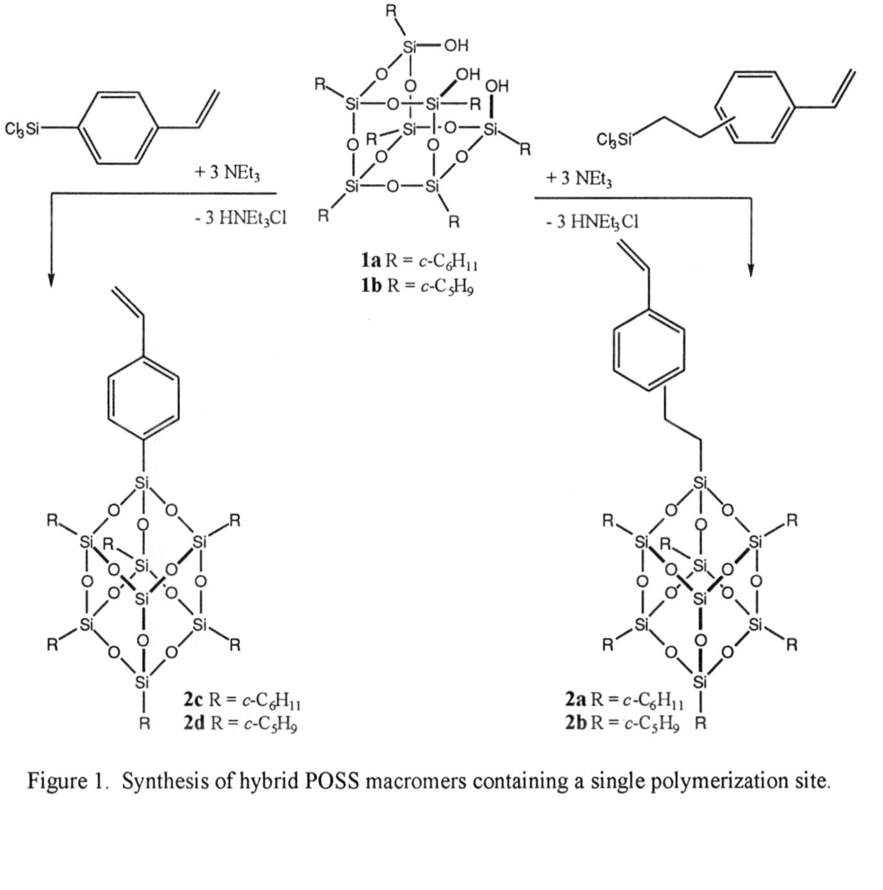

Figure 1. Synthesis of hybrid POSS macromers containing a single polymerization site.

The POSS styryl macromers **2a-2d** are readily characterized by NMR spectroscopy (Figure 2). For example, the ^{1}H NMR spectra of **2c** and **2d** show five distinct resonances for the styryl group and three broad peaks for the cycylohexyls, while the ^{29}Si NMR spectra show 3 resonances in a 4:3:1 ratio (1:3:3:1 ratio predicted) for the cage silicons. The spectra of **2a** and **2b** are more complex as there are two isomers (3:1 ratio) for *para-* and *meta-* substitution at the styrene group and an impurity due to a non-reactive ethylbenzene derivative (The starting trichlorosilane is 86 % $Cl_3SiCH_2CH_2PhCH=CH_2$ and 14 % $Cl_3SiCH_2CH_2PhCH_2CH_3$).

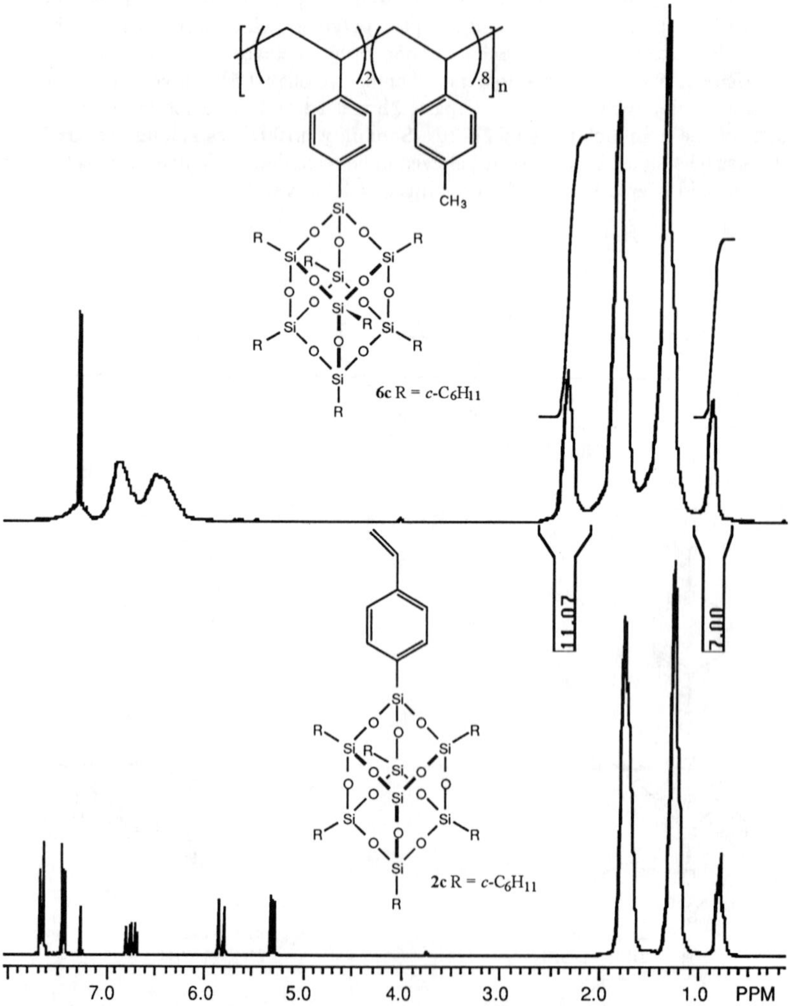

Figure 2. ^{1}H NMR spectra of POSS styryl macromer **2c** (lower trace) and its copolymer with 79 mole % 4-Methylstyrene **6c** (upper trace).

Polymerizations and Characterization The four POSS macromers **2a-2d** have been used to prepare a series of pendent POSS styryl polymers and copolymers by carrying out AIBN-initiated free radical polymerizations. The homopolymers **(3a-3d)** and copolymers with varying proportions of 4-methylstyrene **(4a,b-7a,b)** have been synthesized as shown in Figure 3. For comparison purposes, the homopolymer of 4-methylstyrene, **8**, was also synthesized under the same reaction conditions. These polymers may be viewed as organic polymers filled with monodisperse inorganic particulate which are covalently bound to the polymer backbone.

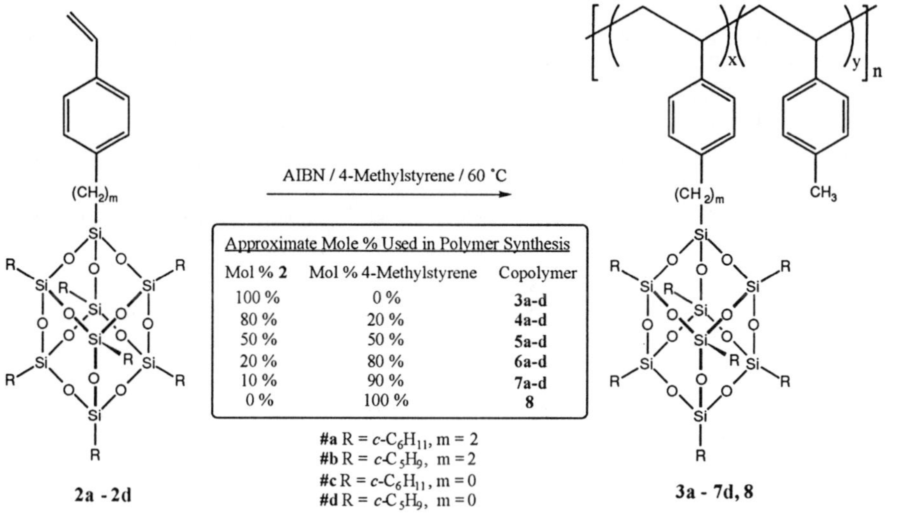

Approximate Mole % Used in Polymer Synthesis

Mol % **2**	Mol % 4-Methylstyrene	Copolymer
100 %	0 %	**3a-d**
80 %	20 %	**4a-d**
50 %	50 %	**5a-d**
20 %	80 %	**6a-d**
10 %	90 %	**7a-d**
0 %	100 %	**8**

Figure 3. Free radical polymerization of hybrid POSS macromers into random copolymers.

All of the polymers are soluble in THF except for the homopolymers **3b** and **3d** (*cyclopentyl*-derivatives) and **4d** (*cyclopentyl*-derivative with 23 mol % 4-methylstyrene). Of these three THF insoluble polymers **3d** is unique as, unlike **3b** and **4d**, it cannot be dissolved in $CHCl_3$. We have previously synthesized POSS methylmethacrylate homopolymers[3d] and found that the *cyclopentyl*-derivatives are also insoluble in THF and $CHCl_3$. This subtle change in inert "R" group which results in drastic solubility differences is perhaps due to intermolecular POSS-POSS interactions forming a type of physical crosslink between polymer chains, with the *cyclopentyl*-derivatives having a stronger interaction than their *cyclohexyl*-counterparts. Above a certain percentage of POSS incorporation, these interactions can result in an insoluble material.

Compositions, molecular weights, and thermal properties of these polymers are summarized in Table 1. ^1H NMR spectra were used to determine the actual compositional % of each comonomer in the copolymer samples. This was achieved by comparison of the integrated signals for the methyl group of the 4-methylstyrene at 2.3 ppm with the signal for ipso-CH resonance of the "R" groups at 0.8 ppm (Figure 2). Interestingly, the mole % POSS and methylstyrene are similar to the % loading of the monomers prior to polymerization. As all of the polymerizations were carried out to about 50 % conversion, this result seems to indicate that the reactivity ratios of these styryl POSS macromers are similar to that of methylstyrene. Further studies are underway to accurately determine reactivity ratios.

Table 1. Compositions, Molecular Weights and Thermal Transitions of POSS Styryl Polymers

Polymer (R, m)[a]	Initial Ratio[b]	Mol % POSS[c]	Wt % POSS	M_n x10^{-3} [d]	M_w x10^{-3} [d]	DP[e]	T_g °C[f]	T_g °C[g]	T_{dec} °C[h]
3a (Cy, 2)	100:0	100	100	150	430	133	396	i	445
3b (Cp, 2)	100:0	100	100	200	930	194	343	i	423
3c (Cy, 0)	100:0	100	100	59	140	54	370	i	443
3d (Cp, 0)	100:0	100	100	insoluble		—	400	i	427
4a (Cy, 2)	77:23	80	97	150	260	162	202	i	434
4b (Cp, 2)	77:23	79	97	87	170	104	272	i	429
4c (Cy, 0)	80:20	60	93	56	180	79	312	i	420
4d (Cp, 0)	80:20	77	97	50	180	62	284	i	421
5a (Cy, 2)	46:54	46	89	210	370	360	179	i	426
5b (Cp, 2)	46:54	44	87	110	200	211	241	i	423
5c (Cy, 0)	50:50	46	89	52	98	91	246	i	427
5d (Cp, 0)	50:50	53	91	60	120	102	232	i	429
6a (Cy, 2)	18:82	16	65	66	110	236	148	115	410
6b (Cp, 2)	18:82	16	62	100	150	378	126, 199	97, 168	407
6c (Cy, 0)	20:80	21	71	35	52	108	182	173	410
6d (Cp, 0)	20:80	24	73	35	53	106	177	159	412
7a (Cy, 2)	9:91	7.8	45	41	71	208	132	110	402
7b (Cp, 2)	9:91	7.8	42	45	72	238	127	95	399
7c (Cy, 0)	10:90	12	56	26	37	110	162	142	397
7d (Cp, 0)	10:90	13	56	28	41	120	143	132	401
8	0:100	0	0	21	34	178	116	93	388

(a) R is the inert alkyl group at silicon (Cy = c-C_6H_{11} Cp = c-C_5H_9) and m is the number of methylene units (CH_2's) between the POSS cage and the styrene group. (b) The mole % POSS to mole % 4-methylstyrene used in the copolymer synthesis. (c) The mole % POSS in the polymer determined from ^{1}H NMR spectra. (d) Molecular weights determined in THF, except for **3b** and **4d** in $CHCl_3$. (e) Degree of polymerization calculated from the number average molcular weights and mole % POSS in each copolymer. (f) T_g determined from the softening point of a pressed pellet on second heating using thermomechanical analysis. Note that **6b** is unique in that it showed two T_g's. The homopolymers **3a-3d** show an initial softening, then harden as decomposition temperatures are reached. (g) T_g determined from powdered polymer on second heating using differential scanning calorimetry. (h) T_{dec} reported as the temperature at which 10 % weight loss has occurred using thermal gravimetric analysis. (i) At or above 50 mole % (90 weight %) POSS loadings, a T_g is not observable by DSC.

Thermal Characterization The thermal characterization of the POSS materials is tabulated in Table 1. The polymers are grouped together for easy comparison as the % POSS and the two variables are changed - variation in inert R group and variation in the distance from the POSS group to the styrene backbone. Each group of four polymers shows the mole % and wt % of POSS styryl comonomer incorporated into each polymer. The disparity between mole % and weight % POSS is, of course, due to the large differences in molecular weight between the two comonomers, 118 g/mole for methylstyrene and over 1000 g/mole for the POSS derivatives.

Perhaps most apparent in Table 1 is the large effect that POSS groups have on thermal transitions. Polymer softening points increase with increasing POSS content, as do the decomposition points. This can be explained by considering the large size[2] of the pendent POSS groups as restricting chain motions and therefore, raising the softening temperatures. The small increase in polymer decomposition point might be a thermoconductive effect resulting from increased free volume of the chains due to the bulk of the POSS cages.

An interesting effect for polymers with high POSS loadings (around 50 mole % or 90 weight %) is that the DSC data do not show a glass transition. If the POSS cages are associating with each other to form physical crosslinks, then a (reversible) network material would result. In a material with 90 weight % POSS groups, the POSS-POSS interactions are so numerous that the material does not soften and flow unless an external force is applied to the polymer. The TMA data for these polymers clearly demonstrates that the polymers do soften, although at these high POSS loadings, some are too viscous to completely flow on the timescale of the experiment. The homopolymers (100 % POSS) show an initial softening, but then harden as decomposition temperatures are reached.

The effect of increasing the distance from the POSS cage to the backbone (by an ethyl linkage) generally results in a lowering of the softening point. This intuitively makes sense, since the extra ethyl-linkage should result in a less restricted polymer chain, and therefore affect the T_g less. However, a more subtle effect due to the inert cyclohexyl and cyclopentyl groups also comes into play. For those polymers where the POSS group is linked to the polystyrene backbone with an ethyl linkage (Table 1, m = 2), the cyclopentyl derivatives generally have higher T_g's than their cyclohexyl counterparts. Conversely, for those polymers without the ethyl linkage (Table 1, m = 0), the opposite is true; the cyclopentyl derivatives generally have lower T_g's than their cyclohexyl counterparts. A working picture of these polymers is that the POSS cages associate with themselves to form a physically crosslinked network. The strength of this association depends both on the R-group and the distance from the POSS cage to the styrene backbone.

A further complication is that one of the copolymers **6b** (cyclopentyl derivative with 16 mole % POSS) is unique in that it shows two distinct T_g's. This would seem to indicate that **6b** is a phase separated material (one much richer in POSS than the other), and that the free radical polymerization process is capable of generating significant amounts of "blockiness" in this particular copolymer.

CONCLUSIONS

The synthesis of soluble, thermoplastic hybrid materials from well-defined inorganic clusters containing only a single polymerization site has been developed. Four styryl-based POSS derivatives were successfully copolymerized with 4-methylstyrene and analysis of the resulting copolymers revealed some interesting trends: 1) Even though the POSS styryl macromers are enormous compared to 4-methylstyrene, these molecules appear to have similar reactivity ratios. 2) Incorporation of the large pendent POSS group results in reduced chain mobility and increased polymer softening points. As a result, T_g's can be raised to the decomposition point of the polystyryl-backbone. 3) Shorter linkages connecting the POSS moiety to the polymer backbone have larger effects on chain mobility. In summary, the large pendent inorganic groups appear to associate with themselves to form (reversible) physical crosslinks between polymer chains, which in turn affect the polymer properties. This suggests the possibility of synthesizing elastomeric hybrid materials wherein POSS-POSS interactions could produce a hard block phase separated from a soft block. We are currently exploring this possibility by developing living polymerization methods for our POSS monomers to enable the preparation of phase separated block copolymers.[6]

ACKNOWLEDGMENTS

This research was supported by the Air Force Office of Scientific Research, Directorate of Chemistry and Life Sciences, and the Phillips Laboratory, Propulsion Directorate.

REFERENCES

1. Reviews and leading references for recent progress in hybrid materials are: (a) Hybrid Organic-Inorganic Composites, edited by J.E. Mark, C.Y-C. Lee and P.A. Bianconi, (ACS Symposium Series **585**, American Chemical Society: Washington DC 1995). (b) L. Mascia, Trends Polym. Sci. **3**, 61 (1995). (c) C. Sanchez and F. Ribot, Nouv. J. Chem. **18**, 1007 (1994).

2. The diameter swept out by a $R_8Si_8O_{12}$ POSS molecule is approximately 15 Å with an inner Si—Si diameter of 5.4 Å. (a) K. Larsson, Ark. Kemi, **16**, 203 (1960). (b) F.J. Feher and T.A. Budzichowski, J. Organomet. Chem. **373**, 153 (1989). (c) T.P.E. Auf der Hyde, H.-B. Burgi, H. Burgy and K.W. Tornroos, Chimia, **45**, 38 (1991)

3. (a) J.D. Lichtenhan, N.Q. Vu, J.A. Carter, J.W. Gilman and F.J. Feher, Macromolecules, **26**, 2141, (1993). (b) J.D. Lichtenhan, Comments Inorg. Chem. **17**, 115 (1995). (c) T.S. Haddad and J.D. Lichtenhan, J. Inorg. Organomet. Polym. **5**, 237 (1995). (d) J.D. Lichtenhan, Y.A. Otonari and M.J. Carr, Macromolecules **28**, 8435 (1995). (e) R. Mantz, P.F. Jones, K.P. Chaffee, J.D. Lichtenhan, J.W. Gilman, I. Ismail and M.J. Burmeister, Chem Mater. in press (1996).

4. (a) F.J. Feher, D.A. Newman and J.F. Walzer, J. Am. Chem. Soc. **111**, 1741 (1989). (b) F.J. Feher, T.A. Budzichowski, R.L. Blanski, K.J. Weller and J.W. Ziller, Organometallics **10**, 2526 (1991). (c) J.F. Brown Jr. and L.H. Vogt Jr. J. Am. Chem. Soc. **87**, 4313 (1965).

5. Isolated yields were: **2a** (97 %), **2b** (95 %), **2c** (78 %), **2d** (87 %).

6. For a discussion on elastomeric phase-separated triblock polymers see G. Odian Principles of Polymerization (J. Wiley & Sons: New York, 1991) pp. 148-149, 426.

INTRAMOLECULAR CONDENSATION REACTIONS OF α, ω-BIS(TRIETHOXY-SILYL)ALKANES. FORMATION OF CYCLIC DISILSESQUIOXANES

DOUGLAS A. LOY*† , JOSEPH P. CARPENTER*, SHARON A. MYERS*, ROGER A. ASSINK*, JAMES H. SMALL**, JOHN GREAVES*** AND KENNETH J. SHEA***
*Properties of Organic Materials and **Organic Materials Processing Departments, Sandia National Laboratories, Albuquerque, NM 87185-1407, daloy@sandia.gov
***Department of Chemistry, University of California Irvine, Irvine, CA 92717-2025

ABSTRACT

Under acidic sol-gel polymerization conditions, 1,3-bis(triethoxysilyl)-propane **1** and 1,4-bis(triethoxysilyl)butane **2** were shown to preferentially form cyclic disilsesquioxanes **3** and **4** rather than the expected 1,3-propylene- and 1,4-butylene-bridged polysilsesquioxane gels. Formation of **3** and **4** is driven by a combination of an intramolecular cyclization to six and seven membered rings, and a pronounced reduction in reactivity under acidic conditions as a function of increasing degree of condensation. The ease with which these relatively unreactive cyclic monomers and dimers are formed (under acidic conditions) helps to explain the difficulties in forming gels from **1** and **2**. The stability of cyclic disilsesquioxanes was confirmed with the synthesis of **3** and **4** in gram quantities; the cyclic disilsesquioxanes react slowly to give tricyclic dimers containing a thermodynamically stable eight membered siloxane ring. Continued reactions were shown to perserve the cyclic structure, opening up the possiblity of utilizing cyclic disilsesquioxanes as sol-gel monomers. Preliminary polymerization studies with these new, carbohydrate-like monomers revealed the formation of network *poly(cyclic disilsesquioxanes)* under acidic conditions and polymerization with ring-opening under basic conditions.

INTRODUCTION

Sol-gel polymerization of α, ω-bis(triethoxysilyl)alkanes normally leads to alkylene-bridged polysilsesquioxanes as insoluble, highly crosslinked gels [1]. Hydrolysis of the six ethoxide groups on each monomer gives silanols that subsequently condense to form a network of siloxane bonds. Unlike most sol-gel monomers that possess a single alkoxysilane functionality, these flexible hydrocarbon-bridged monomers can participate not only in intermolecular condensation reactions that lead to oligomers and polymeric networks, but in *intramolecular* condensation reactions that afford cyclic disilsesquioxanes as well (Scheme 1). Partitioning between intermolecular and intramolecular pathways will be strongly affected by 1) the mechanism of hydrolysis (acid versus base catalysis), 2) the concentration of α, ω-bis(triethoxysilyl)alkane, and 3) the length of the alkylene-bridging group. The degree to which one of these reaction pathways is favored over the other may be an important determinant in how the network is assembled and ultimately of the final morphologies of the network polymers.

The first evidence for the formation of cyclic disilsesquioxanes, rather than polymeric gels, came from the studies of the sol-gel polymerization of α, ω-bis(triethoxysilyl)alkanes, $(EtO)_3Si\text{-}(CH_2)_n\text{-}Si(OEt)_3$ (Figure 1) [2]. Under alkaline conditions (10.8 mol% NaOH), these monomers (0.4 M) reacted rapidly with water (6 equivalents), regardless of the length of the bridging group, to afford viscous solutions of growing, highly branched polysilsesquioxanes that formed gels within a few hours. Under acidic conditions (10.8 mol% HCl), however, it was discovered that solutions of the propylene- (**1**) and butylene- (**2**) bridged monomers required more than 4000 hours for gels to form [2, 3]. In this paper, we describe the discovery that the extraordinarily slow rates of gelation exhibited by **1** and **2** resulting from dominance of the intramolecular reaction pathway leading to the cyclic disilsesquioxanes **3** and **4**, respectively, as the major products. In addition, isolation of these novel siloxane analogs of carbohydrates and their sol-gel polymerization chemistries are described.

Mat. Res. Soc. Symp. Proc. Vol. 435

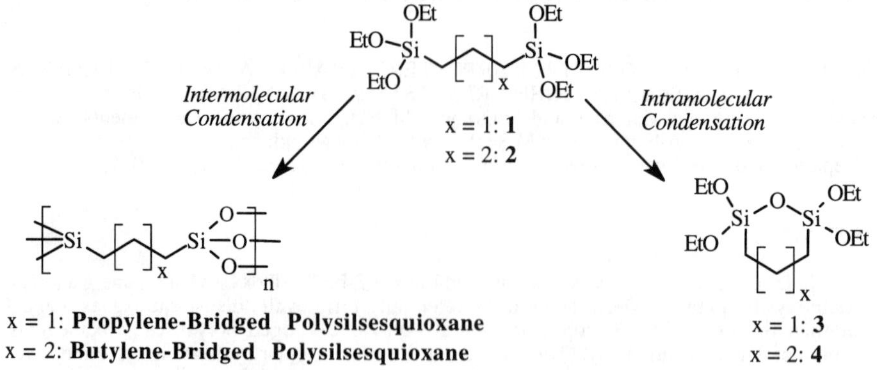

Scheme 1. Inter- and intramolecular condensation pathways for *α, ω*-bis(triethoxysilyl)alkanes.

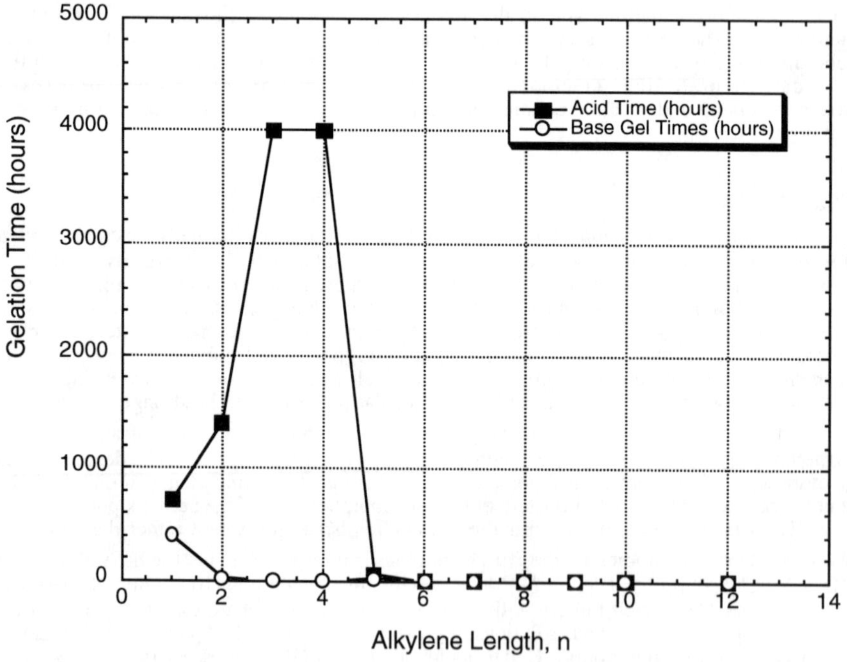

Figure 1. Gelation times for *α, ω*-bis(triethoxysilyl)alkanes, $(EtO)_3Si\text{-}(CH_2)_n\text{-}Si(OEt)_3$ (n = 2-14). Sol-gel polymerizations were carried out at 0.4 M monomer concentration in ethanol with 6 H_2O, 10.8 mol% HCl or NaOH.

EXPERIMENTAL

General Procedures

Reagents were distilled from appropriate drying agents: magnesium (ethanol) and calcium hydride (benzene). Anhydrous tetrahydrofuran (THF, Sure/Seal™), 4-triethoxysilylbutene, chloroplatinic acid, and 3-triethoxysilylpropene was purchased (Aldrich) and used as received. Solution NMR spectra were collected on a Brüker AM-300 spectrometer and referenced to the solvents' residual proton signal (^{1}H, ^{13}C) or internal tetramethylsilane (^{29}Si). The ^{29}Si and ^{13}C CPMAS NMR spectra were acquired on a Brüker AMX-400 spectrometer using $[Si_8O_{12}](OSiMe_3)_8$ (Q_8M_8)[12] and glycine as external references. Infrared spectra were collected on a Perkin-Elmer Model 1750 FTIR. Mass spectra were collected on a VG Autospec using chemical ionization (NH_3 or isobutane).

Monomer Syntheses

Monomer 1

A flask was charged with 3-triethoxysilylpropene (50 g, 245 mmol), triethoxysilane (48.3 g, 294 mmol), benzene (100 mL), and chloroplatinic acid (0.100 g, 0.24 mmol). While stirring for 24 h, the solution gradually turned yellow-brown. Additional chloroplatinic acid (0.100 g, 0.24 mmol) was added to the reaction mixture, which was stirred for another 24 h. Benzene was removed *in vacuo* leaving a dark liquid that was distilled (0.025 Torr, 85-90 °C) to yield a clear colorless liquid (40.5 g, 45%). ^{1}H NMR (300 MHz, $CDCl_3$): δ = 3.78 (q, 12H, O**CH_2**CH_3, J = 7.0 Hz), 1.84 (m, 2H, SiCH$_2$**CH_2**CH$_2$Si), 1.15 (t, 18H, OCH$_2$**CH_3**, J = 7.0 Hz), 0.85 (t, 4H, Si**CH_2**CH$_2$**CH_2**Si, J = 8.2 Hz). ^{13}C NMR (75 MHz, $CDCl_3$): δ = 58.40 (O**C**H$_2$CH$_3$), 18.54 (OCH$_2$**C**H$_3$), 17.20 (SiCH$_2$**C**H$_2$CH$_2$Si), 14.99 (Si**C**H$_2$CH$_2$**C**H$_2$Si). IR (neat): ν = 2975, 2928, 2884, 1484, 1444, 1391, 1366, 1341, 1295, 1229, 1168, 1078, 960, 900, 820, 774, 710 cm^{-1}. ^{29}Si NMR (99 MHz, EtOH): δ = -45.76. HRMS (CI, isobutane): Calcd for $C_{15}H_{37}O_6Si_2$ $[M + H]^+$: 369.2129. Found: 369.2113.

Monomer 2

To a solution of 4-triethoxysilylbutene (25 g, 114 mmol), triethoxysilane (24.3 g, 148 mmol), and benzene (100 mL) was added chloroplatinic acid (0.10 g, 0.24 mmol). Within 10 min, the solution turned yellow and an exothermic reaction occurred. After stirring for 48 h, chloroplatinic acid (0.05 g, 0.12 mmol) was again added and the mixture was refluxed for 20 h. The benzene was removed *in vacuo*. Distillation (95-105 °C, 0.050 Torr) afforded a clear colorless liquid (21.1 g, 48%). ^{1}H NMR (300 MHz, $CDCl_3$): δ = 3.76 (q, 12H), 1.41 (m, 4H), 1.18 (t, 18H), 0.59 (m, 4H). ^{13}C NMR (75.48 MHz, $CDCl_3$): δ = 58.24, 26.26, 18.24, 10.06. ^{29}Si NMR (99 MHz, EtOH): δ = -45.36. HRMS (CI, isobutane): Calcd for $C_{16}H_{39}O_6Si_2$ $[M + H]^+$: 383.2285. Found: 383.2276.

Cyclic Disilsesquioxane 3

To a solution of **1** (15 g, 40.7 mmol) in ethanol (150 mL) was added 1 N HCl (0.733 mL) dissolved in ethanol (50 mL). After stirring for 24 h, the ethanol was removed *in vacuo*. The remaining liquid was distilled (0.05 Torr, 47-55 °C) yielding a clear colorless liquid (**3**, 9.67 g, 81%). ^{1}H NMR (300 MHz, $CDCl_3$): δ = 3.85 (q, 8H, O**CH_2**CH_3, J = 7.0 Hz), 1.78 (m, 2H, SiCH$_2$**CH_2**CH$_2$Si), 1.23 (t, 12H, OCH$_2$**CH_3**, J = 7.0 Hz), 0.71 (t, 4H, Si**CH_2**CH$_2$**CH_2**Si, J = 3.8 Hz). ^{13}C NMR (75 MHz, $CDCl_3$): δ = 58.34 (O**C**H$_2$CH$_3$), 18.23 (OCH$_2$**C**H$_3$), 17.17 (SiCH$_2$**C**H$_2$CH$_2$Si), 11.23 (Si**C**H$_2$CH$_2$**C**H$_2$Si). ^{29}Si NMR (99 MHz, EtOH): δ = -47.25 (T^{1}_{cyclic}). IR (neat): ν = 2977, 2923, 2889, 1442, 1392, 1246, 1169, 1107, 1084, 1007, 968, 899, 787, 756 cm^{-1}. HRMS (CI, ammonia): Calcd for $[C_{11}H_{26}O_5Si_2]^+$: 294.1319. Found: 294.1322.

Cyclic Disilsesquioxane 4

To a solution of **2** (15 g, 39.2 mmol) in ethanol (185 mL) was added 1 N HCl (0.705 mL) in ethanol (10 mL). After stirring for 48 h, the ethanol was removed *in vacuo*. The remaining liquid was distilled (0.05 Torr, 65-70 °C) to give a clear colorless liquid (**4**, 8.22 g, 68%). ^{1}H NMR (300 MHz, $CDCl_3$): δ = 3.83 (q, 8H, $OC\mathbf{H}_2CH_3$, J = 7.0 Hz), 1.64 (m, 4H, $SiCH_2\mathbf{CH_2CH_2}CH_2Si$), 1.16 (t, 12H, $OCH_2C\mathbf{H}_3$, J = 7.0 Hz), 0.76 (m, 4H, $SiC\mathbf{H_2}CH_2CH_2C\mathbf{H}_2Si$). ^{13}C NMR (75 MHz, $CDCl_3$): δ = 58.39 ($O\mathbf{C}H_2CH_3$), 26.67 ($SiCH_2\mathbf{CH_2CH_2}CH_2Si$), 18.58 ($OCH_2\mathbf{C}H_3$), 10.86 ($Si\mathbf{C}H_2CH_2CH_2\mathbf{C}H_2Si$). ^{29}Si NMR (79.5 MHz, $CDCl_3$): δ = -49.06 (T^1_{cyclic}). IR (neat): v = 2975, 2927, 2885, 1484, 1444, 1392, 1307, 1296, 1212, 1169, 1106, 1083, 1034, 961, 884, 849, 800, 762, 737, 652 cm^{-1}. HRMS (CI, isobutane): Calcd for $[C_{12}H_{28}O_5Si_2]^+$: 308.1475. Found: 308.1473.

Sol-Gel Polymerization of Cyclic Monomers.

General Procedure.

All polymerizations were performed at 0.4 M monomer concentration in ethanol. The solutions were prepared in 5 mL volumetric flasks before being transferred to polyethylene bottles. After gelation, the samples were allowed to age for 2 weeks. The gels were then fractured and washed with 3 x 100 mL of H_2O and 2 x 50 mL of ether. The materials were then dried in vacuo at 100 °C for 24 h.

Acid-Catalyzed Polymerizations of 3

To a solution of **3** (0.589 g, 2.00 mmol) in ethanol (2.0 mL) was added 1 N HCl (0.72 mL) dissolved in ethanol (2.0 mL). The volume of the solution was then brought to exactly 5.0 mL with ethanol. A rigid gel formed after 30 days. After drying, an opaque white gel was obtained (268 mg, 91%). ^{13}C CP MAS-NMR (50.20 MHz) δ = 16.5 ($SiCH_2\mathbf{CH_2}CH_2Si$), 13.4 ($Si\mathbf{CH_2}CH_2\mathbf{CH_2}Si$); ^{29}Si CP MAS-NMR (39.65 MHz) δ = -49.74 (T^2_{cyclic}), -58.72 (T^3_{cyclic}); IR (KBr) 2927, 1638, 1413, 1253, 1116, 1018, 902, 754 cm^{-1}.

Base-Catalyzed Polymerizations of 3

To a solution of **3** (0.589 g, 2.00 mmol) in ethanol (2.0 mL) was added 1 N NaOH (0.144 mL) dissolved in ethanol (2.0 mL). The volume of the solution was then brought to exactly 5.0 mL with additional ethanol. Gelation occurred within 1 h. After drying, white monolithic gel fragments were obtained (303 mg, 104%). ^{29}Si CP MAS-NMR (39.65 MHz) δ = -60.78 ($T^3_{acyclic}$); IR (KBr) 2927, 1251, 1111, 1019, 903, 759 cm^{-1}.

Acid-Catalyzed Polymerizations of 4.

To a solution of **4** (0.617 g, 2.00 mmol) in ethanol (2.0 mL) was added 1 N HCl (0.72 mL) dissolved in ethanol (2.0 mL). Ethanol was added to bring the volume to exactly 5.0 mL. After 16 days a rigid gel formed. The final gel product was an opaque white solid (312 mg, 98%). ^{13}C CP MAS-NMR (50.20 MHz) δ = 58.0 ($O\mathbf{C}H_2CH_3$), 25.0 ($SiCH_2\mathbf{CH_2CH_2}CH_2Si$), 13.3 ($Si\mathbf{CH_2}CH_2CH_2\mathbf{CH_2}Si$); ^{29}Si CP MAS-NMR (39.65 MHz) δ = -58.81 (T^3_{cyclic}), -62.75 ($T^3_{acyclic}$); IR (KBr) 2931, 1655, 1638, 1459, 1407, 1029, 852, 795 cm^{-1}.

Base-Catalyzed Polymerizations of 4.

To a solution of **4** (0.617 g, 2.00 mmol) in ethanol (2.0 mL) was added 1 N NaOH (0.144 mL) dissolved in ethanol (2.0 mL). The volume of the solution was brought to 5.0 mL with ethanol. After 2 hours, a rigid gel had formed. The dried gel was an opaque white solid (344 mg, 108%). ^{29}Si CP MAS-NMR (39.65 MHz) δ = -63.74 ($T^3_{acyclic}$); IR (KBr) 2932, 1719, 1510, 1459, 1408, 1311, 1127, 1039, 853, 797 cm^{-1}.

RESULTS AND DISCUSSION

Identification of the gelation "bottleneck" generated during the acid-catalyzed hydrolysis and condensation reactions of **1** and **2** was accomplished by both solution ^{29}Si NMR spectroscopy and mass spectrometry of the reaction solutions. Because no significant changes in viscosity were observed over periods of weeks, we felt that the reaction was most likely forming soluble oligomeric products that would be easily characterized by these solution techniques. One possible structure proposed as the hydrolysis products of **1** (or **2**), based on the precedents for simple organotrialkoxysilanes forming polyhedral oligosilsesquioxanes [4], was the cubic tetramer (Scheme 2).

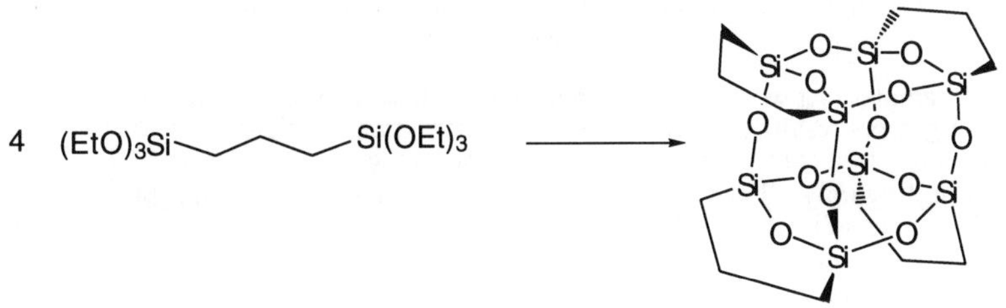

Scheme 2. A proposed, but not realized, structure for the product of the acid-catalyzed oligomerization of **1**.

As polyhedral oligosilsesquioxanes are known to retard or prevent gelation of organotrialkoxysilanes, this fully condensed (T^3) silsesquioxane structure would easily explain anomalous polymerization chemistry of **1** and **2**. The presence of this or any other cyclic species that was preferentially formed would be readily apparent in both NMR and mass spectrometric techniques.

Solution ^{29}Si NMR Spectrscopy.

^{29}Si NMR spectra of the sol-gel reactions (1 M monomer, 1 H_2O, 1.8 mol% HCl) were collected at regular intervals to permit monitoring of the evolution of early hydrolysis and condensation products. The spectra obtained from the hydrolysis of the propylene- and butylene-bridged monomers were dominated by a single resonance at -47.25 ppm for **1** and -48.25 ppm for **2**. The location of these resonances was unusual in that they lie upfield of the expected ^{29}Si NMR chemical shifts of unreacted monomer (-45 ppm; T^0) and significantly downfield of a silsesquioxane with a single, *acyclic* siloxane bond (-52 ppm; T^1). The chemical shifts are *not* consistent with a cyclic (T^3) tetramer (scheme 2). As ^{29}Si NMR chemical shifts are particularly sensitive to ring strain and Si-O-Si bond angles [5], the resonances are in accord with cyclic disilsesquioxanes **3** and **4**, with single siloxane bonds (T^1) linking the silicon atoms [6]. The T^1_{cyclic} resonances persisted for hours even in the presence of excess H_2O indicating that the cyclic species were relatively stable towards ring opening. These spectra are simpler than, and should be contrasted with, those normally obtained from the hydrolysis and condensation of organotrialkoxysilanes R-Si(OEt)$_3$ [7], where, within minutes of the addition of aqueous acid to propyltriethoxysilane, there were more than eight distinct $T^1_{acyclic}$ peaks between -50 ppm and -54 ppm and $T^2_{acyclic}$ peaks near -60 ppm.

Mass Spectrometric Analysis.

Mass spectrometric analysis of the acid-catalyzed reaction solutions obtained from **1** and **2** confirmed that the soluble products were the cyclic disilsesquioxanes **3** and **4**. For example, within one minute of the addition of aqueous HCl to **1**, the mass peaks characteristic of **1** were completely replaced by peaks at m/z 295 and m/z 312 due to $[M+H]^+$ and $[M+NH_4]^+$ ions of the cyclic disilsesquioxane **3**. As in the NMR study, none of the hydrolysis products of **1** are observed in the mass spectra suggesting that cyclization occurs rapidly relative to the rate of further hydrolysis (Scheme 3).

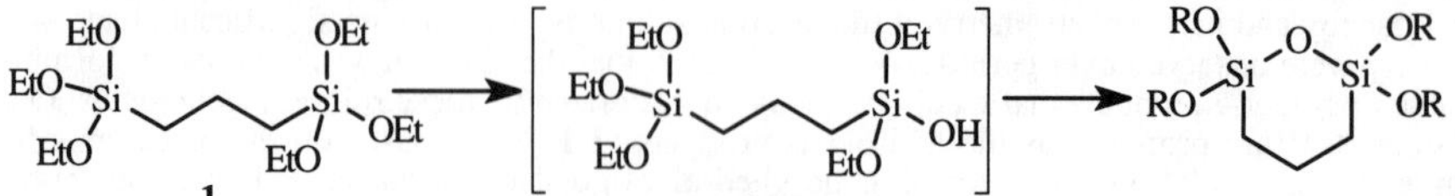

1 **1a** **3**

Scheme 3. Hydrolysis and cyclization of 1,3-bis(triethoxysilyl)propane **1** to give six-membered cyclic disilsesquioxane **3**.

For samples that were allowed to react longer than one minute before mass spectrometric analysis, oligomers resulting from continuing condensation reactions were observed. The first oligomers to be observed were the acyclic (**5a-c**) and cyclic (**6a-c**) dimers of **3**. Presumably, two molecules of **3** hydrolyzed and condensed to form the dimer **5a** (m/z 532, $[M + NH_4]^+$) and its hydrolysis products **5b** and **5c**. Intramolecular cyclization of **5a**, driven by the formation of a tetrasiloxane ring, affords the tricyclic dimer **6a** (m/z 458, $[M + NH_4]^+$) (Scheme 4). After 29 hours of reaction time, the mass spectrum was dominated by the cyclic disilsesquioxane dimer **6a** and its hydrolysis products **6b** and **6c**.

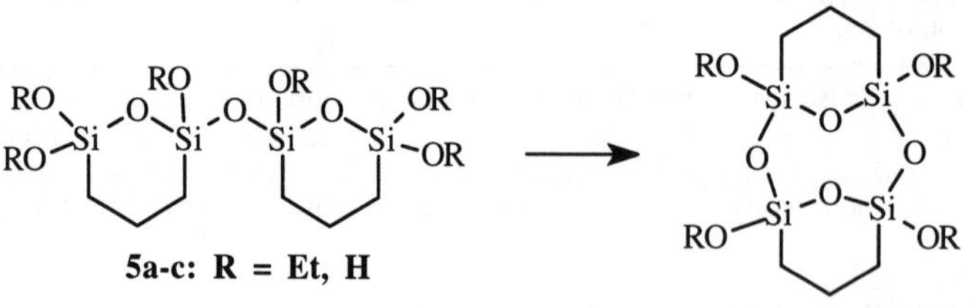

5a-c: R = Et, H

6a-c: R = Et, H

Scheme 4. Intramolecular cyclization of dimer(s) **5a-c** to give tricyclic dimer(s) **6a-c** (Note: thermodynamically favored eight-membered tetrasiloxane ring.).

Two additional groups of peaks corresponding to trimeric (m/z 532-678) and tetrameric (m/z 710-870) species were also observed, indicating that continued oligomerization *without* ring opening is possible.

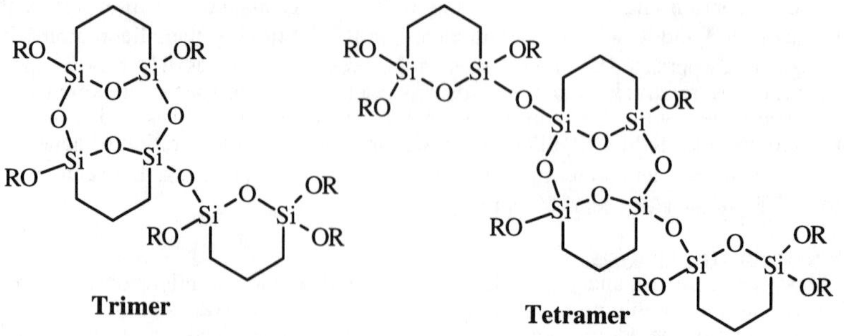

Trimer **Tetramer**

Similarly, mass spectrometry revealed that hydrolysis and condensation of the butylene-bridged monomer **2** also generated cyclic disilsesquioxanes and cyclic oligomers in abundance (Scheme 5). The same pattern of cyclization followed by oligomerization seen with **1** was also observed for **2**. Hydrolysis and condensation of **2** with one equivalent of H_2O in the presence of aqueous acid led to the preferential formation of the seven membered cyclic disilsesquioxane **4**, its cyclic dimer, and trimer along with their hydrolysis products. Formation of **4** was slow compared with **3**, presumably due to greater transannular interactions in the former.

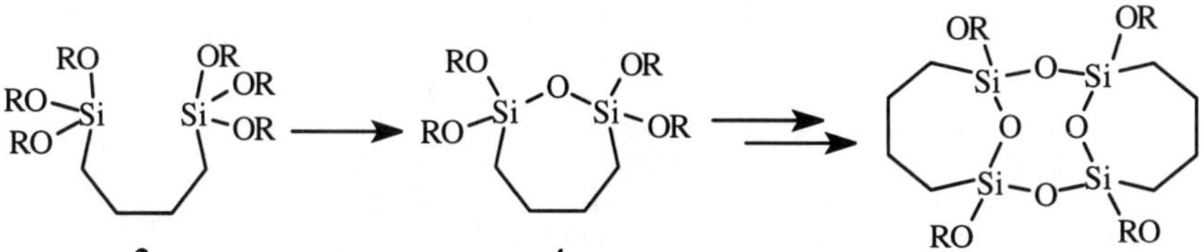

2 **4**

Scheme 5. Intramolecular cyclization pathway for the formation of seven-membered cyclic disilsesquioxane **4** and its dimer.

Synthesis of 3 and 4.

Conclusive proof for the existence of the cyclic disilsesquioxanes was obtained by the deliberate synthesis and isolation of gram quantities of **3** and **4**. When one equivalent of H_2O (1 N HCl) was used in the hydrolysis experiments instead of the excess needed for polymerization, **1** and **2** were *quantitatively* converted into the monocyclic disilsesquioxanes within a few hours. The cyclic monomers were isolated in excellent yields by distillation as clear colorless oils. ^{29}Si NMR spectra for **3** and **4** showed the same downfield shifted T^1 resonances observed in the reaction mixtures.

Polymerizations of Cyclic Disilsesquioxanes.

Despite their reduced reactivity under the acidic conditions normally used to polymerize bridged silsesquioxane monomers, the cyclic disilsesquioxanes are still tetrafunctional sol-gel monomers with intriguing similarities to carbohydrate building blocks (Scheme 6).

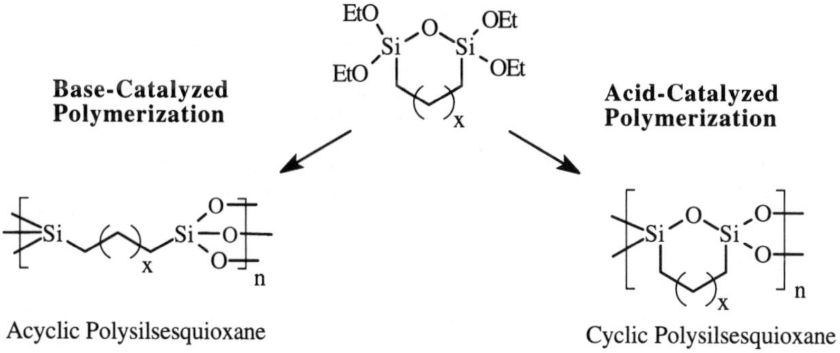

Scheme 6. Possible polymerization pathways for cyclic disilsesquioxane monomers.

Polymerizations of **3** and **4** were performed under basic and acidic conditions (Table 1). With the introduction of NaOH and excess (4 equivalents) H_2O, both monomers reacted within two hours to form transparent, colorless gels. ^{29}Si CP MAS NMR of the dried gels revealed near-quantitative ring opening to afford the acyclic propylene- and butylene-bridged polysilsesquioxanes with $T^2_{acyclic}$ resonances at -60.78 ppm and $T^3_{acyclic}$ resonances at -63.74 ppm [8]. In contrast, aqueous acid-catalyzed polymerization of **3** and **4** occurred much more slowly than under basic conditions, requiring weeks before gels were formed. ^{29}Si CP MAS NMR of the dried polymer derived from **3** revealed a single T^3_{cyclic} peak at -58.7 ppm, with a slight shoulder at -49.7 ppm assigned as the T^2_{cyclic} silicon. Clearly, hydrolysis of the four ethoxy substituents and condensation to give the cyclic polysilsesquioxane occurred faster than ring opening to the acyclic propylene-bridged polysilsesquioxane. Acid-catalyzed polymerization of **4** revealed both cyclic (-58.7 ppm) and acyclic (-62.8 ppm) T^3 resonances in the solid state ^{29}Si NMR spectrum indicating that ring opening was competitive with hydrolysis and condensation to afford the cyclic polysilsesquioxane. The impact of the cyclic disilsesquioxane building block on the properties of the resulting network polymers' physical properties can be seen by the results of nitrogen sorption porosimetry. The gels prepared under acidic conditions from the cyclic monomers were porous with high surface areas (267-969 m^2g^{-1}) and smaller mean pore diameters (2.5-2.8 nm) than observed in the acyclic polysilsesquioxanes prepared under basic conditions (4.2-5.1 nm).

Table 1. Gelation times and porosity data for acid- and base-catalyzed cyclic disilsesquioxane gels.

Monomer (catalyst)	**Gelation Times** (hours)	**BET** ($m^2\ g^{-1}$)	**Pore Diameter** (nm)
3 (acid)	720	471	2.5
3 (base)	< 2	979	4.2
4 (acid)	384	267	2.8
4 (base)	< 2	858	5.1

CONCLUSION

Intramolecular cyclizations of propylene- and butylene-bridged triethoxysilanes **1** and **2** were demonstrated to be the dominant reactions under acidic sol-gel polymerization conditions. Isolation of the six and seven membered cyclic disilsesquioxanes **3** and **4** has not only demonstrated the relative ease of the cyclization reaction, but has provided a new class of siloxane monomers with intriguing similarities to carbohydrates. We are currently investigating further the polymerization chemistry of these new molecules and are preparing new functionalized cyclic disilsesquioxanes from substituted propylene- and butylene-bridged monomers in an effort to control the resulting polymers' stereochemistries.

REFERENCES

[1] Review on hydrocarbon-bridged polysilsesquioxanes: D. A. Loy; K. J. Shea, Chem. Rev. **95**, 1431, (1995).
[2] a) J. H. Small; K. J. Shea; D. A. Loy, J. Non Cryst. Solids **160**, 234 (1993). b) H. W. Oviatt, Jr.; K. J. Shea; J. H. Small, Chem. Mater. **5**, 943 (1993). c) D. A. Loy; G. M. Jamison; B. M. Baugher; E. M. Russick; R. A. Assink; S. Prabakar; K. J. Shea, J. Non Cryst. Solids **186**, 44 (1995).
[3] Sol-gel polymerizations were carried out at 0.4 M monomer concentration. It is significant to note that while bridged triethoxysilanes generally gel with a few hours at this concentration, tetraethoxysilane will not gel until the monomer concentrations is above 1.8 M.
[4] M. G. Voronokov, V. I. Lavrent'yev, Top. Curr. Chem. **102**, 199 (1982).

[5] a) G. Englehardt; H. Jancke; M. Magi; T. Pehk; E. Lippmaa, J. Organomet. Chem. **28**, 293 (1971). b) D. J. Burton; R. K. Harris; K. Dodgson; C. J. Pellow; J. A. Semlyen, Polymer Commun. **24**, 278 (1983).
[6] Siloxane (Si-O-Si) bond angles were calculated to be 131.86° for **3** and 141.02° for **4** in comparison with a value of 145° for acyclic siloxanes. Calculations were performed using MM2 based models in Chemdraw 3D Plus, Cambridge Scientific Computing.
[7] Hydrolysis of organotriethoxysilanes under identical conditions gave rise to three discrete peaks *downfield* from the monomer (T^0) resonance due to the three possible hydrolysis products. Condensation reactions generate three groups of peaks *upfield* relative to T^0 representing silsesquioxanes with one (T^1), two (T^2) and three (T^3) siloxane bonds.
[8] ^{29}Si CP MAS NMR T^1-T^3 signals are broader and farther downfield than the corresponding solution ^{29}Si NMR signals due to contributions from hydrolyzed and cyclic siloxanes. See: L. W. Kelts; N. J. Effinger; S. M. Melpolder, J. Non Cryst. Solids **83**, 353 (1986).

HYBRID ORGANIC-INORGANIC SYSTEMS DERIVED FROM ORGANOTIN NANOBUILDING BLOCKS

F.O. RIBOT, C. EYCHENNE-BARON, F. BANSE, C. SANCHEZ
Laboratoire de Chimie de la Matière Condensée - Université P. et M. Curie / CNRS, 4 Place Jussieu, 75252 Paris, France

ABSTRACT

A new synthesis of the macro-cation $\{(BuSn)_{12}O_{14}(OH)_6\}^{2+}$ is described from commercially available and easy to handle BuSnO(OH) and p-toluene sulfonic acid. A crystalline compound, $\{(BuSn)_{12}O_{14}(OH)_6\}(O_3SC_6H_4CH_3)_2{\bullet}C_4H_8O_2$, is obtained. It can react with tetramethylammonium hydroxide to yield the more versatile composition : $\{(BuSn)_{12}O_{14}(OH)_6\}(OH)_2$. The macro-cations are then used as nanobuilding blocks and assembled in rosary-like structures by carboxy-terminated poly(ethyleneglycol). The resulting system can be pictured as hybrid organic-inorganic alternated block co-polymers. Characterizations were performed with ^{119}Sn NMR (solution and solid state), ^{13}C CP-MAS NMR and FT-IR.

INTRODUCTION

Sol-gel chemistry offers the possibility to blend organic and inorganic components in a single material [1-5]. Depending on the process, the blending scale can range from submicronic to almost molecular. A very large variety of hybrid organic-inorganic materials, mostly based on silicon, has been prepared and characterized in the last years [2-5]. One usual problem of many hybrid materials synthesized by sol-gel process is the size polydispersity that exhibits the inorganic or organic polymer components. Moreover, the interface between the inorganic and organic components, especially when they exchange weak interactions such as hydrogen bonds, is also often poorly defined. These two features can be major drawbacks for the understanding of the relations between structures and properties. A partial answer to the size polydispersity and interface problems would be to synthesize hybrid systems by assembling via known interactions well defined nanobuilding blocks. These tailored systems could be used as models to ease the comprehension and improve the elaboration of more complex but practically relevant systems.

For such an approach, monoorganotin compounds are interesting. They exhibit a rich oxo clusters chemistry with many different structures [6-9], and most of them can be considered as potential inorganic nanobuilding blocks. Moreover, the chemistry of organotin compounds offers several strategies to turn these discrete oxo species into hybrid systems. The stability of Sn-C bond provides a first way to attach the inorganic component to the organic one [10]. Unfortunately, only few $RSnX_3$ precursors, in which the R group bears an organic functionality ($-HC{=}CH_2$, methacrylate, OH, NH_2, etc…), are available so far [11]. However, monoorganotin oxo clusters offer other assembling strategies. Many of them contain strongly bonded ligands such as carboxylate or phosphinate. These ligands can be used to anchor the organic component to the inorganic cluster. A similar strategy is also developed for transition metals [12]. Finally, some

Mat. Res. Soc. Symp. Proc. Vol. 435 © 1996 Materials Research Society

monoorganotin oxo clusters are macro-cations and their charges can be taken advantage of to attach them by means of telechelic dianions or polymers bearing pending anions.

The research dealing with monoorganotin oxo clusters in hybrid systems has been mainly focused on the $\{(RSn)_{12}O_{14}(OH)_6\}^{2+}$ macro-cations [8,13,14]. The assembling strategy based on the Sn-C bond has been partially investigated with $\{(ButenylSn)_{12}O_{14}(OH)_6\}^{2+}$, which was prepared by hydrolysis of ButenylSn$(OAm^t)_3$ [11,15] (Butenyl : $H_2C{=}CHCH_2CH_2$-). However, the very low reactivity of butenyl groups toward radical initiated poly-addition led to poor assembling of the nanobuilding blocks. This strategy, based on unsaturated functionalities born by a chain covalently bonded to tin, is still under investigation. Attractive precursors for such a goal are $H_2C{=}CHC_6H_4$-$(CH_2)_n$-Sn$(OR)_3$ (functionality=styryl) or $H_2C{=}C(CH_3)C(O)O$-$(CH_2)_n$-Sn$(OR)_3$ (functionality=methacrylate). An alkyl spacer (n=3 or 4) seems necessary in such precursors to ease the formation of well-defined nanobuilding blocks such as $\{(RSn)_{12}O_{14}(OH)_6\}^{2+}$. Indeed, preliminary works on PhSn$(OR)_3$ showed that hydrolysis-condensation, of phenyltin tri-alkoxides yielded only small amounts of $\{(PhSn)_{12}O_{14}(OH)_6\}^{2+}$, but mostly ill-defined oxo-polymeric species [16].

Attempts to assemble the Butenyl-functionalized macro-cations by hydrosylilation with a poly(hydrogenomethylsiloxane) were also performed, but did not succeed. Actually, reactions between the Si-H and the Sn-O-Sn framework of the oxo clusters led to the destruction of the $\{(ButenylSn)_{12}O_{14}(OH)_6\}^{2+}$ macro-cations [16].

An other assembling strategy relying on the positive charges borne by $\{(BuSn)_{12}O_{14}(OH)_6\}^{2+}$ was also studied. The reaction between $\{(BuSn)_{12}O_{14}(OH)_6\}(OH)_2$ and different α,ω-dicarboxylic acids based on alkyl or aryl spacers, in ratios such as $-CO_2H/Sn=2/12$, led to the exchange of the hydroxyl counter-ions by carboxylate groups, and resulted in the formation of hybrid organic-inorganic alternated block co-polymers, $[\{(BuSn)_{12}O_{14}(OH)_6\}(O_2C\text{-}R\text{-}CO_2)]_n$ [17]. These systems were obtained as precipitates, and no organic solvent was found to solubilize them. With the same strategy, $\{(BuSn)_{12}O_{14}(OH)_6\}(OH)_2$ was also functionalized with methacrylate, yielding $\{(BuSn)_{12}O_{14}(OH)_6\}(O_2C(CH_3){=}CH_2)_2$, which can then undergo co-polymerization with methylmethacrylate [18].

To study all these systems based on $\{(BuSn)_{12}O_{14}(OH)_6\}^{2+}$, an important issue is the preparation in large quantities of this nanobuilding block. So far, this organotin oxo cluster has been prepared with a yield of 50% by hydrolysis-condensation of butyltin tri-isopropoxyde [14]. With this method, the macro-cation is obtained with hydroxyl counter-ions. This is a versatile composition, but it requires to synthesize BuSn$(OPR^i)_3$, under moisture free conditions. Controlled base hydrolysis of $BuSnCl_3$ also provides the macro-cation with a good yield, but with chloride counter-ions [13]. These counter-ions are less attractive, and moreover the compound is contaminated with undesired butyltin oxo polymer species [16].

This paper presents a new and easier synthesis of the $\{(BuSn)_{12}O_{14}(OH)_6\}^{2+}$ nano-building block and describes hybrid organic-inorganic systems obtained by reacting $\{(BuSn)_{12}O_{14}(OH)_6\}(OH)_2$ with carboxymethyl-terminated poly(ethyleneglycol).

EXPERIMENT

Characterization techniques

Solution ^{119}Sn NMR experiments were performed on a Bruker AM250 spectrometer (93.27 MHz for ^{119}Sn). Solid state ^{119}Sn and ^{13}C NMR experiments were carried out on a Bruker MSL300 spectrometer (111.92 MHz for ^{119}Sn and 75.47 MHz for ^{13}C). Magic angle spinning (MAS) was used for both nuclei. ^{119}Sn NMR spectra were obtained using direct polarization, while cross-polarization (CP) with ^{1}H was used for ^{13}C NMR spectra. For ^{13}C CP-MAS NMR spectra a contact time of 2 ms was used. Chemical shifts are referenced (0 ppm) versus external tetramethylsilane for ^{13}C and external tetramethyltin for ^{119}Sn. ^{119}Sn MAS NMR spectra generally include many spinning side bands even at high rotor speeds (≈13 KHz). Therefore, at least two different spinning rates were used to assign unambiguously the isotropic resonances. To determine the relative proportions in multi sites compounds, ^{119}Sn MAS NMR spectra were fitted with WINFIT [19]. This software was also used to compute the ^{119}Sn shielding tensorial values, using the Herzfeld and Berger's approach. They are reported as the isotropic chemical shift ($\delta_{iso}=-\sigma_{iso}=-(\sigma_{11}+\sigma_{22}+\sigma_{33})/3$; σ_{ii} being the tensorial values in the principal axis system), the anisotropy ($\delta_A=\sigma_{33}-\sigma_{iso}$), and the asymmetry ($\eta=|\sigma_{11}-\sigma_{22}|/|\sigma_{33}-\sigma_{iso}|$), according to Haeberlen's notation [20].

FT-IR experiments were run on a Nicolet Magna 550 spectrometer. Samples were mixed with KBr and pressed into pellet.

Synthesis

$\{(BuSn)_{12}O_{14}(OH)_6\}(O_3SC_6H_4CH_3)_2{\bullet}C_4H_8O_2$ was prepared as follows. 20 g of butylstannonic acid (BuSnO(OH), Strem Chemicals), 6 g of p-toluene sulfonic acid ($HO_3SC_6H_4CH_3{\bullet}H_2O$, Fluka), and 500 ml of toluene were placed in a one liter round-bottom flasks equipped with a Dean-Stark and a condenser. The mixture was refluxed for 48 h. The system was then cooled down and filtered on a glass frit to removed unreacted materials. The so-obtained clear solution was evaporated under reduced pressure to yield around 23 g of solid (chemical analysis : S/"$BuSn_{12}$"=3.1). This crude material was dissolved, by heating, in 50 ml of 1,4-dioxane for crystallization. About 15 g (67% yield) of crystals were obtained within few days at room temperature. Chem. Anal. : Sn=50.7% (49.7), S=2.3% (2.3), C=26.9% (27.6) and H=4.6% (4.7), percentages calculated for $\{(BuSn)_{12}O_{14}(OH)_6\}(O_3SC_6H_4CH_3)_2{\bullet}C_4H_8O_2$ are given between parentheses. The presence of 1,4-dioxane in the final crystals was observed by ^{1}H and ^{13}C NMR [21].

$\{(BuSn)_{12}O_{14}(OH)_6\}(OH)_2$ was prepared by hydrolysis of butyltin tri-isopropoxide following a procedure previously described [14]. This dihydroxy compound was also prepared with a new procedure. 5 g of $\{(BuSn)_{12}O_{14}(OH)_6\}(O_3SC_6H_4CH_3)_2{\bullet}C_4H_8O_2$ was dissolved in 40 ml of isopropanol. To this solution, was slowly added a mixture made of 3.3 g of 10 weight-% aqueous solution of tetramethylammonium hydroxyde ($(CH_3)_4NOH$, Fluka) and 20 ml of isopropanol. The quantities of reagents were such that the ratio $OH^-/\{(BuSn)_{12}O_{14}(OH)_6\}^{2+}$ was equal to two. From this solution, 4 g (80% yield) of crystals were obtained within few days at -15°C. In both preparations, the solids were dried under vacuum to remove isopropanol.

Two poly(ethyleneglycol) bis(carboxymethyl) ethers, $HO_2CCH_2O(CH_2CH_2O)_nCH_2CO_2H$ (Aldrich) were used in this study. They are both soluble in water, and their molecular weights have been determined by titration of their acidity. PEG-1 (mw=255 g/mol, <n>=2.75) is a waxy solid and PEG-2 (mw=630 g/mol, <n>=11.3) is a highly viscous liquid. Reactions between $\{(BuSn)_{12}O_{14}(OH)_6\}(OH)_2$ and carboxymethyl-terminated PEG were carried out as follows. To a 5 weight-% solution of $\{(BuSn)_{12}O_{14}(OH)_6\}(OH)_2$ in tetrahydrofuran (THF), was added drop by drop a 10 weight-% solution of a carboxymethyl-terminated PEG, or a mixture of both, in THF. The amounts of solution added were adjusted to tailor the final $-CO_2H/\{(BuSn)_{12}O_{14}(OH)_6\}(OH)_2$ ratios (Φ). Ratios of 2 or 12 were prepared. Depending on Φ, solutions or precipitates were observed after addition. In both cases, the systems were dried under vacuum to remove the THF, and samples ranging from powders to waxy solids were obtained.

RESULTS AND DISCUSSION

Synthesis of the nanobuilding blocks

The structure of $\{(BuSn)_{12}O_{14}(OH)_6\}(O_3SC_6H_4CH_3)_2 \bullet C_4H_8O_2$ was resolved by X-ray diffraction on single crystal [21]. This compound is based on $\{(BuSn)_{12}O_{14}(OH)_6\}^{2+}$ units, linked in chains by p-toluene sulfonate. The molecular structure of $\{(BuSn)_{12}(\mu_3\text{-}O)_{14}(\mu_2\text{-}OH)_6\}^{2+}$ is shown in figure 1. In this, almost spherical, centrosymmetric, closo-type entity, tin atoms are linked by μ_3-O and μ_2-OH bridges. Every tin bears a butyl chain which spreads outside of the closo Sn-O-Sn framework. Six, of the twelve tin atoms, are five-coordinate ($C\underline{Sn}O_4$) and exhibit a square pyramidal environment. The other six are six-coordinate ($C\underline{Sn}O_5$) and have a pseudo octahedral environment. The bridging hydroxy groups (μ_2-OH) are only involved in the environment of the six-coordinate tin atoms. To describe the connections inside the macro-cation, one can consider three sub-units. One is a closed belt formed by the six square pyramids, each sharing opposite edges of its basal plan. The two other are trimeric sub-units which are based on pseudo octahedra sharing adjoining edges and a common vertex (μ_3-O). The complete structure of the macro-cation is obtained by capping each sides of the closed belt by a trimeric sub-unit (common vertex toward the inside). One positive charge is located at each pole, in the trimeric sub-units.

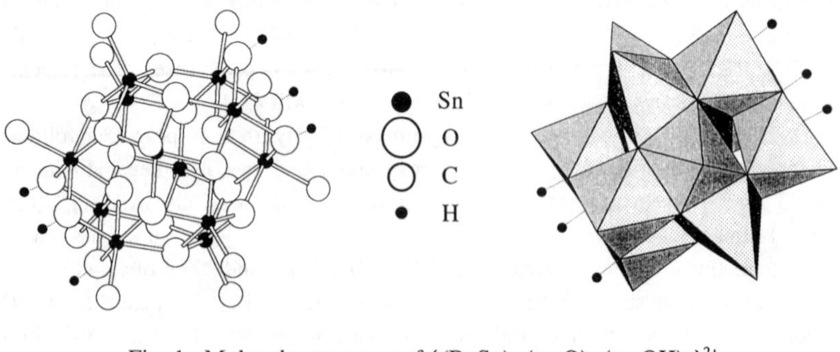

Fig. 1 : Molecular structure of $\{(BuSn)_{12}(\mu_3\text{-}O)_{14}(\mu_2\text{-}OH)_6\}^{2+}$
Balls and sticks (left) - Polyhedra (right)

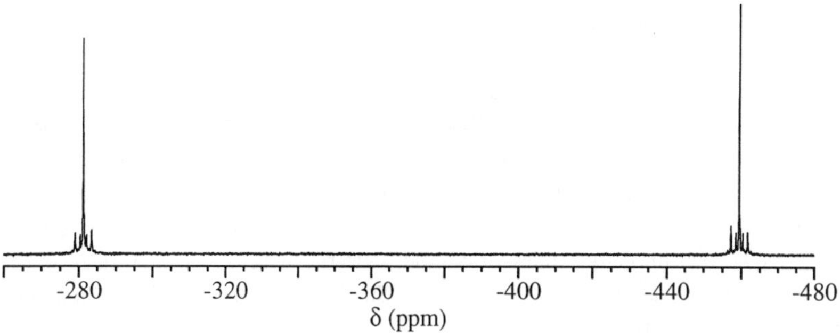

Fig. 2 : ^{119}Sn-{^{1}H} spectrum of $\{(BuSn)_{12}O_{14}(OH)_6\}(O_3SC_6H_4CH_3)_2{\bullet}C_4H_8O_2$ in C_6D_6.

The ^{119}Sn-{^{1}H} NMR spectrum of $\{(BuSn)_{12}O_{14}(OH)_6\}(O_3SC_6H_4CH_3)_2{\bullet}C_4H_8O_2$ dissolved in C_6D_6 is presented in figure 2. It is composed of two sharp resonances, located at -281 and -459 ppm, which are related to five-coordinate ($C\underline{Sn}O_4$) and six-coordinates tin atoms ($C\underline{Sn}O_5$) respectively. Several satellites, associated to two bonds scalar couplings, are also observed at the feet of both resonances [13,14]. Equal amounts of tin was found in each coordination by non proton decoupled ^{119}Sn NMR. The spectrum in figure 2 is characteristic of a compound based on $\{(BuSn)_{12}O_{14}(OH)_6\}^{2+}$ [13,14], indicating therefore that the butyltin oxo cluster is preserved in solution.

Actually, the molecular entity which is characterized by the ^{119}Sn NMR spectrum in figure 2 is the ions triplet $\{(BuSn)_{12}O_{14}(OH)_6\}(O_3SC_6H_4CH_3)_2$ which is, as previously shown for $\{(BuSn)_{12}O_{14}(OH)_6\}(OH)_2$ [14] and $\{(BuSn)_{12}O_{14}(OH)_6\}(O_2CCH_3)_2$ [17], not dissociated in organic solvents with low dielectric constant. The electrostatic interactions which exist in solution between the macro-cation and the anions explain the dependency of the ^{119}Sn NMR chemical shifts of $\{(BuSn)_{12}O_{14}(OH)_6\}(X)_2$ with the anion X. Moreover, these interactions involve the μ_2-OH which are only bonded to the six-coordinate tin atoms [13,14,17]. Therefore, this chemical shift dependency is mostly observed for the six-coordinate tin atoms. This effect is clearly evidenced in the following examples : $\delta(^{119}Sn)$=-280 and -447 ppm for $X=OH^-$ [14], $\delta(^{119}Sn)$=-283 and -464 ppm for $X=Cl^-$ [13], $\delta(^{119}Sn)$=-281 and -458 ppm for $X=CH_3CO_2^-$ [17].

The ^{119}Sn-{^{1}H} NMR spectrum of the crude material, before crystallization in 1,4-dioxane, is identical to the one presented in figure 2. Therefore, only one soluble butyltin compound seems to be formed by the reaction of butylstannonic acid and p-toluene sufonic acid. Moreover, the compound formed is stable in the presence of an excess of acid. Indeed, from chemical analysis, ^{1}H and ^{13}C NMR, the formula of the crude material can be written : $\{(BuSn)_{12}O_{14}(OH)_6\}(O_3SC_6H_4CH_3)_2 + (HO_3SC_6H_4CH_3)_{1.1}$.

This new procedure to synthesize $\{(BuSn)_{12}O_{14}(OH)_6\}^{2+}$ is fairly different from the ones previously reported, which were based on the hydrolysis-condensation of $RSnCl_3$ [13], $BuSn(OPr^i)_3$ or $BuSn(OAm^t)_3$ [14]. The synthesis described here was inspired by the works of R.R. Holmes [6] who prepared numerous oxo-butyltin clusters by reacting butylstannonic acid

(BuSnO(OH)) with various weak acids (RCO_2H, R_2PO_2H, $(RO)_2PO_2H$) to which correspond strong bidentate complexing anions. Indeed, in the oxo-butyltin clusters so-obtained, these ligands were always bonded to tin, and generally bridging them. This method was modified by using a strong acid ($CH_3C_6H_4SO_3H$) to which corresponds a poorly complexing anion. It succeeded to quantitatively prepare large amounts of $\{(BuSn)_{12}O_{14}(OH)_6\}^{2+}$ (95% yield on the crude product and about 70% after crystallization in 1,4-dioxane) when 4 equivalents of acid were used for 12 equivalents of butyltin. The need for an excess of acid, compared to the stoichiometry in the final compound (2:12), is not clear. But, attempts to use only 2 moles of acid for 12 moles of BuSnO(OH) yielded lower quantities of a crude material which contained other species than $\{(BuSn)_{12}O_{14}(OH)_6\}^{2+}$, as evidenced by solution ^{119}Sn NMR. It is interesting to mention that both reagents are poorly soluble in toluene.

The reaction mechanism which transforms BuSnO(OH) and p-toluene sulfonic acid in $\{(BuSn)_{12}O_{14}(OH)_6\}(O_3SC_6H_4CH_3)_2$ is not known, but it can be regarded as an acid catalyzed condensation reaction yielding water :

$$12\ BuSnO(OH) + 2\ HO_3SC_6H_4CH_3 \longrightarrow \{(BuSn)_{12}O_{14}(OH)_6\}(O_3SC_6H_4CH_3)_2 + 4\ H_2O$$

The structure of butylstannonic acid is not known. It likely exists as an oxo-hydroxo organotin polymers. ^{119}Sn MAS NMR shows mainly six-coordinate tin atoms (more than 85%) in BuSnO(OH). From the last statement, it seems that the formation of $\{(BuSn)_{12}O_{14}(OH)_6\}^{2+}$ is stable enough to allow for the reduced coordination of almost half of the tin atoms.

The possibility to directly use $\{(BuSn)_{12}O_{14}(OH)_6\}(O_3SC_6H_4CH_3)_2{\bullet}C_4H_8O_2$ in counter-ions exchange reactions [17,18] was checked. The reaction, in THF, between $\{(BuSn)_{12}O_{14}(OH)_6\}(O_3SC_6H_4CH_3)_2{\bullet}C_4H_8O_2$ and 2 carboxylic acids did not lead to the replacement of the counter anions by carboxylates, on the contrary to what was previously observed for $\{(BuSn)_{12}O_{14}(OH)_6\}(OH)_2$ [17,18]. Instead of a simple substitution of the counter-anions, ^{119}Sn NMR showed a destruction of the macro-cation and the formation of unknown products.

More interestingly, the reaction between $\{(BuSn)_{12}O_{14}(OH)_6\}(O_3SC_6H_4CH_3)_2{\bullet}C_4H_8O_2$ and 2 equivalents of tetramethylammonium hydroxide, in isopropanol, yielded crystals. The solution ^{119}Sn NMR spectrum of these crystals resembles the one presented in figure 2. Yet, the chemical shifts are -281 and -447 ppm indicating the replacement of the p-toluene sulfonate anions by hydroxyl anions [14].

Figure 3 presents the ^{119}Sn MAS NMR spectrum of the dihydroxy compound, obtained by reaction with $(CH_3)_4NOH$, and dried under vacuum. It exhibits three isotropic resonances and their associated spinning side bands. The fitting of this spectrum allowed us to extract the ^{119}Sn NMR characteristics of each site. They are reported in table I

The isotropic chemical shift assigned to five-coordinate tin atoms (-280 ppm) agrees perfectly with the chemical shift determined in solution for the same tin atoms. The average (-453 ppm) of the two $C\underline{Sn}O_5$ isotropic chemical shifts, weighted by their relative population, is also in good agreement with the solution ^{119}Sn NMR data. The splitting in two resonances for six-coordinate tin atoms, compared to solution NMR, is likely related to a peculiar localization of the hydroxyl counter-anions relatively tc the μ_2-OH of the macro-cations. Such a splitting effect has already been evidenced for $\{(BuSn)_{12}O_{14}(OH)_6\}(OH)_2\bullet(HOPr^i)_4$ in which the OH^- is hydrogen bond to only one bridging hydroxy at each pole of the macro-cation [14]. In the isopropanol containing compound, the splitting (28 ppm) is higher than the one observed for the compound dried under vacuum (11 ppm). Therefore, one can speculate that in $\{(BuSn)_{12}O_{14}(OH)_6\}(OH)_2$, the hydroxyl counter-anions are more shared by all the bridging hydroxy groups.

Table I : ^{119}Sn NMR parameters for $\{(BuSn)_{12}O_{14}(OH)_6\}(OH)_2$, extracted from the spectrum presented in Fig. 3. (ΣPopulation is normalized to12)

Label	Coordination	Population	δ_{iso} (ppm)	δ_A (ppm)	η
#1	5	6.1	-280	365	0.25
#2	6	2.0	-446	247	0.00
#3	6	3.9	-457	290	0.30

Because of the little literature on the relation between the geometry of a tin site and its tensorial properties, precise discussion of δ_A and η is delicate. Yet, one can write that, as expected from simple geometrical consideration, the anisotropies (δ_A) of five-coordinate tin atoms are larger than those of six-coordinate ones.

The spectrum in figure 3, and the parameters extracted from it (Table I), are identical, within the analysis accuracy, to the ones obtained for $\{(BuSn)_{12}O_{14}(OH)_6\}(OH)_2$ prepared by hydrolysis of butyltin tri-isopropoxide. Moreover, the yield by exchange of counter-anions is higher than the one by hydrolysis of alkoxide (80% compared to 70% at best). Finally, this new method avoids the synthesis in moisture free conditions of $BuSn(OPr^i)_3$. Therefore, the preparation of the dihydroxy compound by reacting $\{(BuSn)_{12}O_{14}(OH)_6\}(O_3SC_6H_4CH_3)_2\bullet C_4H_8O_2$ with $(CH_3)_4NOH$ appears to be valuable.

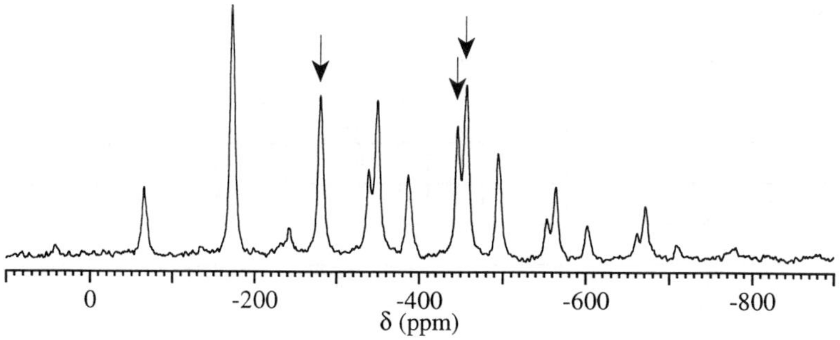

Fig. 3 : ^{119}Sn MAS NMR of $\{(BuSn)_{12}O_{14}(OH)_6\}(OH)_2$ dried under vaccum (isotropic chemical shifts are pointed with arrows, spinning rate = 12 KHz).

Assembling of the nanobuilding blocks

Several samples were prepared from $\{(BuSn)_{12}O_{14}(OH)_6\}(OH)_2$ and carboxymethyl-terminated poly(ethyleneglycol). They are listed in Table II.

Table II : Samples prepared from $\{(BuSn)_{12}O_{14}(OH)_6\}(OH)_2$ and $HO_2CCH_2\text{-}(OCH_2CH_2)_n\text{-}OCH_2CO_2H$ ($<n>=2.75$ for PEG-1 and $<n>=11.3$ for PEG-2).

Sample	$-CO_2H/(BuSn)_{12}$	$HO_2C\text{-}R\text{-}CO_2H$	Aspect	Solubility
P1	2	PEG-1	powder	$THF\text{-}CH_2Cl_2$
P2	2	0.5 PEG-1 + 0.5 PEG-2	soft polymer	$THF\text{-}CH_2Cl_2$
P3	2	0.2 PEG-1 + 0.8 PEG-2	tacky polymer	$THF\text{-}CH_2Cl_2$
P4	2	PEG-2	tacky wax	$THF\text{-}CH_2Cl_2$
S5	12	PEG-1	powder	Insoluble

All the samples (Px) prepared with a $-CO_2H/(BuSn)_{12}$ ratio equal to 2 remained in solution. After removal under vacuum of the THF and the generated water, the samples had various physical appearance ranging from that of a powder to a tacky wax. After drying all these samples could be solubilized (swelled) in THF or dichloromethane. This solubility makes the samples Px very different from the ones prepared analogously with $HO_2C\text{-}R\text{-}CO_2H$ (R=alkyl or aryl) [17]. The sample prepared with 12 acid functions per macro-cation (S5) behaved in a complete different way. As soon as the acids were added, precipitation was observed.

These samples were characterized by ^{119}Sn MAS NMR, ^{13}C CP-MAS NMR, and FT-IR. The ^{119}Sn MAS spectra of P1 (representative of P2 to P4) and S5 are presented in figure 4. They were both analyzed with WINFIT software [19]. Results are reported in Table III. The ^{13}C CP-MAS NMR spectra of P1 and S5 are presented in figure 5.

Table III : ^{119}Sn NMR analysis results for systems P1 and S5.
(ΣPopulation is normalized to12)

Label	Coord.	Population	δ_{iso} (ppm)	δ_A (ppm)	η
			P1		
#1	5	5.8	-282	372	0.30
#2	6	2.3	-463	302	0.35
#3	6	3.9	-474	314	0.30
			S5		
#4	6	"12"	-514	385	0.45

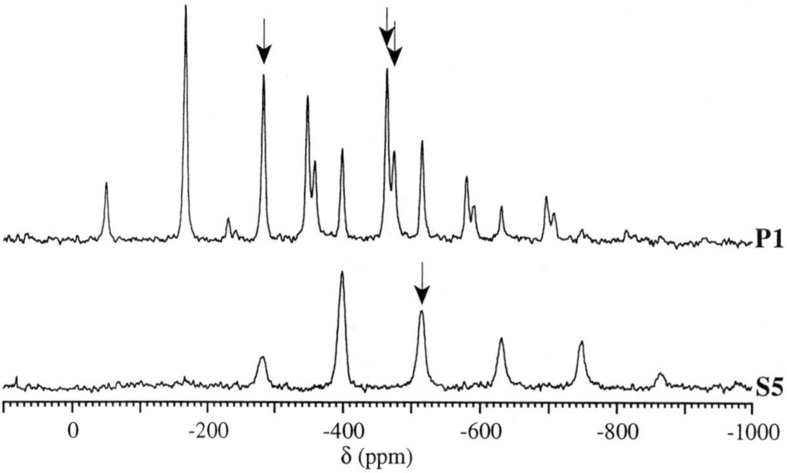

Fig. 4 : ^{119}Sn MAS NMR of sample P1 and S5
(isotropic chemical shifts are pointed with arrows, spinning rate = 13 KHz).

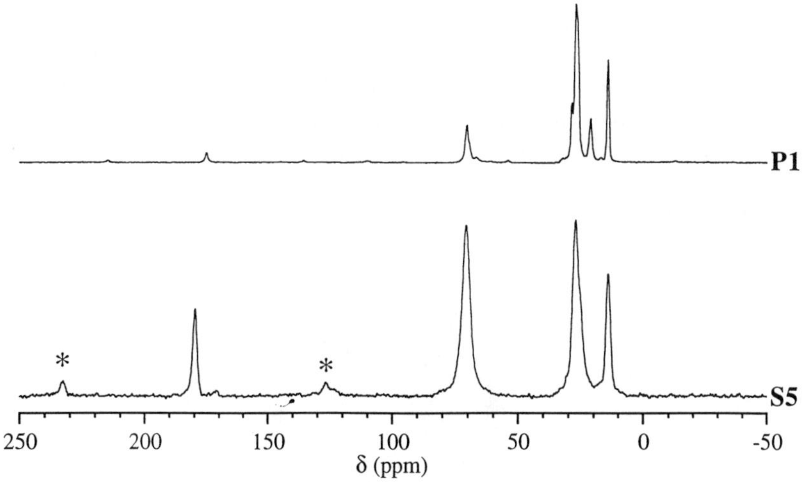

Fig. 5 : ^{13}C CP-MAS NMR of sample P1 and S5
(spinning side bands are pointed with asterisks, spinning rate = 4 KHz).

In the ^{119}Sn MAS NMR spectrum of P1, three different environment for tin atoms are observed. The less shielded (-282 ppm) corresponds to five-coordinate tin atoms ($C\underline{Sn}O_4$). The two others (-463 and -474 ppm) correspond to six-coordinate tin atoms ($C\underline{Sn}O_5$). The same amount of tin is found in both coordination. The ^{119}Sn MAS NMR spectrum of P1 is identical, except for slightly larger line broadening, to the one previously obtained for $\{(BuSn)_{12}O_{14}(OH)_6\}(O_2CCH_3)_2$ [17].

The FT-IR spectrum of P1 exhibits also many similarities with the one of $\{(BuSn)_{12}O_{14}(OH)_6\}(O_2CCH_3)_2$ [18]. A broad absorption between 3500 and 2700 cm^{-1}, related to the stretching mode of hydroxy groups (in SnOH) experiencing strong hydrogen bonding. No absorption at 3650 cm^{-1} which would indicate SnOH groups not involved in hydrogen bonds [22]. Strong absorptions at 1580 and 1410 cm^{-1} which correspond respectively to the antisym. and sym. stretching modes of an ionic carboxylate groups [23]. In $\{(BuSn)_{12}O_{14}(OH)_6\}(O_2CCH_3)_2$ [17] these values were reported at 1540 and 1400 cm^{-1}. The higher frequencies observed in P1 are likely related to the different nature of the group bearing the carboxylate (CH_2O- vs. CH_3). Finally, the absorptions in the low wavenumbers domain (670, 625, 545 and 430 cm^{-1}), related to the Sn-O-Sn stretching modes, show the same pattern in $\{(BuSn)_{12}O_{14}(OH)_6\}(O_2CCH_3)_2$ and P1. This pattern is actually very close to the one observed for $\{(BuSn)_{12}O_{14}(OH)_6\}(OH)_2$ (665, 625, 535, 420 cm^{-1}).

The ^{119}Sn spectra of P1 solubilized in CD_2Cl_2 exhibits two resonances located at -281 and -457 ppm, assigned respectively to $C\underline{Sn}O_4$ and $C\underline{Sn}O_5$ sites. These values agree almost perfectly with the ^{119}Sn chemical shifts reported for $\{(BuSn)_{12}O_{14}(OH)_6\}(O_2CCH_3)_2$ [17]. Therefore solution ^{119}Sn NMR shows that the interaction between the macro-cations and the carboxy-terminated PEG are retained when these hybrid systems are solubilized in organic solvents with low dielectric constant.

Therefore, one can conclude that sample P1, as well as P2 to P4, are alternated block co-polymer made of $\{(BuSn)_{12}O_{14}(OH)_6\}^{2+}$ macro-cations assembled by carboxymethyl-terminated poly(ethyleneglycol). A schematic drawing of the rosary-like structure of such hybrid systems is presented in figure 6. The interactions that hold the macro-cations and the carboxymethyl-terminated poly(ethyleneglycol) together have an electrostatic nature, but also involve a set of hydrogen bonds. They are exchanged between the two oxygen atoms of the carboxylates and the bridging hydroxy groups (μ_2-OH) belonging to the macro-cations [17].

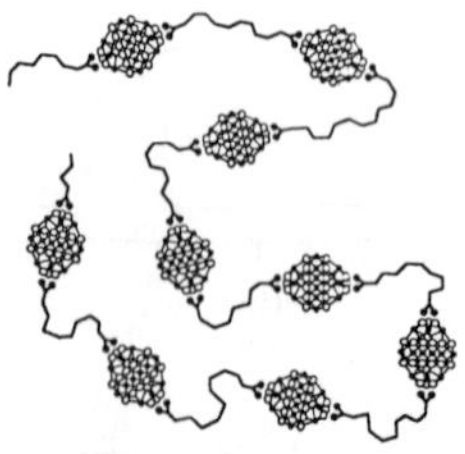

Fig. 6 : Schematic drawing of the rosary-like structure obtained by reacting $\{(BuSn)_{12}O_{14}(OH)_6\}(OH)_2$ and carboxy-terminated poly(ethyleneglycol).

Sample S5 exhibits a completely different ^{119}Sn MAS spectrum. Only one six-coordinate tin environment ($C\underline{Sn}O_5$) is observed at -514 ppm. The ^{119}Sn tensorial parameters for this site (Table III) exhibits a fairly high anisotropy and asymmetry compared to the ones observed for the pseudo ocatedral sites of $\{(BuSn)_{12}O_{14}(OH)_6\}(OH)_2$ and $\{(BuSn)_{12}O_{14}(OH)_6\}(O_2CCH_3)_2$ [17]. The anisotropy is even higher than the ones observed in $[BuSnO(O_2CCH_3)]_6$ [6] which were found around 355 ppm . This seems to indicate the coordination of tin atoms by a carboxylate ligand in S5.

The high wavenumbers domain (around 3200 cm^{-1}) of the FT-IR spectrum of S5 indicates the absence of SnOH groups. They have probably been consumed by the nucleophilic substitution of the carboxylate ligands. In the intermediate wavenumbers domain (1800-1000 cm^{-1}), several antisym. and sym. stretching modes of the carboxylate are observed (ν_{AS}=1600 and 1570 cm^{-1}, ν_S=1440 and 1415 cm^{-1}). In the low wavenumber domain, only one broad absorption related to Sn-O-Sn vibrations appears (540 cm^{-1}).

The ^{13}C CP-MAS spectra (figure 5) are also indicative of changes between S5 and P1. At high field (10-30 ppm) are found the resonances of the carbon atoms belonging to the butyl groups. The ^{13}C resonance related to the poly(ethyleneglycol), $-(\underline{C}H_2\underline{C}H_2O)-$, is observed at 71 ppm. Finally, the carboxylate resonance is found at low field (180-170 ppm). The higher amount of PEG-1, per tin atom, in S5 explains the increase of the intensities of the carbonyl and poly(ethyleneglycol) resonances, compared to butyl resonances. Except from those intensity variations, the main changes between the ^{13}C CP-MAS spectra of P1 and S5 are the shift of the carbonyl resonance, from 175 ppm for P1 to 179 ppm for S5, and the disappearance in S5 of the resonance located at 21 ppm. This resonance was assigned to the first carbon of a butyl chain borne by five-coordinate tin in $\{(BuSn)_{12}O_{14}(OH)_6\}(OH)_2$ [14]. The shift of the carbonyl resonance is probably explained by the change in the carboxylate bonding mode, going from ionic in P1 to directly bridging tin atoms in S5. Among the oxo butyltin carboxylate clusters described in the literature none was reported to contain chelating carboxylate [6]. The same kind of shift is observed between $\{(BuSn)_{12}O_{14}(OH)_6\}(O_2CCH_3)_2$ ($\delta_{carbonyl}$=178 ppm) and $[BuSnO(O_2CCH_3)]_6$ ($\delta_{carbonyl}$=181 ppm) for which the acetato ligand bonding mode goes from ionic to bridging.

All these characterizations show the disappearance of $\{(BuSn)_{12}O_{14}(OH)_6\}^{2+}$ in S5. Previous works have shown that the reaction of $\{(BuSn)_{12}O_{14}(OH)_6\}X_2$ ($X=OH^-$ or Cl^-) with 12 equivalents of acetic acid led to the destruction of the macro-cation and the formation of a hexameric cluster corresponding to the formula $[BuSnO(O_2CCH_3)]_6$ [13,17]. This entity belongs to the drum-shape family [6], named from the disposition of the two superimposed and connected $(SnO)_3$ stannoxane rings. This species is characterized in ^{119}Sn solution NMR by a chemical shift of -485 ppm, and in ^{119}Sn MAS NMR by three isotropic chemical shifts at -484, -488 and -490 ppm. Obviously, the oxo butyltin entity formed in S5 does not correspond to this drum-shaped cluster. Geometrical constraints and limitations of the diffusion, related to the telechelic nature of the carboxymethyl-terminated poly(ethyleneglycol), probably prevent the butyltin moieties to reach the drum-shape configuration.

So far, no relevant proposition could be made for the oxo butyltin entity present in S5. However, the following features are observed : a unique type of six-coordinate tin atoms, no hydroxy groups, and probably carboxylate groups bridging tin atoms.

CONCLUSION

A new synthesis, starting from commercially available BuSnO(OH), has been proposed for the oxo-hydroxo butyltin macro-cation $\{(BuSn)_{12}O_{14}(OH)_6\}^{2+}$. This method yields $\{(BuSn)_{12}O_{14}(OH)_6\}(O_3SC_6H_4CH_3)_2{\bullet}C_4H_8O_2$, which can then be turned into the more versatile $\{(BuSn)_{12}O_{14}(OH)_6\}(OH)_2$ by reaction with tetramethylammonium hydroxyde.

$\{(BuSn)_{12}O_{14}(OH)_6\}^{2+}$ can be considered as a nanobuilding block that can be assembled (co-polymerized) by telechelic di-anions. Reaction between $\{(BuSn)_{12}O_{14}(OH)_6\}(OH)_2$ and one molar equivalent of bis(carboxymethyl) poly(ethyleneglycol) yields hybrid organic-inorganic alternated block copolymers of general formula $(\{(BuSn)_{12}O_{14}(OH)_6\}[O_2CCH_2(OCH_2CH_2)_nOCH_2CO_2])_m$. The inorganic entities and the telechelic dicarboxylates are held together by electrostatic interactions. These co-polymers are soluble in tetrahydrofurane.

REFERENCES

1. C.J. Brinker and G.W. Scherrer, Sol-Gel Science, the Physics and Chemistry of Sol-Gel Processing, Academic Press, San-Diego, CA, 1990.
2. B.M. Novak, Adv. Mater. **5**, 422 (1993).
3. C. Sanchez and F. Ribot, New J. Chem. **18**, 1007 (1994).
4. U. Schubert, N. Hüsing and A. Lorenz, Chem. Mater. **7**, 2010 (1995).
5. D.A. Loy and K.J. Shea, Chem. Rev. **95**, 1431 (1995).
6. R.R. Holmes, Acc. Chem. Res. **22**, 190 (1989).
7. H. Puff and H. Reuter, J. Organomet. Chem. **368**, 173 (1989).
8. H. Puff and H. Reuter, J. Organomet. Chem. **373**, 173 (1989).
9. H. Reuter, Angew. Chem. Int. Ed. Eng. **30**, 1482 (1991).
10. D. Hoebbel, I. Pitsch and D. Heidemann, Z. Anorg. Allg. Chem. **592**, 207 (1991).
11. B. Jousseaume, M. Lahcini, M-P. Rascle, F. Ribot and C. Sanchez, Organometallics **14**, 685 (1995).
12. M. In, C. Gérardin, J. Lambard and C. Sanchez, J. S. S. T. **5**, 101 (1995).
13. D. Dakternieks, H. Zhu, E.R.T. Tiekink and R.J. Colton, J. Organomet. Chem. **476**, 33 (1994).
14. F. Banse, F. Ribot, P. Tolédano and C. Sanchez, Inorg. Chem. **34**, 6371 (1995).
15. F. Ribot, F. Banse and C. Sanchez, Mater. Res. Soc. Symp. Proc. **346**, 121 (1994).
16. F. Ribot, F. Banse, G. Kehr and E. Dien, unpublished results.
17. F. Ribot, F. Banse, F. Diter and C. Sanchez, New J. Chem. **19**, 1163 (1995).
18. F. Ribot, F. Banse, C. Sanchez, M. Lahcini and B. Jousseaume, J. S. S. T. in press (1996).
19. D. Massiot, H. Thiele and A. Germanus, Bruker Rep. **140**, 43 (1994).
20. U. Haeberlen, Adv. Magn. Reson. Suppl. **1** (1976).
21. C. Eychenne-Baron, F. Ribot, F. Robert and C. Sanchez, forth coming paper.
22. H. Reuter and A. Sebald, Z. Naturforsh. **48b**, 195 (1993).
23. K. Nakamoto, Infrared and Raman Spectra of Inorganic and Coordination Compounds, Wiley, New York, NY, 1978.

STUDIES ON THE STRUCTURE AND PROPERTIES OF CERAMIC/POLYMER NANOCOMPOSITES

Kenneth E. Gonsalves*, Xiaohe Chen
Polymer Science Program at the Institute of Materials Science and Department of Chemistry, University of Connecticut, Storrs, CT 06269, USA

ABSTRACT

Unique synthetic approaches for the synthesis of homogeneously dispersed and highly loaded aluminum nitride (AlN) / polyimide (PI) nanocomposites have been developed. The effective interactions at the solid-liquid interface during the preparation of stable dispersions of ceramic/polymers have been investigated. In particular, the surface chemical composition of the nanoparticles has been analyzed. Characteristic model reactions on the surface have been carried out, which revealed the mechanisms for the deagglomeration and stabilization of nanoparticles via chemisorption reactions. Moreover, compared to other synthetic approaches, this method demonstrates the capability of preparing extremely highly loaded nanocomposites and being applicable to a wide range of materials. The thermal and mechanical properties of the AlN/PI nanocomposite have also been studied.

INTRODUCTION

Nanostructured (also referred to as nanocrystalline, nanophase, nanosized or nanoscale) materials are defined as materials having an average phase or grain size of less than 100 nm [1]. In a broader sense, any material that contains grains or clusters below 100 nm, or layers or filaments of that thickness, can be considered to be nanostructured [2]. Due to the small size of the structural unit (particle, grain, or phase) and the high surface to volume ratio, nanostructured materials are anticipated to exhibit unique behavior compared to conventional materials with micron-scale structures [3]. Extensive fundamental and applied investigations have been conducted on nanoscale materials in the last decade [4-9]. The efforts and successes in synthesizing and processing a variety of nanostructured materials have provided a new degree of freedom for the development of advanced materials with enhanced or novel properties.

Nanocomposites represent the current trend in developing novel nanostructured materials. They can be defined as a combination of two or more phases containing different compositions or structures, where at least one of the phases is in the nanoscale regime [10]. Examples of inorganic-organic hybrid nanocomposites include biomimetic ceramic/polymer composites [11], superparamagnetic ceramic/ceramic nanocomposites [12], non-linear optical metal colloid polymer nanocomposites [13], and intercalated polymer-clay nanocomposites [14]. The synthesis of nanocomposites usually involves a multistep process, which may comprise several basic preparation techniques, such as sol-gel/polymerization, sol-gel/thermal pyrolysis, and intercalation/solidification. The critical issue for making nanocomposites with a homogeneous dispersion of the involved phases lies in an understanding of the surface/interface chemistry [15].

Ceramic/polymer nanocomposites provide a new flexibility in designing advanced materials. However, the nanometer-level dispersion of ceramic phases in a polymer matrix is difficult to achieve, because of the agglomeration of the ceramic particles and the small entropy of mixing of polymer-ceramic systems [16]. To prevent agglomeration and to stabilize the surface of ultrafine particles, repulsive interparticle forces are essential. One way to achieve this is through electrostatic repulsion by using polar solvent media that basically provide Van der Waals forces between particles with electric double layers surrounding them. Ionic surfactants are also frequently used for the purpose of dispersing fine particles in aqueous or polar solutions by electrostatic stabilization [17]. In electrolyte macromolecular media, such as poly(acrylic acid), poly(methacrylic acid), or poly(amic acid), the mechanisms for stabilizing a suspension are considered [16] to be mainly (i) steric stabilization, where the macromolecules are attached to the particle surface; (ii) depletion stabilization, in which the macromolecules are free in suspension, and (iii) electrosteric stabilization, the combination of both electrostatic and steric mechanisms. The electrostatic component may originate from a net charge on the particle surface and/or charges associated with the anchored polyelectrolyte. Polyelectrolyte solutions are considered to be particularly effective since they offer the opportunity to introduce both types of interactions. In summary, the success of the stabilization depends on the surface coverage, the type and the configuration of the adsorbed polymers, and the thickness of the adsorbed layer.

Mat. Res. Soc. Symp. Proc. Vol. 435 © 1996 Materials Research Society

An example of a polymer/ceramic nanocomposite has been reported by us in which a homogeneous dispersion of AlN in a polyimide matrix has been achieved [18]. Nanosized AlN particles fixed in the polymer matrix were obtained by utilizing the electrostatic and steric interaction in the non-aqueous system of the chemically synthesized nanostructured AlN powders, the solvent NMP (N-methylpyrrolidinone), and poly(amic acid) (which is a polyelectrolyte and the precursor to polyimide). The agglomeration of the AlN particles had been reduced significantly, as evidenced by morphology studies. (Fig. 1) The average size of the "soft agglomerates" of AlN particles decreased from micron size (observed in SEM) to below 10 nm.

Figure 1. TEM micrograph of AlN/polyimide nanocomposite

Here we present mechanistic details of dispersing the nanostructured AlN in the polymer via model systems studied by FTIR spectroscopy.

RESULTS AND DISCUSSION

Due to the intensive surface tension, the chemical bonds on the first atomic layer of the nanoparticle surface tend to be broken and to form various surface species such as dangling functional groups. Thus, nanoparticles often have a very different surface chemical composition from that of the bulk. We previously reported a comprehensive surface characterization of a chemically synthesized nanostructured AlN powder [19]. It revealed a surface chemical composition that consisted of -OH, $-NH_2$ groups, as well as physisorbed H_2O, CO_2, and NH_3 molecules. Acidic and weak basic surface sites were also evident. Figure 2 illustrates these surface chemical compositions. The reactivities of these surface functional groups have been thoroughly investigated with probe molecules including deuterium, ammonia, pyridine, and methanol [19].

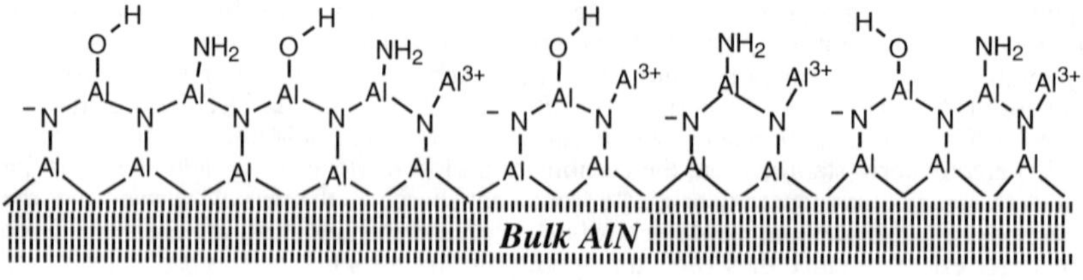

Figure 2. Surface functional groups, and Lewis acidic or basic sites of the nanostructured AlN

As mentioned above [18], the nanostructured AlN powders showed a great tendency to de-agglomerate in NMP (1) and resulted in a homogeneously dispersed system. NMP is a useful, high boiling point, dipolar aprotic solvent. The mechanism of this de-agglomeration process was considered to be from a Lewis acid-base interaction between the solvent NMP molecules and those of chemisorption on the particle surface as proposed by Fowkes [20]. The chemisorption was evident from an NMP-AlN surface reaction experiment monitored by FT-IR spectroscopy. Figure 3(a-d) shows the FTIR spectra following the adsorption of NMP on the activated AlN surface at room temperature and at increasing temperatures under continuous evacuation. All these spectra are depicted by subtracting the spectrum of the activated AlN sample, so that all the vibrational bands are attributed only to the surface adsorbed species and/or to the possibly modified surface groups.

The adsorption of NMP on the nanostructured AlN surface appeared to result in strong hydrogen bonding with the OH groups on the surface as shown in Figure 3a. The N-alkylation promotes a stronger tendency to form a hydrogen bond due to the greater electron richness of its carbonyl oxygen atom. Intermolecular hydrogen bonding caused a broad band in the 3000~3500 cm^{-1} region due to the ν(OH) stretching vibration of the perturbed OH groups, while the negative band at ~3740 cm^{-1} confirmed the decrease of free OH groups.

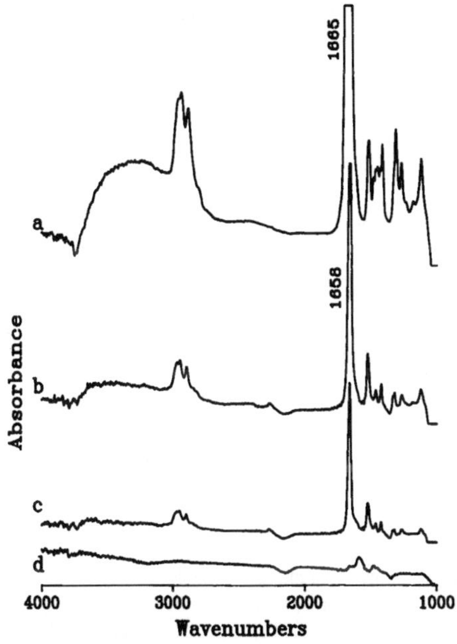

Figure 3. FTIR transmission spectra for the addition of NMP at (a) room temperature; (b) 423K; (c) 573K; (d) 873K. (All the spectra subtracted from the surface activation spectrum)

Associated with the surface hydrogen bonding, a shift to lower frequency of the carbonyl band from 1686 cm^{-1} (unperturbed group in liquid phase) to 1665 cm^{-1} was observed, which proved that the carbonyl oxygen atom was the acceptor site for the hydrogen bonding with the surface OH groups. Considering the overall vibrational bands of the NMP adsorption spectrum, no other apparent shift was shown in such a great extent from its free form spectrum. This suggested that the hydrogen bonding was the main interaction occurring upon NMP adsorption onto the AlN surface at room temperature.

The broad hydrogen bonding band decreased to a great extent upon heating to 423 K (Figure 3b). The difference spectra between the two steps (not shown) also proved this desorption process. Meanwhile, the appearance of a positive band at 3740 cm^{-1} corresponds to the recovery of free surface OH groups. These data showed that the majority of the surface hydrogen bonds were dissociated at this condition. As the temperature was further increased to 573 K, the hydrogen bond was completely deformed (Figure 3c).

At 423 K, however, the broad peak at ~3200 cm^{-1} was still evident (Fig. 3b). Moreover, all of the bands corresponding to NMP molecule were evident, while only the intensity ratio varied. This indicated an unusual strength of the surface hydrogen bond, as compared to the case of methanol addition where the surface hydrogen bond was completely removed at 423 K [19]. A careful observation of the carbonyl band also revealed a further shift to 1658 cm^{-1} (Figure 3b), which is even lower than that at room temperature (1665 cm^{-1}, Figure 3a). Although the stronger intermolecular interactions with carbonyl oxygen will shift the C=O band to the lower frequency, the exact bonding structure for this effect is not fully understood. A rationale for this phenomenon may be a secondary interaction between the adsorbed NMP molecules and the surface as observed in alumina [19]. This secondary interaction could happen because the increasing temperature not only causes the breaking of normal hydrogen bond, but also increases the reactivity between the surface reactive sites (such as the Al^{3+} acidic sites) and the adsorbed NMP molecules. As a result, a portion of the hydrogen bonded carbonyl oxygen was further associated with the surface acidic Al^{3+} sites. Therefore, this part of the adsorbed NMP molecules achieved a stronger association force and remained on the surface. A proposed mechanism for the secondary interactions is described in Figure 4.

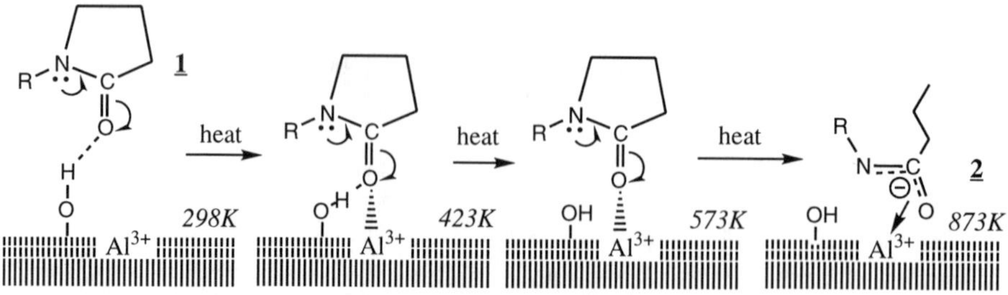

Figure 4. Surface adsorption mechanism of NMP

Although hydrogen bonding is completely deformed at 573 K and more NMP molecules have been removed (Figure 3c), NMP is still detectable on the surface. The bands corresponding to NMP molecule persist, albeit their peak intensities have decreased dramatically and the precise fingerprint structure below 1600 cm^{-1} has changed, probably because of the mixing effects such as the tortured conformation. Because the carbonyl shift (1658 cm^{-1}) at this temperature has not changed compared to that of 423 K, it seems that the secondary interaction still existed and became the main type of association between the surface and the adsorbed molecules.

It is interesting that the surface was not completely reversible upon heating to 873 K (Figure 3d) as in the case of pyridine adsorption [19]. The disappearing characteristic bands of NMP include most of the ν(C=O) band and the bands at 1652 , 1582 , and 1472 cm^{-1} (Figure 3d). A weak aliphatic C-H stretching vibration was also visible in the range of 2800-3000 cm^{-1}. Considering the chemical reactivity at such a high temperature and the affinity of amide resonance structure to the surface acidic Al^{3+} sites, the cleavage of the ring C-N bond, along with a complex coordination between the amide N-C-O bond and Al^{3+} sites, seems plausible. From the residual spectrum (Figure 3d), a partial, i.e., decomposed structure (2) is proposed.

The strength and thermal stability of the hydrogen bond of NMP are of crucial importance to the properties of AlN suspension, since the processing of the AlN powders involve a sonication step that creates an extensive amount of heat during the treatment. The hydrogen bonding and the interactions of NMP with surface acidic Al^{3+} sites on the AlN powder surface provided an adsorbed layer of NMP with decreased basicity compared to unadsorbed NMP in the media. Therefore this adsorption and the surrounded Lewis base would promote dynamic adsorption and desorption [20]. In addition, the formation of a high temperature residual, which could provide steric repulsion, is non-negligible due to the extensive amount of heat during the sonication process. Considering the high surface area of the system, the good deagglomeration force and relatively high stability of the suspension appear to be synergistic.

On addition of poly(amic acid) (**3**), a precursor to polyimide, further deagglomeration was achieved. The surface adsorption of poly(amic acid) can be partially mimicked via a model reaction of 2-pyrrolidinone (NHP) (**4**, Fig.5) on the surface of AlN by FT-IR spectroscopy, as the NHP molecule contains a similar amide structure as poly(amic acid).

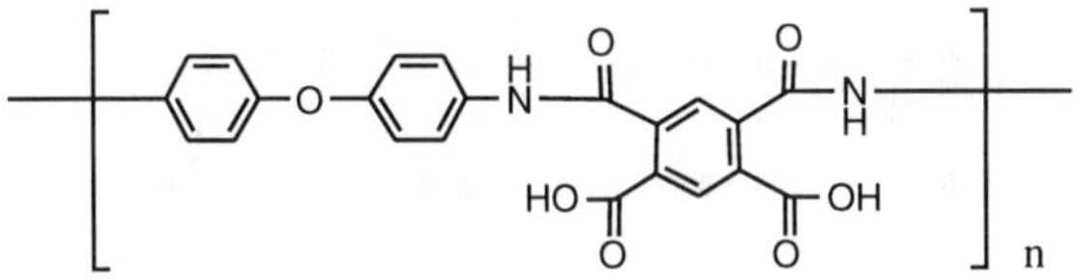

Poly(amic acid) **3**

NHP has an increased acidic character, or the tendency to lose a proton, compared to amines, and the basic character is weakened [21,22]. The N-coordination to metal cations has been observed and utilized for the study of anionic polymerization of lactams [21]. The by-products of the coordination reaction were of proton acidic character [21]. In the case of NHP adsorption at room temperature, careful analyses revealed that the overall surface association originates from both the surface hydrogen bond and the amide nitrogen complex coordination (Fig. 5). The latter involved a proton transfer mechanism that is of significance to the deagglomeration and stabilization process.

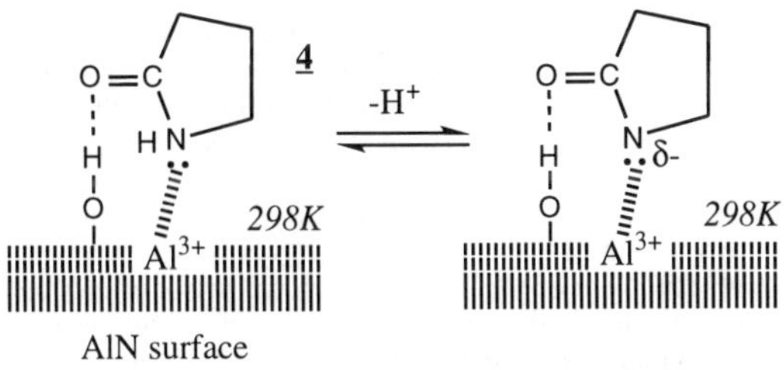

Figure 5. Surface adsorption of NHP

The adsorption spectra of NHP at various temperatures, by subtracting a background spectrum (not shown) of the activated surface, are shown in Figure 6. The formation of a hydrogen bond with the surface OH groups at room temperature is indicated as a broad band in the range of 3000-3500 cm^{-1} and a weak negative band around 3740 cm^{-1} (Figure 6a). A shift for the carbonyl band to a lower frequency was observed from 1686 cm^{-1} (for the free form of NHP [23]) to 1667 cm^{-1} (Figure 6a), which is also associated with hydrogen bonding. Since the lower intensity of the negative 3740 cm^{-1} band is an indication of less "free" surface OH groups in association with NHP, it seemed that the adsorption of NHP molecules through hydrogen bonding was less than that of NMP. Considering the weakened basicity of NHP, this decreased tendency to form hydrogen bonds is reasonable.

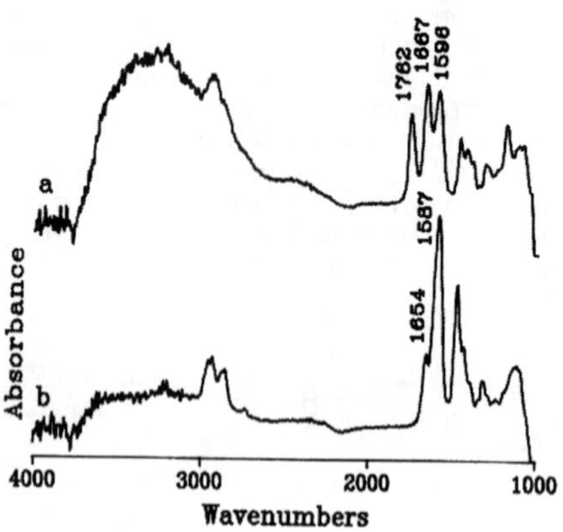

Figure 6. FTIR transmission spectra for the addition of NHP at (a) room temperature; (b) 423K. (All the spectra subtracted from the surface activation spectrum)

The striking difference between the adsorption of NMP and NHP was that, upon the adsorption of NHP at room temperature, the spectrum showed an interesting group of appearing bands (Fig. 6a) compared to the free form spectrum of NHP. The most remarkable new band appeared at 1762 cm^{-1}, which should be assigned to the carbonyl C=O stretching vibration. Compared to the free NHP spectrum, a total 100 cm^{-1} shift toward the higher wavenumber region was attained. The increase of the bond energy is associated with the strengthening of the double bond feature of the amide carbonyl group. It can be achieved by the complex coordination of the amide N atom adjacent to the carbonyl bond. The nitrogen coordination was verified by another frequency shift of the characteristic amide band from 1492 cm^{-1} (free form of NHP) to 1596 cm^{-1}, corresponding to the strengthening of C'-N (carbonyl carbon-nitrogen) stretching vibration. This coordination favors surface acidic sites such as Al^{3+}, as shown in Fig. 5.

On the other hand, the adsorption of carboxylic acid on AlN powder was reportedly through the esterification reaction with the surface hydroxyl groups [24]. In the case of the nanostructured AlN, however, the esterification reaction only occurred upon heating to 423 K, as studied by FT-IR spectroscopy using acetic acid as the probe molecule. Room temperature chemisorption was observed mainly through the hydrogen bonding between the surface hydroxyl groups and the carbonyl oxygen atom of acetic acid.

Consequently, the strong surface chemisorption of the amide and carboxylic acid species implied a strong affinity between poly(amic acid) and the AlN surface. It promoted an adsorption and desorption equilibria that could eventually lead to the diffusion of a part of the polymer chain into the porous structures of the agglomerated nanopowder. Then, a deformation mechanism similar to the exfoliation of *smectite* clays could occur [25] upon continuous external agitation, such as mechanical stirring, resulting in fully deagglomerated nanoparticles in the suspension [18].

Figure 7 summarizes the experimental results and the theoretical interpretations with respect to the mechanism of deagglomeration and stabilization of the AlN nanoparticle suspension in NMP with the addition of poly(amic acid).

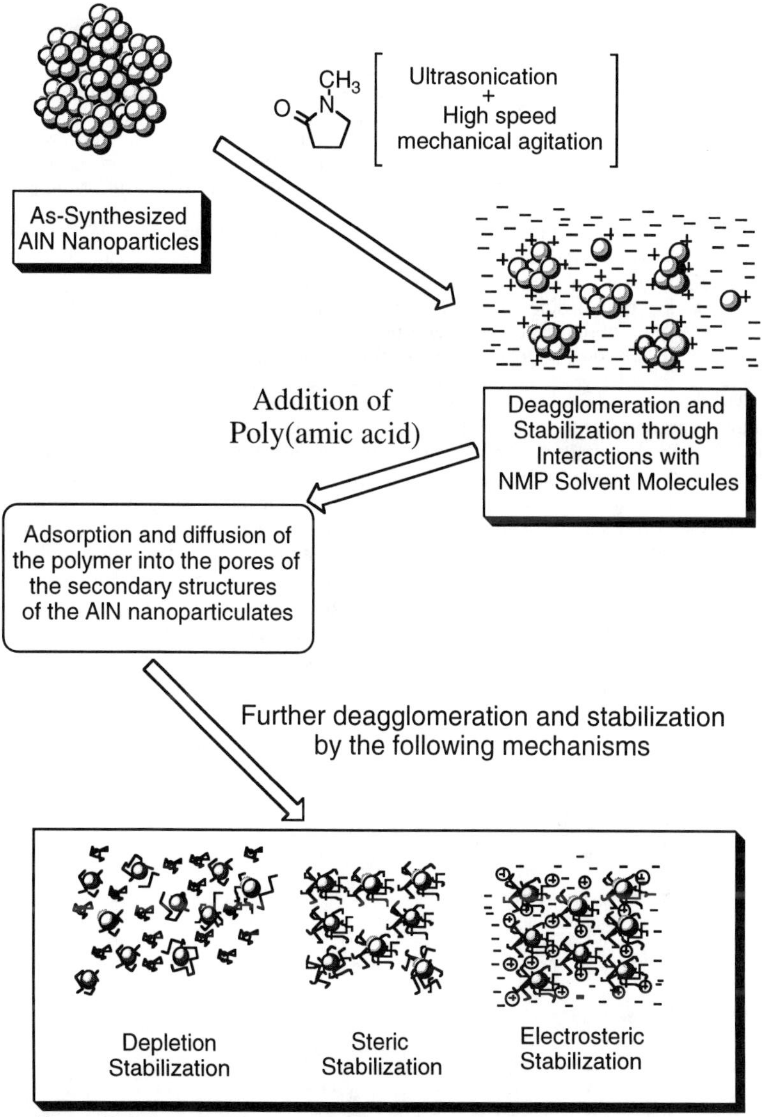

Figure 7. Schematic diagram of the interactions for AlN particle stabilization

The stabilization, however, is usually a dynamic equilibrium process [17]. The magnitudes of the stabilization interaction strongly depend on the particles' transient mutual distances. To prevent the possible destabilization of the homogeneous suspension system, we report here a rapid precipitation method to obtain fine, polymer-coated composite particles. In this process, at least two important factors need to be considered: the type of non-solvent and suspension / non-solvent ratio. H_2O is a common non-solvent for polyimide/NMP system [26], but it is not favorable in this case due to the possible surface oxidation of AlN nanoparticles. Various non-solvents were tested, i.e., methylene chloride, methanol, THF, and triethylamine. Triethylamine appears to be the best choice because it completely precipitated out the composite phase from the suspension. Another advantage of using triethylamine is its catalytic function for poly(amic acid) cyclization [27]. The suspension / non-solvent ratio during the precipitation determines the rate of solidification of the polymer. In order to obtain small particles with a uniform coating of the polymers, the nucleation rate should be controlled and the extent of particle's coalescence should be minimized. In the experiment, several different ratios of suspension vs. solvent were compared. As a result, a typical ratio of 60 ml of the suspension to 300 ml of the non-solvent turned out to be ideal for this system.

The XRD data of the polymer coated composite shows only the characteristic diffraction pattern of polycrystalline AlN nanoparticles, which along with the characteristic AlN band at 690 cm^{-1} in the FTIR spectrum, verifies the purity of the ceramic phase after the composite preparation process. This is expected, since AlN is a thermochemically stable material and the non-aqueous process was very carefully handled under a protective environment to prevent oxidation. As a result, this synthesis route allowed us to obtain a nanocomposite with 65 % AlN loading by volume.

The thermogravimetric measurements were carried out at a rate of 5°C/min. This measured the loading percentage of the composite, as well as the thermal decomposition temperature of the polymer. The polymer starts to degrade from 450°C and is fully decomposed at 700°C. The TGA residue at 700°C, a white substance, was proved to be AlN by FTIR and XRD. The TMA measurements showed an obvious trend in the reduction of the thermal expansion coefficient of the AlN/PI samples from 8 $x10^{-5}$/°C of pure polyimide to 6.21 x 10^{-6}/°C of the 65 vol.% composite (see Fig. 8). The thermal conductivity test also reveals an increase for the nanocomposite sample (1.84 W/m•°C), while the typical thermal conductivity of polyimide is reported to be 0.128 W/m°C [28]. These improved thermal properties demonstrate the effect of the second phase material, AlN, a highly thermally conducting (theoretical value 320 W/m•°C [29,30]) and low thermal expansion ($3.5x10^{-6}$/°C, below 200°C [31,32]) ceramic, as was anticipated [18].

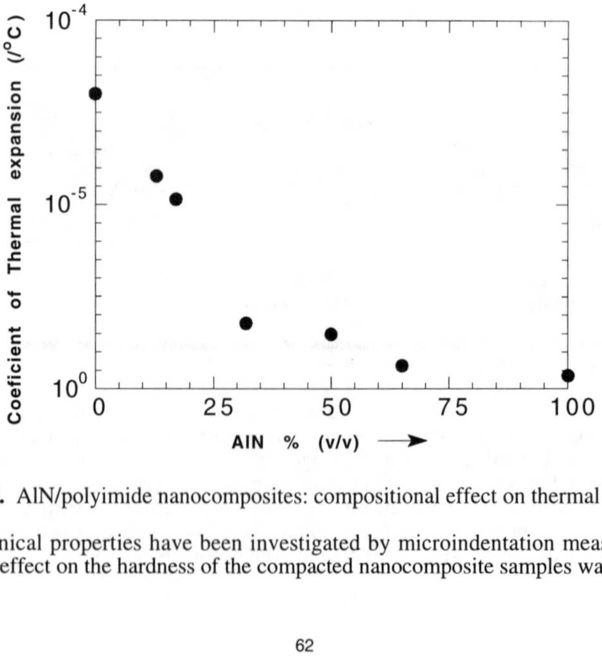

Figure 8. AlN/polyimide nanocomposites: compositional effect on thermal expansion

Mechanical properties have been investigated by microindentation measurements. The compositional effect on the hardness of the compacted nanocomposite samples was evident, which showed an increasing hardness with the increasing ceramic content. (Fig.9) As a result, the Vickers hardness of a 65 vol.% nanocomposite sample was approximately seven fold that of the pure polyimide sample.

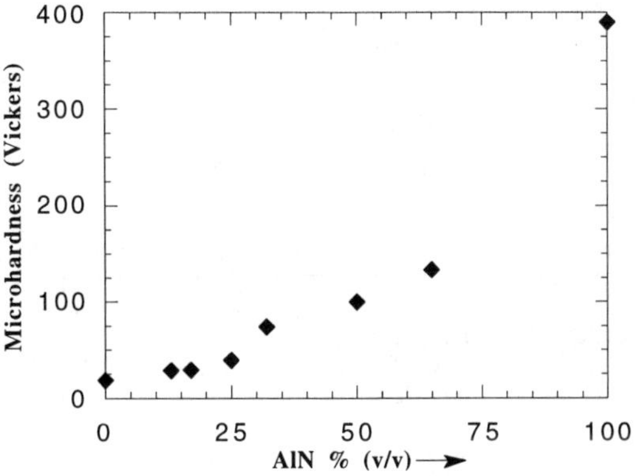

Figure 9. Microindentation measurement of AlN/PI nanocomposites

CONCLUSIONS

A series of polyimide/AlN nanocomposites with highly loaded ceramic content were prepared. Unique preparation methods were developed to efficiently prevent the agglomeration of the nanoparticles. By FTIR spectroscopy, the probe molecule reactions using NMP and NHP were carried out, in order to elucidate the interaction forces and mechanisms involved in the deagglomeration and stabilization process. Moreover, the compositional effects on thermal and mechanical properties of AlN/polyimide nanocomposites have been investigated.

ACKNOWLEDGMENT

The authors wish to express their thanks for partial support by the U.S. Office of Naval Research. Thanks to Dr.M-I. Baraton of the University of Limoges (France) for access to the surface FT-IR spectroscopy facilities at Limoges. Thanks also to Prof.D.D.L. Chung, of the Composite Materials Research Laboratory, State University of New York at Buffalo, for the thermal conductivity measurements.

REFERENCES

1. (a)H. Gleiter, Adv. Mater. **4**, 474 (1992); (b)H. Gleiter, Nanostruct. Mater. **1**, 1 (1992); (c) R. Birringer, H. Gleiter, Encyclopedia of Materials Science & Engineering, Suppl. Vol. 1, edited by R.W. Cahn, (Pergamon Press, Cambridge, 1988), p.339.
2. (a)R.W. Siegel, Nanostruct. Mater. **3**, 1 (1993); (b)R. Dagani, Chem. Eng. News **72**(47), 18 (1992).
3 (a)R.P. Andres, R.S. Averback, W.L. Brown, L. E. Brus, W.A. Goddard, III, A. Kaldor, S.G. Louie, M. Moscovits, P.S. Peercy, S.J. Riley, R.W. Siegel, F. Spaepen, Y. Wang, J. Mater. Res. **4**(3), 704 (1989); (b)H. Gleiter, Prog. Mater. Sci. **33**, 223 (1989); (c)R. Birringer, U. Herr, and H. Gleiter, Suppl. Trans. Jpn. Inst. met. **27**, 43 (1986); (d)F. Feynman, Miniaturization, edited by H.D. Gilbert, (Reinhold, New York, 1961).
4. (a) R.A. Andrievski, J. Mater. Sci. **29**, 614 (1994); (b)R.W. Siegel, Mater. Sci. and Eng. **B19**, 37-43 (1993); (c)R.D. Shull, L.E. Bennett, Nanostruct. Mater. **2**, 1, 53 (1992); (d) C.

Suryanaraya, F.H. Froes, Metall. Trans. **23A**, 1071 (1992); (e)G.M. Whitesides, J.P. Mathias, C.T. Seto, Science **254**, 1312 (1991).
5. K.E. Gonsalves, G.M. Chow, T.D. Xiao, R.C. Cammarata, Molecularly Designed Ultrafine/Nanostructured Materials, MRS Symposium Proceedings, Vol.351, (1994).
6. M.J. Yacaman, T. Tsakalakos, B.H. Kear, "Proceedings of the First International Conference on Nanostructured Materials", Special Proceedings Vol., Nanostruct. Mater., **3**(1-6), (1993).
7. N. Ichinose, Y. Ozaki, S. Kashu, Superfine Particle Technology, (Springer-Verlag, London, 1992). (Translated from Japanese)
8. K.E. Gonsalves, G.M. Chow, "Particle Synthesis by Chemical Route", Nanostructured Materials: Synthesis, Properties, and Uses, (Inst. of Physics, UK 1996), Chapt. 3.
9. D. Chakravorty, A. K. Giri, "Nanomaterials", chapter in Chemistry for the 21st Century: Chemistry of Advanced Materials, edited C.N.R. Rao, (Blackwell Scientific Publications, London, 1993), p.217.
10. C.M. Lukehart, J.P. Carpenter, S.B. Milne, K.J. Burnam, CHEMTECH **8**, 29 (1993).
11. (a)P. Calvert, Mater. Sci. Engi. **C1**, 69 (1994); (b)J. Burdon, J. Szmania, P. Calvert, in Molecularly Designed Ultrafine/Nanostructured Materials, edited by K.E. Gonsalves, G.M. Chow, T.D. Xiao, R.C. Cammarata, MRS Symposium Proceedings, Vol 351, 1994, p.103; (c) Sukun Zhang, K.E. Gonsalves, ibid, p. 245.
12. (a)K.E. Gonsalves, G.M. Chow, Y. Zhang, J.I. Budnick, T.D. Xiao, Adv. Mater. **6**, 4, 291 (1994); (b)R.D. Shull, R.D. McMichael, J.J. Ritter, Nanostruct. Mater. **2**, 205 (1993); (c)R.F. Ziolo, E. P. Giannelis, B.A. Weinstein, M. P. O'Horo, B.N. Ganguly, V. Mehrotra, M. W. Russell, D. R. Huffman, Science **257**, 219 (1992).
13. (a)M.J. Bloemer, J.W. Haus, P.R. Ashley, J. Opt. Soc. Am., **7B**, 5, 790 (1990); (b)A.W. Olsen, Z. H. Kafafi, J. Am. Chem. Soc. **113**, 20, 7758 (1991); (c) K.E. Gonsalves, G. Carlson, J. Kumar, F. Avanda, M. Jose-Yacaman, Nanotechnology: Molecularly Designed Materials, ACS Symposium Series, Vol 622, edited by K. Gonsalves, G.M. Chow, (ACS, D.C., 1996), p.151-161.
14. P.B. Messersmith, E.P. Giannelis, Chem. Mater. **5**, 1064 (1993); ibid, **5**, 1694 (1993).
15. R.J. Pugh, L.Bergstrom, Surface and Colloid Chemistry in Advanced Ceramics Processing, Surfactant Science Ser., Vol. 51, (Marcel Dekker, 1994).
16. (a)D.H. Napper, Polymer Stabilization of Colloidal Dispersions, (Academic Press, 1983); (b)P.Dubin, P. Tong, Colloid-Polymer Interactions: Particulate, Amphiphilic, and Biological Surfaces, ACS Symp. Ser., Vol. 532, (American Chemical Society, 1993).
17. T.F. Tadros (Ed), Solid/Liquid Dispersions, (Academic Press, 1987), p.91.
18. X. Chen, K.E. Gonsalves, G.M. Chow, T.D. Xiao, Adv. Mater. **6**, 6, 481 (1994).
19. M.I. Baraton, X. Chen, K.E. Gonsalves, Nanotechnology: Molecularly Designed Materials, ACS Symposium Series, Vol 622, edited by K. Gonsalves, G.M. Chow, (ACS, D.C., 1996), p.312.
20. F.M. Fowkes, D.W. Dwight, J.A. Manson, T.B. Lloyd, D.O. Tishler, B.A. Shah, Mater. Res. Soc. Symp. Proc. Vol 119, 223, (1988).
21. H. Sekiguchi, in *Ring-Opening Polymerization, Vol 2*, Ed. by K.J. Ivin, T. Saegusa, (Elsevier Applied Science Publishers, 1984), chapter 12.
22. T. Tahara, H. Imazaki, K. Aoki, H. Yamazaki, J. Organometa. Chem. **327,** 157 (1987).
23. (a)D.P. McDermott, J. Phys. Chem. **90**, 2569 (1986); (b)P.S. Peek, D.P. McDermott, Spectrochimica Acta **44** A , 4, 371 (1988).
24. M. Egashira, Y. Shimizu, Y. Takao, R. Yamaguchi, Y. Ishikawa, J. Am. Ceram. Soc., **77**(7), 1793 (1994)
25. (a) P.B. Messersmith, E.P. Giannelis, Chem. Mater. **6**, 1719 (1994); (b)T.J. Pinnavaia, T. Lan, Z. Wang, H. Shi, P.D. Kaviratna, Nanotechnology: Molecularly Designed Materials, ACS Symposium Series, Vol 622, edited by K. Gonsalves, G.M. Chow, (ACS, D.C., 1996), p.250.
26. T. Lin, K.W. Stickney, M. Rogers, J.S. Riffle, J.E. McGrath, H. Marand, Polymer **34**(4), 772 (1993).
27. M.I. Bessonov, V.A. Zubkov, Polyamic acids and Polyimides: Synthesis, Transformations, and Structure, (CRC Press, Inc., 1993), p.38.
28. L. Li, D.D.L. Chung, J. Elect. Mater. **23**(6), 557 (1994).

29. L. M. Sheppard, Ceramic Bulletin **69**(11), 1801 (1990).
30. T. J. Mroz, Jr., Ceramic Bulletin **75**(5), 782 (1992).
31. A.V. Dobrynin, Inorg. Mater. **28**(7), 1063 (1992).
32. M. Hirano, N. Yamauchi, J. Mater. Sci. **28**, 5737 (1993).

NOVEL CARBOXY FUNCTIONALIZED SOL-GEL PRECURSORS

H. WOLTER, W. STORCH, AND C. GELLERMANN,
Fraunhofer-Institut für Silicatforschung, Neunerplatz 2, D-97082 Würzburg, Germany

ABSTRACT

A novel family of inorganic-organic copolymers (ORMOCER*s) derived from urethane- and thioether(meth)acrylate alkoxysilanes has been successfully exploited for a variety of diverse applications. In order to widen the range of applications an additional functionality (carboxy group) has been incorporated in this silane type. Conventional sol-gel processing facilitates the formation of an inorganic "Si-O-Si"-network via hydrolysis and polycondensation reactions of alkoxysilyl moieties and in addition, the (meth)acrylate groups are available for radically induced polymerization to obtain a complementary organic polymeric structure. The presence of a carboxy group would appear to have great potential for a range of diverse areas of application, such as an internal catalyst for the sol-gel process, complexation of elements such as Zr and Ti, increasing the adhesion to various substrates and modification of solubility. A number of novel silanes and their syntheses will be described in this paper.

INTRODUCTION

In addition to organic polymers, glasses and ceramics, inorganic-organic copolymers (ORMOCERs) now constitute an important class of materials. The importance of this development is evident in the success of applications within a variety of areas, for example as coatings possessing desirable properties such as scratch and abrasion resistance [1], corrosion resistance [2], low optical loss factor [3,4], low permittivity constant and high dielectric strength [4,5]. The synthesis of bulk-ORMOCERs [6,7,8] based upon oligo(meth)acrylate alkoxysilanes [6,7] would appear to offer a promising alternative to the outlined "classical" materials. Modification of the structure and composition of these multifunctional silanes allow the synthesis of bulk materials with adjustable properties [6,7,8], such as Youngs modulus, thermal expansion coefficient and refractive index which are important for special applications e. g. soft lenses, optical wave guides [3] and dental filling materials [9,10]. These examples serve to illustrate the importance of special functionalized sol-gel precursors. In particular, the development of novel oligo(meth)acrylcarboxy silanes which possess an additional carboxy group may have potential in the field of ORMOCER-coatings and bulk materials.

GENERAL ASPECTS

A general structural scheme of the new type of sol-gel precursor including the manifold of possible variations (e. g. number of (meth)acrylate, carboxy and alkoxy groups and choice of connecting unit) is shown in Fig. 1. The (meth)acrylate and alkoxysilyl groups of the oligo(meth)acrylate alkoxysilanes allow the construction of an inorganic-organic copolymer network and in addition the modification of final properties. The choice of the connecting unit can also influence the final properties of the material, via variation in chain length and chemical nature (e. g. presence of aliphatic or aromatic moieties) of the unit.

* registered trademark of Fraunhofer-Gesellschaft in Germany

Mat. Res. Soc. Symp. Proc. Vol. 435

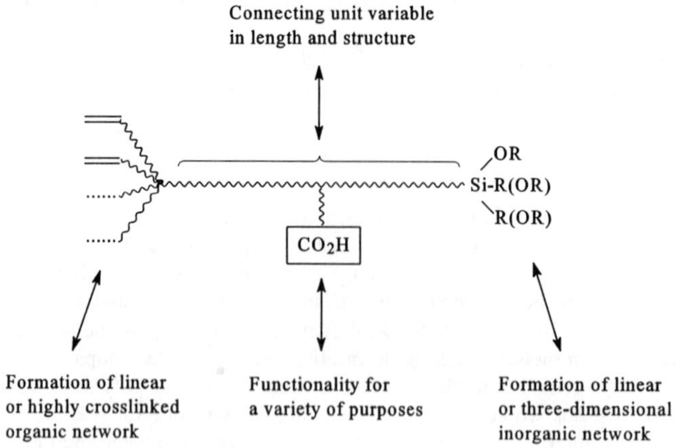

Figure 1. Schematic formula of the novel carboxy functionalized silanes.

EXPERIMENTAL

Infrared spectra were recorded using a FTIR (Bio-Rad FTS-25) in order to monitor the course of the synthesis and characterize the carboxy functionalized silane product. ^{1}H-NMR measurements were performed at 400.14 MHz with a Bruker WM400 Spectrometer using $CDCl_3$ as solvent and internal standard.

Synthesis of silane (2)

19.5 g (0.15 mole) of hydroxyethylmethacrylate (HEMA, Fluka Chemie) were added dropwise to 41.2 g (0.15 mole) of 3-(methyldiethoxysilyl)propylsuccinic anhydride (1) (Wacker-Chemie) under inert atmosphere. The reaction time varied over several hours or days, dependent upon the reaction temperature and specific catalyst employed. The product was obtained as a colourless liquid.

A similar strategy was employed for the synthesis of silane (3). In contrast it was necessary to use a solvent (THF) during the synthesis of silane (4) due to the solid nature of 2.2-bis-[4-(2-hydroxy-3-methacryloxypropoxy)phenyl]propane (BISGMA).

RESULTS AND DISCUSSIONS

The new silane type can be prepared by addition of an OH-substituted (meth)acrylate to an anhydride-substituted silane. For silane (2) (s. Fig. 2) HEMA and silane (1) are the reactants for the esterification. The progress of this reaction can be monitored by IR-spectroscopy. During the course of the reaction, the $\nu_{(OH)}$-absorption in HEMA at 3430 cm^{-1} in addition to the $\nu_{as(C=O)}/\nu_{s(C=O)}$-absorptions in silane (1) at 1786/1864 cm^{-1} decrease, whereas a characteristic $\nu_{(OH)}$-absorption at 3600 - 2500 cm^{-1} of the resulting carboxy group emerges. The corresponding two carbonyl groups result in an enhancement of the $\nu_{(C=O)}$-absorption at 1721 cm^{-1}. In addition, the presence of the $\nu_{(C=C)}$-absorption at 1638 cm^{-1} is consistent with the structure of silane (2) in Fig. 2.

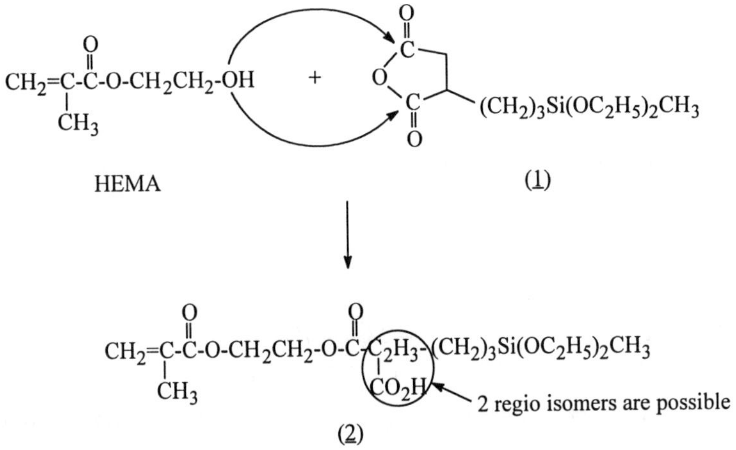

Figure 2. Reaction scheme of the synthesis of silane (2).

Inspection of the reaction scheme reveals that in principle two isomers (circled in Fig. 2) are possible. Actually, the formation of two regio isomers may be evident in the number of ^{1}H-NMR signals. Integration of the signals reveals a product purity of ca. 70 % according to the particular reaction temperature and catalyst employed. By-products originate from transesterification reactions at the silyl moiety. ^{1}H-NMR data in addition to IR investigations are shown in Table I.

Table I. Characterization of carboxy silane (2) by IR and ^{1}H-NMR-spectroscopy.

IR (film on KBr):	ν = 3600 - 2500 cm^{-1}, 1721, 1638, 1455, 1408, 1376, 1320, 1298, 1259, 1167, 1104, 1080, 1043, 949, 815, 801, 768.
^{1}H-NMR ($CDCl_3$):	δ = 0.03 ppm (s, 3H, Si-CH_3), 0.48-0.58 (m, 2H, $CH_2CH_2\underline{CH_2}Si$), 1.12 (t, 6H, 2 $OCH_2\underline{CH_3}$), 1.28-1.41 (m, 2H, $CH_2\underline{CH_2}CH_2Si$), 1.44-1.56, 1.58-1.71 (2m, 2H, $\underline{CH_2}CH_2CH_2Si$), 1.86 (s, 3H, $C(\underline{CH_3})=CH_2$), 2.34-2.44, 2.60-2.84 (2m, 3H, $O_2CC_2\underline{H_3}CO_2H$), 3.67 (q, 4H, 2 $O\underline{CH_2}CH_3$), 4.26 (s, 4H, $O\underline{CH_2}\underline{CH_2}O$), 5.51, 6.04 (2s, 2H, $C(CH_3)=\underline{CH_2}$), 9.13 (s, 1H, COOH).

Silanes (3) and (4) show characteristic $\nu_{(C=O)}$-absorptions in the IR region at 1724 and 1726 cm^{-1} (Table II). Broad $\nu_{(OH)}$-absorptions at 3600 - 2500 cm^{-1} and $\nu_{(C=C)}$-absorptions at 1638 cm^{-1} confirm the presence of carboxy and methacryl functions in both structures. In addition, the two characteristic $\nu_{(C=C)}$-absorptions at 1609 and 1583 cm^{-1} of aryl moieties support the structure of silane (4).

Table II. Characteristic IR absorptions of silanes (2)-(4).

Silane\\ν [cm^{-1}]	COOH	C=O	C=C
(2)	3600 - 2500	1721	1638
(3)	3600 - 2500	1724	1638
(4)	3600 - 2500	1726	1638, 1609, 1583

On account of the presence of two OR-groups and one C=C-group, a two-dimensional inorganic structure in combination with a linear organic polymer structure is expected using silane (2) in the preparation of an inorganic-organic copolymer. Silane (3) (see Fig. 3) has a structure which allows in addition to the linear inorganic structure a three-dimensional organic network; it is expected therefore a copolymer with a higher Youngs modulus as a result. A further increase in Youngs modulus is expected as a consequence of four OR-groups on silane (4) (see Fig. 4), effecting a possibly compact inorganic polymeric structure in combination with the rigid aromatic connecting unit. The additional carboxy group in this particular silane allows further modification in contrast to the single moiety found for silanes (2) and (3).

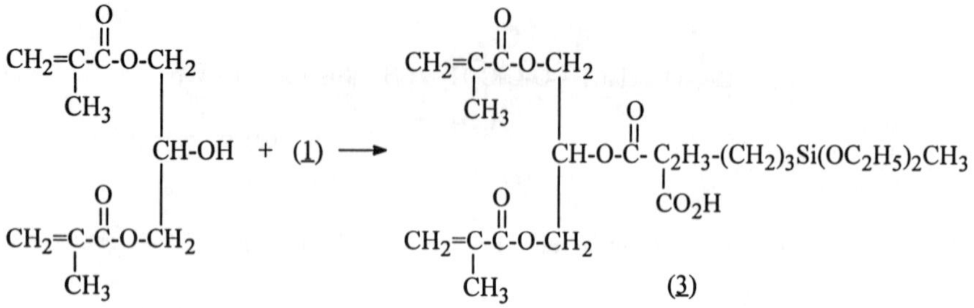

Figure 3. Reaction scheme for the synthesis of silane (3).

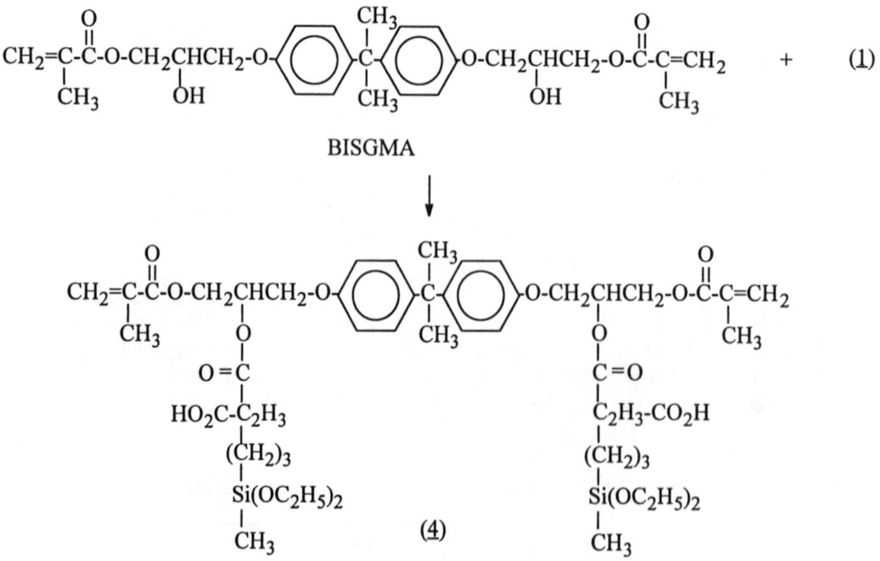

Figure 4. Reaction scheme for the synthesis of silane (4).

CONCLUSION

Three novel methacryl carboxy alkoxysilanes have been synthesized and exemplify a new class of silane. In comparison with the established oligo(meth)acrylate alkoxysilanes, the presence of an additional carboxy functionality would appear to have great potential for a range of diverse areas of application, such as an internal catalyst for the sol-gel process, adhesion promoter, complexation moiety and solubility modifier, etc. Future work will address the synthesis and characterization of further structural modifications of the outlined alkoxysilanes, with subsequent preparation of inorganic-organic copolymers for various applications.

ACKNOWLEDGEMENT

We are very grateful to Mrs. H. Bäuerlein, Mrs. K. Langguth and Mrs. D. Hanselmann for their technical assistance.

REFERENCES

1. S. Amberg-Schwab, E. Arpac, W. Glaubitt, K. Rose, G. Schottner, U. Schubert in High Performance Films & Coatings, edited by P. Vincenzini (Elsevier Science Publ., Amsterdam, 1991), pp. 203-210.

2. K. Greiwe, Farbe+Lack **97**, 968 (1991).

3. P. Dannberg, A. Bräuer, W. Karthe, R. Waldhäusl, H. Wolter in Micro system technologies `94, edited by H. Reichl and A. Heuberger (vde-verlag GmbH, Berlin, 1994), pp. 281-287.

4. M. Popall, J. Kappel, J. Schulz, H. Wolter in Micro system technologies `94, edited by H. Reichl and A. Heuberger (vde-verlag GmbH, Berlin, 1994), pp. 271-280.

5. H. Wolter, H. Schmidt, DVS-Berichte 129 (1990).

6. H. Wolter, W. Glaubitt, K. Rose in Better Ceramics Through Chemistry V, edited by M.J. Hampden-Smith, W.G. Klemperer, C.J. Brinker (Mater. Res. Soc. Symp. Proc. 271, Pittsburgh, PA, 1992), pp. 719-724.

7. H. Wolter, W. Storch in Functional and Structural Materials (Polymer & Mater. Res. Symp. Proc., Bayreuth, 1993), pp. 14-17.

8. H. Wolter, W. Storch in 4th European Polymer Federation Symposium on Polymeric Materials (Baden-Baden, 1992), p. 79.

9. H. Wolter, W. Storch, H. Ott in Better Ceramics Through Chemistry VI, edited by A.K. Chestham, C.J. Brinker, M.L. Mecartney, C. Sanchez (Mater. Res. Soc. Symp. Proc. 346, Pittsburgh, PA, 1994), pp. 143-149.

10. H. Wolter, W. Storch in MakroAkron 94 (35th IUPAC International Symposium On Macromolecules Proc., Akron, Ohio, 1994), p. 509.

THERMAL CHARACTERISTICS OF LAYERED DOUBLE HYDROXIDE INTERCALATES - COMPARISON OF EXPERIMENT AND COMPUTER SIMULATION

I. S. Bell[1,2], F. Kooli[1], W. Jones[1] and P. V. Coveney[2]
[1]Department of Chemistry, University of Cambridge, Lensfield Road, Cambridge. CB2 1EW, UK.
[2]Schlumberger Cambridge Research, High Cross, Madingley Road, Cambridge. CB3 0EL, UK.

ABSTRACT

Terephthalate anions have been incorporated as charge-balancing species inside the layers of Mg-Al containing layered double hydroxides (LDHs). The ease of incorporation as well as the orientation of the guest species has been studied as a function of the Mg/Al ratio. The orientation of the guest as a function of temperature and degree of hydration has also been studied using powder X-ray diffraction. Molecular dynamics simulations, using an NPT ensemble, have been performed and good agreement between experimental and simulated data has been obtained.

INTRODUCTION

In a series of publications we have reported on the synthesis of organic intercalates of LDHs [1-5]. A variety of organic anions have been introduced including cinnamate, benzoate and terephthalate. In particular we have described how the method of preparation (for example, pH of precipitation) as well as the charge on the sheets is important in controlling the incorporation of the organic anion [5]. Here we discuss in more detail how the orientation of the terephthalate anion varies with the Mg/Al ratio, temperature/degree of hydration of the sample and in particular show how it is possible using computer simulation to model the observed changes.

EXPERIMENTAL

Sample synthesis

Full details of the preparative procedures used have been given elsewhere [4,5]. Our approach involves the co-precipitation (at pH 10) of Mg and Al nitrate salts in the presence of the terephthalate anion.

Simulation details

Full details of the simulation procedures and underlying theory will be presented elsewhere [6]. We created a supercell consisting of 4 unit cells in the *a* and *b* directions and either 3 or 2 hydroxide layers in the *c* direction. These are referred to as the 4 x 4 x 3 or 4 x 4 x 2 supercells, respectively. The particular Mg:Al ratio chosen

Mat. Res. Soc. Symp. Proc. Vol. 435

was 3 and was created by appropriate substitution of Mg by Al. In previous clay simulations, the Dreiding force field [7] has been found to be effective [8]. A particular difficulty associated with using the Dreiding force field in the case of LDHs, however, is that it does not contain parameterisation for Mg. In order to overcome this difficulty, we made the approximation of parameterising Mg in exactly the same way as Al (for which Dreiding does have parameters), with the exception of charges (+2 for Mg and +3 for Al). A justification for this approach is that the metal ions are located within the metal-hydroxyl sheets, and that there fore their exact identity should not have too large an effect on species in the interlayer region; their dominant interaction with the interlayers will be Coulombic, which should be parameterised accurately. (The excellent agreement between our simulated and experiment results would appear to vindicate the use of this approximation.)

For the water we used the water geometry and charges (i.e. charge of -0.85 on the O atom) derived from the dedicated TIP3P water force field of Jorgensen et al [9], together with the van der Waals and bonded parameters from the Dreiding force field. The charges on the atoms in the brucite-like sheets were calculated using the Charge Equilibration (QEq) method [10]. The geometry and atomic charges of the intercalated anions were optimized using the MOPAC semi-empirical method, employing the PM3 Hamiltonian [11]. The long range Coulombic interactions were computed using the Ewald summation technique. Van der Waals interactions were treated with a cut-off radius of 14 Å. Constant pressure molecular dynamics requires the use of 3-dimensional periodic boundary conditions, implying that none of the atoms in the supercell were clamped but instead all were free to move.

Most of our computational work was performed using the molecular modeling package CERIUS^2 v1.6.2 from Molecular Simulations Inc. 1995. For example, for the 523K simulation the terephthalate anions were first positioned randomly in the interlayers and then energy minimization was performed on the system, allowing it to relax to a minimum - subject to the constraint of fixed lattice parameters. Molecular dynamics was performed using a constant NPT ensemble (fixed number of atoms, pressure and temperature) which allows the lattice parameters of the supercell to change in response to the externally applied pressure. The equivalent hydrostatic pressure was set to 0.0001GPa (1 atmosphere). The dynamics time step was set to 0.001ps in order to ensure reliable results (the time step must be an order of magnitude smaller than the fastest local motion for the numerical integration of Newton's equations of motion to be stable). The Hoover thermostat [12] was used for thermal coupling with a relaxation time of 0.1ps and a cell mass prefactor of 1. Simulations were typically performed for a simulated time of 20ps, a CPU time of about 36 hours on an R4400 processor, by which time equilibration was attained - as judged by the simulation temperature and constancy of the lattice parameters.

RESULTS

Experimental data

Figure 1 shows the variation in PXRD for a series of terephthalate intercalates obtained for different Mg:Al ratios. For the intercalate formed with a Mg:Al ratio of 1 the d_{003} value close to 14.4 Å is associated with the incorporation of the terephthalate anion into the gallery region in a vertical orientation (approximate size of the anion is 9 Å and a hydroxide layer thickness of 4.8 Å [4,13]. At ratios 2 and 3 similarly expanded strctures are observed. At ratio 5, however, the observed value of d_{003} has been interpreted by us as indicating preferential incorporation of the counter inorganic anion (i.e. nitrate) during synthesis.

Figure 2 shows how the PXRD patterns vary as a function of temperature for the terephthalate intercalate for Mg:Al = 2. At 200 °C the principal d_{003} is found close to 8.8 Å with a shoulder at lower two-theta. At an intermediate temperature of 75 °C it can be seen that this shoulder is resolved (at 11.4 Å) and is associated (indexed as d_{006}) with an additional peak at 23.2 Å (indexed as d_{003}). This has been interpreted by us as an ordered interstratified arrangement of the collapsed (8.8 Å) and expanded (14.4 Å) structures [4,5].

Clearly therefore the experimental data points to a dependence of orientation on both Mg:Al ratio and temperature/degree of hydration.

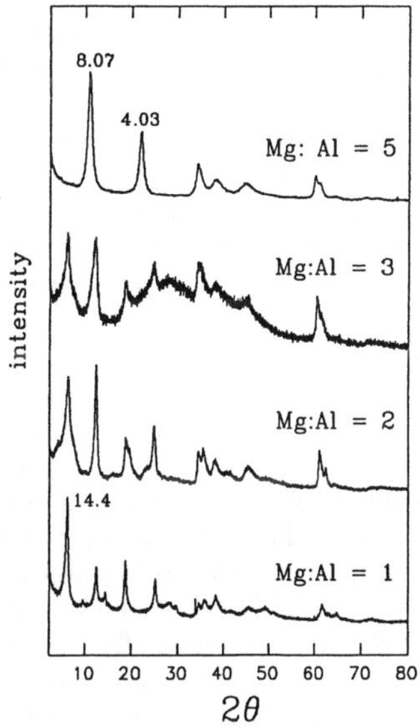

Figure 1. PXRD data for Mg:Al terephthalate intercalates as a function of Mg:Al ratio. The broad background observed at ratio 3 is due to the relatively large amount of water in this sample.

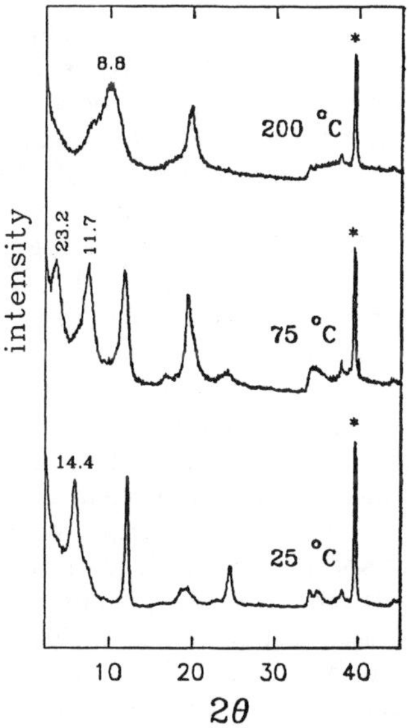

Figure 2. Variation in PXRD as a function of temperature for the Mg:Al ratio = 2 terephthalate intercalate. (* = reflections from Pt holder)

Simulation results.

The interlayer spacing for the terephthalate intercalate has been shown above to be temperature-dependent. For the high temperature case, when it can be assumed that all of the interlayer water has been lost, we initially set the system up such that the anions were at approximately 45° to the brucite-like sheets, with an initial layer spacing of 11.0 Å - a value intermediate between the expanded and collapsed values. After 20ps of molecular dynamics at a simulated temperature of 523K had been performed, the anion was found to adopt an orientation approximately coplanar with the sheets (Figure 3). In this orientation, the simulated interlayer spacing had decreased to 8.9 A - to be compared with the experimental value of 8.8 Å.

At room temperature (300K), it is known that a considerable amount of water is present in the interlayer region of the terephthalate intercalate, but experimental results as to the exact amount vary depending upon preparation procedure. We therefore employed a grand canonical Monte Carlo sorption simulation [14] (fixed μ, V and T), with the initially anhydrous intercalate in contact with a fixed pressure of 100 kPa (1 atmosphere) of water. The lattice parameters of the simulation cell (at the experimental layer spacing of 14.4 Å) were held constant. The contents of the original supercell containing the terephthalate anions were held fixed as was the geometry of the adsorbing water molecules. 20 water molecules were found to adsorb per interlayer per supercell after 5 million Monte Carlo steps, by which time equilibration was attained as judged by the constancy of the system loading and energy. The resulting system was energy minimized, and NPT molecular dynamics performed as described above. In this case, the terephthalate anions adopted an orientation perpendicular to the brucite-like sheets (Figure 4), and the simulated interlayer spacing was 14.1 Å - in good agreement with the experimental value.

To model the intermediate temperature regime (348K) where the experimental interlayer spacing was found to be the sum of that at room temperature and that at 200 °C we removed all of the water from one layer of the result of the 300K simulation described above, while the other retained its full complement of 20 water molecules. After 20ps of NPT molecular dynamics, we found that the anions in the hydrated gallery remained in the perpendicular orientation, whilst those in the anhydrous interlayer formed a collapsed gallery (Figure 5). The resulting layer spacings were 14.0 and 9.3 Å, giving an overall repeat distance of 23.3 A in excellent agreement with the experimental value of 23.2 Å.

The water molecules were then removed from the hydrated layer, but the layer spacings determined above retained, and a grand canonical Monte Carlo simulation employed to allow water to adsorb from a 100 kPa reservoir into this asymmetric supercell. 19 molecules were adsorbed into the originally hydrated layer, with only 2 adsorbed into the originally anhydrous layer, demonstrating that indeed a collapsed gallery would be essentially anhydrous but an expanded gallery would remain hydrated. Whilst this result supports our model of alternating layers of hydrated (perpendicular) and anhydrous (coplanar) layers of terephthalate anions, it does not provide a mechanism by means of which this can be achieved.

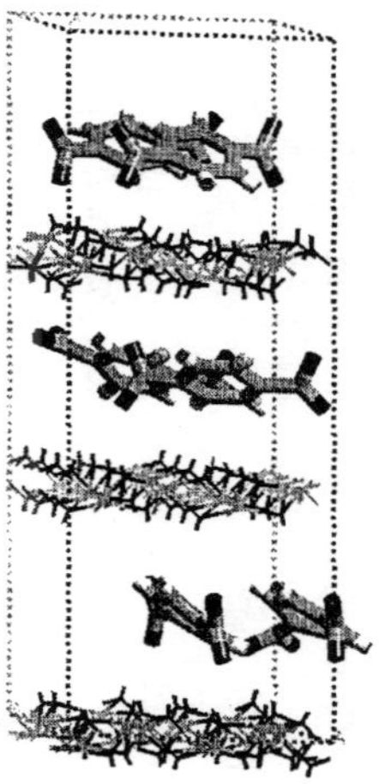

Figure 3. 523K simulation in which the anions adopt a horizontal arrangement

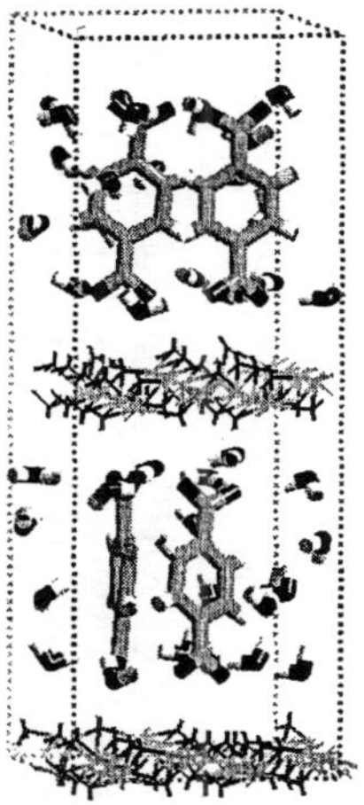

Figure 4. Vertical orientation of anions at 300K

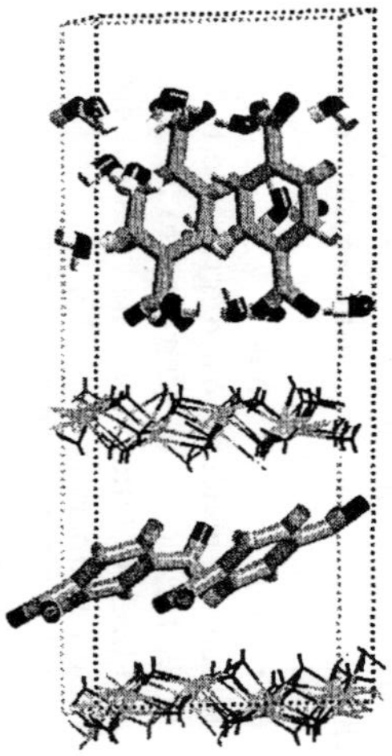

Figure 5. Supercell arrangement at 348K consisting of horizontal and vertical arrangements in alternating layers

CONCLUSIONS

Computational simulations, although based on relatively simple models and simulation procedures, have been demonstrated to be an effective procedure for reproducing the experimental features of the orientation of guest species within layered double hydroxides. The results are in good agreement with experiment despite the simplicity of the approach, and suggest that further modeling of these materials would be beneficial. Simulations involving other Mg:Al ratios are planned as well as with other intercalated anions.

The mechanism by which the water is lost to give rise to the superstructure is as yet unclear. It appears, however, not to be restricted to terephthalate intercalates. Comparable results have also been obtained for a benzoate intercalate. Rehydration also appears to proceed through a superstructure [15].

ACKNOWLEDGMENTS

We are grateful to Biosym/Molecular Simulations Inc. for facilitating access to some of the software necessary for this work. We also wish to thank the European Union and Schlumberger Cambridge Research for financial assistance.

REFERENCES

[1] J. Valim, B. M. Kariuki, J. King and W. Jones, Mol. Cryst. Liq. Cryst., **211**, 271 (1992).

[2] M. Vucelic and W. Jones, Proc. NATO-ASI on Multifunctional Mesoporous Inorganic Solids, Kluwer Academic Publishers, 373 (1993).

[3] M. Chibwe, W. Kagunya and W. Jones, Mol. Cryst. Liq. Cryst., **244**, 155 (1994).

[4] M. Vucelic, G. D. Moggridge and W. Jones, J. Phys. Chem., **99**, 8328 (1995).

[5] F. Kooli and W. Jones, in Synthesis of Microporous Materials: Zeolites, Clays and Nanostructures, American Chemical Society, 1996, in press.

[6] I. S. Bell, P. V. Coveney and W. Jones, in preparation.

[7] S. L. Mayo, B. D. Olafson and W. A. Goddard III, J. Phys. Chem., **94**, 8897 (1990)

[8] A. Bains, E. S. Boek and P. V. Coveney, submitted for publication (1996).

[9] W. L. Jorgensen, J. Chandraskhar, J. D. Madura, R. W. Impey and M. L. Klein, J. Chem. Phys., **79**, 926 (1983).

[10] A. K. Rappe and W. A. Goddard III, J. Phys. Chem., **95**, 3358 (1991).

[11] J. J. P. Stewart, MOPAC Version 6.0, QCPE No. 455, Department of Chemistry, Indiana University (1990).

[12] W. H. Hoover, Phys. Rev. A., **31**, 1695 (1985).

[13] F. Cavani, F. Trifiro and A. Vaccari, Catalysis Today, **11**, 173 (1991).

[14] N. Metropolis, A. W. Rosenbluth, M. N. Rosenbluth, A. H. Teller and E. Teller, J. Chem. Phys., **21**, 1087 (1953).

[15] F. Kooli et al, in preparation (1996).

NANOLAYER ORDERING IN AN EPOXY-EXFOLIATED CLAY HYBRID COMPOSITE

Tie Lan and Thomas J. Pinnavaia
Department of Chemistry and Center for Fundamental Materials Research,
Michigan State University, East Lansing, MI 48824

ABSTRACT

A new class of epoxy-clay nanocomposites have been prepared from alkylammonium exchanged fluorohectorite clays. These nanocomposites have usually large and regular basal spacings up to 105 Å platelet separation. The extra large separation of the clay nanolayers results in reinforcement of the polymer matrix, the reinforcing effect being comparable to that of disordered phased nanocomposites. It is now possible to define two general classes of exfoliated clay nanocomposites, namely those with long range nanolayer ordering, and those with little or no long range ordering. Long range ordering is indicated by the XRD patterns that have multiple *00l* harmonies whereas disordered composites exhibit little or no *00l* diffraction. Regardless of the presence of X-ray Bragg diffraction, exfoliated polymer-clay nanocomposites form when adjacent clay nanolayers are separated by a distance that precludes the possibility of interlayer communication through the association of galleries cations.

INTRODUCTION

The design, synthesis, and characterization of nanostructured inorganic-organic hybrid materials are attracting great attention.[1-3] Smectite clays consisting of tactoids which contain 10-Å thick nanolayers are very important components in the nanostructured materials family,[4] especially for polymer-clay nanocomposites. The well defined structure, composition, and dimensions of the clay nanolayers are usually superior to the nanophase prepared from the hydrolysis and condensation of mononuclear precursors, such as tetraethoxysilane (TEOS) via the sol-gel process.[5,6] Polymer-clay nanocomposites have been widely studied and are beginning to have a great technical impact on materials science.[7-9] Researchers at Toyota Research Center have demonstrated that nylon 6-clay hybrids exhibit substantial improvements in mechanical, thermal and rheological properties, making possible new materials applications of the pristine nylon polymer, particularly in hostile environments such as these encountered in under-the-hood automotive applications.[10] Also, epoxy-clay nanocomposites have been shown to have greatly improved tensile modulus and strength,[11-13] especially when the polymer matrix has a sub-ambient glass transition temperatures.[14] However, the structures of the polymer-clay nanocomposites have not been well defined.

Based on the dispersion of clay nanolayers in the polymer matrix, polymer-clay nanocomposites normally have been classified as "intercalated" and "exfoliated" nanocomposites.[13,14] Intercalated nanocomposites are intercalation compounds of definite structure formed by the insertion of one or more molecular layers of polymer into the clay host galleries. The clay content of the intercalates is typically high (> 50 wt%), and the properties usually resemble those of the ceramic host. In contrast, exfoliated polymer-clay nanocomposites have clay nanolayers dispersed in a continuous polymer matrix uniformly and the exfoliated nanocomposites have a low clay content, a monolithic structure, a separation between nanolayers that depends on the polymer content of the composite, and properties that reflect those of the nano-confined polymer. In general, the distinction between intercalated and exfoliated nanocomposites is indicated by the presence or absence, respectively, of the Bragg X-ray diffraction. However, it is conceivable that even an exfoliated nanolayer composite could exhibit 001 Bragg scattering provided that the layers are regularly ordered on a nanometer length scale. In the present work, we describe the preparation and characterization of an exfoliated polymer-clay nanocomposite system which has a well defined clay nanolayer stacking order of up to 100 Å. This long range ordering between nanolayers allows for the definition of a new class of exfoliated polymer-clay nanocomposites.

Mat. Res. Soc. Symp. Proc. Vol. 435 © 1996 Materials Research Society

EXPERIMENTAL

Synthetic fluorohectorite with a unit cell formula of $Li_{1.6}[Li_{1.6}Mg_{4.4}(Si_{8.0})O_{20}F_4]$ and a particle size ≥ 2 μm. was used to form epoxy-based nanocomposite composition. The large particle size gives us a high aspect ratio for the clay nanolayers, ≥ 2000. Also, the uniform charge distribution of the fluorohectorite layer provides a constant population density of the gallery onium ions, which helps to control the intragallery catalytic polymerization reaction. The evenly distributed catalytic centers in the fluorohectorite galleries provide for a uniform polymerization rate and a uniform gallery expansion upon curing of the matrix. Homoionic $CH_3(CH_2)_{17}NH_3^+$- and $CH_3(CH_2)_{17}N(CH_3)_3^+$-fluorohectorites were prepared from Li^+-fluorohectorite (CEC = 122 meq/100g) by ion exchange reaction with alkylammonium chloride or bromide salts in H_2O/EtOH solution.[18] The epoxide resin selected for this work was the diglycidyl ether of bisphenol A (Epon-828, Shell). Polyetheramine (JEFFAMINE D2000, Huntsman) was used as the curing agent to achieve sub-ambient glass transition temperatures.

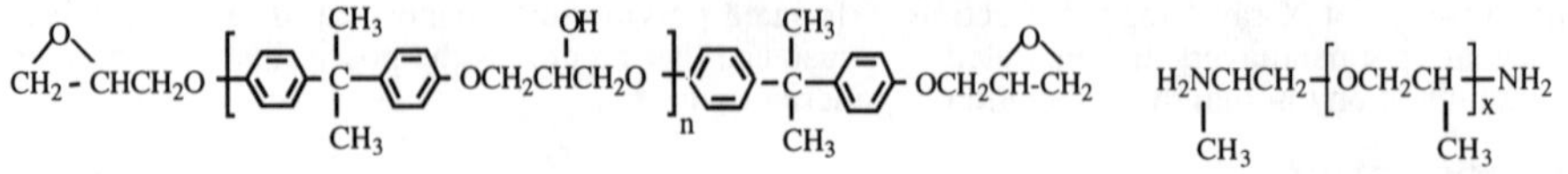

n = 0 (88%); n = 1 (10%); n = 2 (2%).

x = 33.1, MW = 2000.

The epoxy-clay nanocomposites were prepared as follows:[14] Equivalent amounts of the epoxide resin (27.5 wt%) and the polyetheramine (72.5 wt%) were mixed at 75 °C for 30 min. The desired amount of organoclay, from 5 to 10 wt%, was added to the epoxide-amine mixture and stirred for another 30 min. The clay-epoxide-amine complex was then outgassed in vacuum for 10 min and transferred into an aluminum mold for curing at 75 °C for 3 h and then at 125 °C for an additional 3 h. Powder X-ray diffraction, transmission electron micrograph (TEM) were carried out to characterize the epoxy-fluorohectorite nanocomposite. Tensile properties were measured to evaluate the reinforcing effect from the fluorohectorite nanolayer in the epoxy matrix.

RESULTS AND DISCUSSION

The X-ray powder diffraction pattern of the epoxy-clay composite containing 10 wt% of $CH_3(CH_2)_{17}NH_3^+$-fluorohectorite is shown in Figure 1. Included for comparison purpose is the pattern for the pristine organoclay. In contrast to the X-ray diffraction pattern of the epoxy-clay composites containing $CH_3(CH_2)_{17}NH_3^+$-montmorillonite,[14] the composite containing 10wt% $CH_3(CH_2)_{17}NH_3^+$-fluorohectorite gives multiple orders of d_{001} diffraction in the epoxy matrix. The well-defined orders of 00*l* diffraction nanolayer stacking in the matrix reveal that the clay nanolayers retain their stacking registry in the polymer matrix, even though the average separation caused by epoxy polymerization between the layers is in excess of 100 Å.

The XRD pattern for the composite formed from $CH_3(CH_2)_{17}N(CH_3)_3^+$-fluorohectorite gives us a well-defined diffraction peak at 38.8 Å, indicating the clay experiences only a 7.6 Å gallery expansion upon nanocomposite formation. Therefore, the composite prepared from $CH_3(CH_2)_{17}N(CH_3)_3^+$-fluorohectorite is an intercalated nanocomposite. As expected, the clay intercalated with acidic primary alkylammonium ions in their gallery, catalyze the intragallery polymerization process and form exfoliated nanocomposites. In contrast, the clay interlayered by non-acidic quaternary ammonium simply gives an intercalated nanocomposite.

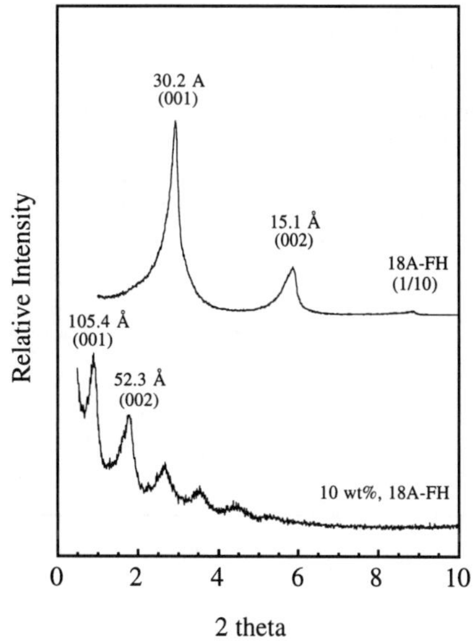

Figure 1. XRD patterns of $CH_3(CH_2)_{17}$-NH_3^+-fluorohectorite clay (upper) and the epoxy-clay nanocomposite prepared with 10 wt% of $CH_3(CH_2)_{17}$-NH_3^+-fluorohectorite clay (lower).

For comparison, the basal spacings of the original clays are given in Table I, the change in basal spacing caused by nanocomposite formation is also listed. It is noteworthy that upon nanocomposite formation, the primary ammonium exchanged clay experiences a much larger increase in basal spacing than the quaternary ammonium exchanged analogue. For the primary ammonium exchanged clay, the basal spacing of the clay in the nanocomposite state (105 Å) is much larger than the chain length of the alkylammonium. The basal spacing for the quaternary ammonium exchanged fluorohectorite composite is the value expected for a lipid-like arrangement of onium ions in the gallery.[11-14] The lipid-like structure in the latter case limits the amount of polymer that can be accommodated in the clay galleries. The absence of acidic protons on the quaternary onium ion ensures that only the monomer initially intercalated in the gallery are converted to cured polymer. In contrast, the catalytic primary onium ion facilitate further incorporation of polymer in the gallery until the curing process is complete at a basal spacing of ~ 105 Å.

It is also very interesting to note the changes in the basal spacing of $CH_3(CH_2)_{17}NH_3^+$-fluorohectorite in the nanocomposite formation process (Figure 2). At short reaction times (e.g., 10 min, curve A), two intercalated phases are detected. One is the $CH_3(CH_2)_{17}NH_3^+$-fluorohectorite slightly solvated by the monomer, with d_{001} at 38.7 Å and d_{002} at 19.4 Å. The second phase is a more solvated phase, with d_{001} at 56.0 Å and d_{002} at 28.1 Å. On the basis of broadness of the diffraction peaks, the solvated intercalates are somewhat disordered. After reaction at 75 °C for 1 h, the X-ray diffraction pattern (curve B) indicates the presence of only one fluorohectorite phase in the composite system with d_{00l} diffraction peaks (l = 1 - 4) at 57.7 Å, 28.8 Å, 19.3 Å and 14.4 Å respectively. After reaction at 75 °C for 3 h, the XRD pattern is indicative of a fully exfoliated, but regularly ordered array of nanolayers with a basal spacing of ~ 110 Å. The peaks at 55.5 and 34.9 and 26.5 Å are assigned to the d_{002}, d_{003}, and d_{004} reflections, respectively.

TEM image of Epoxy-D2000 composite containing 10 wt% $CH_3(CH_2)_{17}NH_3^+$-fluorohectorite indicates that the layer separation is in the range around 100 Å, which is in the agreement with XRD result. The average layer separation is similar to the nanolayer separation in the nanocomposite containing $CH_3(CH_2)_{17}NH_3^+$-montmorillonite.[14] But the $CH_3(CH_2)_{17}$-NH_3^+-fluorohectorite composite is characterized by long range order between the platelets, which gives rise to a 00*l* series of XRD lines, whereas the corresponding montmorillonite composite is disordered and gives an amorphous XRD pattern.

Table I. Comparison of Physical Properties of Organo-cation Exchanged Fluorohectorites as Pristine Intercalates and as Epoxy Nanocomposite Components.

Gallery Onium Ion	$CH_3(CH_2)_{17}NH_3^+$	$CH_3(CH_2)_{17}N(CH_3)_3^+$
d_{001} (Å), Air Dried Clay	29.4	31.2
d_{001} (Å), Clay Composite	105.4	38.8
Δd (Å)	76.0	7.6
Type of Composite	exfoliated	intercalated
Tensile Strength (MPa):		
a. 10 wt% Clay Composite	2.6	1.4
b. Pristine Polymer	0.5	0.5
Tensile Modulus (MPa):		
a. 10 wt% Clay Composite	14.3	11.3
b. Pristine Polymer	3.7	3.7

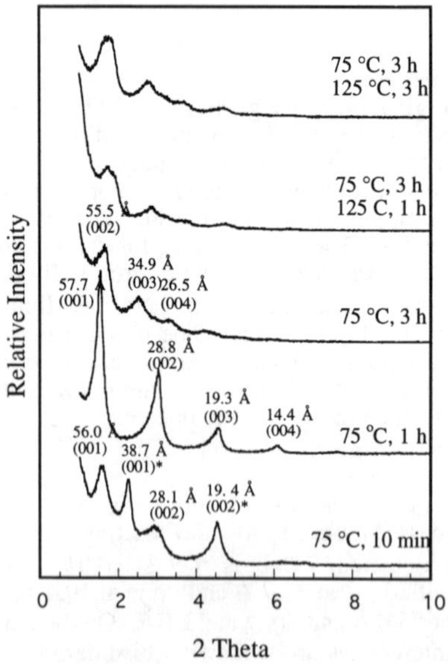

Figure 2. X-ray diffraction patterns of $CH_3(CH_2)_{17}NH_3^+$-fluorohectorite in stoichiometric mixtures of epoxide resin and polyetheramine curing agent after reaction under the following conditions: A: 75 °C, 10 min; B: 75 °C, 1h; C: 75 °C, 3 h; D: 75 °C, 3 h and 125 °C, 1 h; E: 75 °C, 3 h and 125 °C, 3 h. The clay loading was 10 wt%.

The tensile properties of the nanocomposite containing 10 wt% $CH_3(CH_2)_{17}NH_3^+$- and $CH_3(CH_2)_{17}N(CH_3)_3^+$- fluorohectorite are given in Table 1. The nanocomposite containing ordered $CH_3(CH_2)_{17}NH_3^+$-fluorohectorite performs much like the nanocomposite containing exfoliated, but disordered, $CH_3(CH_2)_{17}NH_3^+$-montmorillonite. The reinforcement of the matrix by the clay nanolayers is demonstrated by the increased strength and modulus with increasing incorporation of clay in the matrix.[14] In contrast, for the nanocomposite containing intercalated $CH_3(CH_2)_{17}N(CH_3)_3^+$-fluorohectorite, the reinforcing effect is much less than for exfoliated $CH_3(CH_2)_{17}NH_3^+$-fluorohectorite composites. The clay layer separation (basal spacing) may contribute to the shear deformation and stress transfer to the platelet particles when the test sample is under load. The average clay layer separation caused by the intercalation of cured epoxy matrix (76 Å) in the galleries of $CH_3(CH_2)_{17}NH_3^+$-fluorohectorite composites makes it a much more effective reinforcement agent compared with the composites containing $CH_3(CH_2)_{17}N(CH_3)_3^+$-fluorohectorite, where an average clay layer separation in the epoxy matrix is only 7.6 Å. Therefore, more effective stress transfer between the matrix and the reinforcing elements can be obtained from the more fully exfoliated higher clay layer separation. Further, the composite containing $CH_3(CH_2)_{17}NH_3^+$-fluorohectorite, even though it exhibits a well-defined X-ray diffraction, should be classified as an exfoliated nanocomposite with long range nanolayer ordering.

From our previous discussion, we clearly had demonstrated the existence of an exfoliated polymer-clay nanocomposite with a well-defined long range nanolayer stacking pattern in the polymer matrix. Thus, it is necessary to extend the definition of exfoliated nanocomposites to include these with and without long range ordering between nanolayers. For a polymer-clay nanocomposite prepared from long chain alkylammonium exchanged clays, the intercalated and exfoliated states of the clay may be distinguished based on the ratio of the observed gallery height (h_o) to the gallery height expected for a lipid-like bilayer of onium ions in the gallery (h_e). The value of h_e can be easily computed based on the van der Waals length of the onium ion *(l)*. Thus, $h_e = 2l$. The observed gallery height (h_o) can be determined by subtracting the thickness of a smectite clay (9.6 Å) from the basal spacing obtained by X-ray diffraction measurement (d_{001}). As illustrated in Parts A-D of Figure 3, if $h_o \leq h_e$, then the spacing between the clay layers allows for van der Waals interactions between the onium ions on adjacent layers. These van der Waals interactions between onium ion chains link adjacent clay layers. Under these conditions, the intercalated polymer (polym.) resides in the gallery space between the associated onium ions. The polymer-clay composites in these cases can be regarded as an intercalated clay composites. However, if $h_o > h_e$, as in Part E of Figure 3, then the adjacent layers of the clay are no longer linked through the onium ion interactions. There are two possibilities regarding the arrangement of the adjacent clay layers. As shown in Figure 3E, the clay layers are not parallel to each other due to the non-uniform polymerization reaction rates within the galleries. No long range ordering is realized, and the exfoliated nanocomposite gives no observable X-ray diffraction. In the case of Figure 3F, the intragallery polymerization rate is uniform due to the even charge distribution on the clay layers. The gallery expansion during the network formation process is equitable and the adjacent clay layers remain parallel to each other. Therefore, the exfoliated nanocomposite exhibit long range ordering, as judged by XRD. Consequently, the resulting state of the clay is best described as being exfoliated but with long range order.

According to the work of Lagaly,[15] the onium ions in an organoclay can adopt various orientations depending on the clay layer charge density and the chain length of the onium ions. The observed orientation of onium ions in smectite clays include lateral monolayers, lateral bilayers, pseudotrimolecular layers, paraffin-like layers and lipid-like bilayers. However, the onium ion chains can re-orient in the gallery to accommodate interaction of the polymer. Thus, as the amount of intercalated polymer is increased, the initial lateral bilayer structure of the organoclay may re-oriented to a paraffin structure and to a perpendicular bilayer, or a lipid-like structure and finally to an exfoliated structure. Thus, when the adjacent clay layers lose correlation (communication) through the cations in the clays gallery, an exfoliated nanocomposite will be formed. However, the exfoliated state may be of two types, namely, with or without long range nanolayer ordering, depending on the uniformity of polymerization rates within the clay galleries.

ACKNOWLEDGMENTS

This research has been supported by the Michigan State University Center for Fundamental Materials Research and, in part, by NSF under grant CHE-92241023 and Amcol International Co. We also thank the MSU Composite Materials and Structures Center for the use of the UTS system for mechanical testing.

The onium ion is represented by 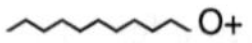 and the polymer chain by (Polym.).

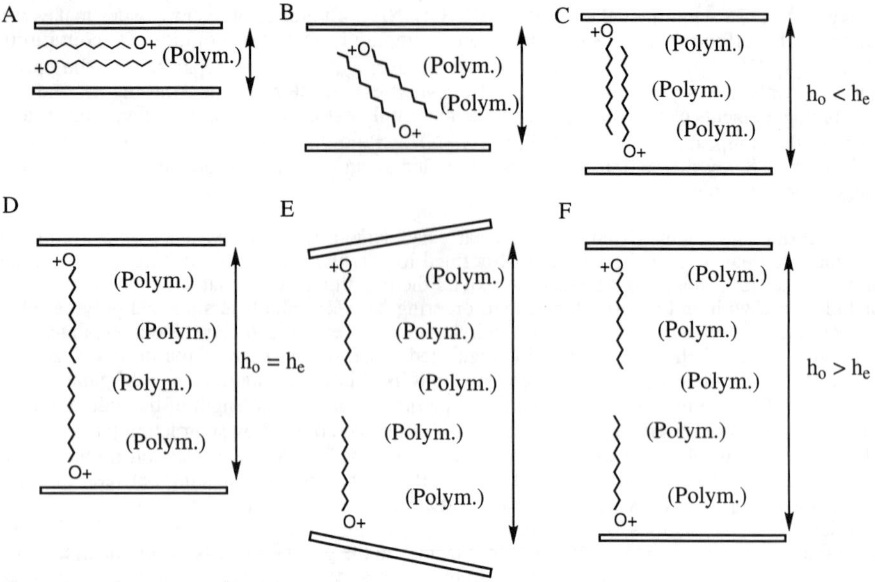

Figure 3. Schematic illustration of intercalated and exfoliated polymer-clay nanocomposites. Parts A-D illustrate intercalated clay-polymer nanocomposites in which the gallery onium ion on adjacent clay layers link the layers through van der Waals interaction. Part E and F illustrate exfoliated clay-polymer nanocomposite in which the onium ions on adjacent layers are separated by the guest polymer chains and the clay layers are no longer linked through van der Waals interactions between the onium ions.

REFERENCES

1. G.M. Whitesides, T.P. Mathias, C.T. Seto, Science, **254**, 1312 (1991).
2. G. Philipp, H. Schimdt, J. Non-Crystalline Solids, **80**, 283, (1984).
3. B. Novak, Adv. Mater., **5**, 422 (1993).
4. T.J. Pinnavaia, Science, **220**, 365 (1983).
5. N.R. Langley, G.C. Mbah, H.A. Freeman, H. Huang, E.J. Siochi, T.C. Ward, G. Wilkes, J. Colloid Interface Sci., **143**, 309 (1991).
6. S. Wang, Z. Ahmad, J.E. Mark, Preceedings of ACS, Div. of Polymeric Materials: Science and Engineering (PMSE), **70**, 305 (1994).
7. E.P. Giannelis, JOM, **44**, 28 (1992).
8. H. Gleiter, Adv. Mater., **4**, 474 (1992).
9. T.J. Pinnavaia, T. Lan, P.D. Kaviratna, M. Wang, "Clay-Polymer Nanocomposites: Polyether and Polyimide Systems". MRS Symposium Proceeding, **348**, 88 (1994).
10. Y. Kojima, A. Usuki, M. Kawasumi, A. Okada, Y. Fukushima, T. Kurauchi, O. Kamigaito, J. Mater. Res., **8**, 1185 (1993).
11. T. Lan, P.D. Kaviratna, T.J. Pinnavaia, Proceedings of ACS, Div. of Polymeric Materials: Science and Engineering (PMSE), **71**, 528 (1994).
12. P.B. Messersmith, E.P. Giannelis, Chem. Mater., **5**, 1064 (1993).
13. T. Lan, P.D. Kaviratna, T.J. Pinnavaia, Chem. Mater., **7**, 2144, (1995).
14. T. Lan, T.J. Pinnavaia, Chem. Mater., **6**, 2216 (1994).
15. G. Lagaly, K. Beneke, A. Weiss, American Mineralogist, **60**, 642 (1975).

SILICATE CLAY PLATELET DISPERSION IN A POLYMER MATRIX[1]

C.D. MUZNY*, B.D. BUTLER**, H.J.M. HANLEY**, F. TSVETKOV†, D.G. PEIFFER††
*University of Colorado, Dept. of Physics, Boulder, CO
**Physical and Chemical Properties Division, NIST, Boulder, CO
†Solar Dynamics, Ashod, Israel
††Exxon Research and Engineering, Annandale, NJ

ABSTRACT

The use of dynamic light scattering to monitor the dispersion of clay platelets in a polymer matrix is described. The dispersed clay/polymer composite was prepared by joining surfactant monomers to synthetic hectorite clay platelets by a cationic exchange reaction and subsequently polymerizing with acrylamide. Dispersion of the organically modified clay in aqueous solution was facilitated by preparing a clay surfactant precursor suspension with a surfactant concentration significantly above the clay cation exchange capacity. At lower surfactant concentrations the clay does not disperse but instead forms large polydisperse aggregates unsuitable for making homogeneous composites. The presence of these aggregates could not be found in an x-ray powder diffraction scan but were clearly detected using dynamic light scattering.

INTRODUCTION

A primary consideration in the design of clay-polymer composite materials is the way in which the clay is distributed in the polymer matrix. Generally, a homogeneous material is desired, so it is important that the clay be dispersed completely. An ideal dispersion is one where the clay mineral delaminates (exfoliates) into discrete platelets — sheets about 1 nm thick — and where these platelets are in turn homogeneously distributed in the polymer. This configuration maximizes the surface contact between the clay and the polymer, minimizes the amount of clay mineral needed to affect material properties, and ensures that the material is homogeneous at length scales only slightly larger than a clay platelet. The resulting material is thus a type of nanocomposite.

Several groups have recently reported syntheses of clay-polymer composite materials including Kamigaito and co-workers [1-3], Giannelis and co-workers [4-5], Pinnavaia and co-workers [6-7], and Moet and Akelah [8]. Typically, montmorillonite clays (dioctahedral, aluminosilicate smectites) are employed, and a variety of polymers, including nylon [1], polyimide [2,7], polystyrene [5,8], and various epoxies [4,6] have been tried. Although many routes have been used in the syntheses, most are based on the following method: The clay mineral is modified by exchanging the surface ions with cationic monomers, the modified clay is mixed in a suitable matrix monomer, and then the polymerization reaction is initiated. The organization of the clay mineral is often monitored by following the x-ray diffraction *00l* (inter-layer) peak which detects swelling of the clay layers. The absence of this diffraction peak is generally taken to indicate that delamination of the clay into individual platelets has occurred and, by inference, that it is likely that these platelets are no longer associated. Electron microscopy of the polymer composite is sometimes used to further investigate the degree of dispersion of the clay platelets.

While x-ray diffraction is a useful tool for monitoring the dispersion of the platelets

[1] Work of NIST, not subject to copyright in the USA.

during and after the synthesis, it does not provide a complete picture of their distribution. The presence of an *00l* diffraction peak reveals clay platelets associated face-to-face and can be used to establish a characteristic repeat distance provided that this distance is less than about 5 nm (the resolution limit of standard x-ray diffractometers). Other associations, such as edge-to-face aggregation, or large and/or irregular lamellar spacings cannot be detected this way. Imaging of the platelets by electron microscopy can be used to detect the distribution directly. Unfortunately, electron microscopy cannot be used conveniently to monitor the distribution of platelets during synthesis, is difficult to use with polymeric systems, only views materials in projection through a thin slice of the material, and there is often the question of how representative a particular image is of the sample as a whole.

In this paper we discuss an alternative — using dynamic (time-correlation) light scattering to monitor, in situ, the dispersion of clay platelets. We demonstrate the technique using a synthetic form of the clay mineral hectorite, organically modified using a monomer similar to the common surfactant CTAB, and dispersed in a polyacrylamide matrix.

EXPERIMENTAL

Dynamic Light Scattering

Diffraction patterns from all disordered systems display a characteristic pattern of 'speckles'; small bright and dark regions in the scattering pattern superimposed on a continuous diffuse background that result from the particular arrangement of scattering centers present in the disordered system. The angular separation of these speckles is inversely related to the coherence length of the optical system employed. In colloidal systems, where particles are continuously diffusing, the arrangement of scatterers (and thus the speckle pattern generated by diffraction of laser light from the material) undergoes reorganization over a time scale characterized by the diffusion rate. This time scale can be most simply determined from a measurement of the intensity-intensity time correlation. For a monodisperse system composed of freely diffusing particles, the correlation function decays as a pure exponential function with a decay time τ that is related to the size of the diffusing particles and the viscosity of the surrounding fluid [9]. A polydisperse distribution of particle sizes can be detected by observing deviations from this simple exponential; for many systems the actual distribution of particle sizes can be extracted by various least-squares fitting procedures [10].

Dynamic light scattering is thus well suited for use as a tool to monitor the degree to which a suspension of clay particles is homogeneous. That is, a monodisperse, homogeneous solution of individual clay platelets (the ideal we are striving for when making a clay-polymer composite) will display a clean exponential decay of the correlation function. The existence of larger clay aggregates can be detected by the presence of long time tails in the decay function which deviate significantly from a pure exponential form.

Materials Used

The synthetic clay mineral used in this study was a synthetic lithium substituted hectorite[11]. This material is a trioctahedral magnesiasilicate smectite that can be readily dispersed in aqueous solution into individual platelets 1 nm thick and approximately 30 nm in diameter [12,13]. It has a cation exchange capacity near 75 meq/100 g and a chemical formula approximated by [14]

$$Si_8[Mg_{5.43}Li^-_{0.57}H_4O_{24}]Na^+_{0.57}$$

The surfactant monomer used to exchange with the Na ions on the surfaces of the clay platelets (and which later forms a bridge between the clay mineral and the polyacrylamide matrix) was hexadecylallyldimethyl ammonium bromide [15]. This compound is a variant of the common surfactant hexadecyltrimethyl ammonium bromide (CTAB) in which an allyl group has replaced one N-methyl. The double bond permits subsequent polymerization [16]:

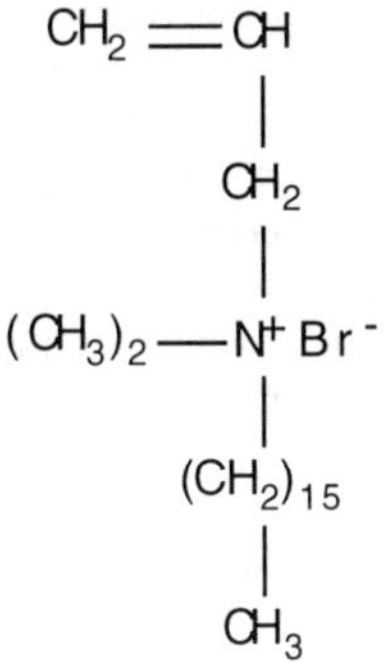

Preparation and Analysis of Suspensions

Several mixtures were prepared for study: suspensions of pure hectorite in water, and aqueous mixtures of hectorite with various concentrations of surfactant monomer. A solution of 1% by mass hectorite in water which has been stirred for several hours is optically clear indicating that the pure clay mineral forms a homogeneous dispersion in water. Addition of cationic monomer to the solution, however, clouds the mixture, indicating that aggregation of the clay takes place in the presence of this organic. (This was not unexpected because the organophilic nature of a surfactant can induce reaggregation of previously dispersed clay platelets [17].) At a surfactant concentration equal to that needed to exchange all of the clay's surface sodium ions (the cation exchange capacity, CEC) we observe large flocs in the mixture that settle to the bottom of a vial after a few hours. When surfactant monomer is added in concentrations above about 5 times the CEC the clouded mixtures become stable with respect to precipitation and the mixture appears to be more homogeneous but still not transparent.

Two particular mixtures were studied in some detail. (1) A mixture prepared by combining 20 ml of 2% by mass clay solution with 20 ml of a 3.9% by mass (0.1 molar) solution of the surfactant. This mixture contains a surfactant excess over the CEC by a factor of about 7. (2) A mixture prepared from (1) by the following procedure: 40 ml of mixture (1) was centrifuged at 19 600 m/s^2 for 30 min. The mixture separated into a dense precipitate and a supernatant liquid which contained most of the NaBr salt produced in the cation exchange reaction. This supernatant was discarded. 20 ml of 3.9% by mass (0.1 molar) surfactant monomer solution was added to the precipitate along with enough deionized water to restore the total volume to the original 40 ml. This solution was stirred for 16 h and centrifuged for 30 min. At this point solids precipitated but the supernatant was mildly cloudy, suggesting that some clay had been promoted to the suspension This supernatant, therefore, was extracted and used as the subsequent clay/surfactant complex precursor. Dry mass analysis showed that the composition of the supernatant was 0.1% by mass clay and 2.5% by mass surfactant (0.06 molar), corresponding to a cationic surfactant excess over the CEC of about 100.

All solutions and mixtures at various stages of preparation were investigated by dynamic light scattering using an Ar-ion laser at a wavelength of 488 nm and a scattering angle of 90°. Figure 1 displays the correlation functions for a 0.1% hectorite solution and for mixtures (1) and (2). The decay of the intensity correlation function for the pure clay mineral dispersed in water shows nearly a pure exponential decay with a characteristic decay time of approximately 50 μs. A least-squares fit of a particle size distribution (using the NNLS algorithm [10]) showed a relatively narrow distribution centered at a hydrodynamic size of about 25 nm. In contrast to this pure clay sample, the measured intensity correlation of the untreated mixture (1) shows very long time tails and does not appear to approximate an exponential decay. This indicates that there is a wide range of particle (cluster) sizes present in the mixture, including some large clay aggregates, as evidenced by the fact that the correlation still has not decayed completely to 0 even after 1 s. Distribution analyses of samples like this one do not yield meaningful results. The washed and centrifuged mixture (2), however, does not show these long time decays but instead, like the clay suspension, closely follows a pure exponential. The characteristic decay time is near 200 μs and a distribution analysis confirmed a monodisperse system with particles of hydrodynamic size approximately 150 nm. This is much larger than for the pure clay suspension presumably because of friction due to the large organic molecules on the surface of the clay plates and the increased viscosity of the solution due to the large excess of surfactant slowing the diffusion of the particles. Dynamic light scattering has thus provided strong evidence that a homogeneous monodisperse suspension of organically modified clay has been produced.

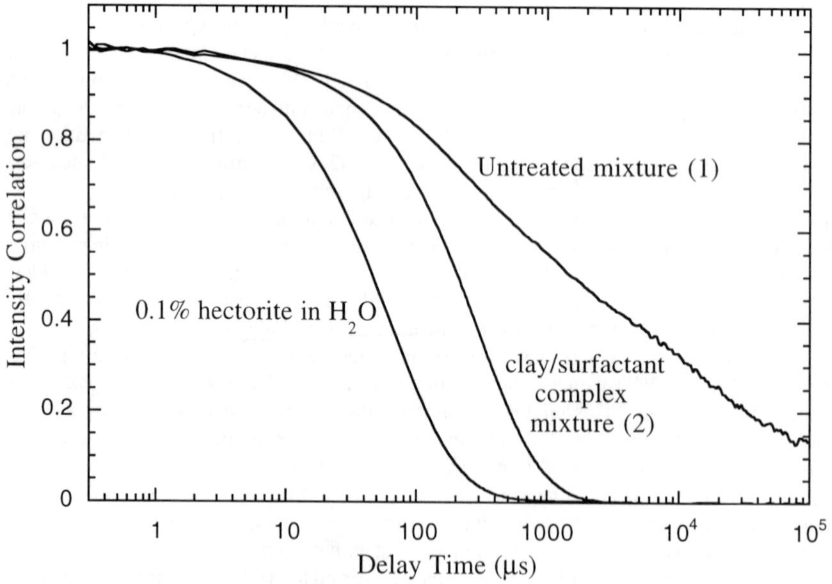

Figure 1: Intensity-intensity time-correlation function for three mixtures: a 0.1% by mass solution of hectorite in water, the untreated mixture (1) of surfactant and clay, and the clay/surfactant complex (mixture (2)).

X-ray powder diffraction scans performed on the residue of mixture (2) dried on a microscope slide show an *00l* diffraction peak at an angle corresponding to a lamellar spacing of 2.4 nm. This can be compared to a spacing of 1.2 nm that is obtained upon drying unexchanged hectorite thus confirming that the surfactant has indeed formed a complex with the clay.

Preparation and analysis of composite

The clay/surfactant complex from mixture (2) was polymerized with acrylamide as follows: 20 ml of mixture (2) was deoxygenated by stirring under nitrogen gas for 3 h; 2 g of acrylamide was added and the mixture stirred at 323 K for 5 min; then 1.55 ml of a deoxygenated water solution of 0.8% by mass potassium persulfate initiator was slowly added. At this point the time-correlation function of the polymer precursor was measured. No significant change in the dynamic light scattering signal was detected from that of mixture (2) confirming that these additions did not cause aggregation of the clay platelets. After 15 min it became clear that polymerization was taking place, so the mixture was compacted between microscope slides and the polymerization allowed to proceed for a further 10 h at 323 K in a nitrogen atmosphere. The resulting composite – designated material-2 – was washed, and rewashed, with acetone and then dried at 323 K for 10 h. The same procedure was carried out using the 0.1% clay/H_2O suspension and the acrylamide/surfactant mixture (1) as precursors.

Scanning electron microscopy (operated in a backscattering x-ray fluorescence mode) was used to detect whether the materials were homogeneous on length scales of 0.4 μm or larger. No inhomogeneity in the clay distribution was detected in the polymers made with mixture (2) and with the clay/H_2O solution. In the polymer prepared from mixture (1), however, clay clusters from 1 μm to 30 μm were clearly seen. X-ray powder diffraction scans showed no evidence of a *00l* diffraction peak in any of the three composites. We can thus assume that the clay in all samples, including the one made from mixture (1) which shows large aggregates in dynamic light scattering and the SEM, contain delaminated clay. The large aggregates observed with dynamic light scattering must, therefore, be composed of clay platelets either associated edge-to face or with large and/or irregular *00l* spacings. The absence of an *00l* interlayer diffraction peak is thus not sufficient evidence to conclude that the clay in a polymer composite material is homogeneously distributed.

CONCLUSION

In this paper we have described a method utilizing dynamic light scattering whereby the dispersion of clay in a polymer can be monitored during all process steps in which the sample is fluid. The method is demonstrated on a system containing a dispersible hectorite clay mineral modified with an organic monomer similar to the surfactant CTAB and polymerized in a polyacrylamide matrix. Dynamic light scattering data demonstrate that a homogeneous dispersion of the clay in solution prior to polymerization is best achieved by saturating the clay with the surfactant monomer at concentrations well above the CEC. In a material where aggregation of the clay was verified by dynamic light scattering and electron microscopy, we found, nevertheless, that the x-ray powder diffraction pattern had no measurable *00l* reflections. We thus conclude that dynamic light scattering is better able to distinguish (and monitor) homogeneous dispersions than commonly employed x-ray powder methods.

ACKNOWLEDGEMENTS

This work was supported in part by a grant from the Office of Basic Energy Sciences, Division of Engineering and Geosciences, US Department of Energy.

REFERENCES

1. Y. Kojima, A. Usuki, M. Kawasumi, A. Okada, T. Kurauchi, O. Kamigaito, *J. Poly. Sci. A* **31**, 983 (1993); A. Usuki, M. Kawasumi, Y. Kojima, A. Okada, T. Kurauchi, O. Kamigaito, *J. Mater. Res.* **8**, 1174 (1993); A. Usuki, Y. Kojima, M. Kawasumi, A. Okada, Y. Fukushima, T. Kurauchi, O. Kamigaito, *J. Mater. Res.* **8**, 1179 (1993).

2. K. Yano, A. Usuki, A. Okada, T. Kurauchi, O. Kamigaito, *J. Poly. Sci. A* **31**, 2493 (1993);

3. US Patents: Kamigaito et al., 4 472 538 (1984); Kawasumi et al., 4 810 734 (1989); Usuki et al., 4 889 885 (1989).

4. P.B. Messersmith, E.P Giannelis, *Chem. Mater.* **6**, 1719 (1994); P.B. Messersmith, E.P Giannelis, *Chem. Mater.* **5**, 1064 (1993).

5. R. A. Vaia, H. Ishii, E.P. Giannelis, *Chem. Mater.* **5**, 1694 (1993).

6. M. S. Wang, T. J. Pinnavaia, *Chem. Mater.* **6**, 468 (1994).

7. T. Lan, P. D. Kaviratna, T.J. Pinnavaia, *Chem. Mater.* **6**, 573 (1994).

8. A. S. Moet, A. Akelah, *Mater. Lett.* **18**, 97 (1993).

9. B.J. Berne, R. Pecora, *Dynamic Light Scattering*. Wiley , New York (1976).

10. E.F. Grabowski, I.D. Morrison, Chap. 7 in: *Measurement of Suspended Particles by Quasi-Elastic Light Scattering*, Ed. B.E. Dahneke, Wiley & Sons, New York (1983).

11. Trade name Laponite RD; Laporte Industries, UK. Other materials may work as well or better. Endorsement by NIST is not implied.

12. M. Morvan, D. Espinat, J. Lambard, T. Zemb, *Colloids and Surfaces Pt. A* **82**, 193 (1994).

13. See, for example: B. S. Neumann, K.G. Sansom, *Israel J. Chem.* **8**, 315 (1970); J.D.F. Ramsay, S. W. Swanton, J. Bunce, *J. Chem. Soc. Faraday Trans.* **86**, 3919 (1990); R.G. Avery, J.D.F. Ramsey, *J. Colloid and Interface Sci.* **109**, 448 (1986); L. Rosta, H.R. von Gunten, *J. Colloid and Interface Sci.* **134**, 397 (1990).

14. Laponite Technical Bulletin, L104/90/A, Laporte Industries, UK.

15. D. G. Peiffer, *J. Poly. Sci. Pt. A* **28**, 619 (1990).

16. D. G. Peiffer, US Patent 4 853 447 (1989).

17. B.K.G. Theng, *The Chemistry of Clay-Organic Reactions*, Wiley & Sons, New York (1974); H. van Olphen, *An Introduction to Clay Colloid Chemistry*, Wiley & Sons, New York (1963).

Part II

Synthesis, Characterization, and Processing of Organic/Inorganic Hybrid Materials

Part II

THE USE OF FUNCTIONALIZED POLYBENZOXAZOLES AND POLYBENZOBISTHIAZOLES IN POLYMER-SILICA HYBRID MATERIALS

J. E. MARK,* J. PREMACHANDRA,* C. KUMUDINIE,* W. ZHAO,* T. D. DANG,** J. P. CHEN,** AND F. E. ARNOLD**

*Department of Chemistry and the Polymer Research Center, University of Cincinnati, Cincinnati, OH 45221-0172

**Wright Laboratory, Materials Directorate, Wright-Patterson Air Force Base, Dayton, OH 45433

ABSTRACT

Hybrid organic-inorganic composites were prepared by precipitating silica into hydroxypolybenzoxazole (HPBO) and sulfopolybenzobisthiazole (SPBT) polymers, with interfacial bonding between the phases improved by use of isocyanato-propyltriethoxy silane and N,N-diethylaminopropyltrimethoxy silane, respectively. The resulting materials were characterized with regard to their transparency, silica particle distribution, tensile modulus and tensile strength, thermal stability, and tendency to absorb water.

INTRODUCTION

The idea of in-situ generation of hybrid inorganic-organic composite materials is based on an important recent advance in the ceramics area. It involves the use of preparative techniques heavily based on chemical reactions, for example the generation of ceramic-type materials by the hydrolysis of an organometallic compound [1-14]. In the case of composite materials, the goal is to carry out reactions of this type in the presence of organic molecules which are typically polymeric and contain functional groups to improve their bonding to the ceramic-like phase. Parts of this research on the sol-gel approach to hybrid composites is described in sections of some of the books and review articles cited above for the sol-gel process in general. In addition, there have been two books devoted entirely to the topic [15, 16], a number of additional recent review articles [17, 17-40].

High-temperature polymers [41-43] are very difficult to treat in the usual sol-gel technique, but some studies have been carried out on a few aromatic polyamides [38, 44-47], a number of polyimides [58, 48-62], and several benzoxazole and benzobisthiazole polymers [63-68]. The unreactivity that many such polymers must have to qualify as high-temperature, high-performance materials is the main disadvantage in this application, since it generally causes poor interfacial bonding between the polymers and the ceramic phases. This problem can be solved, at least in part either by functionalizing the polymer, by adding a bonding agent, or by doing both.

Typically, the in-situ precipitated silica does give some improvements in mechanical properties, at least when an effective bonding agent is present. More specifically, there is an increase in toughness due to increased values of the modulus and ultimate strength at low silica contents; the toughness subsequently decreases, however, due to large decreases in extensibility caused by the presence of the silica at higher silica contents. In some cases, there are increases in thermal

Mat. Res. Soc. Symp. Proc. Vol. 435

stability and hardness, and decreases in water absorption. Effective bonding agents generally have the additional advantage of increasing the transparency of the composite.

One of the challenges in this area of high-performance polymers is to obtain significant improvements in these classes of polymers, since many are relatively resistant to bonding to reinforcing phases and already have superb thermal and mechanical properties.

Two important classes of heat-resistant heterocyclics are the polybenzoxazoles (PBO) and polybenzobisthiazoles (PBT), largely because of their outstanding retention of desirable properties at elevated temperatures. The early PBO and PBT rigid-rod polymers were found to be intractable, and their inertness to solvents resulted in difficulties in fabricating them into useful objects. Introduction of specific functional groups into these polymers and preparation of functionalized copolymers to make them somewhat more tractable have become a main synthetic activity in this area.

As part of continuing efforts to circumvent the lack of solubility as well as to provide bonding sites to the inorganic phase, hydroxy-functionalized copolymers of PBO (HPBO) and sulfo-functionalized polymers of PBT (SPBT) were synthesized for use in organic-inorganic, polymer-ceramic composites [66-68]. In the present study, high-temperature HPBO and SPBT composites were prepared, using isocyanatopropyltriethoxy silane and N,N-diethylaminopropyltrimethoxy silane as bonding agents, respectively. Their morphology, mechanical properties, thermal stability, and water absorption behavior were characterized with regard to the silica content. The results were compared to those of the pure polymers, to identify improvements in properties from the in-situ generated silica.

EXPERIMENT

Polymers and Reagents

The HPBO polymer was prepared by polycondensation of 4,4'-[2,2,2-trifluoro-1-(trifluoromethyl)ethylidene]bis[2-amino-phenol], 4,4'-oxybis(benzoic acid) and 5-hydroxyisophthalic acid in polyphosphoric acid. Similarly, the SPBT polymer was prepared by polycondensation of 2,5-diamino-1,4-benzenedithiol dihydrochloride and 2-sulfo-terephthalic acid in polyphosphoric acid. The detailed syntheses are reported elsewhere [66-68]. The coupling agents used were isocyanatopropyltriethoxy silane (95%) and N,N-diethylaminopropyltrimethoxy silane.

Preparation of the HPBO-Silica Composites

The desired amount of the PBO copolymer was dissolved in dry THF and the bonding agent isocyanatopropyltriethoxy silane was added in the stoichiometric amount, relative to the OH groups in the copolymer; a catalytic amount of triethylamine was also included [68]. The mixture was stirred overnight at room temperature, and tetramethyl orthosilicate (diluted in THF) and a diethylamine aqueous solution (also diluted in THF) were then incorporated into portions of the mixture, to give various silica amounts. The portions were then stirred overnight to ensure homogeneous mixing. The resulting viscous solutions were transferred to a

Petri dish for further hydrolysis and condensation. Pieces of film were obtained after the solvent was evaporated at room temperature, and were further dried *in vacuo.*

Preparation of the SPBT-Silica Composites

A desired amount of SPBT polymer was dissolved in anhydrous methanol, with the help of a stoichiometric amount of tributylamine [68]. The resulting solution was cooled to room temperature and the bonding agent N,N-diethylaminopropyl-trimethoxy silane was added in a fixed ratio relative to the amount of polymer used. Different amounts of TMOS were then incorporated into portions of the mixture to give a series of composites of various silica contents, as already described.

Characterization of the Composite Films

Infrared spectra were recorded on a Perkin-Elmer 1600 Series Fourier Transform Infrared Spectrometer [68].

Thermogravimetric analyses (TGA) were performed using a Perkin Elmer model TAS-7 system at a heating rate of 20 °C/min, under a nitrogen atmosphere.

The tensile strengths and moduli of the dried samples were measured using an Instron mechanical tester at room temperature.

Scanning electron microscopy (SEM) was used to characterize both the film surfaces and their cross-sectional areas, using a Model 90 Cambridge instrument. Energy dispersive x-ray analysis (EDAX) measurements were conducted on a EDS system (Princeton Gamma Tech) with the same samples.

The extents of water absorption by the films under saturated conditions were measured by the method described in ASTM-D570-81 [68].

RESULTS AND DISCUSSION

Interfacial Bonding and Appearances

Solubility measurements indicated good bonding between the two phases, presumably with the isocyanate moiety undergoing an addition reaction with the hydroxy group on the PBO backbone (Scheme 1). The infrared spectra indicated the presence of the -C=O stretching vibration band in the carbamate linkages at 1750 cm^{-1}, and the disappearance of the -N=C=O stretching band of isocyanato groups on the bonding agent at 2269 cm^{-1}. This confirmed the desired covalent bonding. It was remarkable that all the hybrid films were transparent and uniform, even at higher silica contents (up to 30 wt %), whereas the HPBO hybrid materials prepared without bonding agent were opaque when the silica content exceeded 15 wt % [66, 68].

The SPBT pure polymer was a dark reddish solid and insoluble in common organic solvents. The polymer had to be converted to a salt (Scheme 2) before it could be dissolved in methanol – a common solvent used in sol-gel technology. Various amounts of tetramethyl orthosilicate and the bonding agent were subsequently added to the methanol solution of the SPBT-$SO_3^{\ominus}$ $^{\oplus}NHBu^t_3$ (where Bu^t is tertiary butyl). Ionic associations between the -$SO_3^{\ominus}$ groups on the polymer backbone and the ammonium ions of the bonding agent N,N-diethylamino-propyltriethoxy silane facilitated strong interactions between the two phases. The

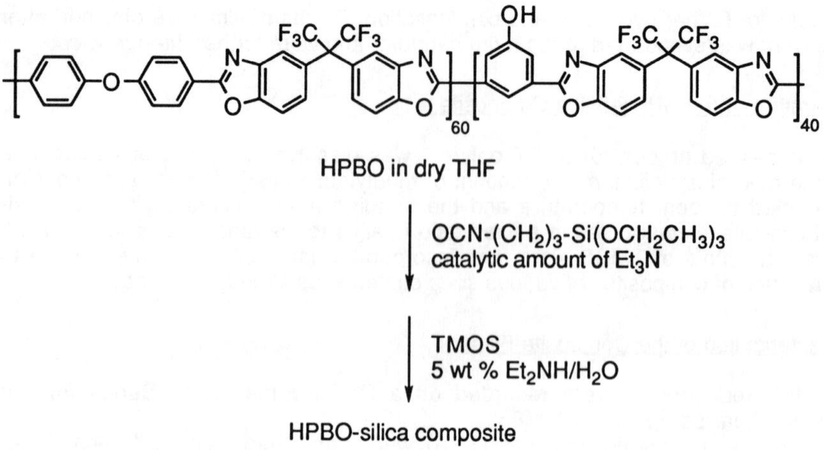

Scheme 1. Synthesis of HPBO-silica composites.

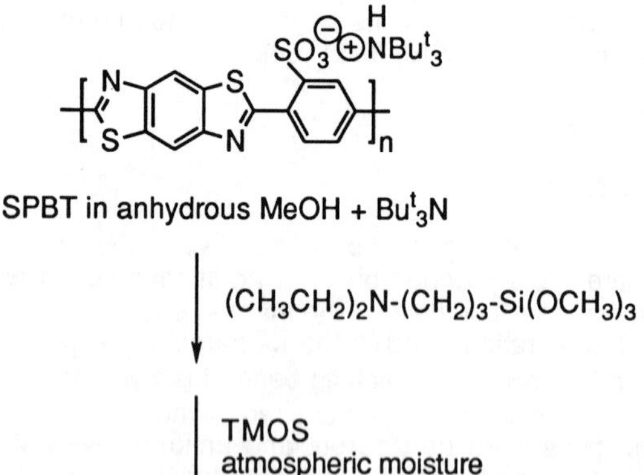

Scheme 2. Synthesis of SPBT-silica composites.

SPBT materials prepared in this manner were highly transparent but reddish. In addition, they were insoluble even after refluxing in methanol for 24 hr. This further suggested the formation of silica networks from hydrolysis and condensation of the inorganic components, with good bonding to the rigid-rod polymer phase [68].

Morphology

The clarity of both the HPBO and SPBT composites indicated good interfacial bonding between the polymer and silica phases, as well as good dispersion of the silica in the polymer matrix. This conclusion was supported by the SEM microphotos, and by the silicon distribution map corresponding to the EDAX spectrum [68]. The EDAX spectrum clearly identified the existence of silicon atoms on the surfaces of both unfractured and fractured surfaces; the distribution map further suggested the uniform dispersion of the silica present. No silica particles were visible in the SEM microphotos, indicating they had sizes below 1 μm. Compared to the previous unbonded films, the particles in these films were much smaller and more uniform. The improved interfacial bonding from the covalent linkages apparently suppressed the agglomeration of silicate particles, thus reducing the sizes of the clusters.

For the SPBT composites, the unfractured smooth surface and the cross-sectional area of the fractured surface were also examined. No silica particles could be observed in the SPBT-30 transparent film [68]. Only a homogeneous dispersion of silicon atoms could be identified by silicon mapping *via* EDAX. It can again be concluded that the improved interfacial bonding facilitated uniformly distributed silicon atoms in the polymer matrix. The particle sizes were apparently much smaller than 1 μm [68].

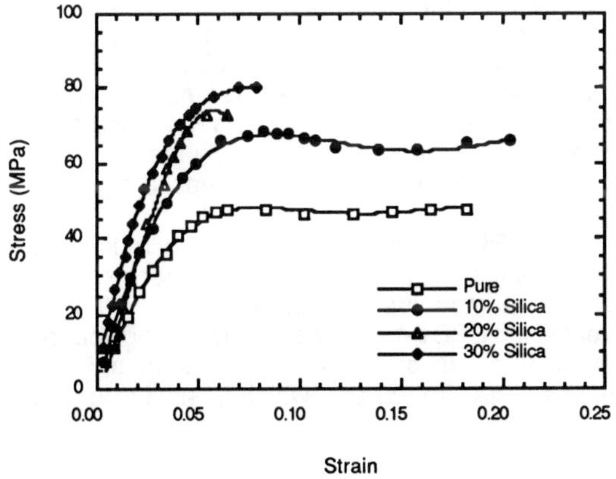

Figure 1. Stress-strain curves for HPBO-0 (□), HPBO-10 (●), HPBO-20 (Δ), and HPBO-30 (◆) [68].

Mechanical Properties

In the case of HPBO-silica hybrid films, both tensile moduli and tensile strengths increased with increase of silica level in the composites (Figure 1). As illustrated in Table 1, 40, 70, and 140 % increases in modulus were obtained for

Table 1. Mechanical Properties of the HPBO Composites

Hybrid Material	Modulus[a] (GPa)	Ultimate Strength[b] (MPa)	Ultimate Elongation[c] (%)	Toughness (MPa)
HPBO-0	1.19	47.0	18.3	14.2
HPBO-10	1.73	65.4	20.5	22.5
HPBO-20	2.06	72.4	6.5	5.9
HPBO-30	2.89	80.2	7.9	8.8

a Initial slope of the stress-strain curve.
b Stress at break.
c Elongation at break.

hybrids HPBO-10, HPBO-20 and HPBO-30, respectively, where the numbers specify the wt % silica. Similar increases were also observed for the ultimate strength, with an almost two-fold increase in strength obtained for the HPBO-30 film. These substantial improvements can be explained by the successful use of the bonding agent to enhance the connectivity of the organic and inorganic phases. Also important are the strong interactions between the silanol groups on the silica surface and the hydrogen bonding acceptors on the polymer backbone (such as nitrogen, oxygen of the oxazole rings in HPBO). The elongation at break was found to be large at low silica contents, and relatively small at higher contents. At low levels of silica, the material seemed to be more flexible and tougher without loss of high strength and high modulus. On the other hand, as the silica amount increased, the materials became harder and stronger, but inevitably less flexible and less tough. As can be seen from Table 1, toughness actually decreased significantly at higher silica levels (above 20 wt %).

In the case of the PBT-silica composites, the mechanical properties also showed substantial increases relative to that of the pure polymer. For example, a nearly two-fold increase in modulus was obtained for SPBT-30. Tensile strength showed a 40% increase for SPBT-10, without diminishing toughness and flexibility. Only when substantial amounts of silica were present did the characteristics of ceramics (*e.g.* brittleness) start to appear, which is similar to what was found in the HPBO series [68].

Thermal Stability

The TGA curves for the HPBO pure polymer and its composites showed slow but gradual weight losses beginning around 200 °C. These materials started to exhibit drastic weight losses at approximately 500 °C, due to the decomposition of the benzoxazole rings. The thermal transition temperatures of the HPBO composites were raised about 20 - 35 °C compared with that of the pure polymer, however, depending on the silica content.

For SPBT pure polymer, two distinct degradations were observed. The first significant weight loss started at 350 °C, probably due to the loss of $-SO_3H$ side groups and the breaking down of the S-C single bonds in the thiazole rings, giving off small molecules. The second significant weight loss started at 570 °C, with the aromatic backbones being destroyed at this stage. Virtually no residues were observed above 690 °C. For the SPBT composites, slight weight losses at the initial stage were observed for all the three hybrid films tested. This initial weight loss can be attributed to the presence of small amount of tributylamine in the composites. The similar two-stage degradation pattern still remained, but the thermal transition temperatures had been raised about 20 - 40 °C, drastic weight losses between 350 - 400 °C in the pure polymer were prevented by the incorporation of the larger amount of silica [68].

Water Absorption

Figure 2 shows the extents of water absorption in the HPBO-silica composites.

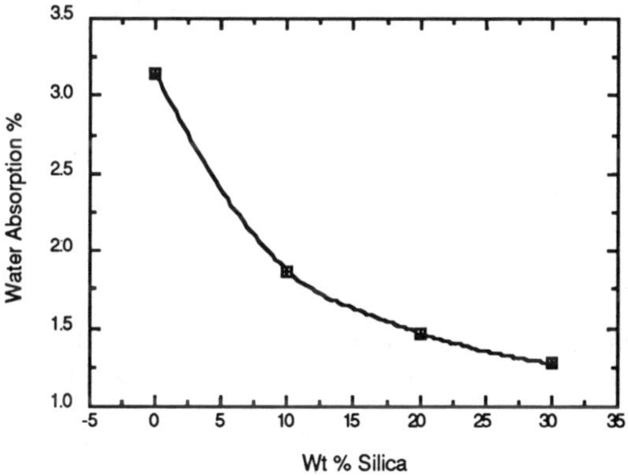

Figure 2. Water absorption for the HPBO composites, as a function of silica concentration [68].

Water absorption was much higher in the pure SPBT (about 400 wt %) than in the pure HPBO (only 3.2 wt %), presumably because of the presence of the very polar SO_3H side groups in SPBT [68]. These acidic groups facilitated strong hydrogen bonding to water. In both composites, however, there was a considerable decrease in the water absorption due to the presence of the silica. In the SPBT series, a 350 wt % decrease in water absorption was observed for SPBT-10. It was most pronounced up to approximately 10 wt % silica, with a slight additional decrease for higher silica contents. These decreases in water absorption are important since water can influence many physical properties of a material, such as its electrical insulating ability, dielectric loss, dimensional stability, and appearance.

ACKNOWLEDGMENT

It is a pleasure to acknowledge the financial support provided by the Air Force Office of Scientific Research (Directorate of Chemistry and Materials Science) through Grant F49620-92-J-0322.

REFERENCES

1. C. J. Brinker, G. W. Scherer, *Sol-Gel Science* (Academic Press, New York, 1990).

2. L. L. Hench, J. K. West, *Chem. Rev.* **90**, 33 (1990).

3. L. L. Hench, W. Vasconcelos, *Annu. Rev. Mater. Sci.* **20**, 269 (1990).

4. B. J. J. Zelinski, C. J. Brinker, D. E. Clark, D. R. Ulrich, Eds., *Better Ceramics Through Chemistry IV*, vol. 180 (Materials Research Society, Pittsburgh, 1990).

5. D. W. Schaefer, J. E. Mark, Eds., *Polymer-Based Molecular Composites*, vol. 171 (Materials Research Society, Pittsburgh, 1990).

6. D. R. Uhlmann, D. R. Ulrich, Eds., *Ultrastructure Processing of Advanced Materials* (Wiley, New York, 1992).

7. R. H. Baney, L. R. Gilliom, S.-I. Hirano, H. K. Schmidt, Eds., *Submicron Multiphase Materials*, vol. 274 (Materials Research Society, Pittsburgh, PA, 1992).

8. M. J. Hampden-Smith, W. G. Klemperer, C. J. Brinker, Eds., *Better Ceramics Through Chemistry V*, vol. 271 (Materials Research Society, Pittsburgh, 1992).

9. L. L. Hench, J. K. West, Eds., *Chemical Processing of Advanced Materials* (Wiley, New York, 1992).

10. J. E. Mark, in *Physical Properties of Polymers* J. E. Mark, et al., Eds. (American Chemical Society, Washington, DC, 1993) p. 3.

11. A. Cheetham, C. J. Brinker, M. L. Mecartney, C. Sanchez, Eds., *Better Ceramics Through Chemistry VI*, vol. 346 (Materials Research Society, Pittsburgh, 1994).

12. L. C. Klein, Ed., *Sol-Gel Optics* (Kluwer Academic Publishers, Boston, 1994).

13. J. Gopalakrishnan, *Chem. Mater.* **7**, 1265 (1995).

14. J. E. Mark, *Hetero. Chem. Rev.* **2**, 000 (1996).

15. C. Sanchez, F. Ribot, Eds., *Proceedings of the First European Workshop on Hybrid Organic-Inorganic Materials* (Chimie de la Matiere Condensee, Chateau de Bierville, France, 1993).

16. J. E. Mark, C. Y.-C. Lee, P. A. Bianconi, Eds., *Hybrid Organic-Inorganic Composites* , vol. 585 (American Chemical Society, Washington, 1995).

17. H. Schmidt, H. Wolter, *J. Non-Cryst. Solids* **121**, 428 (1990).

18. G. L. Wilkes, H.-H. Huang, R. H. Glaser, in *Silicon-Based Polymer Science* J. M. Zeigler, F. W. G. Fearon, Eds. (American Chemical Society, Washington, DC, 1990), vol. 224, p. 207.

19. R. Nass, E. Arpac, W. Glaubitt, H. Schmidt, *J. Non-Cryst. Solids* **121**, 370 (1990).

20. J. E. Mark, in *Silicon-Based Polymer Science. A Comprehensive Resource* J. M. Zeigler, F. W. G. Fearon, Eds. (American Chemical Society, Washington, DC, 1990) p. 47.

21. D. W. Schaefer, J. E. Mark, D. McCarthy, L. Jian, C.-C. Sun, B. Farago, in *Polymer-Based Molecular Composites* D. W. Schaefer, J. E. Mark, Eds. (Materials Research Society, Pittsburgh, 1990), vol. 171, p. 57.

22. H. Schmidt, in *Better Ceramics Through Chemistry IV* B. J. J. Zelinski, C. J. Brinker, D. E. Clark, D. R. Ulrich, Eds. (Materials Research Society, Pittsburgh, 1990), vol. 180, p. 961.

23. J. E. Mark, D. W. Schaefer, in *Polymer-Based Molecular Composites* D. W. Schaefer, J. E. Mark, Eds. (Materials Research Society, Pittsburgh, 1990), vol. 171, p. 51.

24. J. E. Mark, *J. Inorg. Organomet. Polym.* **1**, 431 (1991).

25. J. E. Mark, *J. Appl. Polym. Sci., Appl. Polym. Symp.* **50**, 273 (1992).

26. H. Schmidt, in *Chemical Processing of Advanced Materials* L. L. Hench, J. K. West, Eds. (Wiley, New York, 1992) p. 727.

27. H. Schmidt, in *Ultrastructure Processing of Advanced Materials* D. R. Uhlmann, D. R. Ulrich, Eds. (Wiley, New York, 1992) p. 409.

28. J. E. Mark, *Angew. Makromol. Chemie* **202/203**, 1 (1992).

29. B. M. Novak, *Adv. Mats.* **5**, 422 (1993).

30. S. J. Clarson, J. E. Mark, in *Siloxane Polymers* S. J. Clarson, J. A. Semlyen, Eds. (Prentice Hall, Englewood Cliffs, 1993) p. 616.

31. J. E. Mark, in *Frontiers of Polymers and Advanced Materials* P. N. Prasad, Ed. (Plenum, New York, 1994) p. 403.

32. H. Schmidt, H. Krug, in *Inorganic and Organometallic Polymers II* P. Wisian-Neilson, H. R. Allcock, K. J. Wynne, Eds. (American Chemical Society, Washington, 1994), vol. 572, p. 183.

33. J. E. Mark, P. D. Calvert, *J. Mats. Sci., Part C* **1**, 159 (1994).

34. J. E. Mark, in *Diversity into the Next Century* R. J. Martinez, H. Arris, J. A. Emerson, G. Pike, Eds. (SAMPE, Covina, CA, 1995), vol. 27,.

35. L. Mascia, *Trends in Polymer Science* **3 (2)**, 61 (1995).

36. J. E. Mark, *Macromol. Symp.* **93**, 89 (1995).

37. J. D. Mackenzie, in *Hybrid Organic-Inorganic Composites* J. E. Mark, C. Y.-C. Lee, P. A. Bianconi, Eds. (American Chemical Society, Washington, 1995), vol. 585, p. 226.

38. J. E. Mark, S. Wang, Z. Ahmad, *Macromol. Symp.* **98**, 731 (1995).

39. J. E. Mark, in *Hybrid Organic-Inorganic Composites* J. E. Mark, C. Y.-C. Lee, P. A. Bianconi, Eds. (American Chemical Society, Washington, 1995), vol. 585, p. 1.

40. J. Wen, G. L. Wilkes, in *Polymeric Materials Encyclopedia: Synthesis, Properties, and Applications* J. C. Salamone, Ed. (CRC Press, Boca Raton, 1996).

41. P. E. Cassidy, *Thermally Stable Polymers* (Marcel Dekker, New York, 1980).

42. J. P. Crichley, G. J. Knight, W. W. Wright, *Heat Resistant Polymers* (Plenum Press, New York, 1983).

43. J. F. Wolfe, in *Encyclopedia of Polymer Science and Engineering* H. F. Mark, N. M. Bikales, C. G. Overberger, G. Menges, Eds. (Wiley-Interscience, New York, 1987) p. 635.

44. C. J. T. Landry, B. K. Coltrain, J. A. Wesson, N. Zumbulyadis, J. L. Lippert, *Polymer* **33**, 1496 (1992).

45. S. Wang, Z. Ahmad, J. E. Mark, *Polym. Bulletin* **31**, 323 (1993).

46. Z. Ahmad, S. Wang, J. E. Mark, in *Better Ceramics Through Chemistry VI* A. Cheetham, C. J. Brinker, M. L. Mecartney, C. Sanchez, Eds. (Materials Research Society, Pittsburgh, 1994), vol. 346, p. 127.

47. Z. Ahmad, S. Wang, J. E. Mark, in *Hybrid Organic-Inorganic Composites* J. E. Mark, C. Y.-C. Lee, P. A. Bianconi, Eds. (American Chemical Society, Washington, 1995), vol. 585, p. 291.

48. M. Spinu, A. Brennan, J. Rancourt, G. L. Wilkes, J. E. McGrath, in *Multi-Functional Materials* D. R. Ulrich, F. E. Karasz, A. J. Buckley, G. Gallagher-Daggitt, Eds. (Materials Research Society, Pittsburgh, 1990), vol. 175, p. 179.

49. M. Nandi, J. A. Conklin, J. L. Salvati, A. Sen, *Chem. Mater.* **3**, 201 (1991).

50. K. Yano, A. Usuki, A. Okada, T. Kurauchi, O. Kamigaito, *Preprints, Div. Polym. Chem., Inc., Am. Chem. Soc.* **32(1)**, 65 (1991).

51. F. Breval, M. L. Mulvihill, J. P. Dougherty, R. E. Newham, *J. Mats. Sci.* **27**, 3297 (1992).

52. A. Morikawa, Y. Iyoku, M. Kakimoto, Y. Imai, *J. Mater. Chem.* **2**, 679 (1992).

53. A. Morikawa, Y. Iyoku, M. Kakimoto, Y. Imai, *Polym. J.* **24**, 107 (1992).

54. R. J. Jeng, Y. M. Chen, A. K. Jain, J. Kumar, S. K. Tripathy, *Chem. Mater.* **4**, 1141 (1992).

55. J. J. Burgmeister, L. T. Taylor, *Chem. Mater.* **4**, 729 (1992).

56. S. Marturunkakul, J. I. Chen, R. J. Jeng, S. Sengupta, J. Kumar, S. K. Tripathy, *Chem. Mater.* **6**, 743 (1993).

57. K. Yano, A. Usuki, A. Okada, T. Kurauchi, O. Kamigaito, *J. Polym. Sci., Polym. Chem. Ed.* **32**, 625 (1993).

58. L. Mascia, A. Kioul, *J. Mater. Sci. Lett.* **13**, 641 (1994).

59. X. Chen, K. E. Gonsalves, G.-M. Chow, T. D. Xiao, *Adv. Mater.* **6**, 481 (1994).

60. S. Wang, Z. Ahmad, J. E. Mark, *J. Macromol. Sci., Macromol. Reports* **31**, 411 (1994).

61. S. Wang, Z. Ahmad, J. E. Mark, *Chem. Mats.* **6**, 943 (1994).

62. L. Mascia, A. Kioul, *Polymer* **36**, 3649 (1995).

63. R. F. Kovar, R. W. Lusignea, in *Ultrastructure Processing of Advanced Ceramics* J. D. Mackenzie, D. R. Ulrich, Eds. (Wiley-Interscience, New York, 1988) p. 715.

64. R. F. Kovar, R. W. Lusignea, R. A. Griffiths, E. L. Thomas, in *Chemical Processing of Advanced Materials* L. L. Hench, J. K. West, Eds. (Wiley, New York, 1992) p. 685.

65. S. A. Jenekhe, J. A. Osaheni, *Chem. Mater.* **6**, 1906 (1994).

66. J. P. Chen, Z. Ahmad, S. Wang, J. E. Mark, F. E. Arnold, in *Hybrid Organic-Inorganic Composites* J. E. Mark, C. Y.-C. Lee, P. A. Bianconi, Eds. (American Chemical Society, Washington, 1995), vol. 585, p. 297.

67. T. D. Dang, J. P. Chen, F. E. Arnold, in *Hybrid Organic-Inorganic Composites* J. E. Mark, C. Y.-C. Lee, P. A. Bianconi, Eds. (American Chemical Society, Washington, 1995), vol. 585, p. 280.

68. J. Premachandra, C. Kumudinie, W. Zhao, J. E. Mark, T. D. Dang, J. P. Chen, F. E. Arnold, *J. Sol-Gel Sci. Technol.* **6**, 000 (1996).

STAR GELS. NEW HYBRID NETWORK MATERIALS FROM POLYFUNCTIONAL SINGLE COMPONENT PRECURSORS

KENNETH G. SHARP AND MICHAEL J. MICHALCZYK
Central Research, DuPont Co., Wilmington, DE, 19880-0323 U.S.A.

ABSTRACT

A new family of hybrid inorganic/organic network materials -- the star gels -- has been synthesized from single component molecular precursors. The starting materials comprise an organic core with multiple flexible arms which terminate in network-forming trialkoxysilane groups. The core can be a single silicon atom, linear disiloxane segment or ring system. With at least 12 alkoxysilane groups per molecule, gelation rates in aqueous or formic acid media can be extremely high, but can be attenuated several orders of magnitude by choice of solvent system. The degree of intramolecular condensation of these molecules has been assessed via mass spectrometric techniques. Transparent glasses which show brittle fracture but high levels of toughness have been generated from this family of precursors. The materials do not show plastic deformation even under compressive stress of 350 MPa. No evidence for open porosity in the glasses has been obtained. The organic content of the networks can be increased by lengthening the arms; the inorganic content can be increased via co-hydrolysis with simple tetraalkoxysilanes such as TEOS. Species with longer arms lead to glasses with higher coefficients of thermal expansion.

INTRODUCTION

Sol-gel chemistry has become uniquely useful as a low temperature route to inorganic network materials. Especially when hybrid organic/inorganic materials are sought, sol-gel generation of the inorganic phase is the most commonly practiced route [1]. For purely inorganic materials, however, the convenience of gel synthesis under mild conditions can be compromised by the need for extended drying times and the resultant brittleness of monolithic structures or thick films.

Our objective in the present work was to incorporate compliance in glassy solids at the molecular level while maintaining the general hydrolytic and condensation chemistry characteristic of silica-based sol-gel. Our approach was to generate a family of polyfunctional network precursors with a number of flexible arms radiating from a common atomic, linear or cyclic core and terminating in trialkoxysilane groups. A few structural and synthetic assembly examples of "star gel" precursors are shown in Fig. 1. More specific molecular examples are included in Fig. 2. What results from gelation of these precursors (with or without added tetraalkoxysilane) is a single network which comprises both rigid inorganic and flexible organic segments. Clear glasses can be generated from the wet gels, but their mechanical behavior lies somewhere between conventional glasses and cross-linked rubbers.

A number of desirable attributes would plausibly result from incorporation of compliant polyfunctional network formers. These include high rates of network formation, relatively fast drying rates for wet gels, possible pore collapse on drying and toughness or impact resistance for the resultant dry glasses.

Several features of chemical interest relate to the polyfunctionality of the molecules: the gelation rates, Arrhenius behavior and percolation threshold might be expected to be unusual. Moreover, because of the length of the arms, *intra-* as well as intermolecular condensation reactions should be possible. Finally, the stars might substantially alter the properties of a conventional silicate network when incorporated in relatively small amounts.

Shea, Loy and co-workers have described several families of hexafunctional alkoxysilanes. Where rigid (e.g., arylene) spacers are used, materials of extremely high surface area are possible [2-5]; however, the dried glasses are as brittle as conventional gels derived from tetraalkoxysilanes. Other rigidly separated polyalkoxysilanes have been used in hybrid material systems [6] . More recently, the Shea and Loy groups have extended the chemistry to include

Mat. Res. Soc. Symp. Proc. Vol. 435 © 1996 Materials Research Society

alkylene spacers [7,8]. In some instances gels derived from these precursors have been non-porous.

Si-vinyl functional core:

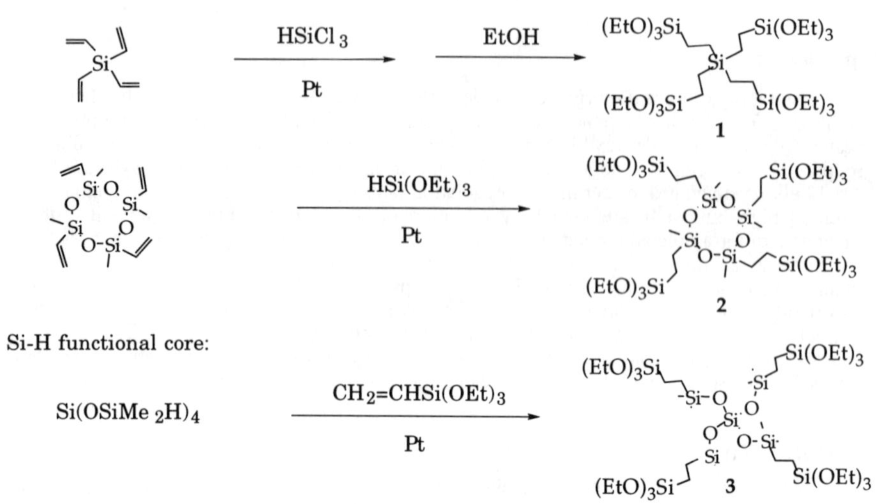

Figure 1. Example syntheses and structures of Star Gel precursors

EXPERIMENTAL

The star silanes can be synthesized in one- or two-step [9] high yield hydrosilylation reactions. The core or arm segments contain -CH=CH_2 groups; the mating segment contains terminal Si-H bonds which add to the double bonds in anti-Markovnikov fashion, producing linear arms. Synthetic examples for Stars **1** and **2** are shown in Figure 1. The reaction occurs smoothly between 60°-110 °C in the presence of a Pt(0) catalyst [10]. Co-solvents can be used but are generally unnecessary. Proton and ^{13}C NMR and K^+IDS mass spectrometry [11] have been used to characterize product purities. Typically, yields of completely substituted material exceed 95% and only for the most crowded molecules were some bent arms (Markovnikov addition) observed. In these cases (e.g., with **1**), fully linear and fully substituted products could be obtained from use of trichlorosilane in the hydrosilylation step followed by alkoxylation. Water used in the uncatalyzed gelations had a resistivity greater than 10^8 ohm-cm, was stored in fluoropolymer containers and was deaerated with nitrogen prior to use.

Solid state ^{29}Si spectra were run on an MSL 200 at 39.73 MHz with cross polarization on samples spun at 3.5 kHz. Spectra were taken with a 20 second recycle delay. The optimum contact time for the cross polarization was determined to be 2 msec.

Bulk densities were determined from low pressure mercury porosimetry or from mass and dimensions of right cylinders grown in fluoropolymer tubes. Skeletal densities were determined from He pycnometry [12] . Nitrogen adsorption isotherms were run on Micromoretics instruments on samples which had been outgassed under high vacuum at a minimum of 150°C.

Compressive stress-strain measurements were conducted on cylindrical disks grown in fluoropolymer containers. Disk faces were polished to improve flatness. Strain rates were about 0.12 mm/min for a typical sample 3 mm thick.

Coefficients of thermal expansion were determined in a thermomechanical analyzer with a stylus under the minimum load necessary to maintain contact with the surface of a star disk.

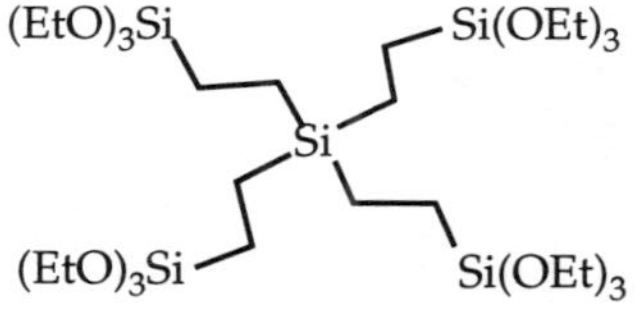

1

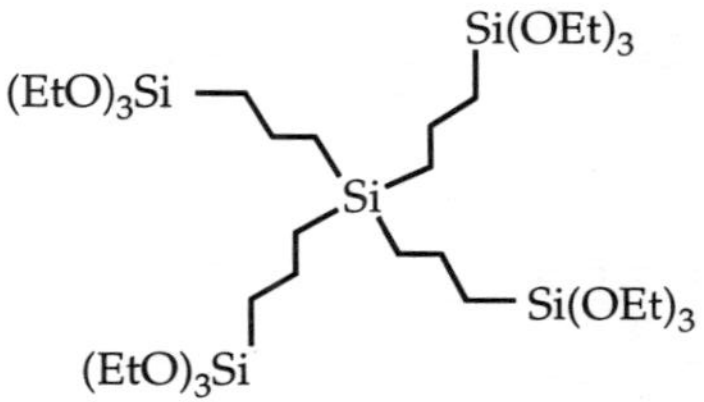

4

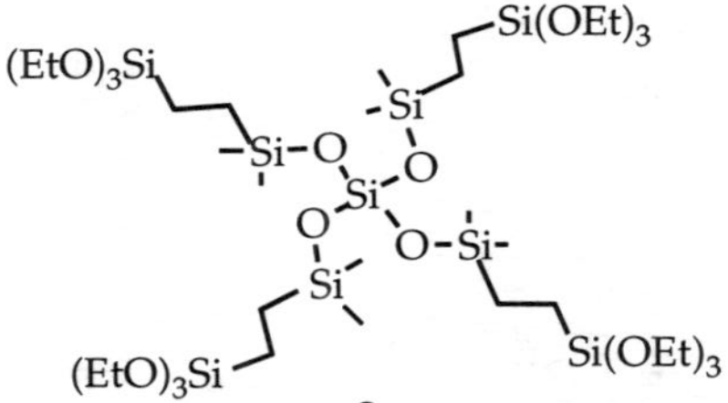

3

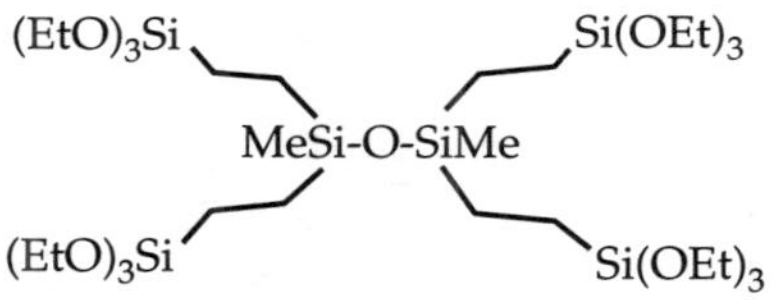

5

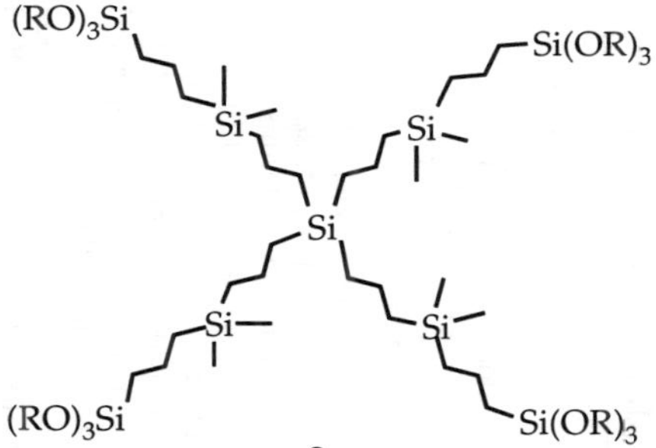

8

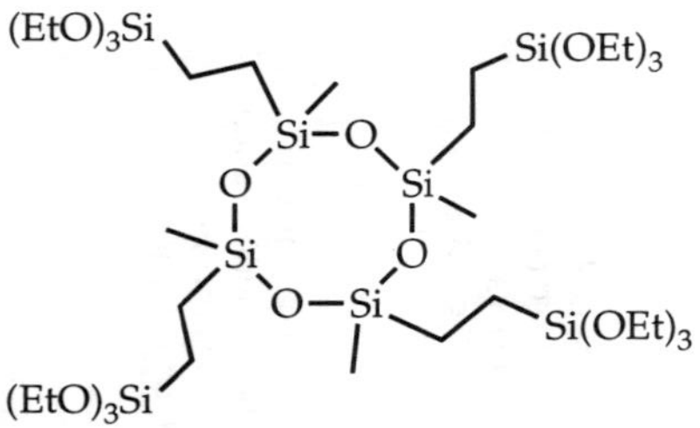

4

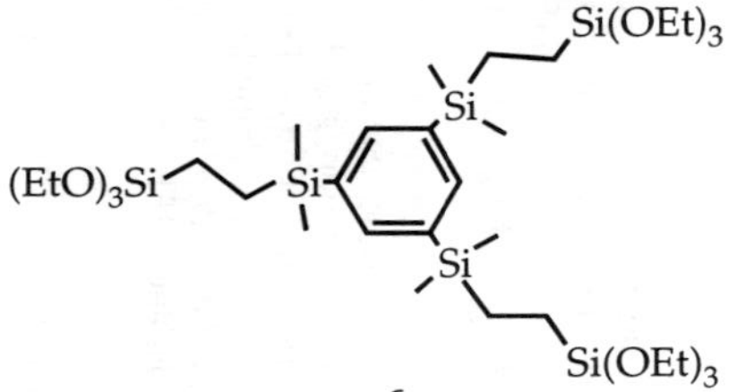

6

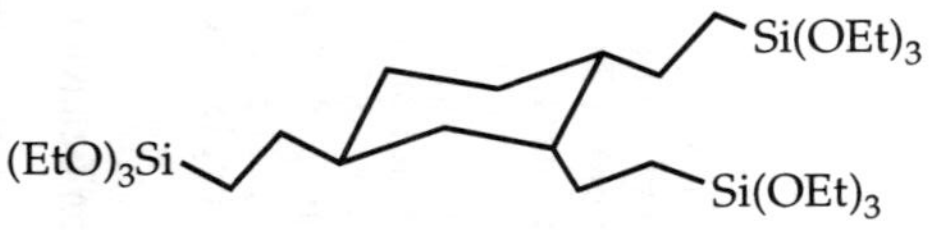

7

Figure 2
Families of Star Gel Precursors

RESULTS AND DISCUSSION

Sol-gel Chemistry

Both traditional and novel sol-gel chemistry can be used to convert the star molecules into gels. In conventional aqueous-based acid- or base-catalyzed systems [13] , the stars gel much more rapidly than TEOS or alkyltrialkoxysilanes such as $C_3H_7Si(OEt)_3$. At pH values near 2 (slowest condensation) or 7 (slowest hydrolysis) the gelation rates of the stars can be several orders of magnitude greater than that of TEOS.

Also useful is a newly developed [14] non-aqueous process using formic acid or non-catalyzed conditions using water at pH 7. Gel times can be varied from about one second to several weeks by choice of silane and its concentration, gelation agent, pH and co-solvent. Table I lists some representative gelation rates of $Si(OEt)_4$ [TEOS] and Star **1** under a variety of conditions. The star reacts much faster in acid media (including the anhydrous formic acid system) and at pH 7. The rate differential is considerably less at high pH, consistent with observations from simple alkyl(trialkoxy)silanes [15].

Table I
Gelation Behavior in Acidic and Neutral Media for Star **1** and TEOS

	pH 2 (H_2O/HCl)	**Neat formic acid**	**uncatalyzed H_2O**
EtOH/Si (molar)	20.0	----	4.0
Gel agent/Si-OR	0.63	0.97	1.00
TEOS t_{gel}	11,000 hrs [16]	450 min [14]	130 hrs pH 5 [17]
Star 1 t_{gel}	3 min.	< 4 min	8 hrs pH 7

Along with the usual network-forming condensation reactions eliminating water, alcohol or formic acid [14], the condensation pathway for the stars can also include *intra*molecular reactions. The distinction between intra- and intermolecular condensation pathways is minimal in silicon NMR due to overlapping chemical shifts of their characteristic resonances. However, the use of K^+IDS mass spectra has been very useful in analyzing partially reacted systems. Figure 3 shows a mass spectrum of partially condensed **1** in which both intra- and intermolecular condensations are apparent. Although stars with relatively short arms (e.g., **1**) are more susceptible to intramolecular condensation they can be converted into gels very rapidly.

The gelation rate of TEOS by formic acid can be profoundly influenced by addition of small amounts of star precursors. A mixture of TEOS and HCOOH at a molar ratio of 1:3 normally requires 18 hours to gel at room temperature. Substitution of 10 mole% **1** for TEOS led to a gelation time of 8 *minutes* under comparable conditions, despite the lowering of the ratio of acid to total alkoxy groups. The percolation threshold for gel formation appears to be quite low: gels have been made from **4** with only 1.36 molar equivalents of formic acid.

The gelation behaviors of Stars **1**, **4** and TEOS in uncatalyzed aqueous media are compared via the Arrhenius plot shown in Fig. 4. The stars gel about 20 times faster than TEOS, but the activation energy for TEOS is lower. Accordingly, there is a very substantial reduction in the pre-exponential factor for the stars. The TEOS data are taken from the literature [17], from a system in which the pH of the uncatalyzed water was reported as 5. It seems likely that the water used contained adventitious quantities of impurities such as dissolved CO_2 or Fe(III) and that the gelation rate of TEOS in highly purified water would be slower yet.

Star Glasses

The wet gels can be dried relatively rapidly to produce transparent, compliant films or glasses which comprise a single organic/inorganic network. The star glasses show intriguing physical behavior somewhere between conventional glasses and rubbery elastomers. Although they typically show brittle fracture under very high levels of stress, their bulk flexibility is readily detected by fingertip pressure. Unlike conventional sol-gel generated glasses, they show no

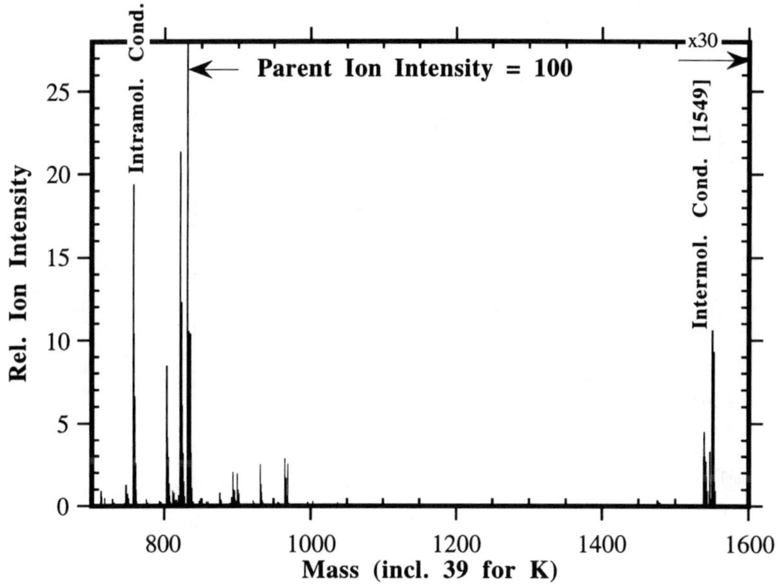

Figure 3. KID$^+$S Mass Spectrum of Early Products of Star 1 Reaction. Peaks above mass 1500 are shown 30x actual intensity.

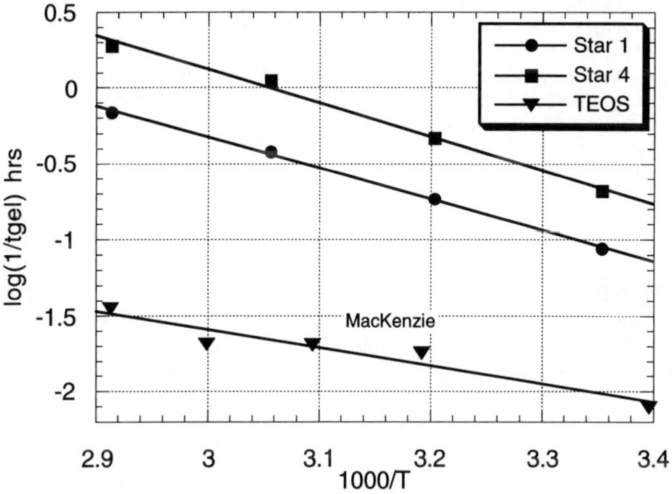

Figure 4. Gelation Arrhenius Plots for Star **1, 4** and TEOS. [Log A, E_{act} (kcal/mole)] values : Star **1** [5.82, 9.40]; Star **4** [6.78, 10.17]; TEOS (Ref. 17) [2.03,4.6]

molecular precursors. Unlike thermoplastic organic materials, they show almost no plastic (irreversible) deformation under stress.

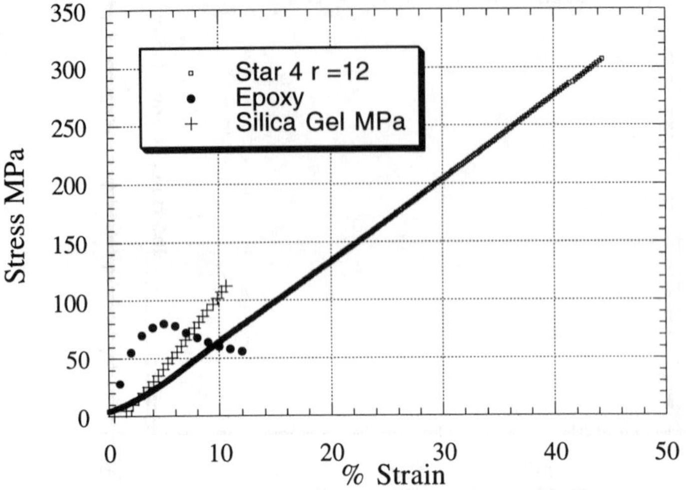

Figure 5. Compressive Stress-Strain Curves for Star Glasses and Related Materials

The dried star glasses show much higher levels of toughness and impact resistance than do conventional sol-gel glasses. Figure 5 shows stress-strain curves of two star glasses, two TEOS-derived gels and a conventional organic network material, an epoxy resin.[18] The moduli of the conventional glasses are higher, but the star glasses show a substantially greater strain (and energy) to break and tensile strength. They are also considerably tougher in compression than the organic network materials. The prominent yield and plastic deformation behavior evident in the epoxy has not detected for the stars during or after compression stress/strain runs.

Attempts to determine toughness by examining crack propagation in an indented sample were thwarted by the failure of a Vicker's diamond (roughly 100 microns in diameter) to produce a permanent indentation in a sample under a load of 400 g.

The coefficients of thermal expansion (CTE) for the Stars thus far examined are significantly greater than that for a TEOS-derived gel (Figure 6). There appears to be a relationship between arm length (and corresponding ring size) and CTE; the highest value is for Star **3**, whose smallest ring size is 12 atoms.

Solid state ^{29}Si CP MAS NMR of the formic acid-catalyzed condensation of **1** shows about 66% of the silicon atoms are T^2 (two network links, one uncondensed Si-OR (R=Et or H)) with about equal amounts (~ 17 % each) of T^1 and fully condensed T^3 silicons. This distribution of network bonds is consistent with NMR observations made for hydrolysis and condensation reactions of $RSi(OEt)_3$ [19]. Gelation in water/alcohol mixtures with NH_4OH as a catalyst leads to a higher proportion of T^3 structures. This behavior has been observed by Shea and co-workers for a series of silsesquioxanes in which the silicon atoms are separated by alkylene spacers [8].

The star molecules can also readily be co-networked with simple tetraalkoxysilanes. Solid state ^{29}Si NMR indicates that both **1** and TEOS are effectively incorporated into the network.

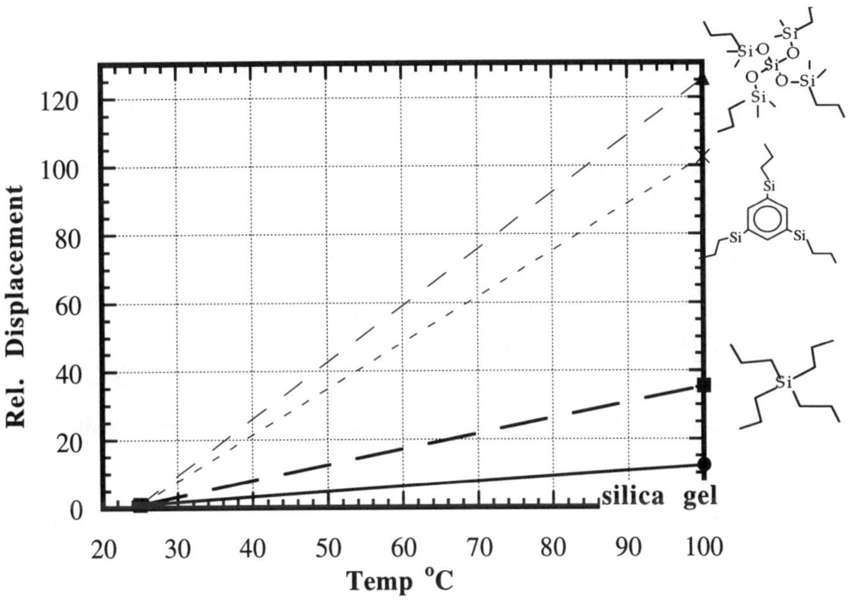

Figure 6. Coefficients of Thermal Expansion for Silica Gel and Stars (cores shown)

The dried pure star glasses from the condensation of **1** by formic acid show no significant surface area by nitrogen adsorption isotherm measurements. The *apparent*[20] lack of open porosity suggested by this experiment was confirmed by comparison of bulk density and skeletal density from He pycnometry [12]. The two densities were identical, indicating an inability of He to penetrate into the interior of the solid. The reason for the lack of measurable surface porosity is presumably related to the flexibility of the network, which may lead to pore collapse upon drying. However star-gel modified silica glasses made from the addition of 10 % of **1** to TEOS have shown surface areas of 625 m^2/g.

Applications for the star glasses or their molecular precursors currently under consideration include abrasion resistant coatings, glass adhesives, hyper cross-linking agents, impact resistant glasses and tough aerogels. The latter low density materials would be made via supercritical fluid drying processes, possibly leading to networks with open porosity.

REFERENCES AND NOTES

1. J. E. Mark, Heterog. Chem. Rev., in press
2. K. J. Shea, D. A. Loy, O. W. Webster, Chem. Mater **1,** 572-4 (1989).
3. K. J. Shea, D. A. Loy, O. W. Webster, Polym. Mater. Sci. Eng **63,** 281-5 (1990).
4. D. A. Loy, K. J. Shea, E. M. Russick. Preparation of aryl-bridged polysilsesquioxane aerogels, in *Mater. Res. Soc. Symp. Proc;* M. Hampden-Smith, W. Klemperer, C. J. Brinker, Eds., 1992; Vol. 271; pp. 699-704.
5. K. J. Shea, D. A. Loy, O. Webster, J. Am. Chem. Soc **114,** 6700-10 (1992).
6. R. J. P. Corriu, J. J. E. Moreau, P. Thepot, M. Wong Chi Man, C. Chorro, J.-P. Lere-Porte, J.-L. Sauvajol, Chem. Mater. **6,** 640-9 (1994).

7. J. H. Small, K. J. Shea, D. A. Loy, J. Non Cryst. Solids **160,** 234-46 (1993).
8. H. W. Oviatt, Jr., K. J. Shea, J. H. Small, Chem. Mater **5,** 943-50 (1993).
9. In some cases, the yield is somewhat improved if trichlorosilane is used in the hydrosilylation, then the chlorosilane is alkoxylated in a second step.
10. The Pt catalyst of choice was $Pt((ViMe_2Si)_2O)$ [Karstedt's catalyst]; concentrations of about 20 ppm by weight were found effective. Caution should be used since hydrosilylation conducted without solvent can be highly exothermic. Soluble catalyst residues were typically removed with activated charcoal.
11. W. J. Simonsick. Characterization of Polymer Building Blocks by K+ Ionization of Desorbed Species. In *Structure-Property Relationships in Polymers;* Urban, M. W.; Craver, C. D. Eds.; American Chemical Society: Washington, D.C., 1993; Vol. 236.
12. T. Woignier, J. Phalippou, J. Non-cryst. Solids **93,** 17 (1987).
13. C. J. Brinker, G. W. Scherer *Sol-gel Science*; Academic Press: San Diego, CA, 1990.
14. K. G. Sharp, J. Sol-gel Sci. Tech. **2,** 35 (1994).
15. H. Schmidt, J. Non Cryst. Solids **73,** 681-91 (1985).
16. V. Gottardi, M. Guglielmi, A. Bertoluzza, C. Fagnano, M. A. Morelli, J. Non-Cryst. Solids **63,** 71-80 (1984).
17. M. W. Colby, A. Osaka, J. D. Mackenzie, J. Non-Cryst. Solids **99,** 129-39 (1988).
18. A. J. Kinloch, R. J. Young *Fracture Behavior of Polymers*; Applied Science Publishers: Essex, England, 1983.
19. W. G. Fahrenholtz, D. M. Smith, D. W. Hua, J. Non-Cryst. Solids **144,** 45-52 (1992).
20. Several reports now exist of silica gel which shows no appreciable nitrogen adsorption at 77 ^{o}K over several hours but does adsorb CO_2 at higher temperatures.

Sol-Gel Kinetics for the Preparation of Inorganic/Organic Siloxane Copolymers

Stephen E. Rankin, Christopher W. Macosko, and Alon V. McCormick
Department of Chemical Engineering and Materials Science
University of Minnesota, Minneapolis, MN 55455

Abstract

A model is described which incorporates hydrolysis pseudo-equilibrium to quantify the evolution of silicon site distributions in sol-gel polymerization up to higher conversions than previously possible. ^{29}Si nuclear magnetic resonance spectroscopy data are used 1) to provide a means of recognizing hydrolysis pseudo-equilibrium and 2) to provide examples where different substitution effects on condensation rate coefficients caused by branching are observed. Extension of the model to copolymerization is discussed.

Introduction

The sol-gel method involves the formation of inorganic oxides from metal alkoxides in solution [1]. It is therefore necessary to approach the formation of these materials by studying the thermodynamics and kinetics of the reactions that lead to the final material. The most studied set of sol-gel monomers are tetraethoxysilane (TEOS) and organically substituted ethoxysilanes. The polymerization process that leads from monomers to inorganic oxides in these systems (assuming that water-producing condensation is dominant [2]) may be written:

$$\equiv SiOC_2H_5 + H_2O \rightleftharpoons \equiv SiOH + C_2H_5OH \quad (1)$$

$$\equiv SiOH + \equiv SiOH \rightarrow \equiv SiOSi \equiv + H_2O \quad (2)$$

where only one of four ligands attached to the silicon site undergoes exchange at a given reaction step.

When hydrolytic polycondensation of TEOS occurs at moderately low pH (pH $\sim$ 2–4), it is known that homogeneous gels are formed and that the species present in solution may be regarded as siloxane oligomers [1]. It would be advantageous, therefore, to regard sol-gel polymerization as ideal (meaning that the reactivity of a silanol is not a function of the types of ligands already attached to that site, cyclization does not occur, and diffusion plays no role). Previous studies of alkoxysilane polymerization kinetics [3, 4] have shown, however, that this condition usually does not hold and that there appear to be substitution effects in the condensation reactivity of alkoxysilanes.

If we consider only first shell substitution effects (caused by ligands directly attached to the reacting sites), then an intractable number of rate coefficients is required to fully characterize the condensation rate between silicon sites of arbitrary degrees of connectivity and of hydrolysis [3]. Here, we overcome this limitation by using the *observed* pseudo-equilibrium hydrolysis state of the system to reduce the number of condensation rate coefficients needed to characterize a given polymerizing system. This hydrolysis pseudo-equilibrium simplification has been explicitly employed by Assink [5], Rankin *et al.* [6, 7], and Sanchez *et al.* [8] and appears to be applicable to many systems (see references 1 and 9). This model is equally applicable to homopolymerizations and copolymerizations.

^{29}Si NMR spectroscopy

^{29}Si NMR has been shown to be an excellent tool for monitoring the evolution of silicon sites according to their first shell environment [2–12]. For ethoxysilanes, the technique is sensitive to both the degrees of hydrolysis and condensation of the observed nucleus. Advanced multipulse techniques are also available to verify assignments of peaks, as well as to provide information about the structure of the oligomer in which a site is found [4].

Mat. Res. Soc. Symp. Proc. Vol. 435 © 1996 Materials Research Society

Table 1: Initial compositions of sol-gel systems

Name	Monomer	[Si] (M)	$(H_2O/Si)_0$	-log[H^+]	Ref.
Brunet TMOS	Tetramethoxysilane	2.27	8	3.0	10
Devreux MTEOS	Methyltriethoxysilane	1.37	10	3.1	11
Devreux VTEOS	Vinyltriethoxysilane	1.35	10	3.1	11
Devreux	Tetraethoxysilane	1.32	10	3.1	11
Rankin MTEOS#1	Methyltriethoxysilane	2.61	4	2.6	6
Rankin MTEOS#2	Methyltriethoxysilane	2.19	2	2.6	6
Sanchez ETEOS	Ethyltriethoxysilane	2.16	2	2.6	8
Sanchez	Tetraethoxysilane	2.07	2	2.6	8

All data discussed here were gathered using ^{29}Si NMR spectroscopy. The conditions used in the systems to be discussed are summarized in Table 1. These systems were chosen for this study because the reported data provide a good indication of the trajectory of the site concentrations in phase space (see below). In keeping with previous notation, tetrafunctional sites of connectivity i (i siloxane ligands attached) are denoted by Q_i, trifunctional by T_i, difunctional by D_i, and monofunctional by M_i [8].

HYDROLYSIS PSEUDO-EQUILIBRIUM

The basis of the simplification we have introduced into the first-shell substitution model of alkoxysilane polymerization is a hydrolysis pseudo-equilibrium. The hydrolysis pseudo-equilibrium condition means that hydrolysis and reesterification both proceed so quickly that the set of reagents and products in equation 1 are able to approach their most favorable free energy state at all times, even as condensation reactions proceed. A mathematical expression of this is just given by:

$$K_h = e^{-\Delta G_{hydrolysis}/RT} = \frac{[\equiv SiOH][C_2H_5OH]}{[\equiv SiOC_2H_5][H_2O]} \quad (3)$$

where G is Gibbs free energy, brackets denote activities and K_h is the hydrolysis equilibrium constant. That hydrolysis proceeds on a shorter time scale than condensation can be inferred from studies where hydrolysis kinetics have been studied before significant condensation begins (see reference 9 for a review). Note that nothing has been stated about the other ligands attached to a silicon site in this expression. It is easily possible that hydrolysis equilibrium coefficient (K_h) values vary with the degree of hydrolysis and of condensation of the site involved.

In many observed cases, this pseudo-equilibrium manifests itself as a sharp change in water and alcohol concentrations followed by relative constancy [8,10,11]. In these cases, the *extents* of hydrolysis (ξ_i), defined as the fraction of non-condensed groups which are silanols for sites of connectivity i, become approximately constant (or evolve very slowly) after an initial sharp transient. When a constant extent is the pseudo-equilibrium hydrolysis state, the rate of reaching that state compared to the rate of condensation is what determines when hydrolysis pseudo-equilibrium is applicable. To see when this is the case, it is sufficient to compare the evolution of the extent of hydrolysis of sites of connectivity i to the concentration of sites of connectivity $i+1$. Because sites appear in the order $(X_0, X_1, \ldots)$, this condition is sufficient to show that hydrolysis pseudo-equilibrium has been reached for all condensations producing sites of connectivity $i+1$.

As an example of a case in which this type of pseudo-equilibrium is reached, Figure 1 shows the evolution of extents of hydrolysis of sites of connectivity i (ξ_i) with the concentration of sites of connectivity $i+1$ for TEOS and ethyltriethoxysilane (ETEOS) [8]. These data were collected using ^{29}Si NMR. In order to compare different sites, the figure shows the extents on a dimensionless scale of the extent divided by the extent reached at pseudo-equilibrium. For TEOS, these pseudo-equilibrium values are 0.35, 0.35, and 0.38 for i = 0, 1, and 2, respectively and for ETEOS, these values are 0.57 and 0.65 for i = 0 and 1.

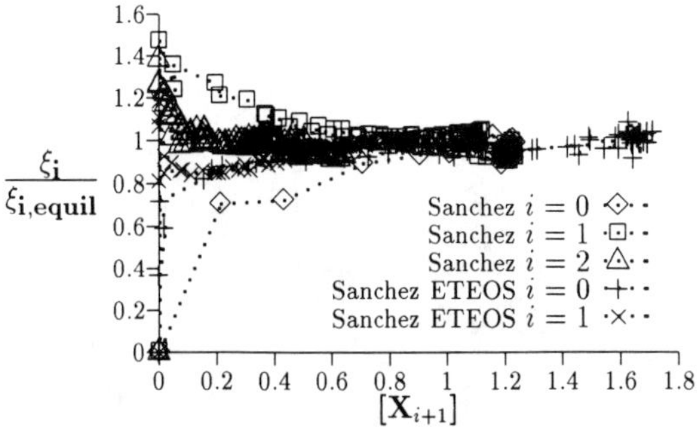

Figure 1: Evolution of dimensionless extent of hydrolysis of sites of connectivity i with concentration of sites of connectivity $i + 1$. Data are adopted from Sanchez *et al.* [8].

The similarity of these pseudo-equilibrium values indicates that K_h is a very weak function of i. Figure 1 shows that these systems approach pseudo-equilibrium before a great deal of condensation begins. For these systems, a condensation model using hydrolysis pseudo-equilibrium was found to be appropriate.

CONDENSATION MODEL

When hydrolysis reaches a pseudo-equilibrium state, it is possible to treat condensation as a network of bimolecular reactions with rate coefficients $k_{ii'}$ occurring between sites X_i of connectivity i and sites $X_{i'}$ of connectivity i'. If condensation is presumed to occur solely by a water-producing route [2], then the set of differential equations describing the evolution of sites is [8]:

$$\frac{d[X_i]}{dt} = \sum_{i'=0}^{f-1} \left\{ k_{(i-1)i'} \langle j \rangle_i \langle j \rangle_{i'} [X_{i-1}][X_{i'}] - k_{ii'} \langle j \rangle_i \langle j \rangle_{i'} [X_i][X_{i'}] \right\} \tag{4}$$

where $\langle j \rangle_i = (f - i)\xi_i$ is the average degree of hydrolysis of sites of connectivity i, f is the functionality of the monomer, and $[X_i]$ is the concentration of sites X_i.

Concentrations from equation 4 were fit to site concentration data from ^{29}Si NMR for TEOS and ethylethoxysilanes at low pH [8]. The set of rate coefficients found corresponds to a strong negative substitution effect on condensation reactivity caused by branching. The trend in rate coefficients with branching of the sites involved can not be expressed using a linear free energy relationship. That is,

$$k_{ii'} \neq k_{00}(e^{-\Delta\Delta G_1^\ddagger/RT})^i (e^{-\Delta\Delta G_2^\ddagger/RT})^{i'} \tag{5}$$

By comparing the trajectories of site concentrations for different systems, we can see how far different sol-gel systems are from ideality and how close they are to satisfying this linear free energy relationship.

SUBSTITUTION EFFECTS

To assess substitution effects, it is fruitful to compare, in phase space, the site concentration trajectories of systems with simple substitution effects and those of real sol-gel systems. This phase space representation is valuable for two reasons: 1) the positions of the points along a trajectory are independent of time, so changes in systems which change the rate of *all* reactions are filtered out, and 2) the region where monomers and fully condensed sites coexist is small enough that all degrees of freedom of the system can be understood by following only the concentrations of intermediates ($\{[T_1], [T_2]\}$ for trifunctional and $\{[Q_1], [Q_2], [Q_3]\}$

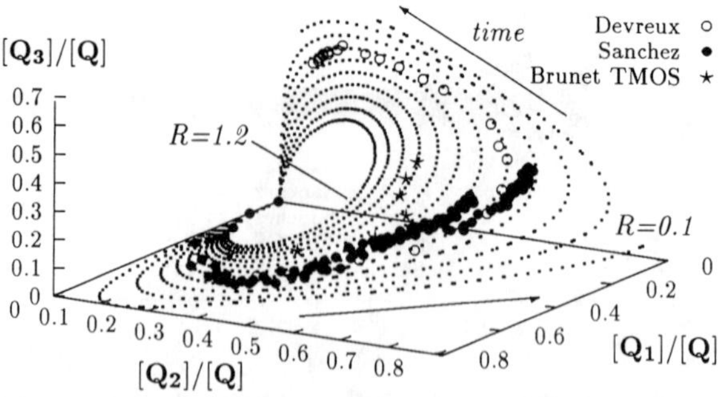

Figure 2: Phase portraits for tetrafunctional systems. Dotted lines correspond to simulations using the substitution effect in equation 6 with R varying in units of 0.1 from R=0.1 (outermost) to R=1.2 (innermost). Points are experimental data. Trajectories start at (0,0,0) and move counterclockwise with time.

for tetrafunctional monomers). One form for the substitution effect with branching that has been proposed [13] is:

$$k_{ii'} = k_{00} R^{(i+i')} \tag{6}$$

In Figure 2, the results of the integration of equation 4 with R varying from 0.1 to 1.2 have been plotted along with experimental data for tetrafunctional monomers. In the calculations, it was assumed that $K_h = 10$ for all sites, in keeping with typical values for ethoxysilane hydrolysis [9]. Notice that the calculated data form a manifold in phase space; this results primarily from material balance considerations and the presence of limited sets of sites at certain conversions.

The first notable feature of Figure 2 is that the experimental trajectories for the TEOS systems do not appear to correspond to a single R value. This agrees with what was found by least-squares fitting of the condensation rate coefficients for sample "Sanchez" – there is no linear free energy relationship. However, the TMOS sample from Brunet *et al.* [10] does stay fairly close to the trajectory with $R = 0.9$. Thus, there may be more of an additive free energy relationship for TMOS than for TEOS under the conditions of the two systems studied. In addition, the trajectories of both TEOS samples match each other exactly up to the end of the Sanchez data. This is somewhat surprising considering that Devreux *et al.* used much more water and a slightly higher pH than Sanchez *et al.* (see Table 1). Thus, the substitution effects appear to agree for these two systems.

Figure 3 shows the corresponding trajectories in phase space for trifunctional systems. The dotted lines correspond to calculated trajectories with R varying from 0.1 to 1.2 and the points are the indicated data. Again, none of the experimental trajectories quite match up with a line, so additive free energy relationships fail. It is again interesting that the trajectories of the methyl and vinyl systems match very closely, indicating a similarity in substitution effects in spite of the differences in the organic group on the monomer and in initial conditions.

For the "Sanchez ETEOS" sample, the substitution effect is even farther from an additive free energy relationship than for the methyl- substituted monomers. There also appears to be a more severe negative substitution effect (the trajectory is near the lines with lower R values). So in addition to reducing in the values of rate coefficients [8], it is clear by comparing Figures 2 and 3 that substitution of one ethoxy group with an ethyl group on TEOS promotes a more strongly negative substitution effect when polymerizations are carried out under similar conditions.

On the other hand, methyl substitution appears to not only accelerate condensation [6] but also to reduce the degree of the negative substitution effect caused by connectivity (comparing sample "Rankin MTEOS#2" to "Sanchez"). Thus for the systems studied, alkyl

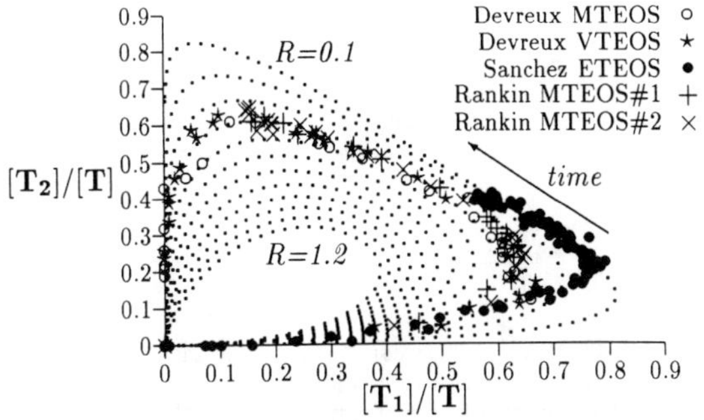

Figure 3: Phase portraits for trifunctional systems. Dotted lines correspond to trajectories using the substitution effect in equation 6 with R varying in increments of 0.1 from R=0.1 (outermost trajectory) to R=1.2 (innermost). The points are the indicated experimental data. The trajectories start at (0,0) and move counterclockwise with time.

substitution appears to have a similar effect on substitution effects caused by branching that it has on the values of the rate coefficients.

COPOLYMERIZATION

In addition to homopolymerization, this model can easily be extended to model copolymerization of monomers of differing functionality or different types of organic groups. If, for instance, sites are defined according to the number of bonds they form with identical monomers and with differing monomers, an equation analogous to equation 4 can be written as long as hydrolysis reaches a pseudo-equilibrium [7].

Such a model was fit to ^{29}Si NMR data for polymerizing trimethylethoxysilane (M) and dimethyldiethoxysilane (D). This copolymerization system is a suitable model because the set of rate coefficients required is tractable and because the phase space required to capture all degrees of freedom is only three-dimensional. As reported by Prabakar *et al.* [12], for copolymerization of monomers with differing numbers of methyl groups, it is possible using ^{29}Si NMR to distinguish between sites involved in different numbers of homo- and hetero-linkages. We employed this ability to determine homo- and co-polymerization rate coefficients from a single set of transient NMR spectra for an MD system at low conversion [7]. The copolymerization rate coefficients were found to be intermediate between the homopolymerization rate coefficients of the condensing sites.

While it is by far preferable to apply this model to situations where copolymerization bonds can be distinguished from homolinkages (for instance, by using ^{17}O NMR at high conversions [14]), it is not necessary in all cases. For instance, if the copolymerization coefficients from reference 7 are used along with the D condensation substitution effect from reference 6, trajectories in $M_1/D_1/D_2$ phase space can be generated for differing initial M/D ratios to get a feeling for how sensitive the overall condensed species population is to the initial composition. Figure 4 shows that the trajectory is quite sensitive to the fraction of M monomer. In the calculations, it was assumed that the conditions could be kept constant enough that the rate coefficients do not change and that $K_h = 10$. Both the qualitative shape and the actual values of the trajectories change as M is added. This indicates that the rate coefficients of the model can be determined by performing several experiments under identical conditions and varying the M/D ratio. This technique may be of use in systems where copolymerization can not directly be quantified.

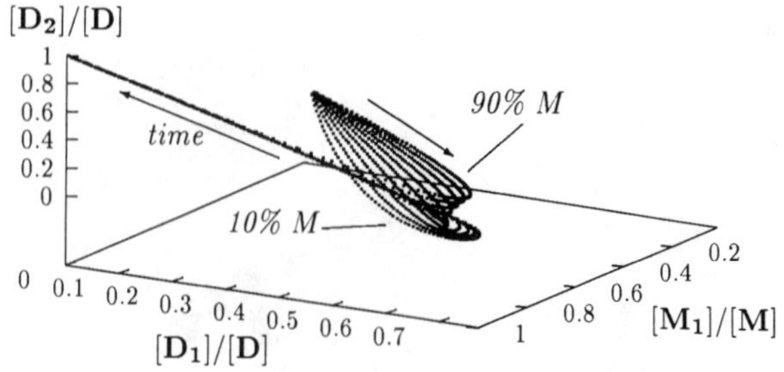

Figure 4: Trajectories in MD phase space using rate coefficients from references 6 and 7, $(H_2O/Si)_0 = 2$, and the fraction of monofunctional monomer varying in increments of 10% from 10% to 90%. Trajectories begin at (0,0,0) and move in the indicated direction.

CONCLUSIONS

A model for hydrolytic polycondensation of sol-gel monomers with hydrolysis pseudo-equilibrium has been shown to be applicable to several alkoxysilane and organically modified alkoxysilane systems. By comparing experimentally measured trajectories to predictions from this model using simplified substitution effects, certain conclusions could be drawn. All systems studied were found not to follow trajectories consistent with additive free energy relationships. It appears as if substitution of ethoxy groups with ethyl groups on the monomer increases the severity of the negative substitution effect on condensation caused by branching, while methyl substitution has an opposite effect. Analysis of data under several conditions using these phase portraits reveals an unexpected constancy in the trajectories, even as conditions are changed. The model has also been shown to be suitable for analysis of copolymerization in situations where heterolinkages and homolinkages can not be distinguished by ordinary NMR techniques.

ACKNOWLEDGMENTS

The authors would like to thank Dr. Gary Wieber of Dow Corning Corporation for helpful discussions, the National Science Foundation for a fellowship to SER, and Dow Corning Corporation for their financial support.

REFERENCES

[1] C. J. Brinker and G. W. Scherer. *Sol Gel Science*. Academic, Boston, 1990.
[2] R. A. Assink and B. D. Kay. *Coll. Surf. A*, 74:1–5, 1993.
[3] R. A. Assink and B. D. Kay. *J. Non-Cryst. Solids*, 107:35–40, 1988.
[4] F. Brunet and B. Cabane. *J. Non-Cryst. Solids*, 163:211–225, 1993.
[5] R. A. Assink and S. Prabakar. ACS National Meeting, New Orleans, LA. March 1996.
[6] S. E. Rankin, C. W. Macosko, and A. V. McCormick. Kinetics of methyl-modified sol-gel processes. *Inorg. Chem.*, to be submitted, 1996.
[7] S. E. Rankin, C. W. Macosko, and A. V. McCormick. Methylethoxysilane copolymerization kinetics in batch and semibatch reactors. *Chem. Mater.*, to be submitted, 1996.
[8] J. Sanchez, S. E. Rankin, and A. V. McCormick. *Ind. Eng. Chem. Res.*, 35:117–129, 1996.
[9] J. Šefčík and A. V. McCormick. *Catal. Today*, to appear, 1996.
[10] F. Brunet, B. Cabane, M. Dubois, and B. Perly. *J. Phys. Chem.*, 95:945–951, 1991.
[11] J. P. Devreux, F.and Boilot and F. Chaput. *Phys. Rev. A*, 41:6901–6909, 1990.
[12] S. Prabakar, R. A. Assink, N. K. Raman, and C. J. Brinker. *Mat. Res. Soc. Symp. Proc.*, 346:979–984, 1994.
[13] C. J. Brinker and R. A. Assink. *J. Non-Cryst. Solids*, 111:48–54, 1989.
[14] F. Babonneau, J. Maquet, and J. Livage. *Chem. Mater.*, 7:1050–1052, 1995.

NMR CHARACTERIZATION OF THE CHEMICAL HOMOGENEITY IN SOL-GEL DERIVED SILOXANE-SILICA MATERIALS

Florence BABONNEAU*, Virginie GUALANDRIS, Monique PAUTHE
Chimie de la Matière Condensée, Université Pierre et Marie Curie / CNRS, 4 place Jussieu, 75005 Paris, France.

ABSTRACT

^{17}O NMR in solution using enriched water as reactant has been used to investigate co-hydrolysis of dimethyldiethoxysilane and tetraethoxysilane, as well as methyltriethoxysilane and tetraethoxysilane. Co-condensation reactions were clearly identified in both systems : the amount of oxo bridges between the two kinds of Si units is rather large, and those bridges are stable during aging, which favor a good chemical homogeneity in the final gels. This last point was investigated on the related dried gels by ^{29}Si Magic Angle Spinning NMR, using Cross-Polarization (CP) technique that allows to probe the local environment of the various Si sites through ^{29}Si-^{1}H dipolar interactions.

INTRODUCTION

Numerous hybrid siloxane-oxide systems have been prepared by sol-gel techniques, using substituted silicon alkoxides, $R'_xSi(OR)_{4-x}$, crosslinked by metal alkoxides, $M(OR)_n$ with M = Si, Ti or Zr [1-4]. The chemical reactions occurring in solution (hydrolysis and condensation), during the aging process before gelation, and then during drying, will strongly influence the chemical homogeneity of the final materials, and thus their properties. It is of extreme importance for the synthetic chemist to be able to characterize all the preparation steps, in order to design hybrid materials with given architecture.

It is of great importance to be able to differentiate self-condensation from co-condensation reactions between two different precursors.

Self-condensation : M-OH + XO-M -> M-O-M + XOH (X = R or H) (1)

Co-condensation : M-OH + XO-M' -> M-O-M' + XOH (2)

M and M' can be two different metal atoms such as Si, Ti or Zr, but can also be the same metal atom in different environment.

Identification and moreover quantification of the various oxo-bridges are essential to characterize the degree of chemical homogeneity of the final network. This problem is not specific to hybrid systems, and can be extended to all sol-gel derived materials. ^{29}Si NMR has been extensively used to follow hydrolysis and condensation reactions of different organosilanes, and more specifically of tetraalkoxysilanes $Si(OR)_4$ (R = Me or Et) [5-9] and methyl substituted alkoxysilanes, $Me_2Si(OR)_2$ [10,11] and $MeSi(OR)_3$ [12,13].

Investigations on the condensation reactions between several precursors such as DMDES ($Me_2Si(OEt)_2$) and TEOS ($Si(OEt)_4$) [14,15], MTES ($MeSi(OEt)_3$) and TEOS [16-18] have clearly identified the formation of co-condensed species. The task is quite easy when low hy-

Mat. Res. Soc. Symp. Proc. Vol. 435

drolysis ratios are used, which just generate short oligomers. The problem is more difficult when higher hydrolysis ratios like those used for sol-gel preparations, are used : oligomers of various size and composition are formed, that will give a distribution of resonance peaks that usually strongly overlap. It is thus possible to get direct evidence for co-condensation, but very difficult to quantify the various species. Identification by ^{29}Si NMR of co-condensation reactions between organosilanes and titanium or zirconium alkoxides was not clearly demonstrated [19-21] mainly because of a lack of references.

^{17}O NMR should appear as the best technique to identify the oxo-bridges formed during condensation. Due to the low sensitivity of this isotope, enriched water is needed especially to follow kinetics which requires to record spectra in short time. Beside the tremendous increase in the quality of the spectra, the use of enriched water in sol-gel reactions has another advantage, to label specifically hydroxyl groups, M-OH and oxo bridges, M-O-M', and thus to minimize the resonance signals due to residual alkoxy groups and to alcohol. The first paper was published on the hydrolysis of TEOS [22] and have concluded quite rightly that the chemical shift dispersion is too small relative to the peak width to provide detailed information about hydrolyzed and condensed species. However the results show clearly that this technique allows to identify and perhaps even quantify the amount of water, hydroxyl groups and oxo bridges. More recently, this technique has been successfully applied to identify directly co-condensation reactions between various purely inorganic as well as hybrid organic-inorganic sol-gel derived systems: DMDES/$Ti(OPr^i)_4$ and DMDES/$Zr(OPr^n)_4$ [23], TEOS/$Ti(OR)_4$ [24], DMDES/TEOS and MTES/TEOS [25,26] and MPS/$Zr(OPr^n)_4$ (MPS : methacryloxypropyltrimethoxysilane) [27]. Besides the clear identification of co-condensation reactions between the various precursors, one important information coming from these experiments was the poor stability of the Si-O-Ti and Si-O-Zr bonds during the aging process that leads to a degree of phase separation in the final materials [23].

This paper is focused on two model systems, precursors for siloxane-silica hybrid systems : DMDES/TEOS and MTES/TEOS. Condensation reactions in solution are characterized by ^{17}O NMR to identify and quantify the extent of co-condensation reactions, and the stability of the as-formed co-condensed species. Then the dried gels are investigated by ^{29}Si CP-MAS NMR techniques to probe the local environment of the Si sites, and thus the degree of chemical homogeneity of the final network.

EXPERIMENTAL SECTION

All the chemicals were used as received. 10 at.% enriched water was purchased from Isotec. For the ^{17}O NMR experiments, the pH of the hydrolysis water was adjusted by mixing in suitable proportions, enriched water and an aqueous solution of normal water acidified with HCl. ^{17}O NMR experiments were performed on a Bruker MSL 400 spectrometer at 54.2 MHz. Samples were held in 8 mm tubes that were in turn placed in 10 mm tubes containing C_6D_6 as lock solvent. Tap water was used as external reference (δ = 0 ppm). Spectra were recorded with 15 μs pulsewidth and 200 ms recycle delays. Simulations of the spectra were done with the Winfit program [28]. The usual D_n, T_n and Q_n units will be used to describe difunctional, trifunctional and tetrafunctional Si units. n is the number of oxo-bridges. The subscript x in Si_x indicates the nature of the Si site, D, T or Q. ^{29}Si MAS-NMR spectra were recorded on a MSL300 Bruker spectrometer. For single pulse experiment, short pulse angles ($\theta \approx 30°$) and recycle delays from 10 to 60 s depending on samples were used. Cross-polarization sequence was applied with variable contact times and proton decoupling during acquisition.

RESULTS AND DISCUSSION

• ***Solution investigation***

Methyl substituted alkoxysilanes, $Me_xSi(OEt)_{4-x}$ (x=1 and 2) are simple models to study the reactivity of difunctional and trifunctional organosilane precursors. However, the nature of the alkyl group directly bonded to the Si units is known to influence the kinetics of hydrolysis and condensation reactions [29] and this should be keep in mind before extending the present results to other hybrid systems.

A study has been performed on 1/1 mixtures of DMDES/TEOS and MTES/TEOS hydrolyzed by low hydrolysis ratio, $H_2O/OEt = 0.1$, to clearly identify co-condensed species [25]. The chemical shift values found for the various oxo-bridges are summarized in Table 1. This preliminary study shows that in both DMDES/TEOS and MTES/TEOS systems, self condensed species can be differentiated from co-condensed species based on chemical shift values. However, it has to be pointed out that under these low hydrolysis ratios, only short oligomers, dimers and trimers, are formed, which correspond to a discrete number of oxygen environments. Under usual sol-gel conditions, $H_2O/OEt \geq 0.5$, a distribution of oligomers is formed, associated with a large number of oxygen environments, which will naturally broaden the resonance signals. Thus it is not obvious to be able to distinguish the various oxo bridges peaks that extend only over few tens of ppm.

^{17}O species	Chemical shift (ppm)	
	$H_2O/OEt = 0.1$	$H_2O/OEt = 0.5$
Si_D-$\underline{O}$H	34	35;39
Si_T-$\underline{O}$H	30	32-36
Si_Q-$\underline{O}$H	15	15-20
Si_D-$\underline{O}$-Si_D	64;67	63; 67; 70
Si_T-$\underline{O}$-Si_T	57	60-70
Si_Q-$\underline{O}$-Si_Q	31	33-48
Si_D-$\underline{O}$-Si_Q	47	50-56
Si_T-$\underline{O}$-Si_Q	45	45-46

Table 1 :

Chemical shift values for Si_X-$\underline{O}$H and Si_X-$\underline{O}$-Si_Y species generated during the hydrolysis of DMDES, MTES, TEOS and DMDES/TEOS and MTES/TEOS mixtures, for two hydrolysis ratios.

An investigation was performed on DMDES/TEOS and MTES/TEOS systems using $H_2O/OEt = 0.5$ to show if co-condensation also occur in such experimental conditions. Obviously the extent of co-condensation will depend on the respective condensation rates of the different organosilanes. For this reason, the hydrolysis of the pure alkoxides in the same experimental conditions, was investigated not only by ^{17}O NMR, but also by ^{29}Si NMR. As previously mentioned, the ^{29}Si NMR data on such organosilanes are quite well documented in the literature. Once the ^{29}Si resonance signals assigned to specific Si sites, the amount of water, of hydroxyl and oxo groups present in the solution can be estimated at a given time of the kinetics, and compared with the evolution of the ^{17}O NMR signals versus time. This comparison is of great help to verify the assignments of the ^{17}O signals.

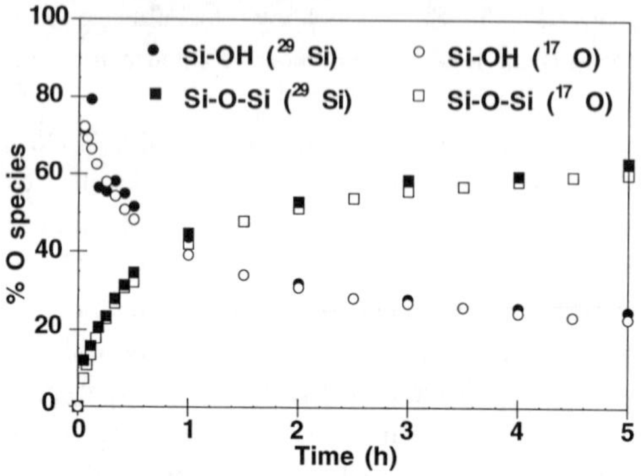

Figure 1 : Comparison between ^{17}O and ^{29}Si NMR results for the hydrolysis of DMDES (DMDES/EtOH/H_2O 1/4/1; pH = 2)

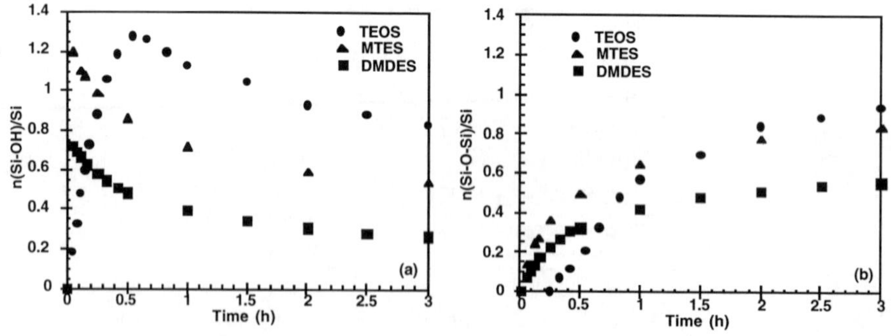

Figure 2 : Evolution of the number of (a) Si-OH and (b) Si-O-Si bonds per Si during the hydrolysis of DMDES, MTES, TEOS ($H_2O/OEt = 0.5$; EtOH/Si = 4; pH = 2). Results from ^{17}O NMR study

To illustrate this purpose, this comparison is done in Figure 1 for the DMDES system. The agreement is quite good and fully confirm the assignments done for the ^{17}O NMR signals. The same comparison between ^{29}Si and ^{17}O NMR data have been done for MTES and TEOS, and all the assignments are reported in Table 1. All the peaks present lowfield shift compared to the previous investigation done with low hydrolysis ratio, due to an increase of the degree of condensation of the various Si sites versus time.

From the ^{17}O NMR spectra, it is thus possible to follow directly the evolution of the number of silanol groups and oxo-bridges per silicon unit for the three systems (Figure 2). It can be verified that DMDES and MTES hydrolyzed faster than TEOS [30]. But more interestingly, it shows that the number of silanol groups for the various precursors is quite high during the first 30 minutes of hydrolysis (Figure 2a), while the number of oxo-bridges due to self-condensation reactions is limited (Figure 2b). The following co-condensation reactions for the respective DMDES/TEOS and MTES/TEOS systems can thus be expected :

$$Si_X\text{-}\underline{O}H + EtO\text{-}Si_Q \; \text{-->} \; Si_X\text{-}\underline{O}\text{-}Si_Q + EtOH \; (x = D, T) \quad (3)$$
$$Si_X\text{-}\underline{O}H + HO\text{-}Si_Q \; \text{-->} \; Si_X\text{-}\underline{O}\text{-}Si_Q + EtOH \; (x = D, T) \quad (4)$$

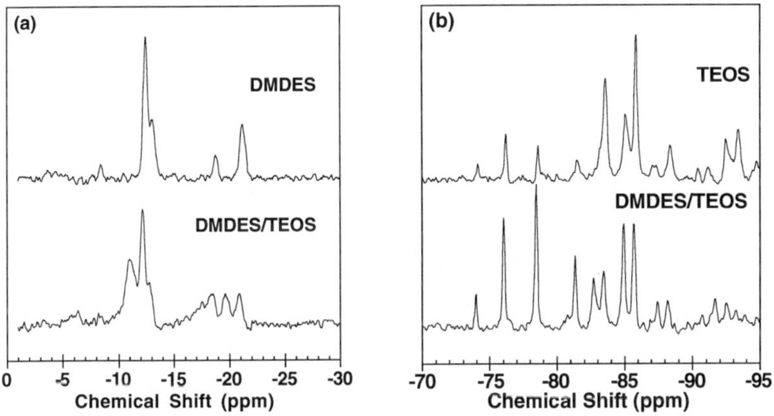

Figure 3 : ^{29}Si NMR spectra of hydrolyzed solutions of DMDES, TEOS and DMDES/TEOS mixture after 5 hours of hydrolysis ($H_2O/OEt = 0.5$; $EtOH/Si = 4$; $pH = 2$)
(a) - D unit region and (b) - Q unit region

The co-hydrolysis of DMDES and TEOS has first been followed by ^{29}Si NMR. Evidence for co-condensation can be clearly seen from the D-unit region of the spectrum, by comparison with the spectrum of an hydrolyzed solution of pure DMDES (Figure 3). New peaks at low field are present that can be assigned to D_1 unit bonded to Q unit (δ = -11.0 ppm) and D_2 units bonded to Q units (δ = -17.5; -19.6 ppm). However, a precise assignment of these peaks, and thus quantification of co-condensation reactions is difficult. It is even more difficult by looking at the Q-unit region : some new peaks appear or increase in intensity (-82.7 ppm; -84.9 ppm ; -87.4 ppm). They could be due to Q_1 units bonded to D units, but also to Q_2 units in tricyclic species [31]. If the last assumption is true, it indicates that these cyclic species are favored in presence of D units. However, the intensity of the new peaks in the D unit region suggests a rather large amount of D units bonded to Q units, and this should also be found in the Q unit region. This shows that the assignments of the ^{29}Si resonance peaks and thus, the identification of co-condensation reactions is difficult with this technique.

The same experiment was thus followed by ^{17}O NMR (Figure 4) and the various resonances are reported in Table 1 with their assignments. The evolution of the various species versus time has already been discussed in a previous paper [26]. The peaks are broader than for low hydrolysis ratios since larger oligomers with distribution of oxygen environments are formed under these experimental conditions (H_2O/OEt = 0.5). Additionally, the presence of hydrogen bonds between the O sites and the solvent (ethanol) can also cause some broadening. This is confirmed by the difference in linewidth between peaks due to Si_D-$\underline{O}$H (≈ 300-400Hz) and Si_D-$\underline{O}$-Si_D (≈ 150-200 Hz) groups; the silanol groups are more subject to hydrogen bonding than the oxo bridges between Si units, with hydrophobic character such as D units.

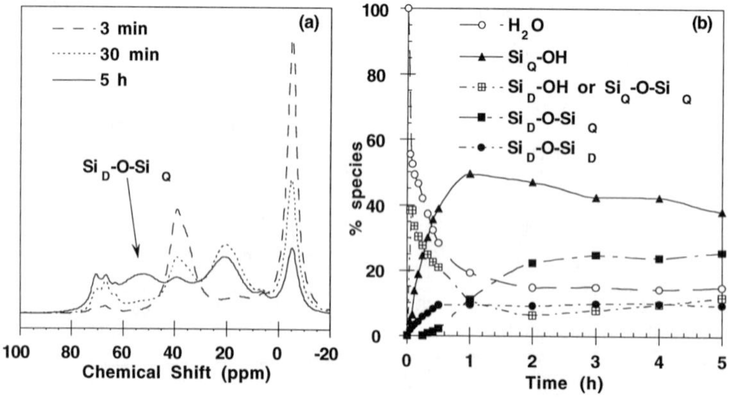

Figure 4 : Hydrolysis of a mixture of DMDES/TEOS followed by ^{17}O NMR (H_2O/OEt = 0.5; EtOH/Si = 4; pH = 2; (a) Spectra recorded 3 min, 30 min and 5 h after hydrolysis; (b) Evolution of the various O species versus time.

However, besides the broadness of the peaks that causes overlapping, a new peak around 50 ppm can be distinguished due to oxo-bridges between D and Q units, Si_D-$\underline{O}$-Si_Q, a clear signature for co-condensation reactions. This peak does not appear immediately, even if silanol groups are present on D units. It seems that at that time, these groups are involved in self-condensation, rather than in co-condensation reactions. Indeed, co-condensation starts when the amount of silanol groups on Q units becomes significant, and then the decrease of these groups occurs simultaneously with the increase of $Si_D\text{-}\underline{O}\text{-}Si_Q$ bridges suggesting that eq. (4) occurs rather than eq. (3). It has also to be noticed that these bridges are formed in rather large quantities; unfortunately, the peaks due to $Si_D\text{-}\underline{O}H$ and $Si_Q\text{-}\underline{O}\text{-}Si_Q$ species overlap and the exact total number of oxo bridges can not be calculated. However, assuming that this peak is exclusively due either to $Si_D\text{-}\underline{O}H$ or to $Si_Q\text{-}\underline{O}\text{-}Si_Q$, the percentage of $Si_D\text{-}\underline{O}\text{-}Si_Q$ ranges between 55 and 75 % of the total number of oxo bridges, after 15 hours of hydrolysis, which shows clearly the large extent of co-condensation reactions in this system. Moreover these bonds are stable in solution, at least during this period of time, which was definitely not the case for the $Si\text{-}\underline{O}\text{-}Ti$ bonds formed in the DMDES/$Ti(OPr^i)_4$ system [23].

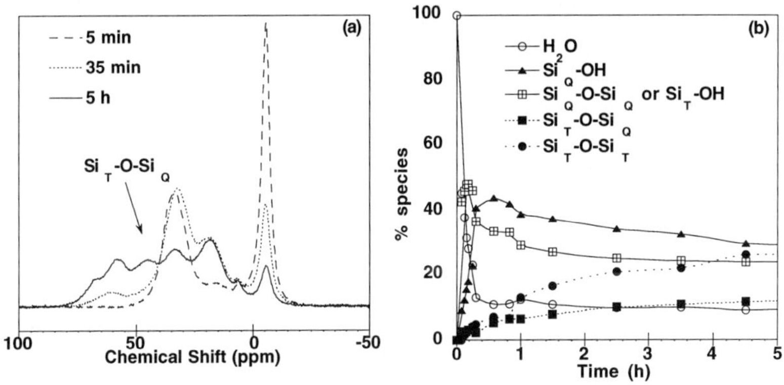

Figure 5 : Hydrolysis of a mixture of MTES/TEOS followed by ^{17}O NMR ($H_2O/OEt = 0.5$; $EtOH/Si = 4$; $pH = 2$ (a) Spectra recorded 3 min, 35 min and 5 h after hydrolysis; (b) Evolution of the various O species versus time.

Similar ^{17}O NMR experiments have been done on the MTES/TEOS system (Figure 5). The peaks are broader (400-700 Hz) due to a higher average functionality of the precursors which causes a larger distribution of as-formed species. However, all the expected species can be detected, with once more a small difficulty coming from similar chemical shift values for $Si_T\text{-}\underline{O}H$ and $Si_Q\text{-}\underline{O}\text{-}Si_Q$ units. The water is consumed during the first 30 minutes to reach a residual content of $\approx$ 10%. Hydrolysis groups are generated during the first minutes, faster as expected on T unit than on Q units. The $Si_T\text{-}\underline{O}H$ and $Si_Q\text{-}\underline{O}H$ units reach maximum values after 10 and 30 minutes, respectively. Condensation reactions start quite rapidly, during the first 30 minutes, and increase continuously versus time. Two resonances can be clearly identified around 60 ppm due to $Si_T\text{-}\underline{O}\text{-}Si_T$ and around 45 ppm due to $Si_T\text{-}\underline{O}\text{-}Si_Q$ units, showing that self-condensation of the most reactive precursor, MTES, but also co-condensation between MTES and TEOS occur. Unlike the DMDES/TEOS system, co-condensation reactions do not seem to occur predominantly in the MTES/TEOS system, if the amount of Si_T-O-Si_Q units is compared to that of Si_T-O-Si_T units (Figure 5b). However, quantitative analysis of these spectra should be considered with caution : the peak due to Si_T-O-Si_Q units that should shift to low field during the course of condensation reactions and thus strongly overlaps with peaks due to Si_T-O-Si_T units, is certainly underestimated.

• ***Solid state investigation***

For both systems investigated, co-condensation reactions occur to a large extent in solution and the as-formed bonds seem stable versus time, which should ensure a good chemical homogeneity in the final gels.

To illustrate this point, the ^{29}Si spectrum of an hydrolyzed solution of DMDES/TEOS in a 1/1 ratio, aged for 15 hours has been compared with the ^{29}Si MAS-NMR spectrum of a dried gel corresponding to the same composition (Figure 6).

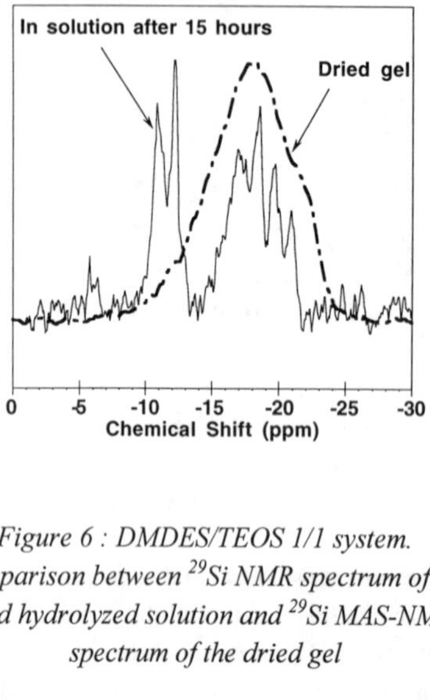

Figure 6 : DMDES/TEOS 1/1 system. Comparison between ^{29}Si NMR spectrum of an aged hydrolyzed solution and ^{29}Si MAS-NMR spectrum of the dried gel

As previously mentioned, in the D unit region of the solution spectrum, various peaks are present at -17.5 ppm, -18.5 ppm, -19.6 ppm and -20.9 ppm due to D_2 units bonded to Q units. For reference, the peak characteristic of D_2 units in polydimethylsiloxanes is around -22 ppm [32]. The shift to low field is due to an increased number of neighboring Q units [33]. Interestingly, the ^{29}Si resonance in the dried gel due to D units corresponds to the envelop of the signals due to D_2 units in solution, indicating that all D units are fully condensed, and that their distribution is close to that observed in solution after aging.

The CP technique is a useful technique to probe the local environment of Si sites in such systems. Indeed the efficiency of the polarization transfer between the protons and the ^{29}Si nuclei depends strongly on the ^{1}H-^{29}Si dipolar coupling, and thus on the distance between protons and ^{29}Si nuclei as well as on the mobility of the corresponding groups.

Two dried gels of respective compositions, DMDES/TEOS = 1/1 and 4/1 have been studied (Figure 7). The spectra recorded with single pulse technique (SPE) show clearly differences between the two gels : the spectrum of the 4/1 composition shows in the D unit region, sharp peaks at -22.3, -21.7, -20.7, -19.3 and -18.3 ppm. The first one is due to D_2 units bonded to D units, while the others are due to co-condensed species. The sharpness of the lines reveals a certain mobility of the network due to its low theoretical functionality (f = 2.4). The gel with the 1/1 composition as already mentioned, presents a broad signal centered at -18.2 ppm, so at a lower field compared to the previous one, suggesting a higher number of co-condensed species. The increase in linewidth indicates a higher rigidity of the network, due to an increase in its functionality (f = 3).

The CP spectra were recorded with variable contact times, and the spectra corresponding to 1 and 10 ms are presented in Figure 7. The peaks due to D units for the 4/1 composition present different polarization dynamics. The peaks at low field show a more efficient polarization transfer than the peaks at higher field. This can be directly related to differences in mobility of such groups : the peaks at high field are related to D units bonded primarily to D units and thus in rather mobile environments compared to the peaks at low field due to D units bonded to Q units, and thus in more rigid environments. Such effect is not present for the 1/1 composition, indicating that all D units experience similar motions, in perfect agreement with the presence of an homogeneous network.

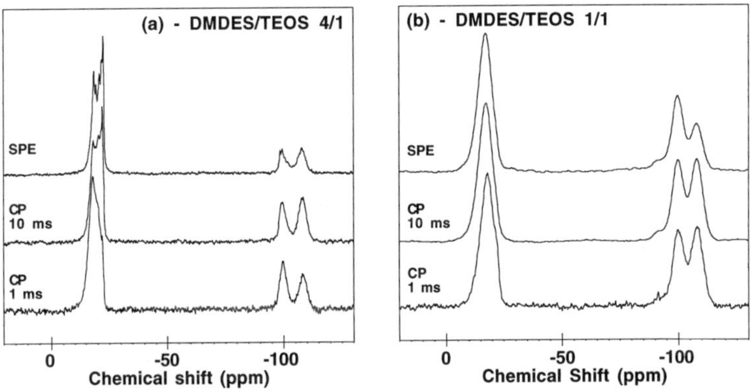

Figure 7 : ^{29}Si MAS-NMR spectra recorded on dried gels using single pulse (SPE) or cross-polarization (CP) sequences. For CP sequence, the contact time is indicated in the figure.

The polarization dynamics of Q_4 units can give information on the proximity of D and T units in DMDES/TEOS and MTES/TEOS derived samples. Cross-polarization dynamics is usually described within the framework of spin thermodynamics considering two reservoirs corresponding to the abundant *I*-spin (^{1}H) and the rare *S*-spin (^{29}Si). This model predicts an exponential build-up of *S*-spin magnetization in the rotating frame, characterized by $1/T_{IS}$. If the loss of spin-locked proton magnetization in the rotating frame is taken into account by a relaxation rate $1/T_{1\rho H}$, the time dependence of the *S*-spin magnetization is [34] :

$$M_S(t_c) = \frac{\gamma_I}{\gamma_S} M_{0S} \frac{1}{1-\lambda} [1 - \exp(-(1-\lambda)\frac{t_c}{T_{IS}})]\exp(-\frac{t_c}{T_{1\rho}^I}) \quad (6)$$

$\lambda = T_{IS}/T_{1\rho}^I$ and M_{0S} is the maximum magnetization in the absence of spin-lattice relaxation.

Typical curves obtained for the peak due to Q_4 units in DMDES/TEOS derived samples are presented in Figure 8, and the corresponding parameters are summarized in Table 2, together with parameters corresponding to TEOS and MTES/TEOS derived samples.

T_{SiH} is related to the dipolar coupling and thus to the ^1H-^{29}Si distance. For Q_4 units, the influence of mobility can be neglected as a first approximation. For TEOS derived silica gel, reported T_{SiH} values are long due to poorly protonated environments [35]. In the gels containing D or T units, the T_{SiH} values are reduced indicating an increase number of protons in the environment of Q_4 units. In those systems, the degree of condensation of the siloxane network, and thus the number of silanol groups are lower than for a pure silica gel [36]; the decrease of T_{SiH} values is thus due to D and T units, containing methyl groups, close to Q_4 units. Indeed, the decrease of T_{SiH} is even more important when the number of D units increases.

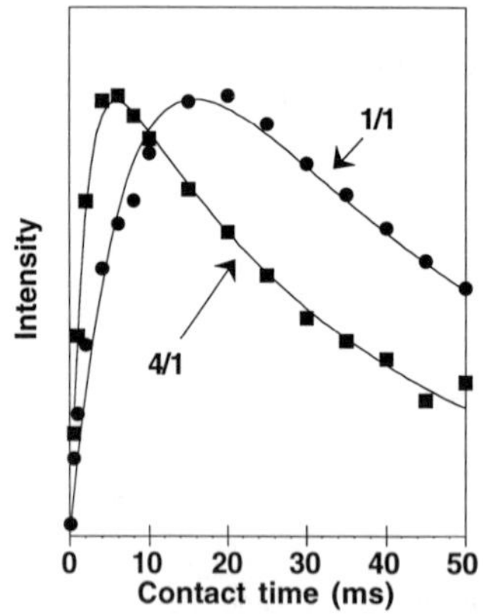

Figure 8 : evolution of the magnetization versus contact time of the peak due to Q_4 units in the DMDES/TEOS dried gels.

Another parameter that can be extracted from eq. (6) is M_{0S}, that leads to an estimation of the amount of spins polarized during the CP sequence. This can be compared with the total number of Q_4 units obtained in a single pulse experiment (SPE) (Table 2). For a silica gel, the Q_4 units are largely underestimated in a CP technique. Zumbulayadis et al. [34] suggests that the analysis of CP dynamics according to eq. (6) measures the amount of Q_4 units directly bonded to either Q_3 or Q_2 units.

In the samples derived from DMDES/TEOS or MTES/TEOS mixtures, the quantitative results extracted from the CP analysis agree well with the total number of Q_4 units, clearly indicating a close proximity between Q_4 units, and D or T units.

These results agree with those reported by Fyfe et al. [37] who investigated MTES/TEOS derived gels by two dimensional $^1H/^{29}Si$ correlation experiment, which clearly demonstrated that the two components are not phase separated.

Sample	$T_{Si\text{-}H}$ (ms)	$T_{1\rho}$ (ms)	Magnetization %*	
			SPE	CP
DMDES/TEOS 4/1	1.9	34	51	50
DMDES/TEOS 1/1	7.2	46	50	50
MTES/TEOS 1/1	5.1	-	42	42
TEOS	8.7	-	49	25

Table 2 : ^{29}Si CP MAS-NMR experiments
Characteristic parameters of the peak due to Q_4 units in various dried gels.
**The percentages are related to the total number of Q units*

CONCLUSIONS

This paper clearly illustrates the usefulness of ^{17}O NMR experiments using enriched water to investigate co-condensation reactions between various organosilanes, precursors for siloxane-silica hybrid systems. Hydrolysis and condensation reactions were followed for two systems, DMDES/TEOS and MTES/TEOS indicating that co-condensation occurs to a large extent, and that the oxo bridges between different units are stable during aging. This leads to homogeneous final networks that do not present phase separation. Their homogeneity was investigated by ^{29}Si MAS-NMR using CP technique. Analysis of the CP dynamics can probe the local environments of Si sites via ^{1}H-^{29}Si dipolar couplings, showing clearly the spatial proximity between D (respectively T) units and Q units, in samples derived from DMDES/TEOS and MTES/TEOS samples.

REFERENCES

1 B.M. Novak, Adv. Mater. **5**, 422 (1993).
2 C. Sanchez, F. Ribot, New J. Chem. **18**, 1007 (1994).
3 H. Schmidt, J. Sol-Gel Sci. and Techn. **1**, 217 (1994).
4 U. Schubert, N. Hüsing, A. Lorenz, Chem. Mater. **7**, 2010 (1995).
5 I. Artaki, M. Bradley, T.W. Zerda, J. Jonas, J. Phys. Chem. **89**, 4399 (1985).
6 J.C. Pouxviel, J.P. Boilot, J.C. Beloeil, J.Y. Lallemand, J. Non Cryst. Solids **89**, 345 (1987).
7 L.W. Kelts, N.J. Amstrong, J. Mater. Res. **4**, 423 (1989).
8 R.A. Assink, B.D. Kay, Annu. Rev. Mater. Sci. **21**, 491 (1991).
9 F. Brunet, B. Cabane, M. Dubois, B. Perly, J. Phys. Chem. **95**, 945 (1991).
10 Y. Sugahara, S. Okada, K. Kuroda, C. Kato, J. Non Cryst. Solids **139**, 25 (1992).
11 P. Lux, F. Brunet, H. Desvaux, J. Virlet, Magn. Reson. Chem. **31**, 623 (1993)
12 F. Devreux, J.P. Boilot, F. Chaput, A. Lecomte, Phys. Rev. A **41**, 6901 (1990).
13 Y. Sugahara, S. Okada, S. Sato, K. Kuroda, C. Kato, J. Non Cryst. Solids **167**, 21 (1994).
14 F. Babonneau, L. Bois, J. Livage, Mat. Res. Soc. Symp. Proc. **271**, 237 (1992).

15 S.K. Mah, I.J. Chung, J. Non Cryst. Solids **183**, 252 (1995).
16 Y. Sugahara, Y. Tanaka, S. Sato, K. Kuroda, C. Kato, Mat. Res. Soc. Symp. Proc. **271**, 231 (1992).
17 S. Prabakar, R.A. Assink, N.K. Raman, C.J. Brinker, Mat. Res. Soc. Symp. Proc. **346**, 979 (1994).
18 I. Hasegawa, J. Sol-Gel Sci. Technol. **1**, 57 (1993).
19 J.D. Basil, C.-C. Lin, in Ultrastructure Processing of Advanced Ceramics, edited by J.D. Mackenzie, D.R. Ulrich, John Wiley & Sons (1988) pp. 783.
20 W. Beier, A.A. Göktas, G.H. Frischat, Ch. Wies, K. Meise-Gresch, W.Müller-Warmuth, Phys. Chem. Glasses, **30**, 69 (1989).
21 S. Diré, F. Babonneau, G. Carturan, J. Livage, J. Non Cryst. Solids, **147 & 148**, 62 (1992).
22 C.W. Turner, K.J. Franklin, , J. Non Cryst. Solids, **91**, 402 (1987).
23 F. Babonneau, Mat. Res. Soc. Symp. Proc. **346**, 949 (1994).
24 F. Babonneau, J. Maquet, J. Livage in Sol-Gel Science and Technology Ceramics Transactions, Vol. 55, Am. Ceram. Soc. Ed. (1995) pp. 53.
25 F. Babonneau, J. Maquet, J. Livage, Chem. Mater. **7**, 1050 (1995).
26 F. Babonneau, C. Toutou, S. Gavériaux, J. Sol-Gel Sci. Technol. (in press)
27 L. Delattre, M. Roy, F. Babonneau, J. Sol-Gel Sci. Technol. (in press)
28 D. Massiot, H. Thiele, A. Germanus, Bruker Report, **140**, 43 (1994).
29 L. Delattre, F. Babonneau, Mat. Res. Soc. Symp. Proc. **346**, 365 (1994).
30 H. Schmidt, H. Scholze, A. Kaiser, J. Non Cryst. Solids, **63**, 1 (1984).
31 L.W. Kelts, N.J. Armstrong, J. Mater. Res. **4**, 423 (1989).
32 G.W. Gray, W.D. Hawthorne, D. Lacey, M.S. White, J.A. Semlyen, Liquid Crystals, **6**, 503 (1989).
33 F. Babonneau, Polyhedron, **13**, 1123 (1994).
34 A. Pines, M.G. Gibby, J.S. Waugh, J. Chem. Phys., **59**, 569 (1973)
35 N. Zumbulayadis, J.M. O'Reilly, Macromolecules **24**, 5294 (1991)
36 R.H. Glaser, G.L. Wilkes, C.E. Bronnimann, J. Non Cryst. Solids, **113**, 73 (1985).
37 C.A. Fyfe, Y. Zhang, P. Aroca, J. Am. Chem. Soc. **114**, 3252 (1992)

SYNTHESIS OF MESOPOROUS ZIRCONIA USING AN AMPHOTERIC SURFACTANT

A. Y. KIM, P. J. BRUINSMA, Y. L. CHEN, AND J. LIU
Pacific Northwest National Laboratory,* Battelle Boulevard, P.O. Box 999, Richland, WA 99352

ABSTRACT

An amphoteric surfactant, cocamidopropyl betaine, was used for the synthesis of mesoporous zirconia. The carboxylate functionality of the surfactant permitted strong bonding with soluble zirconium species, while the quaternary ammonium group ensured large headgroup area and high solubility under acidic conditions. An amphoteric co-template [betaine, or (carboxymethyl)trimethylammonium hydroxide] improved uniformity of the hexagonal mesophase. Transmission electron microscopy (TEM) of the as-synthesized zirconium sulfate mesophase indicated hexagonal mesostructure, and low-angle X-ray diffraction (XRD) showed a 41 Å primary d-spacing and two higher order reflections of a hexagonal lattice. High surface area zirconia was produced by controlled base treatment of the hexagonal mesophase with sodium hydroxide, followed by calcination. TEM and XRD indicated that the mesostructure was stable to 350°C.

INTRODUCTION

The discovery of mesoporous silica molecular sieves [1, 2] has generated wide interest in developing new synthesis routes for silica and non-silica based mesoporous materials [3-7]. Several templating methods for silica have been demonstrated, including ionic condensation of oppositely charged surfactant molecules and inorganic species [1,2], counterion mediated assembly of similarly charged surfactant and inorganic species [3], and hydrogen bonding of nonionic or neutral surfactant and inorganic species [4,5]. In comparison, synthesis of ordered, mesoporous non-silicates is difficult. In the past, a thermally stable titanium hexagonal mesophase has been produced using phosphate surfactant as template [6]. A cationic surfactant was used to produce mesoporous zirconia [7], but this preparation does not appear to involve a templating mechanism based on attractive interactions between the surfactant and the inorganic species. In this paper, we report the synthesis of hexagonal mesophase zirconium salt using an amphoteric surfactant as template. The amphoteric surfactant, cocamidopropyl betaine (CAPB), has the following advantages for mesophase formation: (1) ability to bond with zirconium species (from carboxylate functionality [8-9]), (2) high solubility under acidic conditions (from quaternary ammonium group), and (3) large headgroup area in high salt (from bulky quaternary ammonium group). An amphoteric co-template, betaine (B), increases the effective headgroup area, promoting the formation of hexagonal mesophase. Hexagonal mesophase zirconium salt was converted to high surface area zirconia by controlled condensation, followed by calcination.

EXPERIMENT

For templating of hexagonal mesophase zirconium compounds, the surfactant should have high solubility under reaction conditions, an appropriate packing parameter (headgroup area), and the ability to bond with metal species. Since zirconium shows the highest solubility under acidic conditions with the formation of cationic species [10], the mesophase syntheses were performed at low pH with surfactants having anionic functionalities. Fatty acids were unsuitable because of their low solubility in acid. Solubility was improved with sodium dodecylsulfate (SDS) as template, but SDS produced thermally unstable lamellar mesophases.

Mat. Res. Soc. Symp. Proc. Vol. 435

In this study hexagonal mesophase zirconium salts were prepared under acidic conditions using the following components: CAPB, B, Zr, and H_2O. CAPB is the amphoteric surfactant. B is betaine [(carboxymethyl)trimethylammonium hydroxide], an optional co-template having the same amphoteric structure as the headgroup of CAPB. Zr is one of the following salts: zirconium acetate, zirconyl chloride, zirconyl nitrate, and zirconium sulfate. A typical zirconium salt mesophase was prepared as follows. First, surfactant template 30 wt% CAPB was contacted with strong acid/strong base ion exchange media to remove NaCl impurities. CAPB was mixed with co-template B, if desired, in amounts such that B/CAPB mole ratio was about 0.5. Aqueous solution of zirconium salt (0.25M $ZrOCl_2$, for example) was then added dropwise to the surfactant solution with stirring, until Zr/CAPB mole ratio was in the range 0.8-3. The reaction mixture was sealed in a vessel and hydrothermally treated for three to five days at 70°C. Product was recovered and rinsed with D-I water by filtration. The resulting zirconium salt mesophase was converted to hydrous zirconium oxide mesophase by controlled addition of NaOH at 25-70°C. 1N NaOH was added dropwise to the zirconium salt mesophase dispersed in water to maintain pH = 6-8 for six days. The hydrous zirconium oxide mesophase was calcined in air for four hours to produce high surface area zirconia.

Experiments were designed to determine (1) the range of reaction compositions and temperatures in which amphoteric surfactant acts as a template, (2) zirconium precursors which can be used, (3) effect of co-template on mesophase formation, and (4) effect of base treatment on thermal stability and porosity.

Powder samples were analyzed by transmission electron microscopy (TEM), X-ray diffraction (XRD), electron energy dispersive spectroscopy (EDS), and surface area (BET).

RESULTS

The interaction between surfactant headgroup and inorganic precursor is important to mesophase formation [1-6]. In this work, carboxylate groups of CAPB micelles bond with the soluble zirconium species. Initially, this bonding is probably electrostatic in nature, but covalent bonding may also occur as the reaction proceeds [8-9]. Another consideration is the surfactant packing parameter (v/a_0l_c), which dictates the type of micellar aggregates that form in solution [11]. The surfactant used in this study has a large head group size, or low packing parameter, which promotes formation of rod-shaped micelles over lamellar aggregates. Formation of rod-shaped micelles is important, because their organization during bonding reactions between the zirconium species and the surfactant comprises the hexagonal mesophase.

TEM indicated that amphoteric surfactant CAPB templated the formation of zirconium salt hexagonal mesophase within a window of reaction temperatures and compositions and from a variety of zirconium salts (zirconium acetate, zirconyl chloride, zirconyl nitrate, and zirconium sulfate). Samples prepared with zirconium/surfactant mole ratios in the range 0.8-3 and reacted at 70°C showed hexagonal ordering. Below 0.8 Zr/CAPB, no ordering was observed. At 0.8 Zr/CAPB, increasing the temperature from 70 to 105°C caused a transition from hexagonal to lamellar mesophase (XRD not shown). No templating was observed at higher temperature (175°C). Unfortunately, samples prepared with CAPB alone were non-uniform, as indicated by XRD (Fig. 1A).

Sample uniformity was improved by the addition of co-template B. The co-template increased the effective headgroup area, thereby promoting formation of the rod-like micellar subunits of the hexagonal phase. Low-angle XRD of zirconium sulfate mesophase prepared from a B/CAPB mixture showed 100, 110, and 200 reflections (41.0, 23.8, and 20.8 Å d-spacings) of a hexagonal lattice (Fig. 1B). Mole ratios of the reaction mixture were: 480 H_2O : 1.7 Zr : 0.56 B : 1 CAPB. Under TEM, this sample showed large regions of hexagonal mesophase (Fig. 2).

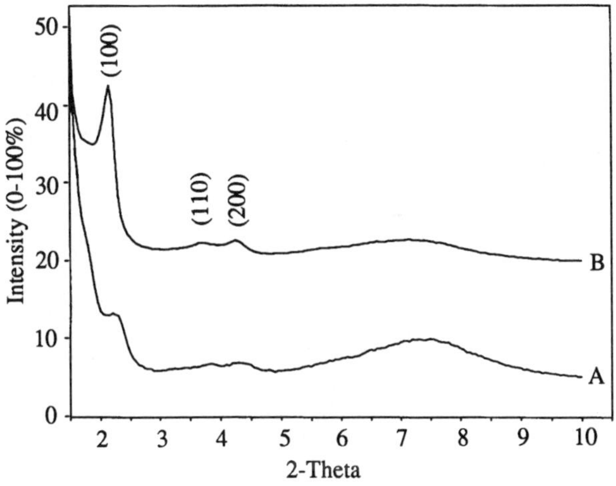

Figure 1. Low-angle XRD of zirconium sulfate mesophases prepared using amphoteric surfactant CAPB as template (Fig. 1A) and using a mixture of co-template B and CAPB (Fig. 1B). Addition of the co-template improved uniformity of the hexagonal mesophase.

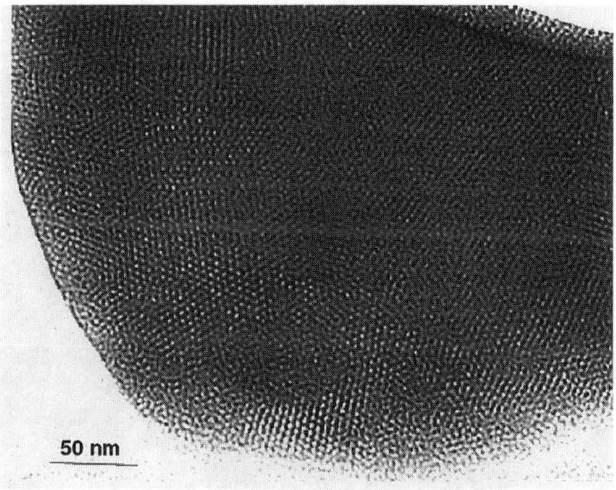

Figure 2. Zirconium sulfate hexagonal mesophase under TEM.

Thermal stability of zirconium salt hexagonal mesophase was investigated. Calcined zirconium sulfate samples showed ordering under TEM and XRD (Fig. 3). However, BET surface areas were less than 5 m^2/g, indicating nonporous materials. The lack of porosity may have been related to structural changes in the zirconium salt mesophase as large amounts of structural water and counterions were rapidly removed during calcination.

Controlled base treatment of hexagonal mesophase zirconium salt with NaOH was studied as a means to improve surface area. EDS suggested that NaOH treatment was efficient at removing counterions (SO_4^{2-} or Cl^-) from the mesophase. Base treatment also facilitated condensation of the zirconium species and removal of structural water. Thermal analysis showed that base-treated samples underwent 40 wt.% weight loss, as compared to 80 wt.% weight loss for untreated sample. Therefore, base treatment may have reduced structural rearrangement during the calcination process. TEM (Fig. 4) indicated that the mesostructure remained in tact throughout the different stages of processing.

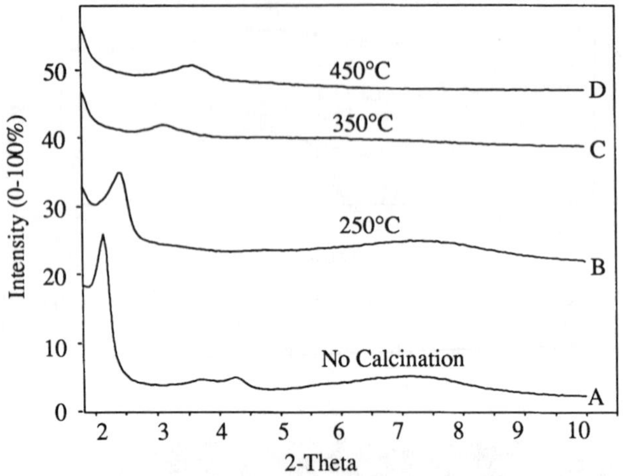

Figure 3. XRD of as-synthesized $Zr(SO_4)_2$ hexagonal mesophase as a function of calcination temperature: no calcination (A), 250°C (B), 350°C (C), and 450°C (D).

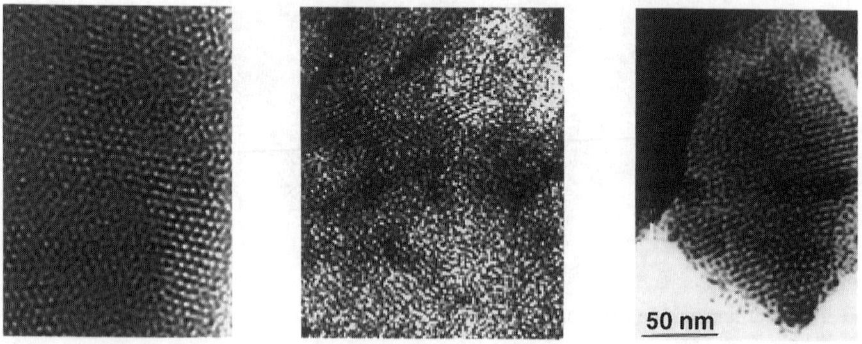

A. As-synthesized hexagonal mesophase

B. After 70°C NaOH treatment.

C. After 250°C calcination.

Figure 4: TEMs of hexagonal zirconium sulfate at different stages of processing: as-synthesized $Zr(SO_4)_2$ salt mesophase (A), after 70°C NaOH treatment (B), and after 250°C calcination (C).

The mesostructure of base-treated $Zr(SO_4)_2$ was stable to 350°C (Fig. 5). Above 350°C, the material began to crystallize, as indicated by high angle XRD (not shown). Most significantly, base treatment of $Zr(SO_4)_2$ mesophase improved porosity and produced high surface area zirconia materials (Fig. 6). Low surface areas at calcination temperatures higher than 350°C were consistent with the observed onset of crystallization.

Thermal stability of zirconium mesophase depended on the zirconium salt precursor. Base-treated mesophases derived from zirconium sulfate showed better surface areas than those derived from zirconyl chloride (Fig. 6). This might be due to the better ability of sulfate relative to chloride to complex zirconium [12].

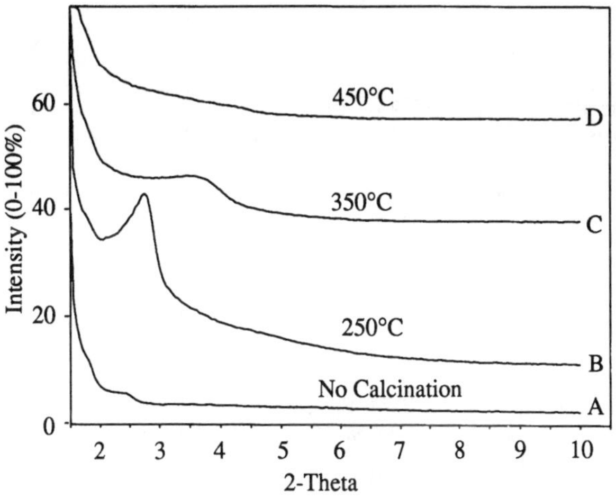

Figure 5: XRD of condensed $Zr(SO_4)_2$ hexagonal mesophase as a function of calcination temperature: no calcination (A), 250°C (B), 350°C (C), and 450°C (D).

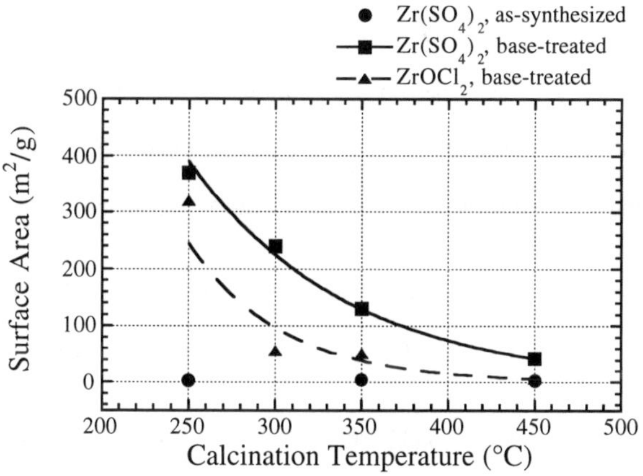

Figure 6: BET surface area of zirconium mesophase materials versus calcination temperature. The base-treated $Zr(SO_4)_2$ mesophase produced the highest surface areas.

CONCLUSIONS

Ordered, mesoporous zirconia was synthesized by an amphoteric surfactant route. Cocamidopropyl betaine has the appropriate headgroup area, solubility, and anionic functionality to template formation of hexagonal mesophase zirconium salts under acidic conditions. An amphoteric co-template [(carboxymethyl)trimethylammonium hydroxide] improved uniformity of the mesophase. As-synthesized $Zr(SO_4)_2$ mesostructures were nonporous after calcination. Controlled base treatment of $Zr(SO_4)_2$ mesophase produced a high surface area product stable to 350°C. Amphoteric surfactant as template broadens the range of compositions which can be fabricated into hexagonal mesophase material.

ACKNOWLEDGEMENT

The authors greatly appreciate the X-ray diffraction work performed by D. E. McCready.

This research is supported by the U.S. Department of Energy under contract DE-AC06-76RLO 1830.

* Pacific Northwest National Laboratory is operated by Battelle Memorial Institute for the U.S. Department of Energy under contract DE-AC06-76RLO 1830.

REFERENCES

1. C. T. Kresge, M. E. Leonowicz, W. J. Roth , J. C. Vartuli and J. S. Beck, Nature **359**, p. 710 (1992).

2. J. S. Beck, J. C. Vartuli, W. J. Roth, M. E. Leonowicz, C. T. Kresge, K. D. Schmitt, C. T-W. Chu, D. H. Olson, E. W. Sheppard, S. B. McCullen, J. B. Higgins and J. L. Schlenker, J. Am. Chem. Soc. **114**, p. 10834 (1992).

3. Q. Huo, D. I. Margolese, U. Ciesla, D. G. Demuth, P. Feng, T. E. Gier, P. Sieger, A. Firouzi, B. F. Chmelka, F. Schüth, and G. D. Stucky, Chem. Mater. **6**, p 1176 (1994).

4. P. T. Tanev and T. J. Pinnavaia, Science **267**, p. 865 (1995).

5. S. A. Bagshaw, E. Prouzet, and T. J. Pinnavaia, Science **269**, p. 1242 (1995).

6. D. M. Antonelli and J. Y. Ying, Angew. Chem. Int. Ed. Engl. **34**, p. 2014 (1995).

7. J. A. Knowles and M. J. Hudson, J. Chem. Soc., Chem. Commun., p. 2083 (1995).

8. G. S. Kharitonova, V. I. Nefedov, L. N. Pankratova and V. L. Pershin., Russ. J. Inorg. Chem. **19**, p. 469 (1974).

9. G. S. Kharitonova and L. N. Pankratova, Russ. J. Inorg. Chem. **22**, p. 379 (1977).

10. C. F. Baes, Jr. and R. E. Mesmer, The Hydrolysis of Cations, John Wiley and Sons, New York, 1976, pp. 152-9.

11. J. N. Israelachvili, Intermolecular and Surface Forces, Academic Press, San Diego, 1992, pp. 380-2.

12. A. Clearfield, Rev. Pure and Appl. Chem. **14**, p. 91 (1964).

MOLECULES WITH POLYMERIZABLE LIGANDS AS PRECURSORS TO POROUS DOPED MATERIALS

L.G. HUBERT-PFALZGRAF*, N. PAJOT*, R. PAPIERNIK*, S. PARRAUD**
* Université de Nice-Sophia-Antipolis, 06108 Nice Cédex (France) hubert@hermes.unice.fr
** CEA, Centre d'Etudes de Limeil-Valenton, 94195 Villeneuve-St-Georges Cédex (France)

ABSTRACT

Titanium and aluminium alkoxide derivatives with polymerizable ligands such as 2-(methacryloyloxy)ethylacetoacetate (HAAEMA), oleic acid and geraniol (HOGE) have been obtained. The various compounds have been characterized by FT-IR and NMR ^{1}H. Copolymerization with styrene and divinylbenzene affords porous doped organic materials which have been characterized by scanning electron microscopy (SEM), elemental analysis, density measurements.

INTRODUCTION

Porous materials based on ballon microspheres are of interest for a broad range of applications such as catalyst supports, thermal insulation, gas or chemical storage, light weight composites and fusion targets, to name just as few [1]. Various methods such as high-intensity ultrasound and a combination of colloidal and emulsion techniques have been used [2] for access to such materials [3].

We report here the synthesis and characterization of titanium and aluminum precursors with polymerizable ligands and their use in copolymerization reactions for the development of porous spherical materials.

EXPERIMENTAL

The synthesis and characterization of the titanium and aluminium derivatives have been achieved under inert atmosphere using standard Schlenk tubes and vacuum line techniques [4]. $Ti(O^iPr)_4$ and $Al(O^iPr)_3$ were distilled before use. HAAEMA and the other ligands were stored over molecular sieves. Solution NMR spectra were obtained with a AC-200 Bruker spectrometer. FT-IR spectra were obtained as Nujol mulls for the air sensitive compounds, or as KBr pellets after the polymerization reactions. Microanalyses were done at the Centre de Microanalyses du CNRS. Scanning electron micrographs were obtained with a SEM 505 Philips microscope. Density measurements were achieved with a picnometer by flotation in water. The samples have been varnished in order to close the peripheral pores.

Typical formulation for the formation of the emulsion

A toluene solution of precursors (1ml-12ml, 0.08 - 0.3M.l^{-1}) was added to 4.4 ml of a solution composed of 47.6 weight % of a surfactant, SPAN 80, 23.8 wt % of styrene, 28.6 wt % of divinybenzene. An aqueous solution containing the polymerization initiator, $Na_2S_2O_8$, (1.84 g.l^{-1}) was used to complete the medium to 40ml. The emulsion was obtained by stirring the mixture in a double-barreled syringe pump device. The polymerization was achieved at 60°C in a drying oven during 12h.

Mat. Res. Soc. Symp. Proc. Vol. 435 © 1996 Materials Research Society

RESULTS

Synthesis and characterization of precursors

Various organic moieties having polymerizable functionalities (chart 1) namely 2-hydroxyethylmethacrylate, 2-(acetatoacetoxy)ethylmethacrylate, oleic acid, and geraniol have been considered as ligands toward titanium and aluminium alkoxides.

2-hydroxyethylmethacrylate (HEMA)

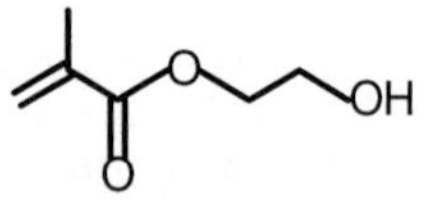

2-(acetatoacetoxy)ethylmethacrylate (HAAEMA)

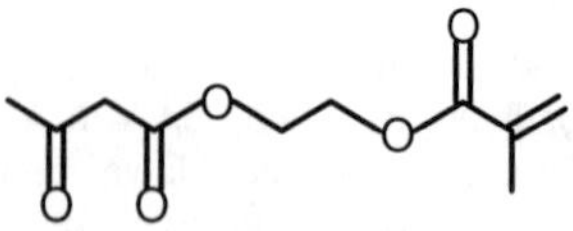

oleic acid (OLEA):

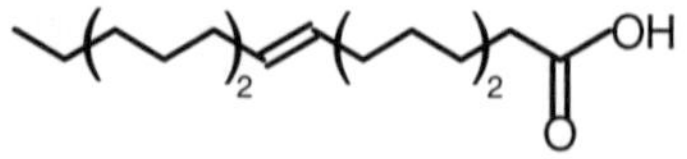

geraniol (HOGE):

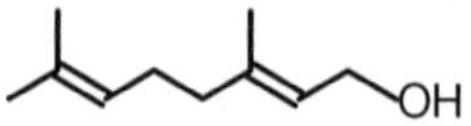

Chart 1: Unsatured organic ligands

Heteroleptic titanium alkoxides such as $Ti(O^iPr)_2(AAEMA)_2$, $Ti(O^iPr)_3(OLEA)$, $Ti(O^iPr)_3(OGE)$ and $Ti(O^iPr)_2(OGE)_2$ have been obtained in high yield ($\approx$ 90%) by reaction between $Ti(OiPr)_4$ and the corresponding ligands in toluene at 20°C for HAAEMA and by azeotropic distillation at 100°C for the other compounds in 1:1 and 1:2 stochiometry. The reactions between aluminium isopropoxide and HAAEMA in 1:3 stochiometry afforded $Al(AAEMA)_3$ purified by recrystallization in hexane. The spectroscopic characteristics of the various compounds are summarized in Table 1. The presence of the β-ketoesterate AAEMA and oleate ligands in the metal coordination spheres is evidenced by absorption bands around 1560 cm^{-1} in the FT-IR spectra corresponding to the C=O streching vibrations. They are shifted to lower frequencies for the β-ketoesterate and carboxylate ligands with respect to the free ligand (for instance 1750 cm^{-1} for HAAEMA) [5]. The presence of the C=C functionality is evidenced by absorption bands around 1650 cm^{-1}. The frequency of νC=C adsorption bands are essentially unaffected by coordination of the ligand to the metal, titanium or aluminium, thus suggesting that they will be available for cross-liking reactions with unsaturated organic molecules.

Methacrylate groups can generally be copolymerized quite easily [6]. Methacrylate, HAAEMA and 2-hydroxyethylmethacrylate are good candidates for the introduction of the functional group in a metal coordination sphere. However, oxophilic metals such as titanium or aluminium can promote side reactions namely the cleavage of the C–O acyl bond. For instance the reaction between $Ti(O^iPr)_4$ and 2-hydroxyethylmethacrylate (1:1 stoichiometry) in toluene at room temperature is characterized by a progressive diminution of the νC=C vibration at 1637 cm^{-1} in the FT-IR spectra. The compound isolated after a reaction time of 20 h corresponds actually to an ethyleneglycolate derivative, $Ti_5(O^iPr)_{10}(OC_2H_4O)_5$, as shown by X-ray diffraction studies. [7] Similar C–O cleavage reactions have been observed when 2-hydroxyethylmethacrylate is reacted with titanium ethoxide or aluminium isopropoxide. These observations require that polymerization reactions are achieved rapidly after the mixing of oxophilic metal alkoxides and 2-hydroxyethylmethacrylate if one wants to take advantage of homogeneity at a molecular level in the subsequent cross-linking reactions.

Compound	IR (cm^{-1})		^{1}H NMR ($CDCl_3$, ppm)	
	νCO	νC=C	OiPr (CH)	Ligand
$Ti(O^iPr)_3(OLEA)$	1561s	1653s	4.5-5 (ovm, 3H)	1.2 m, 22H, CH_2 1.6 m, 3H, CH_3 2 m, 4H, CH_2 2.2 t, J=8.7Hz, 2H, $CH_2C(O)$ 5.4 t, J=4.7Hz, 2H, CH=CH
$Ti(O^iPr)_3(OGE)$	—	1667w 1636w	4.6 (m, 3H)	1.65 d, 4Hz, 6H, $(CH_3)_2C=C$ 1.7 s, 3H, $CH_3C=C$ 2 m, 4H, C_2H_4 4.9 d, J=5.2Hz, 2H, CH_2O 5.1 and 5.4 br, 1H, C=CH
$Ti(O^iPr)_2(OGE)_2$	—	1669w 1638w	4.65 (m, 2H)	1.65 d, 2.6Hz 12H, $(CH_3)_2C=CH$ 1.7 s, 6H, $(CH_3)C=CH$ 2.1 m, 8H, C_2H_4 4.9 d, J=4.7Hz, 4H, CH_2O 5.1 and 5.4 br, 2H, C=CH
$Al(AAEMA)_3$	1576s	1639s 1607s	—	1.9 ovm, 18H, $CH_3C=C$; $CH_3C(O)$ 4-4.4 ovm, 12H, C_2H_4 5 m, 3H, (O)CCHC(O) 5.6 and 6.1 br, 3H, C=CH
$Ti(O^iPr)_2(AAEMA)_2$	1529s	1638s 1615s	5 (sp, 6.1Hz, 2H)	1.89 s, 6H, $CH_3C=C$ 1.92 s, 6H, $CH_3C(O)$ 4-4.4 ovm, 8H, C_2H_4 5 s, 2H, (O)CCHC(O) 5.4 and 6.1 br, 3H, C=CH

Table 1: Spectroscopic characteristics of titanium and aluminium alkoxides with polymerizable ligands (s: singulet; d: doublet; m: multiplet; br: broad; ovm: overlapping of multiplet, sp: septuplet).

Aldolic condensations are a means to achieve non-hydrolytic condensation [8], but also form enolates and thus metal alkoxides with unsaturated ligands.

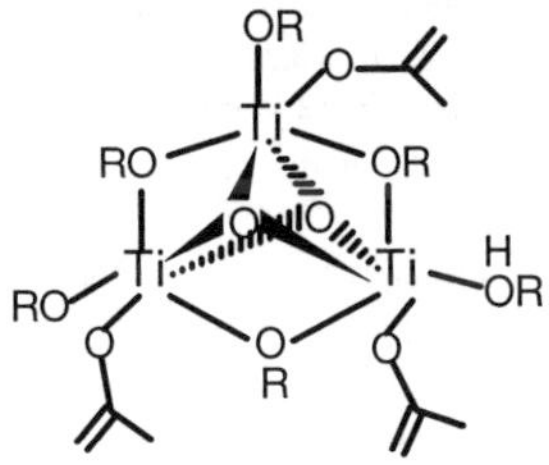

Figure 1: Structure of $Ti_3(\mu_3\text{-}O)_2(O^iPr)_5(OCMe=CH_2)_3(HO^iPr)$

The trinuclear oxoenolate $Ti_3O_2(O^iPr)_5(OCMe=CH_2)_3(HO^iPr)$ (fig 1) has been obtained from condensation between titanium isopropoxide and acetone. The presence of unsatured carbon-carbon bonds is evidenced by a νC=C absorption band at 1655 cm^{-1} in the FT-IR [9]. The compound was also used for copolymerization reactions.

Synthesis and characterization of foams

The synthesis of polystyrene foams doped with titanium or aluminium was achieved following the process summarized in chart 2. The alkoxide derivatives were dissolved in toluene (0.08 - $0.3 M.l^{-1}$). The formation of water-oil emulsion using SPAN-80 (essentially sorbitan monooleate) as a surfactant is achieved. The emulsion is obtained by stirring in a double-barreled syringe pump device. Styrene (St) and divinylbenzene (DVB) readily undergo copolymerization reactions, the use of a mixture of monomers favors the formation of a branched polymer rather than a linear one [10]. Polymerization reactions are anionic, cationic or can proceed by free radicals using initiators such as benzoyl peroxide, azobis(isobutyrolinitrile) (AIBN), or inorganic derivatives.

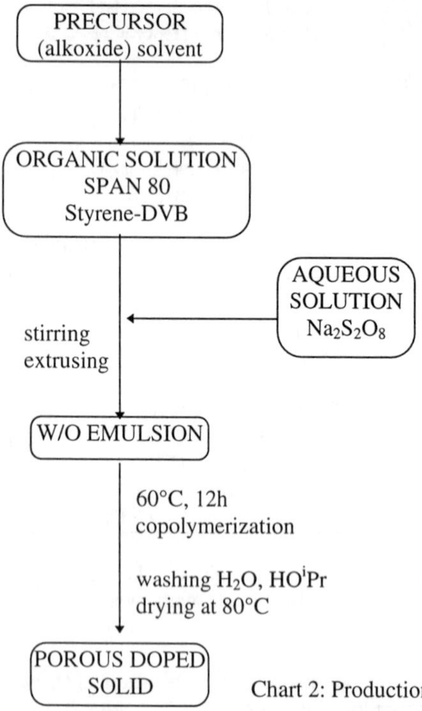

Chart 2: Production of the porous doped organic foams.

During the second step, the polymerization in the organic solvent induced by sodium persulfate $Na_2S_2O_8$, holes are formed through the thin membranes, which separate the water droplets, thus ensuring the formation of an open microstructure. The elimination of the water which acts as a temporary pore former during drying, allows to obtain a porous structure of low density. Extended wash-soak times with water to eliminate inorganics and with isopropanol to remove residual surfactant were used. The obtaining of microcellular foams with different metals shows that the approach is general.

Different experimental conditions have been investigated. The foams have been characterized by elemental analysis for the amount of metal, density measurements, FT-IR and scanning electron microscopy. FT-IR spectra establish that the foams are devoid of residual surfactant. Microanalyses of the metals showed a weight loss (0%-55%) which can probably in part be related to partial hydrolysis of the metal precursors. Table 2 summarizes the characteristics of representative samples.

The physical properties of the foams, especially their mechanical strength can be dramatically changed. The aspect, morphology and microstructure of the doped foams depend on numerous factors, concentration of the precursors, nature of the ligand, functionality of the precursor (number of polymerizable sites), stability of the water-oil emulsion and amount of initiator.

PRECURSOR	%M	C ($M.l^{-1}$)	d±5 (mg/cm^3)	φ (μm)	Aspect
$Ti(O^iPr)_3(OLEA)$ PSTi4	0.36	0.3	_	_	soft
$Ti(OR)_3(OGE)$ PSTi5	0.42	0.3	_	8-12	soft
$Ti(OR)_2(OGE)_2$ PSTi7	0.51	0.3	_	_	soft
$Ti(OR)_2(AAEMA)_2$ PSTi6	0.51	0.3	100	≈10	rigid
$Ti(OR)_2(AAEMA)_2$ PSTi8	1.10	0.3	115	≈24	rigid
$Ti_3O_2(OR)_5(OCH_2{=}CH_2)_3(HO^iPr)$ PSTi9	0.44	0.115	84	8-12	rigid
$Al(AAEMA)_3$ PSAl2	0.78	0.08	_	_	soft
$Al(AAEMA)_3$ PSAl3	0.69	_	_	_	soft
$Al(AAEMA)_3$ PSAl4	0.65	0.3	180	4-6	rigid

Table 2: Properties of the doped foams.

As expected, the foams exhibit a low macroscopic density, the values varying from 80 to 180 mg/cm^3. The difference in density can be observed on the micrographs. They show nearly spherical pores with diameters over the range 4-24 microns and interconnecting walls of various thicknesses as shown on fig. 2. Figure 2 represents the micrographs showing the microstructure morphology of the material identified as PSTi5, PSTi9, PSAl4.

The morphology of the material is dictated primarily at the emulsification stage. The effect of the concentration of the precursor can be illustrated on the Al doped polystyrene foams (0.65 % Al) derived from $Al(AAEMA)_3$. The foam obtained from diluted precursor solution (0.08 M) displays an irregular porous structure with ill-defined cells while the use of higher concentrations (0.3 M) afforded foams with nearly spherical pores.

For instance, in the foams derived from $Ti(O^iPr)_2(AAEMA)_2$, the diameter of the pores increases with the percentage of titanium but the microstructure remains the same.

(a)

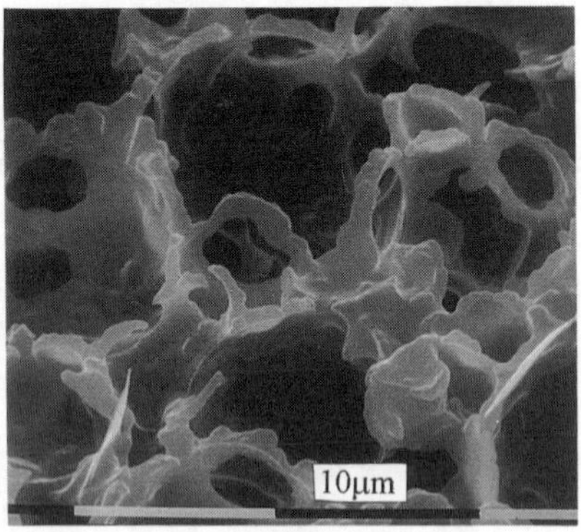

(b)

(c)

Figure 2: SEM micrographs of PSTi5 (a), PSTi9 (b), PSAl4 (c).

CONCLUSION

Microcellular foams based on polystyrene and doped with aluminium or titanium (0.36-1.10 wt %) have been obtained. The technique for making low density doped foams (84-180 mg / cm^3) is based on the use of molecular precursors with polymerizable ligands in toluene with a surfactant SPAN 80, styrene and divinylbenzene. The water phase is a solution of a polymerization initiator, $Na_2S_2O_8$. The stability of the water-oil emulsion and the morphology of the foams depend on the concentration of the precursors and of the nature of the ligand. The microanalyses show a weight loss of the metal depending on the ligand.

Our results suggest a dependence of the ligand on the doping and the stability of the emulsion, but more experiments are necessary in order to determine its effect on the microstructure.The porous doped solids were characterized by FT-IR and SEM. This technique of generating doped foams based on polystyrene can be extended to others metals (Y, Yb, Ge...).

REFERENCES

1. R. Dagani, C&EN News, 33, (1995).
2. J.G. Liu and D.L.Wilcox, Better Ceramics Through Chemistry VI (Mater. Res. Soc. Symp. Proc. 346, Pittsburgh, PA 1994), p. 201.
3. W.R. Even and D.P. Gregory, MRS Bull, XIX, 14 (1994).
4. V.G. Kessler, L.G. Hubert-Plazgraf, S. Daniele, A. Gleizes, Chem. Mater, **6**, 2336 (1994).
5. P. Mastrorilli and C.F. Nobil, J. Mol. Cat, **94**, 217 (1994).
6. U. Schubert, N. Hüsing, A. Lorenz, Chem.Mater, 7, 2010 (1995).
7. N. Pajot, R. Papiernik, L.G. Hubert-Pfalzgraf, J. Vaissermann and S. Parraud, J. Chem. Soc., Chem. Commun, 1817 (1995).
8. S.C. Goel, P.C. Chiang, A.T. Gibbons and W.E. Buhro, Better Ceramics Through Chemistry V (Mater. Res. Soc. Proc. 271, Pittsburgh, PA 1992), p. 3.
9. L.G. Hubert-Pfalzgraf, M. Decams, S. Daniele, J. Vaissermann, J.Sol-gel Technol., in press.
10. P, Cook, G.E. Overtuff III and B.L. Haendler, S.A.Setts LLNL report (1992).

ACKNOWLEDGMENTS

This work was supported by grant N° DAM/CEL V/W003878.

ORGANIC-RICH HYBRID O/I SYSTEMS BASED ON ISOCYANATE CHEMISTRY

S. CUNEY*,**, J.F. GERARD*, J.P. PASCAULT*, G. VIGIER***
*UMR 5627, Laboratoire des Matériaux Macromoléculaires, Bâtiment 403, INSA, 20, Avenue A. Einstein, 69621 Villeurbanne Cedex, France, pascault insa@insa-lyon.fr
**BSN, Le Clairin, BP 16, Saint-Romain, 69702 Givors, France
***UMR 5510, GEMPPM, Bâtiment 502, INSA, 20, Avenue A. Einstein, 69621 Villeurbanne Cedex, France

ABSTRACT

The isocyanate chemistry has been used to prepare, without adding any solvent, organic-rich hybrid O/I systems. α–ω hydroxy-terminated prepolymers (soft segments, denoted SS) can be end-capped with γ-isocyanato propyl triethoxy silane (γ-IPS), or previously reacted with a diisocyanate (DI) and then end-capped with γ-amino propyl triethoxy silane (γ-APS), or with γ-amino propyl methyl diethoxy silane (γ-APMDES). With this second pathway, a double distribution of molecules is present. The aim of this work is to investigate the morphologies and structural properties of different organic-rich hybrid organic/inorganic materials. Two types of α–ω hydroxy prepolymers have been used : hydrogenated polybutadiene, HPBD, and polycaprolactone, PCL.

The inorganic phase is obtained through the hydrolysis and condensation of the silane groups under acidic conditions and with $[H_2O]/Si = 3$. Assuming a complete conversion of SiOH groups, the SiO_2 content never exceed 10% wt. The extent of crosslinking is estimated by the soluble fraction able to be extracted by tetrahydrofurane. Times for gelation were obtained by rheological measurements and/or from the appearance of insoluble fractions. *In-situ* small angle X-ray scattering (SAXS) measurements show that phase separation between an organic-rich phase and an inorganic-rich one can appear before or after gelation depending on the acid catalyst concentration and the nature of SS.

The nanometric silica-rich particles in the final morphologies, after a post-cure at 150°C, were also studied by means of SAXS measurements. Viscoelastic measurements show one or two main relaxation peaks depending on the phase separation process during reaction and the polarity of the initial SS. However for full crosslinked SS/γ-IPS hybrid networks, the relaxed modulus does not depend on SS nature and it is well described by the affine network theory. Silica particles have a small size and probably a high functionality in elastically active network chains (EANC). The fully cured SS/DI/silane hybrid networks equilibrium moduli are higher than SS/γ-IPS one. This can be understood if some silane end-linked DI participate to the EANC.

KEY-WORDS

isocyanate; silane prepolymers; hybrids; phase separation; gelation; SAXS; viscoelastic measurements.

INTRODUCTION

The unique properties of organic and inorganic materials can be synergically combined in a number of different ways [1][2]. The hydrolysis and condensation reactions of alkoxy silanes, specifically tetraethoxysilane (TEOS) and tetramethoxysilane (TMOS), lead to the formation of an oxide network, suggesting a possible way to form glass and ceramic materials. More recently (since a decade ago) investigations on this type of chemistry focused on the use of silicon alkoxides mixed with alkoxysilane-terminated organic monomers or M-Si-$(OR)_3$ prepolymers for different applications such as coupling agents for composites [3], coatings [4-6], electrolytes [7], or adhesives [8].

Mat. Res. Soc. Symp. Proc. Vol. 435

Reactions of tetraalkoxysilanes and trialkoxysilane-terminated prepolymers are similar and give siloxane ≡Si-O-Si≡ bonds. These compounds are also useful model systems to study the chemistry and the processing of such materials [9-17]. The temperature, the water concentration, the nature (basic or acidic medium) and concentration of the catalyst are the most important parameters and their influence in the silicon tetraalkoxide sol-gel process are well documented.

Chemistry has also to be related to the rheological changes occuring during the thermoset process. The hydrolysis and condensation of silane-terminated organic monomers or prepolymers, with a ≡SiOR functionality. f higher than 2 leads to crosslinked materials. As a homogeneous thermosetting system cures, two principal structural transformations may occur : gelation and vitrification [18,19]. Gelation is the transformation from a liquid to a rubbery state and corresponds to the appearance of a cross-linked network displaying with elastic properties which lacked in the low molar mass, linear or branched oligomers. Vitrification involves a transformation from the liquid or rubbery state into the glassy state as a consequence of either the increase in molar mass before gelation or the increase in cross-link density after gelation. After vitrification, reactions become diffusion controlled.

The morphologies, mechanical and physical properties of hybrid materials depend on their thermal history [8], particularly they have to be related to the different transformations occuring during the curing process [20].

In two previous studies [21,22], synthesis of alkoxysilane-terminated oligomers from α–ω hydroxy-terminated oligomers, hydrogenated polybutadiene and polydimethylsiloxane were described. Hybrid organic-inorganic, O/I materials were prepared from these reactive oligomers in bulk. It was found that the two phase model proposed by Wilkes *et al.* [8] could give a correct description of the materials but the effects of different parameters were not investigated.

The aim of this work is to investigate more deeply the morphologies and structural properties of different organic-rich hybrid organic/inorganic materials with different polarity of the organic phase. The isocyanate chemistry was used to prepare different silane-terminated oligomers with different functionality (f = 6 or 4) and different molar mass distributions (MMD). The network formation has been studied by *in-situ* small angle X-ray scattering (SAXS) measurements associated with insoluble determinations. The final morphologies were also characterized by SAXS and dynamic mechanical spectroscopy.

EXPERIMENTAL

Reagents

All the characteristics of the reagents are given in Table I. The α–ω hydroxy-terminated oligomers were two polycaprolactone (PCL) of number average molar mass, $\overline{M_n}$ = 550 g.mol^{-1} and $\overline{M_n}$ = 2000 g.mol^{-1} and a hydrogenated polybutadiene (HPBD) with $\overline{M_n}$ = 2000 g.mol^{-1}. The diisocyanates (DI) were the 3,5,5-trimethyl 3-isocyanatomethyl cyclohexane isocyanate (IPDI) and the 4,4'-dicyclohexylmethane diisocyanate (H_{12}MDI). The crosslinking agents were the γ-isocyanato-propyltriethoxysilane (γ-IPS), the γ-amino-propyltriethoxysilane (γ-APS), or the γ-aminopropylmethyl diethoxysilane (γ-APMDES).

Synthesis

Two types of silane-terminated oligomers have been prepared :

i) the first type consists in reacting 2 moles of γ-IPS with 1 mole of macrodiol (SS) at 80°C for 24 h under an inert atmosphere. The designation is SS/γ-IPS

ii) for the second type, one mole of macrodiol was reacted first at 80°C with 2 moles of diisocyanate (DI). After complete reaction, the prepolymer was dissolved in tetrahydrofurane (THF) and the amino-terminated organosilane was added drop by drop at room temperature. After one hour of reaction the solvent was removed under vaccum. These silane terminated oligomers are denoted SS/DI/γ-APS or SS/DI/γ-APMDES.

Table I Characteristics of reactants

Abbreviation	Supplier	Chemical structure	$\overline{M_n}$ ($g.mol^{-1}$)	Tg (°C)	Tm (°C)
PCL 20	Union Carbide (CAPA 2000)	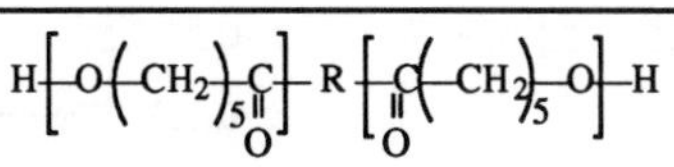	2000	-72	+45
PCL 6	(CAPA 200)		550	-78	+2
HPBD	Nippon Soda (GI 2000)	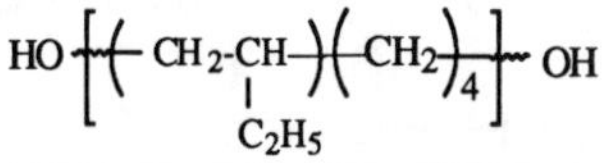	2100	-45	/
IPDI	Hüls	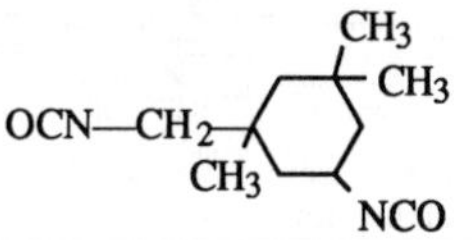	208		
$H_{12}MDI$	Bayer (Desmodur W)	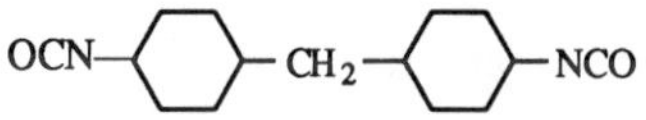	262	-73	
γ-IPS	Roth-Sochiel	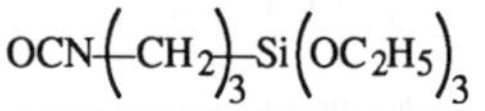	248		
γ-APS	Union Carbide (A1100)	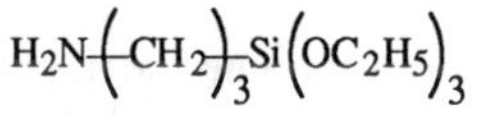	221		
γ-APMDES	Gelest	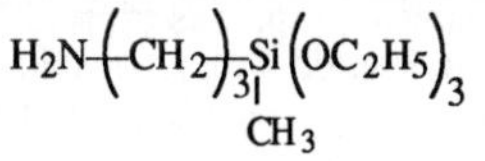	191		

The scheme given in Figure 1 explains the main difference between the two procedures. In the first case the polydispersity of the SS/γ-IPS oligomer equals the polydispersity of the initial macrodiol. In the second case the polydispersity of the final oligomer is controlled mainly by the reactivity ratio of the two isocyanate group, k_1/k_2. As k_1/k_2 never tends toward zero, a certain quantity of silane-terminated diisocyanate is present.

The final network is obtained without solvent through hydrolysis of the ≡SiOR group and condensation of the ≡SiOR and ≡Si-OH group, by addition of deionized water (3 H_2O/Si) and trifluoromethane sulfonic (triflic) acid as catalyst. Different acid-to-alkoxide ratios, $[H^+]$/Si were used. The films were casted from the reactive mixture after stirring at room temperature, and cured 24 h at 100°C under pressure. When specified, they were post-cured at 150°C during 12 h.

Techniques

Gelation times were determined at 40°C from the appearance of insoluble fractions in THF, and from dynamic rheological measurements at various frequencies (from $0.1/2\pi$ to $10/2\pi$ Hz), using a Rheometrics RDA rheometer equipped with parallel plates.

Sol fraction content in our cured networks was determined by swelling (at room temperature) samples in THF solutions containing 1 % polymer by weight. After one week, solid residue was dried and weighted.

First step : silane terminated oligomer synthesis

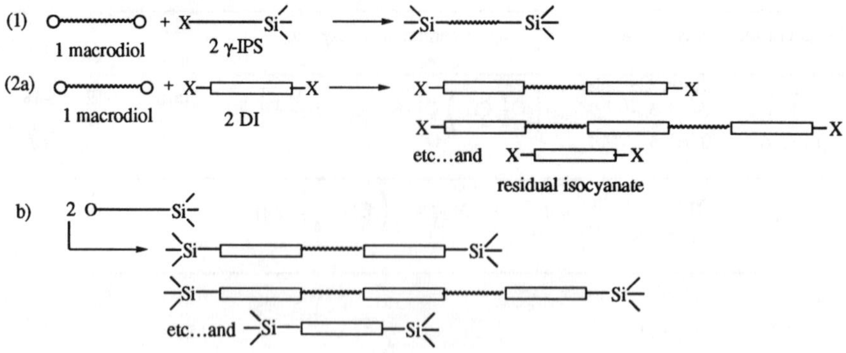

Second step : hydrolysis and condensation

$$\equiv Si\text{-}OR + H_2O \xrightarrow{H_2O} \equiv Si\text{-}OH + ROH$$

$$\equiv Si\text{-}OH + \equiv Si\text{-}OH \longrightarrow \equiv Si\text{-}O\text{-}Si\equiv + H_2O$$

or

$$\equiv Si\text{-}OH + \equiv Si\text{-}OR \longrightarrow \equiv Si\text{-}O\text{-}Si\equiv + ROH$$

Figure 1 Organic-inorganic network synthesis

Dynamic mechanical properties were measured on films by the means of the dynamic analyser RSAII (from Rheometrics Inc.). A frequency of $10/2\pi$ Hz was used to study the temperature dependence of the dynamic mechanical properties from -50°C to 200°C. The equilibrium moduli in the relaxation mode were also measured with a constant deformation of 2%. The temperature was chosen in order to perform the experiment on the rubbery plateau, out of the relaxation region.

Small angle X-ray Scattering (SAXS) experiments were performed with a set-up including a rotating anode X-ray generator with copper target and nickel filter (λ = 1.54 Å), a point collimation produced mainly by two orthogonal mirrors and a line position sensitive proportional counter connected to a computer. The scattered intensity was obtained as a function of the scattering vector q : $q = \frac{4\pi}{\lambda} \sin \theta$. 2θ being the scattering angle, between 2.10^{-3} Å^{-1} and 0.2 Å^{-1}. The correlation length d_c and Porod R_P radius were obtained from the scattering curves. Using Beaucage and Schaefer [23] formalism, we can calculate three other parameters : Guinier radius R_G, a theoretical correlation length ξ and a correlation intensity factor k ($0<k<6$), which is a measure of the degree of particle order in the material. The higher k is, the better ordered the particle dispersion is.

RESULTS AND DISCUSSION

Synthesis and transformations

The gelation times determined from the appearance of a non soluble fraction in THF for the various systems reacted at a low isothermal temperature, T_i = 40°C are reported in Table II. The values were in most cases in a good agreement with those obtained from dynamic rheological measurements except for the HPBD-based sample with the lowest $[H^+]/Si$. t_{gel} from dynamic rheology was chosen as the time at which the loss factor tan δ is independent on the testing frequency [23-24] (Fig. 2) or at which the system displays an elastic response [20].

Table II Gelation times for the various alkoxy-terminated prepolymers at $T_i = 40°C$

SS/DI/γ-APS	H^+/Si	t_{gel} (min) insoluble	t_{gel} (min) rheology
PCL6/H_{12}MDI/γ-APS	10^{-3}	< 3	< 3
PCL20/H_{12}MDI/γ-APS	10^{-3}	< 3	
	10^{-4}	< 3	
HPBD/H_{12}MDI/γ-APS	10^{-3}	20	
	10^{-4}	170	
HPBD/IPDI/γ-APS	$2.5.10^{-3}$	0	11
	10^{-3}	26	30
	$7.5.10^{-4}$	85	90
	5.10^{-4}	165	190

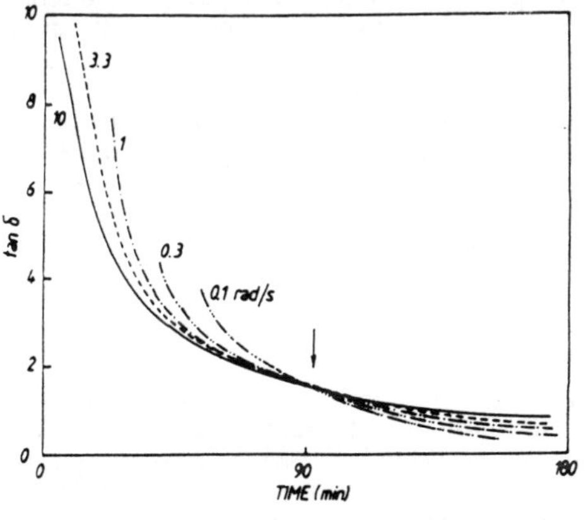

Figure 2 Loss factor, tan δ, as a function of time for various frequencies (system based on H-PBD/IPDI/γ-APS at 40°C; $H^+/Si = 7.5.10^{-4}$)

As expected, the gelation time decreases as the acid-to-alkoxide ratio increases. This effect is in agreement with the conclusions of Glaser and Wilkes [26] on PDMS-modified TEOS glasses. It was shown that increasing the amount of catalyst leads to an increase of the hydrolysis and condensation rates, but the most important effect concerns the condensation reaction rate. The gelation times obtained by changing the hard segment, i.e. the diisocyanate nature, keeping the soft segment (HPBD) and the $[H^+]/Si$ ratio unchanged, are similar. In the opposite, for the same hard segment (H_{12}MDI) and $[H^+]/Si$ ratio, the gelation phenomena is greatly influenced by the nature of the soft segment. In fact, the gelation occurs more rapidly in the case of polycaprolactone-based reactive systems than for the HPBD-based ones (Table II), even if the molar mass of the initial oligomers are very close. This effect could be explained by the fact that the reaction proceeds in a more polar medium in the case of PCL-based alkoxysilane prepolymers.

Gelation is not the one and only transformation occuring during the isothermal polycondensation of the initially miscible alkoxysilane-terminated prepolymers. Beside vitrification, another transformation is expected to occur : phase separation must be present since the final morphology is generally heterogeneous, with an organic-rich phase and an inorganic rich one.

The occurence of the onset phase separation phenomena can be either before or after the gel point. In the case of these particular systems, after phase separation the probability for the organic-rich phase to vitrify and then, for the reaction to become diffusion controlled is high. Another point is that the probability of intramolecular condensation between ≡SiOH group is certainly favored inside the inorganic-rich phase (just like in microgels in the case of free-radical polymerisation).

Having a clear picture of what happens during the O/I material synthesis is certainly difficult. The fact is that the final structures and morphologies result from the appearance of all these events and more generally, from the competition between the reaction rate (≡Si-OH condensation since the hydrolysis is very fast using an acid catalyst) and the phase separation rate (agregation of ≡Si-OH and ≡Si-O-Si≡ units). Some *in-situ* SAXS measurements have been performed in order to locate the time at which phase separation occurs on the reacting time scale.

In the case of the PCL-based hybrid systems the reaction was too fast to run *in-situ* SAXS measurements. Because the reaction rate is high and according to the fact that PCL displays a high polarity, we assume that in this case phase separation occurs after gelation.

On the opposite, in the case of HPBD/H_{12}MDI/γ-APS hybrid systems the reaction rate allows *in-situ* measurements. Samples taken upon curing at T_i = 100°C were cooled down to room temperature for SAXS measurements. No further evolution during measurement was assumed. The results are presented in Figure 3. No correlation peak was observed before t_i = 3 min for $[H^+]/Si = 10^3$ and before t_i = 10 min for $[H^+]/Si = 5.10^{-2}$. The measurements are not precise enough to calculate radius and correlation lengths evolutions. But comparing the phase separation time to the gel time, it appears that phase separation occurs just before gelation for the reaction with the lower acid catalyst concentration ($[H^+]/Si = 10^{-3}$). On the contrary at a higher concentration ($[H^+]/Si = 5.10^{-2}$) gelation appears long before the formation of the inorganic-rich particles. In this latter case the final morphology comprises logically particles with smaller diameters and interparticle distances.

Morphologies and properties of the cured materials

Films have been prepared as described in the experimental part. The degree of crosslinking was estimated by the sol fraction in THF (Table III). In all cases, the sol fraction was higher for the HPBD-based hybrid samples. In some cases, the post cure increased the crosslink density and for the three types of PCL-based hybrids the sol fraction was within the same range ± 3 %.

The scattering curves of the three types of networks are presented in Figure 4.

Table III Analysis of the extent of crosslinking by the determination of the sol fraction in THF (% bw)

Network	pre-cured (100°C/24 h)	post-cured (150°C/12 h)
HPBD/γ-IPS	5	5
PCL20/γ-IPS	2	2
HPBD/H_{12}MDI/γ-APS	23	23
PCL20/H_{12}MDI/γ-APS	12	3
HPBD/H_{12}MDI/γ-APMDES	23	8
PCL20/H_{12}MDI/γ-APMDES	3	3

According to Wilkes model the morphology can be described as an organic-rich matrix in which silica-rich particles are dispersed. The particle radius and the correlation length are obtained from the scattering curves and given in Table IV. For the SS/γ-IPS hybrid networks the particle radius varies from 5 to 10 Å and the correlation length is about 45 Å, i.e.; close to the end-to-end distance of the macrodiol [2]. The macrodiol nature seems to have an influence only on the correlation intensity factor : the ordered particle dispersion in the resulting material. For these materials it is difficult to conclude about a sharp or a diffuse interface between particles and organic-rich matrix because the asymptotic behaviour of the diffusion does not tend clearly to a constant value. This can be explained by the fact that the particles are very small.

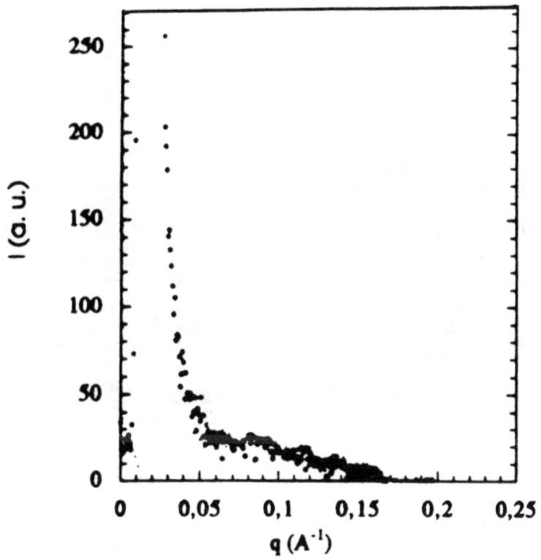

a) scattering curve just after hydrolysis

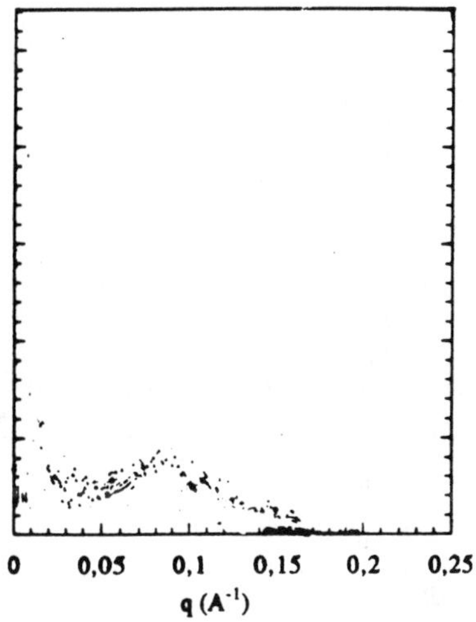

b) $[H^+]/Si = 10^{-3}$ scattering curve at 3 min. No peak before 3 min

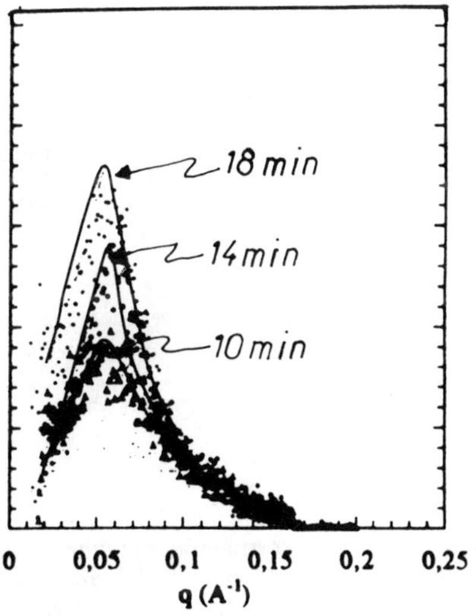

c) $[H^+]/Si = 5.10^{-2}$ scattering curve at different times. No peak before 10 min

Figure 3 *in-situ* scattering measurements for HPBD/H_{12}MDI/γ-APS hybrid system, $3[H_2O]/Si$

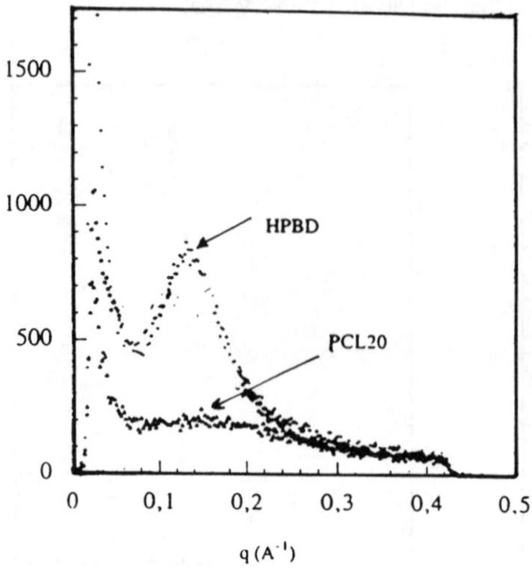

a) SS/γ-IPS

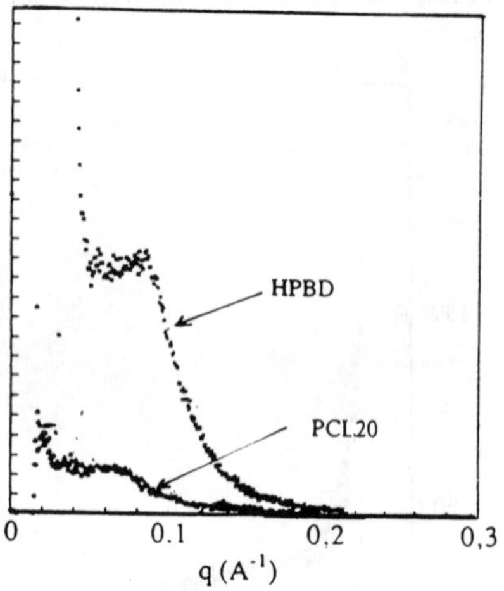

b) SS/H12MDI/γ-APS

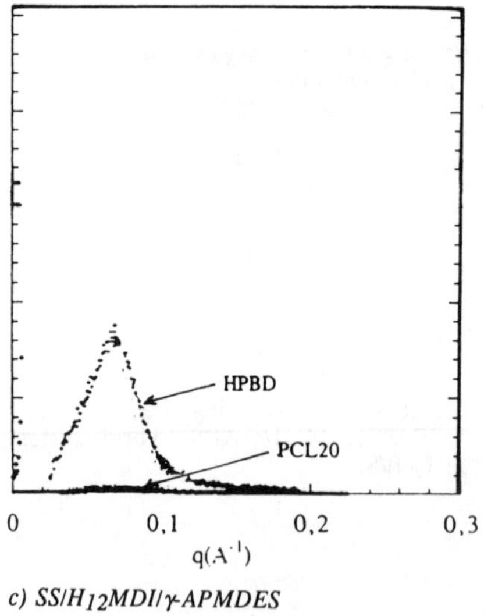

c) SS/H12MDI/γ-APMDES

Figure 4 Scattering curves of hybrid materials cured at T_i = 100°C for 24 h and post-cured at T_i = 150°C/12 h ; 3 $[H_2O]$/Si = 1 ; $[H^+]$/Si = 5.10^{-2}

SS/H_{12}MDI/γ-APS hybrid networks exhibit a quite similar microstructure, but with a sharp interface between particles and matrix, higher particle radius (about 20 Å), and higher correlation lengths (about 100 Å). The correlation length is higher than the end-to-end distance of the SS/H_{12}MDI prepolymer. We suspect the alcoxy silane end-capped residual diisocyanate molecules to disturb the building-up of the network. Due to steric hindrance, the extent of reaction of these rigid molecules is low and they may lead to some chain extension.

Table IV Analysis of the final morphologies by SAXS measurements ($[H^+]/Si = 5.10^{-2}$ $3[H_2O]/Si$ and T_i = 100°C/24 h plus 150°C/12 h

Network	Particle size		Correlation length		Correlation intensity
	R_P (Å)	R_G (Å)	d_c (Å)	ξ (Å)	k
HPBD/γ-IPS	7	9	48	39	3.1
PCL20/γ-IPS	/	5	42	30	0.7
HPBD/H_{12}MDI/γ-APS	16	21	100	82	1.8
PCL20/H_{12}MDI/γ-APS	17	22	84	70	1.5
HPBD/H_{12}MDI/γ-APMDES	16	24	92	74	4.4
PCL20/H_{12}MDI/γ-APMDES			no diffusion		

For SS/H_{12}MDI/silane networks, the macrodiol nature, but also the SiOR functionality of the organosilane have an influence on the particle dispersion in the matrix. SAXS analysis of SS/H_{12}MDI/γ-APMDES hybrid networks indicates that the material containing a PCL macrodiol is homogeneous. Nevertheless, with the non polar polymer matrix, γ-APS and γ-APMDES crosslinked materials have similar microstructures.

Viscoelastic properties and equilibrium moduli are given in Table V and Figure 5. Viscoelastic properties and equilibrium moduli of SS/γ-IPS networks are very interesting to discuss first, in the sense that the type of SS has no influence on the modulus at the rubbery plateau and on the equilibrium one. These networks can be considered as model networks. In this case, the concentration in elastically active network chains (EANC) can be considered to be proportional to $\overline{M_c}^{-1}$, i.e.; to $\overline{M_n}^{-1}$, the average molar mass of the initial macrodiol. The fact that the experimental equilibrium moduli are slightly lower than the theoretical values obtained for an affine network containing pointlike crosslinkings (E_{aff} = 4.4 MPa) can certainly be explained by the combination of a small cluster size with a high functionnality in EANC.

The dissymetry of tanδ peak can be explained either by the PCL rich-phase crystallization during measurement or by a large relaxation time distribution, due to hydrogen interactions between silica-rich particles (urethane units and unreacted SiOH groups) and the polar polymer matrix.

Table V Modulus analysis for hybrid networks (10 rad/s)

Network	Dynamic modulus at 150°C (MPa)	Measurement temperature for relaxed modulus	Relaxed modulus (MPa)
HPBD/γ-IPS	6	60	3.8
PCL20/γ-IPS	6	60	3.8
HPBD/H_{12}MDI/γ-APS	2.2	130	1.5
PCL20/H_{12}MDI/γ-APS	8	130	6.5
HPBD/H_{12}MDI/γ-APMDES	2.5	130	1.6
PCL20/H_{12}MDI/γ-APMDES	6	/	4

The equilibrium and rubbery moduli of the HPBD/H_{12}MDI/silane hybrid materials are drastically lower than the HPBD/γ-IPS one. Due to the immiscibility between soft, silane-terminated HPBD, and hard, silane-terminated DI, segments and premature phase separation, the morphology of these networks is complex. Two relaxations can be observed on tanδ curves (Fig. 5b). The relaxation at low temperature is attributed to the SS, HPBD, relaxation and the relaxation at higher temperature, T ≈ 60°C can be explained by the presence of lightly crosslinked and mixed domains, containing HPBD, H_{12}MDI, and silicate.

As shown in Figure 6, this second relaxation is very sensitive to the acid catalyst concentration i.e.; to the location of the phase separation process relative to the gel point.

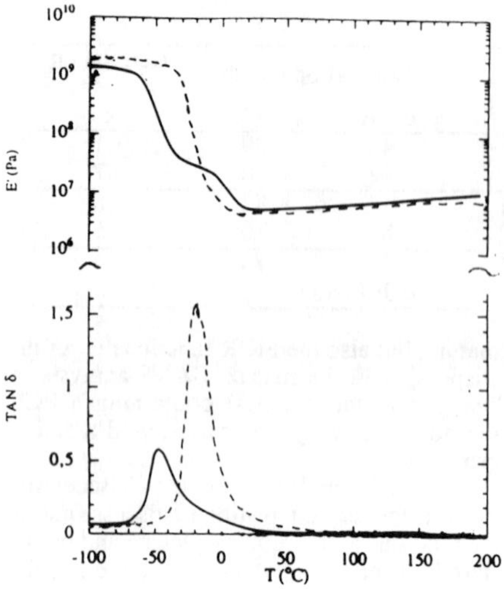

a) SS/γ-IPS

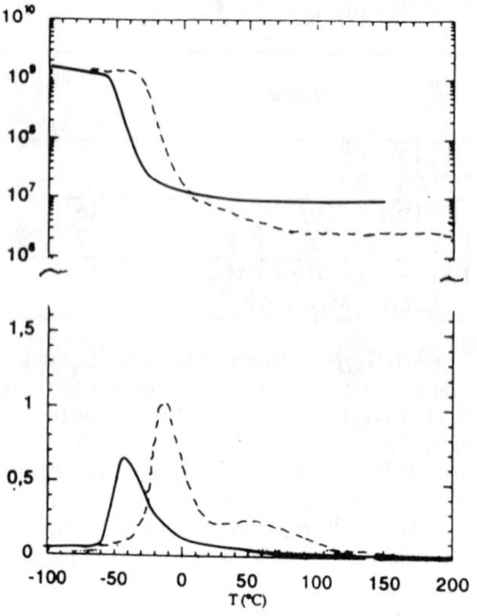

b) SS/H_{12}MDI/γ-APS

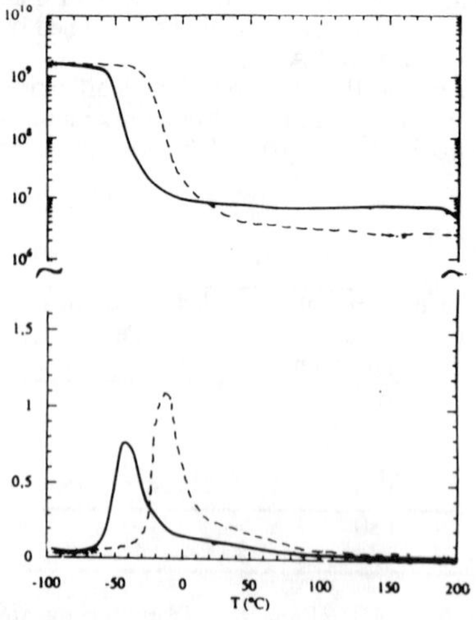

c) SS/H_{12}MDI/γ-APMDES

Figure 5 Dynamic mechanical properties of hybrid materials cured at $T_i = 100°C$ for 24 h and post-cured at T_i = 150°C/12 h ; 3 $[H_2O]/Si = 1$; $[H^+]/Si =$

The low moduli of these materials, in comparison with the SS/γ-IPS model ones, can be explained by a lower concentration in EANC. This is due to i) the introduction of silane-end-capped diisocyanate which microphase separate during reaction and thus decreases the EANC concentration, or ii) some trapped SS in a mixed interfacial region [8]. The main reason is certainly the incomplete condensation of SiOH group. The sol fraction in these networks is high, and in these conditions no further modelling is possible.

The PCL20/H_{12}MDI/γ-APMDES hybrid network is still homogeneous from SAXS measurements and exhibits a higher modulus than the SS/γ-IPS system (Fig. 5c and Table V).

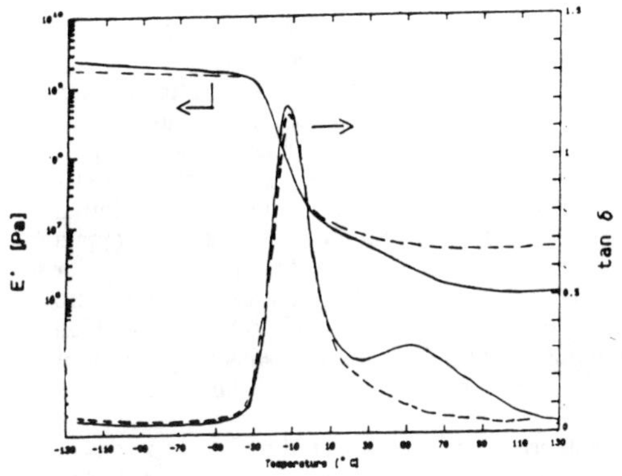

Figure 6 Dynamic mechanical spectra at 10 Hz (storage young's modulus and tanδ vs. temperature) for hybrid films based on H-PBD: $[H^+]/Si$ = (------) : 5.10^{-2} ; (——) : 5.10^{-4}

The double-distribution of miscible molecules at the end of the first stage of the synthesis : silane end-capped PCL 20 and silane end-capped DI resulted in a binodal distribution of EANC in the final homogeneous network. As a part of EANC contains SS and the other one is only the short and hard DI units, the EANC average length decreases leading to increase the modulus. The value calculated from the affine network theory assuming point like crosslinks is lower, 2.6 MPa.

Although a two-phase structure has been revealed by SAXS measurements. The PCL20/H_{12}MDI/γ-APS network seems to be «slightly heterogeneous» because of the low scattered intensity and the low correlation intensity factor, k (Table IV). As in the case of the PCL20/H_{12}MDI/γ-APMDES sample, some hard silane terminated DI units act certainly as EANC, but compared to the γ-APMDES sample some silica-rich clusters with high functionnality are still present.

CONCLUSION

The isocyanate chemistry has been used to prepare solvent free, organic-rich hybrid, O/I systems. α–ω hydroxy-terminated prepolymers (soft segments, SS) can be end-capped with γ-isocyanatopropyltriethoxysilane (γ-IPS), or previously reacted with a diisocyanate (DI) and then end-capped with γ-aminopropyltriethoxysilane (γ-APS), or with γ-aminopropylmethyl diethoxysilane (γ-APMDES). With this second pathway, a double distribution of molecules is present, one based on the end-capped prepolymer, and the other one based on the silane end-capped diisocyanate. The samples are denoted as: SS/γ-IPS; SS/DI/γ-APS and SS/DI/γ-APMDES.

During the hydrolysis and condensation of the silane groups, done in bulk under acidic conditions and with 3 $[H_2O]/Si$, three transformations are expected to occur : gelation and vitrification as usually in thermoset cure, but also phase separation between an organic-rich phase and an inorganic-rich one. For SS/DI/γ-APS systems, *in-situ* SAXS measurements associated with insoluble determinations show that phase separation can occur before or after gelation of the organic matrix depending on the silane condensation rate. For the lower acid catalyst concentration $[H^+]/Si = 10^{-3}$ and considering HPBD soft segment, phase separation appears just before gelation. On the opposite for a higher concentration ($[H^+]/Si = 5.10^{-2}$), gelation appears well before phase separation. In the case of the polar PCL matrix the reaction rate increases strongly, *in-situ* SAXS measurements are not possible and it can be assumed in this case that phase separation occurs after gelation.

The SS/γ-IPS samples can be considered as model systems. For similar synthesis conditions the morphologies characterized by SAXS (radius of particles, R_G, and distances between particles, d_c) or the equilibrium elastic response, E'_c of the networks do not depend on the type of soft segments (for the same $\overline{M_n}$). The E'_c value can be modelled by the affine network theory in which the concentration of elastically active network chains (EANC) corresponds to the initial SS concentration. The connected crosslinks are the inorganic-rich clusters.

Due to the initial bimodal distribution of the prepolymer and diisocyanate molecules, the structures of the SS/DI/γ-APS or γ-APDMES samples are more complex. The final silane conversion can reach full conversion with the γ-APMDES end groups whereas it is never higher than 90 % with the γ-APS end groups. Compared to SS/γ-IPS samples, R_G and d_c measured by SAXS increase and d_c values can not be correlated to the end-to-end distance of the initial hydroxy-terminated prepolymers. Some chain extension reactions of silane end-capped prepolymers through silane-end-capped diisocyanate are assumed. In the case of non-polar HPBD soft segments, the poor miscibility between HPBD and the reacted DI leads to premature phase separation which increases the sol fraction and to a mixed interfacial O/I region which could explain a second relaxation observed at about 50 to 60°C on the viscoelastic spectrum of this material. For all these reasons, the equilibrium elastic response, E'_c, of the networks are difficult to model.

In future studies the properties of hybrid O/I materials will be compared to the properties of crosslinked polyurethanes. Properties introduced by different concentrations of organic and inorganic clusters will be compared.

REFERENCES

1. J.E. J.E. Mark, C.Y-C Lee, and P.A. Bianconi, Hybrid Organic-Inorganic Composites and ref. therein, ACS 585, Am. Chem. Soc., Washington DC, 1995.
2. C. Sanchez and F. Ribot, New J. Chem., p. 1007 (1994).
3. A. Serier, J.P. Pascault, and T.M. Lam, J. Polym. Sci. Chem. **29**, p. 1125 (1991).
4. H. Schmidt and B. Seiferling, Mater. Res. Soc. Symp. Proc. **73**, p. 739 (1986).
5. H. Schmidt, J. of Sol-Gel Sci. and Tech. **1**, p. 217 (1994).
6. B. Tamani, C. Betrabet, and G.L. Wilkes, Polym. Bull. **30**, p. 393 (1993).
7. D. Ravaine, A. Seminel, Y. Charbouillot, and M. Vincens, J. Non-Cryst. Solids **82**, p. 210 (1986).
8. H.H. Huang, G.L. Wilkes, and J.G. Carlson, Polymer 30, p. 2001 (1989).
9. J. Livage, M. Henry, and C. Sanchez, Prog. Solid State Chem. 18, p. 259 (1988).
10. J.K. Bailey, C. Macosko, and M.L. McCartney, J. Non-Cryst. Solids 125, p.208 (1990).
11. K.D. Keefer, Adv. Chem. Ser. 224, p. 227 (1990).
12. C.J. Brinker and G.W. Scherer, J. Non-Cryst. Solids 70, p. 301 (1985).
13. K.D. Keefer, Mater. Res. Soc. Proc. 32, p. 15 (1984).
14. I. Strawbridge, A.F. Craievich, and P.F. James, J. Non-Cryst. Solids 72, p. 139 (1985).
15. B.E. Yoldas, J. Non-Cryst. Solids 83, p. 375 (1986)
16. I. Artaki, T.W. Zerda, and J. Jonas, J. Non-Cryst. Solids 81, p. 381 (1986).
17. M.W. Colby, A. Osaka and J.D. MacKenzie, J. Non-Cryst. Solids 99, p. 129 n(1988).
18. P.J. Flory, Principles of Polymer Chemistry, Cornell University Press, Ithaca, New York, 1953.
19. H. Kaddami, F. Surivet, J.F. Gérard, T.M. Lam, and J.P. Pascault, J. Inorg. and Organometal. Polym. **4** (2), p. 183 (1993).
20. J.B. Enns and J.K. Gillham, J. Appl. Polym. Sci. **28**, p. 2567 (1983).
21. F. Surivet, T.M. Lam, J.P. Pascault, and Q.T. Pham, Macromolecules **25**, p. 4309 (1992).
22. F. Surivet, T.M. Lam, J.P. Pascault, and C. Maï, Macromolecules **25**, p. 5742 (1992).
23. C. Beaucage and D.W. Schaefer, J. Non-Cryst. Solids **173**, p. 797 (1994).
24. D. Adoff, J.E. Martin, and J.P. Wilcoxon, Macromolecules **23**, p. 527 (1990).
25. D. Adoff and J.E. Martin, Macromolecules **23**, p. 3700 (1990).
26. R.H. Glaser and G.L. Wilkes, ACS Polym. Prepr., New Orleans **28** (2), p. 236 (1985).
27. D.E. Rodrigues, A.B. Brennan, B. Wang, and G.L. Wilkes, Chem. Mater. **4**, p. 1437 (1992).

STRUCTURE/PROPERTY BEHAVIOR OF ORGANIC-INORGANIC SEMI-IPNS

A.B. BRENNAN, T.M. MILLER
Department of Materials Science and Engineering, University of Florida, Gainesville, FL 32611

ABSTRACT

Ongoing work in our laboratories continues to demonstrate the unique properties that can be developed by the generation of a glassy semi-interpenetrating network (SIPN) within an elastomeric-based hybrid organic-inorganic composite. The hybrids are synthesized by the acid catalyzed sol-gel-processing of triethoxysilane end-capped poly(tetramethylene oxide) (PTMO) and tetraethoxysilane (TEOS) in an alcoholic solution. The SIPNs are then created by swelling these hybrids with methacrylic acid (MAA) and then gamma-polymerizing the monomer in-situ. The resulting materials exhibit excellent optical transparency and significantly improved toughness. This paper reports on the effect of inorganic loading upon the mechanical properties of acid catalyzed hybrid composites. The results demonstrate that as the amount of TEOS added to the sol prior to gelation increases the elongation-to-break decreases and the stress-to-break increases. Similarly, the post-yield stress drop and post-yield elongation prior to strain hardening increase with increasing inorganic loading. Atomic force microscopy and small angle X-ray scattering data is presented which substantiates the morphological model developed to explain the mechanical response.

INTRODUCTION

Several approaches have been taken in the synthesis of hybrid organic-inorganic composites. Typically, the approaches differ in the type of organic employed. High inorganic hybrids incorporate mixtures of tetraalkoxysilanes and organically substituted alkoxysilanes. These materials find applications in areas such as microelectronic and abrasion resistant coatings [1-3]. Less highly loaded composites often employ organic oligomers and polymers. These systems exhibit much more flexibility and may or may not employ functionalization, i.e., incorporation of reactive endgroups on the polymer [4-7]. Other approaches involving the transformation of an existing membrane to a hybrid composite have been accomplished by swelling the membrane with metal alkoxides and then converting these absorbed species to an inorganic network using sol-gel-processing [8,9]. Regardless of the processing route, the resulting material is a mixture of both organic and inorganic phases. The morphology and hence the degree of interaction occurring between the phases is a function of the sol-gel-processing conditions: water-to-alkoxy ratio, type of solvent and catalyst used, molar mass of the organic species, functionality of precursors and solubility of the sol constituents [10-15]. Therefore, a wide range of morphologies can be engineered into hybrid composites to produce tailored responses.

The organic-inorganic hybrids utilized in our investigations are synthesized by the acid catalyzed, simultaneous sol-gel-processing of triethoxysilane end-capped poly(tetramethylene oxide) and tetraethoxysilane. Small angle X-ray scattering and dynamic mechanical spectrometry experiments indicate that a microphase separated morphology exists consisting of both organic and inorganic rich domains, as well as mixed regions of both [6,16,17]. Our earliest investigations focused on the development of SIPNs generated via gamma polymerization of methacrylic acid, n-vinylpyrrolidone and cyclohexylmethacrylate [18]. The results showed that

Mat. Res. Soc. Symp. Proc. Vol. 435 © 1996 Materials Research Society

the poly(methacrylic acid) (PMAA) - poly(tetramethylene oxide) (PTMO) SIPNs exhibited exceptionally high stress and elongation at break, as well as yielding. Additionally, a 3000% increase in the tensile elastic modulus was observed over that of the benchmark hybrid. These significant changes in the mechanical response of the organic-inorganic hybrids were attributed to the formation of a PTMO-PMAA copolymer upon exposure of the MAA swollen PTMO-silica gel to high energy gamma radiation and/or weak bonding forces between the less than fully developed, acid catalyzed polysilicate network and the polar moities of the PMAA. It is likely that both mechanisms are at work. Considering the inorganic loading as a design variable, research was begun to determine the effect of the polysilicate loading on the MAA swellability and subsequent PMAA loading. This study reports on the tensile mechanical properties of these polysilicate containing PMAA-PTMO SIPNs as a function of the polysilicate loading. Additionally, atomic force microscopy (AFM) and small angle X-ray scattering (SAXS) were utilized to probe the morphology of these mixed-phase systems.

EXPERIMENTAL

Triethoxysilane functionalized PTMO was prepared by the reaction of 2000 g/mole hydroxy terminated poly(tetramethylene ether) glycol (Polysciences, Inc.) and isocyanatopropyl-triethoxysilane (IPTS) (United Chemical Technologies, Inc.). Prior to the end-capping reaction the PTMO was vacuum dried overnight at 40°C and 20 Torr. Additionally, the IPTS was vacuum distilled and the middle 80% used for end-capping. The purity of the middle 80 of IPTS was verified at 99+% using gas chromatography. The end-capping reaction involving a 1:2.09 molar ratio of PTMO:IPTS was carried out at 70°C with constant stirring under nitrogen until infrared spectroscopy indicated no further changes in the ratio of intensities of the diminishing isocyanate peak at 2250 cm^{-1} and developing carbonyl peak at 1700 cm^{-1}, ca. 5 days.

The hybrid composites were synthesized by dissolving masses of 10, 8, 6 and 4 g of the end-capped PTMO each into a mixture of 8 ml of isopropyl alcohol and 2 ml of tetrahydrofuran (THF). Both solvents were used as received (Fisher Scientific-HPLC grade). Continuous stirring for 15 minutes in covered 25 ml polypropylene Erlenmeyer flasks resulted in a clear solution to which 0, 2, 4 and 6 g of TEOS were added, respectively. After 5 minutes had elapsed, the stoichiometric amount of water required for the hydrolysis of every ethoxy group was added along with 0.0143 equivalents of 10N HCl. The 4 resulting sols contained 0, 20, 40 and 60 mass percent TEOS, henceforth abbreviated TEOS(0), TEOS(20), etc., relative to PTMO. The contents of each beaker were then cast into 4 polystyrene petri dishes, which were then covered for 4 days during which the sols gelled to transparent films. The covers were removed for 2 days to evaporate any residual alcohol, water and THF, after which the gels were covered again and stored for later use. These thin films thus produced were used as the benchmark against which the effects of radiation and SIPN formation are compared.

The synthesis of SIPNs was accomplished by cutting the benchmark gels into rectangular pieces approximately 1.1 cm wide and 5 to 7 cm long. Typical sample thickness for each of the inorganic loadings evaluated were 0.40 ± 0.035 mm, 0.35 ± 0.10 mm, 0.25 ± 0.059 mm and 0.14 ± 0.061 mm for the 0, 20, 40 and 60 mass percent TEOS loadings, respectively. The differences in thickness are attributable to increasing polymerization shrinkage with increasing TEOS loading. The strips were cut to characteristic shapes and the initial weights recorded so that mass increases of each could be monitored throughout the processing. The rectangular samples from each TEOS loading were placed in Pyrex petri dishes containing 45 ml of distilled MAA. After 2 hours the monomer swollen strips were removed, blotted dry and their mass uptake of monomer recorded. Each was immediately placed into a prefilled test tube containing 10% (mass/mass) MAA in deionized water. Water was chosen because of its poor ability to swell the benchmark gels [19]. However, MAA is soluble in water. Therefore, a 10% (mass/mass) monomer in water solution was used to reduce the concentration gradient and hence the driving force for monomer desorption.

The test tubes containing the samples were then purged by bubbling ultrahigh purity nitrogen through each tube for ca. 15 seconds. Additionally, test tubes containing samples of the benchmark gels suspended in deionized water only were also purged with nitrogen for 15 seconds. The tubes were then sealed with polyethylene caps and were positioned radially 4 inches from a ^{60}Co gamma radiation source for 3 hrs and 18 minutes. The dose rate was approximately 350 rads/min for a total dose of 0.07 Mrads. This total dose is in keeping with earlier experiments [18]. Upon removal from the source, the MAA solutions in the test tubes had gelled to a transparent, water swollen network which adhered to the surfaces the samples when removed from the tubes. To facilitate removal of the clinging polymer all of the samples were swollen in water for 24 hrs with multiple washings. Vacuum drying at 40°C for 24 hrs at ca. 20 Torr followed by another 24 hrs of swelling in THF and identical vacuum drying completed the processing of the polysilicate containing PMAA-PTMO SIPNs. These SIPNs were stored under vacuum for several days until tested.

The tensile mechanical properties of the benchmark gels, gamma-irradiated gels and PMAA-PTMO SIPNs were all evaluated using an Instron Model 1122 equipped with a 200 lb load cell at ambient conditions (23°C±1°C, ≈50%RH). The strain rate was 2.5 mm/min. Dog-bone shape samples were cut using the Type III ASTM die described in ASTM test D638M-84. The grip-to-grip distance was 25mm, thus the strain rate was 10%/min. Between 4 and 6 samples were tested for the benchmark gels, the gamma-irradiated gels and the PMAA-PTMO SIPNs at the 0, 20, 40 and 60% (mass/mass) TEOS loadings. However, only 2 samples were tested for the TEOS(0) benchmark gel and TEOS(0) PMAA-PTMO SIPNs. Densities were obtained using Archimedes' principle and water with between 8 and 10 samples per datum. The error bars shown on all figures represent 1 standard deviation from the average. Atomic force microscopy was performed in tapping mode using a Nanoscope III instrument manufactured by Digital Instruments. Small angle X-ray scattering was acquired using a Siemens Kratky camera employing a M. Braun position-sensitive detector from Innovative Technologies.

RESULTS

To facilitate the understanding of the effect of polysilicate loading upon SIPN formation the experimental results obtained will be presented as a function of the estimated volume of polysilicate present within the gels. Table I lists the densities of both the benchmark hybrids and the γ-irradiated gels as a function of TEOS loading. The estimated polysilicate volume percentage is based upon the benchmark gel density measurements and the calculations of Huang et al. assuming ca. 75% conversion of the TEOS into an oxygen bridging network of SiO_2 [16].

The effect of high energy gamma radiation upon gels synthesized using 40% (mass/mass) TEOS was previously shown to increase the network cross-link density thereby increasing the elastic modulus and decreasing the swelling in a good solvent, e.g., THF [18]. To more completely evaluate the influence of the glassy phase upon the radiation induced cross-linking of the benchmark gels the elastic modulus, stress at failure and elongation at failure as a function of the polysilicate loadings were determined and appear in Figure 1. Typical stress-elongation responses are not shown as the gels exhibit linearly increasing stress with increasing elongation up to the points of failure given in Figure 1, i.e., no yielding was observed for any of the inorganic loadings evaluated either before or after γ-irradiation. Consistent with our earlier reports, the irradiation of the benchmark gels results in an increase in the elastic modulus and stress at failure with a corresponding decrease in elongation at failure. The magnitude of these changes increases as the inorganic loading increases suggesting that the gamma radiation interacts with the polysilicate phase leading to its densification as a result of localized heating. Increased consolidation of the polysilicate phase would produce an increase in modulus and stress at failure with a corresponding decrease in elongation at failure. The overall magnitude of

Table I. Densities of the benchmark and γ-irradiated gels as well as the estimated volume fraction of polysilicate based upon the benchmark PTMO-silica gel densities and the work of Huang et al. [16].

TEOS Loading (%)	Benchmark Gel Density (g/cm^3)	γ-Irradiated Gel Density (g/cm^3)	Estimated Volume of Polysilicate (%)
0	1.028 ± 3.12E-3	1.029 ± 3.68E-3	3.2
20	1.071 ± 6.95E-3	1.074 ± 2.34E-3	12
40	1.147 ± 2.97E-3	1.150 ± 2.48E-3	21
60	1.261 ± 6.76E-3	1.273 ± 1.16E-2	32

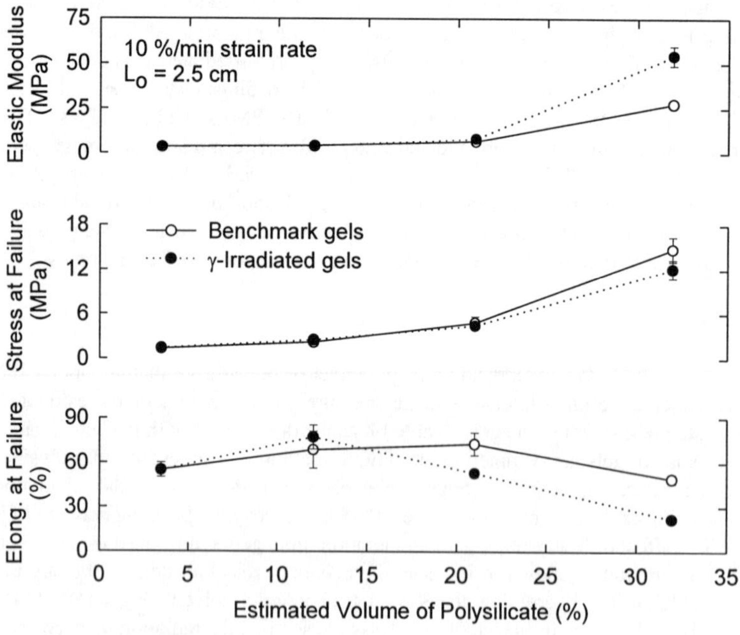

Figure 1. Effect of inorganic loading upon the tensile mechanical response of the benchmark and gamma-irradiated PTMO-silica hybrids

the changes for the TEOS(40) gel observed in this study are less than those observed in previous investigations [18]. Although the total dose is the same, the dose rate in prior investigations was 580 rads/min. This increased dose rate could lead to more heat being generated and hence a higher extent of silicate consolidation exhibited as even larger increases in elastic modulus and stress at failure. Any interaction between the polysilicate phase and the gamma radiation producing heat would lead to increased consolidation of the polysilicate. This increased degree of condensation should manifest itself in an increased gel density. As the data in Table I indicates, there is no apparent increase in overall density of the hybrid gels, suggesting that minimal additional consolidation occurs as a result of irradiation. However, further studies involving FTIR and NMR are underway to more completely understand these effects.

The effect of the inorganic loading upon the mass uptake of MAA and subsequent polymerization to PMAA as a function of the estimated volume percentage of polysilicate present within the swollen gels is graphically represented in Figure 2. It is evident that as the volume of polysilicate increases, the swelling of the benchmark hybrid decreases. However, even though less monomer is absorbed as the volume of polysilicate increases, more is retained within the hybrids structure during gamma polymerization. This behavior suggests that the PTMO phase absorbs the MAA but that both the monomer and subsequent polymer possess an affinity for the polysilicate phase. The swelling response of these hybrids can be explained in terms of two competing mechanisms, i.e., the kinetics of monomer desorption versus polymer formation.

At low polysilicate loadings, and hence minimal restriction of the PTMO chains by the vitreous polysilicate chains, both the concentration gradient and hydrophobic nature of PTMO ensure that the rate of MAA desorption will be great. As the inorganic volume increases, the

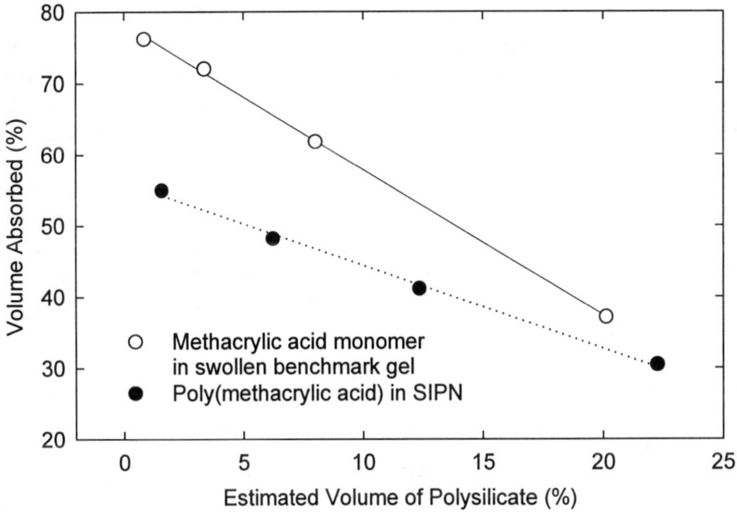

Figure 2. The effect of inorganic loading upon the equilibrium MAA absorption and PMAA SIPN formation for the benchmark PTMO-silica hybrids

vitreous chains restrict the mobility of the PTMO chains and hinder their ability to collapse in the aqueous medium. Consequently, the magnitude of this contribution to monomer desorption is diminished. At the highest polysilicate loading investigated extensive interaction and restriction of the mobility of the PTMO chains occurs thereby resulting in the slowest rate of MAA desorption. The limiting case would be a 100% microporous silica gel in which the only mechanism for monomer desorption would be concentration induced Fickian diffusion. The reduced monomer desorption rates with increasing polysilicate loading allow the formation and retention of increasing amounts of the PMAA SIPN. As a consequence of increased poly(methacrylic acid) formation the hydrophobic nature of the PTMO is further reduced, thus lessening the rate of desorption. Additionally, polar interactions between the acid moities of the MAA and the hydroxy containing, less than fully developed polysilicate network diminish the driving force for desorption of MAA from the gels [20]. It seems reasonable that the preferred morphology resulting from the above described swelling and formation kinetics would entail PMAA-PTMO SIPN formation adjacent to polysilicate domains. Another consideration is that the reduced rates of monomer desorption with increasing polysilicate loading and the favorable polar interactions between the polysilicate phase and the adsorbed MAA suggest that for inorganic loadings that meet or exceed the percolation threshold a monomer adsorption-polymer formation inversion takes place, as is suggested in Figure 2.

The typical engineering stress as a function of the percent elongation for each of the polysilicate loadings is shown in Figure 3 while the elastic tensile moduli and stress and elongation at failure are given in Table II. The elongations shown and reported are calculated using the crosshead extension as a measure of elongation up to the yield point. Beyond the yield, a value of 7.5 mm was used as the gauge length as per ASTM D638. This switch in gauge length accounts for the viscous flow that occurs only in the parallel sided test section of the dog bone shaped sample. The result, however, is artificially high elongations. Preliminary results utilizing a video camera and tick marks 7.5 mm apart within the parallel sided test section indicate that the true elongations at failure are approximately half the values shown and tabulated. Efforts are continuing to more accurately measure the true elongations. Nevertheless, the relative differences in elongations for the samples tested at the various volume fractions of glass are valid and comparable to those reported previously for PMAA SIPNs synthesized using higher dose rates of gamma radiation [18]. Examining the data below, all of the inorganic loadings evaluated exhibit significant increases in the mechanical response as compared to the benchmark and gamma-irradiated gels. Indeed greatly increased stress and elongation to failure along with yielding indicate significant changes in the morphology of the hybrids as a result of PMAA-PTMO SIPN formation. Considering the tabulated data (Table II), there is no dependence of the elastic modulus upon the inorganic loading. However, there is the trend of increasing stress at failure, decreasing elongation at failure and, referring to Figure 3, increased strain hardening rate with increasing inorganic loading.

Table II. The effect of TEOS loading in the sol upon the stress and elongation at failure of the resulting PMAA-PTMO SIPNs.

Estimated Volume of Polysilicate (%)	Elastic Modulus (MPa)	Stress at Failure (MPa)	Elong. at Failure (%)
1.6	930, 1170	34, 35	410, 360
6.2	750 ± 130	37 ± 3.7	390 ± 63
12	870 ± 71	39 ± 1.5	300 ± 24
22	920 ± 140	46 ± 1.2	140 ± 17

The stress and elongation at yield values are given in Table III along with the post yield (P.Y.) stress drop and elongation prior to strain hardening, i.e., the percent elongation occurring after yield but before the onset of strain hardening. The data indicates that the yield stress and elongation at which yielding occurs exhibit no correlation with inorganic loading. However, a clear trend of diminishing post yield stress drop and post yield elongation prior to strain hardening with increasing inorganic loading is observable. Both of these post yield phenomena indicate that as the continuity of the polysilicate phase approaches the percolation limit it hinders the viscous motion of the PMAA-PTMO SIPN. The increased stress at failure and decreased elongation to failure can also be explained based on the increasing continuity of the polysilicate phase.

Table III. The effect of polysilicate loading upon the tensile yield stress and elongation of the resulting PMAA-PTMO SIPNs.

Est. Vol. of PS (%)	Stress at Yield (MPa)	Elongation at Yield (%)	P.Y. Stress Drop (MPa)	P.Y. Elong. Prior to Strain Hard. (%)
1.6	24, 26	3.7, 3.6	7.9, 8.6	37, 26
6.2	19 ± 1.4	4.3 ± 0.90	4.7 ± 0.48	31 ± 5.5
12	18 ± 1.1	3.6 ± 0.10	2.7 ± 0.21	17 ± 0.70
22	22 ± 2.5	4.0 ± 0.22	None Observed	None Observed

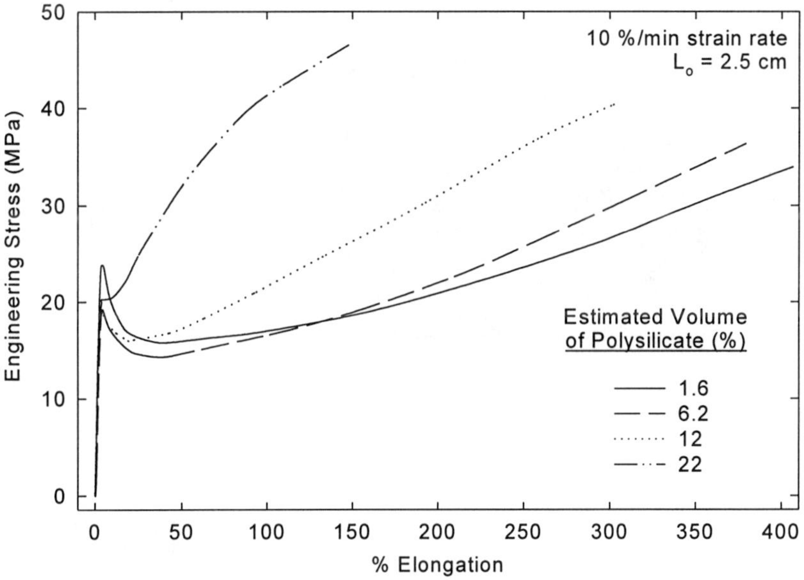

Figure 3. The effect of TEOS loading upon the tensile mechanical response of the PMAA-PTMO SIPNs.

The proposed model of SIPN encapsulated polysilicate domains appears sufficient to explain the observed swelling and mechanical responses. Further evidence of the affinity of the PMAA for the polysilicate phase is provided by the atomic force microscopy images shown in Figures 4 and 5 of the fracture surfaces of both the TEOS (40) benchmark gel and 12 volume percent polysilicate loaded PMAA-PTMO SIPN, respectively. The tapping mode images clearly indicate that the tortuous, microphase separated polysilicate domains clearly observable in the benchmark gel are smoothed over in the SIPN gels. The morphological similarities of this polysilicate containing PMAA-PTMO SIPN to those images collected by Toki et al. on the surface of poly(vinylpyrrolidone) (PVP)-silica hybrids are striking [21]. Their attribution of the homogeneity of the PVP-silica gel to hydrogen-bond formation between the PVP carbonyl groups and residual silanols as measured using NMR lends further support to the proposed poly(methacrylic acid) - polysilicate interactions discussed in this work. Computerized sectional analysis of the SIPN fracture surface indicates an average peak-to-peak distance of 77 ± 36 nm. Small angle X-ray scattering was performed to evaluate the effect of SIPN formation upon the average electron density fluctuations present within the gels. Figure 6 illustrates the changing scattering response observed upon formation of the SIPN within a benchmark 21 volume percent polysilicate containing gel. The scattering profile before swelling and gamma polymerization exhibits a maximum characteristic of these microphase separated gels [6]. An interdomain spacing of ca. 13 nm can be estimated from the reciprocal of the scattering vector, at the maximum scattering intensity, I(s), for the starting gel. Upon polymerization of the adsorbed MAA the overall scattering intensity is reduced due to the lower volume fraction of the relatively electron rich inorganic phase . Additionally, there is no clear maximum in the scattering profile indicating that the PMAA and polysilicate phases have mixed and thereby reduced the electron

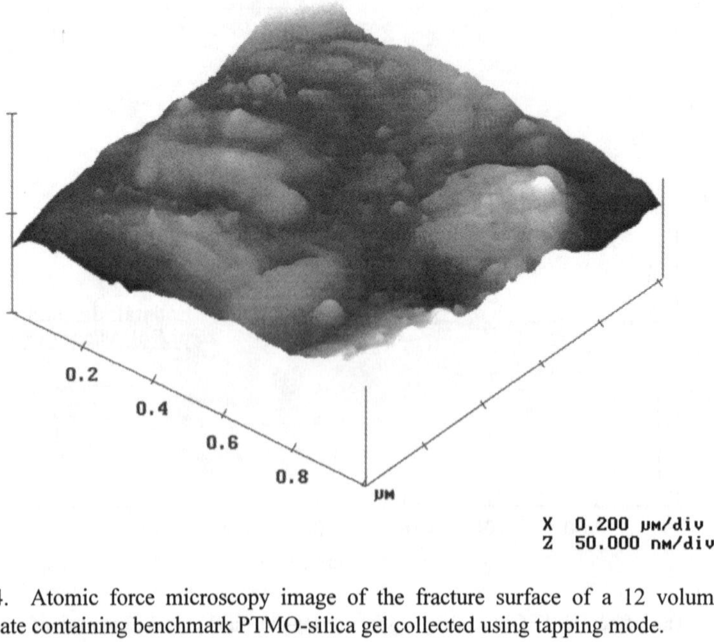

Figure 4. Atomic force microscopy image of the fracture surface of a 12 volume percent polysilicate containing benchmark PTMO-silica gel collected using tapping mode.

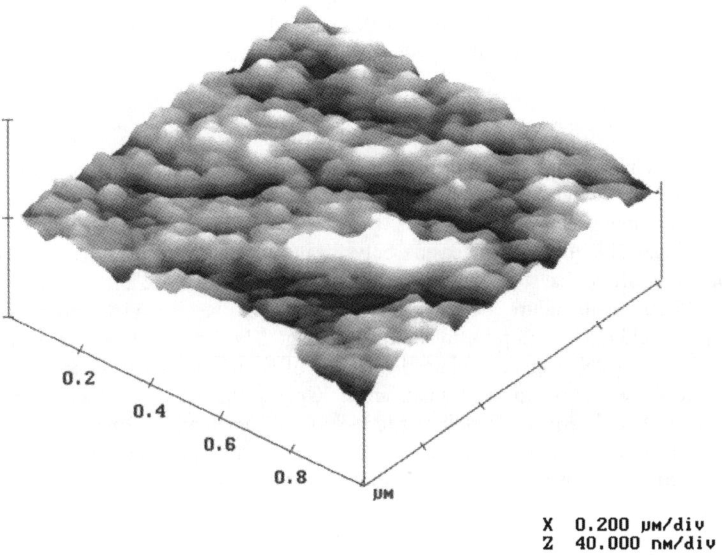

Figure 5. Atomic force microscopy image of the fracture surface of a 12 volume percent polysilicate containing PMAA-PTMO SIPN collected using tapping mode.

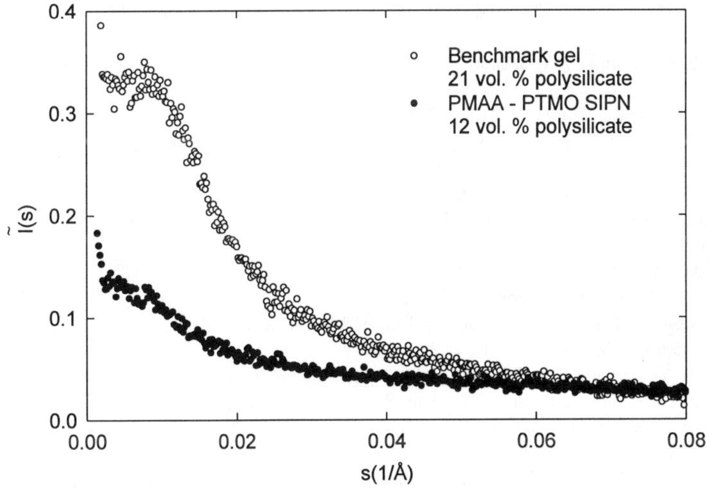

Figure 6. Small angle X-ray scattering profiles of the 21 volume percent polysilicate containing benchmark gel and the subsequent 12 volume percent polysilicate containing PMAA-PTMO SIPN resulting from gamma polymerization of the MAA swollen gel.

density gradient. Dynamic mechanical spectrometry previously reported also indicates that mixing occurs in these PMAA-PTMO-silica SIPNs [18]. This near molecular level of mixing confirms the affinity of the PMAA and polysilicate phases for one another.

CONCLUSIONS

Significant changes in the tensile mechanical response of the benchmark PTMO-silica hybrid composites can be induced by formation of PMAA-PTMO SIPNs. This transformation involves the conversion of elastomeric hybrids to high strength, high elongation organic glasses exhibiting yielding. The effect of increasing polysilicate loading upon the tensile mechanical properties of these SIPNs has been investigated and found to increase the stress at failure, decrease the elongation to failure, decrease the magnitude of the post yield stress drop and decrease the post yield elongation prior to strain hardening. The stress and elongation at failure dependence of these SIPNs can be explained in terms of a morphological model based upon swelling, atomic force microscopy and small angle X-ray scattering data which suggests the formation of the PMAA-PTMO SIPN adjacent to the polysilicate domains. Increased continuity of the polysilicate phase at higher inorganic loadings results in increased stress at break and reduced elongation, due to restricted viscous flow of the SIPN after yielding.

REFERENCES

1. H. Schmidt and H. Wolter, J. Non-Cryst. Solids **121**, p. 428 (1990).
2. M. Popall, M. Meyer, H. Schmidt and J. Schulz, Mater. Res. Soc. Symp. Proc. 180, p. 995 (1990).
3. B. Tamami, C. Betrabet and G.L. Wilkes, Polym. Bull. **30**, p. 39 (1993).
4. C.C. Sun and J.E. Mark, Polymer **30**, p. 104 (1989).
5. A.B. Brennan, B. Wang, D.E. Rodrigues and G.L. Wilkes, J. Inorgan. Organomet. Polym. **1**, p. 167 (1991).
6. D.E. Rodrigues, A.B. Brennan, C. Betrabet, B. Wang, and G.L. Wilkes, Chem. Mater. **4**, p. 1437 (1992).
7. C.J.T. Landry, B.K. Coltrain, J.A. Wesson, N. Zumbulyadis and J.L. Lippert, Polymer **33**, p. 1496 (1992).
8. G.S. Sur and J.E. Mark, Eur. Polym. J. **21**, p. 1051 (1985).
9. Q. Deng, K.A. Mauritz and RB. Moore, Hybrid Organic-Inorganic Composites, edited by J.E. Mark, C.Y-C. Lee and P.A. Bianconi (ACS Symp. Series 585, 1995) p. 66.
10. S. Sakka and K. Kamiya, J. Non-Cryst. Solids **48**, p. 31 (1982).
11. C.J. Brinker, D.K. Keefer, D.W. Schaefer, R.A. Assink, B.D. Kay and C.S. Ashley, J. Non-Cryst. Solids **63**, p. 45 (1984).
12. C.J. Brinker and G.W. Scherer, J. Non-Cryst. Solids **70**, p. 301 (1985).
13. A.B. Brennan and G.L. Wilkes, Polymer **32**, p. 733 (1990).
14. H.H. Huang, B. Orler and G.L. Wilkes, Macromolecules **20**, p. 1322 (1987).
15. T. Saegusa, J. Macromol. Sci. -Chem. **A28**, p. 817 (1991).
16. H.H. Huang, R.H. Glaser and G.L. Wilkes, Inorganic and Organometallic Polymers, Edited by M. Zeldin, K.J. Wynne and H.R. Allcodk (ACS Symp. Ser 360 , 1987) p.354.
17. G.L. Wilkes, A.B. Brennan, H.H. Huang, D.E. Rodrigues and B.W. Wang, Mater. Res. Soc. Symp. Proc. 15, p. 171 (1990).
18. A.B. Brennan, T.M. Miller and R.B. Vinocur, Hybrid Organic-Inorganic Composites, edited by J.E. Mark, C.Y-C. Lee and P.A. Bianconi (ACS Symp. Series 585, 1995) p. 142.
19. A.B. Brennan and T.M. Miller, Chem. Mater **6**, p. 262 (1994).
20. R.H. Glaser, G.L. Wilkes and C.E. Bronniman, J. Non-Cryst. Solids **113**, p. 73 (1989).
21. M. Toki, T,Y. Chow, T. Ohnaka, H. Samura and T. Saegusa, Poly. Bull. **29**, p. 653 (1992).

PREPARATION OF POLYIMIDE-SILICA HYBRID MATERIALS BY HIGH PRESSURE-THERMAL POLYMERIZATION

Kevin Gaw, Hironori Suzuki, Mitsutoshi Jikei, Masa-aki Kakimoto*, and Yoshio Imai
Department of Organic and Polymeric Materials, Tokyo Institute of Technology,
Meguro-ku,Tokyo 152, Japan

Abstract

Polyimide-silica hybrid materials were prepared via a modified sol-gel, high pressure-thermal polymerization procedure. Precursor monomer salts were made from ethanol soluble 2,5-diethoxycarboxyl terephthalic acid (p-PME) and either a disiloxanediamine, an aliphatic diamine (1,9 diaminononane) or combinations of the two. Solutions of tetramethoxysilane (TMOS) and monomer salt were transformed into a gel, dehydrated, and the resulting powders were subjected to high pressure and thermal polymerization and transformed into a polyimide-silica composite. By varying the TMOS content, and/or the siloxane to aliphatic diamine ratio, composites of 0 to 100 wt% SiO_2 were made. The silica morphology changed significantly with siloxane/aliphatic PI ratio. Reaction mechanisms, thermal and physical properties and composite morphologies are discussed.

Introduction

Molecular hybrids formed by a variety of sol-gel reactions between organic polymers and inorganic gels are expected to make composite systems having unique thermal, electrical, and physical properties that are intermediate to the plastic and glass components [1,2]. The sol-gel method is used to make molecularly homogeneous, monolithic bodies at low processing temperatures that allows for uniform composite microstructures to form via covalent bonds between the organic and an inorganic phase eliminating macroscopic phase boundaries[3,4].

Previously, we examined the interactions of aliphatic diamine polyimides and sol-gel derived silica synthesized under high pressure[5]. In the current process, the monomer salt undergoes solid state polycondensation with the evolution and elimination of water and methanol[6] and th siliceous component is transformed from a hydroxide to a glass (Fig. 1a -b).

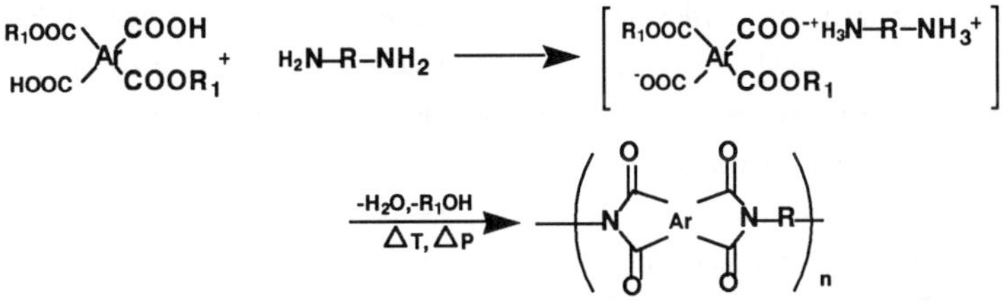

Figure 1a. Polyimide formation from its precursor monomer salt

Mat. Res. Soc. Symp. Proc. Vol. 435

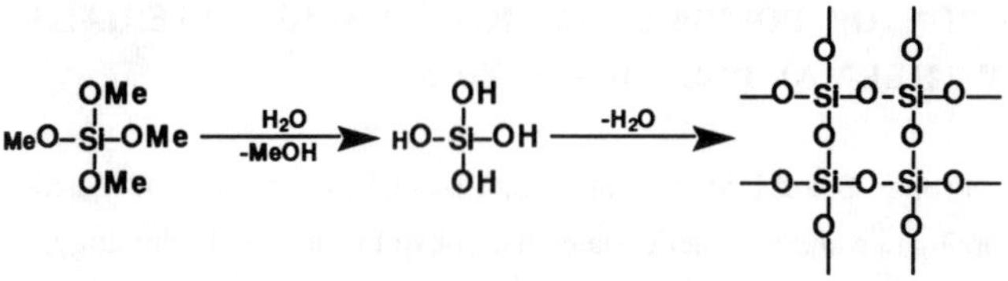

Figure 1b. Sol-gel reaction of tetramethoxysilane (TMOS)

By changing the siloxane/aliphatic amine ratio and TMOS content in the precursor solution, 0 to 100 wt% SiO_2 hybrids were made. In this report, we present the preparation of unique polyimide-silica hybrids by a modified high pressure sol-gel reaction of a metal alkoxide and a monomer salt. The thermal, physical and chemical properties of the composites were determined via thermal analysis (TGA, DTA, and DSC) and micro-hardness measurements. Sol-gel reactions were characterized by FTIR and ^{29}Si NMR measurements. Scanning electron micrographs are used to establish structure-property relationships.

Experimental

Materials and Methods

Preparation of the Monomer Salt - To begin, the pyromellitic diethylester diacid (PME) used for the preparation of the monomer salt was synthesized. A mixture of PMDA (21.81 g, 0.1 mol) in ethanol (500 mL) was heated to 80°C for 15 h , affording a PME solution. After solvent removal, the residue was suspended in ethyl acetate (500 mL) and the soluble material (2,4-diethoxycarboxyl isophthalic acid (m-PME)) was discarded and the insoluble material, 2,5-diethoxycarboxyl terephthalic acid (p-PME) was used in the monomer salt preparation[7]. p-PME was purified with a yield of 14.5 g (46.7 %) by recrystallization from butyl acetate. p-PME (12.41 g, 40 mmol) was dissolved in ethanol (300 mL) under a nitrogen atmosphere and an equimolar amount of either 1,9-diaminononane and/or siloxane diamine with stirring at 60°C was added. After cooling the solution to 0°C, the monomeric salt precipitated was collected by filtration and vacuum dried (yeilds>95%). Structures of reactants in Fig. 1 are given below.

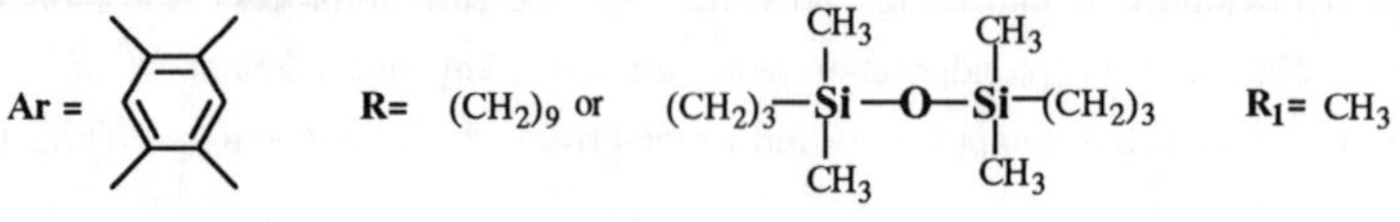

Hybrid precursor powders were made by adding a measure of TMOS to the solution of the monomeric salt with excess water added and allowed to react for 4 hours at r.t. The solvent was then removed under reduced pressure yeilding white precursor powders.

High Pressure-Thermal Polymerization of the Polyimide and Hybrids - The polymerization procedure varied slightly from that used previously described with the precursor powders being compacted at 550MPa for 5 minutes at room temperature followed by thermal polymerization at 200°C for 5 h. The polymer, in the form of a dense, void free, cylindrical pellet was then dried under vacuum at 80 °C to remove the reaction condensates.

Polyimides will be signified as the ratio of siloxane to aliphatic diamine, X/Y. 30/70 means a polyimide containing 30% siloxane amine and 70% aliphatic amine. Composites will also be written as ratios, X/Y/Z, where X and Y are the same as above, and Z is the SiO_2 weight percent (calculated from TMOS) of the total weight of the composite.

Measurements - IR spectra were recorded on a Shimadzu FTIR-8100 spectrophotometer. Differential thermal analysis (DTA), differential scanning calorimetry (DSC), and thermogravimetry analysis (TGA) were performed with Shimadzu thermal analyzers DTA-40, DSC-41, and TGA-40, respectively. Vickers hardness was measured using a Matsuzawaseiki MHT2 (50 g, 15 sec). SEM photographs were taken with a Nippondenshi T-220.

Results and Discussion

FTIR

Monomer salts and the composite precursor powders - IR spectra are shown in Figure 2 (curve a-c). Salt spectra (a-c) show characteristic bands due to the ammonium ion at 2127 and 1632 cm^{-1}, the carboxylate ion at 1564 cm^{-1} and the carboxylic absorption at 1730 cm^{-1} indicating the formation of the monomer salt. For all X>0 samples (1b-1d), there is a characteristic peak at 1250 cm^{-1} indicating the presence of $Si\text{-}CH_3$ linkages.

Polyimide - IR spectrum of polymerized 50/50/50 is shown in Figure 2d. While the absorption bands due to the monomer salt disappeared, bands of the imide ring at 1772, 1711, 1394, and 727 cm^{-1} appeared confirming the formation of polyimide. A broad band associated with SiO_2 complexes is present at 1100 cm^{-1} in all Z>0 samples.

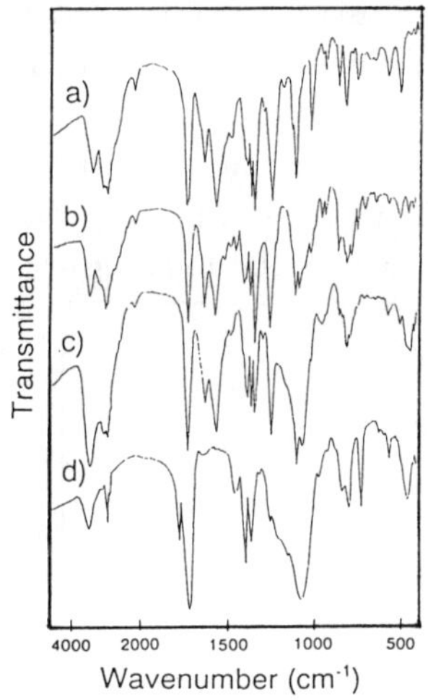

Figure 2 FTIR of monomer salts a) 0/100 aliphatic salt, b) 100/0 siloxane salt, c) precursor 50/50/50, d) polymerized 50/50/50.

From the DTA and TGA curves, the rapid one step imidization of X>0 salts had a split peak at 165°C and 175°C resulting from the partial melting and then polymerization of the molten salt. The weight loss was 23%. The 0/100 monomer salt shows a single, sharp endothermic peak, indicating solid state polymerization, and a stepwise weight loss of 27.0 % at about 178°C . These weight losses closely agree with the calculated values of 27.3 % and 22.9%, respectfully, based on the loss of two molar equivalents of water and ethanol during polymerization (**Figure 3**).

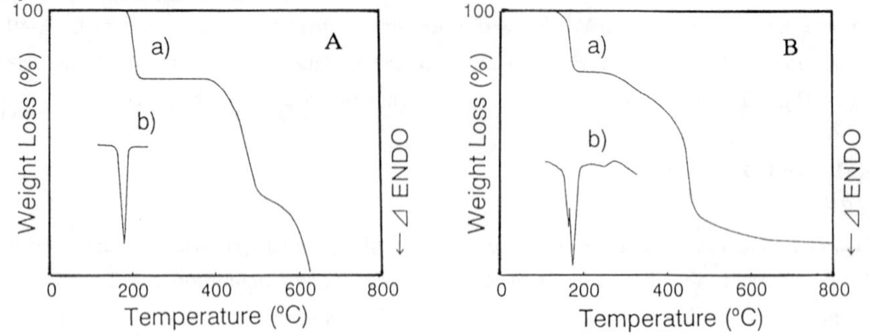

Figure 3 TGA (a) and DTA (b) curves of A) 0/100 B) 100/0 salts. (heating rate = 10°C min.$^{-1}$, in air)

From DSC (not shown) and TGA, the $PI_{0/100}$ had a melting temperature of 310°C but no T_g, began to decompose at 350°C and was completely oxidized by 700°C. $PI_{100/0}$ showed no T_m, decomposed at 260°C and had significantly more ash (11.3%). This amount is about one half the theoretically calculated 22% if all the Si in the polymer was transformed into SiO_2. The inherent viscosity of the $PI_{0/100}$ was 1.30 dL g-1, measured at a concentration of 0.5 g dL^{-1} in H_2SO_4 at 30°C. $PI_{100/0}$ had a lower viscosity and degradation point than $PI_{0/100}$ indicating that it was of lower molecular weight. TGA of the 50/50/50 precursor (4a) shows a weight loss starting at 140°C due to hydroxyl-SiO_2 transformation and ending at 190°C upon polyimide formation. The degradation temperature of the mixed polymer was around 300°C while the silica phase was stable up to the maximum analysis temperature of 800°C. An explanation of the seemingly ambiguous 63% ash content of the 50/50/50 composite is given below.

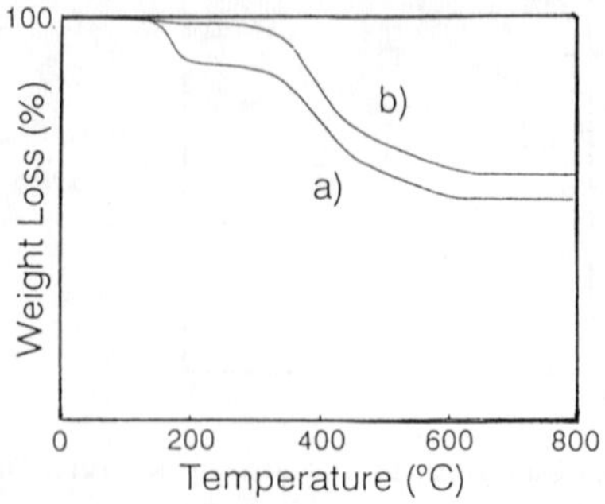

FIGURE 4 TGA a) precursor 50/50/50 and b) polymerized 50/50/50 (heating rate = 10°C min.$^{-1}$, in air)

After burnout, all the X>0, Z>0 composites had final SiO_2 content that was greater than expected from TMOS contributions alone due to the transformation of salt siloxane units into silica or siliceous carbon. This generation of silica or homologues of silica from the polyimide occurs only when TMOS silica is already present. This is explained in the next section. SiO_2 content did not effect the polymer decomposition temperatures.

Reaction Mechanisms

According to widely accepted reaction schemes for the hydrolysis of organic siloxanes, dehydration polymerization follows the reaction sequence: $Si(OR)_4 + xH_2O \rightarrow Si(OR)_{4-x}(OH)_x + xROH$, with the reaction then proceeding with the extension of the silica network by hydrogen bonding or by dehydration again to, $(OH)_{x-1}(OR)_{4-x}$-Si-O-Si-$(OR)_{4-x}(OH)_{x-1}$ $+H_2O$ and finally with the removal of water, to SiO_2, where R is a methyl group[4]. This reaction produced the hydroxide-rich silica-monomer salt precursor powders used here.

The FTIR trace of polymerized 50/50/50 (Figure 2d), shows a broad hydroxide band at 1100 cm^{-1} indicating the incomplete conversion of organic Si to SiO_2. Since the Si-OH and SiO_2 bands overlap in theh FTIR, ^{29}Si-NMR is needed to determine a more detailed Si reaction description. Under high pressure polymerization conditions, the silica hydroxides not only come into intimate contact with one another and fuse but they were also in contact with the polyimide salt and formed interfacial bonds.

It has been noted that organosilaxane derived silicon covalently bonds to the carbon of polymers creating a Si-O-C bond with an absorption at 802 cm^{-1}. This band is seen in all of theX>0, Z>0 composites. Thus, the above reaction scheme can be slightly altered for X>0 salts, where not only Si_{silica}-O-Si_{silica} bonds form but also Si_{silica}-O-$Si_{siloxaneimide}$ bonds[8]. ^{29}Si-NMR shows that there is actually a cleaving of the siloxane bond within the polymer chain resulting in the incorporation of the polymer fragments into the growing silicon network. Thus, if there were more Si present, or more siloxane linkages in the polymer, molecular interaction would increase and a more uniform silica/polyimide material with no phase boundaries would form. This is indirectly confirmed with the SEM.

^{29}Si-NMR Studies

The ^{29}Si-NMR plots (figure 5) show that the Si-O-Si bonds in the salt are cleaved via disproportionation under the acidic conditions of our system. When X>0,Y=0 and Z=0 , there is a single peak located at -6.97pmm. When X>0 and Z>0 negatively shifted peaks centered around -11-13ppm appear indicating a local environment change for the silicon in the polymer. These two peaks either as distinct entities or as a large peak with a shoulder are present only in X>0, Z>0 samples and are attributed to the acid catalyzed disproportionation of the Si-O-Si bonds in the siloxane amine to form a silanol, which reacts with the silica network (Figure 1b). Thus, Pi_{silica} fragments are incorporated into the silica network.

The silica-hydroxide peaks at 100 ppm follow the progress of the inorganic precursor conversion to silica. The three peaks present at 110, 100, and 90 represent singly, doubly and triply hydroxyl substituted silica, respectively. The peak at 110 is largest after the polymerization process and the other peaks diminish with reaction progress.

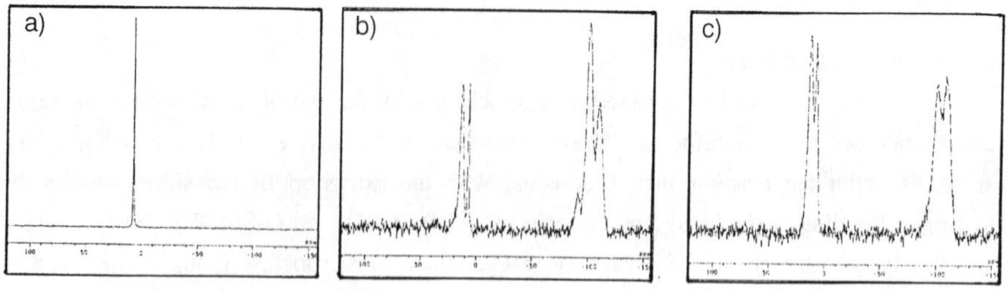

Figure 5 ^{29}Si-NMR of a) precursor 100/0 salt b) precursor 100/0/50 and c) polymerized 100/0/50

Physical Properties

Hardness

Evidence for the establishment of an interconnected silica phase, can be seen in the Vickers hardness data for the 0/100/Z composites (**Figure 6**). The hardness of the hybrids, an indicator of silica monolithicity, increases in two stages as the SiO_2 content increases. When Z<50, the hardness values are similar to the polyimide matrix. Whereas, when Z>50, the silica becomes a continuous matrix and cannot be deformed easily rapidly increasing hardness. Hence, the hardness of the composites is a soft polyimide continuous phase dominated regime (Z<50) and a continuous silica phase dominated one (Z>50). In X>0 composites, hardness increases with X. Thus, the polyimide increases network, monolithicity and brings about the change in hardness at lower Z values.

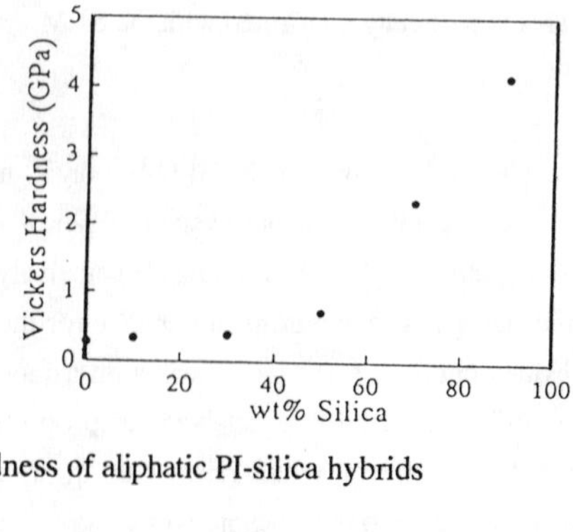

Figure 6 Vickers hardness of aliphatic PI-silica hybrids

Morphology

SEM of the X/0/Z composites show no phase separation of the silica from the polyimide indicating the samples formed nanophase domains of silica and polyimide. In the X/Y/Z samples where Y>0, Z>0 this is not seen. In these specimens, due to the presence of the incompatible aliphatic chain, the silica phase separates and forms 0.1-1mm spherical domains that increase in size with Y.

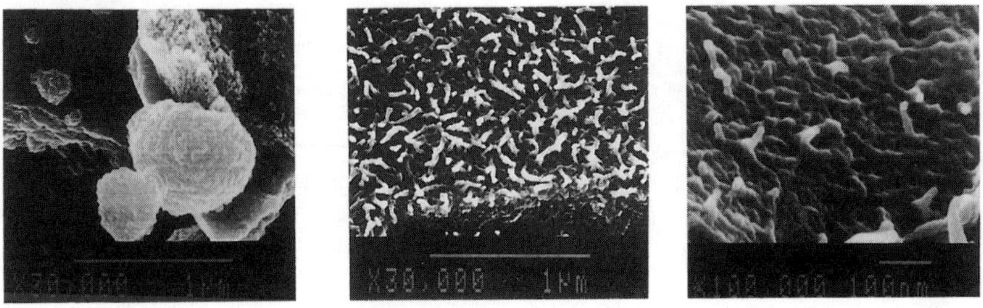

(a) (b) (c)

Figure 7 SEM of composites a) 0/100/50, mag. 30000x, b) 50/50/50, mag. 30000x, c) 0/100/30 mag. 100000x

Conclusion

We prepared novel hybrid materials incorporating polyimide and silica via a modified high pressure sol-gel polymerization process. Aliphatic and siloxane containing diamines, and p-PME were mixed with TMOS in solution, dehydrated and polymerized. Depending on polyimide composition and the amount of TMOS used, the SiO_2 content ranged from 0-100 wt%. These materials were characterized using TGA, DSC, FTIR ^{29}Si-NMR and SEM. The hardness of the hybrids showed two distinct regimes. The incorporation of siloxane amine into the polyimide backbone homogenized the composites. The polymerization process also concurrently completed the condensation reactions of both the inorganic and the organic phase resulting in completely reacted hybrid materials.

References

1. B.Wang , J.C.Hedric, J.E.McGrath, Macromolecules, **24**, 3449 (1991)
2. A.Morikawa, Y.Iyoku, M.Kakimoto, Y.Imai, Poly. J., **24**, 107 (1992)
3. Y.Wai, D.Yang, L.Tang, M.K.Hutchins, J. Mater. Res., **8**, 1151 (1993)
4. H.Schmidt, H.Wolter, J. Non-Cryst. Sol., **121**, 428 (1990)
5. K. Gaw, H. Suzuki, , M. Kakimoto, J. of Photopoly. Sci. and Tech., **8**, 2, 144 (1995)
6. C.D.Papaspyrides, Polymer, **31**, 490 (1990)
7. N.C.Stoffel, E.J.Kramer, W.Wolksen, T.P.Russell, Polymer, **34**, **21**, 4524 (1993)
8. E. Colvin, Silicon in Organic Synthesis (Butterworths Publ., London, 1981), p. 54

POLY(METHYL METHACRYLATE)-TITANIA HYBRID MATERIALS BY SOL-GEL PROCESSING

JUN ZHANG, SHENGCHENG LUO, LINLIN GUI and YOUQI TANG
Institute of Physical Chemistry, Peking University, Beijing 100871, P.R.China

ABSTRACT

Sol-gel derived Poly(methyl methacrylate)-titania hybrid materials were synthesized by using acrylic acid or allyl acetylacetone (3-allyl-2,4-pentanedione) as coupling agents. Titanium butoxide modified with acrylic acid (or titanium isopropoxide modified with allyl acetylacetone) was hydrolyzed to produce a titania network, and then poly (methyl methacrylate) (PMMA) chains formed *in situ* through a radical polymerization were chemically bonded to the forming titania network to synthesize a hybrid material. Transparent hybrid materials with different contents of titania were achieved. With the increase of the titania content, the colors of the products changed from yellow to dark red. The synthesis process was investigated step by step by using FTIR spectroscopy, and the experimental results demonstrated that acrylate or acetylacetonato groups bound to titanium remain in the final hybrid materials. The thermal stability of the hybrid materials was considerably improved relative to pure PMMA. Field emission scanning electron microscopy (FE-SEM) analyses showed the hybrid materials are porous and pore diameters vary from 10nm to 100nm. The hybrid materials using allyl acetylacetone as the coupling agent exhibited thermochromic effects that both pure PMMA and titania do not have.

INTRODUCTION

Recently, there has been increasing interest in the development of organic/inorganic hybrid materials, which exhibit some characteristics and properties of both organic polymers and ceramics[1-9]. They are mostly produced by a sol-gel process, which can synthesize the inorganic network by wet chemical routes. And the low processing temperature allows one to prepare organic/inorganic hybrid materials.

Carbon family elements such as Si, Sn and Pb can usually be chemically linked to the organic component through M-C covalent bonds(M= Si, Sn or Pb), which are chemically stable and do not undergo hydrolysis during the sol-gel process. Transition metals such as Ti, Zr, whose M-C covalent bonds(M= Ti, Zr et al) are not stable in the chemical condition of the sol-gel process, can be bonded to the organic component through coordinative or ionic bonds[1], and most of them can maintain their structure during the hydrolysis and condensation. In the present work, we prepared organic/inorganic hybrid materials by copolymerising methyl methacrylate with titanium alkoxides modified with acrylic

Mat. Res. Soc. Symp. Proc. Vol. 435

acid or allyl acetylacetone. The synthesis process was investigated by FTIR spectroscopy, and the final hybrid materials were characterized by thermogravimetric analysis (TGA), field emission scanning electron microscopy (FE-SEM) and UV-VIS spectroscopy.

EXPERIMENT

Preparation of Hybrid Materials

Acrylic acid and allyl acetylacetone were chosen as coupling agents to connect PMMA chains and the titania network. Allyl acetylacetone was synthesized according to[11]. Using acrylic acid as the coupling agent, the starting metal alkoxide was titanium butoxide. When allyl acetylacetone was used as the coupling agent, the starting metal alkoxide was titanium isopropoxide, because it is difficult to obtain transparent hybrid materials using titanium butoxide under these conditions. In a typical formulation, titanium butoxide was mixed with acrylic acid (or titanium isopropoxide was mixed with allyl acetylacetone) in a molar ratio of 1:2. After standing at room temperature for half an hour, the mixture was hydrolyzed using a solution of water, ethanol and nitrate. Molar ratios of these to titanium were 2, 7 and 0.08. After 24 hours, the organic monomer, methyl methacrylate, was added along with the small amount of benzoyl peroxide (BPO). After half an hour, glass tubes were filled with the solution and sealed. The tubes were heated to 60°C . Usually after a few days, the content of the tubes gelled and heating was continued for about one week. Then the caps on the tubes were slightly loosened to allow the solvent to evaporate slowly. The samples were dried at 60°c for one more week. The synthesis procedure is illustrated in Figure 1.

Characterization

FTIR spectra were obtained using a Nicolet Magna-IR™ 750 spectrometer. Thermogravimetric analyses on the samples were performed under argon atmosphere with a DuPont 1090 System at a heating rate of 15°C/min. Optical transmission was measured on a General TU-1221 UV-VIS spectrometer. Field emission scanning electron microscopy (FE SEM) was carried out using a AMRY 1910FE microscope.

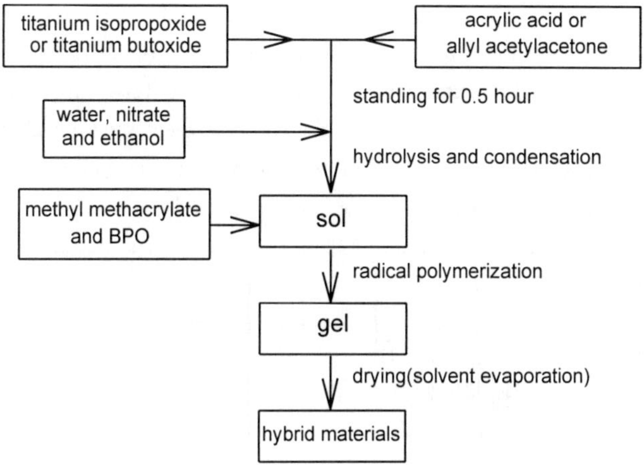

Figure 1. Sol-gel process for the production of PMMA-titania hybrid materials

RESULTS AND DISCUSSION

Figure 2A gives the FTIR spectrum of the products of $Ti(OBu^n)_4$ reacting with acrylic acid. The broad band at 3334cm^{-1} is assigned to stretching vibrations of OH groups. Sharp bands around 2873-2959 cm^{-1} correspond to the stretching vibrations of aliphatic CH_2 and CH_3 groups. The sharp bands at 1730cm^{-1} and 1637cm^{-1} are assigned to C=0 and C=C stretching vibrations respectively. The bands at 1529-1562cm^{-1} correspond to the acrylate groups bonded to titanium[12]. They could be assigned to the asymmetric stretching vibration of the carboxyl group. The symmetric stretching vibration of the carboxyl group is around 1441cm^{-1} and might be superposed with bending vibrations of aliphatic CH_2 groups. Figure 2B shows the FTIR spectrum of the final hybrid materials using acrylic acid as the coupling agent. The bands in the 1400-1600cm^{-1} range characteristic of acrylate groups bonded to titanium still remain intense. This indicates that acrylate groups remain bound to titanium during the synthesis process and in the final hybrid materials.

Figure 3A shows the FTIR spectrum of the products of $Ti(OPr^i)_4$ reacting with allyl acetylacetone. The broad band at 3423cm^{-1} is assigned to i-PrOH formed by the substitution of an O-Pr^i group in $Ti(OPr^i)_4$ by allyl acetylacetone[13]. The sharp bands around 2863-2968cm^{-1} and 1357-1464cm^{-1}

correspond to the stretching and bending vibrations of aliphatic CH_2 and CH_3 groups respectively. The bands around 1718cm^{-1} could be attributed to ν (C=O) vibration of free allyl acetylacetone (keto-forrn). The intense bands around 1580-1602cm^{-1} could be assigned to ν (C-O) and ν (C-C) vibrations of acetylacetonato groups bonded to titanium (enol form)[13]. The FTIR spectrum (Figure 3B) of the final hybrid materials using allyl acetylacetone as the coupling agent shows characteristic bands of acetylacetonato groups bonded to titanium in 1500-1600cm range. This indicates that acetylacetonato groups remain bound to titanium during the synthesis process.

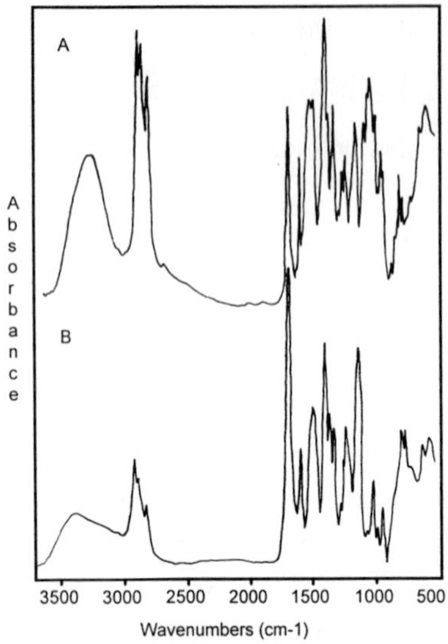

Figure 2. FTIR spectra of the samples using acrylic acid as coupling agent
A) Titanium butoxide-acrylic acid
B) The final hybrid materials

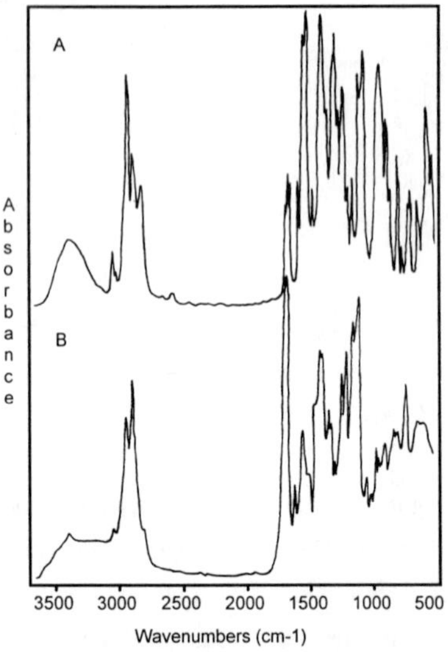

Figure 3. FTIR spectra of the samples using allyl acetylacetone as coupling agent
A) Titanium isopropoxide-allyl acetylacetone
B) The final hybrid materials

Transparent PMMA-titania hybrid materials with titania contents ranging from 6.4% to 53.6% by weight have been prepared when acrylic acid was used as the coupling agent. Transparent PMMA-titania hybrid materials with titania contents ranging from 15.7% to 38.1% have been prepared when allyl acetylacetone was used as the coupling agent. The TGA curves of the two systems of PMMA-titania hybrid materials and pure PMMA are shown in Figure 4. The hybrid materials and pure PMMA begin to lose weight around 200°C, but the pyrolysis temperatures of organic components in hybrid materials, which are around 400°C, are higher than that of PMMA.

The microstructures of the PMMA-titania hybrid materials were characterized by FE-SEM. All the hybrid materials are porous. The microstructures of hybrid materials are different when different compounds are used as the coupling agent. With similar titania contents, the hybrid material using acrylic acid as the coupling agent has a more homogenous distribution of pores and smaller pore sizes. The results indicate that in the hybrid materials using acrylic acid as the coupling agent organic and inorganic components have a higher degree of interpenetration.

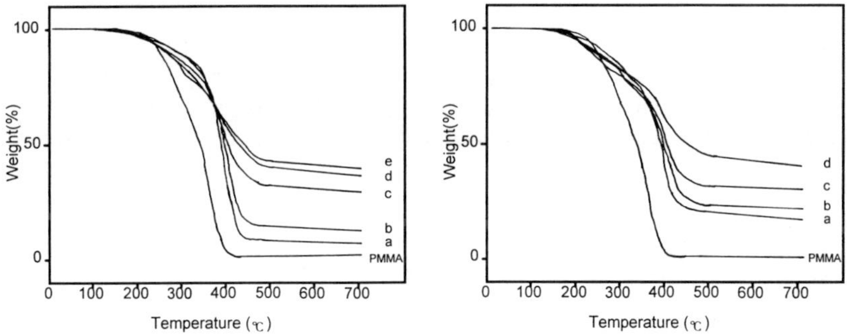

A) Using acrylic acid as coupling agent.
Titania contents(%): a)6.4 b)11.3 c)28.4 d)36.8 e)39.3

B) Using allyl acetylacetone as coupling agent.
Titania contents(%): a) 15.7 b)21.6 c)29.9 d)38.1

Figure 4. TGA curves of the PMMA-titania hybrid materials and PMMA
*Titania content is the content of TiO_2 in the final hybrid materials.

The hybrid materials are transparent and are colored. With the increase of titania content , the colors of the hybrid materials vary from yellow to dark red. The hybrid materials using allyl acetylacetone as the coupling agent show thermochromic effects that both PMMA and titania do not have. The hybrid materials are red and transparent at room temperature and change to yellow and opaque when they are cooled. The reverse transformation is observed as the temperature is raised.

CONCLUSIONS

Using acrylic acid as the coupling agent, transparent PMMA-titania hybrid materials with the titania contents ranging from 6.4 to 53.6% by weight have been prepared. Using allyl acetylacetone as the coupling agent, transparent PMMA-titania hybrid materials with titania contents ranging from 15.7 to 38.1% have been prepared. FTIR spectroscopy result demonstrated that acrylic acid or allyl acetylacetone can introduce efficient chemical bonding between PMMA chains and titania network in a sol-gel process. The thermal stability of the hybrid materials was considerably improved relative to pure PMMA. All the hybrid material were porous. Using acrylic acid as the coupling agent, the hybrid materials had more homogenous distribution of pores and smaller pore sizes. The hybrid materials using allyl acetylacetone as the coupling agent exhibited thermochromic effects that both PMMA and titania do not have.

REFERENCES

1. H. K. Schmidt, in Inorganic and Organometallic Polymers with Special Properties, edited by R. M. Laine (Kluwer Academic Publisher, 1992) p.297

2. H. Schmidt and G. Philipp, J. Non-Cryst. Solids **63**, p. 283 (1984)

3. J. Hu and J. D. Mackenzie, in Better Ceramics Through Chemistry, edited by M. J. Hampden-Smith, W. G. Klemperer and C. J. Brinker (Mater. Res. Soc. Proc. 271, Pittsburgh, PA, 1992) p. 681

4. H. Huang, B. Orler and G. L. Wilkes, Polym. Bull. **14**, p. 557 (1985)

5. H. Huang, B. Orler and G. L. Wilkes, Macromolecules **20**, p. 1322 (1987)

6. B. Wang, G. L. Wilkes, C. D. Smith and J. E. McGrath, Polymer Commun. **32**, p. 400 (1991)

7. M. Nandi, J. A. Conklin, L. Salvati Jr. and A. Sen, Chem. Mater. **3**, p. 2021 (1991)

8. R. H. Glaster and G. L. Wilkes, Polym. Bull. **22**, p. 527 (1989)

9. G. Philipp and H. Schmidt, J. Non-Cryst. Solids **82**, p. 31 (1986)

10. B. M. Novak, Adv. Mater. **5**, p. 422 (1993)

11. R. B. Davis and P. Hurd, J. Am. Chem. Soc. **77**, p. 3284 (1955)

12. K. Nakamoto, Infrared and Raman Spectra of Inorganic and Coordination Compounds, 3rd edition, (Wiley, New York, 1978)

13. A. Leaustic, F. Babonneau and J. Livage, Chem. Mater. **1**, p. 240 (1989)

A STUDY OF THE MOBILITY OF POLY(METHYL METHACRYLATE)-SILICATE INTERPENETRATING NETWORKS WITH ^{2}H NMR

CLARE P. GREY,** KENNETH G. SHARP*
*DuPont, CR&D, Wilmington, DE 19880-0323
**Department of Chemistry, SUNY Stony Brook, Stony Brook, NY 11794-3400

ABSTRACT

Poly(methyl methacrylate) (PMMA)/silicate interpenetrating networks (IPNs) have been synthesized in formic acid solutions. The fast gelation times observed in these solutions reduced the phase separation of the polymer, and the 50% by weight PMMA/silicate hybrid showed virtually no evidence of a glass transition (T_g) from thermomechanical data. ^{2}H NMR showed that the PMMA is extremely mobile in the wet gels, and liquid-like ^{2}H NMR spectra are observed even after prolonged aging of the samples. A ^{2}H static NMR lineshape, indicative of the rigid polymer, does not occur until polymer to solvent ratios of approximately 0.2 (weight basis) are attained. Motional narrowing of the ^{2}H resonances is not observed in the vacuum-dried IPNs below 180 °C. In contrast, pure PMMA shows motional narrowing between 150 and 180 °C. There was little evidence of isotropic motion of the polymer chains, for low polymer concentrations, up to the highest temperature studied (200 °C). The percentage of polymer undergoing isotropic motion increased with polymer content, and as a function of the aging time before solvent stripping.

INTRODUCTION

Interpenetrating networks comprising organic and inorganic polymers have generated considerable interest due to the possibility of synthesizing materials that combine the flexibility and toughness of a polymer with the thermal stability, hardness and strength of a ceramic. We have recently devised a new non-aqueous method of synthesizing IPNs in formic acid solutions.[1,2] Formic acid is a good solvent for several classes of polymers including polyamides and methacrylates, and the very small pore sizes of the gels synthesized in the absence of the polymer suggest that these gels may be very effective at trapping the polymer in the IPN. The gelation times are considerably faster than those in comparable aqueous solutions, reducing the tendency for phase separation of the polymer, even at very high polymer concentrations.[1,2]

Excellent mechanical properties were reported for the polymer/silicate IPNs generated in formic acid solutions.[1] A ^{2}H NMR study was, therefore, commenced, in order to probe the local mobility of the polymer, and to assess the effect of the confinement of the polymer in the SiO_2 network on its mobility in both wet and dried gels quantitatively. The PMMA-d_8 IPNs were chosen for investigation, since the variable temperature ^{2}H NMR spectra of bulk PMMA are relatively well understood. ^{2}H NMR was used as a probe of mobility, because the lineshape and width of the ^{2}H resonance can be used to analyze the motion of the deuterons.[3] Changes in the ^{2}H lineshape are observed when the correlation times of the polymer motion, τ_c, are of the same order of magnitude as the ^{2}H quadrupolar frequency $[(1/\tau_c)^2 \approx (\nu_Q)^2$, where ν_Q, the quadrupolar frequency is a function of the quadrupole coupling constant: $\nu_Q = (3/4)QCC]$. This is the so-called intermediate regime and occurs (for ^{2}H) when $10^{-5}<\tau_c<10^{-7}$s, for typical values of ν_Q.[3] When τ_c decreases further $[(1/\tau_c)^2>>(\nu_Q{}^2)$; $\tau_c >10^{-7}$s] the motion is said to be in the fast-regime, and no further changes in lineshape with increased mobility are observed.

The structure of PMMA-d_8 is shown below:

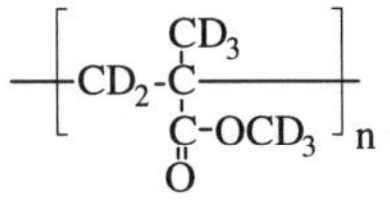

Mat. Res. Soc. Symp. Proc. Vol. 435 © 1996 Materials Research Society

Two distinct Pake doublets from the CD_3 and CD_2 groups are seen at temperatures below T_g.[4] The CD_3 groups rotate rapidly about the C-C bond axis, and a reduced QCC (of approximately 53 kHz) from that of a rigid C-D group (167 kHz) is observed.[4-5] Even above the nominal (low frequency) T_g, the chain motion of bulk PMMA is not sufficiently fast to result in complete averaging of the QCC, although a reduction is observed, which is characteristic of small angular fluctuations of the polymer-backbone and side chains. Within the relatively narrow temperature range of 150 to 180 °C, the Pake doublets from the deuterated backbone (CD_2) and methyl groups of the rigid polymer collapse into to a motionally-narrowed Lorentzian lineshape.[4]

Kodak researchers have prepared PMMA/silicate IPNs in aqueous solutions, with both acid and base-catalyzed silicate condensation reactions.[6] The PMMA/silicate IPNs prepared under acidic conditions were found to exhibit a rubber plateau in the tensile modulus, E', above T_g, which extends to at least 250 °C, if a critical concentration of silica of 16% by weight is used. In contrast, no plateau was observed for the pure polymer or for samples prepared under alkaline conditions.[6] A substantial amount of PMMA remained trapped in the silica matrix after acetonitrile was used to extract the polymer from the samples prepared in acid conditions. PMMA/silicate IPNs have also been synthesized by polymerizing the MMA monomer in situ, in the previously formed SiO_2 host.[7-10] These materials show increased hardness in comparison to pure PMMA,[8,9] but the T_g of the PMMA in some of these composites is lower than that of the bulk polymer.[7,8]

The results of two series of ^2H experiments are presented in this paper. In the first, the immobilization of the polymer in the gels was studied as a function of the solvent content of the wet gel. In the second, the mobility of the PMMA, as a function of the polymer/SiO_2 ratio, the stripping of the solvent, and the aging of the SiO_2 network, was explored at temperatures around the glass transition. The results are compared to thermomechanical measurements for these IPNs.

EXPERIMENTAL

PMMA-d_8 synthesized at DuPont was shown by gel permeation chromatography to have a number average molecular weight of ≈ 835,000. Samples were prepared from separate solutions of PMMA-d_8, dissolved in formic acid, and tetraorthosilicate (TEOS), in the ratios listed in Table I. The TEOS solution was slowly combined with the polymer solution, while stirring to prevent agglomeration of the polymer. After stirring for additional few minutes, the solution was left to gel. The gel was aged for variable time periods, and the solvent was then removed at 60 °C on a vacuum line. Samples are labeled according to their composition (by weight), calculated assuming complete conversion of the TEOS to SiO_2. Actual inorganic content will be somewhat higher due to unreacted SiOEt groups. Samples for the thermomechanical measurements were prepared as cylindrical disks, allowing the wet gels to dry slowly over several weeks at atmospheric pressure.

Table I. The composition of the IPNs. Ratios are given on a weight basis.

Sample	**PMMA/ Total Solids**	**PMMA/HCO_2H**	**HCO_2H/TEOS**
17% PMMA	0.17	0.0795	0.69
26% PMMA	0.26	0.0783	1.19
50% PMMA(a)	0.50	0.0783	3.33
50% PMMA(b)	0.50	0.0783	3.33
70% PMMA	0.70	0.0783	7.94

RESULTS

Thermomechanical Measurements: Unlike the sample of pure PMMA, the PMMA/silicate IPNs show only a small dimensional change above the T_g of pure PMMA, up to the highest temperature studied (200°C). The change observed above T_g decreases with increased SiO_2 content, and at ratios of PMMA/SiO_2 of approximately 1:1 is barely discernible (Figure 1).

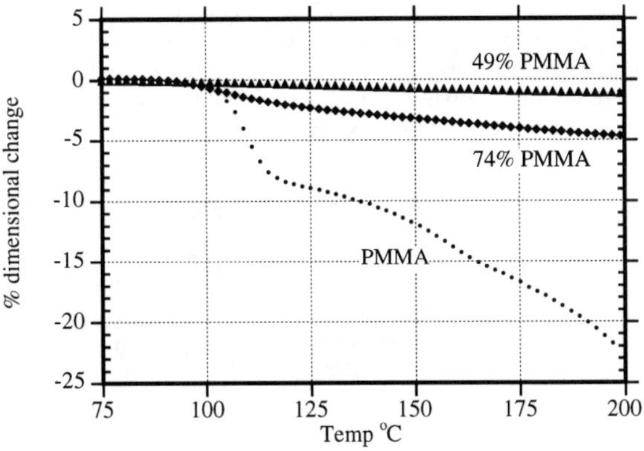

Figure 1. Softening behavior for PMMA and the PMMA/silica IPNs.

^2H NMR: Solvent stripping: A 'liquid-like' spectrum is observed in ^{2}H NMR spectrum of the 70% PMMA sample just after the gel point. The sealed sample was allowed to age for up to four months and was monitored periodically. The spectrum remained liquid-like throughout the entire period; gel-shrinkage did not result in a decrease in polymer mobility. The ^{2}H resonances were observed to broaden, indicating a reduced mobility of the polymer, only when the solvent was allowed to evaporate from the gel. Not until 58% of the solvent is removed, however, was there any evidence of immobilized polymer. This corresponds to a ratio (on a weight basis) of PMMA to solvent of 0.21. At this point a broad resonance from rigid CD_3 groups became visible, in addition to the narrower resonance of the mobile polymer. When 70% of the solvent is removed, virtually no mobile polymer remains, and the Pake doublet from the CD_3 resonances is clearly observed. A weight loss of 88% was obtained after drying at 60 °C under vacuum, at the end of the experiment. Note that a weight loss of 90% corresponds to removal of all solvent from the sample, and a complete conversion of the silicate network to SiO_2. The experiment was repeated for different polymer/SiO_2 ratios, (26 and 17%) and similar spectra were observed. The polymer remained mobile in the wet gels, and was not trapped by the SiO_2 networks prior to drying, even for higher SiO_2/PMMA ratios. On solvent removal, rigid polymer was observed at similar PMMA/solvent concentrations (0.24 and 0.21 for the 17 and 26% samples, respectively). These differences are not significant within the errors associated with the experiment. These samples were then completely stripped of solvent (under vacuum) for the variable temperature studies.

Temperature: ^{2}H NMR were obtained on the dried IPNs as a function of temperature, in order to monitor the mobility of the PMMA-d_8. The onset of isotropic polymer chain motion for bulk PMMA is seen in the ^{2}H NMR spectrum between 150-180 °C. In contrast, a sample with a low concentration of polymer (17%) did not exhibit a glass transition below 180 °C: a small reduction of ν_Q was, however, observed with increasing temperature, decreasing to 34 kHz at 170 °C. This is due to increased fast small angular fluctuations of the CD_3 groups. A spectrum consistent with isotropic motion of the polymer chains was not observed. Assuming a value for ν_Q of 39.75 kHz for the CD_3 groups undergoing C_3 rotations only,[5] this corresponds to a rms angular fluctuation of approximately 12°. Significant angular fluctuations of bulk PMMA are also observed, similar values being observed at 120 - 140 °C. The ^2H variable temperature NMR spectra of the 26% and 70% PMMA samples are shown in Figure 2. Evidence for polymer motion was observed in samples with higher polymer concentrations: at 170 °C, a significant decrease in the splitting (to approximately 30 kHz), was observed, and a component of the CD_3 lineshape with a smaller ν_Q was evident. Spectra were therefore acquired with varying recycle delays and spin-echo evolution times (τ) and two components were separated: one with a shorter recycle time and a splitting of 25 kHz, and another more rigid component with a splitting of 35 kHz. These correspond to angular fluctuations of 20 and 11°, respectively. The more mobile component may be due to the side chains which are freer to undergo more extensive motion, or may result from less tightly

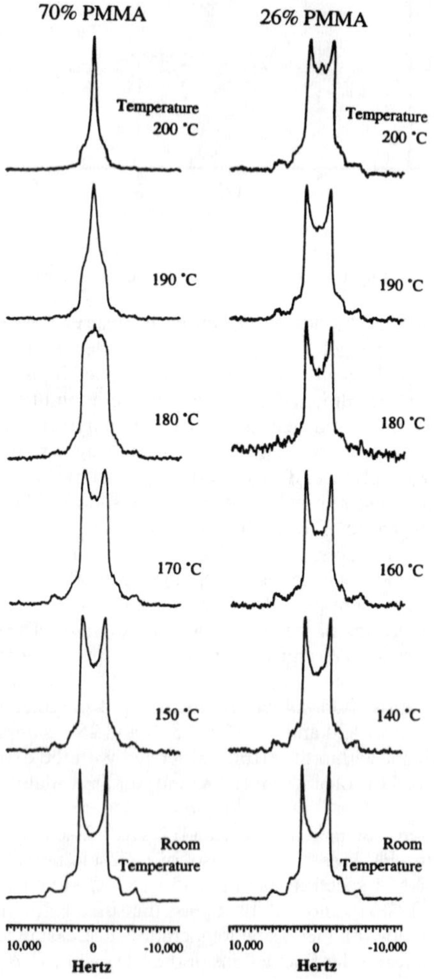

Figure 2. The variable temperature ^2H NMR spectra of 26% and 70% PMMA. NMR spectra were acquired at 46.1 MHz with $\pi/2$ lengths of $\approx$ 1.5 μs, and a quadrupole-echo pulse sequence.

constrained polymer. At higher temperatures, significant changes in the lineshape are observed and a narrower Lorentzian-like lineshape starts to grow in. The Pake doublet from the rigid CD_3 groups is still, however, observed at all the temperatures studied, and ν_Q for this component does not decrease significantly. In contrast, the 26% PMMA sample shows very little change with increasing temperature. A component undergoing extensive angular fluctuations was not observed at higher temperatures, and at 200 °C, only 2% of the polymer is undergoing rapid isotropic motion.

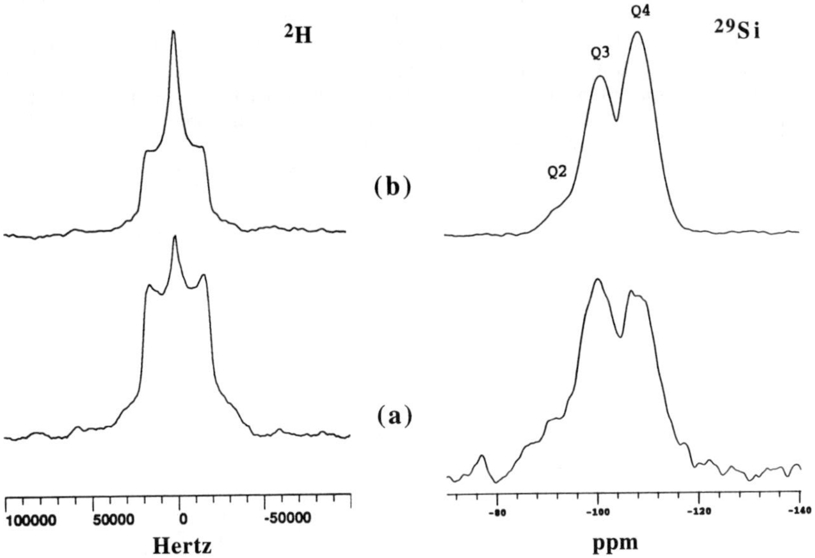

Figure 3. The room temperature ^{29}Si MAS NMR, and ^{2}H NMR quadrupole-echo spectra at 200 °C, for samples of 50% PMMA, aged for (a) 2 hours and (b) 24 hours, before solvent stripping. The ^{29}Si spectra were acquired with $\pi/6$ pulse lengths and recycle times of 60 s.

The work of Pantano et al., has demonstrated that more extensive development of the network in a wet gel (through higher water/silane ratios) leads to less shrinkage when the gel is dried.[11] Consequently, the bulk density and skeletal density of the dried gels are inversely correlated. In order to explore the effect of aging, and thus gel-shrinkage, on the polymer entrainment, two samples of 50% PMMA were prepared where the gel was allowed to age for two and 24 hours, before the solvent was stripped [(a) and (b), respectively]. Dramatically different mobility of the PMMA at 200 °C are seen for the two samples, the sample that has been aged for a longer time showing a higher fraction of mobile polymer (Figure 3). The ^{29}Si NMR of these samples shows three resonances, which can be assigned to the Q^2 (-90 ppm), Q^3 (-100 ppm) and Q^4 (-108 ppm). A significant increase in the percentage of the Q^4 species is seen after aging for 24 hours. Ratios for Q^2:Q^3:Q^4 of 4:53:43 and 4:38:58, for (a) and (b) can be extracted by deconvolution of these spectra. Hence, there are approximately 15% more Q^4 species in the sample that has been aged for longer times.

The variation in the percentage of mobile polymer with temperature, for the 26, 50 and 70% PMMA compositions is compared to that of bulk PMMA in Figure 4. These numbers were obtained by deconvolution of the spectra as follows: a spectrum of the rigid PMMA IPN, just before the onset of any polymer backbone motion was subtracted from the experimentally observed spectrum, and a difference spectrum was obtained. The room temperature ^{2}H spectrum was not used in the subtraction procedure, due to the smaller rms angular fluctuations of the CD_3 groups at lower temperatures. The intensity of the rigid polymer spectrum used in the subtraction was varied until no evidence of the Pake doublet could be observed in the difference spectrum. The percentage of mobile PMMA was then calculated by integrating the "rigid" PMMA spectrum and the difference spectrum, allowing for the fact that both the CD_3 and CD_2 groups are observed in the spectrum of the mobile polymer. The error bars reflect the differences in the percentages estimated when different "rigid polymer" spectra were used in the subtraction. The lower and higher values were obtained when spectra at 165 and 150 °C were used for the 70% sample. The lowest value provides an estimate of the percentage of polymer that is undergoing polymer motion in the fast regime ($\tau_c << 10^{-7}$ s), while the upper value gives an indication of the total percentage of mobile polymer. Comparing the two samples of 50% PMMA, a 15% increase in the concentration of Q^4 species (as seen in the ^{29}Si NMR), results in an increase in the percentage of mobile polymer of from 9 to 28%, at 200 °C. A delay in the onset of mobility by at least 20 °C, in comparison to bulk PMMA, was observed for all samples.

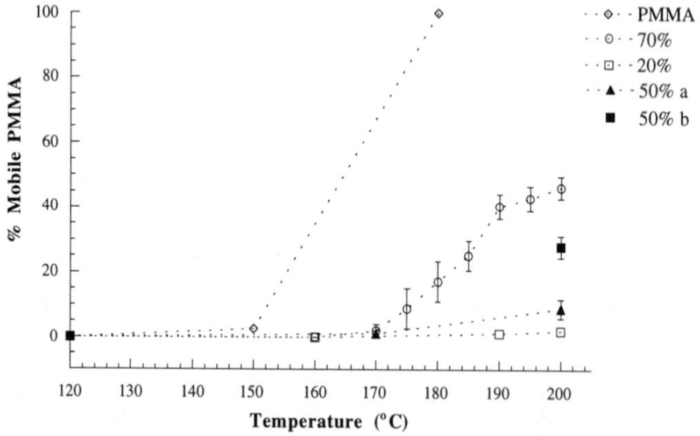

Figure 4. The change in the percentage of mobile PMMA with temperature for different PMMA/SiO_2 compositions

CONCLUSION

The polymer is extremely mobile in the wet gels ($\tau_c << 10^{-7}$s), and not until almost all the solvent is removed is the ^{2}H NMR spectrum of the rigid polymer observed. The majority of the polymer, for the (dry) low polymer-content IPNs, is held rigidly by the SiO_2 network, at temperatures where pure PMMA shows significant isotropic motion of the polymer backbone. The concentration of mobile polymer increases with polymer content, presumably as more polymer rich-domains are formed. However, confined polymer chains are still present for the 70% PMMA IPN. The fraction of mobile polymer at 200 °C varies from 9 to 28% of the total polymer content for the 50% PMMA, for the sample aged for 2 and 24 hours, respectively, before drying. In contrast, a similar composition IPN, aged over a relatively long period, shows only negligible dimensional changes in the TMA experiment at similar temperatures. Hence, at these compositions, enough SiO_2 appears to be present to maintain material integrity and a continuous silica network. The dried gel, however, contains pockets within the silica pores, especially for the gels that have been aged for longer periods, that appear to be sufficiently open to allow some polymer mobility. The resistance of gel to shrinkage on drying increases with Q_4 content, network stiffness and skeletal density, all of which increase as the gel ages. The consequent larger pore size formed after solvent removal logically results in a less confined polymer. Additionally, as the gel ages, the thickening of the SiO_2 matrix may result in a decrease in entanglement of the PMMA and SiO_2 networks, resulting in larger, polymer-rich domains. The presumed reduction in the numbers of silanol groups in the samples that have been aged longer, will also reduce the possibility of hydrogen bonding between the two networks.

The onset of polymer mobility occurs at higher temperatures than for bulk PMMA, even for samples with high polymer concentrations. These results should be contrasted with ^{2}H NMR studies of polymer mobility where the polymer was polymerized inside a preformed silica network.[7] In this case, an isotropic resonance from the monomer was visible even at room temperature, and the T_g of this encapsulated polymer was *lower* than that of bulk PMMA. Finally, this work has demonstrated that ^{2}H NMR is an excellent probe of the confinement of the polymer in these gel systems.

ACKNOWLEDGMENTS

A. J. Vega and Y. Ba are thanked for many stimulating discussions, and help in acquiring the NMR spectra, respectively. Financial support (for CPG) from DuPont and the NSF (DMR-9458017 and CHE-9405436) is gratefully acknowledged.

REFERENCES

1. K. G. Sharp, J. Sol-Gel Sci. Technol. **2**, 35 (1994).
2. K. G. Sharp, in ACS Symp. Ser., edited by J. E. Mark, C. Y. -C. Lee and P. A. Bianconi, Am. Chem. Soc. **585**, Washington, D.C., 1995) p163.
3. H. W. Spiess, Colloid and Polymer Sci., **261**, 193 (1983).
4. A. J. Vega, Polym. Prepr. (Am. Chem. Soc., Div. Polym. Chem.), **22**, 282 (1981).
5. K. Schmidt-Rohr, A. S. Kulik, H. W. Beckham, A. Ohlemacher, U. Pawelzik, C. Boeffel, and H. W. Spiess, Macromolecules, **27**, 4733 (1944).
6. C. J. T. Landry and B. T. Coltrain, Polym. Prepr. (Am. Chem. Soc., Div. Polym. Chem.), 541, (1991).
7. T. M. Che, R. V. Carney, G. Khanarian, R. A. Keosian, and M. Borzo, J. Non-Cryst. Solids, **102**, 280 (1988).
8. E. J. A. Pope, M. Asami and J. D. Mackenzie, J. Mater. Res., **4**, 1018 (1989)
9. L. C. Klein and B. Abramoff, Polym. Prepr. (Am. Chem. Soc., Div. Polym. Chem.), 519, (1991).
10. C. J. T. Landry, B. T. Coltrain, J. A. Wesson, N. Zumbulyadis and J. L. Lippert, Polymer, **33**, 1496 (1992).
11. M. J. Murtagh, E. K. Graham and C. Pantano, J. Am. Ceram. Soc., **69**, 775 (1986).

MOLECULAR PROBES OF PHYSICAL AND CHEMICAL PROPERTIES OF SOL-GEL FILMS

J. I. ZINK *, B. DUNN **, B. DAVE **, F. AKBARIAN *
*Dept. of Chemistry, University of California, Los Angeles, CA 90095, zink@chem.ucla.edu
**Dept. of Materials Science and Engineering, University of California, Los Angeles, CA 90095

ABSTRACT

Two molecular probes, pyranine (8-hydroxy 1,3,6-trisulfonated pyrene) and cytochrome *c*, are used to probe two different aspects of sol-gel SiO_2 films. Pyranine is used as an *in-situ* fluorescence probe to monitor the chemical evolution during sol-gel thin film deposition of silica by the dip coating process. The sensitivity of pyranine luminescence to protonation/deprotonation effects is used to quantify changes in the water/alcohol ratio in real time as the substrate is withdrawn from the sol reservoir. Correlation of the luminescence results with the interference pattern of the depositing film allows the solvent composition to be mapped as a function of film thickness. Cytochrome *c* is used as a probe of the rotational mobility of a large molecule in the pores of sol-gel SiO_2 films. A.C. dipolar relaxation measurements show that the biomolecule remains mobile but experiences a ten-fold increase in local microviscosity.

INTRODUCTION

This paper focuses on the use of two probes, pyranine and cytochrome *c*, of silicate films made by the sol-gel technique. The pyranine probe is chosen to monitor the chemical evolution in-situ during thin film deposition by the dip-coating method. Luminescence spectroscopy enables us to study the chemical changes that occur as the film is formed. The cytochrome *c* probe is chosen to study the interaction of a biomolecule (~17 Å diameter) with a thin-film silicate matrix. AC impedance spectroscopy is used to study the mobility of the protein in the film. The first part of this paper is devoted to film processing and the second part to biomolecular interactions with the film.

PYRANINE AS AN OPTICAL PROBE OF IN-SITU FILM FORMATION

The sol-gel process is a method of preparing inorganic oxide glasses from polymeric or colloidal sols at temperatures well below those used in traditional glass processing techniques.[1-3] Interest in sol-gel thin films has grown in recent years as a number of applications have emerged in areas such as protective and optical coatings, high and low dielectric constant films, electrochromics, waveguides, ferroelectric thin films, sensors, and membranes.[4] The films can be produced readily by several different methods including spin coating and dip coating.[5] Dip coating is perhaps more important technologically since a uniform coating can be deposited onto substrates of large dimensions and complex geometries. In the dip coating process the substrate is immersed into the coating sol and then withdrawn at a constant rate. Brinker, Hurd and co-workers have carried out extensive work, both experimental and theoretical, and have established how various physical and chemical parameters involved in dip coating affect the structure and properties of the final film.[5]

In the dip coating process, a substrate is withdrawn slowly at constant speed (typically 10-20 cm/min) from a sol which contains polymeric or colloidal species in suspension. A liquid film becomes entrained on the surface of the moving substrate. This film thins by solvent evaporation accompanied by gravitational draining and, for multicomponent fluids, possible surface tension gradient driven flows. When the upward moving flux is balanced by that of evaporation, a steady, film profile, 1 to 2 cm in height, is established which terminates at a drying line. The structural evolution in sol-gel thin films is very complex. Unlike the bulk gel system where gelation, aging, and drying occur sequentially over a period of several weeks or longer, all of these processes typically occur within 30 seconds in the thin film. The result is that the drying stage of the sol-gel process overlaps the aggregation/gelation and aging stages. Spatially resolved chemical, structural

Mat. Res. Soc. Symp. Proc. Vol. 435

and rheological characterization of this situation therefore should allow us to understand the genesis of sol-gel derived films as well as gain insight into sol-gel processing.

Determining the composition of the film at different stages during film formation is central to understanding how the changing chemical and physical conditions affect the film's properties and microstructure. The composition of the sol (e.g., ethanol, water and silica precursor) is quite different from that of the final film (silica), and, although models have been developed to consider this evolution, only limited experimental work has addressed this topic. Monitoring the film evolution requires *in-situ* observation of compositional changes. These changes in chemistry, in turn, affect the surface tension, capillary pressure, and condensation rate, and therefore the densification of the network. To date, the characterization of sol-gel thin films has generally been limited to the characterization of the final dried film. Such methods, however, offer only indirect information regarding the physical and chemical evolution during thin film formation. One example of an *in-situ* monitoring technique was the imaging ellipsometry measurements of Hurd and Brinker which measured film thickness and refractive index during film deposition.[6]

The fluorescent probe molecule, trisodiumtrisulfonatehydroxypyrene, or pyranine, is sensitive to proton transfer phenomena. The fluorescence spectrum of the molecule is dependent upon protonation, with the protonated form having an emission maximum at 430 nm and the deprotonated form having a maximum at 515 nm.[7,8] In previous studies with sol-gel materials, pyranine was used successfully to measure the alcohol/water content of aluminosilicate gels during the sol-gel-xerogel transformation and to study the effect of pH on water consumption during the gelation of a silica system.[9,10]

This section describes the use of pyranine luminescent probes to real time measurements within the depositing film. These measurements provide a number of new insights concerning compositional evolution during film drawing including the differential evaporation of alcohol and water, the spatial mapping of the water/ethanol ratio as a function of film thickness, and the determination of the water content of the solvent at the drying line. These experiments demonstrate that luminescent organic molecules may be applied to the processing science of sol-gel thin films.

Materials Preparation

The series of sols used in this study were synthesized with the objective of varying the amount of "excess" water in the sol. A H_2O/Si ratio of 2 is stoichiometrically sufficient to achieve complete hydrolysis and condensation of silicon alkoxides, $Si(OR)_4$, to form SiO_2 because water is produced by condensation. Even if condensation did not occur, $H_2O/Si > 4$ would result in "excess" water in the precursor sol. In this work, careful procedures were used to control the amount of excess water. The mole ratio of H_2O/Si, *r*, was 4 or 2.5. To make the sol, 75 ml of tetraethoxysilane (TEOS) from Fisher, 15.0 or 23.9 ml of deionized water (for $r = 2.5$ or 4.0 sols, respectively), 0.25 ml of 1 N HCL and 75 ml of absolute ethanol (Gold Shield) were refluxed at 65°C for 1.5 hours. The volume of the mixture was then reduced by 50% by rotary evaporation to remove excess water present as the azeotrope, and an equivalent amount of absolute ethanol was added. The evaporation/reconstitution step was repeated 4 times to remove any excess water and acid. This precursor sol was then stored at 4°C until use. Immediately before the film drawing experiments, the sol was combined with an equal volume of absolute ethanol containing pyranine at a concentration of $2x10^{-4}$ M, and then deionized water was added to bring the sol to the desired composition. The films investigated in this study were drawn from the $r = 2.5$ and 4.0 sols with 2.5, 5, 12.5 and 25 volume % of water added. From this series of sols, it was possible to examine the effect of excess water content on film composition.

The films were drawn using the same apparatus as that described by Hurd and Brinker.[6] This approach uses hydraulic motion to produce a steady, vibration-free withdrawal which does not cause optical irregularities in the film. The substrates were polished strips (10 cm x 1 cm x 1 mm) of single crystal <100> silicon cleaned with Chromerge, rinsed with deionized water and then methanol and acetone. These substrates were connected to a weighted float in a water reservoir whose drainage rate was controlled by a flow valve. The substrate was withdrawn in a chamber that was closed to the atmosphere except for a small aperture required for the excitation laser and the collecting lens. The convection-free drying of the film was critical to obtaining high optical quality films. A film drawing rate of 5 cm/min (± 10%) was used for all experiments. Using this drawing rate, steady state conditions for film deposition were established within 15 seconds and maintained for over a minute so that several spectra were obtained for a given substrate withdrawal. After deposition, the refractive index and thickness of the dried films were measured.

Optical Measurements

The primary experiments combined interferometry with luminescence spectroscopy to characterize spatially the chemical composition of the solvent during film pulling. The experimental arrangement is shown in Figure 1.

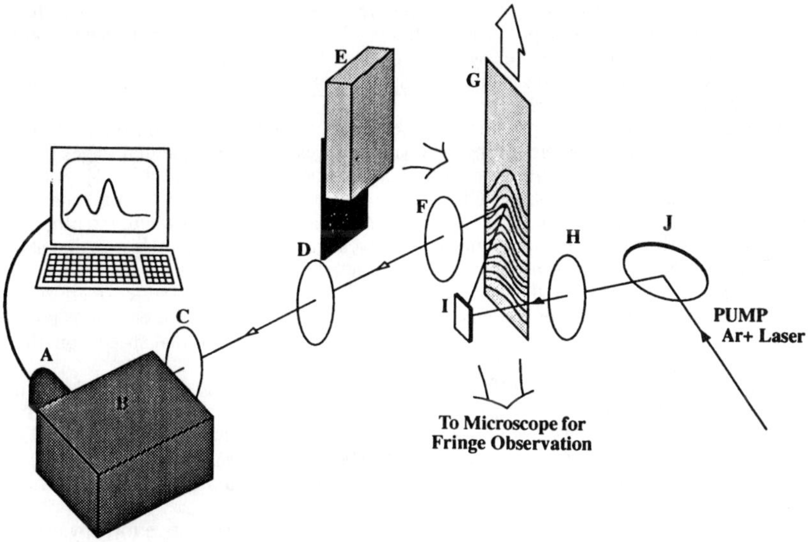

Figure 1. Experimental arrangement combining interferometry with fluorescence spectroscopy. Components are as follows: A. optical multichannel analyzer; B. monochromator; C. focusing lens; D. aperture; E. mercury lamp with filter (546 nm); F. objective lens; G. substrate (silicon wafer); H. focusing lens; I. and J. mirrors.

Interferometry was used to monitor film thickness. A mercury lamp filtered to emit 546 nm light was placed at an angle of 65° to the substrate normal to illuminate the film. A telescopic microscope at an angle of 65° to the substrate normal was used to observe fringes in the film. Interference occurs at thicknesses corresponding to

$$h = \frac{(2\ m+1)\lambda}{4(n^2 - \sin^2\theta_L)^{1/2}} \qquad (1)$$

where h is the film thickness, *m* is the interference fringe number ($m = 0$ corresponds to the drying line) , θ_L is the illumination/viewing angle of the interference pattern and *n* is the refractive index of the solvent. The interference fringes were reproducible with identical pulling conditions. The fringe number not only gives film thickness according to Eqn. 1, but also provides a convenient vertical scale to identify the distance between the drying line and the film reservoir or bath meniscus ($m \approx 10$ to 15 depending upon the film drawing rate).

The fluorescence was excited by the 351 nm line from a coherent Innova 90 Ar^+ laser at a power of approximately 20 mW. The spot size of the excitation beam was less than 50 μm, allowing for excellent spatial resolution. The spectra were recorded using an EG&G model 1420 OMA and a 0.32 m Jobin-Yvon/ISA monochromator for dispersion. The slit width was 200 μm and the integration time was 1 second. The laser spot did not affect the pattern of the interference fringes.

RESULTS

Luminescence spectra of pyranine in ethanol/water solutions of known composition were measured. The emission bands were broad and featureless, with two distinct peaks at 430 nm and 515 nm. The emission intensity at 515 nm increased relative to the intensity at 430 nm as the water increased. The luminescence of a pure water solution of pyranine had almost no peak at 430 nm, while in an absolute ethanol solution of pyranine the peak at 515 nm was not visible. These spectra were used as standards to determine the relative amounts of water and ethanol during the dip-coating process of sol-gel films (*vide infra*).

The characteristic interference pattern for the sol-gel film as it is withdrawn from the reservoir is shown in Figure 2. The dynamics of the dip coating process and the sequential stages of structural development in sol-gel films from this technique have been reviewed.[5,11] This pattern is similar to that obtained in the imaging ellipsometry measurements reported by Hurd and Brinker.[6] The first fringe occurs in the vicinity of the drying line ($m = 0$) where the film attains its final thickness (130 nm to 160 nm in the present investigation). The featureless region above the first fringe, at the top of the substrate, represents a nearly dry film. The fringes offer a convenient vertical scale between the sol reservoir and the drying line, and a series of emission spectra of pyranine in silica sols of known composition were obtained at different fringe positions. The spectra were obtained as the silicon substrate was withdrawn at a constant speed from the reservoir and after the interference pattern had stabilized.

Typical luminescence spectra at different fringe positions (or fringe numbers) are shown in Figure 2 for a sol with a composition of $r = 4.0$. The emission spectrum of the pyranine probe changes with the position of the excitation beam. In the region closest to the reservoir, the 430 nm emission predominates indicating low water content. The 515 nm emission becomes more prominent in the regions closer to the drying line and, just below the drying line, this emission dominates, signaling an increased water content.

The spectral data shown in Figure 2 can be converted to a more quantitative form by using the pyranine reference spectra . It is evident that the film exhibits a rapid increase in water content as the drying line is approached. This sharp gradient is comparable for all the sol/water systems measured: from 2.5% excess water to 25% excess. Preferential evaporation of EtOH has been proposed [12] and now it has been observed directly. The maximum water content in the immediate vicinity of the drying line depends upon the excess water content in the sol. This value ranges from 50% with 2.5% excess water to 85% at 25% excess. Interestingly, there is nearly no change in maximum water content or the composition gradient for the sols prepared with 12.5% and 25% excess water. The water content in the vicinity of the drying line has significant implications for sol-gel films prepared by dip coating. The greater surface tension of water causes a corresponding increase in capillary pressure which is instrumental in the dynamics of film formation and the structural development of sol-gel films.[5]

Discussion

The emission characteristics of pyranine are used to determine the solvent composition in sol-gel thin films as they are withdrawn from the sol reservoir. The emission spectra of pyranine depend on whether or not the molecule is protonated. Adding water to an alcohol solution will increase the number of deprotonated pyranine species and, therefore, the intensity of the luminescence peak at 515 nm. From this protonation/deprotonation behavior, we have been able to quantify chemical changes in the solvent during film pulling, to map spatially the water/ethanol ratio in films in real time, and to investigate how the solvent composition in the film is influenced by the overall composition of the sol.

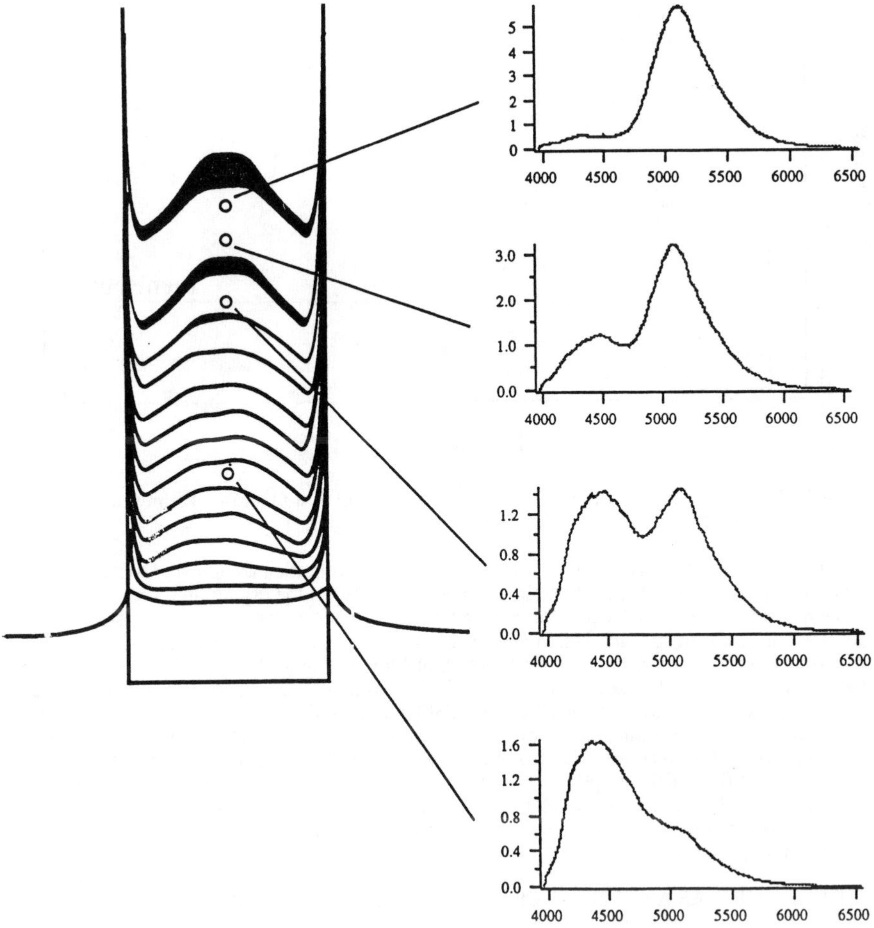

Figure 2. Luminescence spectra at various fringe positions for a film deposited from a sol with *r* = 4.0 and 12.5% excess water added. The dark lines are the interference fringes. The spectra obtained by focussing the laser at the spots represented by the circles are shown at the right.

The increase in the water content of the solvent is due to physical processes (i.e., preferential evaporation of ethanol) rather than chemical processes associated with the production or consumption of water during condensation or hydrolysis. Calculations have shown how preferential evaporation of EtOH leaves behind a water-rich film in the vicinity of the drying line.[12] The results for TEOS/EtOH/H_2O sols are consistent with this model. From the chemical perspective, hydrolysis was complete after the sols were synthesized and any water produced from condensation during synthesis was removed by rotary evaporation. Even if the 4:1 (H_2O:Si) sol had no condensation until film deposition, the water produced by condensation would not exceed

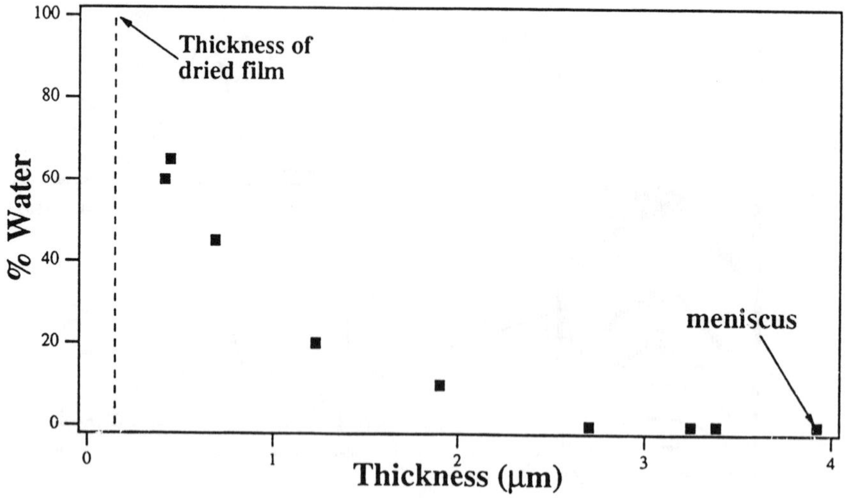

Figure 3. Plot of the water content of the solvent in the film as a function of film thickness for the $r = 4.0$ sol with 5% excess water drawn at a rate of 5 cm/min.

more than a few volume percent in the sol. This amount is insignificant compared to the large increase in water content observed within the first few fringes of the drying line.

The measurements of film water content are quite significant in that the compositional information is obtained spatially and in real time (i.e., as the film is being drawn). The compositional profile can be derived as a function of film thickness. Through the use of Equation 1, the fringe number can be used accurately to calculate film thickness. A compositional profile of the film prepared using the sol with 5% excess H_2O is shown in Figure 3.

The results in Figure 3 underscore an important part of this research: the applicability of this work to thin-film process control. The use of an optical signal to monitor the chemical composition and film thickness provides an opportunity to control rigorously film processing through automated methods. Although the results are obtained for dip coating, it is evident that these methods may be readily adapted to other solution-based materials preparation processes such as fiber spinning.

CYTOCHROME *C* AS A PROBE OF PROTEIN MOBILITY IN A THIN-FILM

Sol-gel thin films containing physically encapsulated enzymes and other proteins have potential applications as biosensors. In the previous section of this paper, we discussed the use of a molecular probe to monitor the chemical composition of a film in real time as it is being formed. The control of film processing to make films with reproducible properties is of obvious importance. Another important property is the interaction of a biomolecule that is the active element of a biosensor with the film in which it is encapsulated. In this section we discuss our studies of cytochrome *c* as a biomolecular probe of protein mobility in films.

Unlike isotropic solution media where the biomolecules have equal mobility and conformational flexibility in all directions, encapsulation constrains them in all three dimensions. The partial order imposed due to presence of a rigid cage may result in unique phenomena usually not feasible in a solution medium [13]. As a result of the confinement of the macrobiomolecules in the nanopores of the gel, significant deviations in structural dynamics and energetics can occur. This section focusses on probing the environment of the protein cytochrome *c* (cyt *c*) in TMOS derived sol-gel films.

Cytochrome *c* (cyt *c*) is a heme-containing electron transfer protein with a molecular weight of ~12 400. The redox properties of this protein are essential for its biological function [14]. To determine the effects of nanoconfinement on the structure and properties of cyt *c*, the protein was encapsulated in silica sol-gel thin films. In our synthesis method, a combination of both strategies was used to obtain longer gelation times without compromising the stability of the protein. One composition which combined long gelation time and protein stability (vide infra), and produced high quality optical thin films was the ratio 40:50:10 (vol percentages) of MeOH, TMOS sol, and buffered solution of cytochrome *c*, respectively (Figure 4). Dip coating techniques can be used to produce films. Excellent quality protein-doped thin-films that were homogeneous, optically transparent, crack-free, and adherent to the substrate were produced by this method [15].

Optical probes of the immobilized ferricytochrome *c* are able to establish the structural integrity of the sol-gel trapped protein. In an idealized D_{4h} symmetry the heme orbitals transform as $a_{2u}(\pi)$, $a_{1u}(\pi)$, and $e_g(\pi^*)$. The absorption spectra of heme moiety derive from electronic transitions from the low lying π orbitals to antibonding π^* orbitals. Cyt-*c*(III) shows a very intense absorption band near 400 nm called the Soret band [16]. The absorption spectrum for a reference solution of the protein dissolved in pH 4.25 acetate buffer (0.1 M) has its Soret maximum at 407 nm, whereas the

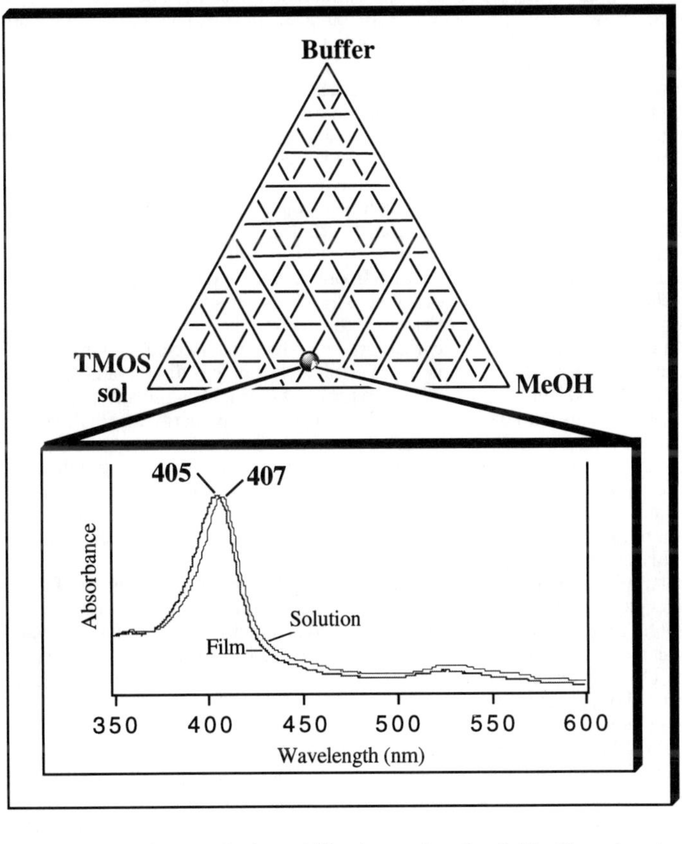

Figure 4. Synthesis space diagram for immobilized cyt-*c* in sol-gel thin films showing the composition used for protein stability.

absorption maximum of the as-prepared film is centered at 405 nm. The blue shifts indicate an increased energy separation between the HOMO-LUMO form the cyt *c* in the sol-gel thin films. The process of encapsulation of the protein within the electrostatically charged pores of the sol-gel thin films and subsequent pore shrinkage as a result of drying can be expected to be the factor responsible for the increased energy of the Soret transition.

A.C. Dipolar Relaxation in a Sol-Gel Thin-film

Nanoconfined biomolecules are characterized by moderate order and mobility. The rate and extent of motion are important in characterizing the degree of confinement. The mobility of the nanoconfined bioparticles is an important issue to be determined. In principle, the translational motions of the particles are expected to be eliminated upon encapsulation, but rotational mobility may be preserved. If so, such site isolation in a porous matrix may induce variations in rotational processes compared to those of the individual biomolecules freely moving in the solution.

In addition to being stiff or rigid the pore walls must be characterized as *active* or *passive* depending on whether the nanocontact interactions of the pore walls with the macromolecules are attractive, repulsive, or negligible. If sufficiently attractive or repulsive, such interactions will influence the rotational dynamics of the nanoconfined biomolecule. The pore in a gel is not a simple cage comprised only of SiO_4 tetrahedra, but also contains a great deal of adsorbed or hydrogen-bonded water. It is also electrostatically negatively charged (pI of silica gel ~2) with an overall negative charge at the pH of the buffer (~7) generally used to encapsulate biomolecules. Such a pore will show affinity towards polar and charged biomolecules. It will also readily alter peripheral hydrogen-bonding interaction of the biomolecules. In addition, due to electrostatic charges inside the porous network, the rotational ground states of hydrogen-bonding and positively charged biomolecules will be stabilized while those of anionic biomolecules will be destabilized with respect to pure solutions. As the overall energetics of the biomolecules are maintained by a large sum of relatively weaker nanocontact interactions, these effects are expected to lead to altered rotational dynamics of trapped particles.

The technique used to probe rotational dynamics is a.c. dipolar relaxation [17-19]. The rotational activation energy associated with a dipolar molecule when placed in an alternating field as it tries to orient its own field is directly related to the characteristics of the immediate microenvironment. Due to a release of electrostatic energy, the impedance of the system drops at the transition frequency (~1 MHz for cytochrome *c*), which is reflected by a maximum in the plot of complex impedance vs. log of frequency. Ferricytochrome *c* is dipolar and shows an effective dipole moment of ~300 Debye units [20]. As a result of physical caging effect and due to noncovalent interactions with the pore walls the total activation energy for rotation is expected to be raised significantly. The overall sum of the effects can be interpreted in terms of assuming that the protein inside the pore of the gel experiences an effective microenvironment with different viscosity. Techniques which are dependent upon the viscosity of the medium can be employed to derive the extent of gel-protein interactions. When applied to sol-gel glasses, the differences are quantitated in terms of *microviscosity* suggesting that the biomolecule is sensing the resistance to movement only in its immediate surroundings.

The dipolar relaxation measurements provide insight concerning the nature of the interaction between molecules and their environment. The rotational activation energy associated with an orienting dipolar molecule when placed in an ac field is directly related to the characteristics of the immediate microenvironment. The cytochrome *c* molecule has an estimated dipole moment of ~300 Debye units. The differences in dipolar relaxation energies in the sol-gel matrix as compared to aqueous media should provide useful information about the extent to which the molecule experiences an altered environment upon encapsulation in silica gel. A plot of the imaginary component of the impedance versus log frequency exhibits a maximum corresponding to the dipolar transition. In general, proteins show a dipolar transition centered at ~10^6 Hz. Using a set of interdigitated gold electrodes on a silicon substrate, complex impedance measurements were performed on the films that were immersed in buffer. Measurements were performed over the frequency range 20 Hz to 1 MHz using a two probe method. Undoped sol-gel films did not show any well-defined transition in this frequency range. Ferricytochrome *c* in 0.01 M acetate buffer

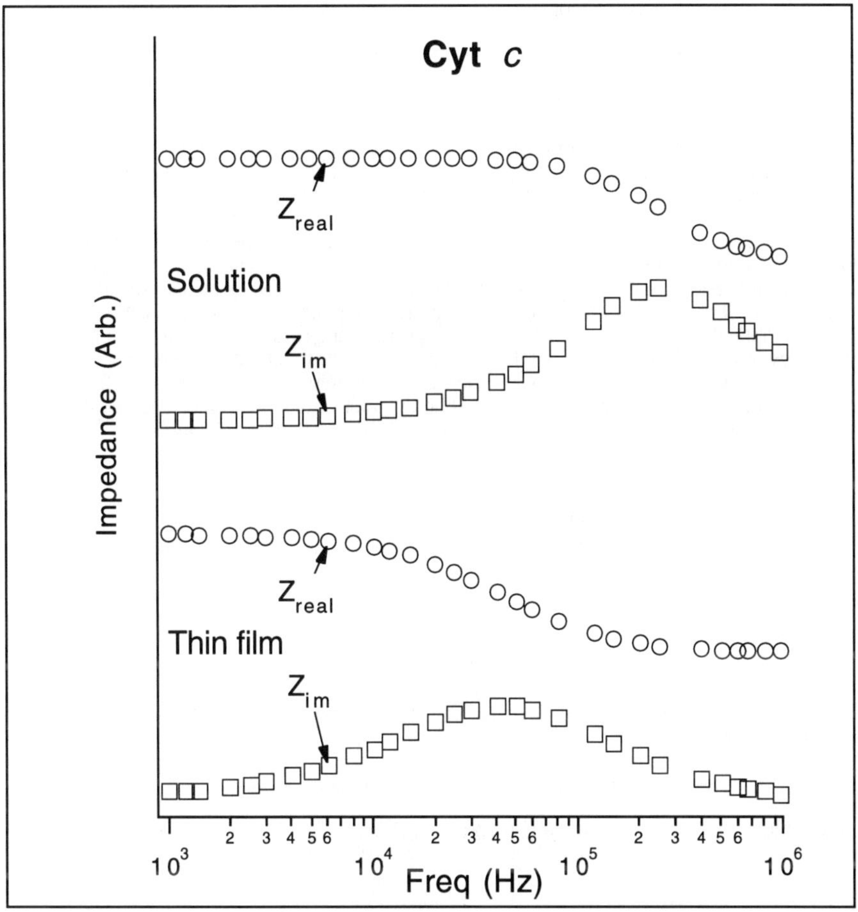

Figure 5. Frequency dependence of the impedance (real and imaginary) for cyt *c* in solution and in SiO_2 thin films.

(pH 4.25) undergoes relaxation at ~$10^{5.5}$ Hz whereas sol-gel thin-film encapsulated ferricytochrome *c* showed a similar transition centered at ~$10^{4.5}$ Hz (Figure 5).

The activation energy (E) for dipolar relaxation process is $E = RT \ln(RT/2\pi hf)$ where R is the gas constant, h is the Planck's constant, f is frequency, and T is the absolute temperature. An additional activation barrier of ~1.1 kcal/mol can therefore be associated with matrix encapsulation. This increase suggests that interaction between the biomolecule and the matrix restricts the rotational movement of the protein to a slightly greater extent inside the sol-gel environment. The small 1.1 kcal/mol difference indicates minimal change in protein noncovalent interaction with the medium and that only a slightly perturbed microenvironment is experienced by the protein upon entrapment in the sol-gel glass film. Additionally, the data can also be used to measure the changes in microviscosity experienced by the protein in the sol-gel medium. For a given relaxation time, $t = (2\pi f)^{-1}$, the viscosity of the medium can be given as $\eta = tkT/4\pi r^3$, where η is the viscosity of the medium, *k* is the Boltzmann constant, T is the absolute temperature, and r is the radius of the bioparticle. Thus, the ratio $\eta_{solution}/\eta_{gel} \sim 0.1$ suggests an approximately ten-fold increase in the microviscosity experienced by the bioparticle upon encapsulation in the sol-gel medium. These results provide a conclusive proof about the confinement effect and its consequences upon the rotational dynamics of cyt *c* biomolecule as a confined bioparticle.

CONCLUSIONS

This paper has demonstrated that pyranine may be used as a fluorescence probe to monitor the chemical evolution *in-situ* during sol-gel thin film deposition of silica by the dip coating process and that cytocrome *c*, in conjunction with a.c. dipolar relaxation measurements provides a probe of biomolecule film interactions. The fluorescence measurements provide a number of new insights concerning the chemical changes which occur during film deposition. The spatially resolved spectral results show that differential evaporation of alcohol and water causes the composition of the entrained sol to become progressively enriched in water with increasing distance from the reservoir surface. These results are significant from the standpoint of automated process control. The use of an optical signal to monitor chemical composition and film thickness in real time offers an opportunity to control film processing through computer-based methods. The relaxation measurements show that cytochrome *c* experiences approximately a ten-fold increase in the microviscosity upon encapsulation in the film. Although translational motion is inhibited, the bimolecule experiences only a 1.1 kcal/mole additional activation barrier to rotation.

ACKNOWLEDGMENTS

We thank Dr. C. J. Brinker and Dr. A. Hurd for their assistance and valuable discussions. We are also grateful for the support of this work by the National Science Foundation (DMR 9003080) and the Department of Energy Basic Energy Sciences Program.

REFERENCES

1. L.L. Hench and J. K. West, Chem. Rev. **90**, 33-72 (1990).

2. C.J. Brinker, G. Scherer, Sol-Gel Science: The Physics and Chemistry of Sol-Gel Processing (Academic Press, San Diego, 1990).

3. L.C. Klein, editor, Sol-Gel Technology (Noyes Publications, Park Ridge, NJ, 1988).

4. D.R. Uhlmann, G.P. Rajendran, in Ultrastructure Processing of Advanced Ceramics, edited by J.D. Mackenzie and D.R. Ulrich (Wiley, New York, 1988), p. 241.

5. C.J. Brinker, A.J. Hurd, G.C. Frye, P.R. Schunk, C.S. Ashley, J. Ceram. Soc. Japan **99**, 862 (1991).

6. A.J. Hurd, C.J. Brinker, Mat. Res. Soc. Symp. **121**, 731 (1988).

7. N.R. Clement, M. Gould, Biochemistry **20**, 1534 (1981).

8. O.S. Wolfbeis, E. Fürlinger, H. Kroneis, H. Marsoner, Fres. Z. Anal. Chem. **314**, 119 (1983).

9. V.R. Kaufman, D. Avnir, D. Pines-Rojanski, D. Huppert, J. Non-Cryst. Solids **99**, 379 (1988).

10. J.C. Pouxviel, B. Dunn and J.I. Zink, J. Phys. Chem. **93**, 2134 (1989).

11. A.J. Hurd, C.J. Brinker, Mat. Res. Soc. Symp. **180**, 575 (1990).

12. P.R. Schunk, A.J. Hurd, C.J. Brinker, in Liquid Film Coating, Scientific Principles and Their Technological Implications, edited by P.M. Schweizer and S.F. Kistler (Chapman and Hall, New York), in press.

13. J. M. Drake, J. Klafter, Physics Today 46 (1990).

14. F. R. Salemme, Ann. Rev. Biochem. **46**, 299 (1977).

15. B. C. Dave, H. Soyez, J. M. Miller, B. Dunn, J. S. Valentine, J. I. Zink, Chem. Mater. **7**, 1431 (1995).

16. B. Cartling, Biological Applications of Raman Spectroscopy, Vol 3. (John Wiley, New York, 1988), pp. 217-248.

17. R. Pethig, Dielectric and Electronic Properties of Biological Materials (John Wiley, Chichester 1979) .

18. J.R. Macdonald, Impedance Spectroscopy (John Wiley, New York, 1987).

19. E.H. Grant, R.J. Sheppard, G.P. South , Dielectric Behaviour of Biological Molecules in Solution (Oxford University Press, Oxford, 1978).

20. W. H. Koppenol, E. Margoliash, J. Biol. Chem. **257**, 4426 (1982).

TAILORING THE MICROSTRUCTURE OF POLYIMIDE-SILICA MATERIALS USING THE SOL-GEL PROCESS.

J.C. SCHROTTER, M. SMAIHI and C. GUIZARD
*LMPM, UMR 5635 CNRS, UMII, ENSCM, 8 rue de l'Ecole Normale, 34053 Montpellier, France
smaihi@crit1.univ-montp2.fr

ABSTRACT

Polyimide-silica materials have been prepared via the sol-gel process by mixing tetramethoxysilane with a polyamic acid. Two polyamic acids have been used. The first is obtained with an equimolar mixture of oxydianiline (ODA) and pyromellitic dianhydride (PMDA) in dimethyacetamide. The second is prepared with a mixture of PMDA with aminopropyltrimethoxysilane (APrTMOS).
The microstructure of the materials obtained with these two polyamic acids are drastically different. The presence of both amino and methoxy side-groups on the APrTMOS enables a chemical bonding between the organic and the inorganic networks resulting in the formation of homogeneous films. On the other side, no chemical bond is provided by the ODA-PMDA polyamic acid resulting in a biphasic microstructure where pure silica particles are embedded in a polyimide matrix.
These two types of materials have been characterized in order to point out the key parameters of their microstructure. ^{29}Si NMR, thermogravimetric analysis, scanning electron microscopy and infra-red spectroscopy have been used to study materials containing various proportions of TMOS and prepared with various hydrolysis ratios.

INTRODUCTION

Because of the combination of high heat resistance, thermal stability, mechanical and electrical properties, polyimides are used in large types of industrial applications such as electrical insulating films, coating compounds and adhesives.
Numerous studies have been carried out on the preparation of polyimide-siloxane hybrid materials. These materials are prepared following different techniques. Co-polymers have been synthesized using co-reaction of polymethyldisiloxane and imide precursors[1-7]. The sol-gel process has been used to prepare polyimide-silica composites where silica particles are dispersed in a polyimide matrix [8-9]. Correlations between the synthesis parameters and the final microstructure have been pointed out[10].
Polyimide-silica materials have been also prepared using precursors which can provide chemical bonding between the organic and the inorganic phase. Spinu et al.[11] presented hexamethoxy functionalized polyimide oligomers which were hydrolyzed and co-condensed with TMOS (tetramethoxysilane). Wang et al.[12] used commercial silicon alkoxides containing amino and chloro groups which provide the bonding with the polyimide. Finally, various bonding agents presenting tailored thermally stable or hydrophobic/hydrophilic functions have been used to prepare polyimide-silica hybrid coatings[13].

EXPERIMENT

Starting materials

The 4-aminophenyl ether (4,4' oxydianiline, ODA, 97%), 1,2,4,5-benzenetetracarboxylic anhydride (pyromellitic dianhydride, PMDA, 97%) and anhydrous n,n-dimethylacetamide

Mat. Res. Soc. Symp. Proc. Vol. 435 © 1996 Materials Research Society

DMAc (Aldrich Chemicals) were stored under argon. The aminopropyltrimethoxysilane (APrTMOS) and the tetramethoxysilane (TMOS) (Hüls America Inc.) have been used as received and stored under argon. Deionized water (18 MΩ) was used for their hydrolysis. The ^{29}Si NMR spectra recorded on TMOS and APrTMOS pure alkoxides contain a single peak at -78.6 and -41.7 ppm respectively showing the absence of any hydrolysed or modified species.

Characterization techniques

Organosiloxane hydrolysis and condensation kinetics have been followed with ^{29}Si NMR spectra recorded on a Bruker 250 spectrometer operating at 49.695 MHz. Molecular structure of the solid materials was determined from 59.62 MHz ^{29}Si NMR spectra recorded on a Bruker MSL 300 spectrometer by using the Magic Angle Spinning (MAS) technique. The samples were put in an alumina rotor which was rotated at 5 kHz. The chemical shifts are given with reference to tetramethylsilane (TMS).
Infrared measurements were obtained on KBr compacted powder or directly on films with a FTIR Nicolet Impact 400D spectrometer.
Thermogravimetric analysis was performed on a Setaram TG 85 apparatus using a 250°C/hour heating rate in air.
Scanning Electron micrographs were obtained from HITACHI S-4500 microscope.

Samples preparation

a-Preparation of PMDA-ODA polyamic acid
0.02 mole of ODA (97%) and 20 ml of DMAc are mixed and stirred at room temperature for one hour under argon atmosphere. A stoichiometric amount of PMDA in 30 ml of DMAc is then added dropwise under stirring and argon atmosphere. This solution is then stirred at room temperature for 18 hours under argon atmosphere giving a viscous yellow solution of polyamic acid in DMAc.
b-Preparation of PMDA-APrTMOS polyamic acid
0.02 mole of PMDA and 50 ml of DMAc are stirred at room temperature for 30 min. under argon atmosphere. Then, 0.04 mole of APrTMOS are added and stirred at room temperature for 3 hours under argon atmosphere. Homogeneous yellow pale amic acid solution is then obtained. The ^{29}Si NMR spectrum obtained on this amic acid solution shows the presence of monomeric species located in the chemical shift range of the pure precursors (-41.7 ppm) and condensed species containing one siloxane bond (-13 ppm).
c- Hybrids preparation and imidization
TMOS and water are added to the polyamic acid solution to form an homogeneous solution after stirring for 60 min. at room temperature. Materials obtained without any addition of TMOS and 77.5% mol. of TMOS have been compared. Two different hydrolysis ratios (h=$[H_2O]/[TMOS]$) molar ratio) have been used : h= 2 and 4.
The PMDA-ODA solutions are then cast onto a glass plate by the tape-casting technique. The PMDA-APrTMOS solutions are cast on alumina substrates.
Polyimide-silica composite materials are obtained by heating the precursor composite using the following heat-treatment : 60°C (6 hours) and 100°C (2 hours) at atmospheric pressure, and then at 230°C (10 min.) and finally 300°C (3 hours) under vacuum.
After cooling to room temperature, films are easily removed by soaking the glass plates in hot distilled water and pealing off.
Typical thicknesses obtained are approximately 15 μm for the films and 2 μm for the coatings.

RESULTS

^{29}Si NMR

Figure 1 presents the ^{29}Si NMR spectra of polyamic acid-silica solutions at different steps of the synthesis.

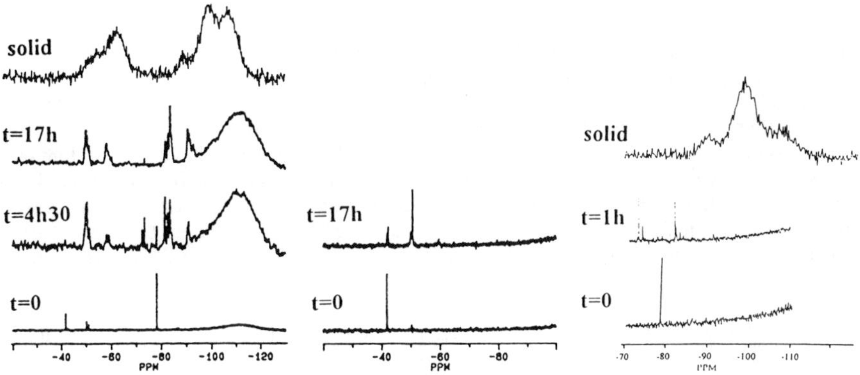

Figure 1 : ^{29}Si NMR spectra of polyamic acid-silica solutions at different steps of their synthesis : PMDA-APrTMOS with 77.5% mol. TMOS (a), PMDA-APrTMOS with no TMOS (b), PMDA-ODA with 77.5% mol. TMOS (c).

When no water has been added, the spectra display peaks which correspond to the pure precursor resonances and to the APrTMOS amic acid condensed species (around -50 ppm). After water addition, the alkoxides undergo hydrolysis. The disapearence of the pure APrTMOS resonance at -41.7 ppm ocurs after 4 hours 30 in the 77.5% solution while it is still present after 17 hours when there is no TMOS in the solution (spectra b). This demonstrates that APrTMOS hydrolysis-condensation is catalyzed by the presence of TMOS. This fact has been previously seen in other systems[14]

Concerning the TMOS, resonances appear in the range of-70/-90 ppm corresponding to its hydrolyzed species. In the PMDA-ODA solution, after 1 hour of reaction, pure TMOS precursor has been completely consumed while its disappearence takes 17 hours in the PMDA-APrTMOS solution.This shows that the presence of APrTMOS in the solution slows down the TMOS reactivity.

Comparison of the chemical shifts in the solid state spectra demonstrates the existence of covalent bonds between the organic and inorganic phase in the PMDA-APrTMOS materials. Table I presents the chemical shifts of TMOS fully condensed species in two types of polyimide-silica materials.

Table I : Chemicals shifts of TMOS (Q4) fully condensed species in two types of polyimide-silica materials.

	PMDA-APrTMOS 77.5%	PMDA-ODA77.5%
Q4 (ppm)	-107.4	-109.7

Standard positions of these species are situated at -110 ppm. When strong interactions between different species occur, large shifts from standard positions are observed[14]. Thus SiO_4 clusters interaction with the polyimide are weak in the PMDA-ODA,77.5% material. This may be due to the large size of the silica clusters which decreases the number of silicon atoms which are neighbors to the polyamic acid. Therefore, this material can be called a composite. On the contrary, materials prepared with APrTMOS show a strong Q4 chemical shift variation from standard Q4 positions. This shows that chemical bonds are created between APrTMOS and TMOS. Therefore, these materials can be called hybrid materials.

Thermogravimetric Analysis

Thermogravimetric analysis of the silica-polyimide films gives informations on the thermal stability and the proportion of silica in the material.
Weight losses are observed below 100°C corresponding to the loss of methanol and water. The polyimide thermal decomposition temperature of approximately 450°C indicates a high thermally stable material. Whatever the composition, all these films have almost the same thermal stability which is equivalent to the resistance of a pure polyimide film. Between 400°C and 500°C, important weight losses are observed for all samples corresponding to the pyrolysis of organic compounds.The final weight residue at 900°C is constituted of silica. Results obtained on three polyimide-silica samples and the influence of the hydrolysis ratio are stated in Table II.

Table II : Experimental and theoretical silica content (wt. %) obtained after 900°C heat-treatment in air on polyimide-silica materials of different compositions obtained with various hydrolysis ratios.

Samples	Theoretical silica content (weight %)	Experimental weight residue (%) h=2	Experimental weight residue (%) h=4
PMDA-ODA-77.5	35	22	32
PMDA-APrTMOS-0	29.8	-	30
PMDA-APrTMOS-77.5	54.3	51.6	53.3

When the hydrolysis ratio is equal to 4, the residue obtained at 900°C is almost proportional to the silica introduced at the beginning of the synthesis. This demonstrates that no pure precursor or hydrolysed species are lost during the cyclization and thus all the silicon alkoxide is transformed in silica.
On the contrary, when the hydrolysis ratio is equal to 2, the silica proportion obtained after the heat-treatment is lower than the amount introduced at the beginning of the synthesis. Thus, when the hydrolysis ratio is lower than the stoichiometric ratio, pure precursor and partial hydrolysed species are lost during the preparation. These results stress the importance of the study of the polycondensation reaction before heat-treatment.

Infra-red spectroscopy

All samples present characteristic bands of imide bonds at 720 cm^{-1}, 1380 cm^{-1} and 1779 cm^{-1}. All hybrid materials exhibit a strong broad absorption band around 1000-1100 cm^{-1} characteristic of Si-O-Si bonds. Absorption bands in the range 3700-3200 cm^{-1} are characterized by OH stretch band of free or hydrogen-bonded OH.

Scanning Electron Microscopy

Figure 2 presents SEM micrographs of different fractured polyimide-silica materials.

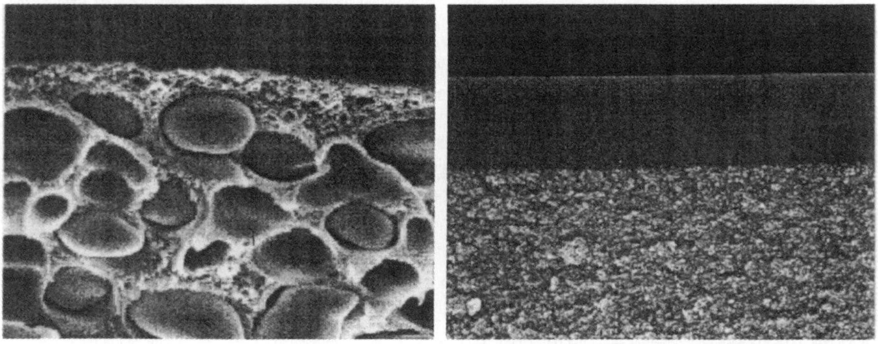

Figure 2 : SEM micrograph of polyimide-silica materials (77.5% mol. TMOS) cross-section : a) PMDA-ODA and b) PMDA-APrTMOS

For PMDA-ODA materials, a homogeneous dispersion of white beads is detected in a polyimide matrix. Electron dispersion spectroscopy (EDS) identified the dispersed particles as silica. The influence of the hydrolysis ratio on the size of the silica particles has been also seen by SEM.The higher the hydrolysis ratio is, the bigger the silica particle size. On the contrary, materials prepared with APrTMOS are completely homogeneous ; no particle can be detected. This observation confirms the ^{29}Si NMR results. As this material is a single phase material, no silica clusters can be seen. These results are presented in Table III.

Table III : Textural characteristics of the polyimide-silica materials (77.5% mol. TMOS) as a function of the hydrolysis ratio and the polyamic acid.

hydrolysis ratio	PMDA-ODA silica particle size (μm)	PMDA-APrTMOS silica particle size (μm)
h=2	0.8	< 1 nm
h=4	2	< 1 nm

CONCLUSIONS

Hybrid polyimide-silica materials have been prepared via the sol-gel technique by mixing TMOS with a PMDA-ODA or a PMDA-APrTMOS polyimide precursor. Various proportions of TMOS and hydrolysis ratios have been used.

^{29}Si NMR study showed that APrTMOS hydrolysis-condensation reactions are catalyzed by the presence of TMOS while the presence of APrTMOS in the solution slows down the TMOS reactivity.

Solid state NMR showed that no chemical bond exist between silica and the polymer in the PMDA-ODA materials while a chemical bond between the APrTMOS and the TMOS induces a chemical shift of the fully TMOS condensed species.
TGA showed that some silica precursor is lost during the imidisation when the hydrolysis ratio is lower than the stoichiometric ratio (h=4).
SEM micrographs demonstrate that PMDA-ODA materials are constituted of silica particles embedded in a polyimide matrix while PMDA-APrTMOS materials are homogeneous. Moreover, the silica particle size depends on the hydrolysis ratio.

ACKNOWLEDGMENTS

It is a pleasure to aknowledge the financial support provided by the European Commission through a Joule program JOU2-CT92-0084.

REFERENCES

1- C.A. Arnold, Y.P. Chen, M.E. Rogers, J.D. Graybeal and J.E. McGrath, 3rd International SAMPE Electronics Conference, **198** (1989).

2- Y. Nagase, S. Mori, M. Egawa and K. Matsui, Makromol. Chem., **191**, p.2413 (1990).

3- T.H. Yoon, C.A. Arnold-McKenna and J.E. McGrath, J. Adhesion, **39**, p.15 (1992).

4- E. Akiyama, Y. Takamura and Y. Nagase, Makromol. Chem., **193**, p.1509 (1992).

5- M. Itoh and I. Mita, J. Polym; Science : Part A: Polymer Chemistry, **32**, p.1581 (1994).

6- M. Ghadir, E. Zimonyi and J. Nagy, J. Thermal Analysis, **41**, p.1019 (1994).

7- J.M. Kaltenecker-Commerçon, T.C. Ward, A.Gungor and J.E. McGrath,J. Adhesion, **44**, p.85 (1994).

8- A. Morikawa, H. Yamaguchi, M. Kakimoto and Y. Imai, Chem. Mater, **6**, p.913 (1994).

9- L. Mascia and A. Kioul, J. Mater. Sci. Letters, **13**, p.641 (1994).

10- S. Goizet, J.C. Schrotter and M. Smaihi, New J.of Chem., submitted.

11- M. Spinu, A. Brennan, J. Rancourt, G.L. Wilkes and J.E. McGrath, MRS Symposia Proceedings, **175**, p.179 (1990).

12- S. Wang, Z. Ahmad and J.E. Mark, Chem. Mater., **6**, p.943 (1994).

13- J.C. Schrotter, M. Smaihi, C. Guizard, J. Applied Polymer Science, in press (1996).

14- M. Smaihi, T. Jermoumi, J. Marignan, Chem. Mater.,7, p.2293 (1995).

Part III

Organic/Inorganic Hybrid Coatings

HYBRID ORGANIC/INORGANIC COATINGS FOR ABRASION RESISTANCE ON PLASTIC AND METAL SUBSTRATES

J. WEN, K. JORDENS AND G. L. WILKES
Department of Chemical Engineering and Polymer Materials and Interfaces Laboratory, Virginia Polytechnic Institute and State University, Blacksburg, VA 24061

ABSTRACT

Novel abrasion resistant coatings have been successfully prepared by the sol-gel method. These materials are spin coated onto bisphenol-A polycarbonate, diallyl diglycol carbonate resin (CR-39) sheet, aluminum, and steel substrates and are thermally cured to obtain a transparent coating of a few microns in thickness. Following the curing, the abrasion resistance is measured and compared with an uncoated control. It was found that these hybrid organic/inorganic networks partially afford excellent abrasion resistance to the polycarbonate substrates investigated. In addition to having excellent abrasion resistance comparable to current commercial coatings, some newly developed systems are also UV resistant. Similar coating formulations applied to metals can greatly improve the abrasion resistance despite the fact that the coatings are lower in density than their substrates.

INTRODUCTION

Transparent polymeric materials have been utilized in recent years as windows in aircraft and public buildings, and glazing for automobiles, buses and aircraft. However, while these polymeric materials may possess excellent optical clarity and many beneficial bulk mechanical properties, they show poor abrasion resistance which can often greatly limit their applications. This has promoted the need for developing abrasion resistant hard coatings for organic polymeric substrates. Many of these coatings have been developed and are based on the use of metal alkoxides or organosiloxanes and generated by the sol-gel process [1-2]. Such coatings may have utility for the protection of metallic surfaces as well. It is often desirable that such coatings for metals should provide good corrosion resistance in addition to good abrasion resistance because of the often rapid corrosion of metals.

Many inorganic/organic hybrid network materials prepared by a sol-gel approach, have been developed over the last decade and studied in our laboratory [3-5]. These materials were prepared from various metal alkoxides and functionalized organics with initial emphasis on functionalized oligomers such as poly(dimethylsiloxane) (PDMS), poly(tetramethylene oxide) (PTMO) and higher T_g components. Many of these materials exhibit high optical clarity. In light of the success of making inorganic/organic composites, hybrid materials based on low molecular weight organics have been more recently developed in our laboratory and successfully used as optical abrasion resistant coating materials. In the previous studies, a series of inorganic/organic coating materials were synthesized and tested [6-17]. In particular, the effect of reaction conditions, the use of metal alkoxide, and chemical structure of the organic component to the abrasion resistance have been principally examined in detail for the polymeric substrate coatings [14-16]. The adhesion behavior, coating durability in a "hot-wet" condition, microhardness tests, UV absorption behavior, and the observation of abraded surfaces by scanning electron microscopy (SEM) were also undertaken to evaluate these coatings. The Si−O−Si inorganic

Mat. Res. Soc. Symp. Proc. Vol. 435

backbone structure, along with its level of crosslinking, promote the high abrasion resistance. The organics contribute some other advantages to the coatings, such as improved adhesion between coating and polymer substrate, reduced shrinkage upon curing, and flexibility relative to a pure brittle inorganic sol-gel coating.

The research summarized in this paper addresses the recent developments with regard to our studies of abrasion resistant coating materials. We particularly address the inorganic/organic coating materials that are based on 3-isocyanatopropyltriethoxysilane functionalized diethylene triamine (DETA) and 3,3'-iminopropyltriamine (IMPA) as the organic components and tetramethoxysilane (TMOS) and tetraethoxysilane (TEOS) as the inorganic components. The coating materials investigated and discussed in this paper have potential for wide application.

EXPERIMENTAL

Materials

Diethylenetriamine (DETA), 3,3'-iminobispropylamine (IMPA), tetraethoxysilane (TEOS), 3-aminopropyltriethoxysilane (3-APS), 3-(trimethoxysilyl)propylmethacrylate (MSi), benzoyl peroxide (BPO), and isopropanol (IPA) were obtained from the Aldrich Chemical Company and used without further treatment. 3-isocyanatopropyltriethoxysilane and tetramethoxysilane (TMOS) were purchased from the Huls America Company. Bisphenol A polycarbonate (3.5×3.5×1/16 inch Lexan®) sheets were obtained from the General Electric Company while CR-39 sheets (diallyl diglycol carbonate resin) were purchased from the Atlantic Plastic Company. These two sheet stock materials were used as the polymeric substrates. Aluminum (0.025 inch thick) and plain steel (0.020 inch thick, ASTM A-366) substrates were obtained from The Q-Panel Company. Norbloc 7966, which is a UV absorber (UVA) of the hydroxyphenylbenzotriazole class, was obtained from Noramco Inc. and Figure 1 shows its structure.

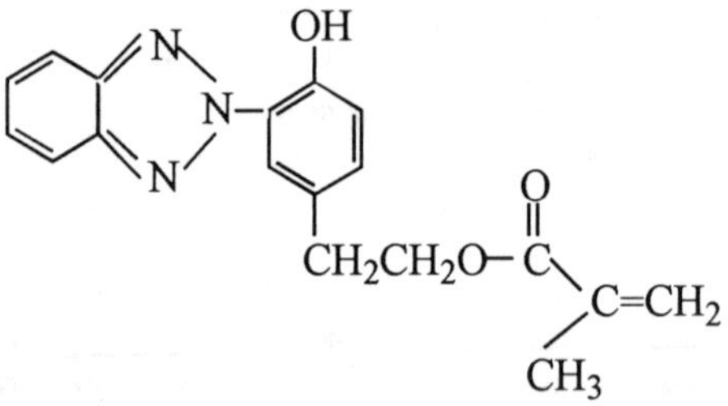

Figure 1. The chemical structure of UV absorber Norbloc 7966.

Coatings for Polymeric Substrates

The synthetic approach for coating preparation is demonstrated in Figure 2. In brief, the synthetic approach involves two major chemical reactions: functionalization of organics followed by the sol-gel reaction. The triethoxysilane functionalized organics in Figure 2 can be prepared through the reaction between the amino groups (or hydroxyl groups) in the organics and the isocyanate groups in 3-isocyanatopropyltriethoxysilane, as reported earlier [14-15]. The loss of isocyanate functionality as the reaction takes place was confirmed by FTIR (2273 cm^{-1}). The reaction was considered complete when this characteristic absorption band disappeared.

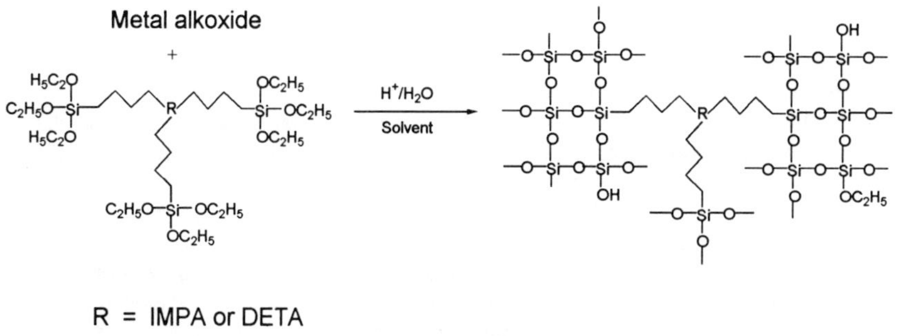

Figure 2. Generalized scheme for generating hybrid inorganic/organic coating materials through the sol-gel process.

During coating preparation, a portion of the resulting solution was mixed with 0.5-1 N aqueous HCl solution and metal alkoxide before being spin-coated at room temperature. The polymeric substrates were pre-treated with an isopropanol solution of a primer, 3-aminopropyltriethoxysilane (3-APS) in order to improve the adhesion between coating and substrate. The hydrolysis and condensation reactions of functionalized organics and metal alkoxides took place when aqueous HCl solution was added. During this procedure, siloxane bonds were formed by the condensation of hydroxyl as well as alkoxy groups which led to the formation of an inorganic backbone incorporated with organics as shown in Figure 2. The extent of reaction was estimated by using the weight loss data and ^{29}Si NMR measurements and found to be around 75% although the value is cure temperature dependent [14,15]. The coatings on the polycarbonate substrates were cured at 120-130°C for three to 12 hours. The coating thickness on the polymeric substrates was found to be either 3-5 microns at higher spin rate (3500 rpm) or 8-12 microns for slower spin rate (1000 rpm) by the surface texture measurement apparatus Talysurf 4 from Rank Talyor Hobson Company. In the case of UVA containing coatings, initiator BPO (1 wt% of MSi) and the desired amount of UVA (see Figures 4 and 5 for detail) were first dissolved in the mixture of MSi, TEOS, and IPA. Then 2 N aqueous HCl solution (H_2O:alkoxy = 2:1) were added to the above solution under brisk stirring for 15 minutes before adding silane functionalized IMPA. The UVA-containing coatings can then be prepared according to the same spin coating and curing procedure described above. The polymerization of the vinyl groups in UVA and MSi has been confirmed by FTIR.

The coated polymeric substrates were then tested for their optical abrasion resistance using the Taber Abraser with CS-10 wheels and 500g loading on each wheel. An index of abrasion was obtained by measuring the intensity of transmitted light through the abraded regions. A monochromatic light source (λ=420 nm) on a Shimadzu-9000 Flyscan spectrometer was used to measure the transmittance through the abraded substrates [6,14-16]. The UV absorption spectra of the coatings were measured by a Hitachi U-2000 Spectrophotometer. The "direct pull method" was used to provide a quantitative index of adhesion of the thermally cured coating on the substrate [14-16].

For the results presented in this report, the sample designation will be denoted as MX-NY where M represents the organic component such as IMPA, and N represents the metal alkoxide such as TMOS. The value of X indicates the weight percent of the precursor functionalized organics in the coating formulation, and Y represents the weight percent of precursor metal alkoxide.

Coatings for Metallic Substrates

Pretreatment of the aluminum and plain steel substrates involved sanding with emery paper followed by washing with IPA. The method of generating coatings on metals is similar to that shown in Figure 2. However, curing has been carried out at a higher temperature of 175°C for 20 minutes to two hours. Using SEM, the coating thickness on aluminum and steel was found to be ca. two microns, but this can be controlled through either turntable spin rate or solution viscosity at the time of spin coating.

Analogously to the coated polymeric substrates, abrasion tests were performed on a Taber Abraser with CS-10 wheels and 500g of load on each arm. Because of the opaqueness of the metals, transmittance methods for evaluating abrasion performance could not be utilized. Instead, coating performance was evaluated visually and with SEM. Corrosion behavior was studied by immersion in 3.5 wt.% NaCl solution. Both full immersion and half immersion tests were performed. Control samples were then compared to coated counterparts. A similar attempt to quantify the adhesive strength between the coatings and metal substrates was made by the "direct pull-off method".

RESULTS AND DISCUSSION

Abrasion Coatings for Polymeric Substrates

The abrasion resistance of uncoated bisphenol-A polycarbonate, uncoated CR-39, and two selected coating systems is shown in Figure 3 as the plot of percent transmission as a function of the number of Taber abrader cycles. As shown in this plot, the abrasion resistance of uncoated bisphenol-A polycarbonate drops quickly at lower number of cycles but levels off at higher number of cycles. On the other hand, the abrasion resistance of CR-39, which is a more highly scratch-resistant crosslinked polymer and has been widely used for eyeglass lenses, decreases in a much slower rate in the beginning although this decrease continues. It was also found that all the coating systems discussed here had much better abrasive resistance than uncoated polycarbonates, and for the coated substrates, all the damage caused by the abrader wheels remained within the coating layer up to at least 500 cycles.

Figure 4 compares the abrasion resistance of two of our coatings with one of the current top commercial abrasion resistant coating Lexan MR5®. It was found that coating IMPA35-TEOS65 has the best performance in terms of abrasion resistance while coating IMPA32-TEOS65-MSi3-UVA1.1wt% has both excellent abrasion resistance and UV-protection.

An important area that has not been addressed in our previous studies is that of promoting ultraviolet stability through incorporation of an appropriate UV absorber. In addressing this issue it may be useful to recall that most commercial organic polymers undergo chemical changes upon exposure to UV light (290 to 400 nm) because they possess chromophoric groups capable of absorbing UV light. The absorption of UV light by many polymers produces noticeable physical and chemical damages. In order to avoid the UV degradation of polymeric materials, the protective coatings utilized should not only offer good abrasion resistance, but also have good durability upon exposure to UV light. Therefore it is a part of our research plan to enhance the UV resistance and sorption characteristics of these coatings without losing optical abrasion resistance and optical clarity. The most common approach is to blend UV absorbers in the coatings. A more desirable route is to covalently bond the UV absorber into the coating which could minimize problems such as incompatibility, migration, volatility, and solvent

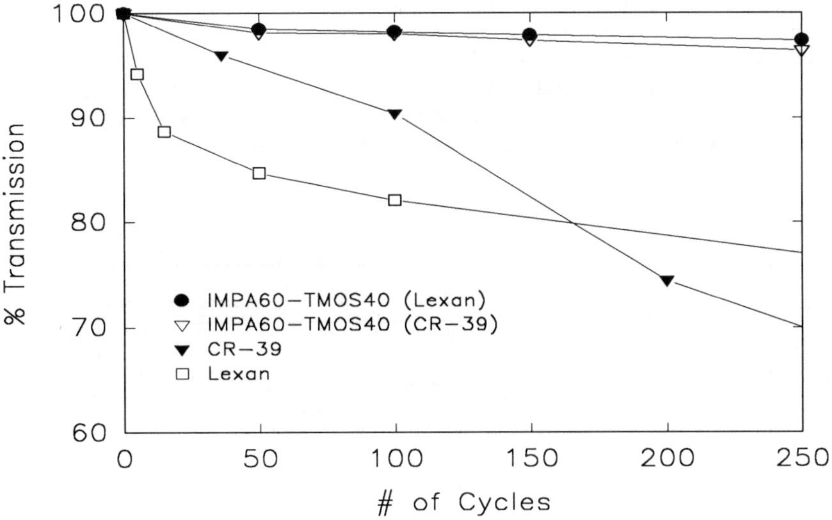

Figure 3. Light transmittance as a function of number of wear cycles for uncoated and coated polycarbonate samples. The cure temperature was 130°C and all the coatings were on a 3-APS primer treated substrate.

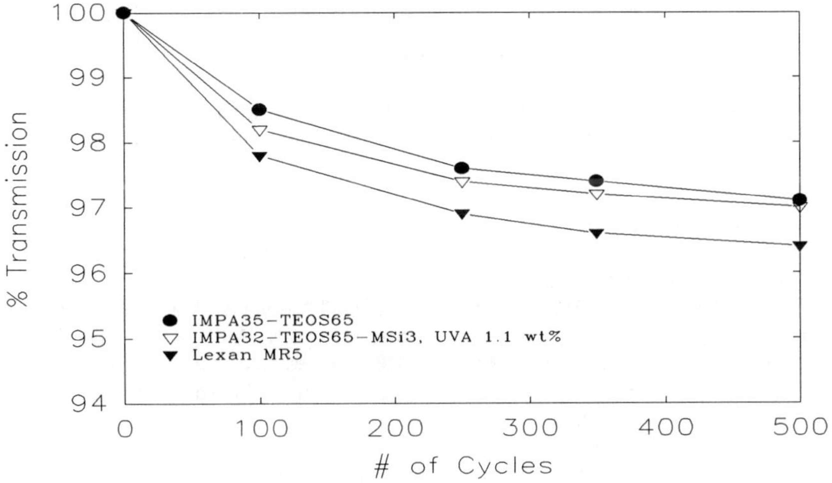

Figure 4. Light transmittance as a function of number of wear cycles for two coated bisphenol-A polycarbonate samples and commercial Lexan MR5® hard coating. Note the scale difference in the % transmission relative to Figure 3.

extraction. In this work, a commercially available UV absorber (Norbloc 7966), was utilized which contains a double bond. It has been successfully bonded into the coating network through the copolymerization with 3-(trimethoxysilyl)propylmethacrylate. Figure 4 clearly shows that the coating containing UV absorber still has excellent abrasion resistance.

Figure 5 shows the UV spectra of the same coating with different amounts of UV absorber on a glass substrate, and shows that the coatings with UVA are capable of absorbing long wavelength UV near 390 nm and the UV absorption ability is dependent, as expected, on the amount of UVA and coating thickness.

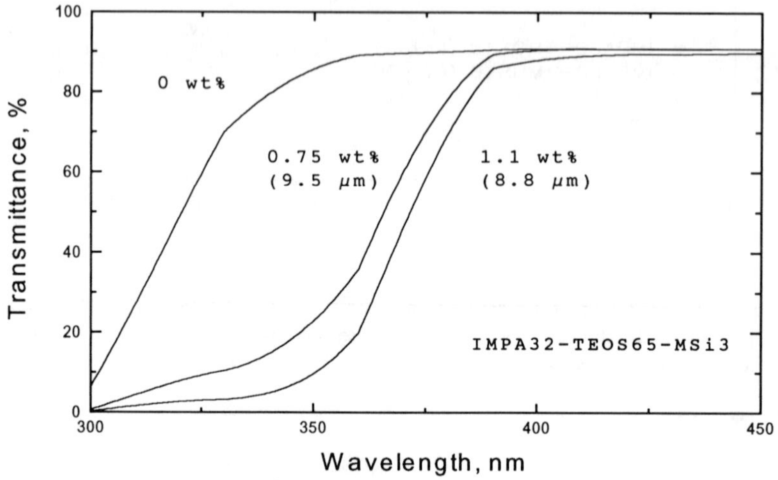

Figure 5. UV spectra of coatings IMPA32-TEOS65-MSi3 with different amounts of UVANorbloc 7966 (wt% in the starting coating solution) coated on glass.

Coatings for Metallic Substrates

One of the best performing abrasion resistant coating on aluminum studied to date is DETA50-TMOS50. After 350 cycles on the Taber abraser, little damage can be discerned visually [15]. A coating produced from neat functionalized DETA is not as abrasion resistant due to the lower content of inorganic material and lower crosslink density. Both coatings have shown good adhesion in the pull-off test (greater than 11.1 MPa). Adhesion of our sol-gel coatings to aluminum substrates may be enhanced by direct covalent bonding between the coating and surface hydroxyls on this metal. Unfortunately, the DETA50-TMOS50 coating performs poorly on steel substrates. Within 350 Taber cycles, pieces of the coating fragment off exposing the bare steel surface. Interestingly, results from the direct pull-off test show that the adhesion between this coating and the steel substrate is significant (greater than 20.9 MPa), and the exact reason for its poor performance is as yet uncertain. Neat DETA performs better in abrasion tests on steel.

Not surprisingly, the coated metal substrates show much better corrosion resistance than uncoated control samples. Both aluminum and steel coated with both neat functionalized DETA and DETA50-TMOS50 display little visual change after 24 hour exposure to salt water, however, the bare metal surfaces show discoloration and pitting. No performance distinction could be made between the neat DETA and DETA50-TMOS50 coatings on either substrate by visual observation.

CONCLUSIONS

Novel transparent abrasion resistant coating materials prepared by the sol-gel method have been developed and applied to bisphenol-A polycarbonate, diallyl diglycol carbonate resin (CR-39), aluminum, and plain steel substrates. It was found that these hybrid organic/inorganic networks afford excellent abrasion resistance to polymeric substrates. In addition to having excellent abrasion resistance comparable to current commercial coatings; some newly developed systems are also UV resistant. Some extent of corrosion inhibition is afforded to aluminum and steel through these coating systems.

ACKNOWLEDGMENTS

The authors acknowledge the financial support of the Air Force Office of Research under grant number F49620-94-1-0149DEF. One of the authors (KJ) greatly appreciates the support of the Adhesive and Sealant Council and the Center of Adhesive and Sealant Science at VPI&SU.

REFERENCES

1. J. Kasi, JPn. Kokai Tokkyo Koho, JP6335, 633 (88,35,633).
2. H. Schmidt, H. Walter, J. Non-Cryst. Solids **121**, 428, (1990).
3. G. L. Wilkes, B. Orler, H. H. Huang, Polymer Preprints **26(2),** 300, (1985).
4. H. Huang, R. H. Glaser, G. L. Wilkes, ACS Symp. Ser. **360**, 354, (1988).
5. H. Huang, G. L. Wilkes, Polymer **30**, 2001, (1989).
6. C. Betrabet, G. L. Wilkes, Polymer Preprints **32(2),** 286, (1992).
7. B. Tamami, C. Betrabet, G. L. Wilkes, Polymer Bulletin **30**, 39, (1993).
8. B. Tamami, C. Betrabet, G. L. Wilkes, Polymer Bulletin **30**, 393, (1993).
9. B.Wang and G.L.Wilkes, J. Macromol. Sci.-Pure Appl. Chem. **A31**, 249, (1994).
10. B. Wang, G. L. Wilkes, C. D., Smith, J. E. McGrath, Polymer Commu. **32**, 400, (1991).
11. C. Betrabet and G. L. Wilkes, J. Inorganic & Organometallic Polymers **4**, 343, (1994).
12. Wang, B; Wilkes, G. L. US patent 5,316,855.
13. Betrabet, C. S.; Wilkes, G. L. J. Inorganic & Organometallic Polym. **4**, 343, (1994).
14. J. Wen and G. L. Wilkes, Journal of Inorganic & Organometallic Polymers **5**, 343, (1995).
15. J. Wen, V. J. Vasudevan, and G. L. Wilkes, J. Sol-Gel Science & Technology **5**, 115, (1995).
16. J. Wen and G. L. Wilkes, PMSE Preprints **73**, 429, (1995).
17. K. Jordens and G. L. Wilkes, PMSE Preprints **73**, 290, (1995).

USE OF AN ORGANOSILANE COUPLING AGENT IN COLLOIDAL SILICA COATINGS

M. W. DANIELS, L. CHU, and L. F. FRANCIS
University of Minnesota, Department of Chemical Engineering and Materials Science, 421 Washington Ave. SE, Minneapolis, MN. 55455

ABSTRACT

Coatings were prepared from suspensions containing colloidal silica (20nm) in isopropanol and 3-glycidoxypropyltrimethoxysilane (GPS). Thicker coatings without cracks were achieved by increased addition of GPS. The microstructure of the resultant coatings contained well packed colloids with some pore space. When high amounts of GPS were added (i.e., GPS:silica weight ratio (R) greater than 0.5), polymeric material appeared and covered the necks between particles and filled in pores. ^{29}Si NMR of coating suspensions (R=1) showed GPS hydrolysis within a few minutes and GPS adsorption on the silica surface within a few hours. The excess silane remained in solution and further polymerized, instead of forming a thick multilayer on the colloids. Diffuse reflectance IR spectroscopy data suggests saturated surface coverage beginning around 2 to 3 molecules/nm^2 based on observed depletion of silica surface hydroxyls.

INTRODUCTION

The hardness and wear resistance of ceramic materials makes them excellent materials to coat softer materials, such as polymers: however, traditional ceramic processing requires high temperatures or expensive deposition systems. An alternative process involves the use of organic binders, such as organosilane coupling agents, in ceramic coatings to improve low temperature strength. Silane coupling agents are widely used in polymer composites containing inorganic particles or fillers [1-3]. These molecules are able to form strong bonds to inorganic particle surfaces and also to the surrounding polymer matrix, linking the different materials together. A similar approach can be used in a "nano-composite" coating prepared with ceramic particles.

In this research, nano-sized silica colloids are mixed with silane coupling agents, which are hydrolyzed and reacted with the silica surface. Coatings are made from these suspensions on polyethyleneteraphthalate (PET) film substrates. The advantages of this approach are the availability and flexibility of the raw materials (silica colloidal suspensions and organosilanes), the ease of processing, and the low temperature processing. In this report, we explore the effects of different amounts of organosilane coupling agent on coating microstructure and cracking behavior. In the context of coating microstructure, we also examine the organosilane adsorption and polymerization reactions which occur in this system.

EXPERIMENTAL PROCEDURE

Coating suspensions were prepared by combining a suspension of colloidal silica in isopropanol (Nalco TX 9476), 3-glycidoxypropyltrimethoxysilane (GPS, Aldrich), and water. Nalco TX 9476 is an experimental product containing 20 nm spherical silica colloids suspended in isopropanol with 6 wt% water and a small amount of acid. Before combining with other ingredients, the suspension was diluted to 10 wt% silica with isopropanol. Water was then added to the suspension, followed by GPS. The mixture was stirred for 24 hours at room temperature. Suspensions with varying amounts of silica and GPS were prepared, with the GPS to silica weight ratio (R) ranging from 0 to 1. In each suspension, the amount of water was at least 30 times that required for complete hydrolysis of the silane.

Coatings were made on polyethyleneteraphthalate (PET) film substrates using a wire-wound rod coater (Mayer rod#6 and #22). The coatings were immediately put in a drying oven

Mat. Res. Soc. Symp. Proc. Vol. 435

1a

1b

1c

Figure 1 SEM micrographs of 20 nm colloidal silica coatings with different GPS to silica weight ratios (R). The fine-grained features on the colloidal surface are from the sputtered platinum coating used for SEM sample preparation.

at 75 °C for 1 hour, and then additionally at 100 °C for 2 hours. Coating microstructure was characterized by optical microscopy and SEM.

A portion of each coating solution was heated in an aluminum pan to produce dry material for examination by diffuse reflectance infrared Fourier transform spectroscopy (DRIFTS). Solutions were heated under identical conditions as those used for the coatings. The dried material was ground and placed in a Spectra Tech Collector sample cup, and analyzed by DRIFTS. In separate experiments, the dried material was heated for 2 hours at different temperatures (75 °C, 120 °C, and 150 °C). Dried specimens were also prepared from suspensions containing added nitric acid (0.001M HNO_3 in the final suspension). Base additives were not investigated, since these destabilized the colloidal solution.

^{29}Si NMR studies were carried out on coating suspensions which contained a relaxation agent (0.03g $Cr(acac)_3$ per 10 ml solution) and reference material (0.3 ml phenyltrimethylsilane (PTMS) per 10 ml solution). Solutions were examined using a Varian 500 at 99.3 MHz and 32 or 64 scans. Spectra were collected over time up to 24 hours. GPS solutions without silica were also examined for comparison.

RESULTS AND DISCUSSION

The relative amount of GPS compared with silica (R) impacted the coating microstructure. Coatings prepared without GPS were comprised of close-packed 20 nm spherical colloids (Fig. 1a). Gaps between colloids due to imperfect packing formed pores 5-10 nm in diameter. SEM micrographs show a grainy structure on the silica surface due to the sputtered platinum coating. Coatings with GPS also generally contained visible 20 nm colloids, often packed fairly tightly (see Fig. 1). As the silane concentration was increased, the presence of polymeric material became evident, forming necks and smoothing over sharp features. When R reached 1.0 (i.e., equal weights of silica and GPS), no colloids were visible. Presumably, particles were completely covered by the silane. The size of the colloids (when they were visible) did not appear to change, regardless of GPS concentration. This observation indicates little multilayered adsorption, which would have increased the colloid size.

BET nitrogen adsorption measurements revealed changes in coating porosity. The coating prepared with no silane had an average pore size of 6.5 nm and a pore volume of 0.37 cm^3/g. Coatings with a relatively small silane content (R=0.05) had similar values, with an average pore size of 7 nm, and a pore volume of 0.23 cm^3/g. The R=0.5 coating had no porosity according to nitrogen adsorption. These results are consistent with SEM observations.

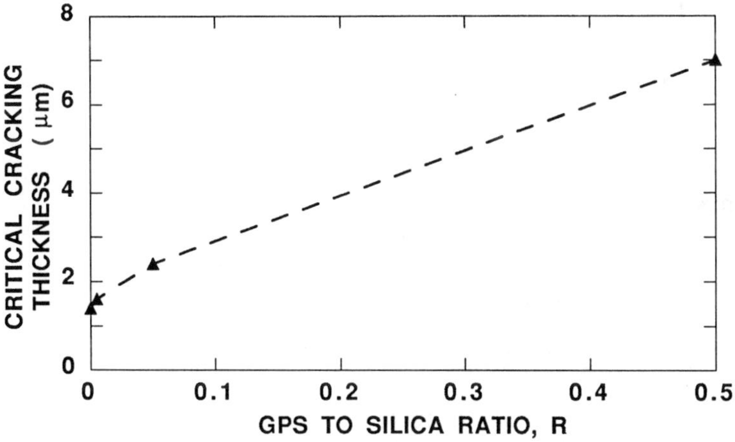

Figure 2 Effect of GPS on critical cracking thickness of dried coatings

Coatings would crack during the drying process, but cracks could be eliminated if the coating is made thin enough. Figure 2 shows the maximum coating thickness achieved for some GPS contents. Silica coatings (R=0) suffered from severe mud-cracking which limited their thickness to ~1.5 μm. As well, these coatings adhered poorly to the PET substrate and were easily rubbed off. Coatings with more GPS could be made thicker, and when cracks did form, they were fewer and more widely spaced as compared with silica coatings. (Future work must be done to fill in the data gap between R=0.05 and 0.5.) The addition of the silane is expected to improve coating toughness by providing an inorganic polymeric matrix which allows some plastic deformation without catastrophic cracking. The inorganic polymer would be made up of siloxane linkages through the silicon end of the silane molecule. Coating strength may also be improved by enhanced bonding between particles due to silane on their surfaces. Another consideration is that silane treatment lowers the surface energy of the silica colloids [1], which reduces capillary stresses during drying.

To better understand the role of GPS in the coatings, DRIFTS experiments were carried out to determine the extent of bonding between GPS and silica. Of primary interest is the surface isolated free silanol concentration. A free silanol is isolated on the surface of the silica colloid and not hydrogen-bonded to any adsorbent or neighboring surface silanols. A silane molecule is most likely to adsorb on this species [4], hence the disappearance of free silanols indicates adsorption. An adsorbed silane can interact with the free silanol by a condensation reaction forming a siloxane bridge, or by hydrogen bonding, which perturbs the surface silanol so that it is no longer "free." Infrared spectroscopy is a useful technique to track adsorption. The free silanol peak is at 3750 cm^{-1} and silanols hydrogen-bonded to their neighboring silanols is at 3660 cm^{-1}[5]. Unreacted silanols belonging to GPS molecules probably give peaks near 3730 cm^{-1}[5]. For comparison between different specimens, we used a normalized free silanol peak, defined as the integrated peak intensity of the free silanol divided by that of the Si-O combination/overtone peak at 1850 cm^{-1}[6].

Increasing the relative concentration of GPS leads to a decrease in the observed free silanol concentration (Fig. 3). The free silanol content decreased until around R=0.1 to 0.5, where it leveled off. An R of 0.5 corresponds to a surface coverage of about 8.9 molecules/nm^2 of silica surface, a value well in excess of the amount needed for complete surface coverage. The total amount of surface silanols is expected to be around 4-5 OH/nm^2, and the amount of free surface silanols is around 1.1 OH/nm^2[4]. Studies with a similar methacrylate silane shows saturated surface adsorption at 2 molecules/nm^2 [7]. Addition of

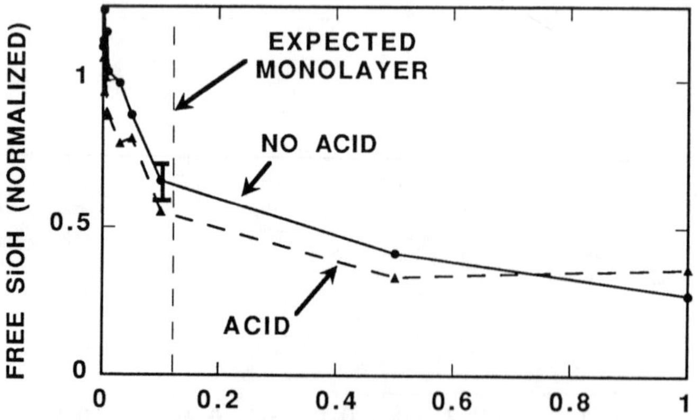

Figure 3 Silica surface depletion in dried solutions as measured by DRIFTS. Acid solutions contain 0.001 M HNO_3.

acid resulted in a small decrease in the amount of free surface silanols, perhaps as a result of decreased condensation between GPS molecules. In this case, the adsorbed silane would have more silanols available for hydrogen bonding or reaction with the silica surface. Increasing the curing temperature resulted in a slight increase in the amount of remaining free silanol. Considering the temperature, this effect is likely due to increased removal of adsorbed water that was hydrogen-bonded to the silica surface.

The IR data for dried material indicates that the surface silanols are removed as silane is added, though it could not be used to clearly distinguish between condensation and hydrogen bonding. Above R=0.1, additional silane had little effect on the surface silanol concentration. If remaining GPS silanols are present, their peak was overwhelmed by the silica free silanol and hydrogen-bonded silanol peaks. GPS silanols were not significant, since increasing the amount of GPS had little effect on the data.

For ^{29}Si NMR studies of suspensions, a high concentration of GPS (R=1.0) was used to ensure an adequate GPS signal. Hydrolysis and condensation were rapid; most of the GPS monomer had disappeared within an hour (Fig. 4). Near -40 ppm are a set of peaks corresponding to GPS monomer (T^0) in different hydrolysis states. Within 15 minutes most of the GPS had fully hydrolyzed. Meanwhile, a set of peaks corresponding to molecules with one siloxane bond (T^1) appeared near -49 ppm and continued to grow. This peak contains dimers and other species. A third type of species (T^2) containing two siloxane bonds began appearing at -58 ppm. This peak appeared after 15 minutes and also grew over time. A GPS solution (without the silica) was prepared with equivalent amounts of isopropanol, water and GPS. This solution showed no hydrolysis even after 30 minutes. When the pH was adjusted to approximately match that of the silica suspension, similar hydrolysis and condensation behavior occurred. Actually, the condensation was somewhat less than that of the silica suspension, probably due to the slightly lower pH of the solution without silica.

The amount of GPS adsorbed on silica colloids could be observed by the reduction of the total of the integrated peaks. In the R=1 suspension, the amount of silane contributing to the NMR signal decreased around 15% over 24 hours; this decrease was not very significant compared to the data scatter. A 13% decrease is expected for monolayer coverage. An R=0.1 solution showed the presence of hydrolyzed monomer for at least the first 10 minutes, and by 4 hours no peaks were observed, indicating adsorption had quickly occurred in solution. Similar behavior was observed with methacrylate silane molecules by Nishiyama, et al[8].

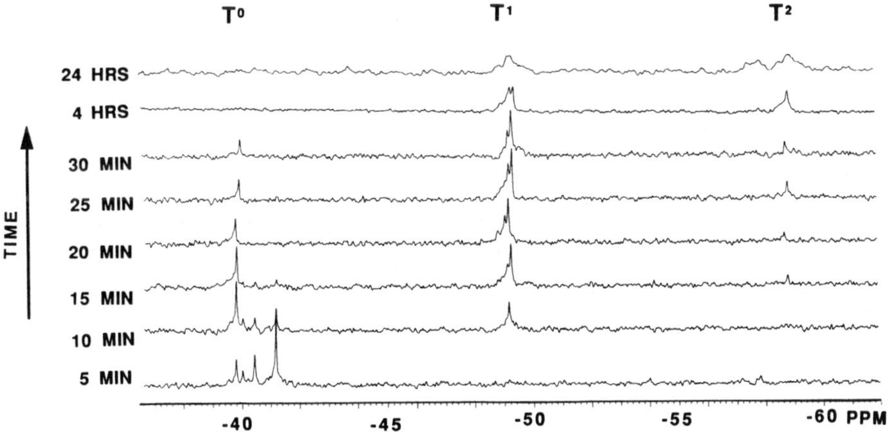

Figure 4 ^{29}Si NMR spectra for a silica suspension (R=1) at various times after initial mixing.

The adsorption behavior observed by ^{29}Si NMR is expected to differ slightly from that observed by DRIFTS, due to effects from increased concentrations during drying. As the volatile solvent is removed, the increased concentration of GPS and silica left behind may increase the chance for contact between them, enhancing adsorption[7]. Drying effects such as this will not occur in ^{29}Si NMR experiments, but will occur in DRIFTS and in coatings.

CONCLUSIONS

The addition of GPS to a colloidal silica suspension in alcohol resulted in coatings with decreased cracking and porosity. GPS hydrolyzed quickly in the silica suspensions, and at low weight ratios of GPS to silica ($R \leq 0.1$) complete adsorption occurred within 4 hours, as determined by ^{29}Si NMR. At a high GPS to silica ratio, adsorption is likely occurring but stops near the expected monolayer coverage. The remaining GPS in the suspension further condenses to form various oligomers. Observations of coating microstructure are consistent with these results. At greater R values, the size of colloidal particles did not increase appreciably as it would if there were extensive multilayer adsorption. Instead, the GPS oligomers in solution were incorporated in the coating during drying and filled in necks and pores. DRIFTS data confirmed adsorption of the silanes on the colloids in the dried state, and suggests little additional GPS adsorption occurs beyond R=0.1, corresponding to 1.8 molecules/nm^2. With increasing amounts of GPS, the final coatings showed less cracking and porosity, and could be prepared to greater thicknesses.

ACKNOWLEDGEMENTS

This work was supported by the NSF Center for Interfacial Engineering. We thank Prof. A. W. McCormick and Mr. S. E. Rankin for useful discussions and assistance with the ^{29}Si NMR.

REFERENCES

1. Jang, J., Ishida, H. and Plueddemann, E.P. *Studies of Methacrylate Functional Silanes Containing Different Spacer Groups; Preparation, Characterization and Properties* p.2-C 1-6 (1986).
2. Nishiyama, N., Shick, R., Horie, K. and Ishida, H. *Influence of Adsorption Behavior of a Silane Coupling Agent onto Silica on Visco-Elastic Properties* p.707-713 (1990).
3. Ikuta, N., Maekawa, Z., Hamada, H., Ichihashi, H., Nishio, E. and Abe, I. *Controlled Interphases in Composite Materials: Proc. of the Third Int. Conf. on Composite Interfaces* p. 757-766 (1990).
4. Morrow, B.A. and McFarlan, A.J. *Langmuir* **7**, p. 1695-1701 (1991).
5. Vrancken, K.C., Coster, L.D., Voort, P.V.D., Grobet, P.J. and Vansant, E.F. *Journal of Colloid and Interface Science* **170**, p. 71-77 (1995).
6. Parrill, T.M. *Journal of Materials Research* **7**, p. 2230-2239 (1992).
7. Nishiyama, N., Shick, R. and Ishida, H. *J. Colloid Interface Sci.* **143**, p. 146-156 (1991).
8. Nishiyama, N., Horie, K. and Asakura, T. *J. Colloid Interface Sci.* **129**, p. 113-119 (1989).

GLYCIDOXYPROPYLTRIMETHOXYSILANE MODIFIED COLLOIDAL SILICA COATINGS

L. CHU, M.W. DANIELS AND L.F. FRANCIS,
Department of Chemical Engineering and Materials Science and Center for Interfacial Engineering, University of Minnesota, Minneapolis, MN 55455

ABSTRACT

Coatings were prepared from a suspension of colloidal silica particles containing glycidoxypropyltrimethoxysilane (GPS) and a polyamine curing agent. GPS was first added to an aqueous silica suspension which contained ethanol (30 wt%) to enhance mixing. The addition of GPS to a basic silica suspension favored condensation among the silane monomers and oligomers, resulting in precipitation. By contrast, acidic conditions resulted in slower condensation which allowed adsorption of the silane on silica, as followed by ATR-FTIR. After GPS addition and aging, the pH of the suspension was increased, a polyamine was added, and coatings were prepared on polyester web. Coatings with GPS modification were denser, adhered better to the polymer substrate, and could be made thicker than unmodified silica coatings.

INTRODUCTION

Colloidal ceramic coatings require high temperature thermal treatments to develop strength through sintering. These thermal requirements preclude the application of ceramic coatings on substrates such as paper and polymers, although the properties of the ceramic (e.g., abrasion resistance, dielectric constant) are often desired for these applications. In addition, ceramic coatings are brittle and often suffer from cracking. Organic binders provide strength at low temperatures and prevent cracking, as those processing bulk ceramics have well documented [1]. Usually organic binders are removed during thermal processing; however, the binder, if properly chosen and processed, can be an integral part of the final material. For coating applications on polymer and paper webs, an organic binder is essential and ceramic coatings prepared with organic binders have been reported [2-4]. Studies on traditional polymer coatings containing minor amounts of ceramic particles as fillers or pigments have shown that the adsorption of the organic onto the particles is crucial to the strength of the coating [5]. Therefore, attention is paid to the characterization of adsorption [6-9].

In this paper, we describe the use of an epoxyfunctional silane to bind colloidal silica particles in a coating. Van Ooij et al [4] described a similar approach using 3-aminopropyl trimethoxysilane. Coatings are prepared from a silica suspension containing glycidoxypropyl trimethoxysilane (GPS) with a polyamine added as a curing agent. GPS adsorption on silica surfaces was followed using ATR-FTIR. Coatings were characterized with optical and electron microscopy. Here, we briefly discuss the effect of suspension pH on processing of coatings; more information on pH effects as well as the role of the relative amount of GPS will be given elsewhere [10].

EXPERIMENTAL PROCEDURE

Coating suspensions were prepared by adding GPS and a polyamine curing agent to a colloidal silica suspension. The aqueous colloidal silica used in this study was Ludox-LS (Aldrich Chemicals) with a silica particle size of 12 nm. This suspension was diluted to 10 wt% silica by adding a combination of water and ethanol to give an aqueous suspension with 30 wt% ethanol. Before adding the GPS, the pH of the silica suspension was adjusted by adding HNO_3 or NH_4OH solution while stirring. Glycidoxypropyltrimethoxysilane (Dow Corning Chemicals) was added dropwise to the suspension while stirring. The weight ratio of GPS to silica (R) was

Mat. Res. Soc. Symp. Proc. Vol. 435 © 1996 Materials Research Society

varied between 0 and 0.33. Two pH conditions were investigated: pH 4 and pH 9.5. At pH 4, the suspension was usually stirred for 24 hours, while at pH 9.5 the mixture was stirred for only several hours. The interaction between GPS and colloidal silica surfaces under acidic conditions was investigated by ATR-FTIR (Nicolet 750). A ZnSe crystal was used in an ATR accessory with an incident angle of 45°. A spectrum for the $H_2O/EtOH/HNO_3$ solution was first collected and then subtracted from the spectra for the silica suspensions containing GPS. Before coating, triethylene tetraamine (TETA, Aldrich Chemicals) was added as a curing agent for the epoxy groups on GPS. The pH of the suspension was adjusted to 8 before the addition. The amount of TETA was kept constant at a TETA:GPS molar ratio of 1:2. Coatings were prepared one to four hours after TETA was added.

Coatings were fabricated by spreading a suspension onto a polyethyleneterephthalate (PET) substrate (ICI-6138) using a glass rod. This substrate has a hydrophilic coating containing amino functional groups. The coating thickness was controlled by the volume of the suspension and the area of the substrate. The wet coating was then put in a drying oven with air convection at 80°C for at least 2 hours. Coating microstructure was characterized by optical microscopy (Olympics BX60) and scanning electron microscopy (Hitachi S-900).

RESULTS AND DISCUSSION

The first step in the preparation of a coating is the addition of GPS to the colloidal dispersion. The commercial Ludox LS suspension is stabilized by a base additive at roughly pH 8 - 8.5; addition of GPS to the commercial suspension resulted in difficulties in achieving a homogeneous mixture due to immiscibility. While this problem was overcome by the addition of a small amount of ethanol, the basic conditions promoted rapid condensation among silane precursors [11]. When the GPS to silica ratio (R) was 0.3 or above, rapid condensation resulted in the formation of particles which settled over time. Particle formation was also evident in GPS solutions (without silica) under identical conditions. For R values less than 0.3, additional problems in coatings were encountered [10]. By contrast, addition of GPS to an acidified silica suspension resulted in a uniform suspension. This suspension clouded over time due to particle aggregation, but became clear again when the pH was raised back to 9. This suspension was stable at pH 9. When an identical sequence was carried out on a GPS solution (without silica), the solution gelled. The fact that the solution with the silica present did not gel indicates that some of the GPS adsorbed on the silica surfaces and therefore would not polymerize in the same way as free silane.

ATR-FTIR results on suspensions support the conclusion that GPS is adsorbing under acidic conditions. Figure 1 shows data for a suspension over the span of several hours after mixing. The adsorption band around 970 cm^{-1} is assigned to the stretching of terminal Si-OH on the surface of silica particles [12]. The intensity of 970 cm^{-1} band decreases when GPS is mixed into the suspension and continues to decrease until about 25 hours after GPS addition. This decrease is due to the adsorption of silane on the particle surfaces. When GPS is added, a band appears at around 915 cm^{-1} which can be assigned to non-bridging Si-OH stretching from GPS [13,14]. This peak also decreases with time due to condensation between GPS monomers or oligomers or adsorption on silica.

Before coating, a curing agent, polyamine (TETA), was added in an attempt to provide chemical bonding between colloids in the coating. A gelation study was carried out on GPS modified silica suspensions with and without polyamine addition. GPS was first added to an acidified silica colloidal suspension and stirred for 24 hours; the pH was then readjusted to 8. At this time, TETA was added to half of the GPS modified suspension (TETA:GPS molar ratio at 1:2). The gelation behavior of both sets were then observed for a range of R values (See Fig. 2). The suspensions without polyamine addition did not gel. Suspensions with TETA added gelled with a gel time that decreased with increasing GPS amount. The gelation is related to the curing of the epoxy functionalized particles and silane monomers and oligomers in solution. The R values chosen for this study are low and hence the majority of the silane is likely adsorbed onto the particle surfaces. These results suggest that for the colloidal suspension with acidic GPS modification, the epoxy-amino bonding between particles is likely dominant.

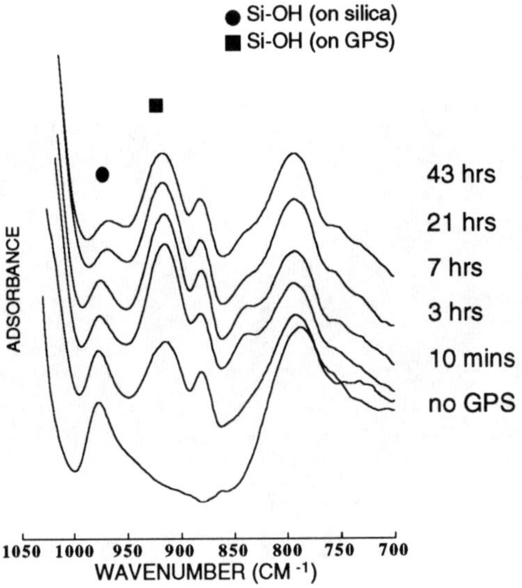

Figure 1 ATR-FTIR data for silica suspensions containing GPS (R = 0.24). Data is shown for increasing time after mixing.

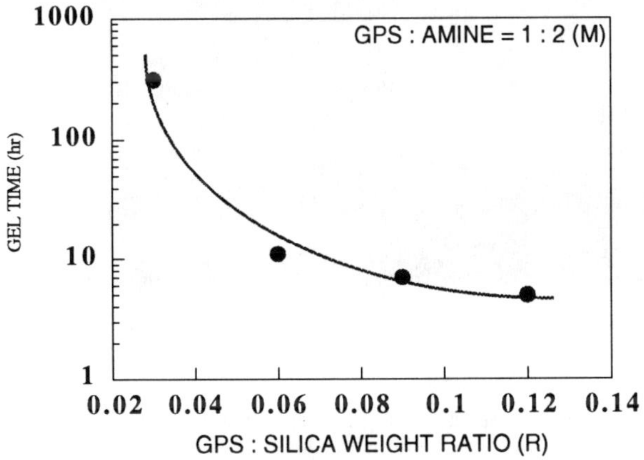

Figure 2 Change in gel time with increasing GPS content for modified silica suspensions containing TETA. Suspensions containing GPS but not TETA did not gel.

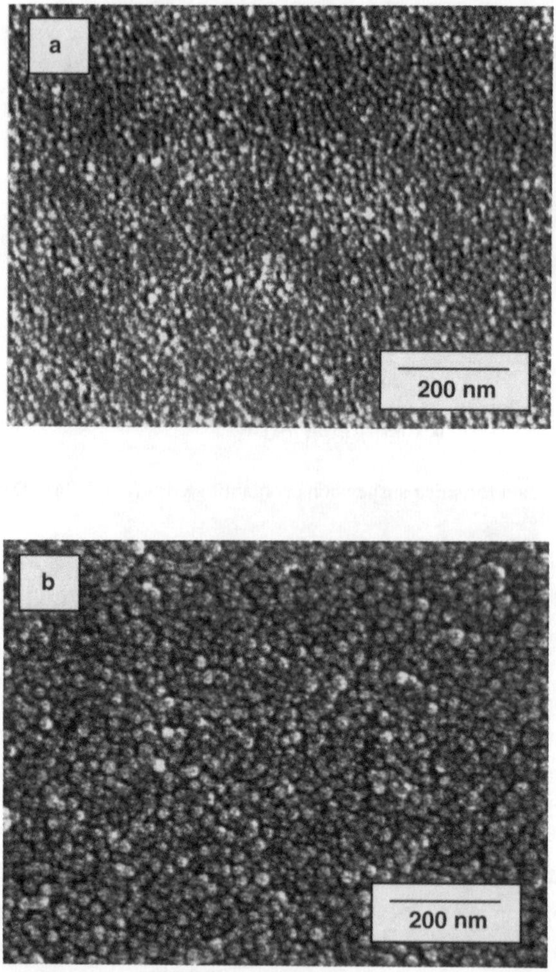

Figure 3 SEM photomicrographs of (a) an unmodified silica coating and (b) a silica coating modified by GPS (R=0.33) under acidic conditions with TETA added as a curing agent

Figure 3 shows the microstructures of modified and unmodified silica coatings. The coating prepared from unmodified silica has a microstructure characterized by packed 12 nm particles. With GPS modification under the acidic conditions, particles are also evident; however, the apparent particle size has increased slightly. The particle size increase is indicative of multilayer adsorption of silane on the particle surfaces and is consistent with other observations. Coatings prepared from modification under basic conditions did not experience the particle size increase, indicating that less GPS adsorbed.

Coating cracking behavior depended on the pH used for modification of the silica with GPS and on the relative amount of GPS (R). Coatings prepared from a colloidal silica suspension without GPS modification adhered poorly to the PET substrate and were prone to cracking. Typically, the maximum coating thickness was only about 1.5 μm. Modification with GPS and addition of a curing agent to create epoxy-amine linkages improves the cracking behavior. Coatings prepared by GPS addition under acidic conditions followed by polyamine addition under basic conditions gave the best results. For these conditions we expect the colloidal particles to have adsorbed GPS on their surfaces and the epoxy-amino bonding between the particles established in the coating. A final thickness of 20 μm was achieved for an R value of 0.33. These coatings also had good optical clarity.

SUMMARY

GPS modified colloidal silica suspensions were used to prepare clear, thick coatings on PET substrates. The modification process involved aging hydrolyzed GPS in an aqueous silica suspension under acidic conditions. After pH adjustment to basic conditions, TETA was added to the suspension and coatings prepared. The TETA was used to cure the GPS modified silica particles in the coating. ATR-FTIR results show GPS adsorption over time and gelation behavior of the coating suspensions supports a curing mechanism of amino-epoxy bonding. Acidic conditions for modification were necessary to avoid rapid condensation among GPS monomers. To achieve a thickness of ~20 μm, a GPS to silica ratio of 0.33 was needed.

ACKNOWLEDGMENTS

The authors gratefully acknowledge research support from the NSF Center for Interfacial Engineering and Rexam Graphics Company.

REFERENCES

1. J. S. Reed, Introduction to the Principles of Ceramic Processing, John Wiley & Sons, New York, 1988, pp. 152-173.
2. R.K. Iler, J. Colloidal and Interface Science, **51** (3), pp 388-393, 1975.
3. Y. Abe and T. Noda. U.S. Patent. #5,372,884, 1994.
4. W.J. Van Ooij, C. Golden, D. R. Boston and E.J. Woo, U.S. Patent. #5204219, 1993.
5. M.A.C. Stuart. Polymer J., **23** (5), pp 669-682. 1991.
6. C.P. Tripp and M.L. Hair. Langmuir, **9**. pp 3523-3529, 1993.
7. H.A. Ketelson, M.A. Brook and R.H. Pelton, Chem. Mater. , **7**, pp 139-187, 1995.
8. S. Brandriss and S. Margel. Langmuir, **9**, pp 1232-1240, 1993.
9. K. Mühle. Colloid & Polymer Sci. **263**, pp 660-672, 1985.
10. L. Chu, M. Daniels and L.F. Francis, in preparation.
11. C.J.Brinker, T.L. Ward, R, Sehgal, N.K. Raman, S.L.Hietala, D.M. Smith, D.W. Hua and T.J.Heedley, J. Membrane Science, **77** (2-3), pp 165-179, 1993.
12. D.l. Wood and E.M. Rabinovich, Applied Spectroscopy, **43** (2), 263-267, 1989.
13. C.A. Capozzi, R.A. Condrate Sr., L.D. Pye and R.P. Hapannowicz. Mater. Letters, **18**, 349-352, 1994.
14. G. Xue, J.L. Koenig, D.D. Wheeler and H. Ishida, J. Applied Polymer Science, **28**, pp 2633-2646, 1983.

Part IV

Mechanical Properties of Selected Organic/Inorganic Materials

EFFECTS OF TEMPERATURE ON PROPERTIES OF ORMOSILS

J.D. MACKENZIE, Q. HUANG, F. RUBIO-ALONSO, AND S.J. KRAMER
Department of Materials Science and Engineering, University of California
Los Angeles, CA 90095-1595, jdmac@seas.ucla.edu

ABSTRACT

Organically modified silicates, the so-called Ormosils, based on silica (SiO_2) and polydimethylsiloxane (PDMS), can have wide variations of mechanical properties depending on their chemical compositions and processing variables. The random network of the SiO_2 is structurally modified by the reactive incorporation of PDMS chains to give materials ranging from hard glassy solids with Vickers number up to 200 Kg/mm^2 to rubbery solids with resilience in excess of 75%. The variations of strength, elastic modulus and mechanical damping with temperature have been studied. The chemical stability of rubbery Ormosils as a function of temperature has been measured in air and in nitrogen. Examples of potential applications are presented.

INTRODUCTION

The first organically modified silicates were independently prepared by Philipp and Schmidt [1] in 1984 and by Wilkes, Orler and Huang [2] in 1985. These new inorganic-organic hybrids were named "Ormosils" by Schmidt and "Ceramers" by Wilkes. Typically, tetraethoxysilane (TEOS) and hydroxyl-terminated PDMS are reacted in alcoholic solutions such that chemical bonds are formed between PDMS chains and SiO_2 clusters [3]. When the PDMS: SiO_2 molar ratios are small, for instance 1:10, the resulting Ormosils are hard and glass-like and have Vickers numbers up to 200 Kg/mm^2 [4]. At higher PDMS: SiO_2 ratios, 3:10 for example, the resultant solid becomes rubbery [5]. The rubbery behavior has been attributed to the ability of the PDMS chain to coil under externally applied pressure since the Si having two CH_3 groups is free to rotate and the Si-O-Si angle is approximately 150°[6]. When the pressure is removed, volumetric recovery is possible via the uncoiling of the chains. The formation of hard Ormosils and rubbery Ormosils as a function of the concentration of PDMS chains in solution can be compared to the preparation of inorganic silicate glasses via the high temperature reactions between SiO_2 and metallic oxides (e.g., Na_2O, CaO). The random network of the SiO_2 structure is modified by the PDMS in the case of Ormosils and by Na_2O and CaO for silicate glasses [7]. Thus, properties such as hardness, brittleness, tensile strength and elastic modulus of Ormosils will decrease as a function of the PDMS concentration. Figure 1 shows how the random network of SiO_2 tetrahedrons, having low expansion coefficients similar to that for silica glass when the PDMS contents are small, increases sharply when the PDMS contents are in excess of about 15%. Figure 2 is an illustration of the typical transitions from rigidity to rubbery behavior of Ormosils.

At present, little information exists on the effects of temperature on the physical and chemical properties of these Ormosils. Because they constitute a new family of hybrid materials containing relatively high concentration of inorganic components, it is possible that they may be thermally more stable than pure organic plastics and

Mat. Res. Soc. Symp. Proc. Vol. 435

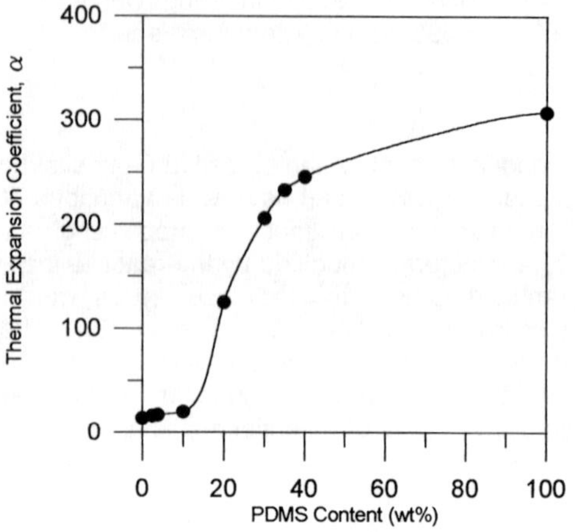

Figure 1. Linear coefficient of thermal expansion vs. PDMS content

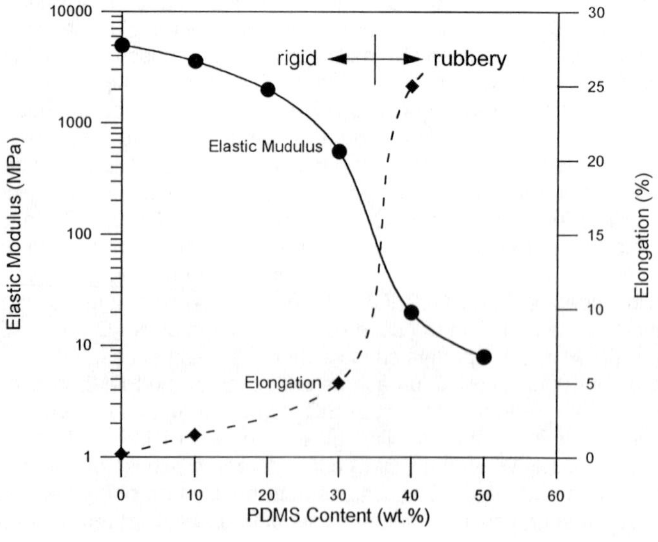

Figure 2. Variation of mechanical properties of Ormosils as a function of PDMS content [7].

rubbers. Thus, a study of the thermal stability of Ormosils is deemed desirable. This report is an up-to-date broad review of the effects of temperature on some mechanical properties of the PDMS-TEOS based Ormosils in general, but primarily on the rubbery solids.

THERMAL STABILITY OF ORMOSILS

For an Ormosil based on PDMS and TEOS, it is likely that in an oxidizing atmosphere, the CH_3 groups will be the most unstable when the temperature is raised. For the silicone rubbers, Nielson has shown that this is indeed the case [8]. Since the decomposition of the CH_3 groups must have an adverse effect on the mechanical properties of materials such as silicone rubbers and Ormosil rubbers, it is important to consider various possibilities of thermal stabilization. It is known that for silicones, the substitution for the CH_3 groups by phenyl or carborane groups can increase thermal stability. However, this study is concerned mainly with additives to the TEOS-PDMS system and not with substitution of the CH_3. The addition of approximately one weight percent of Fe, in the forms of Fe_2O_3, Fe powder and $FeC\ell_3$ are known to have thermal stabilization effects on silicones [8]. Preliminary work in our laboratory had also shown that $FeC\ell_3$ additions to the TEOS-PDMS alcoholic solution were effective in the thermal stabilization of the resultant Ormosil [9]. Recently, we have used iron isopropoxide as an additive in place of $FeC\ell_3$. The TGA results for an Ormosil rubber with a TEOS/PDMS weight ratio of 60/40 containing various amounts of Fe ions are shown in Figure 3. Although trace amounts of Fe does appear to influence the

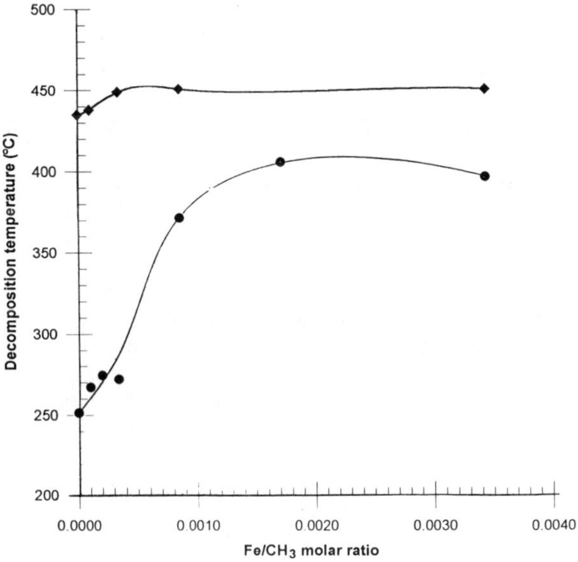

Figure 3. Effect of iron on thermal decomposition temperature, by TGA, of ORMOSILs in air (bottom curve) and in nitrogen

decomposition of Ormosils in an N_2 atmosphere, the effects are seen to be such smaller than that in air. The surprising fact is that less than 1 Fe to more than 1000 CH_3 groups can have much significant effect on the oxidative decomposition of Ormosils. Various mechanisms of the protective role of Fe have been proposed [8]. These involve the formation of free radicals when O_2 reacts with CH_3 groups. The free radicals would react further with O_2 to form volatile products such as CH_2O. However, this latter step can be suppressed if the Fe^{3+} ions present preferentially react with the free-radicals first formed to give Fe^{2+} ions. The Fe^{2+} ions are then reconverted to Fe^{3+} ions by the O_2 present and thus suppress the reaction which produces CH_2O. It is difficult to apply such mechanisms in the case of Ormosils where there are less than one Fe^{3+} ion for more than 1000 CH_3 groups. The most likely coordination number for Fe^{3+} in oxides is six. If one assumes, as for silicate glasses, that three bridging oxygens and three non-bridging oxygens are the nearest neighbors of every Fe^{3+} ion, then there can be no more than twelve CH_3 groups which are nearest neighbors to the Fe^{3+} ion. At temperatures of 300˚, there may be sufficient molecular motion to permit each Fe^{3+}, trapped by ionic attractive forces in the octahedral hole formed by six oxygens, to provide the "protection" suggested above to the twelve nearest CH_3 groups. It is extremely difficult to account for the inertness of the other 1000 CH_3 groups at some distance away. The Ormosil remains highly porous and thus there is no hindrance to the motion of the O_2 molecules in the ambient atmosphere during heating. At present, it appears that a plausible explanation would involve some mechanisms which permit the rapid motion of electrons along the surfaces of the SiO_2 clusters between different PDMS chains. Fe ions are apparently not the only transition metal ions which are effective in minimizing the oxidation of Ormosils. Preliminary experiments in our laboratory have confirmed that vanadium ions are also effective although to a lesser degree than Fe ions. Much research are required to attain understanding of this very interesting mechanism involving Fe and other ions.

MECHANICAL PROPERTIES OF ORMOSILS AS FUNCTION OF TEMPERATURE

Some mechanical properties of Ormosils, with and without the additions of Fe ions, have recently been measured in our laboratory as a function of heat treatment at various temperatures. Because the mechanical properties of Ormosils are dependent on their compositions, processing variables, additives and heat treatment temperatures and times, the objective of the presentation of the results shown below is primarily to illustrate the general trends of Ormosils when temperatures are altered. No attempt is made here to correlate properties with the microstructure of the samples studied. The elastic modulus and logarithmic decrement of two Ormosil samples, one containing Aerosil as a filler are shown in Figures 4 and 5, respectively in comparison with samples of a Neoprene rubber and a silicone rubber (red silicone containing iron oxide additive). The molecular weight of the PDMS was 1700 and the TEOS/PDMS weight ratio was 60/40. Measurements were made at temperatures in air. In general the Ormosils resemble the silicone rubber in elastic modulus, damping and T_g. The Ormosil sample containing Aerosil as a filler appears to have higher temperature stability in damping than the sample without filler. The T_g values of Ormosils are approximately -100˚. The resilience of the 60/40 Ormosil with and without Fe are

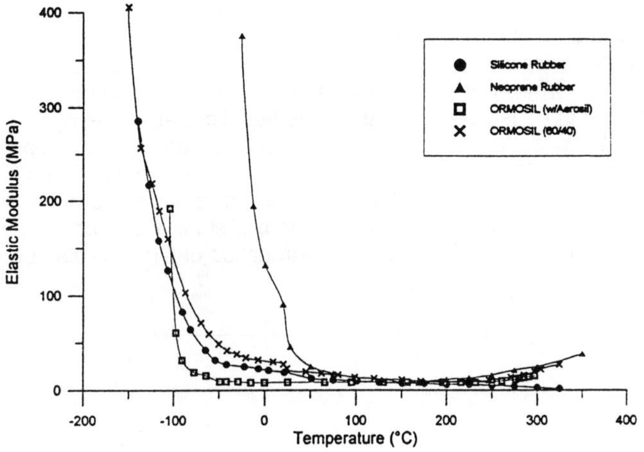

Figure 4 Elastic modulus of ORMOSILs and rubbers

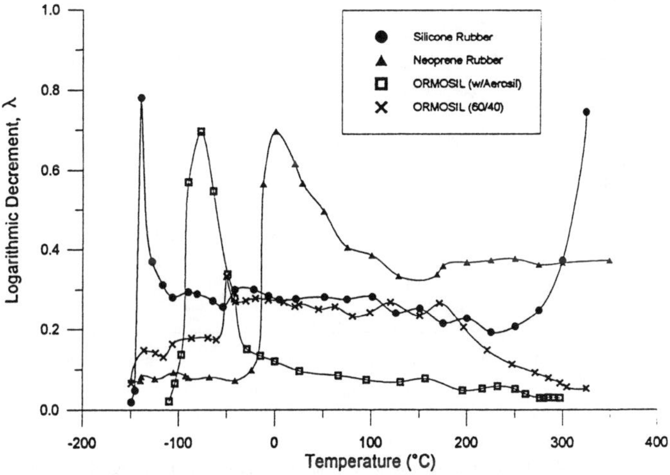

Figure 5 Damping properties of ORMOSILs and rubbers

compared to the behavior of a styrene-butadiene rubber in Figure 6. The samples were all heated in air at the temperatures shown for 24 hrs and resilience measured at 20°. The beneficial effects of the small additions of Fe ions are clearly shown. The tensile strengths of Ormosils with varying amounts of Fe ions are shown in Figure 7. Again, the addition of Fe is seen to have a large beneficial effect despite the fact that allowance must be made for the usual scatter of strength data. The sample containing a Fe/CH_3 ratio of 0.00171 appears to have the highest strength even after heat-treatment at 350° for 24 hrs. Although the reasons why small amounts of Fe ions can have very significant beneficial effects on the thermal stability of rubbery Ormosils are still unknown, it is certain that the oxidative degradation of Ormosils can be minimized.

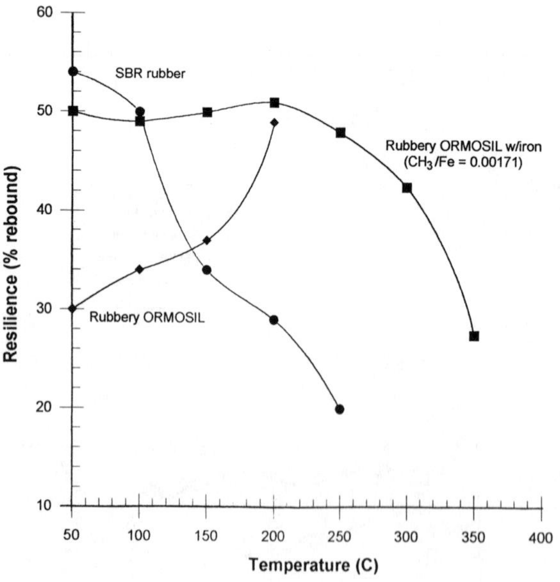

Figure 6 Resilience of rubbery ORMOSIL and an organic rubber with heat treatment temperature (24 hrs.)

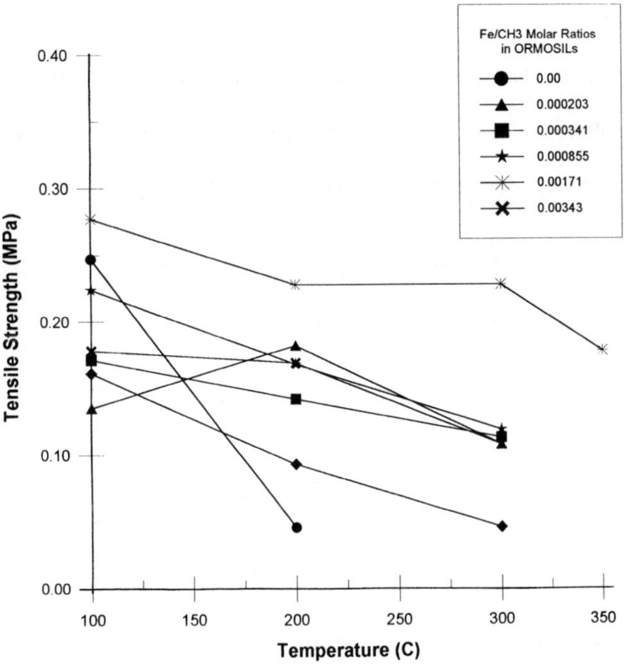

Figure 7 Tensile strengths of ORMOSILs with different iron concentrations as a function of temperature (heated 24 hrs., measurement at room temperature)

CONCLUSIONS

Ormosils based on TEOS-PDMS can be prepared as hard glassy solids with Vickers number up to 200 Kg/mm^2 or as rubbery materials with properties resembling silicone rubbers depending on the SiO_2/PDMS ratio. The addition of very small amounts of iron in the form of Fe^{3+} ions is highly effective in minimizing oxidative degradation and permits the use of Ormosils to temperatures in excess of 325°. Scientific understanding of the protective role of Fe ions is lacking at present. New Ormosils based on the substitution of CH_3 with other organic groups together with Fe additives will further increase the upper use temperatures of these new organic-inorganic hybrids.

ACKNOWLEDGMENTS

We are grateful to the Air Force Office of Scientific Research (Grant F49620-94-1-0071), the Scientific Laboratory of the Ford Motor Company and the Research Division of Hoechst Celanese Corporation for financial support.

REFERENCES

1. G. Philipp and H. Schmidt, J. Non-Cryst. Solids, **63**, p. 283 (1984).

2. G.L. Wilkes, B. Orler and H. Huang, Polym. Prepr., **26**, p. 300 (1985).

3. T. Iwamoto, K. Morita and J.D. Mackenzie, J. Non-Cryst. Solids, **159**, p. 65 (1993).

4. T. Iwamoto and J.D. Mackenzie, J. Sol-Gel Sci. and Tech. **4**, p. 141 (1995).

5. Y.J. Chung, S.J. Ting and J.D. Mackenzie, MRS Symp. Proc., **180**, p. 981 (1990).

6. J.D. Mackenzie, Y.J. Chung and Y. Hu, J. Non-Cryst. Solids, **147/148**, p. 271 (1992).

7. J.D. Mackenzie, Q. Huang and T. Iwamoto, J. Sol-Gel Sci. and Tech., in press (1996).

8. J.M. Nielson, Stabilization of Polymers and Stabilization Processes, Advances in Chemistry Series, 85, (1968), p. 95.

9. S.J. Kramer and J.D. Mackenzie, MRS Symp. Proc. **346**, p. 709 (1994).

TAILORING OF THERMOMECHANICAL PROPERTIES OF THERMOPLASTIC NANOCOMPOSITES BY SURFACE MODIFICATION OF NANOSCALE SILICA PARTICLES

C.Becker*, H.Krug, H.Schmidt

*Composite Technology Group, Institut für Neue Materialien gem. GmbH, Im Stadtwald, Geb.43, D-66123 Saarbrücken, Germany, Tel.: +49/681/302 5000, Fax.: +49/681/302 5223

Keywords: Thermoplastics, nanocomposites, thermomechanical properties

Abstract

Thermoplastic nanocomposites based on linear polymethacrylates as matrix materials and spherical silica particles as fillers have been synthesized using the in situ free radical polymerization technique of methacrylate monomers in presence of specially functionalized SiO_2 nanoparticulate fillers. Uncoated monodisperse silica particles with particle sizes 100 nm and 10 nm were used as reference fillers. For surface modification, the alcoholic dispersions of the fillers were treated with appropriate amounts of methacryloxypropyltrimethoxysilane (MPTS) and acetoxypropyltrimethoxysilane (APTS). Transmission electron microscopy (TEM) was used to investigate dispersion behaviour in dependence on surface modification. Dynamic mechanical properties were measured by dynamic mechanical thermal analysis (DMTA).

1. Introduction

Thermoplastic polymers filled with inorganic spherical particles in the nanoscale range, so called nanocomposites, are expected to show new interesting thermal and mechanical properties compared to thermoplastics filled with microparticles. According to Wu[1,2,3], the interparticulate distance i (or matrix ligament thickness) in a composite is dependent on the filler volume fraction ϕ and on the particle diameter d. It can be calculated by the following equation (1) assuming that monodisperse filler particles are arranged in a simple cubic packing:

$$i = ((\pi / 6\,\phi)^{1/3} - 1) * d \qquad (1)$$

In general, the properties of composites filled with microparticles mainly depend on the filler volume fraction and only slightly on any interfacial phase existing between filler and matrix. The reason for this is the unfavourable surface to volume ratio of microparticles. Polymer segments adsorbed on the filler surface in a definite layer thickness (layer thickness 2 nm for example as shown in the literature[4]) in general possess structures different from the bulk structure. The interfacial phase volume percentage (A) can be calculated by equation (2)

$$A = F / (v(1-F)) * (V-v) * 100 \qquad (2)$$

where F is the filler volume fraction, v is the volume of one particle without interfacial layer and V the volume of one particle having an interfacial layer of definite thickness. According to equation (2), (A) is approximately 30 vol.% for a nanocomposite with 15 vol.% 10 nm particles assuming a layer thickness of 2 nm. From this point of view it can be expected that nanoparticles influence the mechanical and thermal properties of the resulting composites by the in-

Mat. Res. Soc. Symp. Proc. Vol. 435

terfacial phase in addition to the pure "filling" effect. In order to observe the predicted effects, an almost perfect dispersion of the nanoparticles in the matrix is required. For this reason, the particle surface and the matrix have to be "compatiblized", which means the interfacial free energy has to be decreased.

2. Experiment

Monodisperse 10 nm (Nissan) and 100 nm (Merck) Stöber silica particles were coated with methacryloxypropyltrimethoxysilane (MPTS) and acetoxypropyltrimethoxysilane (APTS) in a methanolic dispersion containing approximately 1 vol.% silica. MPTS was used to achieve covalent bonding between filler and matrix by fixation of the polymerizable methacrylate end group along the polymer backbone. APTS was used to realize compatibilization without covalent bonding because of the absence of the polymerizable double bond. The 100 nm particles were used as reference systems for bigger particles with a less favorable surface to volume ratio. Butylamine was used as catalyst to accelerate the coating reaction. In the case of 100 nm particles, the silanes were added in twofold excess compared to the calculated amount of silanole groups. After reaction at 50°C for 12 h the unreacted silane was removed from the silica by centrifugation and subsequent washing with methanol. In the case of 10 nm particles the coating reaction was performed using equal molar amounts of silane compared to the calculated amount of silanole groups. The dispersions with 10 nm particles were used without further centrifugation.

The monomers methylmethacrylate (MMA) and 2-hydroxyethylmethacrylate (HEMA) were distilled in vacuum prior to use. Equal amounts of both monomers were mixed with calculated amounts of the methanolic silica dispersions. After removal of methanole the monomer mixtures contained 2, 5 and 10 vol.% silica. These mixtures were cured in a temperature program until 120°C by free radical polymerization using azobisisobutyronitrile (AIBN) as initiator. The curing was performed in special reaction containers under exclusion of oxygen, resulting in plate like poly(MMA-co-HEMA) nanocomposites with a thickness of 4 mm. The nanocomposites were ultramicrotomed using a Reichert ultracut and their morphology was investigated in a transmission electron microscope (TEM). Dynamic mechanical thermal analysis (DMTA) was performed using samples with rectangular cross section. The samples were measured in single cantilever bending mode at 1 Hz with a deflection of ±8 μm between 40°C and 240°C (heating rate 0,5 K/min).

3. Results and Discussion

For the preparation of hydrophilic polymer matrices, HEMA was added to MMA. The poly(MMA-co-HEMA) nanocomposites containing 2, 5 and 10 vol.% filler, based on pure silica, were synthesized by free radical polymerization in the presence of the silica particles. In the case of 10 nm particles, transparent thermoplastic nanocomposites could be obtained. In contrast to this, the composites with 100 nm silica particles were only translucent. Fig 1 shows the transmission electron micrographs from ultramicrotomed sections of the nanocomposites filled with 2 vol.% 10 nm SiO_2 particles with different particle coatings.

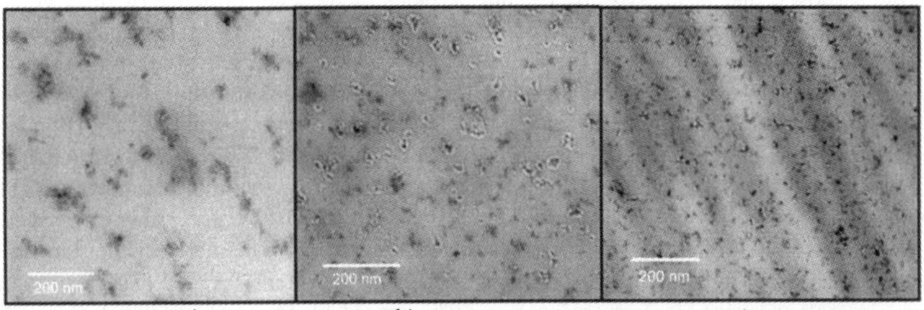

a) b) c)

Fig.1: TEM micrographs of poly(MMA-co-HEMA) nanocomposites with 2 vol.% of a) uncoated, b) APTS coated and c) MPTS coated 10 nm SiO_2 particles.

In figure 1, large aggregates of nearly 100 nm in size are visible in the case of uncoated particles, whereas in the case of particles with APTS coating, small aggregates are visible and in case of MPTS coating the particles are nearly homogeneously distributed in the polymer matrix. It is obvious, that the surface modification of the filler particles by compatibilization agents is an essential step to produce well dispersed agglomeration free composites. This is evident especially for the MPTS-coated particles, which have the same functionality on the surface as the matrix material along the polymer backbone.

Dynamic mechanical thermal analysis (DMTA) was performed in order to investigate the rheological properties, because it offers much information about the influence of the particle coating and the state of particle agglomeration on the thermomechanical behaviour of the resulting composites[5,6,7,8]. The storage modulus E' is equal to the elastic response, in fact the reversible part of the resistance against the applied dynamic strain (stiffness). As shown in literature, especially above the glass transition temperature (Tg), E' will be increased compared to the unfilled matrix by increased filler volume content, by increased phase adhesion between filler and matrix and, in the presence of particle aggregates, by particle/particle friction at the points of contact between the primary filler particles[9,10]. Figure 2 shows the storage modulus E' at 170°C, in the rubbery plateau region above the glass temperature, for poly(MMA-co-HEMA) nanocomposites in dependence on the filler volume content for different filler surface coatings.

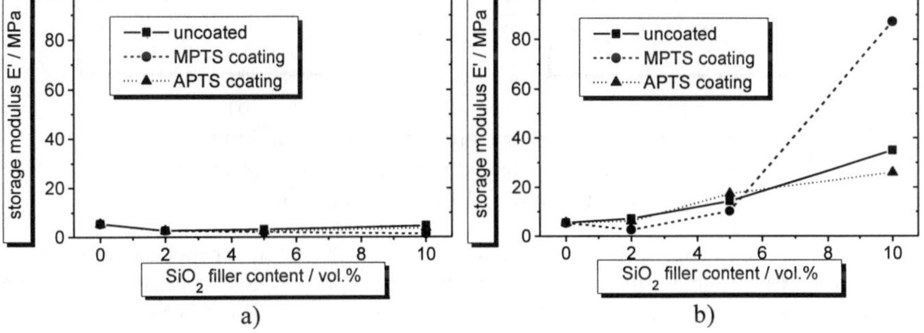

Fig.2: Dependence of the storage modulus E' for poly(MMA-co-HEMA) nanocomposites on the filler volume fraction for different surface coatings of the a) 100 nm and b) 10 nm silica particles; T = 170 °C.

Figure 2 shows, that in the case of 100 nm particles there is no influence of filler volume fraction and filler/matrix adhesion on the storage modulus at 170°C compared to the systems filled with 10 nm silica particles. In the case of 10 nm particles the storage modulus E' increases with filler content for all systems. Up to 5 vol.% particle content, there is no significant influence by the particle coating. At 10 vol.% filler content the MPTS coated particles cause a large increase in E' compared to the uncoated and APTS coated particles. The MPTS coating of the particles allows covalent adhesion of polymer molecules near the filler surface during composite synthesis, resulting in immobilization of these polymer segments. The immobilized shell like polymer increases the apparent particle diameter of the pure silica particles obtained by TEM and therefore increases the filler volume fraction. The result is an increase in melt viscosity, in fact an increase in E' caused by an increased resistance of the greater particles against the applied dynamic strain.

Polymers in general are viscoelastic materials. Usually the applied dynamic deformation and the response of the sample are out of phase. The phase shift δ in a DMTA experiment is expressed as damping tan δ. The peak area under the tan δ curve at the glass transition for example, is a measure of the energy dissipated during the dynamic experiment and gives information about the viscous parts of the sample. The glass transition temperature (Tg) itself is defined as the temperature corresponding to the maximum in the tan δ curve. Changes in Tg can be classified as significant if they show differences of more than ±5°C compared to the Tg of the pure matrix. The Tg is expected to increase compared to the Tg of the pure matrix in the case of covalent filler adhesion, because for the part of polymer segments near the filler surface the chain mobility is restricted[11]. Figure 3 shows the glass transition temperature for all composites with 100 nm and 10 nm particles in dependence on the silica filler content.

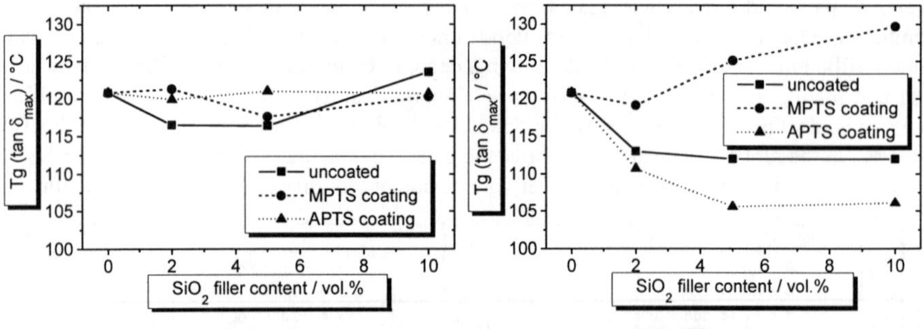

a) b)

Fig.3: Dependence of the glass transition temperature from tan δ maximum for poly(MMA-co-HEMA) nanocomposites on the filler volume fraction for different surface coatings of the a) 100 nm and b) 10 nm silica particles.

For 100 nm particles (fig. 3a)), there is no significant Tg change in dependence on the filler content and the filler coating. In contrast to this, the Tg is changing to a marked extent in the case of 10 nm particles (fig 3b)). The MPTS coating causes immobilization of polymer chains near the filler surface by covalent adhesion with resulting restricted chain mobility. The APTS coating leads to a decrease in Tg compared to the unfilled polymer. It can be assumed, that the covalent bonding of MPTS-coated particles results in a rigid polymer morphology at the particle surface and Tg increases. Although the ATPS-coated particles are less agglomerated in comparison to the uncoated particles by compatibilization, the polymer morphology at the surface is "softened" by ATPS-molecules and Tg decreases caused by increased chain mobility.

According to the above results, the nanocomposites with 10 vol.% 10 nm silica seem to be the interesting ones to be compared over the whole investigated temperature range. Figure 4 shows the temperature dependence of the storage modulus and the tan δ between 40°C and 240°C for all these nanocomposites compared to the unfilled matrix.

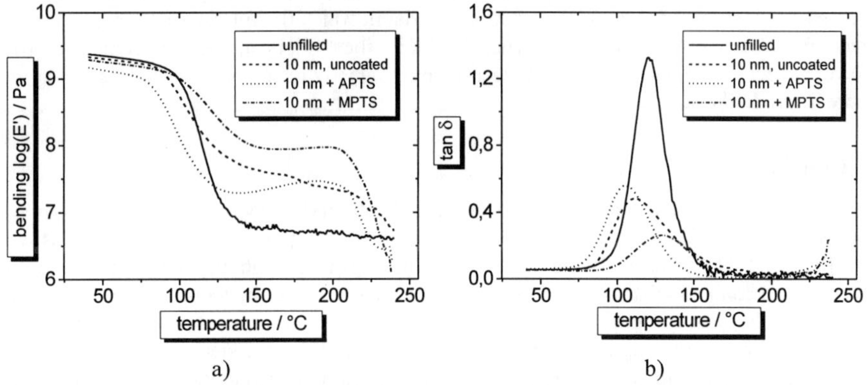

Fig.4: Temperature dependence of a) the storage modulus E' and b) the damping tan δ of poly(MMA-co-HEMA) nanocomposites filled with 10 vol.% 10 nm SiO_2 particles with different coatings compared to the unfilled matrix.

The plateau-like behaviour of the storage modulus between 120°C and 220°C for the unfilled poly(MMA-co-HEMA) shown in figure 4a), indicates that the pure bulk matrix is built up by high molecular weight polymer species ($M_n > 10^6$ g/mol) with a large number of entanglements. The system with 10 vol.% uncoated 10 nm particles therefore shows the expected enhancement in E' value in the same temperature range caused by the filler loading. However, there is a faster decrease in E' value with increasing temperature compared to the pure matrix system, indicating a faster destruction of entanglements between the polymer chains. In comparision with this, the storage modulus shows a maximum with MPTS and APTS coated particles after going through a minimum at a higher level compared to the unfilled system with increasing temperature. It is generally expected, that the storage modulus decreases with increasing temperature. This phenomenon cannot be attributed to residual monomers and additional curing effects, since IR-measurements proved that the residual monomer content is only about 1 %. An explanation for the effect could be that the particle to polymer bond is strengthened by temperature, leading to less flexible interfacial structures. At temperatures above 210°C an additional flow process starts in the systems with APTS and MPTS coated particles, visible in a sharp decrease in E'. This indicates the presence of lower molecular weight polymer chains with less amounts of entanglements, which tend to flow at lower temperatures.

Figure 4b) shows that there is a large decrease in the peak area under the tan δ curve for all filled systems compared to the unfilled polymer. The damping is not only reduced by the presence of 10 vol.% filler but also by covalent filler/matrix adhesion in the case of MPTS coating. The polymer segments near the filler surface are covalently bound and do not seem to contribute to the glass transition of the matrix. Besides this it should also be noted that the particles in a composite normally contribute to the damping behaviour too, especially by particle/particle and particle/matrix friction as indicated by Nielsen[9], a fact which can not be quantified here.

In fact it can be concluded for the system with 10 vol.% MPTS coated 10 nm particles that the interparticulate distance in the nanocomposites, which lies in the range of the silica particle diameter, is just too small to be treated as a material in which the filler particles are well separated from each other. The MPTS coated 10 nm silica fillers in the nanocomposite with 10 vol.% seem to be interconnected by their outer covalently fixed polymer shell, resulting in a more network like structure.

4. Conclusions

Thermoplastic nanocomposites based on methylmethacrylate (MMA) and 2-hydroxyethylmethacrylate (HEMA) as comonomers could be prepared using monodisperse uncoated as well as MPTS and APTS coated 10 nm and 100 nm SiO_2 particles. High dispersity was achievable by increased compatibilization between filler and matrix as shown by transmission electron microscopy. The nanocomposites with MPTS coated 10 nm particles showed the most pronounced effects in rheological and thermal properties. For these systems also an increase in glass transition temperature of about 10°C was detectable compared to the unfilled polymer. In both cases the behaviour could be interpreted by covalent fixation of polymer chain segments at the filler surface produced while synthesis. The results indicate interconnection between the filler particles by entanglements via their outer covalently bound polymer shell.

Acknowledgements

The financial support for the Minister of Research and Development of the state of Saarland is greatfully acknowledged.

5. References

[1] S. Wu, Polymer **26**, 1855 (1985)
[2] S. Wu, A. Margolina, Polymer **29**, 2170 (1988)
[3] S. Wu, J.Appl.Polym.Sci. **35**, 549 (1988)
[4] K. Kendall, F.R. Sherliker, Brit.Polym.J. **12**, 85 (1980)
[5] C.T.J. Landry, B.K. Coltrain, J.J. Fitzgerald, Macromolecules **26**, 3702 (1993)
[6] K. Gandhi, R. Salovey, Polym.Eng.Sci. **28**, 1628 (1988)
[7] P. Cousin, P. Smith, J.Polym.Sci.: Part B: Polym.Phys. **32**, 459 (1994)
[8] M. Sumita, H. Tsukihi, K. Miyasaka, K. Ishikawa, J.Appl.Polym.Sci. **29**,1523 (1984)
[9] B.L. Lee, L.E. Nielsen, J.Polym.Sci.: Part B: Polym.Phys. **15**, 683 (1977)
[10] T.B. Lewis, L.E. Nielsen, J.Appl..Polym.Sci. **14**, 1449 (1970)
[11] G.J. Howard, R.A. Shanks, J.Macromol.Sci.Phys. **19**, 167 (1981)

IMPROVEMENT OF CRYOGENIC FRACTURE TOUGHNESS OF EPOXY BY HYBRIDIZATION

S.Nishijima*, M.Hussain**, A.Nakahira***, T.Okada* and K.Niihara *

*ISIR Osaka Univ., 8-1, Mihogaoka, Ibaraki, Osaka, 567, Japan
**Bangladesh Insulator and Sanitaryware Fact.Ltd,Bux Nagar,Mirpur,Dhaka-1216, Bangladesh
***Dept. Inorganic Mater., Kyoto Institute of Tech., Matsugasaki, Sakyo-ku, Kyoto 606,Japan

ABSTRACT

Improvement of the fracture toughness of epoxy at cryogenic temperatures has been carried out aiming at cryogenic application. The macroscopic molecular state of hybrid material was studied and the possible mechanism to improve the fracture toughness even at cryogenic temperature was discussed.

Based on the mechanism the hybrid materials were prepared by dispersion of silica through hydrolysis of alkoxide. To connect the silica to epoxy molecule a coupling agent was used. The connection was successfully performed and was confirmed by positron annihilation method. The fracture toughness of the developed hybrid material was found to be approximately 2.6 times higher than that of epoxy itself at liquid helium temperature.

INTRODUCTION

Impregnation has been often employed to restrain wire motion in a superconducting magnet which is one of the main causes of instability in the magnets. However crack introduction to the impregnating material epoxy induces heat generation and results in another instability [1,2]. High fracture toughness of epoxy even at cryogenic temperature is inevitable for stable operation of superconducting magnets[3,4].

The macro-molecular state of epoxy resin for cryogenic use was evaluated in terms of positron annihilation lifetime (PAL) method and revealed that the epoxy having larger molecular weight between crosslinks shows higher fracture toughness even at cryogenic temperatures. The plasticizer did not improve or even degraded fracture toughness at cryogenic temperature[5,6]. This means that the three-dimensional epoxy network should not be broken and the local relaxation mode in main chain is important to improve fracture toughness at cryogenic temperatures.

In this work ceramic filler was dispersed through hydrolysis reaction of alkoxide[7] and the improvement of fracture toughness at cryogenic temperatures was attempted.

MOLECULAR DESIGN

Though macroscopic inorganic fillers have often been employed to improve the mechanical properties of polymers [8], an increase in viscosity resulted as well as difficult impregnation processing. In this case the larger sized filler is effective to improve the fracture toughness due to the crack pinning, deflection or bridging. In this sense the size of the filler in hybrid materials is too small.

Other than these factors in the hybrid material the modulation of local relaxation in the epoxy main chain can be expected. When the molecular sized silica is crosslinked to the epoxy main chain, the modulation effects would be expected.

Fig. 1 shows the molecular state model which was assumed. In the hybrid materials without using coupling agent, the three-dimensional epoxy network will be destroyed due to the steric

Mat. Res. Soc. Symp. Proc. Vol. 435 © 1996 Materials Research Society

hindrance around silica and voids will be introduced as shown in Fig.1(a). Two effects of the voids to mechanical properties are expected depending on the size. When the size of the void is large (say larger than 1 micrometer) which will be eliminated with improving the process, the mechanical properties would be decreased markedly. The pressurization during curing (which is often used in the process) will recover the degradation because the void is depressed and size of the void becomes small enough to neglect the stress concentration at the void. On the other hand if the size of the void is small enough (same order of crosslink distance of epoxy main chain or free volume) which cannot be avoided, the hybrid material will become softer at room temperature because the introduced void would act as free volume. In this case the pressurization will not be so effective to improve the mechanical properties because the nano-sized void is just the same order of crosslink distance of epoxy molecules and hence is difficult to eliminate.

By using the coupling agent the epoxy molecules are crosslinked to silica. Addition of coupling agent will be effective to eliminate both the macroscopic and nano-sized voids as shown in Fig.1(b). Furthermore if there exist enough small-sized silica molecules to modulate the local relaxation in the epoxy main chain, the improvement of fracture toughness would be expected even at cryogenic temperatures

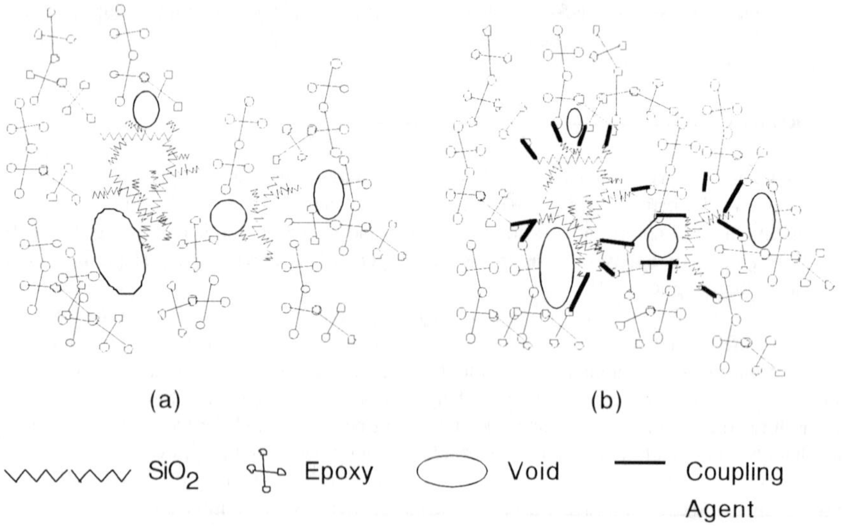

Fig.1 Molecular state model of the system (a) without coupling agent and (b) with coupling agent. Micro void (a) and crosslinks between epoxy and silica (b) are also presented.

EXPERIMENTAL

Specimen

The epoxy used in this work was tetraglycigyl-meta-xylen-diamine (TGMXDA) having four reactive groups and hence high crosslinking density and good cryogenic properties were expected. The curing agent was 1,2 cyclohexane-dicalboxylic anhydride (HHPA). The weight ratio of TGMXDA to HHPA was 100:131. In Fig.2 the chemical structures are presented.

To disperse nanometer scale silica filler in epoxy matrix, the hydrolysis reaction of tetraethoxysilane (TEOS) as shown in Fig.3 was employed. After silica was precipitated, the catalysis and the alcohol produced by the hydrolysis reaction were fully evaporated. The curing agent was mixed and the system was pre-cured at 363K for 2 hours and post-cured at 423K for 2 hours. In order to shrink the macro void introduced by dispersing silica filler the system was also cured at 20 atm. Changing the amount of the silica up to 9.4% by weight the epoxies were cured. The coupling agent was used to crosslink between silica and epoxy molecules. The coupling agent used was N-β aminoethyl-γ–aminopropyl-methyldimethoxy-silane. The chemical structure is also presented in Fig.2. Keeping the amount of the silica at 4.9 % by weight constant, the amount of coupling agent was changed. The amount of coupling agent was small and hence the weight fraction of silica was not affected by that of the coupling agent.

Four types of specimen were fabricated: (i) silica dispersed in the epoxy and (ii) silica crosslinked to epoxy molecules with coupling agent. They were cured at 1 or 20 atm. The pressurization was made in order to examine whether the pressurization improve the mechanical properties of hybrid materials.

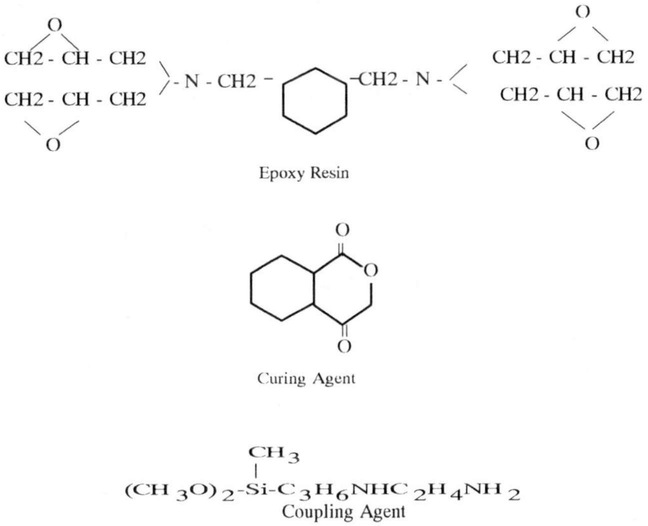

Fig.2 Chemical structure of epoxy resin, curing agent and coupling agent.

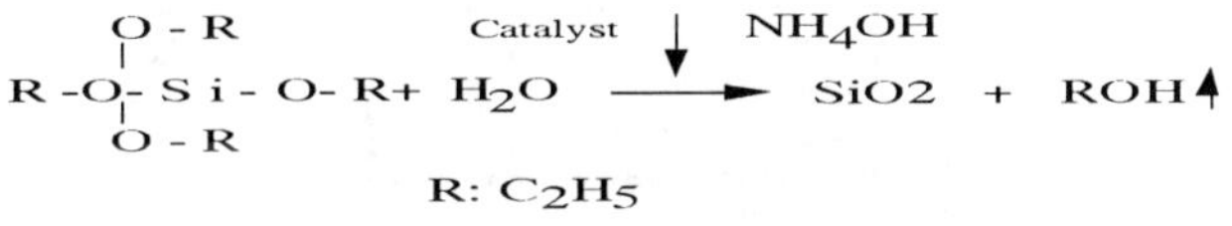

Fig.3 Hydrolysis reaction to form silica filler.

Evaluation

The size of free volume or void was evaluated by means of Positron Annihilation Lifetime Spectroscopy (PAL). PAL was performed by fast-fast coincidence system using 1μCi 22NaCl as a source. The obtained spectra was divided into 3 components and the longest lifetime τ_3 was used to evaluate the size of the free volume. The longer positron lifetime means larger free volume or void. The long-lived lifetime τ_3 is related to the size of the free volume as following formula,

$$1/\tau_3 = 2\{1-R/R_0+1/2\pi \sin(2\pi R/R_0)\}$$

where τ_3 is lifetime in nanoseconds, R is radius of free volume in angstroms and $R=R_0-\Delta R$ where $\Delta R=1.656$ Å. Consequently an increase in τ_3 means an increase in free volume size [9]. The free volume size in epoxy resin which is evaluated by PAL is usually smaller than 1nm . The important fact for applying PAL is that PAL cannot detect larger voids. Hereafter the voids cannot be detected by PAL is called as "macrovoid".

The fracture toughness was measured at room (RT), liquid nitrogen (LNT: 77K) and liquid helium (LHeT: 4.2K) temperature. The single edge notched beam method was used to obtain the fracture toughness. The size of the specimen was 4x8x32 mm and the notch was introduced by knife edge.

RESULTS AND DISCUSSION

In Fig.4 the positron lifetime τ_3 is plotted against weight fraction of coupling agent keeping the amount of silica 4.9 % by weight. To discuss the results it is important to keep in mind that the detectable void size with PAL is smaller than several nanometers (named nanovoid). The lifetime of epoxy itself was low compared with that of silica dispersed epoxy (hybrid material) as shown by closed markings. The addition of silica increased the void size. Especially, systems without coupling agent show longer positron lifetime and indicate that the larger nanovoids were introduced (but the size is smaller than several nanometers).

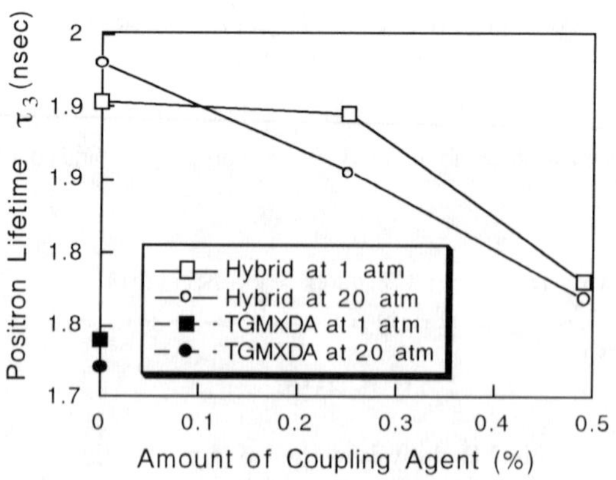

Fig.4. Amount of coupling agent dependency of positron lifetime.

As the amount of coupling agent increases the positron lifetime decreases that is indicating decreases in nanovoid size. This indicates that silica crosslinked with epoxy molecules. The hybrid materials including 0.49% of coupling agent by weight have almost same size of nanovoid as epoxy itself. The introduced nanovoids are eliminated by coupling agent. This suggests that modulation of relaxation of the epoxy main chain can be expected in this system if the size of the crosslinked silica is small enough. The pressurization during curing did not decrease void size markedly.

In Fig.5 the temperature dependency of the fracture toughness of the series without coupling agent is presented. In this figure the amount of silica was changed. The fracture toughness of hybrid materials increased at RT and decreased at LHeT compared with those of epoxy resin. This suggested that dispersed silica act as plasticizer and the nanovoids are certainly introduced.

The existence of macrovoids was not observed by SEM. The effect of pressurization during the curing was also found to be small. These two facts indicates that the number of macrovoid in this system should be very small.

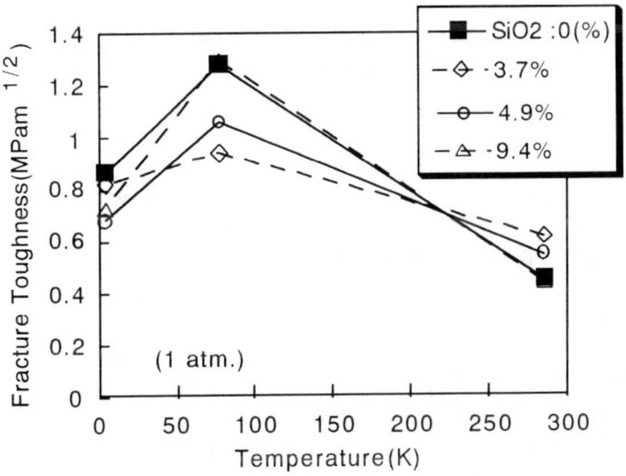

Fig.5 Temperature dependency of fracture toughness of silica dispersed epoxy cured at 1 atm.

Fig.6 shows the temperature dependency of fracture toughness of the coupling agent added system cured at 1 atm. Though the fracture toughness of epoxy usually increases with decreasing temperature down to LNT, it decreases from LNT to LHeT. The silica dispersed without coupling agent specimen (presented open square in the figure) showed the larger fracture toughness at RT, whereas at LNT and LHeT it showed smaller toughness than that of epoxy. The fracture toughness of the coupling agent added system, especially 0.49% added material, did not decrease from LNT to LHeT. The fracture toughness of the system was improved by approximately 2.6 times compared with that of epoxy. This modulation of the local relaxation of the epoxy main chain would be one of the reasons.

CONCLUSIONS

The fracture toughness of epoxy at LHeT can be improved markedly by dispersing silica with coupling agent in epoxy. The pressurization was found not to cause marked improvement of

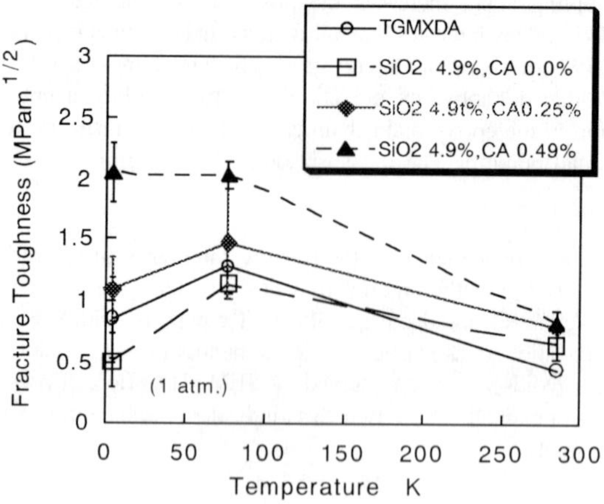

Fig.6 Temperature dependency of fracture toughness of silica dispersed epoxy cured with coupling agent at 1 atm.

fracture toughness. The introduced silica into the hybrid materials without coupling agent acts just as a plasticizer that made epoxy softer at RT and more brittle at LHeT. The PAL method was found to be useful to construct the molecular model of epoxy.

ACKNOWLEDGMENTS

The authors are grateful to Dr. Y.Kobayashi in National Chemical Laboratory and Industry for his discussion of positron lifetime measurement. This work was. partly supported by the cooperative work "Advanced Materials Creation and Their Limit State Protection of Environmental Preservation" between Institute of Scientific and Industrial Research and Welding Research Institute in both of Osaka University.

REFERENCES

1. K.Shibata, S,Nishijma, T.Okada, K.Matsumoto, M.Hamada and T.Horiuchi, Adv. Cryog. Eng. **29** , 175 (1984).
2. A.Iwamoto, S.Nishijima and T.Okada, IEEE Trans. Appl. Supercond. **3**, 269 (1993).
3. F.Sawa, S.Nishijima and T.Okada, Cryogenics **35**, 767 (1995).
4. A.B.Brennan, T.M. Miller, J.J,Arnold, K.V.Huang, N,L.Gephart and W.D. Markewicz, Cryogenics **35**, 783 (1995).
5. T.Okada, S.Nishijima, Y.Honda and Y.Kobayashi,, Suppl. J. Phys.II. **3** , 291 (1993).
6. S.Nishijima, T.Okada and Y.Honda,, Adv. Cryog. Eng. **40** , 1137 (1994).
7. T.Saegusa and Y.Chujo, Macromol. Chem., Macromol. Symp., **64**, 1 (1992).
8. S.Nishijima, K.Nojima, K.Asano, A.Nakahira, T.Okada and K.Niihara,, Adv. Cryog. Eng. **40** , 1051 (1994).
9. Q.Deng, C.S.Sunder and Y.C.Jeans, J.Phys. Chem., **96**, 492 (1988).

DEGREE OF DISPERSION OF LATEX PARTICLES IN CEMENT PASTE, AS ASSESSED BY ELECTRICAL RESISTIVITY MEASUREMENT

XULI FU and D.D.L. CHUNG
Composite Materials Research Laboratory, Furnas Hall, State University of New York at Buffalo, Buffalo, NY 14260-4400

ABSTRACT

The degree of dispersion of latex particles in latex-modified cement paste was assessed by measurement of the volume electrical resistivity and modeling this resistivity in terms of latex and cement phases that are partly in series and partly in parallel. The assessment was best at low values of the latex-cement ratio; it underestimated the degree of latex dispersion when the latex/cement ratio was high, especially > 0.2.

INTRODUCTION

Admixtures in the form of polymers, fine particles and short fibers are used in concrete mixes in order to improve the mechanical properties of the concrete. In particular, polymer admixtures improve both the flexural strength of the concrete and the bond strength between reinforcement (fibers or rebars) and the concrete [1-3]. In addition, polymer admixtures reduce permeability [1] and improve the vibration damping ability [4] and flexural toughness [5]. Among the various types of polymers, latex in the form of styrene butadiene is particularly common for use in cement [1-9].

In spite of the practically valuable properties of latex-modified cement [6-9], the microstructure of this material (an organic/inorganic hybrid material) has not been clarified. Most attention regarding the microstructure has been given to the voids, the content of which decreases with increasing latex/cement ratio [1,5]. The polymer (latex) is a distinct phase, but its observation by microscopy is difficult, partly due to the water in the cement paste and partly due to the small particle size (0.2 μm) of latex. An obvious question regarding the microstructure relates to the degree of dispersion of this phase. The use of microscopy to assess the degree of dispersion globally is difficult and tedious. In this work, through electrical resistivity measurement, we have been able to assess the degree of dispersion of latex in cement paste. This technique is based on the fact that latex is essentially non-conducting compared to cement, so that the degree of latex dispersion strongly affects the resistivity of the latex-modified cement paste. Although the mechanical properties are also affected by the degree of dispersion, they are strongly affected by the polymer/cement bonding and the void content, so they cannot serve as indicators of the degree of dispersion.

EXPERIMENTAL METHODS

The cement was portland cement (Type I) from Lafarge Corp. (Southfield, Michigan). No fine or coarse aggregate was used. The water/cement ratio was 0.23, except that it was 0.45 when latex was absent. The slump was around 160 mm, as determined conventionally using a 77 mm-diameter and 58 mm-high plastic cylinder.

Mat. Res. Soc. Symp. Proc. Vol. 435 © 1996 Materials Research Society

No water reducing agent was used. The latex was a styrenebutadiene dispersion (Latex 460NA from Dow Chemical Corp., Midland, Michigan; 48 wt.% solid, density 1.01 g/cm^3, particle size 0.19-0.21 μm); it was used in the amount of 0.05 - 0.30 of the weight of the cement, and was used along with an antifoam (Dow Corning 2410), which was in the amount of 0.5% of the weight of the latex. The latex and antifoam were first mixed by hand for about 1 min. Then this mixture, cement and water were mixed in a Hobart mixer with a flat beater for 5 min. After pouring the mix into oiled molds, a vibrator was used to decrease the amount of air bubbles. Then specimens were demolded after 1 day and then allowed to cure at room temperature and room relative humidity (33%) in air for 28 days.

The air content was measured using ASTM C185-91a, modified for the absence of sand and the presence of latex. This method also gave the latex volume fraction and the density. Six specimens of each type were tested.

The volume electrical resistivity was measured by the four-probe method, using silver paint for the electrical contacts. The DC current used ranged from 0.1 to 4 A. The specimen size was 160 x 40 x 40 mm. Six specimens of each type were tested.

RESULTS AND DISCUSSION

Table 1 shows that the density decreases with increasing latex/cement ratio, in spite of the fact that the void content decreases with increasing latex/cement ratio (Fig. 1). This is because the density of the latex dispersion (1.01 g/cm^3) is lower than that of cement (3.15 g/cm^3). The latex volume fraction obviously increases with increasing latex/cement weight ratio.

Fig. 2 shows the volume electrical resistivity of cement pastes containing various amounts of latex. The resistivity increases with increasing latex/cement ratio.

Assuming that the latex and cement phases are in parallel (Fig. 3(a)) and neglecting the conductivity of latex compared to that of cement, the volume resistivity of the latex-modified cement $\rho_{||}$ is given by

Table 1. Latex content, void content and density of cement pastes with various values of the latex/cement weight ratio

Latex*/cement weight ratio	Latex* vol.%	Void vol.%	Density (g/cm^3)
0	0	2.32	1.99
0.05	3.64	2.07	1.95
0.10	7.35	1.88	1.90
0.15	11.07	1.70	1.87
0.20	14.91	1.53	1.83
0.25	18.81	1.25	1.79
0.30	22.55	1.10	1.76

*Dispersion.

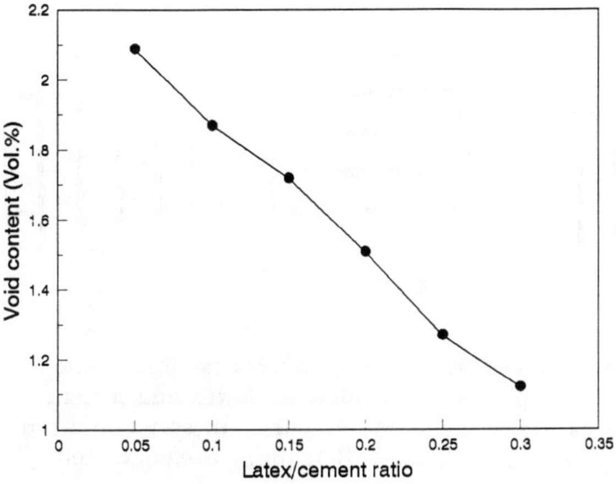

Fig. 1 Effect of the latex/cement ratio on the void content.

$$\rho_{||} = \frac{\rho_c}{V_c} \qquad (1)$$

where ρ_c is the volume resistivity of the cement phase and V_c is the volume fraction of the cement phase. Assuming that the latex and cement phases are in series (Fig. 3(b)) and neglecting the conductivity of latex, the resistivity of the latex-modified

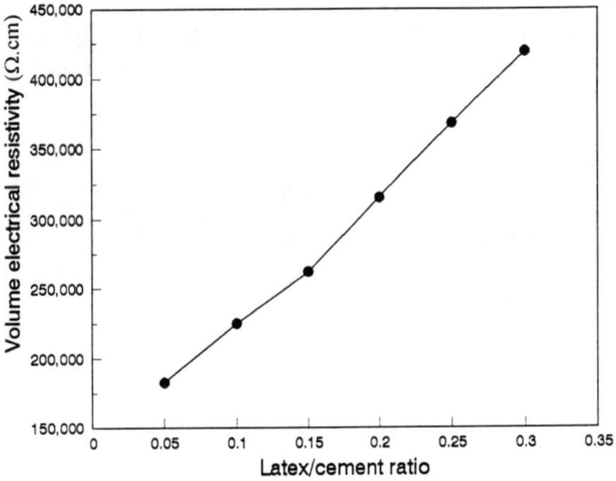

Fig. 2 Effect of the latex/cement ratio on the volume electrical resistivity.

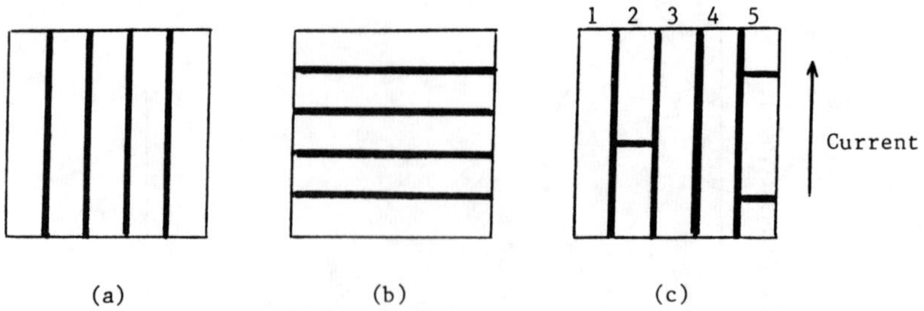

Fig. 3 Electrical conduction model of in latex-modified cement paste. (a) Latex and cement phases in parallel. (b) Latex and cement phases in series. (c) Latex and cement phases partly in series and partly in parallel. Dark bands : latex phase. Remaining areas : cement phase.

cement ρ is ∞. In reality, the situation is in between these two extremes, so that the real resistivity is between ∞ and the value given by Eq. (1). Fig. 3(c) is an electrical conduction model of the real situation. In this model, the latex and cement phases are partly in parallel (e.g., sections 1, 3 and 4 of Fig. 3(c)) and partly in series (e.g., sections 2 and 5 of Fig. 3(c)). The sections which are in series do not contribute to electrical conduction, so the effective volume fraction of the cement phase V_c^{eff} is given by

$$V_c^{eff} = (1-f)\ V_c\,, \tag{2}$$

where f is the volume fraction of the cement phase which belongs to the sections that are in the series configurations. Thus, the resistivity of the latex-modified cement ρ is given by

$$\rho = \frac{\rho_c}{V_c^{eff}}\ . \tag{3}$$

Combining Eq. (2) and (3),

$$\rho = \frac{\rho_c}{(1-f)\,V_c}\ . \tag{4}$$

The quantity f provides a description of the degree of dispersion of the latex phase. The greater is f, the greater is the degree of dispersion, as illustrated in Fig. 4, where

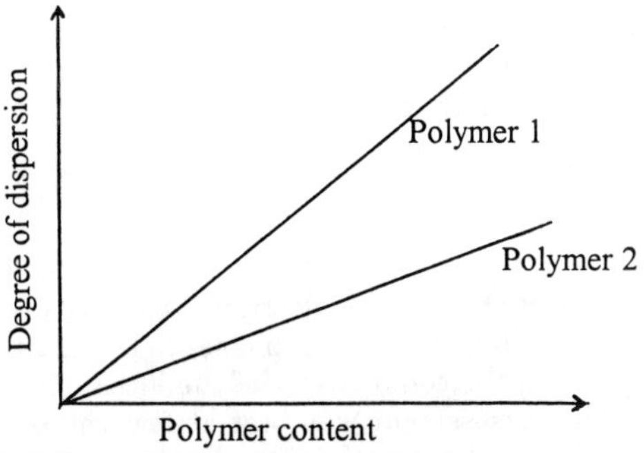

Fig. 4 Schematic illustration of the effect of polymer species and content on the degree of dispersion (f) of polymer in cement paste.

Polymer 1 has a higher degree of dispersion than Polymer 2 and, for a given polymer, f increases with the polymer content. The value of f is obtained from Eq. (4), if ρ, ρ_c and V_c are known. The quantity V_c is obtained by deducting the latex volume fraction and void volume fraction (Table 1) from unity. The quantities ρ and ρ_c are measured (Fig. 2).

Fig. 5 gives the variation of f with the latex/cement ratio; f increases more rapidly with the latex/cement ratio at low values of this ratio than at high values of this ratio. This is because a high latex/cement ratio corresponds to a situation in which some of the sections of Fig. 3(c) (such as Section 5) have more than one

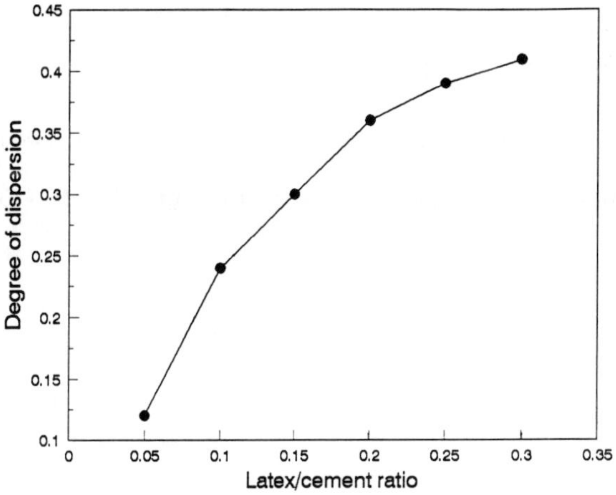

Fig. 5 Degree of dispersion (f) of latex as a function of the latex/cement ratio.

horizontal band of latex. A section does not contribute to conduction as long as it has at least one band. Therefore, f underestimates the true degree of dispersion when the latex/cement ratio is high, particularly > 0.2.

CONCLUSION

The degree of dispersion of latex particles in latex-modified cement paste was assessed by measurement of the volume electrical resistivity. The resistivity was modeled by considering the latex and cement phases to be partly in series and partly in parallel and by neglecting the conductivity of the latex phase compared to that of the cement paste. This assessment was best at low values of the latex-cement ratio; it underestimated the degree of latex dispersion at high values of the latex/cement ratio, especially > 0.2.

REFERENCES

1. ACI Committee 548, ACI Materials J. 91(5), 511 (1994).

2. Parviz Soroushian, Fadhel Aouadi and Mohamad Nagi, ACI Materials J. 88(1), 11 (1991).

3. Xuli Fu and D.D.L. Chung, Cem. Concr. Res. 26(2) (1996).

4. Xuli Fu and D.D.L. Chung, Cem. Concr. Res. 26(1), 69 (1996).

5. P.-W. Chen, Xuli Fu and D.D.L. Chung, ACI Mater. J., in press.

6. L.A. Kuhlmann, Concrete International: Design and Construction 12(10), 59 (1990).

7. D.G. Walters, ACI Materials Journal 87(4), 371 (1990).

8. D.G. Walters, Transportation Research Record 1204, 71 (1988).

9. L.A. Kuhlmann, Transportation Research Record 1204, 52 (1988).

Organic-Inorganic Hybrid Dental Restorative Material

PARTHA P. PAUL, SCOTT F. TIMMONS, WALTER J. MACHOWSKI
Materials Engineering Department, Southwest Research Institute, San Antonio, Texas.

ABSTRACT

A method for the preparation of dental restorative composites based on an organic-inorganic hybrid material is presented. A sol-gel process was used to produce composites comprised of submicron silica in a polymer matrix. Using alcohols such as, 3-allyloxy-1,2-propanediol and 2,3-dihydroxypropyl methacrylate in combination with polysilicic acid or cyclic siloxane precursors, modified siloxanes are produced which can be polymerized photochemically with concurrent hydrolysis and condensation to generate a silica-reinforced polymer with superior mechanical properties and abrasion resistance.

INTRODUCTION

Composites have attracted attention as dental restorative material since 1962 [1]. Though the formulation of the composite materials have changed in the last three decades, the underlying chemical process involved has not changed. An acrylate monomer is used as the source of resin, and silica is used as the filler material. Though these composites have superior esthetic quality, issues of shrinkage, adhesion, toughness and wear resistance have not been solved. These problems are more acute in the case of posterior restoration.

Durability is a major problem with posterior composite restoration [2]. Lifespans of large fillings are usually fewer than five years, which can be attributed to inadequate resistance to wear of composites under masticatory attrition. The insufficient interaction between the reinforcing filler and the resin binder might be responsible for the lack in wear resistance of these composites. It has been demonstrated that an ultrafine compact filled composite has a Young's modulus higher than that of the dentine [3]. It also has a good Vickers hardness and high compressive strength. This improvement in the mechanical property might be the result of better binding of filler particles with the resin.

Two of the other major problems associated with the present day dental restorative materials are lack of adhesion and shrinkage of the resin during polymerization. The lack of adhesion results in micro leakage and formation of secondary cracks along the interface [4]. The shrinkage which occurs during the conversion of monomer to polymer, works against the formation of an adhesive bond between the resin and the dentine. The new fourth generation material invokes the idea of a hydrophilic primer which can penetrate into the dentinal substrate and produces enhanced adhesion [5]. Surface preconditioning of dentine and the use of HEMA as a primer results in shear bond strengths of 17 to 20 MPa [4]. Though these shear bond strengths are acceptable, shrinkage is often a problem.

The shrinkage of commercially available filled composite resin ranges from 2.6 - 7.1% [6]. These shrinkage values differ because of their monomer composition, various degree of polymerization, filler type and filler concentration. Use of oxaspiro monomers have been considered as the precursor for the resin [7]. These monomers expand 3.5 to 3.9% in volume [8] under polymerization, which is not acceptable either.

Mat. Res. Soc. Symp. Proc. Vol. 435 © 1996 Materials Research Society

The properties of a composite material are greatly influenced by the degree of mixing between the inorganic (filler) and the organic (resin) phases [9]. In a molecularly-tailored system, an organically modified ceramic precursor will result in the synchronous formation of inorganic (silica) and organic (resin) components [10]. Such a composite could be obtained by using sol-gel chemistry [11,12]. The mixing of the inorganic and the organic matrix is at the molecular level, and the particles are often nano sized. As expected, these hybrid nano composite materials have toughness three orders of magnitude higher than the ceramic alone [13]. Depending upon the morphology, phase behavior and organic-inorganic ratio, these composite materials span a continuum ranging from glass reinforced organic polymer to polymer modified glass [14,15].

The conventional sol-gel chemistry has a drawback of shrinkage. This problem can be avoided by the use of metal alkoxides (orthosilicate or orthozirconate) obtained through the use of an appropriate alcohol which could be easily polymerized (unsaturated alcohol) [13,16]. The hydrolysis and condensation of these alkoxides occur with the simultaneous polymerization of the eliminated alcohol by a free radical technique. This simultaneous approach circumvents the insolubility problem, while at the same time eliminates the shrinkage problem. It has been shown recently that a hydrophilic group will improve the adhesion of dentine to the resin. Attaching a hydrophilic group like -OH, -SH or $-NH_2$ to the polymerizable alcohol, through a short spacer group, might help in the adhesion property of the resin. Here, we report the development of organic-inorganic hybrid composite materials. Many of these composites were obtained by sol-gel chemistry. We have also used modified clay materials to obtain hybrid materials.

EXPERIMENTAL

All manipulations were carried out under dry nitrogen unless otherwise noted. All solvents used in this study were rigouously dried by refluxing over sodium or potassium with the exception of anhydrous ethanol which was used as received from McCormick Distilling Co., Inc. Tetrahydrofuran (THF) was purchased from J. T. Baker and dried by refluxing in the presence of the potassium ketyl of benzophenone. All reagents were of reagent quality or better and used as received unless otherwise noted. Sodium metasilicate was provided by Diamond Shamrock Corp. Rhone-Poulenc provided the HEMA. Calcium chloride and toluene were purchased from Fisher Scientific. Wollastonite (quartz, Vansil W-30) was provided by the R. T. Vanderbilt Co., Inc. All deuterated solvents and (±) camphorquinone and 2-(dimethylamino)ethyl methacrylate were purchased from Aldrich Chemical Co., Inc. All NMR analyses were carried out with a Varian VXR-300 spectrometer operating at 300 MHz.

Preparation of Compound **A** Polysilicic acid was prepared as reported [16]. To 100 mL of a THF solution of polysilicic acid, 50 mL of 3-allyloxy-1,2-propanediol was added. This solution was heated and the THF/H_2O azeotrope removed until 75 mL was collected. At this point, dry THF was added continuously and distilled, until 500 mL had been collected. The solution was vacuum distilled (0.02 mm of Hg) to remove 30 mL of unreacted diol to produce Compound **A**.

Preparation of $(SiO)_4(OEt)_8$ The titled compound was synthesized by making minor variations to the reported method [17]. $Ca_8(SiO_3)_4Cl_8$ (10 g) was added to a solution containing 180 mL of toluene and 160 mL of anhydrous ethanol. To this suspension, 11 mL of 9M anhydrous, ethanolic HCl was added dropwise over a period of 15 minutes. This solution was distilled until 270 mL of distillate was collected. The remaining suspension was filtered and the filtrate collected. The solid material was washed with 35 mL of dry hexane and the washing was collected. The combined filtrates were concentrated to an oil under vacuum (<0.1 mm of Hg). This material was analyzed by ^{1}H and ^{13}C NMR and judged to be of sufficient purity for further studies.

Preparation of Compound **B** To a solution of $(SiO)_4(OEt)_8$ (3g), HEMA (30 mL) and a catalytic amount of HCl in ethanol were added. This solution was heated to 50°C and was periodically evacuated to remove any ethanol which was released during the reaction. After 1 hour of heat treatment, the solution was cooled to room temperature resulting in a light yellow oil.

Preparation of Compound **C** This compound was prepared in an identical fashion to that of Compound **B** utilizing 2,3-dihydroxypropyl methacrylate as the alcohol.

Preparation of Composite **A** Compound **A** (1.8g) was added to a 1:1 mixture of Bis-GMA and HEMA (1.2g) containing 4 drops of 2-(dimethylamino)ethyl methacrylate and 10 mg of (±) camphorquinone. After several minutes of mixing, 2 drops of water were added and was thoroughly mixed. This mixture was left under a fluorescent lamp for a period of 5 hours until a hard composite had formed.

Preparation of Composite **B** This composite was prepared in an identical fashion to that of Composite **A** using Compound **C**.

RESULTS

Synthesis of Alkoxides To obtain a superior restorative material with high strength, a composite with a high glass content is desired. Inorganic-organic hybrid materials derived from poly or cyclic siloxanes will provide a high silica content. Reaction of an alcohol with polysilicic acid will render an alkoxide through esterification, as shown in Scheme 1.

We have used alcohols that posses a polymerizable group along with two hydroxyl groups. The reaction of both the hydroxyl groups in the formation of the esters might be unlikely due to the steric factors. The unreacted OH group will improve the adhesion to the dentine. It has been observed that 2-hydroxyethylmethacrylate (HEMA), which has only one hydroxyl group, produced an excellent adhesion to the dentine [4]. The two commercially available monomers, 3-allyloxy-1,2-propanediol and 2,3-dihydroxypropyl methacrylate were used in our study. The reaction of 2,3-dihydroxypropylmethacrylate with the polysilicic acid resulted in a gummy solid which was not amenable for further use. The reaction of polysilicic acid with the 3-allyloxy-1,2-propanediol resulted in a liquid (Compound **A**) which was further used to produce Composite **A**.

Scheme 1

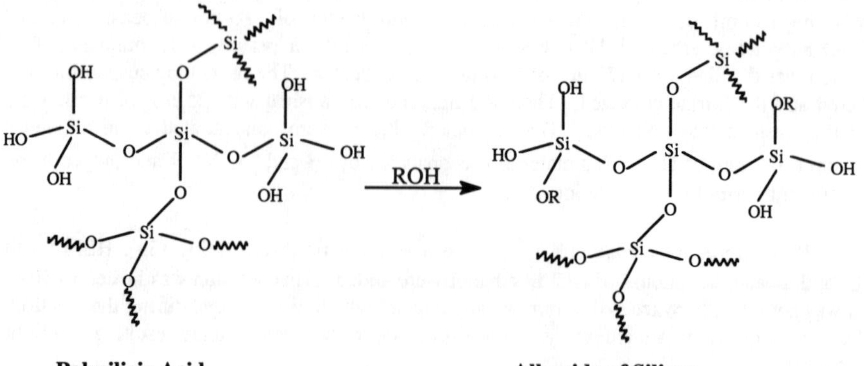

Characterization of Compounds The reaction of polysilicic acid with many desirable alcohols resulted in solid products. A cyclic siloxane (4-6 Si) might be a liquid and suitable for further use. Goodwin and Kenney observed the formation of liquid cyclic siloxanes [17]. $Ca_8(SiO_3)_4Cl_8$ was used as a source of cyclic siloxane. Characterization of $Ca_8(SiO_3)_4Cl_8$ was performed by different techniques, including XRD. The d-values observed for different peaks were identical to the reported values. The cyclic siloxane with ethoxy groups was obtained by treating $Ca_8(SiO_3)_4Cl_8$ with an excess of ethanol saturated with hydrogen chloride gas. The cyclic siloxane, $(SiO)_4(OEt)_8$ was characterized by ^{29}Si and ^{1}H NMR. A single peak at -95.06 ppm was observed which is characteristic of the silicon for such an ester. The transesterification reaction of $(SiO)_4(OEt)_8$ was performed with HEMA and 2,3-dihydroxypropyl methacrylate. The reaction of $(SiO)_4(OEt)_8$ with HEMA to give Compound **B** is shown in Scheme 2 and was studied in detail using ^{1}H NMR spectroscopy. The protons associated with the methylene groups in HEMA have chemical shifts of 4.26 ppm and 4.09 ppm. The transesterification was promoted with the addition of HCl and gently warming the reaction mixture to 50°C. Downfield shifts of the CH_2 peaks of the ethyl group are indicative of the esterification. Under similar conditions, transesterification of $(SiO)_4(OEt)_8$ was also achieved with 2,3-dihydroxypropyl methacrylate to obtain Compound **C**.

Development of Composite The composites were obtained by using sol-gel chemistry. Compounds **A** and **C** are alkoxy-modified silicates where the alkoxy group contains a polymerizable functionality. To obtain a composite with good mechanical properties, the gel formation should be concurrent with the polymerization. We used visible light catalyzed polymerization. The initiator used was (±) camphorquinone (CQ), and 2-(dimethylamino)ethyl methacrylate (NDEM) was the promoter. A stoichiometric amount of water was added to promote the gelation. The composite formation took 5-6 hours under visible light. No special light source was used for this purpose. The overall composite formations are shown in Schemes 3 and 4.

Scheme 2

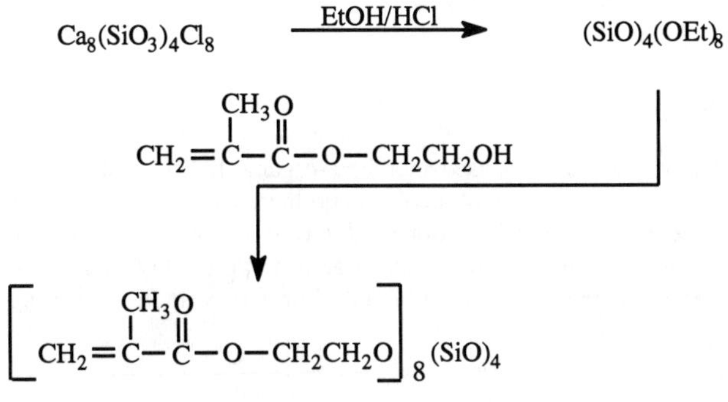

Compound **B**

Scheme 3

Compound **A** (60%) + HEMA + Bis-GMA (40%)

$\xrightarrow[\text{H}_2\text{O (1drop), CQ (10 mg)}]{\text{NDEM (5drops), Light}}$ Composite **A**

Scheme 4

Compound **C** (60%) + HEMA + Bis-GMA (40%)

$\xrightarrow[\text{H}_2\text{O (1drop), CQ (10 mg)}]{\text{NDEM (5drops), Light}}$ Composite **B**

Water Absorption Test A composite specimen of 5 mm thickness was polished, carefully dried to constant weight and placed into deionized water for 6 hours. The specimen was removed and blotted dry, air dried overnight and reweighed. The weight gain was 1.9%. There was no obvious visible change to either the specimen or the water used to immerse the sample.

Toothbrush Abrasion Test Composite specimens were ground to a thickness of 2 mm and polished with 80 and 400 grit sandpaper prior to the test. The test articles were soaked in distilled water for one hour , blotted dry with tissue paper and allowed to air-dry overnight. Each of the test articles were covered with 96 grams of toothpaste (CREST with Fluoristat®) and brushed with an Oral-B *P40* soft-bristle toothbrush containing 47 tufts at a rate of 10,000 strokes/hr. The area brushed by the apparatus was 450 mm^2. Every two hours the sample was rinsed with distilled water, blotted dry, air dried for two hours and weighed. The test was conducted for a total of eight hours and the average weight loss rate determined. For Composite **A** the weight loss rate was 4.7 mg/hr.

CONCLUSION

We have successfully used inorganic-organic hybrid materials as dental restorative composites. Our specific choice of alcohols and the poly or cyclic silicic acid provides easily processable precursors to inorganic-organic hybrid composites. These composites have high silica content, low shrinkage, low abrasion and the possibility of good adhesion due to the available hydrophilic groups in the polymer. Further research is ongoing be achieve better performance.

ACKNOWLEDGMENTS

We thank Mr. Stephen Salazar for his assistance in several experiments. We thank Prof. H. R. Rawls for helping us with toothbrush abrasion test. We thank Dr. Nollie F. Swynnerton for his assistance with the NMR studies. This research was supported with an Internal Research grant by The Southwest Research Institute.

REFERENCES

1. R. L. Bowen, U.S. Patent 3 066 122 (1962).
2. R. L. Bowen, W. A. Marjenhoff, Adv. Dent. Res., **6** , 44-49 (1992).
3. R. G. Craig, J. Dent. Res., **58**, 1544-1550 (1979).
4. J. D. Eick, S. C. Robinson, M. J. Byerley, C. C. Chappelow, Quintessence Int., **24**, 632-640 (1993).
5. D. Bouvier, J. P. Duprez, D. Nguyen, M. Lissac, Dent. Mater., **9**, 365-369 (1993).
6. A. J. Feilzer, A. J. de Gee, C. L. Davidson, J. Prosthet Dent., **59**, 297-300 (1988).
7. T. J. Byerley, J. D. Eick, G. P. Chen, C. C. Chappelow, F. Millich, Dent. Mater., **8**, 345-350 (1992).
8. J. W. Strawberry, J. Dent. Res., **71**, 1408-1412 (1992).
9. L. H. Sperling, Interpenetrating Polymer Network and Related Materials, (Plenum Press: New York, 1981).
10. J. E. Mark, CHEMTECH, **19**, 230 (1989).
11. D. R. Ulrich, CHEMTECH, **18**, 242 (1988).
12. C. J. Brinker, G. W. Scherer, Sol Gel Science, (Academic Press, New York, 1990).
13 B. M. Novak, Adv. Mater., **5**, 422-433 (1993).
14. H. Schmidt, J. Non-Cryst. Solids, **112**, 419 (1989).
15. D. R. Ulrich, J. Non-Cryst. Solids, **121**, 465 (1990).
16. M. W. Ellsworth, B. M. Novak, Chem. Mater., **5**, 839-844 (1993).
17. G. B. Goodwin, M. E. Kenney, Inorg. Chem., **29**, 1216-1220 (1990).

Part V

Organic/Inorganic Hybrids–A Route to Controlled Porosity Materials

SYNTHESIS OF ORDERED SILICATES BY THE USE OF ORGANIC STRUCTURE-DIRECTING AGENTS

YOSHIHIRO KUBOTA, MARK E. DAVIS
Division of Chemistry and Chemical Engineering, California Institute of Technology, Pasadena, CA 91125

ABSTRACT

Microporous, crystalline silicates can be synthesized using organic structure-directing molecules whose function is to organize silicate species into particular arrangements that then spontaneously self-assemble into the final crystalline materials. The porosity is obtained by removal of the organic component. We discuss the molecular properties (size, rigidity, hydrophobicity) necessary for the organic component to interact with aqueous silicate species and prepare microporous silicates. It is shown that a strategy to construct large organic molecules in order to prepare large pore crystalline silicates could involve the use of two charge centers if these functionalities are distributed within the molecules such to prevent aggregation in aqueous media. A charge distribution that allows aggregation of the organic molecules at synthesis conditions directs the formation of a locally amorphous, mesoporous silicate rather than a crystalline, microporous material.

INTRODUCTION

Zeolites and other crystalline, microporous solids have found great utility as catalysts, sorption materials and ion exchangers. To date, the largest proven pore size in crystalline solids is 12-13 Å in the material called VPI-5.[1] In 1992, workers at Mobil prepared the first ordered mesoporous materials.[2] These (alumino)silicate solids contain pores of uniform size in the range of ~ 20-100 Å. Although there is order on the nanometer length-scale, the mesoporous materials are locally amorphous.[3] Because of the amorphous nature of the mesoporous materials, their thermal and hydrothermal stabilities and their acidities (aluminosilicate versions) are far below those of crystalline solids.[3] Thus, the quest for larger pore crystalline materials continues to flourish.

For ordered microporous and mesoporous silicates, the porosity is obtained by removal of the organic fraction from an inorganic-organic hybrid material. The hybrid materials consist of organic species that are non-covalently interacting with the silicate phase via coulombic, van der Waals and hydrogen bonding forces and are formed by hydrothermal synthetic techniques. That is to say that silica and the organics are reacted in aqueous media at moderate temperatures (below 200°C) and autogenous pressures. In the assembly of the crystalline structures that lead to microporous materials, the inorganic-organic interactions occur between inorganic species and single organic molecules.[4] With the locally amorphous, mesoporous materials, the organic molecules must be able to self-assemble into an aggregate, *e.g.*, a micelle, in order to direct the formation of the hybrid material.[5]

Mat. Res. Soc. Symp. Proc. Vol. 435 © 1996 Materials Research Society

In this paper we describe some of our efforts to design larger organic molecules for use in synthetic preparations aimed at crystallizing large pore zeolites. It is shown that the properties of the organic molecules can determine whether crystalline microporous or amorphous mesoporous silicates are formed.

EXPERIMENTAL SECTION

Tetramethyl-, tetraethyl-, tetrapropyl-, tetrabutyl- and tetrapentylammonium halides and hydroxides were used as received. Other organocations used here were synthesized via steps including various amine synthesis followed by the exhaustive alkylation of the amines with alkyl halides. The quaternary ammonium halides were determined to be pure compounds as vrified by ^1H and ^{13}C NMR and elemental analyses. The synthetic procedures used for their preparation have been described in detail elsewhere.[6] All the quaternary ammoniums were placed in their hydroxide form by ion exchange.

The hybrid materials syntheses were accomplished from mixtures of composition $0.3 < OH^-/SiO_2 < 0.5$, $0.1 < R/SiO_2 < 0.3$, $0 < Na^+/SiO_2 < 0.1$ and $50 < H_2O/SiO_2 < 70$ where R is an organic moiety at temperatures of 135-175 °C and autogenous pressure.

The as-synthesized materials were collected by filtration, washed with water and dried at ambient conditions. Phase identification of the collected solids was made by powder X-ray diffraction (XRD) with a Scintag XDS 2000 diffractometer using Cu-Kα radiation.

RESULTS AND DISCUSSION

In order to initiate the structure-direction in the synthesis of microporous, crystalline silicates, individual organic species become enclathrated with silica.[4] These inorganic-organic entities then form nucleation centers from which the crystallization proceeds. Several points concerning the properties of the organic species necessary to elicit these processes have been enumerated by Zones and co-workers based on correlations from numerous experimental observations.[6-8] For example, increasing the size of the organic molecule has been shown to increase the selectivity to the crystalline products.[7] That is to say as the size of the organic molecule increases, the number of different crystalline phases produced from the use of that organic decreases. The selectivity increases also with the rigidity of the organic molecule.[8] These observations are easy to rationalize. With small organic molecules they are able to reside in a wide variety of cage/channel sizes, *e.g.*, with tetraethylammonium (TEA), the zeolites beta, ZSM-12 and ZSM-20 can all be synthesized and each contain different pore architectures of sufficient size to accommodate TEA. Additionally, molecules like TEA and tetrapropylammonium (TPA) can adopt multiple conformations, *e.g.*, TEAI solid, TEA in zeolite beta and TEA in SAPO-34 adopt S_4, D_{2d} and C_2 conformations, respectively.[9] As stated by Brand *et al.*, the ability of the organic to adopt different conformations via low barriers to interconversion during silicate encapsulation makes numerous crystalline structures accessible by hydrothermal synthesis.[9] Thus, larger, more rigid organic molecules are likely candidates for the structure-direction of large pore crystalline materials.[7,8]

The size of the organic molecule can not be increased without regard to another property, *i.e.*, hydrophobicity. Zones, Davis and their co-workers have now clearly established that organic molecules with intermediate hydrophobicity are necessary for the structure-direction of crystalline silicates.[4,6-8,10,11] In order to develop a fast and simple screening experiment for assessing the hydrophobicity of organic cations, Zones investigated the partitioning of the organic species between water and chloroform and showed that only those molecules that had intermediate transfer into the chloroform structure-direct the formation of crystalline silicates.[7] Kubota *et al.* investigated this behavior further.[6]

Kubota *et al.* have shown that the greatest discrimination amongst tetraalkylammonium ions (R_4N^+ : R : methyl, ethyl, propyl, butyl, pentyl) as far as their ability to transfer from water to chloroform was when the counter-anion was iodide.[6] This is not unexpected since it is known that quaternary ammonium salts are soluble in organic solvents with iodides > bromides > chlorides.[12] For larger, carbon-rich cations, iodides tend to form better tight-ion pairs with the polarizabilities of the cation and anion better matched. This leads to more favorable transport into lower dielectric constant media.

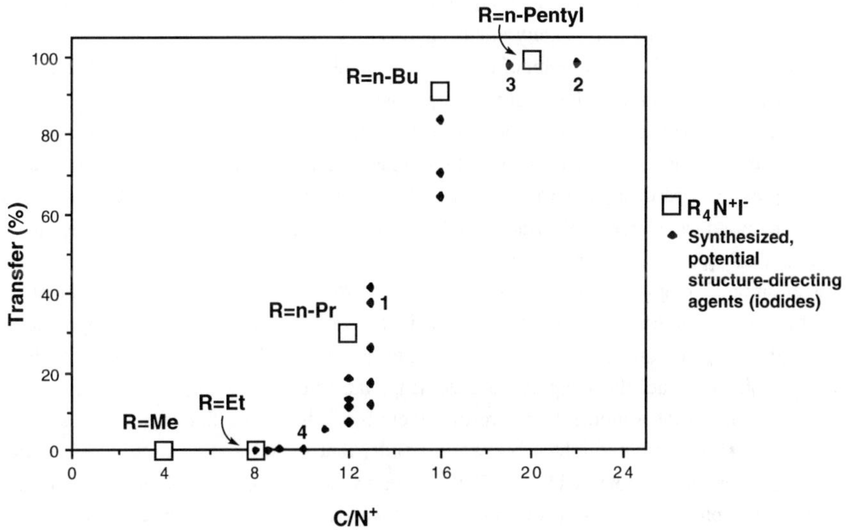

Fig. 1. Transport of simple tetraalkylammonium iodides and synthesized, potential structure-directing agents (SDA) from water to chloroform. C/N^+ is the ratio of the number of carbon to nitrogen atoms in the potential SDA.

The phase transfer behavior of the tetraalkylammonium iodides from water to chloroform is shown in Figure 1. For R=Me and Et, no transfer is observed. At the opposite extreme, when R=n-Bu (TBA) or n-Pentyl (TPentA) the transfer is above 80%. Only TPA shows intermediate transfer. Burkett and Davis have reported that of these tetraalkylammonium ions, only TPA forms a crystalline silicate when combined with aqueous silicate solutions and heated to 110°C.[10,11] Burkett and Davis suggest that ions like TMA and TEA are too hydrophilic to elicit interactions with hydrophobically hydrated silicates in aqueous solution while TBA and TPentA most likely phase separate into aggregates. Clearly, this is true with TPentA as it is visible in the reaction mixture. Additionally, Zones has shown that organic cations with phase transfers between 10-40% structure-direct the synthesis of crystalline silicates.[7] In Figure 1, the points for $8 \leq C/N^+ \leq 16$ with phase transfers of 0-80% can structure-direct the synthesis of crystalline silicates. The difficulty of attempting to specify a range in transfer that is observed for structure-directing agents is that it depends on the temperature of the synthesis. As previously mentioned, TBA can not be used to form a crystalline silicate at 110°C but is able to do so at temperatures greater than 150°C. We believe that this information leads to the conclusion that intermediate hydrophobicity at synthesis temperature is necessary for crystallization to occur. The reason for this is that some hydrophobicity in the organic is required to bring about interactions with the silicate species; however, too much hydrophobicity elicits aggregation. Thus, in the search for structure-directing agents for the preparation of large pore crystalline silicates care must be taken with regard to the organic size, flexibility and hydrophobicity.

In addition to the phase transfer behavior of tetraalkylammonium ions, Figure 1 shows the phase transfer results from a wide variety of organic cations (points on graph). Each datum point represents a particular molecule and the details of the molecules used and the preparation procedures used can be found elsewhere.[6] Of importance is the fact that there does appear to be a "universal" curve for the phase transfer behavior of organic cations when iodide is the counter-anion. Roughly speaking salts that have $8 \leq C/N^+ \leq 17$ appear to structure-direct the assembly of crystalline silicates and are also in the range of intermediate hydrophobicity. Although one can argue about occurrences at the low and high end of this range, there is overwhelming evidence that the previous statement holds true.[6]

In spite of the aforementioned results, there is another factor that is not addressed by previous work and is highlighted below. Consider that molecule **1** is a strong structure-directing agent for the zeolite CIT-1.[13] CIT-1 contains large and medium sized zeolitic pores. In attempts to prepare new zeolites with large pores we synthesized molecules **2**, **3** and **4**.

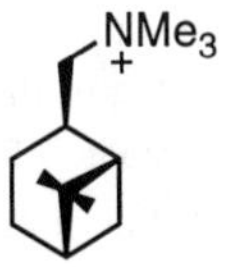

1

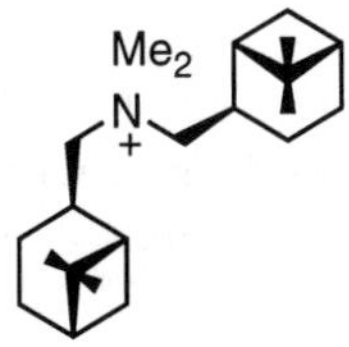

2

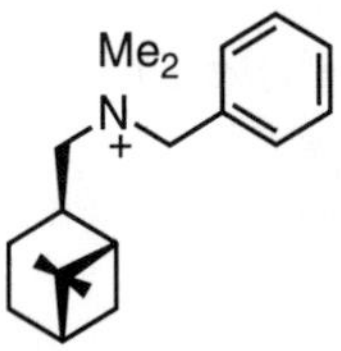

3

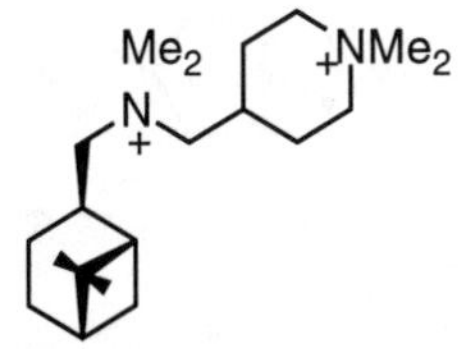

4

Molecules **2** and **3** have C/N$^+$ values above 18 and phase transfers exceeding 80%. Thus, we expected that these molecules would not be able to structure-direct the synthesis of crystalline silicates. This did indeed turn out to be the case. Only amorphous silica is obtained from hydrothermal synthesis involving molecules **2** and **3.** In order to lower both the C/N$^+$ and the % transfer, a second charge was placed into a molecule of type **3** to give molecule **4**. Molecule **4** has a C/N$^+$ of 10 and a % transfer of about 1%. From the data in Figure 1, we would expect that **4** could structure-direct the formation of crystalline silicates. However, the use of **4** yielded a mesoporous silicate with properties indicated in Table I.

Table I. Properties of mesoporous silicate obtained using **4**

Property	*Value*
as-synthesized : d-spacing of XRD peak	35 Å
calcined to 550°C : d-spacing of XRD peak	32 Å
BET surface area	974 m^2/g
pore volume	224 cm^3/g STP
pore diameter	21 Å

From the data listed in Table I, the properties of the mesoporous silica obtained using **4** are fairly typical.[2,3] Clearly, this result shows that in addition to the hydrophobicity of the organic used to organize silica, the distribution of the hydrophobicity within the molecule is also important. It is known that **5** can be used to form pure silica zeolite ZSM-12 at the same conditions that **4** yields the mesoporous silica.[14] It is apparent that **4** must aggregate in aqueous media while **5** does not.

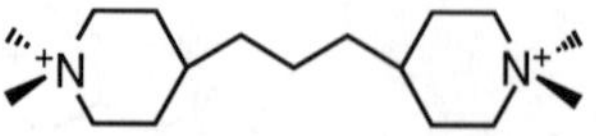

$C/N^+=8$

5

The conclusions from this work are that a strategy to construct large organic molecules could involve the addition of a second charge center if this functionality is distributed within the molecule such to prevent aggregation in aqueous media. Additionally, subtle variations in hydrophobicity can determine whether crystalline microporous or locally amorphous, mesoporous silicates are formed.

ACKNOWLEDGMENTS

Financial support for the zeolite synthesis work was provided by the Chevron Research and Technology Company. We thank Dr. Stacey Zones for many helpful discussions and his continual guidance and Mr. Katsuyuki Tsuji for technical assistance.

REFERENCES

1. M.E. Davis and R.F. Lobo, Chem. Mater. **4**, 756 (1992).

2. C.T. Kresge, M.E. Leonowicz, W.J. Roth, J.C. Vartuli and J.S. Beck, Nature **359**, 710 (1992).

3. C.Y. Chen, H.X. Li and M.E. Davis, Microporous Mater. **2**, 17 (1993).

4. M.E. Davis, Stud. Sur.Sci. Catal., **79**, 35 (1995), and references therein.

5. C.Y. Chen, S.L. Burkett, H.X. Li and M.E. Davis, Microporous Mater. **2**, 27 (1993).

6. Y. Kubota, M.M. Helmkamp, S.I. Zones and M.E. Davis, Microporous Mater., in press.

7. R.F. Lobo, S.I. Zones and M.E. Davis, J. Incl. Phenom. **21**, 47 (1995).

8. S. I. Zones and M. E. Davis, in Synthesis of Microporous Materials: Zeolites, Clays, Nanocomposites, edited by M.Occelli, H. Kessler (Marcel Dekker, New York), in press.

9. H.V. Brand, L.A. Curtiss, L.E. Iton, F.R. Trouw and T.O. Brun, J. Phys. Chem. **98**, 1293 (1994).

10. S.L. Burkett and M. E. Davis, Chem. Mater. **7**, 920 (1995).

11. S.L. Burkett and M. E. Davis, Chem. Mater. **7**, 1453 (1995).

12. H.Z. Sommer, H.I. Lipp and L.L. Jackson, J. Org. Chem. **36**, 824 (1971).

13. R.F. Lobo and M.E. Davis, J. Am. Chem. Soc. **117**, 3766 (1995).

14. M. Goepper, H.X. Li and M.E. Davis, J. Chem. Soc., Chem. Commun., 1665 (1992).

CONTROLLING THE POROSITY OF MICROPOROUS SILICA BY SOL-GEL PROCESSING USING AN ORGANIC TEMPLATE APPROACH

YUNFENG LU[1] G.Z. CAO[1] RAHUL P. KALE[1] L. DELATTRE[1] C. JEFFREY BRINKER[1,2,*] and GABRIEL P. LOPEZ[1*]

[1]The UNM/NSF Center for Micro-Engineered Ceramics, Department of Chemical & Nuclear Engineering, University of New Mexico, Albuquerque, NM 87131.

[2]Ceramics Synthesis and Inorganic Chemistry Department, Organization 1846, Sandia National Laboratories, Albuquerque, NM 87185.

*Authors to whom correspondence should be addressed.

ABSTRACT

We use an organic template approach to prepare microporous silica with controlled pore size and narrow pore size distributions. This was accomplished by fabricating relatively dense hybrid silica matrices incorporating organic template ligands by sol-gel synthesis and then removing the organic ligands to create a microporous silica network. Comparison of computer simulation results and experimental data indicated that using this fugitive template approach, pore volume can be controlled by the amount of organic template added to the system, and pore size can be controlled by the size of the organic ligands.

INTRODUCTION

Porous inorganic membranes have been proposed for a variety of applications such as high temperature gas separation.[1-2] The ideal characteristics of a porous gas separation membrane are a sufficiently small pore size, a narrow pore size distribution and a large pore volume to efficiently separate gases on the basis of size with high flux of the target gas.

We have previously reported the use of an organic templating strategy to fabricate amorphous silica membranes having approximately 5 Å pore size, by co-polymerization of tetraethoxysilane (TEOS) with 3-methacryloxypropyltrimethoxysilane (MPS) followed by oxidative pyrolysis to remove organic constituents.[3] In this approach, organic ligands were dispersed at the molecular level in a relatively dense silica matrix. A microporous silica network was then obtained by removing these organic ligands. Compared to conventional sol-gel techniques used to prepare microporous sol-gel materials, in which the desired small pores are obtained at the expense of pore volume via the promotion of collapse of the silica network upon drying,[4,5] the organic template approach has the advantage of facile, independent control of pore size and pore volume. It was proposed that both pore size and pore volume can be tailored by judicious choice of the size and amount of organic ligands added. However, there is very limited work concerning the relationship of pore size and pore volume to the template ligand size and amount of organic templates added. This information is necessary to tailor the micro-structure of porous silica materials, and correspondingly, the performance of the membrane in gas separation processes. In this paper, we attempt to establish this relationship for the hybrid MPS/TEOS system.

EXPERIMENT

The chemical precursors used were 3-methacryloxypropyltrimethoxysilane, $H_2C=C(CH_3)CO_2(CH_2)_3Si(OCH_3)_3$ (abbreviated as MPS, from PCR, Gainesville, FL), and tetraethylorthosilicate, $Si(OC_2H_5)_4$, (TEOS, from Kodak, Rochester, NY). Each MPS molecule contains one organic ligand, $H_2C=C(CH_3)CO_2(CH_2)_3$ (this organic ligand is also referred to as the organic template). The sol was prepared by mixing different amounts of TEOS and MPS with ethanol, H_2O, and HCl in a molar ratio of 1.0-x : x : 3.8 : 5.1 : 5.3 $\times 10^{-3}$ (x <1) at 60 °C for 90 minutes at approximately 400 rpm stirring rate. The sols were cast into petri dishes and allowed to dry in air to obtain xerogels, which were then ground into a fine powder and calcined for 3 hours in air at 150°C (heating rate =1°C/min.) resulting in materials that allowed exhibited negligible nitrogen adsorption at 77 K. Microporous silica were obtained by further heating the powders at

Mat. Res. Soc. Symp. Proc. Vol. 435 © 1996 Materials Research Society

350°C for 3 hours in an oxygen atmosphere using a heating rate 1 °C/min. The xerogels were characterized by nitrogen adsorption porosimetry at 77 K (Micromeritics ASAP 2010, Norcross, GA).

METHOD OF SIMULATION

Molecular modeling was used to examine the relationship between the molecular size of the template ligand and the pore size and total pore volume of the microporous silica. The average diameter of the template ligand in MPS was estimated by minimizing the configurational energy of a model of the propylmethylacrylate molecule ($H_2C=C(CH_3)CO_2(CH_2)_2CH_3$) using BIOSYM software. We considered different configurations of the organic ligand in the silica matrix by rotating the molecular model in three dimensions in increments of 5 degrees. The average volume occupied by the ligands was obtained by calculating the average dimensions of the rotating molecule model in three dimensions and equating the volume to that of an equivalent sphere. We projected the propylmethylacrylate in the X-Y plane and measured the dimensions, X_i and Y_i. The radii of equivalent spheres were calculated by $R_i = (X_i * Y_i)^{1/2}$. The average diameter, R, of the equivalent sphere was calculated by averaging R_i for all configurations. The volume occupied by a model molecule in silica matrix is then estimated by $4/3\ \pi\ (R/2)^3$.

The pore size was simulated using the above method. By observing the dimensions of the rotating model molecule in X or Y direction (X_i or Y_i), we obtained a frequency distribution of dimensions observed. From the frequency distribution, we assigned the lowest dimensions observed to the smallest projected area of a template ligand (aperture), and compared the aperture area to the pore size of the membranes obtained by the gas permeation experiment.

RESULTS

Figure 1 shows the nitrogen adsorption isotherms of MPS/TEOS (20: 80) xerogels measured at 77 K. It shows that the xerogel is non-porous to nitrogen after heat treatment at 150°C, but become porous after heat treatment at 350 °C for 3 hours. The sharp type I adsorption isotherm for the latter sample indicates microporosity and a narrow pore size distribution.

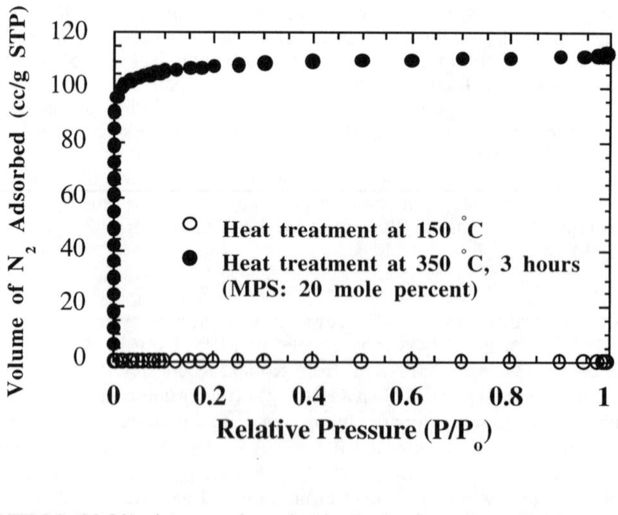

Figure 1. MPS/TEOS (20:80) nitrogen adsorption isotherms heated at 150 °C or 350 °C (holding time 3 hours).

Figure 2 shows the pore volume of hybrid materials (the xerogels heated at 150 °C in air) and microporous silica (the xerogels after heated at 350 °C for 3 hours in an oxygen atmosphere) prepared by reacting different amounts of MPS and TEOS. MPS/TEOS hybrid materials with different MPS ratios after heat treatment at 150 °C are non-porous to nitrogen at 77 K. After heat treatment at 350°C for 3 hours, pore volume increases with the amount of MPS added. However, only when the ratios of MPS are higher than 2%, do the xerogels have the measurable pore volume; when the ratios of MPS are smaller than 2%, the xerogels are non-porous to nitrogen at 77 K.

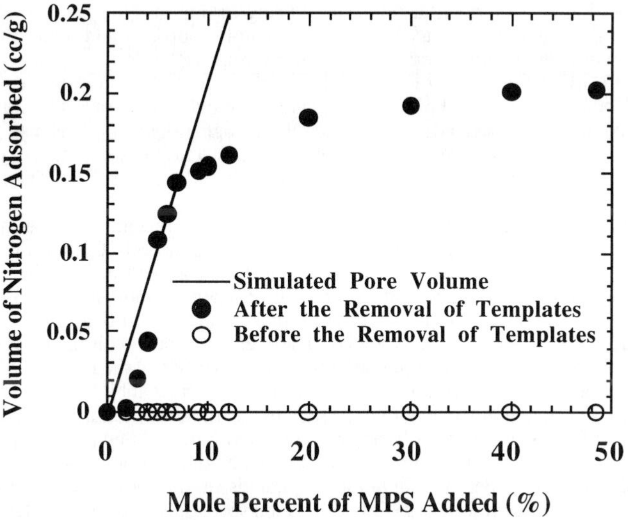

Figure 2. Pore volume of silica xerogels prepared by copolymerizing different mole ratios of MPS with TEOS before and after the removal of organic ligands.

Computer simulation results show that the volume of one propylmethacrylate molecule occupied in the matrix is 208.6 Å^3 and that the aperture size (corresponding to the minimum cross-sectional area) is about 5.2Å. Comparing the size of a propylmethacrylate molecule calculated from a space filling model (150 Å^3),[6] we expect that the approximation of the ligand volume from the simulation is reasonable. The higher estimated pore volume is expected due to contributions from free space around the templates which the silica network cannot access. Assuming that the silica network is fully condensed and does not collapse upon removing the templates, and that all the pores created by the templated ligands can be detected, the volume occupied by organic ligands would be 2.09 X (cc/g), where X is mole fraction of MPS added to the MPS/TEOS hybrid system. So the maximum pore volume created by the template ligands varies with the mole fraction of MPS as illustrated by a straight line passing through the origin with a slope 2.09 cc/g shown by Figure 2.

DISCUSSION

The organic template approach is based on the formation of a relatively non-porous silica matrix incorporating organic ligands as templates and the removal of these templates to create a microporous network. It was proposed that in this way pore volume could be controlled by the amount of templates added, and pore sizes by the ligand size, if there is no phase separation of ligands in the matrix, and if the matrix is non-porous or containing only pores smaller than the templates.

Because pores in the silica formed by this template method can be created both by the free volume of silica network (primary pore) and by the volume occupied by organic template ligands (secondary pore), this requires that the size of the primary pore be smaller than that of the secondary pore or the silica matrix be non-porous. Relatively dense hybrid materials incorporating organic ligands were prepared by minimizing the modulus of the silica network and maximizing the capillary pressure during drying. MPS and TEOS were reacted near the isoelectic point (pH = 2.2) at which the condensation rate is minimized.[4] The slow condensation reaction promotes the formation of a weak inorganic network and favors the creation of hydrophilic silanol groups. This results in the creation of high capillary pressure during drying and the promotion of hydrogen bonding between propylmethylacrylate groups and silanol groups,[7] and therefore, leads to the formation of relatively dense hybrid organic-inorganic xerogels.

As mentioned, MPS/TEOS hybrid materials in all range of MPS ratios are non-porous to nitrogen at 77 K before the removal of organic ligands, while they are porous to nitrogen after the removal of organic templates when the MPS mole ratio is higher than 2%. This result suggests that the primary pores cannot be detected by nitrogen at 77 K, or that the silica matrix is non-porous, while the secondary pores are large enough to be detected by nitrogen at 77 K. Thus nitrogen is an appropriate probe for the secondary pores at 77 K.

It is difficult to determine volume of organic ligands in the silica matrix experimentally; thus, computer simulation is applied in this study as an alternative approach. Comparing the measured pore volume and the simulated pore volume shown in Figure 2, we find that the measured pore volume is always smaller than the simulated pore volume when the mole fraction of ligands added is less than 5%. The measured pore volume increases sharply at 5%, and rises further in proportion to the mole fraction of MPS added. The measured pore volume corresponds to the simulated pore volume very well when the mole fraction of MPS is in the range of 5% to 8%. This implies that in this range the pore volume is contributed to that created by template pyrolysis. Thus it is possible to control pore volume at a limited range by the mole fraction of templates.

The deviation of measured pore volume and simulated pore volume at low MPS mole ratios (<5%) can be explained using percolation theory. When the volume fractions of organic ligands in the matrix are below than the percolation point, the ligands cannot form a percolation network. In this case, the measured pore volume is mainly due to the secondary pores connected to the sample powder surface, and thus it is always smaller than the simulated pore volume. The measured pore volume thus depends both on the connectivity of secondary pores and the size of sample particles. After the percolation point, the removal of a continuous template ligand network creates a continuous pore network permitting the detection of all secondary pores in the network. The measured pore volume increases sharply followed by a linear increase at 5%. These results suggest that the percolation point should be about 5% corresponding to 18% volume percent.

When the mole fraction of MPS is larger than 8%, the pore volume still increases with the amount of MPS added, and almost remains almost constant when MPS ratios are larger than 10%. However, the measured pore volume is smaller than the simulated pore volume. This may be due to continued condensation reactions, structural relaxation, or sintering during heat treatment and removal of templates. With increasing volume fraction of organic templates added, the network becomes weaker and thus collapses more easily when we remove the templates.

By this template approach, we have prepared membranes which have pore size larger than 4.5 Å and smaller than 5.5 Å so that SF_6 (5.5 Å) cannot pass through the membranes while propylene (4.5 Å) can pass through.[3] Computer simulation results, which suggest a pore size of 5.2Å, correspond well to the experimental data results obtained from membrane transport experiments. The simulated pore size corresponds to the smallest projected area of a template ligand (aperture). This result is reasonable because the packing of organic templates in the silica network cannot be completely tight, and thus the pore size created will be larger than the actual minimum cross-sectional area of the template ligands. The pore size achieved is likely to depend both on the minimum cross sectional area of the template ligand and on the packing density. In turn, the packing density will depend on the degree of network collapse during drying and the interaction forces between the silica matrix and the template ligands. The lowest limit of pore size will thus be the aperture size of the templates if the network does not collapse during the removal of these templates. These results demonstrate that it is possible to use computer simulation and experimental methods to predict and control pore sizes. Further work needs to be done to simulate the possible aperture size of different template ligands incorporated in a silica matrix by considering different interactions between matrix, solvent and template ligands.

CONCLUSIONS

We have demonstrated that the porosity (total pore volume) and the pore size of microporous silica can be controlled using the organic template approach. Microporous silica materials with small pore size and narrow pore size distributions were obtained by pyrolysis of propylmethylacrylate ligands from a relatively dense hybrid materials. A new model has been derived using a molecular model to simulate the aperture size of the organic ligands and the volume occupied. Experimental and simulation results suggest that pore size and pore volume can be controlled by the aperture size of the ligands and the amount of ligands added respectively, which demonstrates the possibility of preparing the microporous silica using the organic template approach.

ACKNOWLEDGMENTS

This work was supported by a grant from the Center for Micro-Engineered Ceramics at the University of New Mexico.

REFERENCES

1. R. R. Bhave, Inorganic Membranes Synthesis, Characteristics and Applications, (Van Nostrand Reinhold, New York, NY, 1991).

2. A. B. Shelekhin, A. G. Dixion, Y. H. Ma, Theory of Gas Diffusion and Permeation in Inorganic Molecular-Sieve Membranes, AICHE Journal, **41**, 58-67, 1995.

3. G.Z. Cao, Y.F. Lu, G. P. Lopez, and C. J. Brinker, Amorphous Silica Molecular Sieve, Advanced Materials, in press.

4. C. J. Brinker, G. W. Scherer, Sol Gel Science, The Physics and Chemistry of Sol Gel Processing, (Academic Press, New York, NY, 1990).

5. C. J. Brinker, R. Sehgal, S. L. Hietala, R. Deshpande, D. M. Smith, D. Loy and C. S. Ashley, Sol-Gel Strategies for Controlled Porosity Inorganic Materials, Journal of Membrane Science, **94**, 85-102, 1994.

6. G. L. Slonimskii, A. A. Askadskii and A. I. Kitaigorodskii, Vysokomol Soyed. A12: No. 3, 494-512, 1970.

7. X. Li and T. A. King, Structural and Optical Aspects of Sol-Gel Optical Composites, in Better Ceramics Through Chemistry VI, edited by A. K. Cheetham, C. J. Brinker, M. L. Mecartney and C. Sanchez (Mater. Res. Soc. Proc. 346, Pittsburgh, PA, 1994) pp.541-546.

CONTROLLING POROSITY IN BRIDGED POLYSILSESQUIOXANES THROUGH ELIMINATION REACTIONS

Mark D. McClain†, Douglas A. Loy*,†, and Sheshasayana Prabakar‡
† Org. 1812, Sandia National Laboratories, Albuquerque, NM 87185-1407
‡ Advanced Materials Laboratory, University of New Mexico, Albuquerque, NM 87106

ABSTRACT

The retro Diels-Alder reaction was used to modify porosity in hydrocarbon-bridged polysilsesquioxane gels. Microporous polysilsesquioxanes incorporating a thermally labile Diels-Alder adduct as the hydrocarbon bridging group were prepared by sol-gel polymerization of trans-2,3-bis(triethoxysilyl)norbornene. Upon heating the 2,3-norbornenylene-bridged polymers at temperatures above 250°C, the norbornenylene-bridging group underwent a retro Diels-Alder reaction losing cyclopentadiene and leaving behind a ethenylene-bridged polysilsesquioxane. Less than theoretical quantities of cyclopentadiene were volatilized indicating that some of the diene was either reacting with the silanol and olefinic rich material or undergoing oligomerization. Both scanning electron microscopy and nitrogen sorption porosimetry revealed net coarsening of pores (and reduction of surface area) in the materials with thermolysis.

INTRODUCTION

Hydrocarbon-bridged polysilsesquioxanes are a class of hybrid organic-inorganic materials prepared by sol-gel polymerization of bridged monomers, $(EtO)_3Si\text{-}R\text{-}Si(OEt)_3$, in which siloxane bonds and organic bridging groups are both integral structural components of a highly crosslinked network polymer [1]. Air drying affords micro- and mesoporous xerogels [1-4] in which the surface area and pore sizes can be controlled to some extent through the choice of the bridging group [4,5]. Alternatively, porosity can be created in non-porous gels or modified in porous gels by using the bridging organic functionality as a pore template. Previous efforts have used hydrolysis [2,6] or thermal [7] and low temperature plasma oxidations [8] to completely remove the hydrocarbon groups without any selectivity affording only inorganic silica gels as the final product.

We have prepared and characterized the first polysilsesquioxane utilizing a Diels-Alder adduct as the organic bridging group in a new approach to altering the porosity in a gel: partial decomposition of the 2,3-norbornenylene-bridging group through a *retro* Diels-Alder reaction (**Figure 1**). Volatile cyclopentadiene is evolved, leaving a 1,2-ethenylene-bridged polysilsesquioxane.

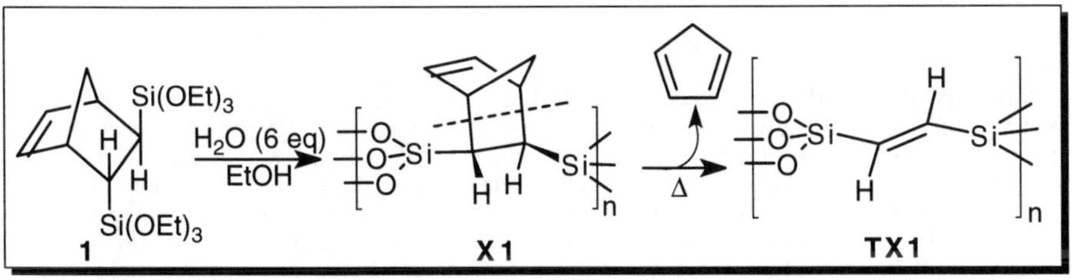

Figure 1. Preparation and retro Diels-Alder reaction of norbornenylene-bridged xerogels.

EXPERIMENTAL SECTION

The norbornenylene-bridged monomer (**1**) was prepared in several steps: (E)-1,2-bis(trichlorosilyl)ethene was reacted with cyclopentadiene to give *trans*-2,3-bis(trichlorosilyl)-5-

Mat. Res. Soc. Symp. Proc. Vol. 435 © 1996 Materials Research Society

norbornene (94% yield), which was subsequently esterified with ethanol in 71% yield. Gels were prepared in ethanol (0.1-0.4 M) using 1.0 M NaOH (10.8 mol %) as catalyst/water source (6 eq. H_2O) and processed with water as described previously [3].

Norbornenylene-Bridged Polysilsesquioxanes: Solutions of **1** and aqueous NaOH (6 eq. H_2O) in ethanol were prepared and transferred to tightly stoppered glass bottles. Cloudy solutions were formed which set to form opaque gels (**Table I**). ^{13}C CP-MAS NMR (50.2 MHz) δ 187 (ssb), 180 (ssb), 137.4, 132.8, 90 (ssb), 83 (ssb), 58.4 (CH_2CH_3), 49.1, 44.8, 25.8, 18.0 (CH_2CH_3).

Thermal Treatment of Xerogels. Weighed samples were placed in a ceramic boat and inserted into a quartz tube containing an in-line trap for collection of volatiles. The apparatus was evacuated (60 μ) and sealed from the vacuum source. The trap was cooled with liquid nitrogen as the tube was heated in a tube furnace. Xerogel **X1A** (0.398 g) was heated at 300°C for 3 h to give thermally treated xerogel **TX1A** as a light yellow powder (0.290 g, 73%). ^{13}C CP-MAS NMR (50.2 MHz) δ 146, 135, 132, 59, 45, 43, 26, 18; ^{29}Si CP-MAS NMR (39.6 MHz) δ -63.7, -72.1. Anal. Calcd for $C_2H_2Si_2O_3$: C, 18.4; H, 1.5. Found: C, 23.69; H, 3.91. Xerogel **X1B** (0.398 g) was heated at 250°C for 3 h to give thermally treated xerogel **TX1B** as a white powder (0.327 g, 82%). ^{13}C CP-MAS NMR (50.2 MHz) δ 146, 140, 138, 132, 59, 49, 44, 26, 18; ^{29}Si CP-MAS NMR (39.6 MHz) δ -63.2, -73.6. Anal. Calcd for $C_2H_2Si_2O_3$: C, 18.4; H, 1.5. Found: C, 27.98; H, 4.52. ^1H NMR(acetone-d_6) analysis of the volatiles from the thermolyses revealed the presence of cyclopentadiene, ethanol, and water.

RESULTS

Monomer **1** is the formal Diels-Alder adduct of (E)-bis(triethoxysilyl)ethene (**2**) and cyclopentadiene. The reaction of **2**, synthesized according to the method of Marciniec, *et al.*, [9] and cyclopentadiene or dicyclopentadiene in a high pressure reactor failed to give **1**, presumably due to the steric hinderance from the triethoxysilyl groups. To generate a more reactive dienophile, **2** was chlorinated using thionyl chloride and DMF catalyst [10], giving (E)-1,2-bis(trichlorosilyl)ethene (**3**) [11-14] in 36% yield. In contrast to **2**, dienophile **3** reacted exothermically with cyclopentadiene in solution (THF or benzene) to quantitatively give the expected Diels-Alder adduct, 2,3-bis(trichlorosilyl)norbornene (**4**). Esterification of **4** with ethanol in the presence of triethylamine readily gave the desired norbornenylene-bridged monomer **1** in good yield (71%).

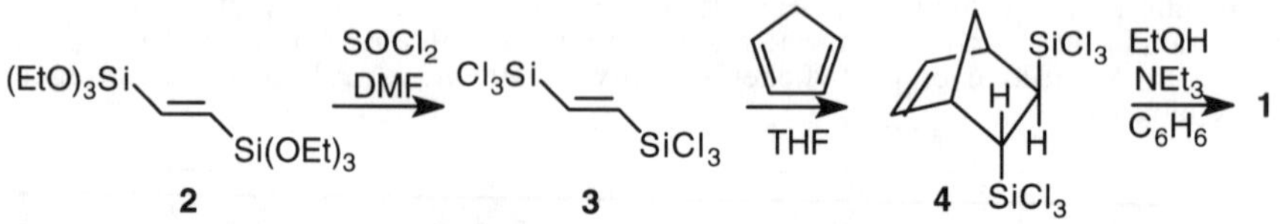

Polymerization and Characterization. Sol-gel polymerization of bridged triethoxysilane monomers involves the hydrolysis of ethoxysilyl groups to silanols, followed by intermolecular condensation reactions to form siloxane linkages in a network polymer [1]. Examination of gel times for **1** gives some indication of the impact of the sterically bulky norbornenylene-bridging group upon intermolecular siloxane bond formation. Polymerization of ethenylene-bridged monomer **2** (0.4 M in ethanol) proceeds to give gels under basic conditions within a few minutes [15]. Solutions (0.4 M) of norbornenylene-bridged monomer **1** with NaOH (10.8 mol %) took six days to form opaque gels. Since **2** has olefinic (sp^2) carbons rather than the saturated (sp^3) carbons of **1**, its polymerization rate would be expected to be enhanced under basic conditions compared with alkyl (sp^3) substituted triethoxysilanes. However, 1,6-bis(triethoxysilyl)hexane, a monomer with an alkyl (sp^3) substituent attached to the triethoxysilyl groups, formed gels within 1 hour under identical conditions. Under acidic conditions, where alkyl substituents should enhance condensation rates, no gels were obtained from **1**, even after several months at 1 M monomer concentration [16] Under identical conditions, **2** formed gels within 6 days [15] and 1,6-bis(triethoxysilyl)hexane formed gels within 12 hours.

Upon processing and drying, xerogels from both **1** and **2** were obtained as insoluble, white powders. The clear colorless solutions obtained from the acid-catalyzed sol-gel polymerization of **1** gave solid residues with removal of solvent *in vacuo*. After drying, white powders were obtained that were soluble in organic solvents indicating that the materials were either oligomers or mostly linear polymers.

Table I. Conditions, Gel Times, and Surface Area Measurements for the Sol-Gel Polymerization of **1** under Base Catalysis.

Xerogel	Conc. (M)	Gel time	Yield (%)	BET Surface Area (m^2/g)	t-Method Micropore Area (m^2/g)	Mesopore Area (m^2/g)	Ave. Pore Diam. (Å)
X1A	0.4	6 d	79	419	316	103	25.3
X1B	0.4	6 d	88	434	329	105	24.6
X1C	0.2	13 d	100	422	311	111	25.2
X1D	0.1	197 d	106				

Nitrogen sorption porosimetry revealed all the xerogels to be porous (**Table I**). Norbornenylene-bridged xerogels **X1A-C** had higher micropore areas and smaller average pore diameters than those prepared from **2**, which were completely mesoporous. Based on plots of the incremental pore volume from the adsorption isotherm versus pore size [4], most of the pore volume in norbornenylene-bridged xerogels **X1A-B** resulted from microporosity (pore diameter <20Å). The soluble materials prepared from **1** under acidic conditions were non-porous.

Cross Polarized-Magic Angle Spinning (CP-MAS) ^{13}C NMR of the xerogels revealed that the hydrocarbon bridges remained intact throughout the polymerization and processing. The norbornenylene-bridged xerogels **X1A-C** gave resonances due to both the hydrocarbon bridge as well as residual ethoxy groups, a consequence of the slow rate of hydrolysis at the sterically hindered silicon atoms. In the ^{29}Si CP-MAS NMR, **X1A-C** gave broad resonances which could not be deconvoluted to quantify T^n distribution (T^n describes a trifunctional silicon containing n siloxane bonds) and calculate the degree of condensation, however the signals were centered in the T^2 region for alkylene-bridged systems (δ -58.8) [17]. The soluble oligomers gave several peaks in the ^{29}Si NMR spectra (acetone-d_6) indicative of partial hydrolysis and condensation: δ -50.0 (T^1), -51.6 (T^1), -54.9 (T^1), -58.6 (T^2), -65.0 (T^3). No Q^n resonances due to silica were observed for any of the gels, indicating no Si-C hydrolysis had occured.

Thermolysis. Gels made from norbornenylene-bridged monomer **1** provide an interesting opportunity for chemical modification *after* polymerization due to the thermal reversibility of the [4 + 2] Diels-Alder cycloaddition reaction. Ideally, an ethenylene-bridge would remain in the polysilsesquioxane network and volatile cyclopentadiene (b.p. 40°C) would be eliminated. Complete elimination of cyclopentadiene from a fully condensed norbornenylene-bridged gel would result in a 34% mass loss.

Thermogravimetric Analysis (TGA) was used to establish the onset temperature of the retro Diels-Alder reaction. Under flowing nitrogen (25 cc/min) at a moderate heating rate (10°C/min), the norbornenylene-bridged gels started losing mass at 160°C with a second onset of mass loss at 360°C. At a slower heating rate (1°C/min), the mass loss onset temperatures were lowered (140 and 330°C, respectively), indicating that diffusion may play a role in the rate of mass loss. The identity of the volatiles during the thermolysis of norbornenylene-bridged gels was established by TGA-MS (heating rate = 60°C/min). Below 340°C, only molecular ions due to water and ethanol from further condensation reactions were observed. Above 340°C, the molecular ion for cyclopentadiene (m/z = 67, $[M+H]^+$) was detected.

Larger scale themolyses (0.4 g) were run in a tube furnace (300 and 250°C) to allow analysis and quantitation of the pyrolysate and volatiles. Judging from the change in mass, residual norbornene groups remained in the thermally treated gels **TX1A** and **TX1B** (mass loss = 27 and 18%, respectively); a somewhat larger mass loss was expected since the precursor gels **X1A-B** should also undergo further hydrolysis and condensation. ^{1}H NMR analysis of the volatiles collected with a cold trap (-197°C) during the thermolyses confirmed that water and ethanol, as well as cyclopentadiene, were evolved. Thermolysis of ethenylene-bridged gels under identical conditions resulted in virtually no mass loss and only traces of volatiles were produced. By TGA (10°C/min), untreated ethenylene-bridged gels heated to 900°C under nitrogen gave a ceramic yield of 96% .

^{13}C CP-MAS NMR spectra of the thermally treated gels (**TX1A** and **TX1B**) along with the unheated norbornenylene-bridged (**X1A**) and ethenylene-bridged xerogels are shown in **Figure 2**. The spectra reveal the decomposition of the norbornenylene group and the emergence of the new ethenylene-bridged structure. The sample heated at 250°C (**TX1B**) exhibited the resonance (δ 146) from the ethenylene carbons and reduced contributions from norbornenylene sp^3 (δ 49.5, 44.8, 26.2) and sp^2 (δ 137.5, 132.7) carbons relative to the ethoxide signals (δ 58.4, 18.0). Greater conversion to the ethenylene-bridged material at 300 °C (**TX1A**) was evidenced by an even greater contribution from ethenylene carbons. However, even after 3 hours at 300°C, attenuated peaks from the norbornenylene group still remained in the gel.

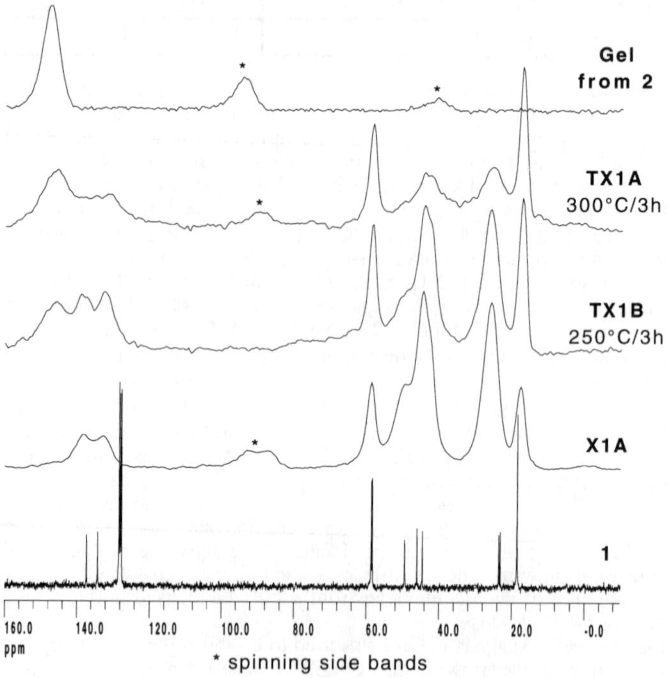

Figure 2. Evolution of ^{13}C CP-MAS NMR Spectra of Norbornenylene-Bridged Gels with Thermal Treatment and Comparision with an Ethenylene-Bridged Gel.

The ^{29}Si CP-MAS NMR spectra also revealed the conversion of the norbornenylene bridge to an ethenylene-bridging group (**Figure 3**). **TX1B** displayed a new shoulder upfield of the norbornenylene-silsesquioxane T^2 at δ -58.8. After thermolysis at 300°C, the shoulder had become a distinct resonance at δ -72, indicative of T^2 silicon in an ethenylene-bridged polysilsesquioxane. The T^2 silicon is consistent with the assignment of the silicon peak in the untreated norborneylene-bridged polymers as T^2 suggesting that the degree of condensation was relatively unperturbed by the thermolysis.

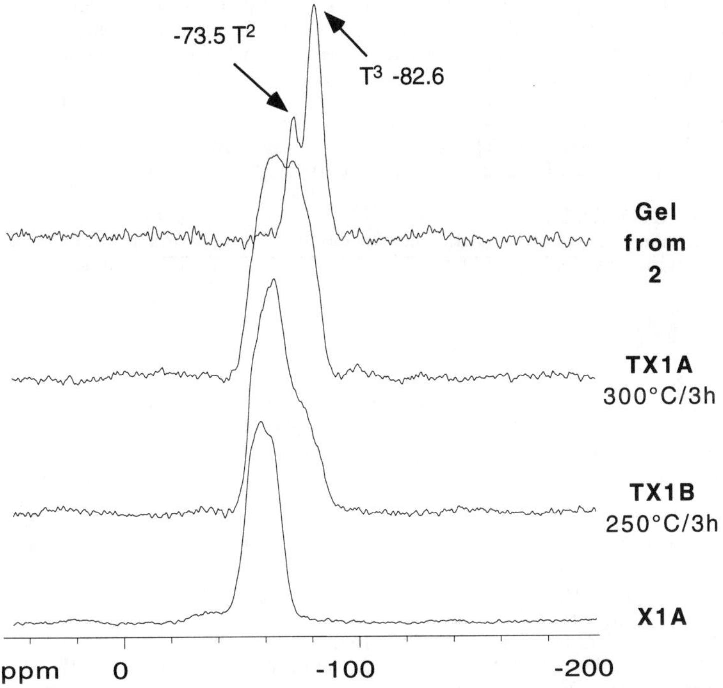

Figure 3. Evolution of ^{29}Si CP-MAS NMR Spectra of Norbornenylene-Bridged Gels with Thermal Treatment and Comparision with an Ethenylene-Bridged Gel.

After thermolysis, the composition [18] of **TX1A** and **TX1B** reflected the loss of hydrocarbon from the polymer, moving considerably closer to the theoretical composition for the ethenylene-bridged polysilsesquioxane (C, 18.4%; H, 1.5%). Incomplete conversion to the ethenylene-bridged material may be due to either oligomerization of cyclopentadiene, irreversible (radical) reactions with the ethenylene-bridging groups, or reactions with silanols. In our hands, thermal polymerization (300 °C) of neat cyclopentadiene was possible in a ethenylene-bridged polysilsesquioxane gel, resulting in a white carbon-rich polymer. Acidic silanol groups in the gels may assist or catalyze polymerization of some of the volatilized cyclopentadiene during the thermolysis of **X1A** as well, however, no resonances in the ^{13}C CP-MAS NMR for polymeric cyclopentadiene were identified possibly as a consequence of overlapping signals.

TX1A and **TX1B** still retained high surface areas (282 and 321 m^2/g, respectively), however, some microporosity and mesoporosity was lost. In contrast, the thermal treatment of ethenylene-bridged gels at 250 or 300 °C resulted in *no change* in surface area or pore size distribution. The change in porosity for the norbornenylene-bridged gels, therefore, was a direct result of the elimination of cyclopentadiene as well as from further condensation. Scanning electron micrographs of a norbornenylene-bridged polysilsesquioxane before and after treatment at 300 °C confirmed the coarsening of the pores for the thermally treated norbornenylene-bridged gels.

CONCLUSIONS

Norbornenylene-bridged polysilsesquioxanes were prepared as the first representatives of sol-gel processed materials that can be structurally modified utilizing the retro Diels Alder reaction. These materials were successfully prepared as gels under alkaline conditions, but under acidic conditions, only soluble oligomers were isolated. The retro Diels Alder reaction was realized with thermolysis of these materials at temperatures over 250 °C. However, complete conversion to an ethenylene-bridged polysilsesquioxane was never observed, even after continued treatment at 300 °C, suggesting that the cyclopentadiene may have undergone side reactions that prevented its egress from the porous material. Thermolysis was accompanied by an apparent sintering of the materials. Significant loss of microporosity was observed by nitrogen sorption porosimetry.

ACKNOWLEDGMENTS

We would like to thank Brigitta Baugher for the nitrogen porosimetry measurements. This work was supported by the United States Department of Energy under contract No. DE-AC04-94AL85000.

REFERENCES

1. D.A. Loy, K.J. Shea, Chem. Rev. **95**, 1431 (1995).
2. K.J. Shea, D.A. Loy, O. Webster, J. Am. Chem. Soc. **114**, 6700 (1992).
3. J.H. Small, K.J. Shea, D.A. Loy, J. Non-Cryst. Solids **160**, 234 (1993).
4. H.W. Oviatt, Jr., K.J. Shea, J.H. Small, Chem. Mater. **5**, 943 (1993).
5. D.A. Loy, G.M. Jamison, R.A. Assink, S. Myers, K.J. Shea, ACS Symp. Ser. **585**, 264 (1995).
6. R.J.P. Corriu, J.J.E. Moreau, P. Thepot, M.W.C. Man, J. Mater. Chem. **4**, 987 (1994).
7. C.J. Brinker, R. Sehgal, S.L. Hietala, R. Deshpande, D.M. Smith, D. Loy, C.S. Ashley, J. Membr. Sci. **94**, 85 (1994).
8. D.A. Loy, K.J. Shea, R. Buss, ACS Symp. Ser. **572**, 122 (1994).
9. B. Marciniec, H. Maciejewski, J. Gulinski, Rzejak, J. Organomet. Chem. **362**, 273 (1989).
10. D.A. Loy, Ph.D. Thesis, University of California, Irvine, 1991.
11. C. Ruedinger, H. Beruda, H. Schmidbaur, Z. Naturforsch., B: Chem. Sci. **49**, 1348 (1994).
12. V.D. Sheludyakov, V.I. Zhun, V.G. Lakhtin, V.N. Bochkarev, T.F. Slyusarenko, V.N. Nosova, A.V. Kisin, Zh. Obshch. Khim. **54**, 640 (1984).
13. V.D. Sheludyakov, V.I. Zhun, V.G. Lakhtin, V.V. Shcherbinin, E.A. Chernyshev, Zh. Obshch. Khim. **53**, 1192 (1983).
14. V.D. Sheludyakov, V.G. Lakhtin, V.I. Zhun, V.V. Shcherbinin, E.A. Chernyshev, Zh. Obshch. Khim. **51**, 1829 (1981).
15. J.P. Carpenter, S.A. Yamanaka, M.D. McClain, D.A. Loy, J. Greaves, S. Hobson, K.J. Shea, Chem. Mater. , submitted for publication
16. A mass spectrometric study has concluded that the slow gelation for acid catalyzed polymerization of **1** results from much slower intermolecular condensation reactions due to the unfavorable steric interactions between norbornenylene groups: M. D. McClain, D. A. Loy, S. Prabakar, J. Greaves, K. J. Shea, to be published.
17. D.A. Loy, G.M. Jamison, B.M. Baugher, E.M. Russick, R.A. Assink, S. Prabakar, K.J. Shea, J. Non Cryst. Solids **186**, 44 (1995).
18. Due to the formation of carbides, elemental analyses were generally low in carbon, even with higher analysis temperature and the addition of combustion aid.

A NEW CATEGORY OF MEMBRANES EXPLOITING POTENTIALITY OF ORGANIC/INORGANIC HYBRID MATERIALS

C. GUIZARD, P. HECKENBENNER, J.C. SCHROTTER, N. HOVNANIAN, M. SMAIHI
Laboratoire des Matériaux et Procédés Membranaires - CNRS UMR 5635
Ecole Nationale Supérieure de Chimie
8, rue de l'Ecole Normale - 34053 Montpellier Cedex (France)

ABSTRACT

Organic/inorganic hybrid materials offer specific advantages for the realization of artificial membranes exhibiting high selectivity and flux as well as good thermal and chemical resistances. Preparations of hybrid membranes able to work in liquid or gas media are described. The influence of physicochemical properties of the membranes on transport parameters was also investigated. Results are discussed related to solution-diffusion theory in dense homogeneous membranes.

INTRODUCTION

Whatever the nature of artificial membranes, organic or inorganic, they can be classified in two main categories based on transport mechanisms. The first category is porous membranes in which transport is essentially due to convection. In the second category membranes exhibit a homogeneous dense structure in which mass transport is described by diffusion. Accordingly, porous or dense is the first characteristic to be defined in the design of new membranes. In the case of a porous membrane, pore size distribution and porous volume have a predominant influence on flux and selectivity while the chemical composition is concerned with polarization phenomena due to interactions between membrane and solutes. These two aspects can be considered independently in the preparation of porous membranes. Things are different with dense membranes for which flux and selectivity are strongly dependent on the chemical and physical structures. A homogeneous membrane merely consists of a dense film through which a mixture of chemical species is transported under the driving force of a pressure, concentration, or electrical potential gradient.

Regarding hybrid materials as potential materials for membrane separation both porous and dense membranes can be envisaged. The main interest of hybrid materials for porous membranes is the ability to tailor porosity through the template effect of the organic part eliminated from the structure in the final step of the preparation [1,2]. Improved membrane performance is expected due to the narrow pore size distribution which results from this method. This paper is concerned with another approach to hybrid membrane synthesis in which the hybrid character is preserved in the final membrane. One characteristic of these membranes is a dense or ultramicroporous (pore size $\leq$ 7 Å) homogeneous structure compared with micro or mesoporous structure generally obtained when the organic part is removed from the hybrid material. With homogeneous membranes, separation of various components from a feed phase is directly related to their transport rate within the membrane phase, which is determined by their diffusivity and concentration in the membrane matrix. For the sake of simplification, discussion concerning potentiality of hybrid materials in membrane separation has been limited to membranes working in water and gas media under concentration gradient and pressure gradient, respectively. Taking into account basic principles of mass transfer in homogeneous membranes, improvements of selectivity and flux are expected with the association of organic and inorganic materials. Two examples are

Mat. Res. Soc. Symp. Proc. Vol. 435 © 1996 Materials Research Society

given of hybrid membrane synthesis specially designed for facilitated diffusion dialysis of metal ions in liquid media and for gas mixture separation.

THEORY OF TRANSPORT IN HOMOGENEOUS MEMBRANES

A general description of mass transfer through dense or ultramicroporous membranes in which convective flux can be neglected is given by the solution-diffusion theory. The differential form for one-dimensional flux of species i within the membrane is given by Fick's law as,

$$J_i = -\frac{D_i}{h}\frac{dc_i}{dx} \tag{1}$$

with D_i the diffusion coefficient, h the membrane thickness and dc_i/dx the gradient of concentration.

In the case of a uniform homogeneous membrane separating two liquids and for steady state transport the flux equation (1) can be rewritten in terms of the external solution interface concentrations c' and c'':

$$J_i = D_i\ k_i\ (c' - c'')/h \tag{2}$$

with k_i being the constant solute equilibrium distribution or partition coefficient. It is customary in the solution-diffusion theory to denote the membrane permeability by:

$$P = D_i\ k_i \tag{3}$$

If the external media are mixtures of gases or vapors instead of liquid solutions, it is often more convenient to employ partial pressures in place of external concentrations. When the solubility of species i in the membrane is linear in external pressure, the steady state flux expression analogous to equation (2) becomes,

$$J_i = D_i\ s_i\ (p'_{\text{int}} - p''_{\text{int}})/h \tag{4}$$

in which p_{int} is the interfacial partial pressure of i and,

$$s_i = {c_{i\,[0]}}/{p'_{\text{int}}} = {c_{i\,[1]}}/{p''_{\text{int}}} \tag{5}$$

is a constant solubility coefficient, $c_{i[0]}$ and $c_{i[1]}$ being the concentration of i on both interfaces of the membrane. The definition of permeability complementary to equation (3) is then,

$$P_i = D_i\ s_i \tag{6}$$

Since k_i is dimensionless, whereas s_i is not, the units of P depends upon which the two definitions, equation (3) or (6) is employed.

One can see that permeability P for a homogeneous membrane, equations (3) and (6), is the product of the diffusion coefficient and the partition or solubility coefficient, respectively depending if liquid phase or gas phase is considered. The diffusion coefficient is assumed to be mainly dependent on the physical structure. In general mass transfer will be greater and selectivity lower in amorphous matrices than in highly crystalline or cross-linked matrices. An important property of homogeneous membranes is that chemical species of similar size, and hence identical diffusivities, may be separated when their concentration, or more appropriately their solubility in the film differs significantly. Regarding the role of chemical composition, the partition and solubility coefficients can be adjusted in order to improve the permeability and the selectivity of a membrane material.

CORRELATION BETWEEN TRANSPORT PARAMETERS AND PHYSICOCHEMICAL PROPERTIES OF HYBRID MEMBRANES

Heteropolysiloxane membranes designed for facilitated diffusion dialysis of metal ions

Interest of facilitated diffusion dialysis for metal ion recovery

Dialysis is a membrane process by which various solutes having widely different molecular weights may be separated by diffusion through semipermeable membranes. Dialysis requires that the membranes separating the two liquids permit diffusionnal exchange under a concentration gradient between at least some of the solutes while effectively preventing any convective mixing between the concentrated and dilute solutions. Apart from artificial kidney applications, dialysis has never become industrially important. The main reason is that processes such as ultrafiltration, reverse osmosis or electrodialysis have been preferred to dialysis because they utilize external driving forces resulting in a higher transmembrane flux. Moreover dialysis membranes are not very selective. Recently industries were urged to develop clean technologies to prevent pollution resulting from manufacturing waste streams. In particular there is an increased need for heavy metal removal from waste water. Silver ions are an example of a high valuable metal which has a wide range of industrial applications in which resulting effluents require silver concentrations as low as 0.1 mg/L (10^{-6} mol/L). Presently, dialysis could reappear as a soft and competitive separation technology for environmental application, provided that selectivity and flux can be improved. That is the reason why incorporation of silver specific complexants in a diffusion dialysis membrane is expected to increase partition coefficient compared to simple diffusion of other penetrants and then to achieve high selectivity and flux. Heteropolysiloxanes can be easily processed to form films and thin layers with physical and chemical structures which answer requirements of facilitated diffusion dialysis.

Influence of membrane structure and composition on mass transfer

When homogeneous membranes are in contact with aqueous phases the balance between hydrophilicity and hydrophobicity is an important parameter influencing volume flux J_V. Membranes in which hydrogen bonding sites are absent are usually hydrophobic and the water content is very low, often less than 1%. Accordingly water transport through the membrane is very low. On the contrary, when hydrogen bonding sites are present, high water sorption is observed. The diffusivity of water in these systems is quite low at low activities but the value increases, however, quite rapidly with increasing water content in the membrane [3]. Increase of water diffusivity is all the more important as swelling of the membrane contributes to free volume. Swelling phenomena and associated increase of free volume can be prevented by higher cross-linking of the polymeric matrix. In dialysis no pressure gradient is applied and only solutes are supposed to diffuse through the membrane. However a concentration gradient of solutes of small molecular weight across hydrophilic membranes results in an osmotic pressure $\Delta\Pi$ responsible for a mass flux of water, J_W. As shown in figure 1, the osmotic flux J_W goes in the opposite direction to the normal volume flux J_V when a pressure gradient Δp is applied to the membrane [3].

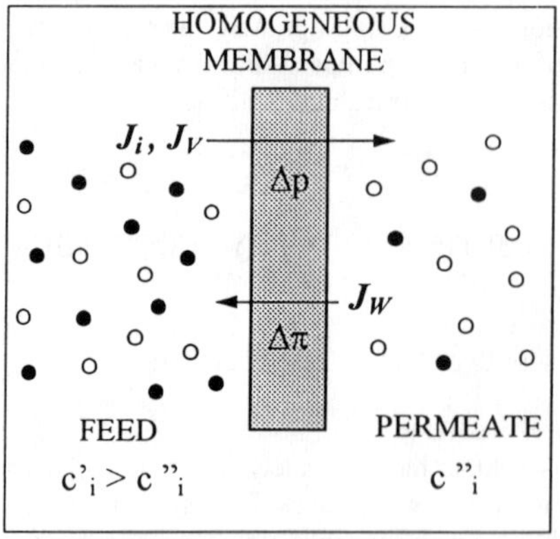

Figure 1. Schematic representation of mass transfer across a homogeneous membrane

Solubility of solutes in the membrane can be described in terms of polarisability of the matrix and related complex formation with the solutes. Solubility in the matrix is associated with an overall reduction in free energy of the system. In the case of ionic species, the enthalpy of solvation in a non-electrically charged matrix is essentially the result of electrostatic interactions between the cation positive charge and the negative charge on the dipolar group of the matrix or a partial sharing of a lone pair of electrons on a coordinating atom in the matrix. Partition coefficient k_i for a cation can be increased through specific complex formation. The stability of complexes is predicted by the hard-soft acid-base principle suggested by Pearson [4]. For hard cations, including alkali and alkaline earth ions, the best electron donor groups follow the trend O>NR, NH>>S, which has been confirmed in stability constant measurements in cyclic polyether complexes [5]. The stability sequence for soft cations as Ag^+ is NH>S>O [6]. Usually, in absence of an electrical field as a driving force, anions are transported across the membrane along with the cations. Differences in the general solvation energies of anions do occur as the dielectric constant of the matrix varies. On passing from a polar, protic medium (water) through to a less polar one (membrane), most anions are destabilized, the destabilization being greatest when the charge density and basicity of the ion are low [7]:

$$F^- >> Cl^- > Br^- > I^- \sim SCN^- > ClO_4^- \sim CF_3SO_3^- > BF_4^- > AsF_6^-$$

The most suitable choices of anions for aprotic, low-dielectric-constant matrices are those to the right of the preceding series.

Membrane preparation

Synthesis and modification of the heteropolysiloxane(HPS) membrane materials, figure 2, were performed according to requirements on mass transfer defined in the previous paragraph. Aminopropylmethyldiethoxy silane (APDS) and terephtaloyl chloride (TPC) were used as commercial reactants for the preparation of the organically modified silica matrix, following the method described in reference [8]. Condensation of TPC with APDS resulted in a matrix precursor (A) containing polar aromatic amide groups and hydrophobic $-CH_2-$ and $-CH_3$ groups. The aim of the TPC bridged monomer is to raise the dielectric constant of the membrane in order to increase ion solubility. Regarding transport of water, the osmotic flux has been prevented by introduction of hydrophobic alkane groups on silicon atoms. Selective solubility of silver in the membrane has been obtained by grafting of a specific complexant, MFA15 thioether (B), in the heteropolysiloxane matrix. Membranes were prepared from a solution of the precursors by film casting on a polyacrylonitrile (PAN) mesoporous support. The description of the casting method can be found in reference [8]. Due to the structure of this heteropolysiloxane membrane high silver partition coefficient compared to other ions present in the feed solution was expected as well as higher flux of silver.

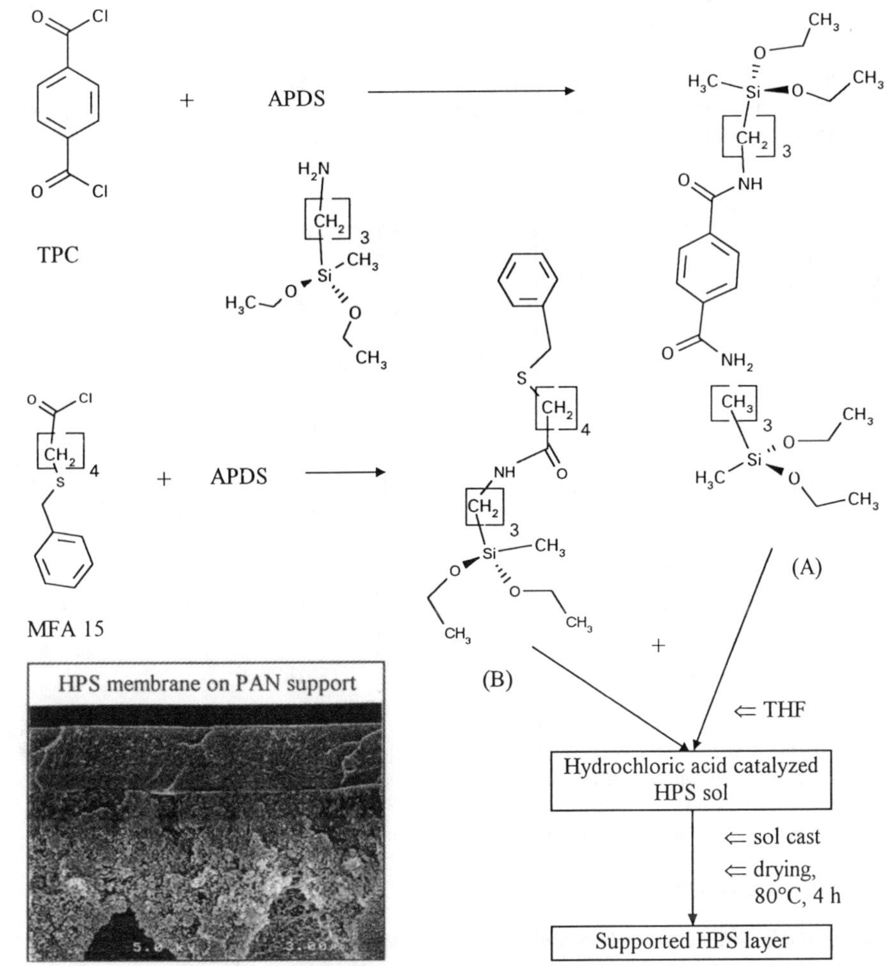

Figure 2. Preparation of a heteropolysiloxane membrane with grafted silver specific complexants.

Membrane performance evaluation

Facilitated diffusion of silver was evidenced by comparing performance of a membrane containing 35% (mole/Si) of MFA 15 complexant (from precursors A + B) with a free complexant membrane (from precursor A). Transport experiments were performed in a two chambers polypropylene cell equipped with a peristaltic pump allowing circulation of the feed and the strip phases on each side of the membrane. In order to accelerate decomplexation of ions at the downstream interface of the membrane in the case of copper/silver competitive transport, a specific extractant, EDTA, was used in the strip compartment.

No water transport was observed due to the cross-linked matrix and its hydrophobic character. In figure 3a, one can see that transport of silver has been enhanced by the introduction of selective complexants in the membrane. Furthermore, silver/copper selectivity was determined for both categories of membranes. Figure 3b shows that no selective transport is achieved with the free complexant membrane; however a selective transport of silver against copper was measured in the case of the membrane containing selective complexants. These results can be taken as the evidence of hybrid material versatility to fulfill requirements of membrane materials designed for selective transport of metal ions.

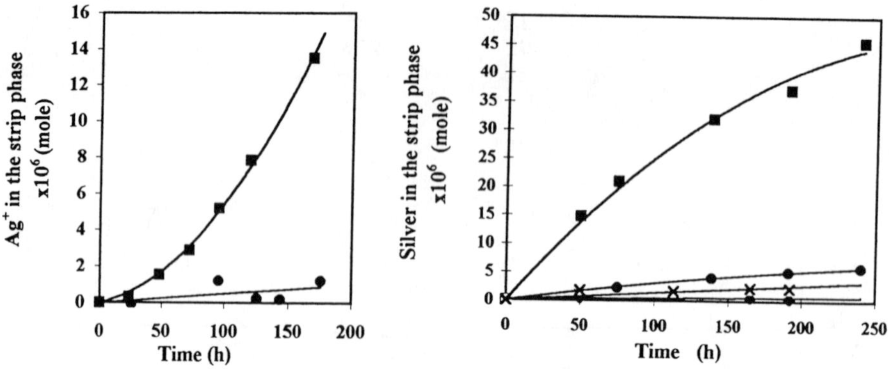

Feed phase : $AgNO_3$ 10^{-3}M
Strip phase: Water

(a)

Feed phase : Ag acetate 10^{-3}M + Cu acetate 10^{-3}M
Strip phase : EDTA 10^{-2} M + NaOH pH=11

(b)

Figure 3. Comparison of silver transport (a) and silver/copper competitive transport in heteropolysiloxane membranes containing or not specific complexant of silver ion.

a) silver transport with a carrier containing membrane (■) and a free carrier membrane (●).

b) competitive silver and copper transport with a 35% wt. carrier containing membrane (silver ■, copper ●) and a free carrier membrane (copper ×, silver ◆).

Polyimide/silica membranes designed for gas separation

Interest of polyimide/silica hybrid materials for hydrogen separation

In addition to their excellent thermal stability and high processability [9], a wide range of chemical and physical properties provided by molecular engineering makes polyimides highly versatile in membrane processes for gas separation. A number of references note interesting results of permeability and selectivity of common gases for this category of polymers [10]. In other respects, ultramicroporous silica membranes have been studied for their selectivity to hydrogen permeation [11]. In the present work, the combination of these two materials has been investigated with the aim to provide membranes with unique hydrogen separation properties. Two sorts of membranes are reported. One is a nanocomposite material made of silica beads dispersed in a polyimide matrix, the other is an organic/inorganic polymer at the molecular level.

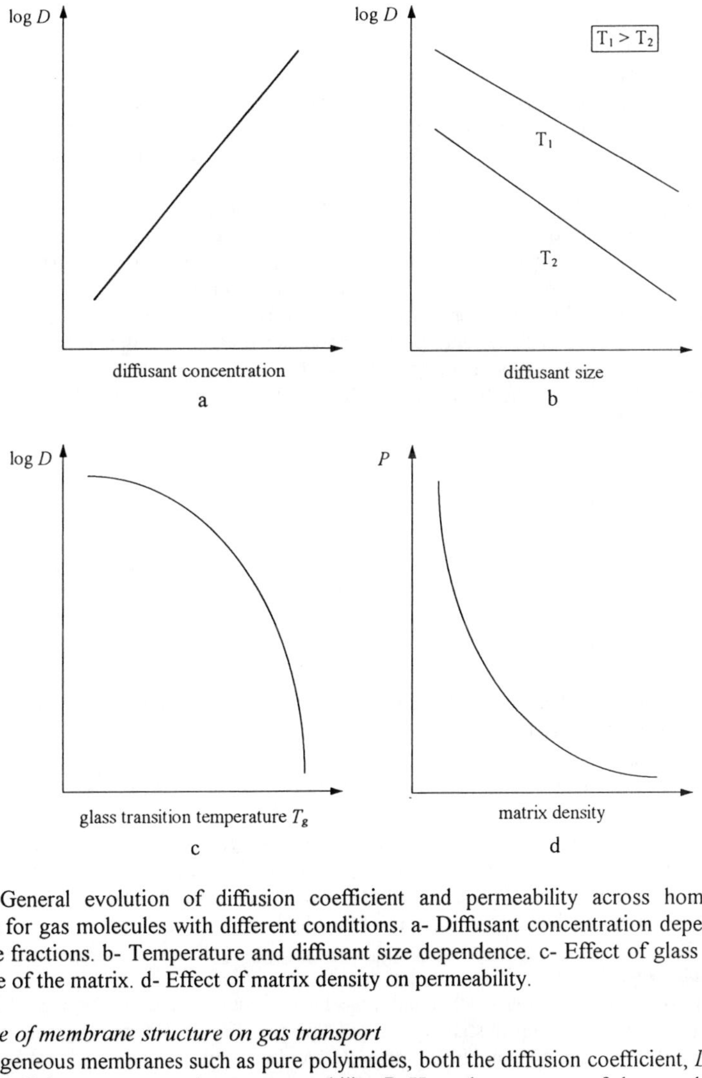

Figure 4. General evolution of diffusion coefficient and permeability across homogeneous membranes for gas molecules with different conditions. a- Diffusant concentration dependence at low volume fractions. b- Temperature and diffusant size dependence. c- Effect of glass transition temperature of the matrix. d- Effect of matrix density on permeability.

Influence of membrane structure on gas transport

In homogeneous membranes such as pure polyimides, both the diffusion coefficient, D, and the solubility coefficient, s, govern the gas permeability P. How the structure of the membrane may affect D and P is summarized in figure 4. Although variations in D are much greater than the variations in s, it is interesting to mention also how s varies with different conditions. First, the concentration of gas molecules in the membrane affects their diffusion coefficient. Linear logarithmic plots are obtained at low volume fractions, figure 4a. For dense or rigid matrices, it appears that the diffusion coefficient at low molecule concentrations is small and the dependence on concentration is great. In other respects, regarding the effect of gas molecular size, solubilities in elastomeric polymers increase with molecular size. However, the increase of s with increasing size is much less than the decrease of D and so permeabilities decrease with increasing size. As shown in figure 4b, with permanent gases, D decreases with increasing penetrant size but more important than actual diffusant size is diffusant shape. For larger organic molecules of the same molar volume, the diffusion coefficient in elastomeric polymers is expected to be higher for a flexible molecule than for a more rigid one [12].

Also the dependence on temperature is generally high whenever the diffusion coefficient is low. It is possible to interpret the temperature dependence in terms of the activation energy. Moreover activation energy increases with penetrant size and tends towards a constant value for large penetrants. As might be expected, an important increase of D with decreasing value of glass transition temperature T_g, figure 4c, is accompanied by an important decrease in the concentration dependence of the diffusion coefficient. Finally, an increase of membrane density due to crystallization or cross-linking of the matrix is responsible for a decrease in permeability. This can be explained by a simultaneous diminution of D and s, figure 4d.

A peculiar case results from dispersion of fillers such as aluminium particles in a polymer matrix [13]. Even though such systems are heterogeneous on a microscopic scale, the membranes can be considered homogeneous on a larger scale and it is possible to discuss the system in terms of D, s and P coefficients. The theoretical problem is to relate these overall coefficients to the values D_A, a_s, P_A of the disperse phase, and d_B, S_{BA}, P_{BS} of the continuous phase. Because the solubility's in the two phases are additive,

$$s = s_A \phi_A + s_B (1 - \phi_A) \qquad (7)$$

with ϕ_A is the volume fraction of the disperse phase A. In many cases of polymer/filler systems it can be assumed that D_A, a_s and P_A are zero and then,

$$s = s_B (1 - \phi_A) \qquad (8)$$

The effect of impermeable spheroids randomly dispersed in a continuous phase on overall permeability has been derived by Fricke [14],

$$P = P_B (1 - \phi_A)/(1 + \phi_A / X) \approx P_B [1 - \phi_A (1 + 1/X)] \qquad (9)$$

Here, the parameter X is a function solely of the spheroid geometry, increasing as a function of asymmetry. However, for a high volume fraction of finely dispersed fillers, adsorption at the filler interface can sometimes occurs so that a filled polymer film actually adsorbs more gas than an unfilled one resulting in a higher permeability of the filled one.

Membrane preparation

Poly(4,4'-oxydiphenylene pyromethylimide) has been chosen for the organic matrix because of its excellent thermal and chemical stability. Preparation was carried out starting from pyromellitic dianhydride (PMDA) and 4-aminophenyl ether (ODA) in dimethyl acetamide (DMAc). Because almost all polyimides are insoluble in organic solvents and infusible, they are fabricated into products such as films and coatings in the form of the soluble polyamic acid.

The preparation of polyimide/silica nanocomposite materials schematized in figure 5 has been described in references [15,16]. Various proportion of tetramethoxysilane (TMOS) were introduced in the polyamic solution yielding polyimide-silica composite films from 0 to 45%wt of silica. Moreover the quantity of water introduced in the mixture allowed modulation of the size of silica beads.

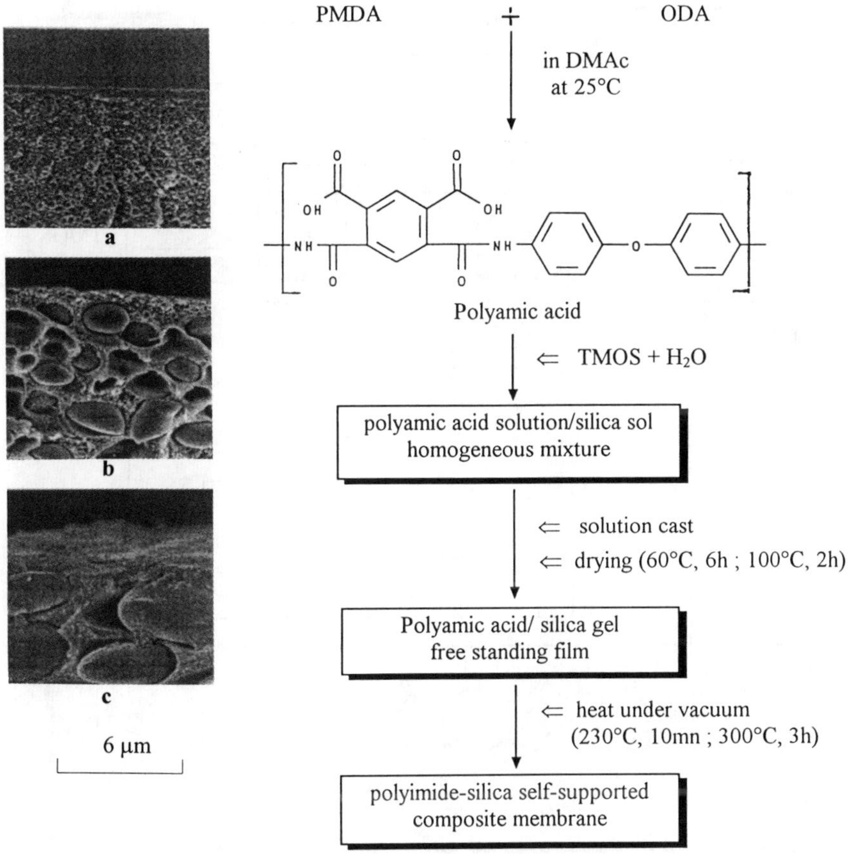

Figure 5. Preparation of composite membranes made of silica beads dispersed in a polyimide matrix. SEM images: a) 14% Wt SiO_2, b) 35% Wt SiO_2, 45% Wt SiO_2.

Polyimide/siloxane polymers at the molecular level have been obtained using organically modified silicon alkoxides for the siloxane part, figure 6. Mixtures of polimide precursors, PMDA and ODA, with siloxane precursors were reacted in DMAc in order to form soluble polymers in a first step. Polyimide formation was performed at 300°C in a second step, after casting of the film. Two kinds of polymer, linear (precursor I) or cross-linked (precursors II, III, IV), have been obtained depending on the nature of organically modified silicon alkoxide. Moreover, TMOS was used as cross-linking agent in preparation II, III and IV. The physicochemical structure of these polymers has been investigated by infrared spectroscopy, NMR and differential thermal and thermogravimetric analysis [17]. A good thermal resistance has been evidenced, the membranes being able to work up to 250°C in a continuous mode.

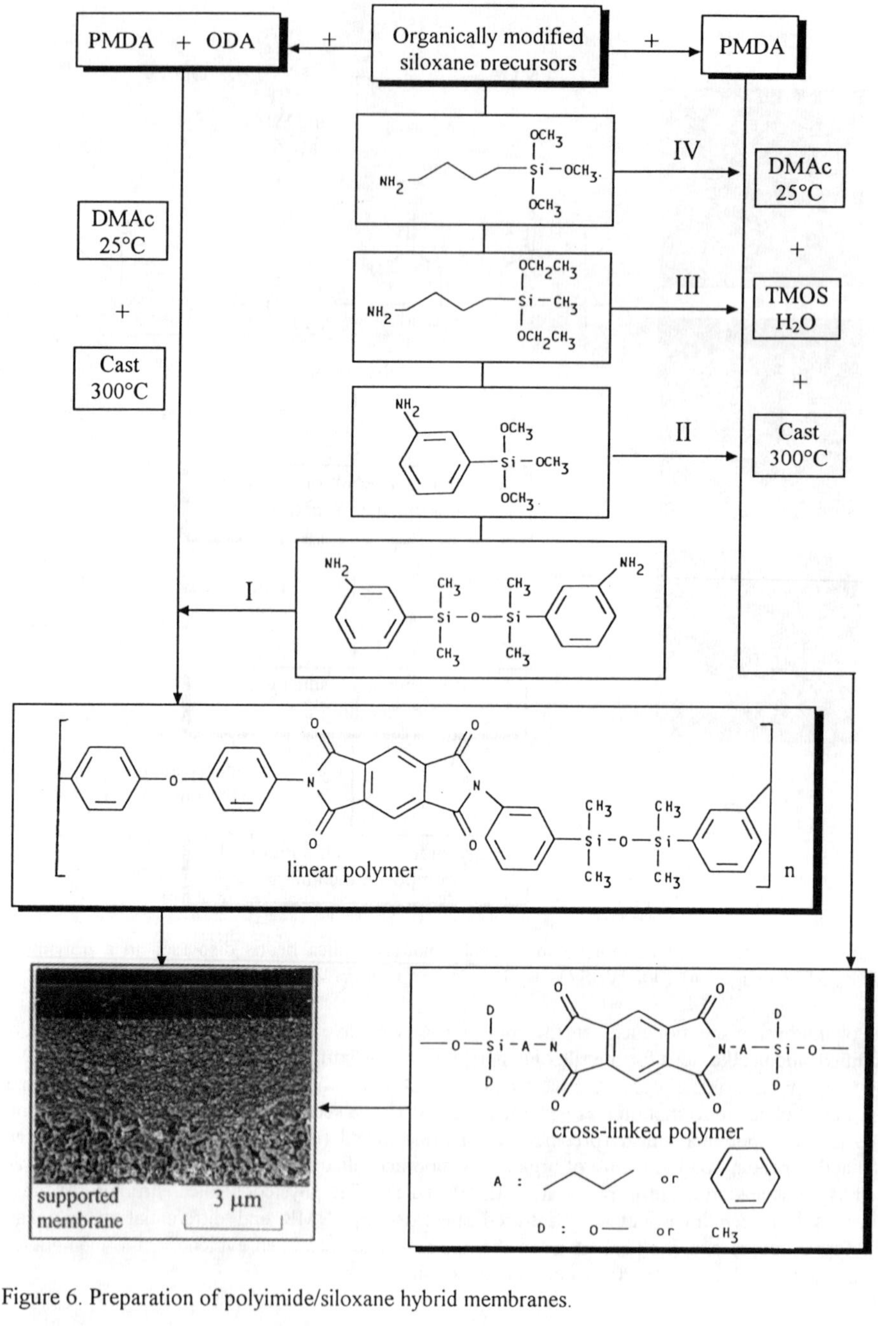

Figure 6. Preparation of polyimide/siloxane hybrid membranes.

Permeability and selectivity measurements

Preliminary gas permeation measurements showed that membrane performance can be varied according as nanocomposite or hybrid polymer membranes are considered.

Table I. Permeability measurement with different gases on a pure polyimide film (Kapton ™) and a polyimide/silica composite membrane (silica content 35%wt).

	H_2	O_2	CO_2	N_2	CH_4	H_2/CO_2	H_2/N_2	H_2/CH_4	O_2/N_2	CO_2/CH_4
PI Kapton	1.6	0.15	0.67	0.06	0.05	2.4	26.7	32	2.5	13.4
PI-SiO_2	7.31	0.45	2.22	0.14	0.17	3.3	52.2	43	3.2	13

Permeability unit: Barrer: 10^{-10} [cm^3 (STP).cm/(cm^2.s.cmHg)]

Permeability to different gases has been measured for a composite membrane with a silica content of 35 %wt. and compared to the value obtained with a film made of Kapton™. Kapton™ is a commercial polyimide synthesized with the same precursors, PMDA and ODA, used in the preparation of the composite films containing dispersed silica beads. The permeabilities are reported in table I. The permeabilities increased for the gases when silica beads are introduced in the polyimide matrix. A higher permeability has been noted for H_2 compared to other gases. Two main phenomena can explain the role of silica fillers. Preferential adsorption of hydrogen at the surface of the silica beads may be assumed to produce better selectivity towards hydrogen whereas the overall increase of permeability is explained by a preferential permeation at the interface between the matrix and the beads.

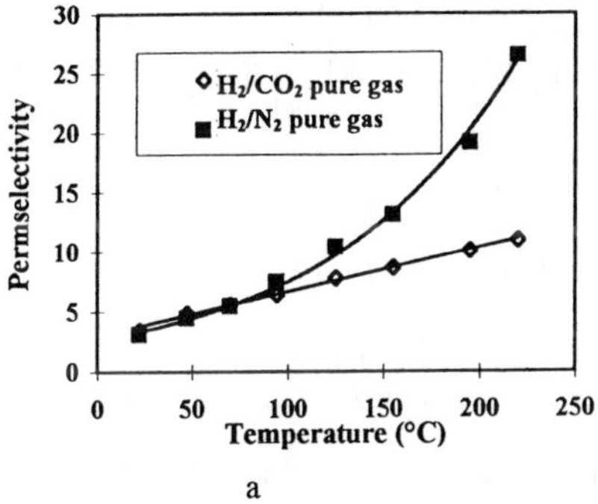

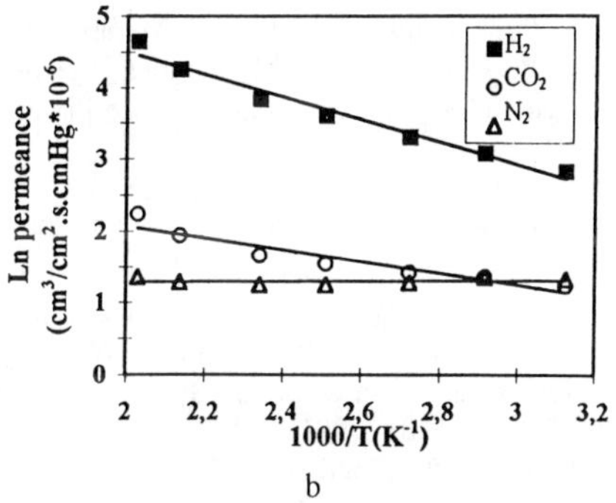

Figure 7. Evolution of selectivity and permeance versus temperature of a polyimide-siloxane hybrid membrane.

A supported membrane with an equivalent silica content (35%wt) but made of an hybrid matrix at the molecular level (polymer III, figure 6) has been tested towards permeance of H_2, CO_2 and N_2. Permeance instead of permeability was chosen in this case because membrane thickness cannot be accurately determined. Contrary to the composite membrane, no potential selectivity between H_2 and CO_2 or N_2 was detected at room temperature. Still an effect of the temperature is evident showing that permeance of H_2 is enhanced with the increase of temperature relative to CO_2 or N_2, figure 7a. This behavior can be explained by the calculation of the apparent activation energy for each gas which is derived from the linear evolution of the logarithm of permeance versus 1/T, figure 7b. The value is 10.6 kJ/mol versus 4.4 kJ/mol for CO_2 and virtually zero for N_2.

CONCLUSION

It has been demonstrated in this work that hybrid materials are good candidates for membrane preparation because of their high processing versatility. They allow to combine in a very efficient way basic properties of organic and inorganic membranes. Membrane synthesis can be directed by solution-diffusion theory applied to dense homogeneous materials. Membrane permeability to a given species is the product of solubility by diffusivity in the membrane phase. These two parameters can be related to the chemical composition (organic part) and the physical structure (inorganic part) of the membranes, respectively. Beyond examples given above, this approach is currently investigated in our laboratory as a generic method to design more specific hybrid membrane materials.

ACKNOWLEDGMENTS

This work has been supported by the Commission of the European Communities under Brite Euram CT 92-0294 and Joule 2 CT 92 -0084 projects.

REFERENCES

1. T. Dabadie, A. Ayral, C. Guizard and L. Cot in Better Ceramics Through Chemistry VI, edited by A.K Cheetham, C.J. Brinker, M.L. Mecartney and C. Sanchez (Mat. Res. Soc. Proc. **346**, Pittsburgh, PA, 1994) pp 849-854.
2. N.K. Raman and C.J. Brinker, J. Membrane Sci. **105**, 273 (1995).
3. G.S. Park in Synthetic Membranes: Science, Engineering and Applications edited by P.M. Bungay, H.K. Lonsdale and M.N. de Pinho (NATO Asi Series, Reidel, 1986) **Vol. C181**, pp 57-108.
4. R.G. Pearson, J. Am. Chem. Soc. **85**, 3533 (1963).
5. H.K. Frensdoff, J. Am. Chem. Soc., **93**, 600 (1971).
6. J.R. Lotz, B.P. Block and W.C. Fermelius, J. Phys. Chem. **63**, 541 (1959).
7. Y. Marcus, Ion Solvation (Wiley, Chichester, 1985).
8. P. Lacan, C. Guizard, P. Le Gall, D. Wettling and L. Cot, J. Membrane Sci. **100**, 99 (1995).
9. M.J.M. Abadie and B. Sillion, Polyimides and other High Temperature Polymers (Elsevier Sc. Pub., Amsterdam, 1991).
10. S.A. Stern, Y. Mi and H. Yamamoto, J. Polymer Sci. Part B: Polymer Phys. **27**, 1887 (1989).
11. T. Ioannides and G.R. Gavalas, J. Membrane Sci. 77, 207 (1993).
12. G.S. Park and M. Saleem, J. Membrane Sci. **18**, 177 (1984).
13. R.L. Hamilton and O.K. Crosser, Ind. Eng. Chem. Fundamentals **1**, 187 (1962).
14. H. Fricke, Physics 1, 106 (1931).
15. J.C. Schrotter, S. Goizet, M. Smaihi and C. Guizard in Proceedings of Euromembrane'95, edited by W.R. Bowen, R.W. Field and J.A. Howell, (European Society for Membrane Science and Technology, Univ. of Bath, England, 1995) **Vol.1** pp 313-318.
16. M. Smaihi, J.C. Schrotter and C. Guizard in Better CeramicsThrough Chemistry VII - Organic/Inorganic Hybrid Materials, edited by B. Coltrain, C. Sanchez, D.W. Schaefer and G.L. Wilkes (Mat. Res. Soc. Proc., Pittsburgh, PA, 1996) to be published.
17. J.C. Schrotter, M. Smaihi and C. Guizard, J. Applied Polymers, in press.

ORGANICALLY MODIFIED SILICATE AEROGELS, "AEROMOSILS"

S.J. KRAMER*, F. RUBIO-ALONSO**, AND J.D. MACKENZIE*
*University of California, Department of Materials Science and Engineering, Los Angeles, CA 90095
**Instituto de Ceramica y Vidrio (CSIC), Metodos Fisico-Qimicos, Madrid, Spain

ABSTRACT

Aerogels derived from sol-gel oxides such as silica have become quite scientifically popular because of their extremely low densities, high surface areas, and their interesting optical, dielectric, thermal and acoustic properties. However, their commercial applicability has thus far been rather limited, due in great part to their brittleness and hydrophilicity. In prior work by our research group, modifying silicate gel structures with flexible, organic containing polymers such as polydimethylsiloxane imparted significant compliance (even rubbery behavior) and hydrophobicity. These materials have been referred to as Ormosils. This study expounds on our current effort to extend these desirable properties to aerogels, and in-so-doing, creating novel "Aeromosils".

Reactive incorporation of hydroxy-terminal polydimethylsiloxane (PDMS) into silica sol-gels was made using both acid and two-step acid/base catalyzed processes. Aerogels were derived by employing the supercritical CO_2 technique. Analyses of microstructure were made using nitrogen adsorption (BET surface area and pore size distribution), and some mechanical strengths were derived from tensile strength testing. Interesting Aeromosil properties obtained include optical transparency, surface areas of up to 1200 m^2/g, rubberiness, and better strength than corresponding silica aerogels with elongations at break exceeding 5% in some cases.

INTRODUCTION

In the aerogel realm of feathery lightness and high porosity, brittleness becomes much more acute, so flexibility from organic modification could result in drastic improvements in mechanical properties. The organic inclusion would also impart a degree of hydrophobicity, which is also beneficial in preserving the aerogel integrity in humid or wet conditions. Improvements in compressive strength and toughness, better optical transparency, and decreased hydrophilicity have already been reported by Novak et al. for silica aerogels with poly(2-vinylpyridine) and polymethylmethacrylate-co-(3-trimethoxysilyl)propylmethacrylate **[1]**. This same study indicated that PDMS aerogels were made, but oddly, no data for which were presented. Better hydrophobicity and elasticity was also reported by Schubert et al. for $RSi(OMe)_3$ incorporations **[2]**. Another interesting consequence of organic modification has been exploited by Brinker and Smith **[3]**. By coating the surface of silica aerogels with with methyl groups, ever-present surface hydroxyls are eliminated or occluded. This prevents pore surfaces from reacting during drying shrinkage, so when evaporation is complete, the gel miraculously springs back to near its original wet volume. Additionally, Loy and Shea have formed novel materials from bridged polysil- and polygerm-sesquioxanes **[4]**. Hunt's group has exploited organic modification in order to dope aerogels with metal ions **[5]**.

Previous work in our laboratory has centered on the reactive incorporation of flexible linkages, such as polydimethylsiloxane (PDMS) **[6,7]**, into sol-gel derived silicate networks to form what Schmidt deemed ORMOSILs, for organically modified silicates **[8]**. Moderate incorporations of such polymers can lead to gels which are much more compliant than silica gels, and can even display rubbery elasticity. Since such interesting properties were forthcoming in xerogels, we applied the same technology to produce more compliant aerogels that we refer to as *Aeromosils* for aerogel, organically modified silicate; or more generally, *Aeromocer*, for any modified ceramic aerogel.

Mat. Res. Soc. Symp. Proc. Vol. 435 © 1996 Materials Research Society

Aeromosils in this study were made through acid and two step acid-base catalysis of the condensation and co-condensation of tetraethoxysilane (TEOS) and PDMS. Very low density, high surface area materials were obtained by both methods.

EXPERIMENTAL

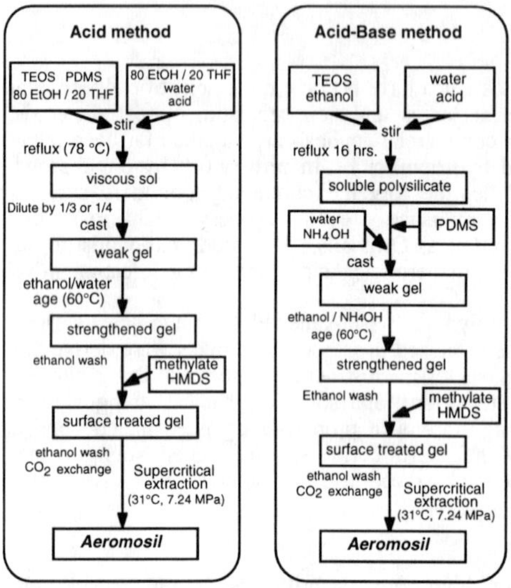

Figure 1. Flow chart schematic of acid and acid-base catalyzed Aeromosil formation.

Two synthetic methods were employed for preparation of gels in this study, an acid (A) and an acid-base (AB) catalyzed hydrolysis/condensation of TEOS with varying amounts of hydroxy-terminal PDMS (M.W~1700), which are represented schematically in figure 1. In A, gel formers TEOS and PDMS were mixed in ratios 0, 10, and 20 wt% (higher proportions have not been successful due to miscibility problems during dilution after reflux) and then diluted 2:1 by volume with an 80:20 solvent mixture of ethanol and tetrahydrofuran (THF). Water and hydrochloric acid (HCl) in molar ratios of 3 and 0.3 to TEOS, respectively, were then added in enough solvent to accomplish a second 2:1 by volume dilution (with respect to the original TEOS+PDMS volume). Under vigorous stirring, the solution was then refluxed for 30 min. Immediately thereafter, portions of the clear solution were diluted by 1/3 and 1/4 with the solvent mixture containing the same proportions of water and HCl, then cast into polyethylene containers, sealed and placed in an oven at 60 °C to gel. Clear gels were obtained within 24 hrs., but they were aged further to allow syneresis to occur, whereupon the gels were removed from their containers and immersed in pure ethanol which was exchanged three times within 24 hrs. Our studies found that pore surface methylation substantially reduced shrinkage during supercritical drying (SCD). Methylation was accomplished by soaking the washed gels in 10 vol% hexamethyldisilazane (HMDS) in ethanol at 60°C for 24 hrs., followed by another triple ethanol wash to eliminate unreacted species. In a jacketed pressure vessel (Polaron E3000), alcohol in the gel was exchanged for liquid CO_2 at 15 °C and 6.20 MPA by purging every 20 min. until no ethanol effluent was apparent. Subsequently, the CO_2 was vented at less than 5 ml/min after slowly taking it above the critical point of CO_2 (31 °C and 7.38 MPa) and allowing time for equilibration.

Synthesis of the AB gels started with a soluble polydiethoxysiloxane (PDEOS) oil, obtained from TEOS, by a method described by Hrubesh et al. **[9]**. 0, 10 and 20 wt% PDMS were mixed with the PDEOS and diluted with 2:3 THF/isopropanol to obtain similar reactant concentrations as the 1/3 and 1/4 solutions described above. The dilution was based on the original TEOS used to prepare the PDEOS. Gelation was accomplished by rapidly stirring in 2.5 mole equivalents water to original TEOS, containing enough NH_4OH to produce a catalyst concentration of 2 ml per liter solution. Clear to slightly hazy solutions resulted which were cast immediately; gelation occurred within one hour. Aging, washing, pore surface modification, and SCD were as described above.

Nitrogen adsorption surface analysis was made using a Quantachrome Autosorb1, and surface areas were calculated utilizing the BET method. From the desorption isotherms, pore size distributions were obtained. Sample densities were calculated using mass and dimension. An Instron model 1122 provided tensile strength measurements of cylindrical samples.

Samples in this study were designated by codes such as 1E33AB, where 1 refers to the TEOS concentration (100 wt% in this case--9 and 8 mean 90 and 80 wt%, respectively), E is for ethanolic solution, the 33 refers to a 0.33 (1/3) final dilution, and the ending letters AB stand for acid-base.

RESULTS and DISCUSSION

Table 1 lists the samples evaluated in this study with their corresponding surface area, density and porosity. Percent porosity was calculated assuming an ideal mixture of fully dense silica glass and solid PDMS, densities 2.2 and 1.1 g/cm^3, respectively. With most samples analyzed, specific surface area decreased with increasing PDMS content. This makes sense in view of the pore size distributions shown in figure 2, which show a shift toward larger pore size and broader pore size distribution with greater PDMS content for acid and acid-base catalysis. Presumably, this is because PDMS is a much larger molecule than TEOS, though it has been shown to break down to some degree during the sol-gel reaction [10], and it can only link via its two end groups whereas TEOS may react at four sites. Therefore, as PDMS increases, the gel would have a less branched and consequently a less

Table 1. Aerogel specific surface area and linear correlation (BET) with corresponding bulk density and porosity.

Sample	SSA (m^2/g)	Corr	Density (g/cm^3)	Porosity (% theory)
1E33A	1169	0.9969	0.071	96.8
1E25A	1369	0.9993	0.061	97.2
9E33A	1273	0.9990	0.059	96.6
9E25A	1193	0.9990	0.057	96.7
8E33A	818	0.9997	0.066	95.6
8E25A	786	0.9993	0.051	96.6
1E33AB	1103	0.9991	0.079	96.4
9E33AB	920	0.9995	0.099	94.3
8E33AB	658	0.9996	0.150	90.1

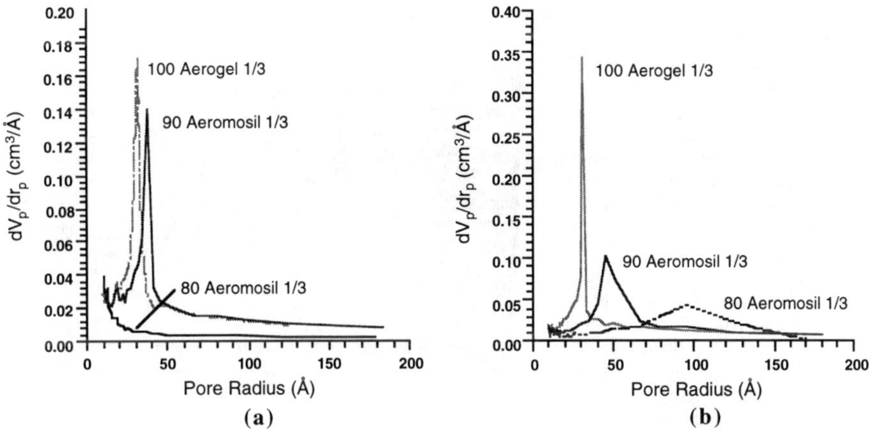

Figure 2. Pore size distributions for (a) acid catalyzed, and (b) acid-base catalyzed Aeromosils.

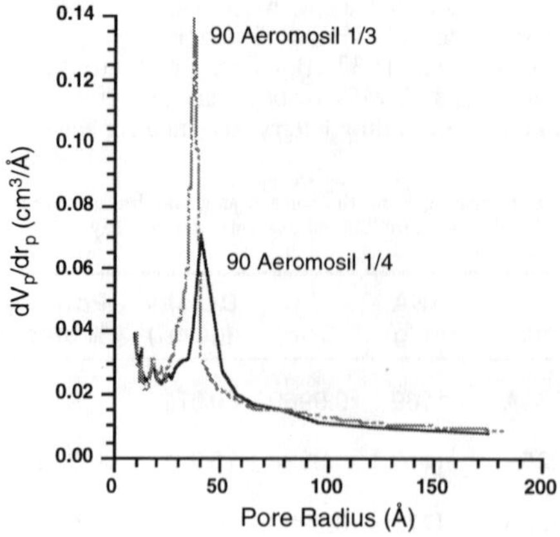

Figure 3. Pore size distributions for acid catalyzed Aeromosils, 1/3 and 1/4 concentration.

convoluted structure. The same trend is not seen, however, in going from 1E33A to 9E33A, which may be due to the significant density difference between these two samples.

In aeromosils containing the same PDMS ratio, surface areas are smaller for the more dilute gels, 1/4. Figure 3 shows the pore size distribution difference between 90 aeromosils of 1/3 and 1/4 dilution. The reduction in surface area is once again resultant from a shift in the pore structure to larger pores and broader pore size distribution, similar to what was observed for increased PDMS content. This dilution phenomenon is not seen in the pure TEOS compositions, however, which may be due to the higher inherent rigidity in these gels. Indeed, the pore size distribution (not shown) for 1/4 dilution shows little difference from that of 1/3. We surmise that the compliance in the gels with PDMS at greater dilution allows more fine structure collapse or rearrangement during stress induced by SCD.

Refocusing on figure 2, though gelation catalysts are different, the two different 100 aerogels have quite similar pore size distributions centered on about 30Å. However, the acid-base aerogel of this composition is quite optically clear, while the acid aerogel is nearly opaque white, as can be seen in figure 4. Presumably, the opacity results from light scattering from a significantly greater percentage or polydispersity of macropores (>50 nm) which are not measurable using this technique. In both systems, as the PDMS content is raised, the distribution peak is moved toward larger pore size and is broader, which is especially true in the acid-base catalyzed material. This broader pore size distribution leads to increased light scattering, visibly apparent in the decreased light transmittance of acid-base 80/20 versus 90/10 in figure 2. The distribution becomes very broad in the acid-base 80 composition.

The nondescript, flat, pore size distribution obtained for the acid 80 aeromosil was somewhat puzzling because BET analysis indicated a surface area of over 800 m^2/g. It is hard to imagine that all of that pore area could be attributed to macropores and/or capillary condensation in the micropores, though some condensation was apparent from the N_2 sorption isotherms (adsorption and desorption isotherms values did not coincide at $P/P_0 = 1$). This may be explained by work of Scherer et al., which indicates that N_2 adsorption can cause near 50% shrinkage in a pure TEOS aerogel **[11]**. Expectedly, this shrinkage would be significantly greater in a very compliant gel, explaining the lack of pore size information in the

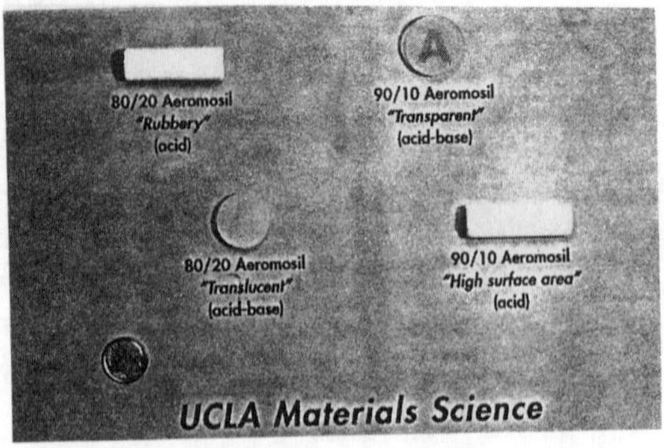

Figure 4. Photographs of Aeromosils from both acid and acid-base catalyzed reactions.

Figure 5. Demonstration of rubbery behavior obtained in acid catalyzed 80/20 Aeromosil.

desorption data. Interestingly, acid catalyzed 80 aeromosil is so compliant, in fact, that it displays rubbery behavior, which is demonstrated in the series of pictures in figure 5--quite surprising for only 20 wt% PDMS. Up to 30% compressive strain was recoverable.

Some preliminary tensile strength measurements were also made on the aeromosils. Acid catalyzed 90 aerogels were found to exhibit strengths between 40 and 70 kPa, and elongations at break from 2 to 4% for densities near 0.1 g/cm^3, which is about four times better than that of a corresponding silica aerogel. Further mechanical characterization is currently in progress.

CONCLUSIONS

High surface area, organically modified silicate aerogels, *Aeromosils*, containing PDMS in the composition range of 0-20 wt% were successfully synthesized utilizing both acid and acid-base catalysis methods. Acid catalyzed materials were more compliant than corresponding silicate aerogels. By tailoring the amount of PDMS inclusion, these aerogels could be made with greater respective strength, or made to exhibit rubbery behavior. Materials with quite narrow pore size distributions were possible through acid-base catalysis, which resulted in modified gels with optical clarity. These unique property possibilities make PDMS exciting for further research and potential applications where silica aerogels fall short in mechanical and environmental stability.

ACKNOWLEDGMENTS

I gratefully acknowledge the Air Force Office of Scientific Research, for financial support of this work. I also wish to thank Dave Lindquist for giving me my start in aerogels; Arlon Hunt and Mike Ayers at LBL for providing more aerogel insight and troubleshooting; Bruce Dunn, UCLA Materials Science, for use of supercritical extractor; Yoram Cohen, UCLA Chemical Engineering, for providing use of surface analysis equipment; and Bill Warrel, USC, for the TEM imaging.

REFERENCES

1. B.M. Novak, D. Auerbach, and C. Verrier, Chem. Mater. **6**, 282-286 (1994).
2. U. Schubert, F. Schwertfeger, N. Husing, and E. Seyfreid, in *Better Ceramics Through Chemistry VI* (Mat. Res. Soc. Symp. Proc. **346**, Pittsburgh, PA 1994) pp. 151-162.
3. (a) D.M. Smith, R. Deshpande, and C.J. Brinker in *Better Ceramics Through Chemistry V*, edited by M.J. Hampden-Smith, W.G. Klemperer and C.J. Brinker (Mat. Res. Soc. Symp. Proc. **271**, Pittsburgh, PA 1992) pp. 567-572; (b) R. Deshpande, D.M. Smith, and C.J.

4. (a) D.A. Loy, K.J. Shea and E. M. Russick *in Better Ceramics Through Chemistry V*, edited by M.J. Hampden-Smith, W.G. Klemperer and C.J. Brinker (Mat. Res. Soc. Proc. **271**, Pittsburgh, PA 1992) pp. 699-704; (b) D.A. Loy G.M. Jamison, R.A. Assink, S. Myers, and K.J. Shea in *Hybrid Organic-Inorganic Composites*, edited by J.E. Mark, C.Y-C. Lee, and P.A. Bianconi (Am. Chem. Soc. Symp. Ser. **585**, Washington D.C. 1993) pp. 265-277.
5. W. Cao, and A.J. Hunt, in *Better Ceramics Through Chemistry VI* edited by (Mat. Res. Soc. Symp. Proc. **346**, Pittsburgh, PA 1994) pp. 631-636.
6. Y.J. Chung and J.D. Mackenzie, in *Better Ceramics Through Chemistry IV*, edited by B.J. Zelinski, C.J. Brinker, D.E. Clark and D.R. Ulrich (Mat. Res. Soc. Symp. Proc. **180**, Pittsburgh, PA 1990) pp. 981-986.
7. Y. Hu and J.D. Mackenzie, J. Mater. Sci. **27,** 4415-4420 (1992).
8. H. Schmidt, J. Non-Cryst Sol. **63,** 283-292 (1984).
9. L.W. Hrubesh, T.M. Tillotson, and J.F. Poco, in Chemical Processing of Advanced Materials, edited by L.L. Hench and J.K. West (John Wiley and Sons, New York 1992) pp. 19-27.
10. T. Iwamoto, K. Morita, and J.D. Mackenzie, J. Non-Cryst. Solids **159**, 65 (1993).
11. G. W. Scherer, D.M. Smith, X. Qiu, and J. M. Anderson, J. Non-Cryst Solids **186**, 316-320 (1995).

ORIGIN OF POROSITY IN ARYLENE-BRIDGED POLYSILSESQUIOXANES

DALE W. SCHAEFER*, GREG B. BEAUCAGE**, DOUGLAS A. LOY*, TAMARA A. ULIBARRI*, ERIC BLACK*, KENNETH J. SHEA***, RICHARD J. BUSS*
*Sandia National Laboratories, Albuquerque, NM 87185-0340, dwschae@sandia.gov
**Materials Science and Engineering, University of Cincinnati, Cincinnati, OH 45221-0012
***Department of Chemistry, University of California Irvine, Irvine, CA 92717

ABSTRACT

We investigate the porosity of a series of xerogels prepared from arylene-bridged silsesquioxane xerogels as a function of organic bridging group, condensation catalyst and post-synthesis plasma treatment to remove the organic functionalities. We conclude that porosity is controlled by polymer-solvent phase separation in the solution with no evidence of organic-inorganic phase separation. As the polymer grows and crosslinks, it becomes increasingly incompatible with the solvent and eventually microphase separates. The domain structure is controlled by a balance of network elasticity and non-bonding polymer-solvent interactions. The bridging organic groups serve to ameliorate polymer-solvent incompatibility. As a result, when the polymer does eventually phase separate, the rather tightly crosslinked network limits domain size to tens of angstroms, substantially smaller than that observed in xerogels obtained from purely inorganic precursors where incompatibility drives phase separation earlier in the gelation sequence.

INTRODUCTION

Hydrocarbon-bridged polysilsesquioxanes [1-6] are network polymers in which the basic building block is two silicons directly attached to an organic bridging group. These materials offer two routes to porous materials: one via sol-gel polymerization followed by drying and another through elimination of the organic bridging group via thermal or plasma treatment of the dried gels. There is evidence that porosity achieved via both routes depends on the nature of the organic spacer, but the relationship has not been systematically studied.

Surface areas (BET) in phenylene-bridged polysilsesquioxane *xerogels* (air-dried) have been measured by gas porosimetry to be as high as 1200 m^2/gram [4]. Surface areas in *aerogels* (supercritically extracted) of the same chemical composition range as high as 1880 m^2/gram [7]. Because these materials are amorphous, surface area measurements that rely on cylindrical-pore models preclude clear interpretation of the measured surface areas. We find no evidence for smooth surfaces at all, further compromising the meaning of BET data. The high surface areas may actually be due to surface roughness.

EXPERIMENTAL

Arylene-bridged polysilsesquioxanes were prepared by sol-gel polymerization bis(triethoxysilyl)aryl monomers under acidic or basic conditions. Arylene-bridged monomers gel rapidly at concentrations as low as 0.01 M, more than two orders of magnitude lower than possible with triethoxysilylbenzene. Since polyarylsilsesquioxanes preferentially form as oligosilsesquioxanes [8], gels can only be prepared at high monomer concentrations with heating and in the presence of strong base. Arylene-bridged materials, however, easily form gels. Alkylene-bridged materials also readily form gels [3] indicating that positioning two or more triethoxysilyl groups about the organic bridging group is the key feature that differentiates these materials from those made by polymerizing simple organotriethoxysilanes. For the gels used here, the solvent was tetrahydrofuran and the monomer concentration was 0.4 M.

Small-angle neutron scattering experiments were carried out with the Low-Q Diffractometer (LQD) at the Manuel Lujan Jr. Neutron Science Center at Los Alamos National Laboratory and 30 meter SANS camera [9] at the High Flux Isotope Reactor at Oak Ridge National Laboratory (ORNL). Small-angle X-ray scattering experiments were accomplished at ORNL's 5-meter SAXS facility [10].

Mat. Res. Soc. Symp. Proc. Vol. 435 © 1996 Materials Research Society

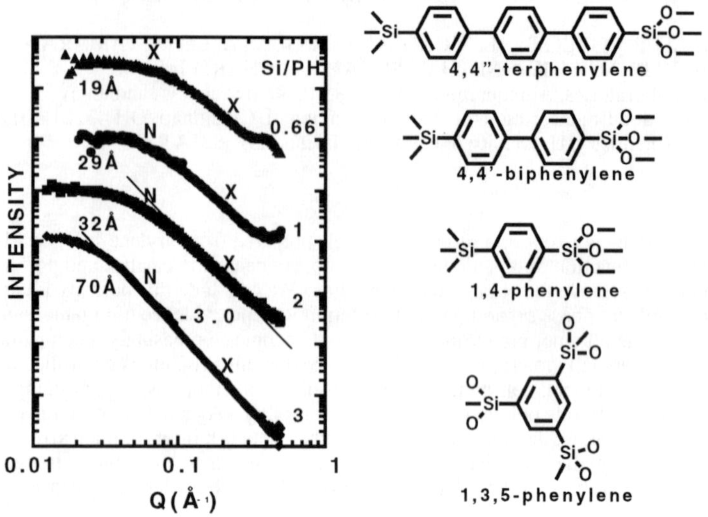

Figure 1. Combined x-ray (high Q) and neutron (low Q) data for arylene-bridged silsesquioxanes. The lengths listed are the Guinier radii obtained from an analysis of the initial curvature. An uninteresting contribution from the large-scale grains of the powder was removed from the neutron data. Details of the data analysis are presented elsewhere [6]. The 15 Å spacing calculated from the peak in the scattering curve is is based on Bragg's law, a rather severe approximation in the present case. The letters X and N stand for x-ray and neutron data.

In most cases, absolute intensity measurements were made by packing powdered samples into quartz cuvettes allowing the density to be directly determined from the cell volume and weight. These samples were suitable for neutron scattering, but not for X-ray scattering due to the absorption of the quartz cells. The hydrocarbon-containing samples show considerable incoherent neutron scattering at large scattering vectors. Since this background compromises Porod analysis, it was necessary to supplement the neutron data with X-ray data in order to establish the so called Porod slopes which characterize the short-scale interfacial structure of the pores. Fortunately, the X-ray and neutron data were identical, apart from a multiplicative factor, in region of overlap allowing the complete scattering curve to be constructed by matching the relative x-ray data to the absolute neutron scattering data. The cross sections reported are neutron cross sections.

Several morphological features can be extracted from the scattering data in Figure 1. From the knee in the curve at small Q we extract the so called Guinier radius which, in the case of low-density samples, is proportional to the domain size of the minority phase; i. e. the mean solid chord.

We use Porod analysis to establish the interface morphology [11, 12]. Although caution is in order when interpreting Porod slopes over a limited regime, in the arylene-bridged materials we consistently observe slopes of ≥ -3, indicative of a stringy domain structure. On plasma treatment, this stringy network evolves somewhat toward smooth interface structure, but even after extensive treatment, the remaining silica is best described as "bushy." Such a structure gives a Porod slope of -3.0, reflecting a unique structure at the crossover from stringy mass-fractal domains to uniformly dense colloidal domains, albeit with very rough, sea-coast-like surfaces.

In the case of the triphenylene-bridged xerogel, we find an unexpected maximum in the scattering curve corresponding to a periodic structure with a spacing of 15 Å. For the biphenyl sample, there is also evidence of a peak, but the maximum is out of range of our instruments. These maxima remain a mystery since the Si-bridge-Si distance is about 11 Å for the longest (4, 4"-terphenylene) bridge, much too small to account for the observations. Since these systems are not ordered lattices, calculating the spacing as $2\pi/Q$ is not strictly justified. It seems unlikely, however, that errors introduced by this "Bragg" assumption would account for the discrepancy between 11 Å and 15 Å.

Finally, the integral under the intensity profile on an absolute scale is related to the contrast which, in turn, depends on the density and composition of the phases. From the contrast change on plasma treatment, we can tell if the organic moieties are uniformly dispersed in the silica matrix, the other possibility being that the moieties themselves are forming the domains via organic-inorganic microphase separation. It turns out that, for the arylene-bridged materials, the former case is realized.

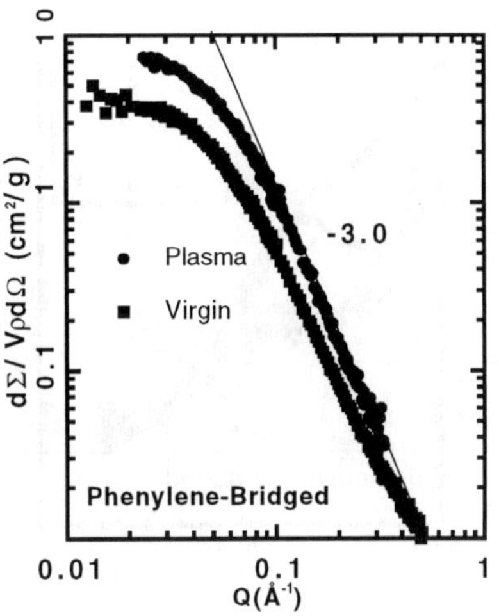

Figure 2. SANS data showing the effect of treatment by a low temperature oxygen plasma on the structure of a 1,4-phyenylene-bridged xerogel. The ordinate is the differential neutron scattering

CONCLUSIONS

Origin of porosity

The scattering curves shown for the arylene-bridged xerogels are quite similar to those of inorganic xerogels made from nonbridged precursors [13]. The difference is primarily one of length-scale as determined by the Guinier radius. Depending on the Si/Phenyl (Si/PH) ratio, the Guinier radius varies from 19 Å to 70 Å (see Figure 1). Only in the latter case does the Guinier radius approach the 100 - 200 Å typical of silica gels prepared conditions from tetraalkoxysilanes.

Perhaps the domains giving rise to the scattering in Figure 1 are due to organic/inorganic phase separation rather than polymer solvent phase separation. Figure 2 helps settle this issue. Little change is observed in scattering profile on removal of the organic phase by an oxygen plasma [14]. The Porod slope steepens slightly indicating emergence of an interface in the stringy domain structure. The increase of intensity is consistent with a change in contrast due to elimination of the organic. In fact, the calculated increase of the Porod invariant (53%) closely matches that calculated (48%) assuming the scattering arises from polymer-void domains in which the organic is uniformly distributed, assuming a skeletal density of 2.0 g/cm^3 for the treated sample and 1.5 g/cm^3 for the untreated. The domains would be expected to collapse somewhat on elimination of the organic bridging groups [14] leading to a decrease in the Guinier radius, which we observe. From the BET surface area, the mean chord decreases by 20% on plasma treatment. We conclude therefore that the domain structure is similar to conventional sol-gel derived silica xerogels, but we must still explain the difference in length scale.

The origin of network structure in aerogels has been the subject of considerable debate. Keefer [15] attributed the fractal nature of silica to inductive effects, an idea that is still widely accepted. Later, Schaefer and Keefer [16] showed that the structural changes with pH could be attributed to kinetic growth processes leading to fractal clusters. Finally, Schaefer [13] and Schaefer and Pekala [17], advanced the idea that the network structure in aerogels was due to phase separation of a crosslinked network.

The arylene-bridged polysilsesquioxane data seem to fit best with the idea that the network structure arises from polymer-solvent microphase separation in which the domain size is fixed by a balance of enthalpic and entropic factors. At low degrees of polymerization, the growing polymers remain soluble since the entropy of mixing overwhelms the attractive forces between like molecules that favor phase separation. Eventually, however, the increasing polymer connectivity decreases the entropy of mixing to the point that phase separation is energetically favored. If the network has percolated, however, macroscopic phase separation is precluded and

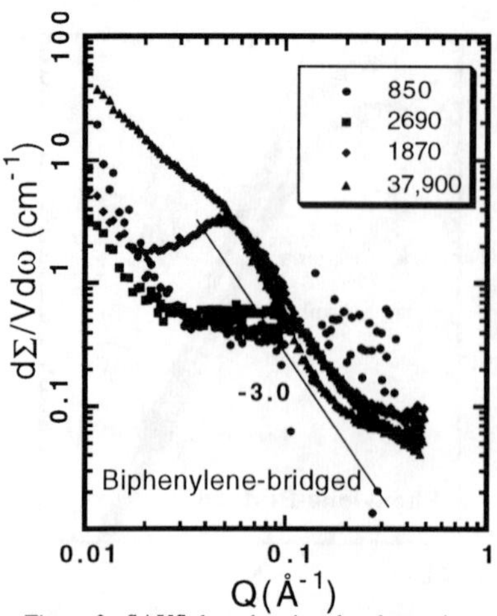

Figure 3. SAXS data showing the change in scattering profile for a series of samples in which the molecular weight of the elastic PDMS varied from 850 to 37,900. The ordinate is the scattering crossection per unit volume.

microscopic domain formation offers the only route to separating the incompatible constituents. Due to the connectivity, however, elastic stresses resist domain formation. A compromise between these competing factors imposes a length scale on the system. The scale is long enough to segregate the polymer from the solvent, but is short enough to not introduce unacceptable stresses. The latter is controlled by network elasticity.

Role of Network Elasticity

If competing interactions control the domain size, one expects that network elasticity plays a significant role. We tested this idea by replacing the solvent with telechelic polydimethylsiloxane (PDMS) polymer. We copolymerized the biphenyl-bridged silsesquioxane (10 wt%) with hydroxy-end-functionalized polydimethylsiloxane of varying chain length. These systems also microphase-separate, presumably into phases dominated by PDMS and silsesquioxanes. Since it is well known from rubber elasticity theory that the elastic modulus goes down with chain length as (molecular weight) $^{-1/2}$, the scale of phase separation should increase with PDMS molecular weight.

To the sympathetic eye, Figure 3 shows the expected trend in domain size with the molecular weight of the elastic chains. The short chain system does not seem phase separated at all but, for the other samples, there is a trend in the broad maximum in the scattering curve to smaller Q with increasing PDMS molecular weight. The presence of the maximum implies that the phase separation is qualitatively different from the bridged polysilsesquioxane xerogels which show no peak. Here, however, we are only concerned with the trend in length scale. Although the data are too limited to establish the expected (molecular weight) $^{-1/2}$ dependence, network elasticity seems to be playing the anticipated role in domain formation since the size scale does increase with molecular weight. Such trends were also seen previously by Ulibarri *et al.* for PDMS-silica hybrids [18], who did find the square-root molecular weight dependence.

Role of Crosslinking

Assuming that elasticity controls phase separation, one is immediately led to ask if the domain size follows catalyst concentration as Schaefer and Pekala [17] found for resorcinol-formaldehyde aerogels. To answer this question, we investigated both the difference between acid and base "catalysis" as well as the effect of varying pH under basic condition. It is known that the degree of crosslinking increases[19] with pH leading us to expect that the Guinier radius should decrease with pH. That is, the increased crosslinking under basic conditions would lead to a stiffer network restricting phase separation to shorter length scales. Figure 4 contrasts gels prepared with acid and base while Figure 5 shows the dependence on the concentration of base. In both cases, slight shifts to larger Q are observed in the Guinier region indicating a decrease of domain size with increased crosslinking. The changes, however, are quite small.

We believe the weak dependence on pH for the bridged polysilsesquioxanes is due to the high reactivity of the bridged-monomers compared to tetraethoxysilane. The bridged monomers generally gel rapidly indicating facile condensation under all conditions. Shea and Loy [20] found condensation around 60% for the arylene-bridged monomers, with little dependence on the nature of the bridge. It appears that although the condensation is facile, the presence of the bridge restricts available configurations sufficiently that a high degree of condensation are precluded.

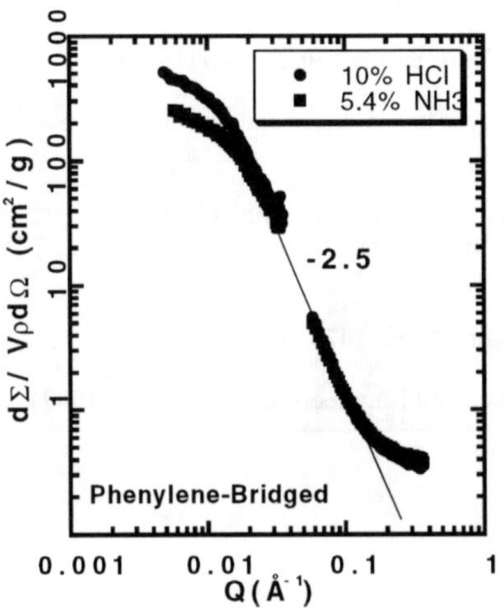

Figure 4. Comparison of SANS profile for acid and base-catalyzed phenyl-bridged aerogels. The catalyst concentration is in mole percent relative to the monomer concentration, so 100%

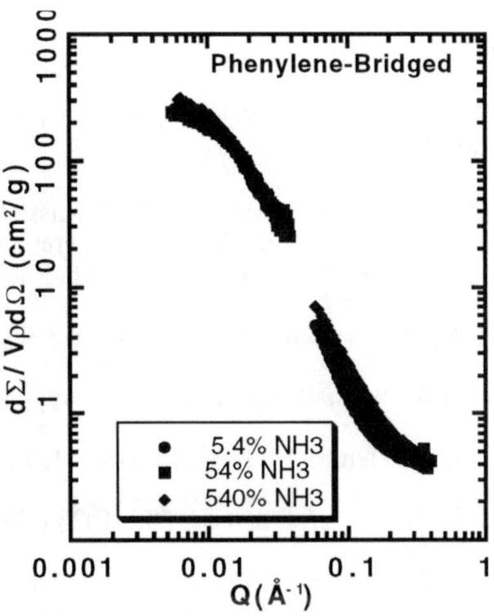

Figure 5. Change in SANS profile with concentration of ammonia measured in mole% relative to the monomer concentration.

Role Of Non Bonding Interactions

If all the arylene-bridged silsesquioxanes enjoy roughly the same degree of crosslinking, it is not obvious why the Guinier radius varies so strongly with Si/PH ratio as shown in Figure 1. Our speculation is that specific interactions between the organic bridges and the solvent lead to enhanced compatibility. In the presence of the organic bridge, the system remains compatible at a higher degrees of crosslinking compared to the unbridged analogue. As a result, when phase separation eventually occurs, the tightly crosslinked network necessarily phase separates on short scales, thus explaining the small Guinier radius of the bridged systems. Furthermore, since the compatabilizing influence of the bridge decreases at high Si/PH ratio, the Guinier radius should vary inversely with this ratio, approaching that of conventional aerogels at low high Si/PH. This trend is indeed observed.

ACKNOWLEDGMENT

We thank Rex Helm, Phil Seegar, George Wignall, and J. S. Lin for assistance in the scattering measurements. This work was supported by the Office of Basic Energy Sciences of the United States Department of Energy under contract #DE-AC04AL85000.

REFERENCES

1. K. J. Shea, D. A. Loy and O. W. Webster, Chem. Mater. **1**, 572-4 (1989).

2. K. J. Shea, D. A. Loy and O. W. Webster, Polym. Mater. Sci. Eng. **63**, 281-5 (1990).

3. K. J. Shea, O. Webster and D. A. Loy, in *Better Ceramics Through Chemistry IV*, edited by (Mat. Res. Soc. Symp. Proc., **180**, 1990) pp. 975-80.

4. K. J. Shea, D. A. Loy and O. W. Webster, J. Am. Chem. Soc. **114**, 6700-10 (1992).

5. J. H. Small, K. J. Shea and D. A. Loy, J. Non Cryst. Solids **160**, 234-46 (1993).

6. D. A. Loy, D. W. Schaefer, G. Beaucage and K. J. Shea, J. Inorg and Orgometallic Pol. **submitted**, (1996).

7. D. A. Loy, K. J. Shea and E. M. Russick, in *Better Ceramics Through Chemistry V*, edited by M. J. Hampden-Smith, W. G. Klemperer and C. J. Brinker (Mat. Res. Soc. Symp. Proc., **271**, Materials Research Society, Pittsburg, PA, 1992) pp. 699.

8. M. G. Voronkov and V. I. Lavrent'yev, Top. Curr. Chem. **102**, 199-236 (1982).

9. G. D. Wignall, Encyclopedia of Polymer Science and Engineering **10**, 112-184 (1987).

10. R. W. Hendricks, J. Appl. Cryst. **11**, 15-30 (1978).

11. D. W. Schaefer, Science **243**, 1023 (1989).

12. P. W. Schmidt, in *The Fractal Approach to Heterogeneous Chemistry*, edited by D. Avnir (John Wiley & Sons, New York, 1989) pp. 67-78.

13. D. W. Schaefer, MRS Bulletin **24** (4), 49-53 (1994).

14. D. A. Loy, K. J. Shea, R. J. Buss and R. A. Assink, ACS Symposium Series **572**, 122 (1994).

15. K. D. Keefer, in *Better Ceramics Through Chemistry II*, edited by C. J. Brinker, D. E. Clark and D. R. Ulrich (Mat. Res. Soc. Symp. Proc., **73**, Mat. Res. Soc., Pittsburgh, PA, 1984) pp. 15-24.

16. D. W. Schaefer and K. D. Keefer, in *Better Ceramics Through Chemistry II*, edited by C. J. Brinker (Mat. Res. Soc. Symp. Proc., **32**, Pittsburgh, PA, 1984) pp. 1-14.

17. D. W. Schaefer, R. Pekala and G. B. Beaucage, Journal of Noncrystalline Solids **186**, 159-167 (1995).

18. T. A. Ulibarri, G. B. Beaucage, D. W. Schaefer, B. J. Olivier and R. A. Assink, in *Submicron Multiphase Materials*, edited by R. H. Baney, L. R. Gilliom, S.-I. Hirano and H. K. Schmidt (Mat. Res. Soc. Symp. Proc., **274**, Mat. Res. Soc., Pittsburgh, 1992) pp. 85-90.

19. S. A. Myers, R. A. Assink, D. A. Loy and K. J. Shea, J. Am. Chem. Soc. **submitted**, (1996).

20. K. J. Shea and D. A. Loy, J. Am. Chem. Soc. **submitted**, (1996).

GENERAL ROUTES TO MICROPOROUS THIN FILMS: FORMATION OF ORGANIC-INORGANIC NETWORK

Y. YAN*, Y. HOSHINO, Z. DUAN, S. RAY CHAUDHURI, AND A. SARKAR.
YTC America Inc. 550 Via Alondra, Camarillo, California 93012.

ABSTRACT

It has been found that inorganic thin films, prepared via sol-gel route, are nonporous or much less porous than bulk gel derived from the same precursor solutions, because of the overlap of rigorous solvent evaporation with continuing condensation during the process of film deposition. The synthetic strategy of this work is to control the condensation reaction by incorporation of terminal ligands through the formation of organic-inorganic network at molecular scales on the thin film coatings. Depending on the nature of the organic groups, the hybrid film can be designed with micropores close to the kinetic diameter of nitrogen as indicated by sorption isotherms and kinetics. The incorporation of the organic groups also renders the porous films with a more hydrophobic surface. AFM, ^{13}C and ^{29}Si MAS/NMR and FTIR data also qualitatively support the proposed microstructures of the hybrid materials.

INTRODUCTION

Porous thin films, especially those with interconnected micropores, have been of great interest in preparation of selective chemical sensors[1-3], membranes[4, 5], and other thin film-based devices[5, 6]. Different methods for the synthesis of powder or bulk porous materials have been developed during the last several decades. These techniques, however, can not be used to prepare continuous, pin-hole-free and crack-free thin film counterparts. Xerogels prepared by sol-gel process and dried by sub-critical drying process[7], for example, show a peak pore radius of 1.1 nm catalyzed by hydrochloric acid or larger than 10.5 nm catalyzed by hydrofluoric acid. Their surface areas can be larger than 1000 m^2/g. Although a microporous xerogel can be easily obtained, thin films derived from the same sol-gel precursor solutions are non-porous[8]. Such a phenomenon is attributed to the overlap of solvent evaporation with condensation film deposition. The fast evaporation may cause the pore collapse and formation of a fully cross-linked, dense film upon thermal treatments.

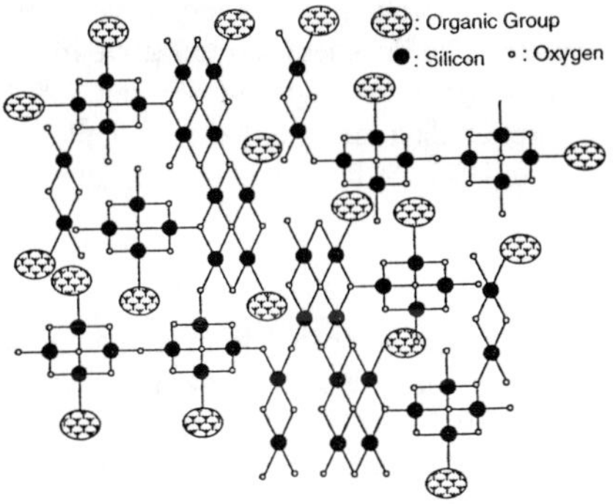

Figure 1. The spacer organic groups reduce the extent of cross-linkage in the network and create a interconnected microporous nanophase in the hybrid thin film at molecular scale, regardless of the processing conditions.

Mat. Res. Soc. Symp. Proc. Vol. 435 © 1996 Materials Research Society

A considerable literature has reported the formation of hybrid inorganic-organic structure to modify the materials at molecular scale[9]. One strategy to design porous thin films based on the sol-gel coating is to reduce the extent of cross-linkage by introducing organic terminal groups into the inorganic network. One possible approach is to replace some of the alkoxide groups by unhydrolyzable organic groups. The organic groups attached to the silicon through a Si-C bonding not only reduce the number of Si-O-Si linkages, but also act as a spacer in the inorganic network by forming an interconnected organic nanophase and hence a micropore phase around the organic groups (Figure 1). In this paper we present a general route to porous thin films by the formation of an organic-inorganic molecular composite. The experimental results demonstrate, for the first time, that a crack-free, well defined and characterized microporous thin films can be obtained from the hybrid materials by using a simple dip-coating or spin-coating process.

EXPERIMENT

The coating solutions were prepared in a similar process as we reported earlier[10]. Tetramethoxysilane (TMOS) was mixed with different portions of methyl trimethoxysilane (MTMS, Aldrich) in ethanol (EtOH). Under stirring the solution was hydrolyzed with water containing hydrochloric acid (pH = 2). The typical concentration of the as prepared solution had a molecular ratio of xTMOS: yMTMS: 2(x + y)EtOH: (4x + 3y)H_2O. After reaction over 24 hr at room temperature the solution was further diluted with 8(x + y)EtOH. Thin films were spin-coated at 2000 rpm for 60 sec on different substrates. The coatings had a typical thickness of 200 nm after cured at 200 °C for 24 hr. Thickness of the coating is also dependent on the ratio of MTMS/(TMOS + MTMS) in the solutions.

Infrared spectra of the thin films coated on KBr substrates were obtained on Nicolet 740 FTIR system. Measurements were performed on the samples thermal treated at different temperatures in air for 4 hr, during which the spectrometer was constantly purged with nitrogen. The morphology and microstructure of the films were characterized by AFM (Nanoscope III, Digital Instruments) operated in a tapping mode. All images (256 X 256 pixels) were obtained in ambient conditions using a lowest auto-flatten and low-pass filtering during the data acquisition. Sorptions of nitrogen were performed in a modified BET apparatus on the film coated quartz crystals at the boiling point of nitrogen. The mass changes on the thin film matrix (<300 nm) caused by adsorption or desorption of the sorbates were in situ monitored by an auto-data acquisition system. The system is sensitive to change of sorption of 0.04 monolayer nitrogen on non-porous surfaces at constant partial vapor pressure. The thickness and refractive index of the coatings was measured by spectroscopic ellipsometer (incident angles of 65° and 70°, MOSS, ES4G, Sopra) at ambient conditions. Thin film sorption data indicate that the samples used for AFM and ellipsometry study may have absorbed with 1.5-3.0% by weight of water in air.

RESULTS AND DISCUSSIONS

Figure 2 shows the FTIR spectra of a hybrid film prepared from a coating solution derived from a 75% MTMS/(TMOS + MTMS) composition. One of the noticeable features is the Si-OH stretching vibration at 930 cm^{-1} shown after 100° and 200°C thermal treatments. The presence of a larger amount of the Si-OH groups at low temperature possibly provides sufficient ligands to bond the coating to the substrate. Indeed, this coating solution can form uniform, continuous thin film coatings on the inorganic, metallic and polymeric surfaces by using dip- or spin-coating techniques. The intensity of this peak is decreased with increasing temperature,

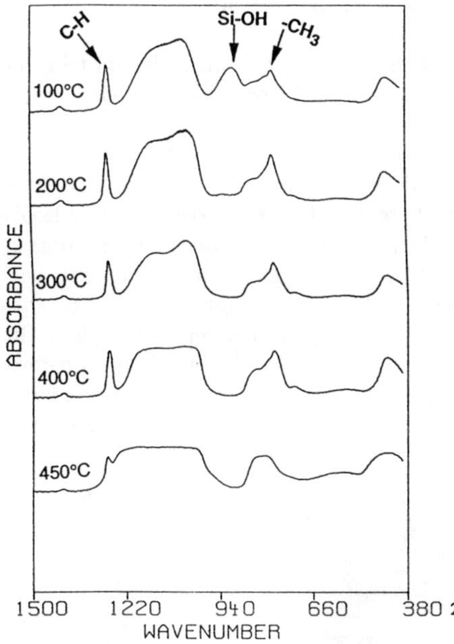

Figure 2. FTIR spectra of the hybrid film thermal treated in air at different temperatures for 4 hr.

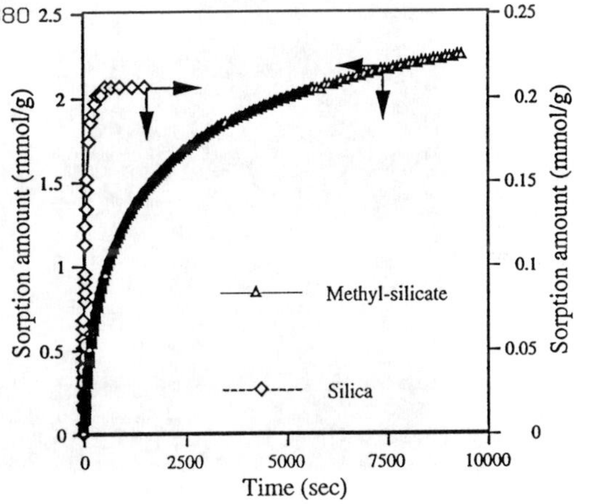

Figure 3. Adsorption kinetics of nitrogen at nitrogen boiling temperature on the inorganic silica and hybrid methyl-silica thin film (thickness < 210 nm).

which indicates the formation of more Si-O-Si linkages in the hybrid network. The presence of the organic terminal groups is reflected by the presence of the symmetric C-H bending vibration at 1277 cm^{-1} and the CH_3 rocking modes at 775 cm^{-1}. The intensity of the organic groups begin to decrease after thermal treated in air between 280-300°C, and the vibrations of those peaks can still be observed up to 420 °C. ^{13}C CP/MAS solid state NMR spectra (not shown) suggesting the presence of two kinds of C groups: the dominant peak is located -2.46 ppm attributed to the Si-CH_3 carbon resonance and a small peak at 51.7 ppm associated to the SiO-CH_3 carbon resonance. The hybrid materials cured at 200 °C are mainly cross-linked by T_2, T_3 and Q_3, Q_4 according to ^{29}Si CP/MAS NMR data, where $T_{2,3}$ and $Q_{3,4}$ denote the coordination numbers of silicon in the network derived respectively from MTMS and TMOS precursors.

The following Table lists some of the properties of the methyl-silicate hybrid thin film compared with those of a silica film. The former was prepared from TMOS derived sol-gel coating solution. The films have a comparable thickness, and the refractive index is 1.44 and 1.41 at the wavelength of 550 nm respectively for the two films cured at 200°C . Those values are consistent with those reported by others[11]. One distinct difference between the films is the sorption capacity of water and nitrogen at 50% partial pressure. It was observed that only 1.50% by weight of water adsorbed on the hybrid film while more than 3% adsorbed on the inorganic silica film dehydrated at 200°C. In contrast the hybrid film has a nitrogen sorption capacity more than 8% which is 12 times higher than that adsorbed on the inorganic film. Sorption isotherms are also shown that sorption of nitrogen is characterized by a type I isotherm on the hybrid film while type II on the silica film. These results suggest that the hybrid film is microporous, whereas the inorganic film is non-porous and hydrophilic materials. The low refractive index of the hybrid film obtained at ambient conditions can be partially attributed to the presence of porosity and hydrophobic surface property in comparison with the inorganic film.

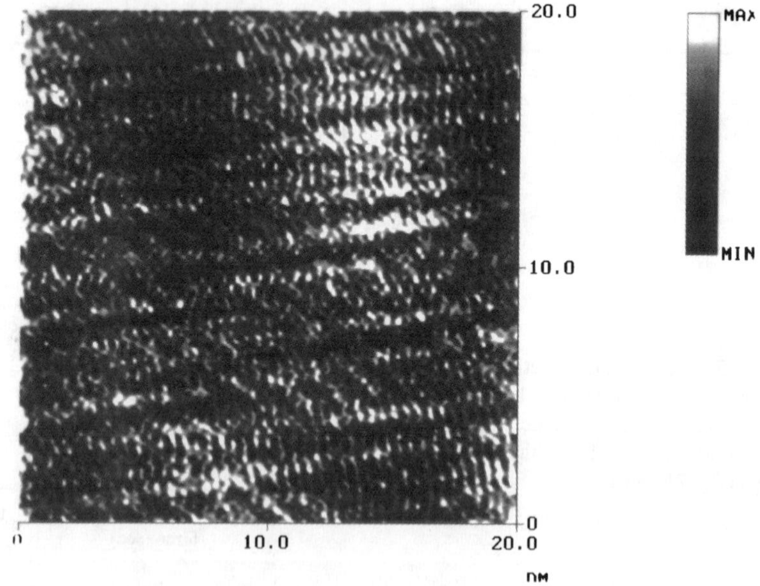

Figure 4. AFM image of the hybrid methyl-silicate thin films. The microporous, nano-feature is completely absent from the inorganic silica film.

Comparison of the properties of the hybrid film with inorganic film cured at 200°C

	Thickness (nm)	Refractive index	Sorption of H_2O by weigh at RT (P/Po = 50%)	Sorption of N_2 by weigh at LN_2 (P/Po = 50%)
Silica film	201.9	1.44	3.01%	0.677%
Hybrid film	184.4	1.41	1.51%	8.40%

The sorption kinetics of nitrogen on the two films also behaved completely different (Figure 3). Sorption equilibrium is reached in a couple of min on the silica film (0.21 mmol/g). On the other hand, no equilibrium is reached on the hybrid film (<200 nm) after 3 hr of sorption (2.4 mmol/g). The activated diffusion sorption kinetics also suggests that the pore size of the amorphous hybrid thin film is close to that of the kinetic diameter of nitrogen at nitrogen boiling temperature. If the sorption of nitrogen on the bulk hybrid gel occurs at same rate as that on the thin film, it might take more than 20 years for the sorption to reach equilibrium on a 8 cm^3 cubic bulk gel. Such an assumption also possibly explains why the bulk hybrid gel have been reported as no-porous materials.

The presence of micropores on the hybrid film is also quantitatively supported by AFM and TEM studies. Figure 4 is the AFM image of the surface morphology of the hybrid film coated on silicon wafer. The presence of about 0.4 nm micropores is consistent with the nitrogen sorption kinetic data. In a comparative study, the nano-features on the hybrid film are absent from the dense, inorganic silica film. The amorphous, hybrid film possibly consists of interconnected porous channels around the organic phase. In contrast to the crystalline zeolitic system, the narrow and tortured hybrid pores may account for the activated diffusion sorption of nitrogen on the film. It is possible to control the size of the micropore by tuning the molecular size and concentration of the organic groups, hence the extent of the network cross-linkage in the hybrid materials. The hybrid microporous thin films reported here can be used as the top layer coating for design separation membranes, which are expected to have a high efficiency for separation nitrogen and oxygen from air. More experiments are in progress.

CONCLUSION

Porous thin film can be prepared by incorporation of stable, terminal organic groups into the inorganic network. The hybrid film can have a micropore size close to the kinetic diameter of nitrogen while maintain a hydrophobic surface. Formation of the organic-inorganic molecular composite can provide a general route to design microporous thin films.

REFERENCES

1. Bein T.; Brown K.; Frye G. C., and Brinker, C. J. J. Am. Chem. Soc. **111,** 7640 (1989).
2. Yan Y., and Bein T. Chem Mater. **4**, 975, (1992) and **5**, 907 (1993).
3. Yan Y., and Bein T. J. Am. Chem. Soc. **117**, 9990 (1995).
4. Lin, Y.S., and Burggraaf A.J., J. Am. Ceram. Soc. **74**, 219 (1991).
5. Yan Y.; Ray Chaudhuri S., and Sarkar, J. Am. Ceram. Soc. in press, (1996).

6. Yan Y.; Ray Chaudhuri S., and Sarkar, A. Chem. Mater. **7**, 2007 (1996) and in proceedings MRS 1996 spring meeting: Micro- & Macroporous Materials.
7. Murata H.; Kirkbir F.; Meyers D.; Ray Chaudhuri S., and Sarkar A. SPIE, Sol-Gel Optics III, **2288**, 709 (1994).
8. Frye G.C.; Ricco A.J.; Martin S.J., and Brinker, C.J. in Better Ceramic Through Chemistry III, Edited by. Brinker, C.J; Clark, D.E. and Ulrich D.R. (Mater. Res. Soc. Proc. **121**, Pittsburgh, PA, 1988) p. 349.
9. For recent review refers Schubert U; Husing N., and Lorenz A. *Chem. Mater* **7**, 2010 (1995)
10. Sarkar A., Yan Y., Fuqua P.D., and Jahn W. in Glass Surface, Sol-gel glasses: Proceedings of XVII international Congress on Glass Vol. 4, 1995, edited by F. Gong, Inter. (Academic Publisher, Beijing), p. 228.
11. Dislich H. in Sol-Gel Technology for Thin Films, Fibers, Preforms, Electronics and Specialty Shapes. Edited by Klein, L. C. (Noyes Publications, Park Ridge, NJ), 1988, pp. 50.

Part VI

Poster Session

SYNTHESIS OF TITANATES $MTiO_3$ (M=Mn,Co,Ni)BY THE SOL-GEL METHOD

J. MORENO*,G, TAVIZON*, L. VICENTE* AND T. VIVEROS**
*Departamento de Física y Química Teórica, Facultad de Química, UNAM, 04510 México, D.F.
**Departamento de Ingeniería Química, UAM-Iztapalapa, Apdo. Postal 55-534, México, D. F.

ABSTRACT

The titanates $MTiO_3$ (M = Ni,Co,Mn) were prepared by a sol-gel method. The reaction between the metal solution and titanium butoxide leads to the formation of gels. The amorphous gels so obtained under different pH conditions, were calcined at temperatures between 750^0C and 1200^0C to obtain in some cases high purity, crystalline $MTiO_3$ powders. The characterization of the samples was done by X-ray powder diffraction, differential thermal analysis (DTA), thermal gravimetry (TG) and BET surface measurements.

INTRODUCTION

Perovskites of metal oxides ceramics are of great research interest due to the wide range of physical properties which include ferroelectrics, pyroelectrics, piezoelectrics and dielectric behaviour that vary with the average valence of the transition metal [1]. This compounds are normally synthesized by solid-state reactions between the metal oxides at temperatures above 1000^0C, and pure phases have been obtained at temperatures of *ca* 1400^0C.

The sol-gel method of hydrolysis and condensation of metal alkoxide precursors is attractive for formation of such materials due to the low processing temperatures and control of stoichiometry [2-3]. A variation of this technique for the synthesis of compounds such as alkaline earth perovskites ceramics [4-7], is one in which titanium alkoxide is hydrolysed by an aqueous solution of the alkaline hydroxide. This method produces ultrafine powders where submicron particles of great uniformity can be produced and larger specific areas than those obtained by conventional routes can be obtained. The sol-gel method involves a large number of possible modifications during hydrolysis, condensation, drying and calcination.

In this work we report results on the synthesis of transition metal perovskite ceramics $MTiO_3$ (M = Mn,Co,Ni) using sol-gel techniques. Synthesis and experimental conditions for the formation of the titanates were studied and their properties were characterized by a multitechnique approach.

EXPERIMENTAL

$MTiO_3$ were prepared by hydrolysis of a dilute alcoholic solution of titanium n-butoxide (TBO), $Ti(OC_4H_9)_4$. The procedure is as follows : the alkoxide was dissolved in isopropyl alcohol and kept stirring for two hours. The metal oxides (MnO/Mn_2O_3, CoO, NiO) were dissolved into distilled water adding HNO_3, The pH measurements were made using a Philips pH meter and depending of this value we worked with conditions of "strong and intermediate acidity". The reactions were also tried in conditions of alkalinity but they were not succesful as is discussed later. The solutions were mixed and stirred at room temperature for 2 hr using a magnetic stirrer. The

Mat. Res. Soc. Symp. Proc. Vol. 435

$H_2O/Ti(OC_4H_9)_4$ molar ratio for the different solutions and pH values presented here are shown in Table I.

TABLE I. $H_2O/Ti(OC_4H_9)_4(TBO)$ molar ratio and pH for the different solutions

solution	water/TBO	pH
Mn solution	292.18	1.25
Co solution	171.90	3.42
Ni solution	171.90	1.23

The resulting gel was evaporated to dryness; the Mn and Ni compound showed a green color while the Co compound showed a red color. The residual solids were ground, pelletized and calcined first in an alumina crucible at temperatures of 700^0C for 72 hrs. Subsequent heat treatments were necessary at 1200^0C for 72 hrs for the Mn and Co compounds in order to obtain pure phases. The $NiTiO_3$ compound was only heated at a temperature of 735^0C for 72 hrs. The $MTiO_3$ phases were obtained at the end of this process.

Powder X-ray diffraction data of calcined samples were obtained on a Sie mens Crystalloflex 810 automatic powder diffractometer with monochromated CuK_α radiation. Intensity data were collected over the range $4^0 \leq 2\theta \leq 70^0$, at room temperature.

Specific surface areas of calcined samples were measured by the BET technique using N_2 adsorption at 75 K on a Micromeritics 2100E apparatus.

TGA and DTA analysis of dried samples were performed using a calibrated DuPont 2000 Thermal Analyser. All TGA measurements were done in platinum containers in air flow. Heating was done from 20^0C to 1000^0C at a rate of $20^0C/min$.

RESULTS AND DISCUSSION

The reaction between the metal solution and titanium(IV) butoxide leads to the formation of the metal titanate gels in acid conditions. When the solution had an alkaline pH (from 9 - 11) the starting materials were $Mn(OH)_2$, $Co(OH)_2$ and NiO and $NaOH$ was added to control the pH value. In this case a precipitate was formed instead of a gel. This is due to the poor solubility of the metal hidroxides in this medium. In this case the X-ray of the products always showed the presence of TiO_2 and the phase $MTiO_3$. The presence of TiO_2 may be due to the high degree of hydrolysis of the colloidal precursor, which may result in the generation of a great diversity of complex compounds.

In the other cases the gels were converted to the metal titanates by drying and calcination, the best results being otained for the "strong acidity" condition. XRD results for the calcined powders prepared under those conditions are given in fig. 1. Sharp diffraction lines indicate the well-crystallized nature of the formed perovskite phase. The products were identified to be $MnTiO_3$, $CoTiO3$ and $NiTiO_3$ (JCPDS No. 23-0902, 15-0866 and 33-960, respectively) which are rhomboheadral. As can be seen in fig. 1a) and 1b), the compounds with Co and Mn are pure but the compound with Ni show the presence of small quantities of TiO_2. For lower temperatures of calcination, the X-ray peaks show the presence of different MO oxides besides TiO_2.

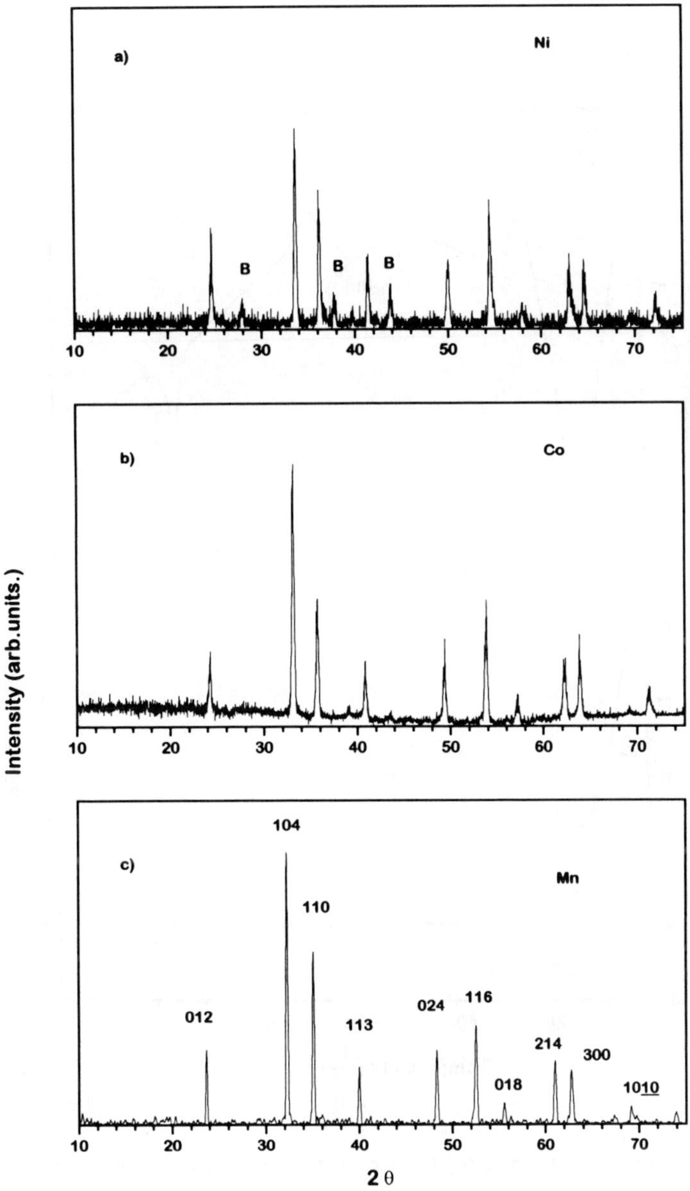

Fig.1. X-Ray diffactograms of : a) $NiTiO_3$, b)$CoTiO_3$, c) $MnTiO_3$. (B= reflection peak from TiO_2).

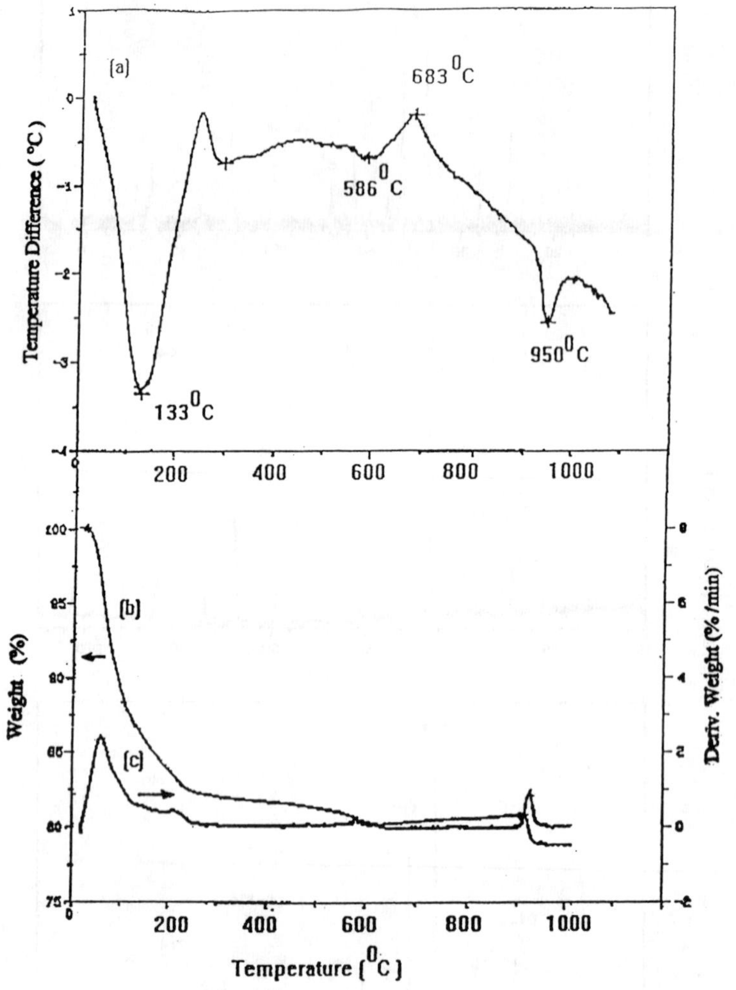

Fig. 2 (a) DTA, (b) TG and (c) DTG curves of the oven-dried gel precursors of $MnTiO_3$.

Figs. 2-3 show the DTA, TG and its derivative (DTG) curves. In all the cases there are three stages of weigth loss. The first weight loss results from the evaporation of alcohol and desorption of the adsorbed moisture. The second is associated with the combustion of the organic compounds formed during the hydrolysis reaction. The third stage corresponds to the pyrolysis of the residual organics.

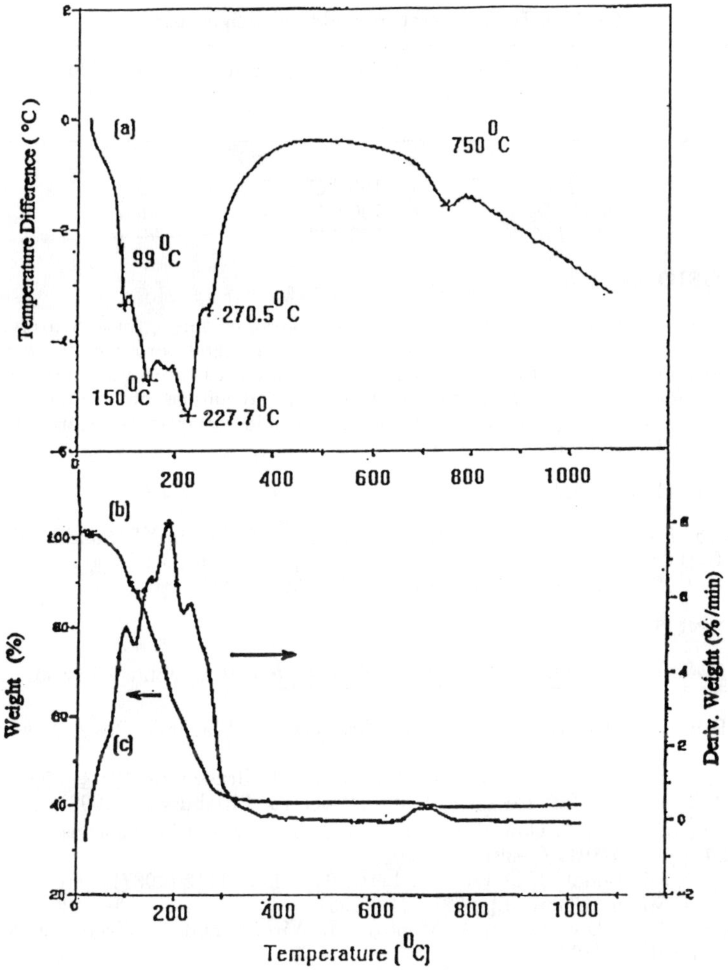

Fig. 3 (a) DTA, (b) TG and (c) DTG curves of the oven-dried gel precursors of $CoTiO_3$.

Table II contains the total weight loss. The three steps are registered as endothermic peaks in the DTA curves. Exotherm peaks around 700^0C in all cases is attributed to the crystallization of the gels.

The BET results showed that the titanates posses small specific surface areas resulting from the relatively high temperature of calcination (1200^0C). In all cases they were less than $2\,m^2g^{-1}$.

TABLE II. Temperature and final weight losses

compound	final temperature (^0C)	weight loss %
$NiTiO_3$	1000^0C	46
$CoTiO_3$	1000^0C	40
$MnTiO_3$	1000^0C	79

CONCLUSIONS

In conclusion, transition metal titanates were obtained through the hydrolyisis of the alkoxides in the presence of an acid catalyst. This method permitted us to obtain homogeneous and high crystalline systems. The specific areas are similar as those obtained with more traditional ceramic methods with the advantage that the temperature of calcination is lower. Applications of these materials in catalytic oxidation reactions will be reported elsewhere.

ACKNOWLEDGMENTS.

We thank the Institute of Materials (IIMUNAM) for the facility in XRD and M. Portilla for the TG-DTA measurements. This work was supported in part by grant N_0.0132-E9106 from CONACYT and IN100293 from DGAPA.

REFERENCES

1. A.R. West in Solid State Chemistry and its Application, (John Wiley and Sons, 1989).
2. C.J. Brinker and G.W. Scherer, Sol – Gel Science, Academic Press, New York, 1990.
3. C. D. Chandler, M. J. Hampden-Smith and C.J. Brinker in Better Ceramics Through Chemistry (Mat. Res. Soc. Symp. Proc. **73**, Pittsburg,PA 1992)p.89.
4. M. I. Diaz-Guemez, T. Gonzales Carreco, C.J. Serna and J.M. Palacios, J. Mater. Science, **24**, pp 1011-1014 (1989).
5. F. Chaput, J.P. Boilot, J. Mater. Sci. Lett., **6**pp. 1110-1112 (1987).
6. G. Pfaff, J. Mater. Chem. **3**,pp.721-724,(1993)
7. J. Moreno,J. M. Dominguez, A. Montoya, L. Vicente and T. Viveros, J. Mater. Chem.,**5**, pp.509-512 (1995).

SYNTHESIS OF INORGANIC-ORGANIC HYBRIDS FROM METAL ALKOXIDES AND SILANOL-TERMINATED POLYDIMETHYLSILOXANE

SHINGO KATAYAMA, IKUKO YOSHINAGA and NORIKO YAMADA
Nippon Steel Corporation, Advanced Technology Research Laboratories,
1618, Ida, Nakahara-Ku, Kawasaki 211, Japan, singok@lab1.nsc.co.jp

ABSTRACT

Inorganic-organic hybrids have been synthesized by reaction of $Ti(OC_2H_5)_4$ and $Ta(OC_2H_5)_5$ with silanol-terminated polydimethylsiloxane (PDMS). The chemical modification of the metal alkoxides with ethyl acetoacetate (EAcAc) was carried out in order to obtain a transparent and uniform hybrid. The hydrolysis behavior of $Ti(OC_2H_5)_4$ modified with EAcAc in the presence of PDMS and the formation of the Ti-O-Si bond in a Ti-O-PDMS hybrid were revealed by FT-IR experiments. Dynamic mechanical measurements showed that a Ta-O-PDMS hybrid was harder than a Ti-O-PDMS hybrid, indicating the effect of metal on the storage modulus of hybrids.

INTRODUCTION

In recent years, a new concept of utilizing the sol-gel process to produce inorganic-organic hybrids has attracted considerable attention [1-3]. This can provide a wide variety of materials with varying properties depending on the constituent inorganic and organic species. One way for preparing inorganic-organic hybrids is to copolymerize an organic-based polymer or oligomer with an inorganic-based alkoxide thereby leading to "Ceramers" [3].

It has been demonstrated that PDMS can be hybridized with the siloxane network prepared by in-situ sol-gel polymerization of tetraethoxysilane (TEOS) [3-6]. The chemical and physical properties of Metal-O-PDMS hybrids are anticipated to be widely altered by choosing other metallic elements than silicon as an inorganic component. However, titanium alkoxide is the only known transition-metal alkoxide with which PDMS can be copolymerized [7-10]. This is because transition-metal alkoxides are difficult to handle owing to their high reactivities. Therefore, it is necessary to stabilize transition-metal alkoxides in order to reduce their reactivities. Chemical modification of metal alkoxides with chelating ligands is commonly employed to retard the hydrolysis and condensation reactions [11, 12]. For example, β-diketones are often used as stabilizing agents in the synthesis of sol-gel ceramics because they react readily with metal alkoxides to yield chelate compounds [13].

In the present work, inorganic-organic hybrids have been synthesized from PDMS and $Ti(OC_2H_5)_4$ or $Ta(OC_2H_5)_5$ chemically modified with EAcAc in order to investigate the effect of Metal-O component on the property of a Metal-O-PDMS hybrid. The structures and properties of these hybrids are discussed based on the results of Fourier transform infrared (FT-IR) analysis and dynamic mechanical measurements.

Mat. Res. Soc. Symp. Proc. Vol. 435

EXPERIMENTAL

$Ti(OC_2H_5)_4$ or $Ta(OC_2H_5)_5$, PDMS, EAcAc and water in the mole ratio of 1 / 0.25 / 2 / 2 were reacted in ethanol. The ratio is one of the representatives to yield homogeneous gel. The solution was allowed to gel in an aluminum dish at 70℃ for 2 to 7 days depending on the sample size. Some of the samples were further heat-treated at 150℃ for 3 days in order to obtain harder hybrids.

FT-IR spectra of samples were recorded on a Perkin-Elmer System 2000 FT-IR Spectrometer. The dynamic mechanical data was obtained with a Rheology DVE-V4 dynamic visco-elastic analyzer. The specimen size was about 5×25×1 mm. Most samples were tested within the temperature range of -150℃ to +300℃ with a heating rate of 3℃/min. A frequency of 110 Hz was selected for all the experiments.

RESULTS AND DISCUSSION

Sample Appearance

Although $Ti(OC_2H_5)_4$ and $Ta(OC_2H_5)_5$ are normally so reactive that their oxides or hydroxides precipitate upon hydrolysis, the solutions of PDMS and alkoxide modified with EAcAc remained clear even after addition of water. While the hybrids heat-treated at 70℃ were fragile and breakable, those heat-treated at 150℃ were flexible and rubbery. The color of hybrids changed from yellow to brown after heating at 150℃. The appearances of the hybrid containing Ti-O component (Ti-O-PDMS hybrid) and that containing Ta-O component (Ta-O-PDMS hybrid) were similar. A typical hybrid after a heat-treatment at 150℃ is shown in Fig. 1.

Figure 1. Appearance of Ti-O-PDMS hybrid.

FT-IR Analysis of the Synthesis of Ti-O-PDMS Hybrids

For a typical synthesis of Ti-O-PDMS hybrids, the behavior of chemical species during reaction was examined by FT-IR spectrometry. Figure 2 shows FT-IR spectra for steps from the starting alkoxide to the final Ti-O-PDMS hybrid. The FT-IR spectrum of $Ti(OC_2H_5)_4$ exhibited absorption peaks assigned to the ν(Ti-O) and ν(C-O) vibrations of bridging ethoxy groups because of the trimeric nature of titanium alkoxides [14]. The absorption peak of the ν(Ti-O) vibration was at 592 cm^{-1} and those of the ν(C-O) vibrations were at 885 and 1044 cm^{-1} as shown in Fig. 2(a). These peaks corresponding to bridging ethoxy groups disappeared in the FT-IR spectrum of $Ti(OC_2H_5)_4$ modified with

2 moles of EAcAc as shown in Fig. 2(b). The modified alkoxide had absorption peaks assigned to the ν(Ti-O) and ν(C-O) vibrations of terminal ethoxy groups. The modified alkoxide also had typical absorption peaks around 1530 and 1620 cm^{-1} corresponding to EAcAc groups bonded to titanium, as described by the previous papers [13,15]. They can be assigned to the ν(C-O) and ν(C=C) stretching vibrations in the enol form of EAcAc, respectively. Therefore, two ethoxy groups of $Ti(OC_2H_5)_4$ are substituted by EAcAc to form a chelate complex with a monomeric structure as $Ti(OC_2H_5)_2(EAcAc)_2$.

The FT-IR bands resulting from the modified alkoxide did not change after PDMS was added to it. The modified alkoxide in the presence of PDMS and ethanol solvent was hydrolyzed with 2 moles of water per mole of $Ti(OC_2H_5)_4$. The FT-IR spectrum of this solution in Fig. 2(c) had no peaks corresponding to ethoxy groups. The absorption peaks

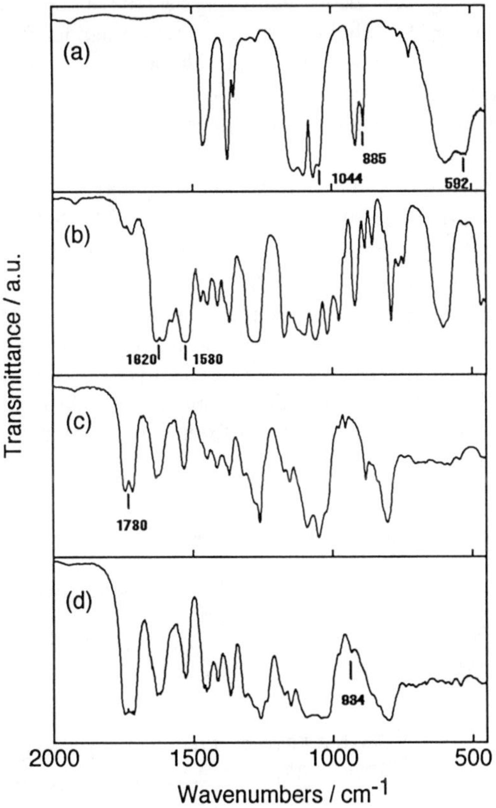

Figure 2. FT-IR spectra during the modification and reaction in the Ti-O-PDMS hybrid system. (a)$Ti(OC_2H_5)_4$, (b)$Ti(OC_2H_5)_4$ modified with EAcAc, (c) hydrolyzed solution of modified $Ti(OC_2H_5)_4$ and PDMS, (d) Ti-O-PDMS hybrid synthesized at 70℃.

corresponding to free EAcAc (keto form) appear around 1730 cm^{-1}, as well as the peaks of chelating EAcAc. In spite of the addition of the stoichiometric amount of water for hydrolysis of the ethoxy groups, all of the ethoxy groups and some of the chelating EAcAc groups were lost. It is assumed that hydroxyl groups of hydrolyzed metal alkoxide react with the hydroxyl groups of another hydrolyzed metal alkoxide and/or the terminal silanol groups of PDMS to produce water. The resulting water leads to hydrolysis of chelating EAcAc.

In the FT-IR spectrum for the Ti-O-PDMS hybrid prepared by drying at 70℃, absorption peaks corresponding to keto and enol forms of EAcAc were still observed although absorption peaks of ethanol disappeared as shown in Fig. 2(d). Thus, the sample contains free EAcAc and chelating EAcAc. However, the absorption peaks of EAcAc disappeared with further heat-treatment at 150℃.

The FT-IR spectrum in Fig. 2(d) also exhibited a characteristic absorption peak at 934 cm^{-1} assigned to the ν(Ti-O-Si) vibration [17]. The formation of Ti-O-Si bonds is consistent with results of ^{29}Si-NMR study reported previously [9]. Therefore, some hydroxyl groups formed by the hydrolysis of the chelated alkoxide are thought to react with silanol groups of PDMS to produce a three-dimensional structure.

Dynamic Mechanical Behavior of Ti-O- and Ta-O-PDMS hybrids

Dynamic mechanical properties of Ti-O- and Ta-O-PDMS hybrids heat-treated at 150℃ are shown in Fig. 3. The storage modulus exhibited a plateau with a magnitude of 10^9 Pa below -120℃. This value is typical for glassy polymers. As the temperature increased, both hybrids exhibited a transition region showing a gradually decreasing modulus and finally reached another plateau with a magnitude of 10^7 Pa for the Ti-O-PDMS hybrid and that of 10^8 Pa for the Ta-O-PDMS hybrid. This suggests that Ta-O component makes the hybrid harder than Ti-O component. The reason can be attributed to the nature of Metal-O component resulting from the difference in valence and electronegativity between Ti and Ta.

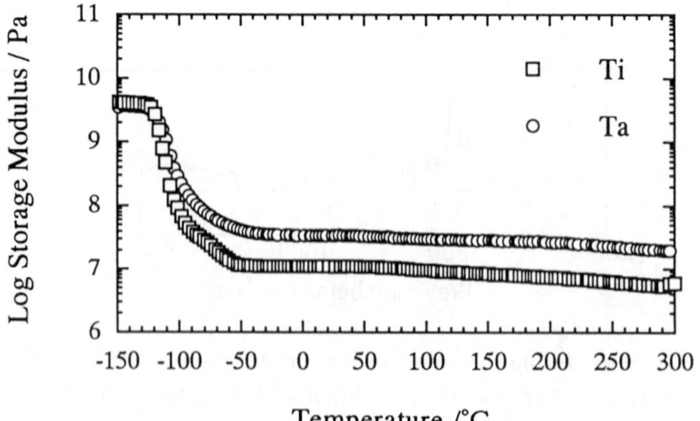

Figure 3. The storage modulus for Ta-O- and Ti-O-PDMS hybrids heat-treated at 150℃.

While Ti-O-PDMS hybrids heat-treated at 70℃ were breakable and the mechanical properties could not be measured, Ta-O-PDMS hybrids were strong enough for the dynamic mechanical measurement. This is considered to be evidence that mechanical properties of Metal-O-PDMS hybrids are influenced by the kind of metal.

Figure 4 shows a comparison of dynamic mechanical properties of Ta-O-PDMS hybrids heat-treated at 70℃ and 150℃. The storage modulus of a hybrid treated at 150℃ is higher than that of a hybrid treated at 70℃. While pure PDMS shows a T_g peak at -120℃, the tan δ curve of the hybrids displayed a T_g peak around -105℃. This shift may be caused by the restriction of micro-Brownian motion of PDMS bonded to an inorganic network because such PDMS will need higher energy to cause the transition [5]. The peak height was lower and the transition region was broader in the hybrid heat-treated at 150℃ than in that heat-treated at 70℃. As a heat-treating temperature becomes higher, the reaction between hydrolyzed alkoxide and PDMS proceeds further. At the same time, hydrolyzed alkoxide forms a more rigid network through self-condensation. It is considered that the progress of these reactions results in the difference in dynamic mechanical behavior between the hybrids heat-treated at 70℃ and 150℃.

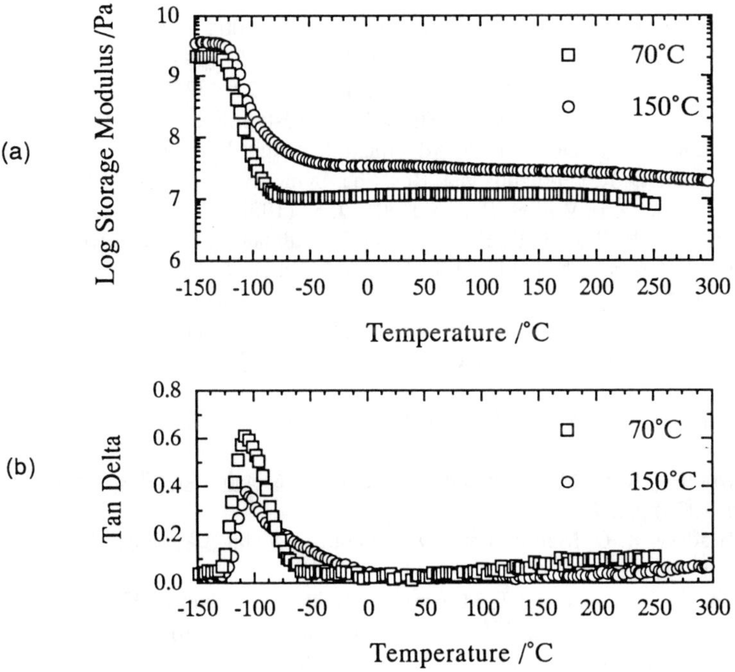

Figure 4. The dynamic mechanical behavior for Ta-O-PDMS hybrids treated at 70℃ and 150℃. (a) storage modulus, (b) tan δ.

CONCLUSIONS

Transparent and flexible inorganic-organic hybrids were successfully synthesized from $Ti(OC_2H_5)_4$ or $Ta(OC_2H_5)_5$ and PDMS by utilizing the chemical modification of metal alkoxide. FT-IR spectroscopic study has revealed the hydrolysis behavior of $Ti(OC_2H_5)_4$ modified with EAcAc in the presence of PDMS and the Ti-O-Si bond formation in the Ti-O-PDMS hybrid. Dynamic mechanical properties were found to be influenced by the kind of metal and the processing conditions. In the Metal-O-PDMS hybrids, Ta-O component made it harder than Ti-O component. The rigidity of the hybrid structure was increased by heat-treating at a higher temperature.

ACKNOWLEDGMENT

This work has been entrusted by NEDO as part of the Synergy Ceramics Project under the Industrial Science and Technology Frontier (ISTF) Program promoted by AIST, MITI, Japan. The authors are members of the Joint Research Consortium of Synergy Ceramics.

REFERENCES

1. G. Philipp and H. Schmidt, J. Non-Cryst. Solids **63**, 283 (1984).
2. H. Schmidt, J. Non-Cryst. Solids **73**, 681 (1985).
3. G. L. Wilkes, B. Orler and H. Huang, Polym. Prep. **26**, 300 (1985).
4. H. Huang, B. Orler and G. L. Wilkes, Polym. Bull. **14**, 557 (1985).
5. H. Huang, B. Orler and G. L. Wilkes, Macromolecules **20**, 1322 (1987).
6. J. D. Mackenzie, Y. J. Chung and Y. Hu, J. Non-Cryst. Solids **147&148**, 271 (1992).
7. C. S. Parkhurst, W. F. Doyle, L. A. Silverman, S. Singh, M. P. Andersen, D. McClurg, G. E. Wnek and D. R. Uhlmann, Mat. Res. Soc. Symp. Proc. **73**, 769 (1986).
8. R. H. Glaser and G. L. Wilkes, Polym. Bull. **19**, 51 (1988).
9. C. L. Schutte, J. R. Fox, R. D. Boyer and D. R. Uhlmann, in Ultrastructure Processing of Advanced Materials, edited by D. R. Uhlmann and D. R. Ulrich (John Willy & Sons, New York, 1992) p.95.
10. S. S. Joardar, M. A. Jones and T. C. Ward, Polym. Mater. Sci. Eng. **67**, 254 (1992).
11. J. Livage, M. Henry and C. Sanchez, in Progress in Solid State Chemistry, **18**, 259 (1988).
12. C. Sanchez, J. Livage, M. Henry and F. Babonneau, J. Non-Cryst. Solids **100**, 65 (1988).
13. D. C. Bradley, R. C. Mehrotra and D. P. Gaur, in Metal Alkoxides (Academic Press, London, 1978) p.209.
14. D. C. Bradley, R. C. Mehrotra and D. P. Gaur, in Metal Alkoxides (Academic Press, London, 1978) p.95.
15. H. Uchihashi, N. Tohge and T. Minami, Seramikkusu Ronbunshi **97**, 396 (1989).
16. C. Sanchez, F. Babonneau, S. Doeuff and A. Leaustic, in Ultrastructure Processing of Advanced Materials, edited by J. D. Mackenzie and D. R. Ulrich (John Willy & Sons, New York,1988) p.77.
17. Z. Liu and R. J. Davis, J. Phys. Chem. **98**, 1253 (1994).

ELECTRICAL PROPERTIES OF TERNARY SI-C-N CERAMICS

Christoph Haluschka, Christine Engel and Ralf Riedel

Fachbereich Materialwissenschaft, Fachgebiet Disperse Feststoffe
TH Darmstadt, Hilpertstraße 31 D, D-64295 Darmstadt (Germany)

ABSTRACT

Ternary Si-C-N ceramics were derived from silicon containing polymers by thermally induced hybrid processing. These silicon carbonitrides were investigated by impedance spectroscopy depending on the synthesis conditions. The electrical behavior correlates with the solid state reactions and phase transformations, which take place during the processing. It has also been shown that the electrical properties can be controlled in a wide range.

INTRODUCTION

During the last decade, silicon nitride (Si_3N_4) has become increasingly of interest as a component for applications in engines and gas turbines [1]. To improve the properties of this material, SiC containing Si_3N_4-based materials have been developed and characterized [2].
Apart from powder technology these composite materials can also be made by thermally-induced ceramization of organometallic polymers [3]. Due to the homogeneous composition of the starting materials, the polymer-derived Si-C-N ceramics show excellent thermomechanical properties in comparison to traditionally processed counterparts.
Beside conventional methods to investigate the amorphous and polycristalline structure of polymer-derived ceramics, structural information can also be obtained by the analysis of electrical properties of these materials. Therefore, the impedance was studied and compared with experimental data derived from traditional characterization methods.

EXPERIMENTAL PROCEDURE

NCP 200, a polyhydridomethylsilazane of the Nichimen Corp. (Japan), was used as starting organometallic polymer. First, the precursor polymer was crosslinked at about 350 °C in Ar, resulting in an unmeltable polymer. After milling and sieving, cold isostatic pressing was used to produce a compact, which was pyrolysed at temperatures of about 1000 °C in Ar. An amorphous ceramic was formed with the chemical composition of $Si_{1.7}C_{1.0}N_{1.5}$ with the outgassing of CH_4 and H_2. Also, traces of H and O were observed in this intermediate material. Carbon free and carbon poor samples can be obtained by using an Ar/NH_3 gas mixture instead of pure Ar during the pyrolysis. At temperatures above 1500 °C, annealing in N_2 leads to a Si_3N_4/SiC composite, which contains a microcrystalline α-Si_3N_4 matrix and nanocrystalline α-SiC inclusions [4]. The complex impedance (Z) was determined in the frequency range between 20 Hz and 1 MHz using the LCR-meter HP 4284 A. The real and imaginary terms of Z are calculated from the applied ac-voltage and the ac-current which flows through the sample (Fig. 1). Silver suspension, as contact layer, was used for a good electrical contact between the sample and the measurement electrodes.

Mat. Res. Soc. Symp. Proc. Vol. 435

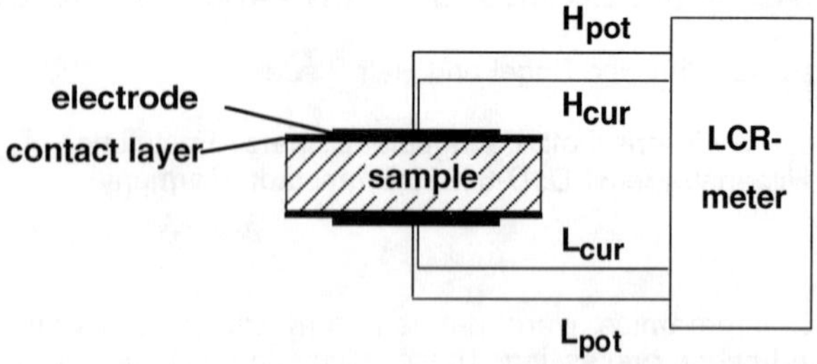

Fig. 1. Measurement configuration

Taking the geometry into account, the complex conductivity $\sigma^*(\omega)$ is obtained:

$$\sigma^*(\omega) = \left(Z \cdot \frac{c}{h}\right)^{-1} \qquad (1)$$

c: cross section
h: height

The complex conductivity can be written as

$$\sigma^*(\omega) = \sigma'(\omega) + i\sigma''(\omega) \qquad (2)$$

$$\sigma'(\omega) = \sigma_{dc} + \sigma'_{ac}(\omega)$$

percolation paths (σ_{dc}) clusters ($\sigma'_{ac}(\omega)$, $\sigma''(\omega)$)

The frequency independent part σ_{dc} is caused by states which are arranged in percolation paths, whereas $\sigma'_{ac}(\omega)$ is determined by states which form clusters and do not constitute a percolation path throughout the sample [5].
In order to obtain information about the transport mechanisms of the charge carriers (transport in the conduction band, hopping mechanisms), which have a different temperature dependence, we carried out impedance measurements in the range between 77 K and 1000 K.

RESULTS AND DISCUSSION

In general, the silicon carbonitrides show semiconducting behavior with a positive temperature coefficient of the electrical conductivity σ_{dc}. Plotted against the reciprocal temperature the electrical conductivity shows nonlinear behavior (Fig. 2), which is typical for amorphous and polycrystalline materials [6]. The reason for this is the high density of states within the so-called "mobility gap". These localized states are responsible for the hopping of charge carriers which was observed at lower temperatures.

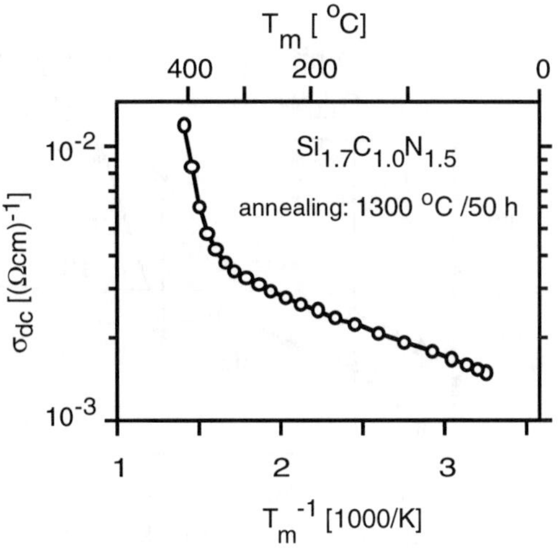

Fig. 2. Electrical conductivity depending on the reciprocal measuring temperature T_m^{-1} (Arrhenius plot)

In order to get some information about the amorphous intermediates and the phase transition to the polycrystalline state σ_{dc} was determined depending on the temperature of the final heat treatment (Fig. 3). A nearly continuous increase of the electrical conductivity was observed and can be divided into several stages:

In the temperature region between 1000 °C and 1100 °C (**I**) a mass loss of about 1 wt.% can be detected, which is mainly caused by outgassing of H_2. It is assumed that similar to amorphous silicon carbide (a-SiC:H), the sp^2-hybridization of the carbon is favored against the sp^3-hybridization with decreasing H-content [7], i.e. the electrical behavior changes from a diamond-like high temperature semiconducting behavior to a graphite-like behavior. Therefore, the electrical conductivity increases, whereas the activation energy for the transport of charge carriers decreases with increasing annealing temperature. This tendency can be observed in the entire amorphous region.
However, in contrast to region **I**, nearly no outgassing of H_2 is detected for temperatures above 1100 °C (**II**). The increase of the conductivity in this region is much smaller and is attributed to a reduction of localized states caused by deviations from the ideal amorphous structure. Due to the change of the local order, the solid state density increases in both regions. At 1400 °C, the formation of α-Si_3N_4 is observed by TEM techniques including ESI (electron spectroscopic imaging) and XRD [4]. The conductivity of the material decreases because the effective cross section for the transport of charge carriers is reduced, whereas the density of this composite material rises due to the higher solid state density of Si_3N_4. At higher temperatures (**III**) the

conductivity increases again by several orders of magnitude due to crystallization of α-SiC.

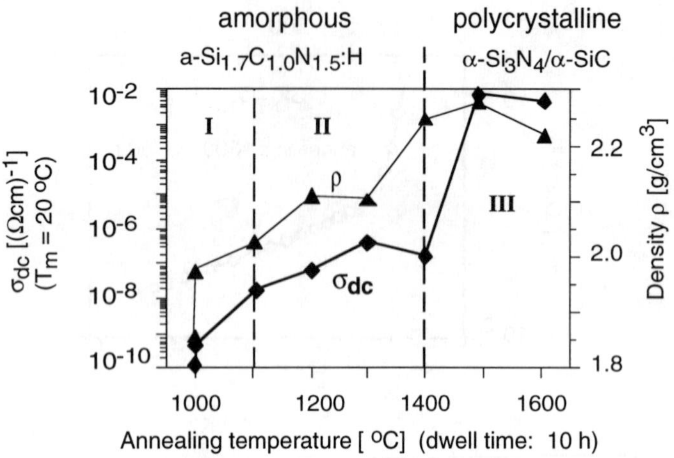

Fig. 3. Influence of the annealing temperature on the electrical conductivity σ_{dc} and the density ρ

Also the ac-conductivity is influenced by the annealing conditions (Fig.4):

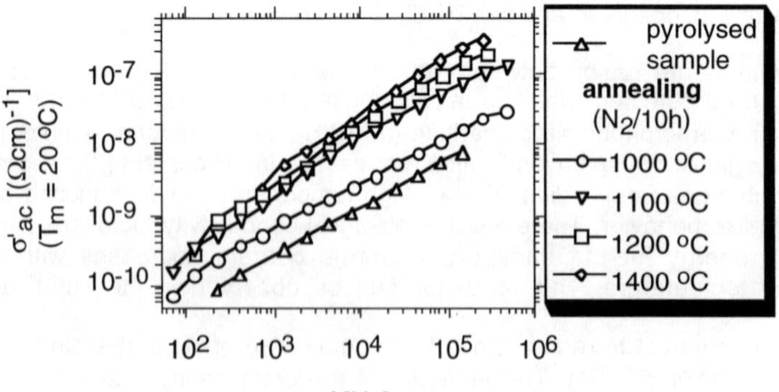

Fig. 4. Influence of the annealing temperature on σ'_{ac}

Apart from the annealing conditions, the pyrolysis conditions have a strong influence on the electrical properties. For this reason it has been shown that the electrical conductivity can be controlled by a variation of the pyrolysis atmosphere. By using NH_3 or an Ar/NH_3-mixture instead of pure argon it is possible to lower the carbon content in the ceramic product due to a higher amount of methane, which is lost during pyrolysis. As a consequence, the electrical conductivity decreases by as much as 4 orders of magnitude (Fig. 5).

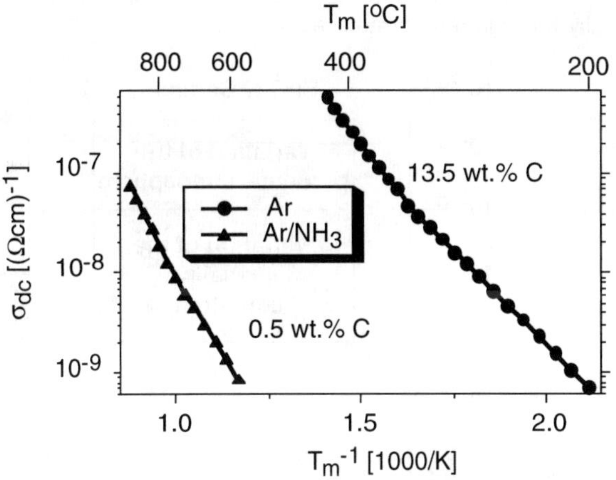

Fig. 5. Influence of the pyrolysis atmosphere on σ_{dc}.

The chemical composition of the pyrolysed samples is in the range between $Si_{1.7}C_{1.0}N_{1.5}$ and Si_3N_4 (Fig.6), i.e. with increasing substitution of carbon by nitrogen during the pyrolysis the chemical composition and therefore the electrical properties are shifted to those of silicon nitride.

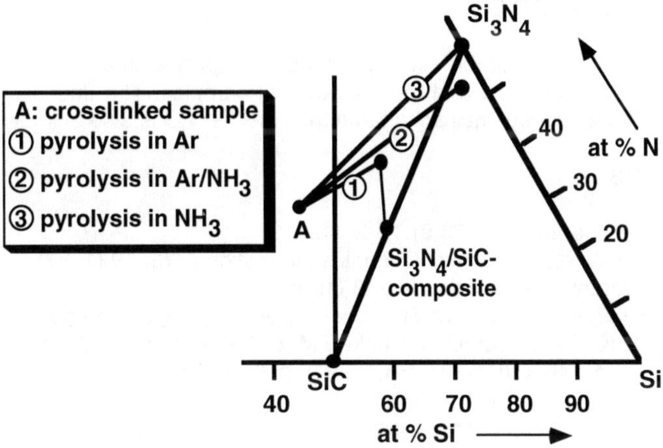

Fig. 6. Part of the Si-C-N phase diagram.

CONCLUSIONS

The investigation of ternary Si-C-N ceramics by impedance spectroscopy leads to a better understanding of the amorphous intermediates and polycrystalline composites and agrees with results derived from conventional characterization methods. It has also been shown that the electrical properties of ternary Si-C-N ceramics made from polymeric precursors can be controlled with changes of the pyrolysis and annealing conditions (Fig. 7), which is very important for the development of new engineering ceramics, especially for high temperature applications.

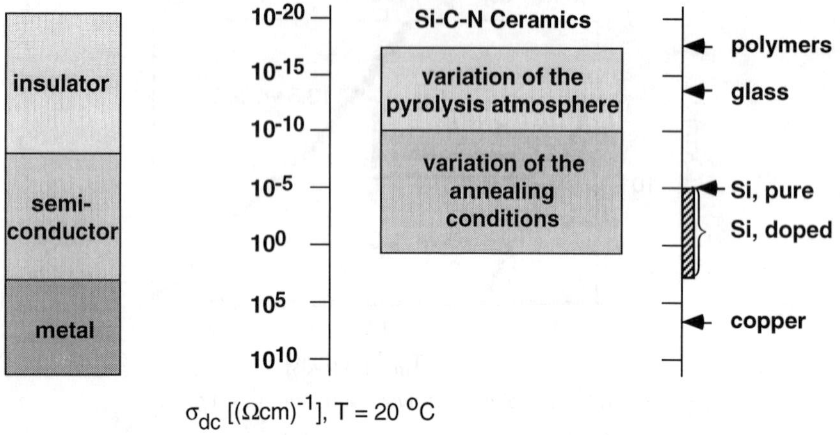

Fig. 7. Electrical conductivity σ_{dc} $[(\Omega cm)^{-1}]$ measured at 20 °C in comparison to conventional materials.

ACKNOWLEDGEMENTS

J. Bill (Max-Planck-Institut für Metallforschung, Stuttgart) is greatfully thanked for the use of measuring equipment and the providing of samples. The financial support of the Deutsche Forschungsgemeinschaft and the Hoechst AG is also acknowledged.

REFERENCES

[1] R. Raj, J. Am. Ceram. Soc. **76**(9), 2147 (1993).
[2] P. Greil, G. Petzow, H. Tanaka, Ceramics International, **13**, 19 (1987).
[3] R. Riedel, Naturwissenschaften **82**, 18 (1995).
[4] J. Bill, M. Frieß, F. Aldinger and R. Riedel in Better Ceramics Through Chemistry VI, edited by A. K. Cheetham, C. J. Brinker, M. L. Mecartney and C. Sanchez, Mater. Res. Soc. Proc. 346, Pittsburgh, PA, (1994), p. 605-610.
[5] S. R. Elliot, Adv. Phys., **36**(2), 135 (1987).
[6] W. Heywang and R. Müller, Amorphe und polykristalline Halbleiter, Halbleiter-Elektronik Bd. 18, Springer-Verlag, Berlin Heidelberg, 1984, p. 13-76.
[7] P. C. Kelires, Europhys. Lett., **14**(1), 43 (1991).

NEW PHOSPHANYL-SUBSTITUTED TITANIUM AND ZIRCONIUM ALKOXIDE PRECURSORS FOR SOL-GEL PROCESSING

ANNE LORENZ AND ULRICH SCHUBERT
Institut für Anorganische Chemie der Technischen Universität Wien, Getreidemarkt 9, A-1060 Wien, Austria, uschuber@fbch.tuwien.ac.at

ABSTRACT

Zirconium and titanium alkoxide derivatives with phosphanyl substituents were synthesized by reaction of the alkoxides with 2-acrylamido-2-methyl-propane sulfonic acid followed by addition of $HPPh_2$ to the acrylic double bond. Alternatively, the phosphinated sulfonic acid was prepared first and then reacted with the alkoxides. The second route is preferred because of the milder reaction conditions, shorter reaction times and avoidance of byproducts. The bifunctional organic ligand is bonded to the titanium or zirconium atom via the sulfonate group, while the PR_2 group is available for further reactions such as the coordination of metal compounds.

INTRODUCTION

One of the possibilities to design new materials by sol-gel processing is the use of organically modified metal or semi-metal alkoxide precursors. Particularly useful are organofunctional derivatives of the type $(RO)_nE\text{-}X\text{-}A$ [1]. The letter A represents the functional organic group, and X is a chemically inert spacer linking A and the alkoxide moiety. Alkoxysilanes of this type, $(RO)_3Si(CH_2)_xA$, are readily available and have found widespread industrial applications [2]. There are also interesting potential applications for compounds $(RO)_nE\text{-}X\text{-}A$ with $E \neq Si$. However, the chemistry of the organofunctional alkoxysilanes cannot be directly extended to metal alkoxides. The more polar metal-carbon bonds are easily cleaved by water, and therefore the functional group A has to be linked to the alkoxide moiety in another manner.

The reactivity of metal alkoxides towards water is much higher than that of alkoxysilanes, and chemical additives are used to moderate their reactivity. When non-silicon alkoxides $E(OR)_n$ are, for example, reacted with carboxylic acids, some of the alkoxide groups are substituted by carboxylate groups, and a new molecular precursor $E(OR)_{n-m}(OOCR')_m$ is formed. Upon addition of water, the alkoxy groups and not the complexing ligands are primarily hydrolyzed [3]. The role of the carboxylate groups in $E(OR)_{n-m}(OOCR')_m$ is similar to that of the groups R' in $(RO)_{4-m}SiR'_m$: they block condensation sites, lower the degree of crosslinking in the oxide materials and allow to introduce functional organic groups into sol-gel materials [4]. The same arguments apply to other anionic complexing ligands.

The chemistry of *metal* alkoxide derivatives substituted by *organofunctional* complexing ligands and their use for materials syntheses is still rather underdeveloped. There are only a few derivatives with $E \neq Si$ carrying unsaturated organic groups (for the preparation of inorganic-organic hybrid polymers) [3,5,6] or amino groups (for coordinating metal ions, for example) [7]. Binding catalytically active *metal complexes* to TiO_2 or ZrO_2 could be attractive not only for heterogenizing these complexes, but possibly also for promoting their catalytic activity. Since many catalytically active metal complexes contain phosphine ligands, phosphanyl-substituted metal alkoxides of the type $(RO)_nE\text{-}X\text{-}PR_2$ are needed to tether the metal complexes. We previously reported the preparation of metal alkoxide derivatives $(RO)_3E(OOC\text{-}X\text{-}PPh_2)$ (E = Ti, Zr), in which a PR_2 group is connected with the metal alkoxide moiety via a carboxylate group

Mat. Res. Soc. Symp. Proc. Vol. 435 © 1996 Materials Research Society

[8]. In this work we report the synthesis and characterization of new organically substituted titanium and zirconium alkoxides with *sulfonate* ligands carrying a dangling phosphanyl group.

EXPERIMENTAL SECTION

All operations were performed in an atmosphere of dry and oxygen-free argon, using dried solvents. All starting compounds were used as received and stored under nitrogen.

Preparation of CH_2=CHC(O)NHC(CH_3)$_2$$CH_2$$SO_3$E(OR)$_3$ (E = Ti: R = Et, ^{n}Pr, ^{i}Pr, ^{n}Bu, ^{t}Bu (**1a-e**); E = Zr: R = ^{n}Pr (**1f**)): 30 mmol of Ti(OR)$_4$ or Zr(O^{n}Pr)$_4$ (70% in n-propanol, 14.04 g) was added to a suspension of 30 mmol (6.21 g) of 2-acrylamido-2-methyl-propane sulfonic acid in 100 ml THF. The resulting pale yellow (E = Ti) or colorless solution (E = Zr) was stirred for 1 h at room temperature. After removal of the solvent in vacuo a yellowish solid was obtained in 87-100% yield (relative to the formulae given in the Results Section), which was washed several times with petroleum ether.

The following bands are observed in the IR spectra (Nujol) of each compound: ν = 3300, 3080 (br, NH), 1660 (s, C=O), 1630 (w, C=C), 1585 (br, C=O), 1550 (sh), 1410 (vw), 1250, 1280 (br), 1140 (s, br, C-O), 1080, 1040.

1a: ^{1}H NMR (DMSO, 250 MHz) δ = 1.08 (m, 7.4 H, HOCH_2*CH_3* and TiOCH_2*CH_3*), 1.40 (s, 6 H, C(*CH_3*)$_2$), 2.67 (s, 2 H, *CH_2*SO_3), 3.41 (quintet, 0.6 H, HO*CH_2*CH_3), 4.33 (q, 4.7 H, TiO*CH_2*CH_3), 4.71 (t, 0.3 H, *H*O$CH_2$$CH_3$), 5.49 (dd, 1 H, trans-*CH_2*=CH), 5.97 (m, 2 H, cis-*CH_2*=CH and CH_2=C*H*), 8.41 (s, 1H, N*H*).

1b: ^{1}H NMR (DMSO) δ = 0.83 (t, 6.7 H, TiO$CH_2$$CH_2$*$CH_3$* and HO$CH_2$$CH_2$*$CH_3$*), 1.41 (m, 10.4 H, C(*$CH_3$*)$_2$, TiOCH_2*CH_2*CH_3 and HOCH_2*CH_2*CH_3), 2.68 (s, 2 H, $CH_2$$SO_3$), 3.31 (m, 0.4 H, HO*$CH_2$*$CH_2$$CH_3$), 4.26 (t, 4.2 H, TiO*$CH_2$*$CH_2$$CH_3$ and *H*O$CH_2$$CH_2$$CH_3$), 5.47 (dd, 1 H, trans-*CH_2*=CH), 5.97 (m, 2 H, cis-*CH_2*=CH and CH_2=C*H*), 8.41 (s, 1H, N*H*).

1c: ^{1}H NMR (DMSO) δ = 1.02 (d, 1.9 H, HOCH(*CH_3*)$_2$), 1.12 (d, 7.4 H, TiOCH(*CH_3*)$_2$), 1.41 (s, 6 H, C(*CH_3*)$_2$), 2.68 (s, 2 H, *CH_2*SO_3), 3.76 (sept, 0.3 H, HOC*H*(CH_3)$_2$), 4.33 (d, 0.3 H, *H*OCH(CH_3)$_2$), 4.77 (sept, 1.2 H, TiOC*H*(CH_3)$_2$), 5.49 (dd, 1 H, trans-*CH_2*=CH), 5.97 (m, 2 H, cis-*CH_2*=CH and CH_2=C*H*), 8.45 (s, 1H, N*H*).

1d: ^{1}H NMR (DMSO) δ = 0.86 (t, 8.3 H, TiO$CH_2$$CH_2$$CH_2$*$CH_3$* and HO$CH_2$$CH_2$$CH_2$*$CH_3$*), 1.34 (m, 17.1 H, TiO$CH_2$$CH_2$*$CH_2$*$CH_2$$CH_3$ and C(*CH_3*)$_2$), 2.68 (s, 2 H, *CH_2*SO_3), 3.37 (q, 2.1 H, HO*CH_2*$CH_2$$CH_2$$CH_3$), 4.30 (t, 4.4 H, TiO*$CH_2$*$CH_2$$CH_2$$CH_3$ and *H*O$CH_2$$CH_2$$CH_2$-$CH_3$), 5.49 (dd, 1 H, trans-*CH_2*=CH), 5.97 (m, 2 H, cis-*CH_2*=CH and CH_2=C*H*), 8.44 (s, 1H, N*H*).

1e: ^{1}H NMR (DMSO) δ = 1.26 (s, 1.3 H, HOC(*CH_3*)$_3$), 1.34 (s, 8.1 H, TiOC(*CH_3*)$_3$), 1.42 (s, 6 H, C(*CH_3*)$_2$), 2.69 (s, 2 H, *CH_2*SO_3), 4.17 (s, 0.1 H, *H*OC(CH_3)$_3$) 5.49 (dd, 1 H, trans-*CH_2*=CH), 5.97 (m, 2 H, cis-*CH_2*=CH and CH_2=C*H*), 8.43 (s, 1H, N*H*).

1f: ^{1}H NMR (DMSO) δ = 0.85 (m, 8.6 H, ZrO$CH_2$$CH_2$*$CH_3$* and HO$CH_2$$CH_2$*$CH_3$*), 1.47 (m, 11.7 H, C(*$CH_3$*)$_2$,ZrO$CH_2$*$CH_2$*$CH_3$ and HOCH_2*CH_2*CH_3), 2.68 (s, 2 H, *CH_2*SO_3), 3.32 (q, 1.4 H, HO*CH_2*$CH_2$$CH_3$), 3.84 (t, 4.3 H, ZrO*$CH_2$*$CH_2$$CH_3$), 4.35 (t, 0.7 H, *H*O$CH_2$$CH_2$-$CH_3$), 5.49 (dd, 1 H, trans-*CH_2*=CH), 5.97 (m, 2 H, cis-*CH_2*=CH and CH_2=C*H*), 8.44 (s, 1H, N*H*).

Preparation of Ph_2P$CH_2$$CH_2$C(O)NHC($CH_3$)$_2$$CH_2$$SO_3$E(OR)$_3$ (E = Ti: R = Et, ^{n}Pr, ^{i}Pr, ^{t}Bu (**2a-e**); E = Zr: R = ^{n}Pr (**2f**))

(a) By radical addition of $HPPh_2$ to **1**: 5.5 mmol $HPPh_2$ (0.95 ml) and 100 mg AIBN were added to a suspension of 5 mmol **1a-e** in 60 ml of benzene. The resulting solution was heated to 70°C for 6 h. After removal of the solvent in vacuo, the obtained pale yellow (**2a-d**) or colorless (**2f**) solid was washed several times with petroleum ether and THF and spectroscopically investigated. There was no formation of **2e** by this route.

(b) By reaction of $Ph_2PCH_2CH_2C(O)NHC(CH_3)_2CH_2SO_3Li$ [9] with $E(OR)_4$ (E = Ti: R = Et, ^nPr, ^iPr, ^tBu. E = Zr: R = ^nPr): 5 mmol $E(OR)_4$ was added to a solution of 5 mmol $Ph_2PCH_2CH_2C(O)NHC(CH_3)_2CH_2SO_3Li$ in 50 ml ROH. The resulting solution was stirred for 1 h at room temperature (in the case of R = ^tBu the mixture was heated to 40 °C). Yellowish solids were obtained after removal of the solvent in vacuo in 70-81% yield (relative to the formulae given in the Results Section), which were washed several times with petroleum ether.

The following bands are observed in the IR spectra (Nujol) of each compound: ν = 3320 (br, NH), 3050 (sh, NH), 3030 (vw), 1655 (br, C=O, Ph), 1580 (sh, Ph), 1550 (br, C=O), 1300 (w), 1230 (br), 1190 (sh), 1160, 1120 (vw), 1060 (br) cm^{-1}.

2a: ^{31}P NMR (DMSO) δ = -15.27 ppm. ^1H NMR (DMSO) δ= 1.04 (t, 5.1 H, $TiOCH_2CH_3$ and $HOCH_2CH_3$), 1.36 (s, 2 H, $C(CH_3)_2$), 2.00 (m, 2 H, $Ph_2PCH_2CH_2CO$), 2.22 (m, 2 H, $PPh_2CH_2CH_2CO$), 2.64 (s, 2 H, CH_2SO_3), 3.42 (q, 3.3 H, $HOCH_2CH_3$), 4.35 (s, 1.6 H, $HOCH_2CH_3$), 4.75 (t, 0.2 H, $TiOCH_2CH_3$), 7.35 (m, 10 H, Ph), 8.14 (s, 1 H, N*H*).

2b: ^{31}P NMR (DMSO) δ = -15.27 ppm. ^1H NMR (DMSO) δ = 0.81 (t, 7.2 H, $TiOCH_2CH_2CH_3$ and $HOCH_2CH_2CH_3$), 1.36, 1.39 (s and m, 10.8 H, $C(CH_3)_2$, $TiOCH_2CH_2CH_3$ and $HOCH_2CH_2CH_3$), 2.00 (m, 2 H, $PPh_2CH_2CH_2CO$), 2.22 (m, 2 H, $PPh_2CH_2CH_2$ CO), 2.64 (s, 2 H, CH_2SO_3), 3.32 (t, 2.9 H, $HOCH_2CH_2CH_3$), 4.36 (m, 4.7 H, $HOCH_2CH_2CH_3$ and $TiOCH_2CH_2CH_3$), 7.35 (m, 10 H, Ph), 8.14 (s, 1 H, N*H*).

2c: ^{31}P NMR (DMSO) δ = -15.27 ppm. ^1H NMR (DMSO) δ = 1.02 (d, 5.9 H, $TiOCH(CH_3)_2$ and $HOCH(CH_3)_2$), 1.36 (s, 6 H, $C(CH_3)_2$), 2.00 (m, 2 H, $Ph_2PCH_2CH_2CO$), 2.22 (m, 2 H, $Ph_2PCH_2CH_2CO$), 2.64 (s, 2 H, CH_2SO_3), 3.75 (m, 0.8 H, $HOCH(CH_3)_2$), 4.34 (d, 0.8 H, $HOCH(CH_3)_2$), 4.77 (m, 0.1 H, $TiOCH(CH_3)_2$), 7.35 (m, 10 H, Ph), 8.14 (s, 1 H, N*H*).

2e: ^{31}P NMR (DMSO) δ = -15.27 ppm. ^1H NMR (DMSO) δ = 1.09 (s, 5.6 H, $TiOC(CH_3)_3$ and $HOC(CH_3)_3$), 1.36 (s, 6 H, $C(CH_3)_2$), 2.00 (m, 2 H, $Ph_2PCH_2CH_2CO$), 2.22 (m, 2 H, $Ph_2PCH_2CH_2CO$); 2.64 (s, 2 H, CH_2SO_3), 4.18 (s, 0.4 H, $HOC(CH_3)_3$), 7.36 (m, 10 H, Ph), 8.14 (s, 1 H, N*H*).

2f: ^{31}P NMR (DMSO) δ = -15.27 ppm. ^1H NMR (DMSO) δ = 0.82 (t, 6 H, $ZrOCH_2CH_2CH_3$ and $HOCH_2CH_2CH_3$), 1.36, 1.38 (s and m, 10.1 H, $C(CH_3)_2$, $ZrOCH_2CH_2CH_3$ and $HOCH_2CH_2CH_3$), 2.00 (m, 2 H, $Ph_2PCH_2CH_2CO$), 2.22 (m, 2 H, $Ph_2PCH_2CH_2CO$), 2.64 (s, 2 H, CH_2SO_3), 3.32 (t, 2.6 H, $HOCH_2CH_2CH_3$), 3.82 (t, 1.4 H $ZrOCH_2CH_2CH_3$), 4.36 (s, 1.3 H, $HOCH_2CH_2CH_3$), 7.35 (m, 10 H, Ph), 8.13 (s, 1 H, N*H*).

RESULTS AND DISCUSSION

2-Acrylamido-2-methyl-propane sulfonic acid reacts with an equimolar amount of the titanium alkoxides $Ti(OR)_4$ (R = Et, ^nPr, ^iPr, ^nBu, ^tBu) or $Zr(O^nPr)_4$ by substitution of one alkoxide group as previously described for $Ti(O^iPr)_4$ [6] (Scheme 1). The sulfonate derivatives are obtained as solid compounds (monomeric formulas are written in this paper for simplicity).

NMR (Table I) and IR spectra clearly show that the SO_3-group is bonded to the titanium or zirconium atom. In the reaction products, the NMR signal for the SO_3H group is no longer observed, and the signals of the α-CH_2 and the NH groups are shifted to higher field relative to the uncoordinated sulfonic acid. The resonances of the methyl groups and the vinyl protons are not significantly changed. The ν(O-H) band in the IR spectrum also disappeared. The two ν(N-H) vibrations at 3000 and 3080 cm^{-1} gain intensity, while the intensity of the ν(S=O) band at 1070 cm^{-1} decreases. It can be excluded from the IR spectra that the NH or CO group interacts with the metal atom.

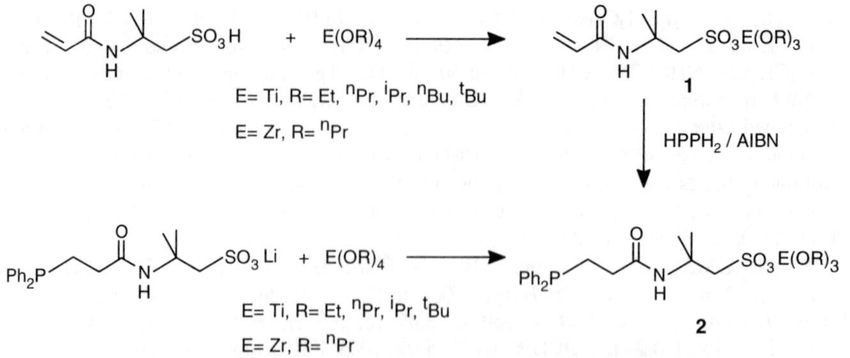

Scheme 1: Reaction of the sulfonic acids with $Ti(OR)_4$ or $Zr(OPr)_4$

Table I: Comparison of the ^{1}H NMR resonances of uncoordinated and coordinated 2-acrylamido-2-methyl-propane sulfonic acid

	CH_2=CHCONHC(Me)$_2$CH$_2$SO$_3$H	CH_2=CHCONHC(Me)$_2$CH$_2$SO$_3$EO$_x$(OR)$_y$
SO_3H	14.16 s	–
$C(CH_3)_2$	1.39 s	1.40-1.41
CH_2SO_3	3.00 s	2.67-2.69
CH_2=CH	5.46 dd 5.94 d 6.11 dd	5.49 dd 5.97 m
NH	8.11 s	8.41 s

Integration of the NMR signals (with the CH_2SO_3 signal [= 2 H] as reference) confirmed the presence of alcohol in the reaction products, which could not be removed in vacuo. In all cases condensation reactions occurred during the reaction or during workup. Integration of the alkoxide signals showed that less than three alkoxide groups were retained in the products. The following composition is proposed based on the NMR spectra.

1a: R`$SO_3TiO_{0.35}(OEt)_{2.3}$ · 0.3 EtOH (R´ = CH_2=CHC(O)NHC(CH_3)$_2$CH$_2$-)
1b: R`$SO_3TiO_{0.5}(O^nPr)_{2.0}$ · 0.2 ^{n}PrOH
1c : R`$SO_3TiO_{0.9}(O^iPr)_{1.2}$ · 0.3 ^{i}PrOH
1d: R`$SO_3TiO_{0.65}(O^nBu)_{1.7}$ · 1.1 ^{n}BuOH
1e : R`$SO_3TiO_{1.05}(O^tBu)_{0.9}$ · 0.1 ^{t}BuOH
1f: R`$SO_3ZrO_{0.4}(O^nPr)_{2.2}$ · 0.7 ^{n}PrOH

With increasing size of the alkoxide group the amount of coordinated alkoxide decreases. The available data do not provide information about the degree of association of the alkoxide derivatives or the bonding mode of the R'SO_3 ligand.

The N and S values determined by elemental analyses (Table II) are in good agreement with the theoretical values calculated from the NMR spectroscopically obtained formulae. The found C and H values are lower than calculated, except for **1e**. We attribute this to the formation of carbides during elemental analysis. Another reason could be a lower sulfonate : titanium ratio than 1:1. However, this is less likely, because we got clear solutions after the reaction (unreacted sulfonic acid would be unsoluble in THF), and the found N and S values then would be smaller.

Table II: Elemental analyses, calculated (found)

	C	H	N	S
1a	38.76 (32.32)	6.74 (5.80)	3.70 (3.47)	8.48 (7.69)
1b	41.63 (35.41)	7.09 (6.26)	3.57 (3.70)	8.17 (8.36)
1c	35.64 (28.42)	6.43 (4.96)	3.92 (3.54)	8.97 (8.12)
1d	46.39 (43.67)	8.19 (7.86)	2.97 (2.88)	6.80 (5.86)
1e	38.48 (37.68)	6.19 (6.69)	4.08 (3.97)	9.33 (8.83)
1f	39.62 (29.42)	6.99 (5.43)	2.94 (2.85)	6.74 (6.32)

The reactive double bond in **1** can, inter alia, be used to introduce PR_2 groups by radical addition of $HPPh_2$. This approach was previously used to prepare ß-phosphanyl propionate derivatives of titanium and zirconium alkoxides by addition of $HPPh_2$ to $(RO)_3E$(acrylate) [8]. When $HPPh_2$ is reacted with **1** in the presence of AIBN as a radical starter, the ^{31}P NMR of the spectra products clearly prove the formation of the phosphinated products **2** (signal at -15.27 ppm). However, another, yet unidentified phosphane (signal at 30.47 ppm) is additionally formed, its portion ranging from about 50% (reaction of **1f**) to 100% reaction of **1e**, i.e., no phosphinated alkoxide derivative is formed with the bulky ^tBu substituents. With increasing size of the alkoxide groups the phosphane : product ratio increases:

$$Zr(O^nPr)_3 > Ti(OEt)_3 > Ti(O^nPr)_3 > Ti(O^iPr)_3 > Ti(O^nBu)_3 > Ti(O^tBu)_3$$

Due to the difficulties to separate the compounds **2** from the byproduct, an alternative method of synthesis was investigated,. The phosphinated sulfonate is easily prepared by Michael addition of $LiPPh_2$ to the acrylic double bond of the sulfonic acid [9]. When the Li-salt is reacted with an equimolar amount of $Ti(OR)_4$ (R = Et, ^nPr, ^iPr, ^tBu) or $Zr(O^nPr)_4$, the solid alkoxide derivatives **2** are formed (Scheme 1). The time required to complete the reaction is much shorter than for the previous method. There is no change in the IR spectra of **2a-f** relative to the phosphinated lithium sulfonate. Coordination of the ligand to titanium or zirconium via the SO_3 group is again shown by the NMR spectra. The resonance of the α-CH_2 group was shifted to higher field, but the shift is much smaller (0.05 ppm). The resonances of the NH and the other groups are not significantly shifted. Integration of the NMR signals (with the $CH_2CH_2PPh_2$ signal [= 4 H] as reference) confirmed again the presence of alcohol in the reaction products. The integration of alkoxide group signals indicated that condensation has occurred to a higher degree than in the previous reaction, i.e. a smaller portion of alkoxide groups is retained in the reaction products. The following compositions are proposed based on the NMR spectra:

2a: $R'SO_3TiO_{1.45}(OEt)_{0.1} \cdot 1.6\ EtOH$ $R' = (Ph_2PCH_2CH_2C(O)NHC(CH_3)_2CH_2\text{-})$

2b: $R'SO_3TiO_{1.15}(O^nPr)_{0.9} \cdot 1.5\ {}^nPrOH$

2c: $R'SO_3TiO_{1.45}(O^iPr)_{0.1} \cdot 0.8\ {}^iPrOH$

2e: $R'SO_3TiO_{1.4}(O^tBu)_{0.2} \cdot 0.4\ {}^tBuOH$

2f: $R'SO_3ZrO_{1.15}(O^nPr)_{0.7} \cdot 1.3\ {}^nPrOH$

The N, P and S values of the elemental analyses (Table III) are in good agreement with the theoretical values calculated from the NMR spectroscopically obtained formulae. The lower C and H values (except for **2f**) are again attributed to the formation of carbides during elemental analysis.

Table III: Elemental analyses, calculated (found)

	C	H	N	P	S
2a	49.03 (43.01)	5.97 (5.70)	2.58 (2.27)	5.72 (5.22)	5.92 (5.07)
2b	52.42 (42.90)	6.76 (5.83)	2.33 (2.06)	5.16 (4.97)	5.34 (4.81)
2c	50.38 (47.24)	5.67 (5.72)	2.70 (2.71)	5.89 (5.87)	6.19 (6.11)
2d	50.79 (49.66)	5.54 (6.15)	2.76 (2.63)	6.33 (6.08)	6.12 (5.57)
2e	48.45 (41.59)	6.06 (5.66)	2.26 (2.12)	4.99 (5.07)	5.17 (4.82)

CONCLUSIONS

We have shown in this paper that reaction of $Ti(OR)_4$ (R = Et, ^nPr, ^iPr, ^nBu, ^tBu) and $Zr(O^nPr)_4$ with functionalized sulfonic acids results in alkoxide sulfonate derivatives. The bifunctional organic ligands are chemically bonded to the titanium or zirconium atom via the sulfonate group. PR_2-substituted derivatives were obtained either by addition of $HPPh_2$ to the double bond of 2-acrylamido-2-methyl-propane sulfonate ligands or by direct reaction of the alkoxides with the phosphinated lithium sulfonates. There is no interaction of the PR_2 group with the titanium or zirconium atom. This group therefore is available to tether metal complex fragments to titanate or zirconate groups and allows to prepare metal complex-containing TiO_2 or ZrO_2 based hybrid materials. Additional preliminary results show that this reaction can also be extended to sulfanilic acid.

ACKNOWLEDGEMENT

This work was supported by the Austrian Fonds zur Förderung der wissenschaftlichen Forschung (FWF).

REFERENCES

1. U. Schubert, N. Hüsing, A. Lorenz, Chem. Mater. **7**, 2010 (1995).
2. U. Deschler, P. Kleinschmitt, P. Panster, Angew. Chem. **98**, 237 (1986); Angew. Chem. Int. Ed. Engl. **25**, 236 (1986).
3. Leading references: C. Sanchez, J. Livage, M. Henry, F. Babonneau, J. Non-Cryst. Solids 1988, **100**, 65. C. Sanchez, J. Livage, New J. Chem. **14**, 513 (1990). J. Livage, M. Henry, C. Sanchez, Prog. Solid State Chem. **18**, 258 (1988).
4. U. Schubert, J. Chem. Soc. Dalton Trans., in press.
5. U. Schubert, E. Arpac, W. Glaubitt, A. Helmerich, C. Chau, Chem. Mater. **4**, 291 (1992). C. Sanchez, M. In, J. Non-Cryst. Solids **147&148**, 1 (1992). C. Sanchez, M. In, P. Toledano, P. Griesmar, Mat. Res. Soc. Symp. Proc. **271**, 669 (1992). M. Chatry, M. Henry, M. In, C. Sanchez, J. Livage, J. Sol-Gel Sci. Technol. **1**, 233 (1994).
6. Ch. Barglik-Chory, U. Schubert, J. Sol-Gel Sci. Technol. **5**, 135 (1995).
7. U. Schubert, S. Tewinkel, F. Möller, Inorg. Chem. **34**, 995 (1995). U.Schubert, S. Tewinkel, R. Lamber, Chem. Mater., in press.
8. H. Buhler, U. Schubert, Chem. Ber. **126**, 405 (1993). A. Lorenz, U. Schubert, unpublished results.
9. G. Fremy, Y. Castanet, R. Grzybek, E. Monflier, A. Mortreux, A. M. Trzeciak, J. J. Ziolkowski, J. Organomet. Chem. **505**, 11 (1995).

CHEMICAL FUNCTIONALIZATION OF SILICA AEROGELS

NICOLA HÜSING*, ULRICH SCHUBERT*, BERNHARD RIEGEL[§], AND WOLFGANG KIEFER[§]
* Institut für Anorganische Chemie der Technischen Universität Wien, Getreidemarkt 9, A-1060 Wien, Austria, uschuber@fbch.tuwien.ac.at
[§] Institut für Physikalische Chemie der Universität Würzburg, Marcusstraße 9-11, D-97070 Würzburg, Germany

ABSTRACT

Organofunctional silica aerogels were prepared by base catalyzed hydrolysis and condensation of $Si(OR')_4$ / $RSi(OMe)_3$ mixtures, followed by supercritical drying. In the trialkoxysilanes, R is of the type $(CH_2)_nA$, the group A being an organic function. Amino, ethylenediamino, mercapto, cyano, vinyl, and methacrylate groups were thus incorporated in the aerogels. The sol-gel reaction of a 4:1 mixture of $Si(OMe)_4$ and (vinyl)$Si(OMe)_3$ was exemplarily monitored by Raman spectroscopy. The Raman spectra show that under basic conditions hydrolysis and condensation of $Si(OMe)_4$ is much faster than that of (vinyl)$Si(OMe)_3$. For the other $RSi(OMe)_3$ precursors the reaction rates strongly depend on the electronic properties and the polarity of the functional organic group. Addition of the aminopropyl and cyanoethyl substituted precursors to $Si(OR)_4$ results in shorter gel times relative to pure $Si(OR')_4$, while the mercaptopropyl, methacryloxypropyl and vinyl substituted precursors retard gelation. Depending on the ratio and the kind of the precursors, 70 - 100 % of the functional organic groups were incorporated into the aerogel network. The hydrophobicity of the aerogels depends on the kind and amount of $RSiO_{1.5}$ units.

INTRODUCTION

The "chemical design" of sol-gel materials has two aspects: the design of their composition and of their micro- and nanostructure [1]. The properties of aerogels [2], for example, largely result from their very porous micro- and nanostructure. While silica aerogels have been intensively investigated for some time, modification of their properties by design of their composition was only recently attempted. Hybrid inorganic-organic silica aerogels were obtained by sol-gel processing of $Si(OR')_4$ / $RSi(OR')_3$ mixtures, followed by supercritical drying [3 - 6]. The goal of any organic modification of silica aerogels is to maintain their unique physical properties originating from their typical structure (high porosities, very low densities, large surface areas), and to supplement new properties arising from the organic groups. For example, aerogels prepared from 4:1 mixtures of $Si(OMe)_4$ and $RSi(OMe)_3$, with R = an alkyl or a phenyl group, are permanently hydrophobic and have a higher compliance than unmodified silica aerogels [3]. In an extension of this work, *functional* organic groups were incorporated by the same approach [4, 5] using organofunctional alkoxysilanes of the type $A(CH_2)_nSi(OR)_3$ [7], with A being the functional group. Such aerogels are promising materials for new types of sensors, catalysts, adsorbents etc. The incorporation of functional organic groups is not trivial, because such groups may influence or even prevent the built-up of the aerogel structure. Furthermore, the supercritical drying process may be hostile to the organofunctional groups. Following our preliminary reports on some organofunctional aerogels [4], these principal issues are the main focus of the present article.

Mat. Res. Soc. Symp. Proc. Vol. 435 © 1996 Materials Research Society

EXPERIMENTAL

$Si(OMe)_4$, $Si(OEt)_4$, and the trialkoxysilanes $RSi(OMe)_3$ (R = vinyl, 3-methacryloxypropyl, 3-mercaptopropyl, 3-aminopropyl, N-(2-aminoethyl)-3-aminopropyl, 2-cyanoethyl) were used as received without further purification (ABCR, purity >97%). The organically modified alcogels were prepared from methanolic solutions of $Si(OR')_4$ and $RSi(OMe)_3$ (in different ratios from 9:1 to 3:2) as described in Ref. [3]. Hydrolysis and condensation reactions were started by adding the calculated amount of 0.01N aqueous NH_4OH solution, corresponding to the amount of water necessary to hydrolyze all methoxy groups. The density of the final materials was adjusted to about 220 - 270 kgm^{-3} by the amount of methanol. In the case of the amino substituted precursors the reaction was initiated by adding only water, and the density was adjusted to 70 - 170 kgm^{-3}.

For the preparation of the corresponding organically modified aerogels, the sol was stirred for 5 min and then transferred into cylindrical polyethylene vessels. After gelation, the gels were aged for 7 days at 30°C. Supercritical drying was performed using either methanol (vinyl samples) or liquid carbon dioxid (other samples) as fluid.

The chemical composition of the resulting aerogels was determined by elemental analyses and the amount of methacrylate and mercapto units by titration [8,9]. The shrinkage was determined by measuring the diameter of the cylindrical samples before and after drying. The density of the aerogels was calculated from the volume and the mass of the samples.

Hydrophobicity tests were performed by measuring the water uptake in 97% humidity at 25°C with a Shimadzu TG-50 analyzer in nitrogen atmosphere.

The Raman spectra were taken with the 1064 nm line of a neodym YAG laser with a power of about 1000 mW. The spectral resolution was 3 cm^{-1}. The experimental setup consisted of a Bruker IFS 120 FT-IR spectrometer equipped with a Raman module and a tempered glass cell. The catalyst was added 30 min after dissolving the silanes in methanol. The spectra were recorded from the solutions in short intervals after starting the reaction until the Raman bands of the precursors disappeared.

RESULTS AND DISCUSSION

Despite the importance and the frequent use of multicomponent systems for the preparation of hybrid materials by sol-gel processing, little is known about the kinetics of such systems, particularly about the mutual influence on their hydrolysis and condensation rates. Due to the complex reactions, the different chemical processes are very difficult to monitor. It is obvious that the physical and chemical properties of the final material are affected by the hydrolysis and condensation behaviour of the different precursors and the resulting distribution of the different structural units.

Table I shows the gel times of selected $RSi(OMe)_3$ / $Si(OR')_4$ mixtures in different ratios, and the appearance of the resulting aerogels. When VINYL, MEMO, or MTMO (abbreviations defined in Table I) is added to $Si(OMe)_4$, the gel time of the mixture is retarded relative to pure $Si(OMe)_4$. It is qualitatively known that under base-catalyzed conditions the reaction rates for the hydrolysis and condensation of $(alkyl)_xSi(OMe)_{4-x}$ are slower than that of $Si(OMe)_4$ [10]. Solutions containing only MEMO, MTMO or VINYL only gel after months, while under the same experimental conditions $Si(OMe)_4$ gels within 20 min. The $RSi(OMe)_3$ / $Si(OR')_4$ mixtures containing these silanes therefore show the expected trends. An increasing amount of $RSi(OMe)_3$ in the mixture increases the gel time.

Raman spectra were recorded to get detailed information on how the precursors mutually influence the sol-gel process. Raman spectroscopy permits following each precursor unequivocally by its characteristic bands during the whole sol-gel process until the precursor is completely consumed, independent of the physical state of the system. This has previously been shown for the gel-to-glass transition of silica [11]. Therefore, this method is particularly useful for monitoring multicomponent systems.

Table I: Precursor silane mixtures, their ratios and abbreviations, gel times and the appearance of the resulting aerogels

precursor mixture		abbreviation	ratio	gel times [min]	appearance
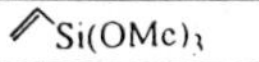	$Si(OMe)_4$	VINYL-20	1:4	55	opaque
	$Si(OMe)_4$	MTMO-10	1:9	25	opaque
		MTMO-20	1:4	65	opaque
		MTMO-40	2:3	180	opaque
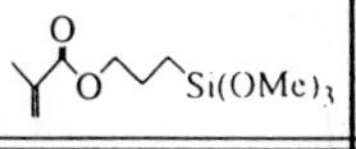	$Si(OMe)_4$	MEMO-10	1:9	38	opaque
		MEMO-20	1:4	80	opaque
		MEMO-40	2:3	~250	opaque
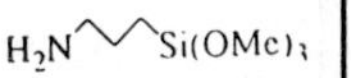	$Si(OEt)_4$*	AMMO-10	1:9	17	white
		AMMO-20	1:4	15	white
		AMMO-40	2:3	15	white
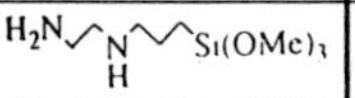	$Si(OEt)_4$*	AEAPS-10	1:9	50	white
		AEAPS-20	1:4	30	white
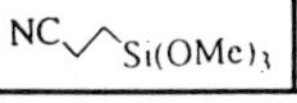	$Si(OMe)_4$	CYMO-10	1:9	15	transparent
		CYMO-20	1:4	7	transparent

*these samples were prepared with a lower density (70-170 kgm^{-3})

The Raman spectra of the liquid, solvent-free precursors $Si(OMe)_4$ and (vinyl)$Si(OMe)_3$ were measured in the range of 150-3500 cm^{-1}. Between 620-650 cm^{-1}, $Si(OMe)_4$ has a strong band at 641 cm^{-1}, and (vinyl)$Si(OMe)_3$ at 622 cm^{-1} . These bands are assigned to the $\nu_s(SiO_4)$ and $\nu_s(SiO_3)$ vibration, respectively. These Raman lines allow easy detection of the individual precursors during the whole sol-gel process. Upon partial or full substitution of the methoxy groups by SiOH or SiOSi groups the symmetric condition is no longer fulfilled and these Raman bands disappear.

Figure 1 shows the Raman spectra of (vinyl)$Si(OMe)_3$ / $Si(OMe)_4$ (ratio 1:4) before and after adding of the water/catalyst mixture. An assignment of the bands is given in Table II. At t = 0 (i.e., before addition of water) the band of (vinyl)$Si(OMe)_3$ at 622 cm^{-1} is observed as a shoulder on the low wavenumber side of the $\nu_s(SiO_4)$ vibration of $Si(OMe)_4$ at 641 cm^{-1}. In the beginning of the reaction, the decrease in the intensity of the $\nu_s(SiO_4)$ vibration is much faster than that of the $\nu_s(SiO_3)$ vibration of the organically substituted silane. That means that initially $Si(OMe)_4$ is preferrentially hydrolyzed. Only after more than 4 h at 50°C the vibrational mode of the substituted precursor also disappeares. The methanol band at 1031 cm^{-1} increases in intensity corresponding to the production of methanol during the progress of the sol-gel reaction. The peak of *iso*-propanol, added as an internal standard, at 817cm^{-1} is nearly unchanged.

The Raman spectra show that hydrolysis of (vinyl)$Si(OMe)_3$ is much slower than that of $Si(OMe)_4$. The solution gels after 40 minutes at 50°C. This means that the major amount of (vinyl)$Si(OMe)_3$ is reacting when the gel network is already built from $Si(OMe)_4$. Therefore, condensation of the $RSiO_{1.5}$ groups will mainly occur with surface silanol groups of the primary and secondary SiO_2 particles of the gel matrix or with non-condensed vinyl substituted siloxane species. The Raman spectra provide no information on homo-condensed molecules. However, species not condensed to the silica network would be leached out during the supercritical drying process due to the extensive rinsing processes. The elemental analysis of the final material clearly shows that there is no loss of vinyl groups (see Table III), which are therefore fully incorporated into the silica network.

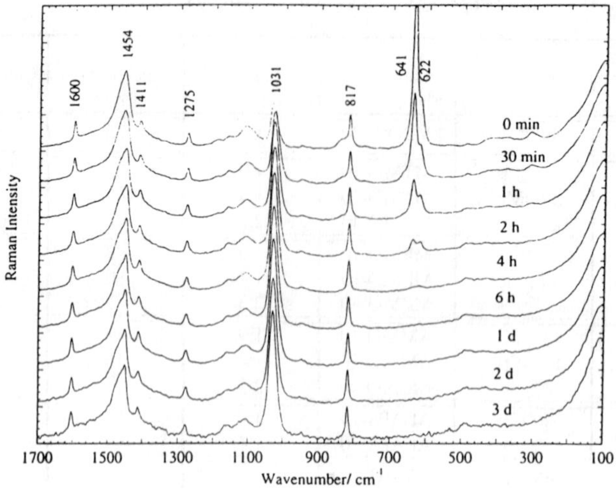

Figure 1: Raman spectra of a (vinyl)RSi$(OMe)_3$ / Si$(OMe)_4$ mixture (4:1)

Table II: Assignments of the Raman modes

622 cm^{-1}	ν(SiO_3) sym. (vinyl)Si$(OMe)_3$	1031 cm^{-1}	ν (CO) sym. (methanol)	1454 cm^{-1}	δ $(CH)_3$ sym.
641 cm^{-1}	ν(SiO_4) sym. Si$(OMe)_4$	1275 cm^{-1}	=CH def. (vinyl)Si$(OMe)_3$	1600 cm^{-1}	ν (C=C) (vinyl)Si$(OMe)_3$
817 cm^{-1}	ν (CO) sym. (iso-propanol)	1411 cm^{-1}	$=CH_2$ def. (vinyl)Si$(OMe)_3$		

3-Chloropropyl-Si$(OMe)_3$ and 3-glycidoxypropyl-Si$(OMe)_3$ were used as modification reagents i previous papers [4]. Because the gel times of RSi$(OR')_3$ / Si$(OR)_4$ mixtures containing one of thes silanes, MEMO or MTMO show the same trends as VINYL / Si$(OR)_4$ mixtures, it is safe to assum that the development of the microstructure is the same, i.e. that in all these cases the surface of th primary and secondary silica particles is covered by the functional $RSiO_{1.5}$ units.

Contrary to that, the gel times of RSi$(OR')_3$ / Si$(OR)_4$ mixtures with RSi$(OR')_3$ = AMMO, AEAP: or CYMO (despite the absence of a strongly basic group) are shorter than that of pure Si$(OR')_4$. Thi behaviour cannot be correlated with the behaviour of solutions containing only one of the component: AMMO and AEAPS do not gel during a period of 3 days, and addition of the water / catalyst mixtur to CYMO results in precipitates. Addition of the water / catalyst mixture to AMMO / Si$(OMe)_4$ o AEAPS / Si$(OMe)_4$ mixtures also results in the spontaneous formation of precipitates. Severa modifications of the system were made to avoid this: (i) replacement of Si$(OMe)_4$ by Si$(OEt)_4$, whicl reacts more slowly, (ii) the aqueous ammonium hydroxide solution was replaced by distilled wateı and (iii) the density of the aerogels was lowered to around 100 kgm^{-3}. Only under these conditions arı stable three-dimensional networks formed. Although the gel network is clearly formed in a differen manner, physical properties, such as density changes or shrinkage during supercritical drying (Tabl III) of these aerogels are in the typical range for organically modified aerogels [3 , 4], and monolithi bodies are obtained. The shrinkage of the gels during supercritical drying shows the expected trends a: discussed in Refs. [3, 4]. Shinkage increases with an increasing amount of $RSiO_{1.5}$ units, probabl because a less stable network is formed due to the larger number of non-condensable and sterically hindered groups.

However, the optical appearance and preliminary BET measurement of the aerogels derived from the mixtures containing AMMO, AEAPS or CYMO indicate that they have a different microstructure. While all samples with retarded gel times are opaque, the amino-substituted samples are white, and the CYMO aerogels are much clearer than unmodified silica aerogels. There is no correlation between the gel times and the optical appearance of the aerogels.

The chemical composition and some properties of the organofunctional aerogels are given in Table III. A very important issue of this work was the question to what extent the functional moieties are incorporated into the aerogels. Elemental analyses show that most of the organic groups are retained. Because of the extensive rinsing processes during supercritical drying, we can exclude the possibility that monomers or oligomers containing these groups are only embedded in the matrix and not chemically bonded to the silica gel skeleton.

Table III. Properties of the functionalized aerogels

Sample	nominal amount of R [mmol/g]	experimental amount of R [mmol/g]	incorporated functional groups [%]	bulk density [kgm^{-3}]	shrinkage [%]	water uptake [%] 40 days/ 97% humidity / 25°C
VINYL-20	2.9	2.9	100	266	5.5	5
MTMO-10	1.5	1.4	93	268	8.1	130
MTMO-20	2.7	2.2	81	225	10.5	23
MTMO-40	4.6	3.2	70	234	16.6	2
MEMO-10	1.3	1.3	100	270	9.1	20
MEMO-20	2.3	1.9	83	259	15.0	10
MEMO-40	3.7	2.4	65	234	22.5	11
AMMO-10	1.6	1.7	106	73	8.2	10
AMMO-20	2.9	2.9	100	118	5.0	-
AMMO-40	5.0	4.5	90	178	17.5	49
AEAPS-10	1.4	1.5	107	113	4.8	19
AEAPS-20	2.6	2.6	100	149	11.5	-
CYMO-10	1.6	1.8	113	261	8.6	105
CYMO-20	2.9	3.0	103	251	8.7	-

In general, the organic groups were quantitatively incorporated into the gels when ≤ 10 mol % of $RSi(OMe)_3$ was initially employed (incorporation of more than 100% $RSiO_{1.5}$ units means that small oligomers formed from $Si(OR')_4$ are removed during supercritical drying). With an increasing amount of $RSi(OMe)_3$ in the starting mixture the percentage of incorporated groups decreases. This decrease is less pronounced for the gels with the shorter gel times, probably because of the different composition and arrangement of the building blocks.

Unmodified silica aerogels have a large number of residual SiOH and SiOMe moieties on their inner surface which are the main source of their hydrophilicity. Replacing these groups by organic groups improves the hydrophobicity. A complete coverage of the inner surface by unpolar organic groups leads to hydrophobic aerogels. Water uptake in humid atmosphere (as a measure of the hydrophobicity) therefore provides additional information on the composition of the inner surface of the aerogels. The water uptake of the vinyl-modified aerogel with 20% of the Si atoms substituted by vinyl groups is strongly reduced relative to unmodified silica aerogels. For the MTMO and MEMO-functionalized samples there is an evident correlation between the hydrophobicity and the amount of organofunctional units corresponding to the degree of coverage of the inner surface by the organic groups. Increasing the amount of $RSi(OMe)_3$ in the starting mixture to 40% (corresponding to about 30% incorporation of $RSiO_{1.5}$ groups) results in reasonably hydrophobic aerogels. (This is not a contradiction of our previous reports on alkyl- and aryl-substituted aerogels, where a 20% substitution was sufficient to achieve hydrophobictity [3]. A different supercritical drying method was employed for the preparation of these aerogels, which also influences the structure). The VINYL, MTMO- and MEMO-functionalized aerogels are inherently less hydrophobic than alkyl- or aryl-modified aerogels, because of the polarity of the organic group. However, these aerogels show a reduced water uptake over a period of 40 days (unmodified aerogels: about 100% under the same conditions).

An inverse trend is observed for amino-modified aerogels. The water uptake increases with an increasing portion of amino groups. This is explained by the strongly hydrophilic character of the amino groups. The CYMO modified aerogels show about the same behaviour as unmodified silica aerogels.

CONCLUSIONS

Monolithic aerogels were prepared from $Si(OR')_4$ / $RSi(OMe)_3$ mixtures, with R = vinyl, methacryloxypropyl, mercaptopropyl, aminopropyl, ethylendiaminopropyl and cyanoethyl. The organofunctional groups are incorporated into the aerogel matrix to 70-100%, depending on the ratio and composition of the starting precursor solution. The gel time and therefore the formation of the silica network is strongly influenced by the organic groups due to the electronic properties and the polarity of the different functional units. Two different cases must be distinguished: an acceleration and a retardation of the gel time relative to pure $Si(OR)_4$. The typical microstructure of unmodified silica aerogels is retained, when the gel time of the mixture is slower. Raman spectra suggest that in this case (under basic conditions) the network is built from $Si(OMe)_4$, while the organofunctional $SiO_{1.5}$ units condense later to the inner surface of this network. This behaviour is also found for MEMO and MTMO. A different structure is obtained when the gel time of the mixture is faster than that of $Si(OR')_4$. Due to the change of the composition and size of the structural building blocks, macroscopically different properties are observed (found for AMMO, AEAPS and CYMO).

ACKNOWLEDGEMENT

Two of us (Bernhard Riegel and Wolfgang Kiefer) acknowledge financial support from the Deutsche Forschungsgemeinschaft.

REFERENCES

1. U. Schubert, J. Chem. Soc. Dalton (1996), in press.
2. J. Fricke, *Aerogels*, Springer Proc. Phys. Vol.6, Heidelberg, 1986. H. D. Gesser, P. C. Goswami, Chem. Rev. **89**, 765 (1989).
3. F. Schwertfeger, W. Glaubitt, U. Schubert, J. Non-Cryst. Solids **145**, 85 (1992). U. Schubert, F. Schwertfeger, N. Hüsing, E. Seyfried, Mat. Res. Soc. Symp. Proc. **346**, 151 (1994). N. Hüsing, F. Schwertfeger, W.Tappert, U. Schubert, J. Non-Cryst. Solids **186**, 1 (1995).
4. F. Schwertfeger, N. Hüsing, U. Schubert, J. Sol-Gel Sci. Technol. **2**, 103 (1994). N. Hüsing, U. Schubert, J. Sol-Gel Sci. Technol. (1996), in press.
5. W. Cao, A. Hunt, Mat. Res. Soc. Symp. Proc. **346**, 631 (1994).
6. D. A. Loy, K. J. Shea, Chem. Rev. **95**, 1431 (1995). D. A. Loy, G. M. Jamison, B. M. Baugher, E. M. Russick, R. A. Assink, S. Prakabar, K. J. Shea, J. Non-Cryst. Solids **186**, 44 (1995).
7. U. Schubert, N. Hüsing, A. Lorenz, Chem. Mater. **7**, 2010 (1995).
8. B. Saville, Analyst, **86**, 29, (1961).
9. R. E. Byrne, Anal. Chem. **28**, 126, (1956). J. B. Johnson, Z. Anal. Chem. **154**, 58 (1957).
10. F. D. Osterholtz, E.R. Pohl, J. Adhesion Sci. Technol. **6** (1), 127 (1992).

CROSS-CONDENSATION KINETICS OF ORGANICALLY MODIFIED SILICA SOLS+

S. PRABAKAR* and R.A. ASSINK, **

*Advanced Materials Laboratory, 1001 University Blvd., SE., #100, Albuquerque, NM 87106.
**Sandia National Laboratories, Albuquerque, NM 87185-1407.

ABSTRACT

The hydrolysis and self- and cross-condensation kinetics of the hybrid sol tetraethoxysilane and ethyltriethoxysilane were investigated by high resolution ^{29}Si NMR spectroscopy. A kinetic model in which hydrolysis is reversible and condensation is irreversible was developed. We found excellent agreement between the product distributions measured by ^{29}Si NMR spectroscopy and calculated by the model. The cross-condensation rates for each of the sols were intermediate to the condensation rates of the individual components. Calculations show that for these sols, the concentration of cross-condensed species is a weak function of the relative rates of self-condensation.

INTRODUCTION

Sol-gel techniques provide a convenient method for the production of high purity, homogeneous materials [1]. The properties of sol-gel derived materials can be tailored by combining an organic component with the conventional inorganic components. Hybrid materials can be prepared from either a single monomer or a mixture of two or more monomers. These hybrid silicate materials are being widely investigated because of their potential application in optical devices, fibers and thin films [2,3]. Some of the factors controlling the homogeneity of multiple component hybrid sol-gels are pH and water to silicon ratio. These factors affect the relative hydrolysis rates and the extent of competition between self- and cross-condensation reactions.

When an inorganic precursor such as tetraethoxysilane (TEOS) and an organically modified triethoxysilane such as ethyltriethoxysilane (ETES) are reacted with water in the presence of an acid the following hydrolysis reactions take place:

$$\text{TOR} + H_2O \underset{k_r}{\overset{k_f}{\rightleftharpoons}} \text{TOH} + \text{ROH} \qquad (1)$$

$$\text{QOR} + H_2O \underset{k'_r}{\overset{k'_f}{\rightleftharpoons}} \text{QOH} + \text{ROH} \qquad (2)$$

+ This work was supported by the United States Department of Energy under contract DE-AC04-94AL8500.

Mat. Res. Soc. Symp. Proc. Vol. 435 © 1996 Materials Research Society

where T represents the trifunctional silicon in ETES and Q represents the tetrafunctional silicon in TEOS . The forward rate constants are denoted k_f and k'_f while the reverse rate constants are denoted k_r and k'_r for the T and Q silicons respectively. Our results will show that reversible reactions are significant. The hydrolyzed species condense by three routes, two due to self-condensation, (3), (5), and one due to cross-condensation, (4), as follows:

$$\mathrm{TOH} + \mathrm{TOH} \xrightarrow{k_{TT}} \mathrm{TOT} + H_2O \quad (3)$$

$$\mathrm{TOH} + \mathrm{QOH} \xrightarrow{2\,k_{TQ}} \mathrm{TOQ} + H_2O \quad (4)$$

$$\mathrm{QOH} + \mathrm{QOH} \xrightarrow{k_{QQ}} \mathrm{QOQ} + H_2O \quad (5)$$

where TOT is the dimer formed by the condensation of two T silanols, QOQ is the dimer formed by the condensation of two Q silanols and TOQ is the dimer formed by the condensation of one T and one Q silanol. k_{TT} is the rate constant associated with the condensation of two TOH species, k_{QQ} is the rate constant for the condensation of two QOH species and $2k_{TQ}$ is the rate constant associated with the condensation of one TOH and one QOH species. A factor of two is associated with the formation of the cross-condensed product because statistically, this species is twice as likely to be formed as either self-condensed species.

^{29}Si NMR spectroscopy is a widely employed technique to study the chemical kinetics of silica sol-gels [1]. The sensitivity of ^{29}Si NMR to nearest neighbors, next-nearest neighbors and cyclic species makes it a convenient probe to study the hydrolyzed and condensed species. In a previous publication [4], we identified the resonances corresponding to hydrolyzed and self- and cross-condensed species in TEOS/ETES hybrid sol-gels. In this work we have investigated the hydrolysis and condensation kinetics of the hybrid sol-gel TEOS/ETES by ^{29}Si NMR and developed a kinetic model for the hydrolysis and condensation reactions.

EXPERIMENTAL

Reagent grade tetraethoxysilane (TEOS) and ethyltriethoxysilane (ETES) were used as received from Hüls America. All the sols were prepared by HCl-catalyzed hydrolysis of an equimolar mixture of TEOS and ETES in ethanol. The total silicon concentration was 2.2M. Two sols were prepared one with water-to-silicon ratio of 0.15 at a pH of 2.2 (designated as TEOS/ETES(I)) and the other with water-to-silicon ratio of 0.3 at a pH of 1.7 (designated as TEOS/ETES(II)). 5 mM chromium acetylacetonate (CrAcAc) was added to reduce the ^{29}Si spin relaxation time. Previous studies have found no effect of CrAcAc on the initial reaction rates of sol-gels [5,6].

The ^{29}Si NMR spectra were recorded at 39.6 MHz (magnetic field 4.7 Tesla) on a Chemagnetics spectrometer described previously [5]. Line widths were typically 0.3 Hz. The spectra were referenced with respect to an external standard tetramethylsilane (0 ppm). The resonances of the observed species were well resolved and could be integrated quantitatively.

RESULTS AND DISCUSSION

When TEOS/ETES in ethanol is reacted with water, the monomers first undergo hydrolysis reactions, eqs. (1) and (2), to give silanol functional groups represented by TOH and QOH. We carried out several experiments to determine if the OH groups in TOH and QOH exchange with each other. This exchange would be transparent when the sol has reached equilibrium because the net exchange between silanols would exactly cancel each other. The exchange can be observed, however, when the sol is in a nonequilibrium state. The nonequilibrium state was prepared by hydrolyzing TEOS and then adding ETES to the hydrolyzed solution. Immediately after mixing, ETES hydrolysis species were observed and the extent of hydrolysis of the TEOS was reduced. This clearly shows that the OH groups are exchanging between TEOS and ETES. Experiments conducted in the reverse order ie. hydrolyzing the ETES and then adding TEOS also showed hydrolyzed species of TEOS. The distribution of hydrolyzed species was calculated from the equation

$$\frac{[TOH]}{[QOH]} = K_R \frac{[TOR]}{[QOR]} \quad (6)$$

where K_R is defined from the appropriate forward and reverse hydrolysis rates [7]. Analogous experiments in which the first monomer was fully condensed and then the second monomer added, showed that the condensation reactions are irreversible on a timescale of 24 h.

Equation (6) enables K_R to be determined during the first 3-5 minutes of the reaction before significant condensation products have formed. The initial distributions of hydrolyzed species and the calculated K_R are shown in Table I for each of the sols. Since the T and Q monomers are equimolar, we expect that the ratio of hydrolyzed species would be 0.75 and K_R = 1.0 if each alkoxy group were equally likely to be hydrolyzed. The concentration of hydrolyzed T silicons for the TEOS/ETES sol is less than that predicted by statistical arguments.

Table I Initial distributions and equilibrium constants of hydrolyzed species

Sol	Initial TOH/QOH	K_R
TEOS/ETES	0.61	0.81
TEOS/ETES	0.54	0.76

Since limited amounts of water are used, and since the early stages of the process are being investigated, the primary condensation reactions observed are those between two monomeric species to form a dimeric species given by equations 3-5. We assume alcohol producing condensation is negligible compared to water producing condensation. Although alcohol producing condensation can be important in tetramethoxysilane sol-gels [6], it was found to be very low in TEOS sol-gel systems [8]. For second order condensation reactions, the three rate equations can be written as:

$$d\,[TOT] / dt = k_{TT}\,[TOH]\,[TOH] \quad (7)$$
$$d\,[TOQ] / dt = 2\,k_{TQ}\,[TOH]\,[QOH] \quad (8)$$
$$d\,[QOQ] / dt = k_{QQ}\,[QOH]\,[QOH] \quad (9)$$

The time evolution for each of the self- and cross-condensed species was calculated by numerically integrating eqs. (7)-(9). The initial values for the concentrations of TOH and QOH were determined from spectra recorded 3 to 5 minutes after the solutions were prepared. When the first spectra were recorded, the water had already been consumed by hydrolysis, and the concentration of hydrolyzed species had reached a state of quasi-equilibrium. During the longer condensation period, the concentrations of TOH and QOH species decrease as the silanols undergo condensation reactions. The condensation reactions yield additional water which immediately participates in hydrolysis reactions to form silanols.

Figs 1(a) and 1(b) show the experimental and calculated time dependencies for the concentrations of TOT, TOQ and QOQ species in the TEOS/ETES(I) and TEOS/ETES(II) sols respectively. The concentrations of TOT, TOQ and QOQ species increase rapidly for the first 300 minutes followed by a slight increase over the next 13 h. We see excellent agreement between the calculated (full curve) and experimental values (symbols). The absolute values of the condensation rate constants depend on fitting the initial slope and are subject to some error because of the difficulty of rapidly recording accurate spectra. The relative values of the condensation rate constants depend on the concentration of dimer species at long times and are quite accurate. The absolute and relative rate constants for the three condensation reactions and the final product distribution for these sols are shown in Table II.

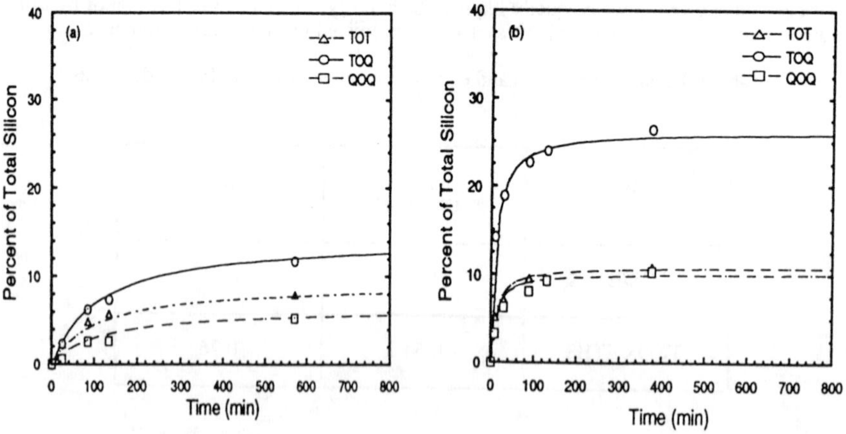

Fig 1 Calculated and experimental distributions of condensed silicon species in (a) ETES/TEOS (I) (b) ETES/TEOS (II) as a function of time.

Table II Absolute and relative condensation rates and final product distributions

Sol	AbsoluteRates $k_{TT} : k_{TQ} : k_{QQ}$ x (l/m) s^{-1} 10^{-4}	Relative Rates $k_{TT}: k_{TQ}: k_{QQ}$	Relative Concentrations TT : TQ : QQ (after 24h)
TEOS/ETES(I)	14 : 6.4 : 3.4	0.59: 0.27 : 0.14	31 : 48 : 21
TEOS/ETES(II)	54 : 36: 16	0.51: 0.34: 0.15	22 : 55 : 23

For each of the TEOS/ETES sols investigated, we found that the self-condensation rate of the organically modified monomer was greater than the cross-condensation rate which in turn was greater than the self-condensation rate of TEOS, $k_{TT} > k_{TQ} > k_{QQ}$. The relative rates for the two sols are comparable, demonstrating the utility of the model. We expect that the rates for sol(II) would be 3.2 times as large as the rates for sol(I) assuming that the condensation rates are proportional to the hydrogen ion concentrations for these acid catalyzed sols. The experimentally derived rates average 4.7 times as large. Although this difference is larger than we expected, the overall trends are reasonable.

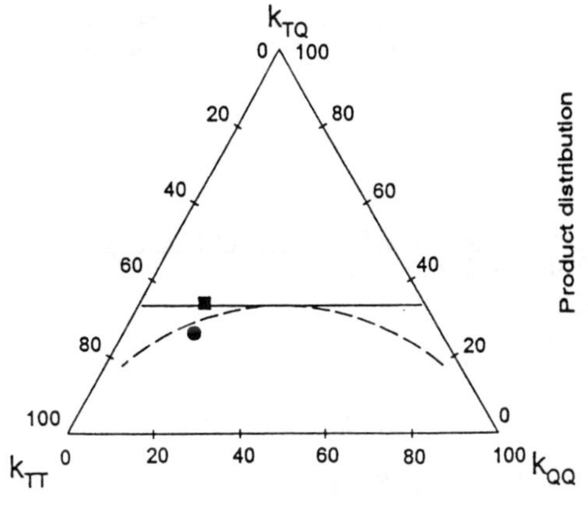

Fig 2. Ternary diagram of rate constants.
The solid line represents $k_{TQ} = 1/2\ (k_{TT} + k_{QQ})$
and the broken line represents $k_{TQ} = (k_{TT}\ k_{QQ})^{1/2}$
● - TEOS/ETES (I) ■ - TEOS/ETES (II)

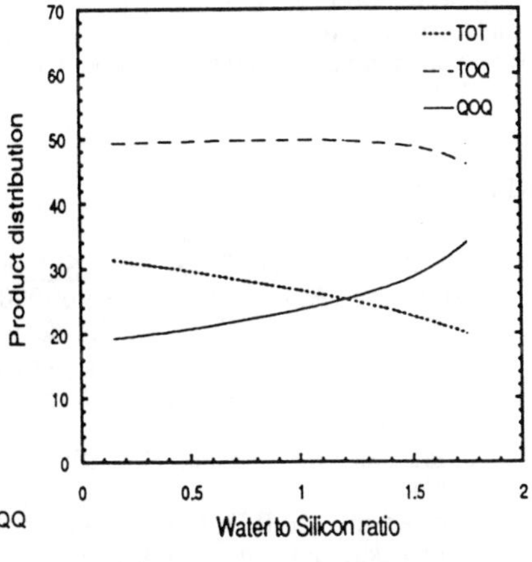

Fig 3. Product distribution as a function of water to silicon ratio (r)

The cross condensation rate constants calculated for the two sols are shown in the form of a ternary diagram (Fig 2). Each of the vertices represents the domination of one of the condensation rates. The calculated ratios of rate constants for the two sols are represented by symbols. The solid line corresponds to a cross-condensation rate equal to the arithmetic mean of the self-condensation rates, $k_{TQ} = 1/2\ (k_{TT} + k_{QQ})$. The broken line corresponds a cross-condensation rate equal to their geometric means, $k_{TQ} = (k_{TT}\ k_{QQ})^{1/2}$. The range of rate constants for these sols is too limited to provide an accurate assessment of the relative merits of the two simple models.

In Fig 3 we have plotted the product distribution of the various silicon-oxygen-silicon condensation bonds as function of water-to-silicon ratio. The product distributions were calculated assuming $K_R = 1$. We find that approximately 50% of the bonds formed are TOQ bonds, independent of water-to-silicon ratio. The relative amounts of TOT vs. QOQ bonds formed are a function of water-to-silicon ratio. For low ratios, more of the TOT bonds form because these silicons compete more successfully for the limited amount of water present. As the water-to-silicon ratio reaches 1.75, sufficient water is present for the reaction to go to completion. Both silicon species react fully and stoichiometric amounts of bonds are formed.

CONCLUSIONS

We have developed a kinetic model for the hydrolysis and self and cross condensation of the hybrid sol TEOS/ETES. The model treated hydrolysis as reversible and condensation as irreversible reactions. The final product distributions for these sols are a function of both the equilibrium between T and Q silanols and the relative rates for the various self- and cross-condsation reactions. The concentrations of the self- and cross-condensed species as a function of reaction time predicted by this model are in good agreement with the concentrations measured by ^{29}Si NMR.

REFERENCES

1 C.J. Brinker and G.W. Scherer, Sol-gel Science: The Physics and Chemistry of Sol-Gel Processing (Academic, San Diego, 1990) p. 166, 840.
2 C. Sanchez and F. Ribot, New J. Chem. **18**, 1007 (1994).
3 F. Babonneau, New J. Chem. **18**, 1065 (1994).
4 S. Prabakar, R.A. Assink, N.K. Raman, S.A. Myers and C.J. Brinker J. Non-Cryst. Solids (In Press).
5 S. Prabakar, R.A. Assink, N.K. Raman and C.J. Brinker. Mat. Res. Symp. Proc. **346**, 979 (1994).
6 R.A. Assink and B.D. Kay, J. Non-Cryst. Solids **99**, 359. (1988)
7 S. Prabakar and R.A. Assink, J. Non-Cryst. Solids (Submitted)
8 R.A. Assink and B.D. Kay, Colloids and Surfaces A: Physicochemical and Engineering Aspects **74,** 1 (1993).

PROCESSING METHYL MODIFIED SILICATE MATERIALS FOR OPTICAL APPLICATIONS: A STRUCTURAL STABILITY STUDY

A. SARKAR, Y. YAN, Z. DUAN, Y. HOSHINO AND S. RAY CHAUDHURI
YTC America Inc., 550 Via Alondra, Camarillo, CA 93012

ABSTRACT

Organic laser dyes in methyl modified silicate hosts have been found to have poor photo stability. This could be caused by the instability of either the laser dyes or the host materials themselves. To get a better understanding of the chemical reactions and the relationship between the thermal stability and the microstructures of the hybrid network, the methyl modified silicate materials were investigated by TGA, DTA, ^{13}C and ^{29}Si solid state NMR, FTIR, and AFM as a function of processing conditions. The structural stability is found to directly relate to the reaction and processing conditions and the results will be discussed.

INTRODUCTION

Organically modified silicates (ORMOSILS) have been the subject of much research interest since the mid-eighties [1-3]. Recently, ORMOSILS have attracted increasing attention as a host for dye doped solid state optical devices [4-9]. Since sol-gel processing is generally done at low temperatures, organic chromophores can be easily incorporated into the solution precursors and trapped by the host material into the interstices of its atomic structure. Methyl modified silicates show many of the characteristics required as host materials for solid state dye lasers. For instance, they can be processed in a relatively short time [4] and polished by conventional glass polishing methods [5]. The light scattering in these materials is reasonably low with no detectable phase separation [7]. However, the laser dyes doped into methyl modified silicate host exhibited poor photo stability, compared with its solution [8]. Although such a phenomenon was assumed to be due to the difference of the dye concentration in solution and solid host [8], it might also result from the instability of the host material. Therefore, it is necessary to investigate the thermal and structural stability of the hybrid material itself. This is the focus of the present paper.

EXPERIMENTAL PROCEDURE

A series of methyl modified silicates with the molar ratios of 50 to 50, 60 to 40, 75 to 25, and 90 to 10 for methyl trimethoxysilane (MTMS) to tetramethoxysilane (TMOS) was synthesized by using normal sol-gel process. Stoichiometric amount of water, which is considered to be in the ratio of 0.5 moles of water to one mole of alkoxy groups in the reactants, was utilized for all of the compositions. For comparison, two and four times of stoichiometric amount of water were also used in preparation for the molar composition of 75% and 25% for MTMS and TMOS, respectively. The general procedure is as follows. MTMS and TMOS of desired composition were mixed in a polymethylpentene (PMP) beaker and stirred for about 10 min. De-ionized water, whose pH value was adjusted to 2.00 with HCl solution, was added drop-wise over a period of five minutes into the precursor solution. Then the beaker was covered with a plastic film and the solution was stirred at room temperature for about one day. The sol was cast into polypropylene containers and allowed to dry open in an oven at 60 or 70 °C for two days. These samples were further heat treated at 100, 200, 300, 400, and 500 °C for different lengths of time as stated in the next section. Unless otherwise stated, characterizations were generally performed on the thermally treated samples which were obtained from the reactions of the corresponding alkoxides and stoichiometric amount of water.

RESULTS AND DISCUSSION

The synthesis of methyl modified silicates, carried out in the absence of solvent, involves the hydrolysis and cross-linking (or condensation) processes, the typical reactions in sol-gel chemistry, and can be illustrated in reaction 1 for the idealized product.

Mat. Res. Soc. Symp. Proc. Vol. 435

$$x\,\text{MTMS} + y\,\text{TMOS} + z\,H_2O \xrightarrow[\Delta]{\text{Catalyst}} (\text{MeSiO}_{1.5})_x(\text{SiO}_2)_y + w\,H_2O \quad (1)$$

The actual product could be rather complicated due to the presence of unreacted alkoxy groups and uncross-linked hydroxyl groups, which will be discussed later. After aging at 70 °C, hard and transparent sample could be obtained except that a white opaque sample was produced by the reactions of MTMS and TMOS (molar ratio = 75 to 25) with four times of stoichiometric water. The bulk density of the materials decreased with increasing MTMS content, changing from 1.44 g/cc for 50 mole% of MTMS to 1.28 g/cc for 90 mole% MTMS, as expected for increasing the lighter organic parts.

The thermal stability of the methyl modified silicates was determined by TGA and DTA methods under flowing air with the heating rate of 10 °C/min and found to depend upon the heat treatment and the reaction conditions. Figure 1 and 2 show the TGA curves of the 75 mole% MTMS material heat treated at 100 and 200 °C for 96 hours and at 200 °C for different times, respectively. It can be seen that the onset of weight loss and temperature increases with increasing the curing temperature and also with increasing the heating time, while the opposite is true for the weight loss of the decomposition process of the material. Normally the total weight loss for the decomposition of the materials is higher than the expected value for the idealized product, due to the presence of unreacted alkoxy and hydroxyl groups as well as trapped volatile components such as water or methanol. The sample treated at 100 °C for 96 hr had an onset temperature of 148 °C and a total weight loss of 19.3%, compared with the theoretical value of 8.07%, while the one heat treated at 200 °C for the same time showed an onset temperature of 285 °C and a weight loss of 12.7% (Fig. 1). The onset temperature of the sample cured at 200 °C varies from 240 °C by heating for 4 hours to 299 °C by heating for 192 hours with a concomitant decrease of the total weight loss from 16.6 to 11.5%, respectively (Figure 2). This indicates that the methyl group bonded to the silicon atom remains thermally stable up to about 300 °C. The thermal stability can also be improved by increasing the water amount in the reactions. For instance, the sample reacted with stoichiometric amount of water and treated at 200 °C for 96 hr exhibited an onset temperature of 285 °C with a total weight loss of 12.7% and the one reacted with two times of stoichiometric amount of water and treated at the same temperature for the same time showed an onset temperature of 317 °C with a weight loss of 11.0%. DTA data were consistent with the TGA results. The 75 mole% MTMS sample heat treated at 100 °C for 4 hours exhibited two exothermic peaks at 292 and 453 °C respectively (Figure 3a) while the same sample heat treated at 300 °C for 4 hours showed only one exothermic peak at 481 °C (Figure 3b). The peak at 292 °C may be caused mainly by the decomposition of unreacted methoxy groups and the elimination of water. The presence of methoxy groups in the sample cured at 100 °C for 96 hours has been confirmed by ^{13}C

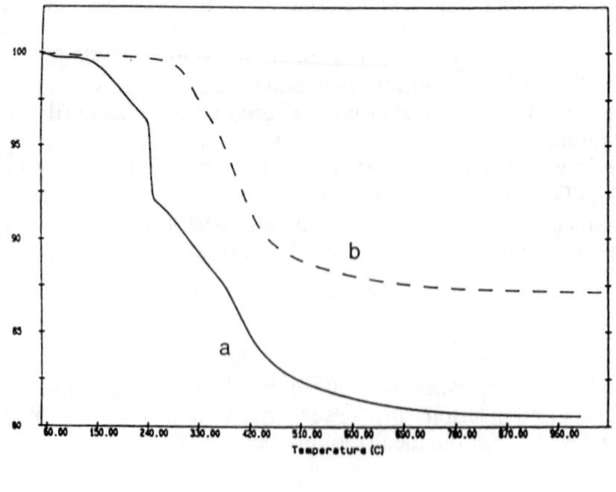

Figure 1. TGA curves of MTMS/TMOS (75/25) ORMOSIL sample cured at 100 (a) and 200°C (b) for 96 h.

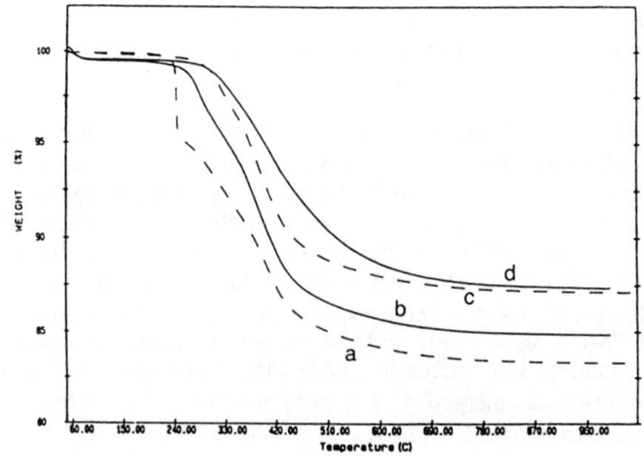

Figure 2. TGA curves of MTMS/TMOS (75/25) ORMOSIL sample heat treated at 200 °C for 4 h (a), 24 h (b), 96 h (c), and 192 h (d).

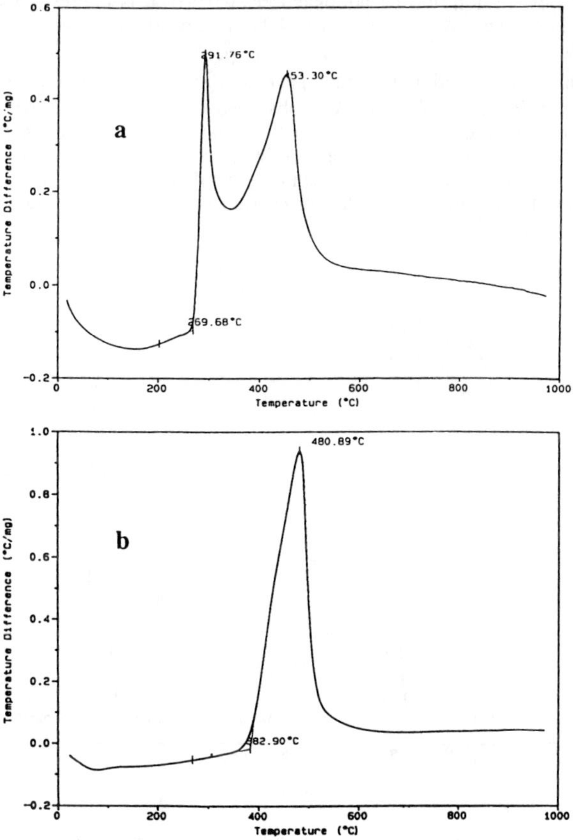

Figure 3. DTA curves of MTMS/TMOS (75/25) ORMOSIL sample heat treated at 100 (a) and 300 °C (b) for 4 h.

CP/MAS solid state NMR experiment. The peak at above 450 °C is owing to the burning out of the methyl group bonded to the silicon atom and the cross linking of the remaining hydroxyl groups.

To get a better understanding of the structural stability of the methyl modified silicate materials, ^{13}C and ^{29}Si CP/MAS solid state NMR experiments were carried out for some of the samples. Figure 4 shows the ^{13}C CP/MAS solid state NMR spectrum of the MTMS/TMOS (75/25) sample cured at 100 °C for 96 hours. Two peaks can be seen clearly. The one at 49.75 ppm is due to the unreacted methoxy carbon resonance while the one at -4.37 ppm is attributable to the silicon-methyl carbon resonance. ^{29}Si CP/MAS NMR spectra of the samples heat treated at 200 °C or below exhibited peaks for T^2, T^3, Q^2, Q^3, and Q^4 species, where T and Q represent the species derived from MTMS and TMOS respectively and the number indicates the condensed siloxane groups bonded to the silicon atom. With increasing the heating temperature, the intensity of the T^2 peak decreased gradually at first and merged with T^3 peak at 300 °C. This is because of the cross-linking of the species derived from MTMS at higher temperature. Both T^2 and T^3 peaks disappeared at 500 °C, as did the Q^2 peak. These results are consistent with the thermal analysis data. The detailed discussion will be published elsewhere. It is interesting to note that TGA analysis of the sample heated at 500 °C for 4 hours reveals a weight loss of 1.2% in the temperature region of 525 to 840 °C. However, no peaks can be identified on the ^{13}C NMR spectrum of the same sample and only a broad peak at -100.54 ppm containing Q^3 and Q^4 species due to the unsymmetric shape of the peak is observed on its ^{29}Si CP/MAS NMR spectrum, consistent with the complete decomposition of the organic parts such as methyl and methoxy groups. This indicates the presence of unreacted hydroxyl groups in this sample. Therefore, it seems likely that it is almost impossible to obtain fully condensed materials by the aforementioned method and thermal treatment without decomposition of the organic groups. New synthetic methods should be exploited.

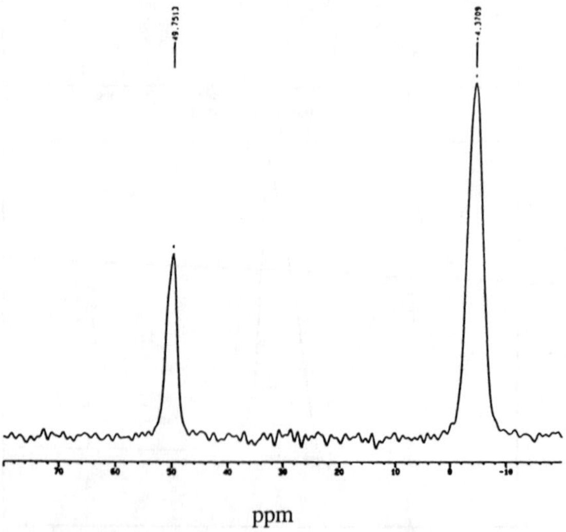

Figure 4. ^{13}C CP/MAS solid state NMR spectrum of MTMS/TMOS (75/25) ORMOSIL sample heat treated at 100 °C for 4 hours.

Fourier transform infrared (FTIR) spectra revealed that the intensity of the peaks at 1274 and 780 cm^{-1} for the methyl rocking and bending modes respectively decreased with increasing temperature and the intensity of the peak at 804 cm^{-1} for SiOSi vibrations increased with increasing temperature. These results are consistent with thermal analysis and solid state NMR data.

The microstructure of some of the samples were measured by atomic force microscopy (AFM). A dramatic change in the microstructure was observed by changing the process temperature. The small cluster units, which appear on the rough surface of the sample cured at lower temperature such as 60 °C, are cross-linked to form a smoother external surface at 100 °C and a much smoother surface at 200 °C. Thus, it is reasonable to assume that, if as-dried methyl modified silicate materials are used as hosts for solid state dye lasers, with slight increase in temperature during laser experiments, the material may structurally relax or thermally release unstable components causing photo instability.

CONCLUSIONS

The thermal stability of methyl modified silicate materials is dependent on the reaction and processing conditions and can be improved by increasing the heating time and the curing temperature as well as using excess water. Small amount of unreacted hydroxyl groups are still present in the materials after heating at 500 °C. Further study is required on how to obtain fully cross-linked materials without decomposition of the organic groups.

ACKNOWLEDGMENTS

The authors wish to acknowledge many scientists at YTCA for their kind assistance and support of this work.

REFERENCES

1. H. Schmidt, Mater. Res. Soc. Symp. Proc., **32**, 327(1984).
2. G. L. Wilkes, B. Orler and H. H. Huang, Polymer Prepr., **26**, 300(1985).
3. U. Schubert, N. Husing and A. Lorenz, Chem. Mater., **7**, 2010(1995) and references therein.
4. Y. Haruvy and S. E. Webber, Chem. Mater. **3**, 501(1991).
5. C. A. Capozzi and L. D. Pye, SPIE, **1513**, 320(1991).
6. Y. Huruvy, A. Heller and S. E. Webber in Supramolecular Architecture, Synthetic Control in Thin Films and Solids, ACS Symposium Series 499 (American Chemical Society, Washington, DC, 1992) p. 405.
7. A. B. Sheldon, C. A. Capozzi, S. L. Lana, T. A. King, X. Li and G. J. Gall, Sol-Gel Sci. Technol., **2**, 181(1994).
8. M. D. Rahn, T. A. King, C. A. Capozzi and A. B. Sheldon, SPIE, **2288**, 364(1994).
9. R. C. Chambers, Y. Haruvy and M. A. Fox, Chem. Mater., **6**, 1351(1994).

[illegible]

[illegible]

CONCLUSIONS

[illegible]

ACKNOWLEDGMENTS

[illegible]

REFERENCES

1. [illegible] Mater. Res. Soc. Symp. Proc. [illegible]
2. [illegible]
3. [illegible]
4. [illegible]
5. [illegible]
6. [illegible] Chem. Mater. [illegible]

SHRINKAGE AND MICROSTRUCTURAL DEVELOPMENT DURING DRYING OF ORGANICALLY MODIFIED SILICA XEROGELS

N. K. RAMAN[1], S. WALLACE[2], AND C. J. BRINKER[1,3]. [1]Center for Micro-Engineered Ceramics, University of New Mexico, Albuquerque, NM 87131. [2]Nanopore Corporation, 2501 Alamo Ave., SE, Albuquerque, NM 87106. [3]Sandia National Laboratories, Advanced Materials Laboratory, 1001 University Blvd., SE, Albuquerque, NM 87106.

ABSTRACT

We have studied the different driving forces behind syneresis in methyltriethoxysilane/tetraethoxysilane (MTES/TEOS) gels by aging them in different H_2O/EtOH pore fluids. We show using shrinkage, density, contact angle, and N_2 sorption measurements, the influence of gel/solvent interactions on the microstructural evolution during drying. Competing effects of syneresis (that occurs during aging) and drying shrinkage resulted in the overall linear shrinkage of the organically modified gels to be constant at ~50%. Increasing the hydrophobicity of the gels caused the driving force for syneresis to change from primarily condensation reactions to a combination of condensation and solid/liquid interfacial energy. In addition the condensation driven shrinkage was observed to be irreversible, whereas the interfacial free energy driven shrinkage was observed to be partially reversible. Nitrogen sorption experiments show that xerogels with the same overall extent of shrinkage can have vastly different microstructures due to the effects of microphase separation.

INTRODUCTION

The extent of gel shrinkage during aging and drying from solvents such as water and ethanol is determined by: (1) condensation or reesterification reactions that strengthen or weaken the gel network, (2) solid/liquid interfacial free energy, (3) coarsening mechanisms that reduce the solid/liquid interfacial area and increase the network modulus, (4) microphase separation that increases the average pore size in gels, and (5) the competition between the capillary pressure exerted by the pore fluid that serves to compact the gel and the network modulus that resists compaction [1,2]. Several applications including gas separation membranes require both small pore sizes and high pore volumes. The organic template approach in which ligands are embedded in a dense inorganic matrix and selectively removed to create pores offers the ability to tailor pore volume independently of pore size [3]. In addition to their role as pore size-directing templates, organic ligands introduced as pendant groups (alkyltrialkoxysilanes, RTES) modify the network by reducing the functionality and therefore the mechanical stiffness of the network, and make the network more hydrophobic. The hydrophobicity of the gel governs the interactions between the gel and the pore fluid during aging and drying, and strongly influences both the tendency for phase separation and the magnitude of the drying stress. A better understanding of these issues would obviously help in the rational control of xerogel microstructures.

Aging

Syneresis is defined as the spontaneous shrinkage of the gel during aging due to the expulsion of solvent from its pores [1,4]. Syneresis is known to occur by different mechanisms in inorganic or organic gels. Syneresis in inorganic gels is attributed mainly to continuing condensation reactions and coarsening, and the syneresis rate is minimized at isoelectric point of silica where the condensation rate is minimized [1]. Syneresis driven by condensation reactions is generally observed to be irreversible. For example, irreversible syneresis was observed in pure TEOS gels to increase from about 1% in pure ethanol to about 16% in pure water [4]. Coarsening occurs due to differences in the solubility between surfaces having different radii of curvature [1]. During the coarsening process, smaller particles dissolve and precipitate on larger particles or on necks between particles that have negative radii, hence, lower solubility. Coarsening does not produce shrinkage because the centers of the particle do not move toward one another, but reduces the interfacial area and increases the strength of the network [1]. Because of the low solubility of silica under acidic conditions, significant coarsening is not expected to occur in acid-catalyzed gels.

Syneresis could also result from the gel network attempting to reduce its huge solid/liquid interfacial area. Scherer [1,4] has derived an expression for the linear strain rate that occurs during the interfacial free energy driven syneresis. However, the shrinkage rate predicted by the equation

Mat. Res. Soc. Symp. Proc. Vol. 435 © 1996 Materials Research Society

was too fast, and based on syneresis studies on acid-catalyzed TEOS gels, it was concluded that the structural rearrangements needed to cause this shrinkage does not occur in inorganic gels. Syneresis driven by interfacial free energies is at least partially reversible. For example, organic gels are known to undergo reversible syneresis [5] with changing temperature, ionic strength etc. The shrinkage and the microstructural changes that occur during syneresis greatly impact the shrinkage and microstructural evolution during drying.

Drying

Syneresis is a self-limiting process. It slows down due to the decreasing network permeability and the stiffness of the gels which increases with shrinkage as a power law [1]:

$$K_p = K_0 (V_0/V)^m \quad (1)$$

where K_0 and K_p are the initial and instantaneous bulk modulus of the gel, V_0 is the shrunken volume, and m =3.0-3.8. To further densify the gel it is necessary to remove the pore fluid by drying. During the initial stages of drying, capillary tension develops in the pore fluid. The gel network, in response, shrinks to support this tension. Drying shrinkage stops when the increasing gel stiffness balances this tension. This balancing point is referred to as the critical point because it establishes the final (dried) pore volume and pore size. The strain at the critical point, ε, is given by [1]:

$$\varepsilon = \{(1-\phi_s)/K_p\} P_c \quad (2)$$

where ϕ_s is the volume fraction of solids and P_c is the maximum capillary stress at the critical point. During drying, if gelation precedes drying and K_0 is low (as in organically modified gels), the network may collapse sufficiently creating pores that are too small to empty at the relevant P/P_0, so P_c is given by a form of the Kelvin-Laplace equation,

$$P_c = (R_g T/V_m) \ln(P/P_0) \quad (3)$$

where, R is the ideal gas constant, T is the temperature, V_m is the molar volume of the pore fluid, and P_0 is the saturation pressure of the pore fluid. Because K_p increases due to both syneresis and drying shrinkage, the extent of drying strain (eq. 2) will vary approximately inversely with the extent of syneresis shrinkage.

For organically modified systems in general, the RTES/TEOS ratio has several effects on syneresis: (1) organic ligands reduce the connectivity of the network and thus its initial modulus K_0, (2) introduction of organic ligands causes the gel network to become progressively hydrophobic, and (3) organic ligands inhibit condensation reactions due to steric and solvation effects. The pore fluid water concentration has two related effects: (1) water is generally observed to promote condensation reactions, and (2) increased water concentration causes the solvent to become more polar reducing the solvent quality for hydrophobic polymers.

In this paper we report on the different driving forces behind syneresis in MTES/TEOS gels aged in different H_2O/EtOH pore fluids, and show using shrinkage, density, contact angle, and N_2 sorption measurements, the influence of gel/solvent interactions on the microstructural evolution during drying.

EXPERIMENTAL PROCEDURE

Sol and Bulk Gel Preparation

Sols were prepared by co-polymerization of methyltriethoxysilane (MTES) and TEOS using a two-step acid-catalyzed procedure, referred to as the A2 process [3]. The MTES/TEOS mole ratio was varied by introducing MTES as a percentage of total silicon in the sol. Xerogel (dry gel) bulk samples for shrinkage measurements were cast in cylindrical molds (length 90 mm, diameter 5 mm), sealed, gelled at 50°C and aged for 24-48 hr. The gel rods were then inserted into 22 mm x 175 mm glass tubes for pore fluid exchange and drying. The original pore fluid was replaced by immersing gels in H_2O/EtOH mixtures in a series of six steps. Seven pore fluid compositions between 0 and 100 vol% H_2O/EtOH were investigated. The change in length of the gels was measured during syneresis and drying at 50°C. The gel rods were subsequently dried in air at 150°C for 8 hr with a heating and cooling rate of 0.1°C/min prior to further characterization.

Bulk and Skeletal Density Measurements

Samples for bulk density measurements were prepared by polishing 15-20 mm cylindrical gel rods to a smooth finish using a Carbimet® 600 Grit Silicon Carbide grinding paper (Buehler, Lake Bluff, IL). The weight of the gel rods measured at 150°C was used in the bulk density calculation to avoid errors due to water condensation in the pores. The length and diameter of the samples were measured using a MAX-CAL digital vernier caliper (Cole-Parmer, Vernon Hills, IL) with an accuracy of ±0.03 mm. An average of three to four measurements was used to calculate the bulk densities. Gel samples for skeletal density measurements were prepared by grinding the gel rods into fine powder using a mortar and pestle. An Accupyc 1330 Helium Pyconometer (Micromeritics, Norcross, GA) was used to measure skeletal densities after outgassing at 150°C for 12 hr. in ambient air. An equilibration rate of 0.01 psig/min was used for the density measurements. An average of 10 measurements with a standard deviation of ±0.005 g/cc was used to calculate the skeletal densities.

Contact Angle Measurements

A VCA-2000 (AST, Bellerica, MA) sessile drop system with a video monitor and a tilting sample stage was used for contact angle measurements. The measurements were made on corresponding thin films prepared with identical compositions and dried at 150°C for 0.5 hr. Advancing contact angles were measured by adding liquids with a syringe. The image of the liquid drop was captured within 2 sec to avoid errors due to its reaction with the film surface. The measurements were performed in a closed environment that was shielded from dust and saturated with the solvent used to make the measurements. An average of six readings with an error of ±3° was used to calculate the contact angles. The polar ($\sim\gamma_p$) and dispersive ($\sim\gamma_d$) components of the interfacial free energies were calculated using the Owens-Wendt geometric mean method [6]. Contact angles of deionized water and 200 proof ethanol were used in the calculations.

Microstructural Characterization

An ASAP 2000 (Micromeritics, Norcross, GA) instrument was used to measure N_2 sorption isotherms of the bulk xerogels after outgassing at 150°C for 12 hr. at 0.01 torr. The apparent surface area (m^2/g) was calculated from the BET equation [7] using N_2 molecular cross sectional area = 0.162 nm^2 and the linear region between $0.05 < P/P_0 < 0.20$ that gave a least square correlation coefficient $R^2 > 0.9999$ for at least 4 adsorption points. The pore volume (cm^3/g) was calculated from the high P/P_0 portion of the isotherm where the volume of N_2 adsorbed was constant.

RESULTS AND DISCUSSIONS

Shrinkage results presented in Fig. 1(a) for MTES/TEOS gels as a function of increasing MTES and water concentration demonstrate the interplay of MTES/TEOS ratio and pore fluid water concentration on syneresis. At constant MTES/TEOS ratios, all the gels exhibited progressively higher syneresis with increasing pore fluid water concentration. For example, syneresis of the 55 mol% MTES/TEOS gels increased from about -1.5% in pure ethanol to about 25% in pure water. Increasing the MTES/TEOS ratio from 0 to 55 mol% caused the shrinkage to increase in pure water from 22% to 29%, while in pure ethanol it caused ~1.5% expansion. About 16% of syneresis was reversible for the 55 mol% MTES/TEOS gels aged in water.

The linear shrinkage of MTES/TEOS gels during drying is shown in Fig. 1(b) as a function of MTES/TEOS ratio and pore fluid water concentration. The linear shrinkage of the gels during drying exhibited an opposite trend compared to syneresis, consistent with the physical and chemical changes that occurred during syneresis. For example, shrinkage of the 55 mol% MTES/TEOS gels decreased from ~50% when dried from pure ethanol to ~20% when dried from pure water. However the overall shrinkage of the gels (including syneresis and drying) remained constant at ~50% (see Fig. 1(c)).

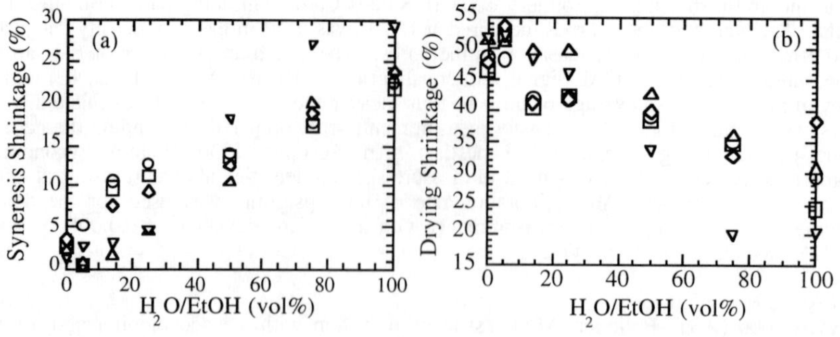

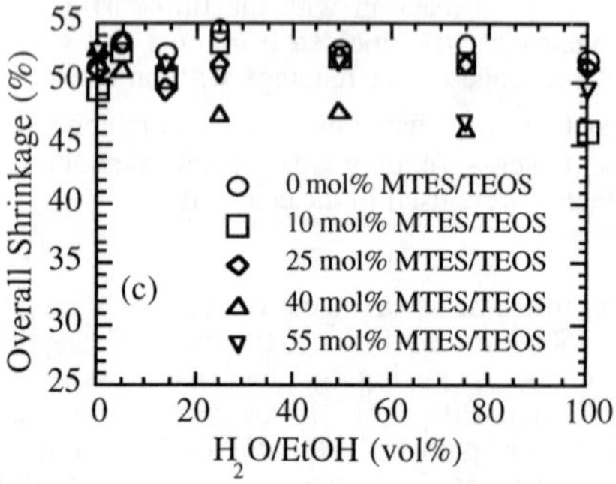

Fig. 1 Syneresis shrinkage (a), drying shrinkage (b), and overall shrinkage (c) of MTES/TEOS gels as a function of MTES/TEOS mole ratio and pore fluid composition, H_2O/EtOH (vol%).

Bulk densities of MTES/TEOS gels as a function of pore fluid water concentration are shown in Fig. 2. The bulk densities of all the gels ranged from 0.9 to 1.1 g/cc showing very little variation with vol% water in the pore fluid from which they were dried. The bulk density of 40 and 55 mol% MTES/TEOS gels decreased somewhat with increasing pore fluid water concentration following the trend in overall shrinkage of these gels. Skeletal densities of MTES/TEOS xerogels as a function of pore fluid water concentration are plotted in Fig. 3. The skeletal density of 0 mol% MTES/TEOS gels increased from 1.7 g/cc to 2.185 g/cc as the water concentration in the pore fluid increased from 0 to 100 vol%. However, as the MTES content increased the variation (increase) in skeletal density with vol% water progressively decreased until at 55 mol% MTES/TEOS there was almost no variation in skeletal density. These results imply that with increasing MTES/TEOS ratios the driving force behind syneresis changes from condensation reactions which should result in an increase in skeletal density to a combination of condensation and interfacial free energies. This is consistent with the observation that with increasing MTES/TEOS mole ratios the syneresis shrinkage in 100% water became progressively more reversible.

Advancing contact angles for various EtOH/water mixtures are plotted in Fig. 3, as a function of MTES/TEOS mole ratios. As seen in Fig. 3 all the films exhibited a zero advancing contact angle for pure ethanol. The contact angles for the films increased with increasing H_2O/EtOH ratios with the 55 mol% MTES/TEOS showing the highest water contact angles (~80°). γ_s was very similar for all the samples at about 10 dynes/cm whereas γ_p decreased from about 49 dynes/cm for the 0 mol% MTES films to about 20 dyne/cm for the 55 mol% MTES film. The decrease in γ_p with increasing MTES content is consistent with shrinkage and contact angle

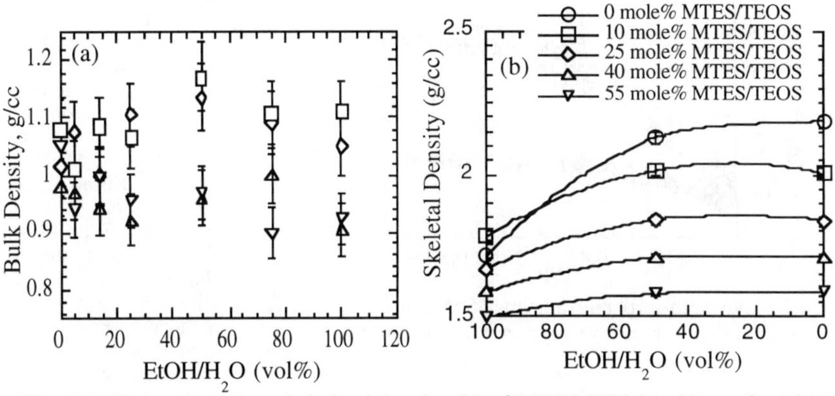

Fig. 2 Bulk density (a), and skeletal density (b) of MTES/TEOS gels as a function of pore fluid water concentration. The line is a visual guide.

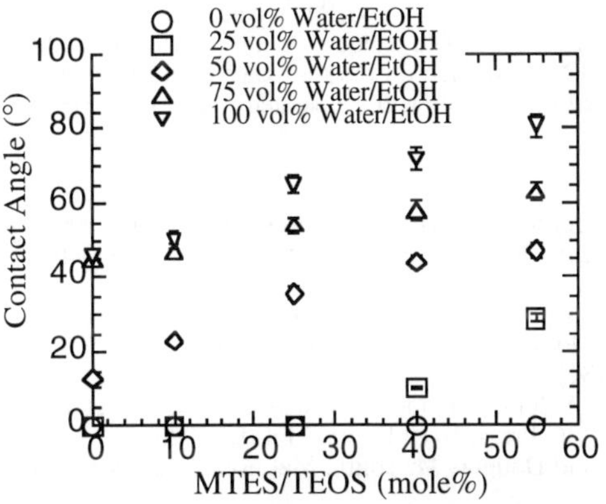

Fig. 3 Advancing contact angles of MTES/TEOS films as a function of MTES/TEOS mole ratio.

measurements which show the xerogels becoming increasingly hydrophobic. The total interfacial free energy of about 30 dynes/cm calculated for the 55 mol% MTES film is higher compared to a closed packed surface of methyl groups (21 dynes/cm [8]) indicating the presence of low levels of silanol groups in these films.

The effect of polymer/solvent interactions on xerogel microstructures is dramatically illustrated by the N_2 sorption isotherms shown in Fig. 4 for 55 mol% MTES/TEOS gels dried from EtOH, 50 vol% H_2O/EtOH and H_2O. Despite the fact that these gels exhibited about the same overall extent of drying shrinkage (Fig. 1(c)) they have considerably different microstructures. The isotherm of the gel dried from pure ethanol is of Type I, characteristic of microporous materials. In addition, for a 5 sec equilibration interval, the adsorption and desorption branch of the isotherm did not converge due to the slow kinetics of adsorption and desorption in the very small pores [9]. On increasing the pore fluid water concentration from 0 to 50 to 100 vol%, the isotherms change from Type I to Type IV that is characteristic of mesoporous materials, showing a net increase in the average pore size of the gels. The increase in average pore size is consistent with microphase separation similar to that illustrated in reference [1].

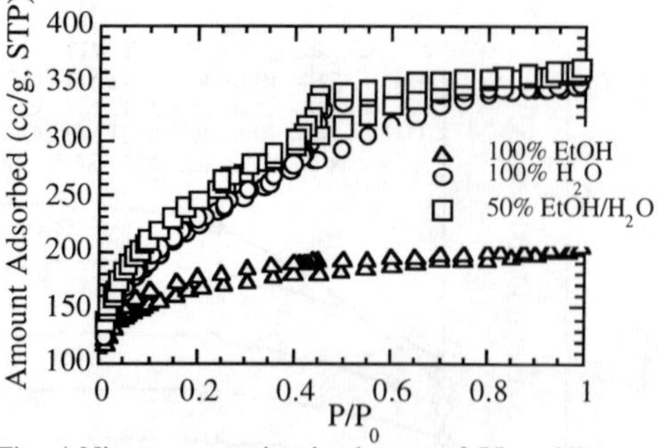

Fig. 4 Nitrogen sorption isotherms of 55 mol% MTES/TEOS xerogels versus pore fluid composition.

CONCLUSIONS

Syneresis in organic-inorganic gels results from both condensation and gel/solvent interactions and is sensitive to the organic/inorganic ratio and the pore fluid composition. The overall extent of shrinkage of MTES/TEOS gels dried from H_2O/EtOH compositions was constant at ~50%. With increasing hydrophobicity the driving force for syneresis changed from primarily condensation to a combination of condensation and interfacial free energies. For the same overall extent of shrinkage, the increase in average pore size with increasing pore fluid water concentration was attributed to microphase separation.

ACKNOWLEDGMENTS

Portions of this work were supported by the Electric Power research Institute, the National Science Foundation (CTS 9101658), the Gas Research Institute, and the Department of Energy - Morgantown Energy Technology Center. One author (NKR) wishes to acknowledge the University of New Mexico for the award of RPT grant to attend this conference. In addition, we are grateful for H. Naraghi of University of New Mexico for the shrinkage Measurements. Sandia National Laboratories is a U.S. Department of Energy facility operated under Contract No. DE-AC04-94AL85000.

REFERENCES

1) C. J. Brinker and G. W. Scherer, Sol-Gel Science (Academic Press, New York, 1990).
2) W. G. Fahrenholtz and D. M. Smith, Mat. Res. Soc. Symp. Proc. Vol. 271, 705-710, (1992).
3) N. K. Raman and C. J. Brinker, J. Memb. Sci., 105, 273, (1995).
4) G. W. Scherer, Mat. Res. Soc. Symp. Proc. Vol. 121, 179-197, (1988).
5) T. Tanaka, Sci. Amer., 244, 124, (1981).
6) Allan F. M. Barton, CRC Handbook of Solubility Parameters and Other CohesiveParameters, chapter 8, pages 159-161 and 430-432, (CRC Press, Boca Raton, Florida, 1983).
7) Stephen Wallace, C. Jeffrey Brinker, and Douglas M. Smith, Mat. Res. Soc. Symp. Proc.vol. 371, 241-246, (1995).
8) M. J. Owen, Silicon-Based Polymer Science: A Comprehensive Resource, Chapter 40, 705-739, (1990).
9) Y. Polevaya, J. Samuel, M. Ottolenghi, and D. Avnir, J. Sol-Gel Sci. Tech., 5, 65, (1995).

PREPARATION, CHARACTERIZATION AND PROPERTIES OF NEW ION-CONDUCTING ORMOLYTES

K. DAHMOUCHE, M. ATIK, N.C. MELLO, T.J. BONAGAMBA, H. PANEPUCCI, M. AEGERTER[1] and P. JUDEINSTEIN[2]

Instituto de Física de São Carlos, Universidade de São Paulo,CP 369, 13560-970, São Carlos-SP (Brazil)
[1]Institut für Neue Materialen (INM), Im Stadtwald, Gebäude 43, D 66123, Saarbrücken (Germany)
[2]Laboratoire de Chimie Structurale Organique, URA CNRS 1384, Université Paris-sud, 91405 Orsay (France)

ABSTRACT

Two families of hybrid organic-inorganic composites exhibiting ionic conduction properties, so called ORMOLYTES (organically modified electrolytes), have been prepared by the sol-gel process. The first family has been prepared from a mixture of 3-isocyanatopropyltriethoxysilane (IsoTrEOS), O,O' Bis (2-aminopropyl)polyethyleneglycol and lithium salt. These materials present chemical bonds between the organic (polymer) and the inorganic (silica) phases. The second family has been prepared by an ultrasonic method from a mixture of tetraethoxysilane (TEOS), polyethyleneglycol and lithium salt. The organic and inorganic phases are not chemically bonded. The Li^+ ionic conductivity has been studied by AC impedance spectroscopy up to 100 ^{0}C. Values of σ up to 10^{-4} Scm^{-1} have been found at room temperature. The conduction properties have been related to the materials structure using linewidth and relaxation times NMR measurements of ^7Li between -100 ^{0}C and 90 ^{0}C. A systematic study has been done changing the lithium concentration, the polymer chain length and the polymer to silica weight ratio. The structures and the ionic conduction properties of both families are compared with emphasis on the nature of the bonds between the organic and inorganic components.

INTRODUCTION

Many studies have been reported in the field of solid electrolytes [1] due to their potential for various applications such as batteries, energy and data storage, sensors, electrochromic and photochromic devices [2]. Systems showing conductivity induced by the addition of lithium offer the most favorable properties [3]. Recently, studies have been reported on hybrid materials obtained by reacting silicon alkoxydes with polyethers [4,5] showing that poly(ethylene)oxyde could act as a "solid" solvent for numerous chemical species, while the structural silica network reinforces mechanically the final material. Specific physical properties can be obtained by dissolving suitable doping agents within such network, for example lithium salt for ionic conductivity. The tailoring of the properties is strongly related to the connectivity of the two phases and the mobility of both the structural network and the active species.

This work reports on the preparation and structural properties of new hybrid ionic conductor SiO_2-Polyethyleneglycol with and without covalent organic-inorganic chemical bonds. The electrical properties of the materials and the mobility of the polymer chains and Li^+ ions have been determined by complex impedance spectroscopy and solid state NMR measurements.

EXPERIMENTAL

All chemical reagents are comercially available (Fluka, Aldrich). For the first family (type I) equimolar amounts of 3-isocyanatopropyltriethoxysilane (IsoTrEOS) and O,O' Bis (2-aminopropyl)polyethyleneglycol (i.e NH_2---PEG---NH_2) were stirred together in tetrahydrofuran (THF) under reflux for 6 hours. THF was evaporated and a pure hybrid precursor $_3$(OEt)Si---PEG---Si(OEt)$_3$ was obtained. 0.5 g of this precursor was mixed with 1ml of ethanol containing NH_4F (NH_4F/Si=0.005) to which was added desired quantities of the lithium salt ($LiClO_4$). Finally 0.2 ml of water was added under stirring. A monolithic gel was obtained in 4h. Ethanol was then slowly removed to give a piece of rubbery material. The existence of covalent chemical bonds between the silica network and the polymer chains has been reported [6,7].

The materials of the second family (type II) were prepared by an ultrasonic method: 12.5 ml of tetraethoxysilane (TEOS) and 4 ml of water were stirred together under ultrasound to hydrolyse the TEOS. Then the desired quantities of PEG_n (n=molecular weight of the PEG) were added in neutral PH conditions. The lithium salt ($LiClO_4$) was then introduced and dissolved under ultrasound in order to obtain a transparent monophasic liquid. Gelation occured in a few minutes and the samples were allowed to dry slowly as a monolithic piece. Studies of similar samples prepared by the classic Sol-Gel method have shown that only weak physical bonds exist in these materials [4,6].

The electrical properties of the samples were determined by complex impedance spectroscopy between 20 and 100 ^{0}C with a Solartron 1260 apparatus, in the range 1Hz to 10 MHz, with an applied voltage amplitude of 5 mV. The samples were pieces of monolith about 0.5 mm thick, with surfaces as flat as possible. The contacts were obtained with plasticized conductive probes (Altoflex) pressed on the sample. Stable and reproductive values of the ionic conductivity σ were obtained by conditioning the dried samples under vacuum at 90 ^{0}C for 24h.

^{7}Li solid state NMR spectra, consisting of only the central transition, were recorded between -100 to 90 ^{0}C at 32.9855 MHz using a TECMAG LIBRA system and a variable temperature double resonance Doty probe. The linewidths were measured from the Fourier transform of the Free Induction Decays (FID) obtained by a simple 10 μs $\pi/2$ excitation. The ^{7}Li spin-lattice relaxation time T_1 was measured by the inversion-recovery method. When necessary, ^{1}H decoupling was used during the FID acquisition to improve the signal to noise ratio, allowing ^{7}Li T_1 measurements over large temperature range. Full widths at half maxima (FWHM) were obtained by fitting the lineshape by Lorentzian or Gaussian functions, depending on the temperature ranges.

RESULTS AND DISCUSSION

Type I materials with covalent bonds between the organic and inorganic network

a) *Effect of the lithium concentration*: Figure 1 presents the variation of room temperature conductivity σ_{amb} for the chemically bonded hybrid silica-PEG_{800} as a function of the ratio [O]/[Li] (where the oxygens considered are only those of the ether type). The maximum (1.5 10^{-6} Scm^{-1}) is observed for [O]/[Li] $\cong$15, in contrast with the behavior observed in most polymeric systems in which σ_{max} ocurrs for [O]/[Li] = 8 [8,9].

For pure polymers [10], the increase of conductivity with the salt concentration is attributed to long-range coulomb interactions. The decrease after the maximum is a consequence of the immobilization of the polymer chains by interactions with the Li^+ ions. In our material, the saturation occurs at a lower concentration, showing that the PEG chains mobility is smaller when both networks are chemically bonded. The temperature variation of the conductivity follows strictly an Arrhenius law $\sigma = \sigma_o \exp(-E_a/RT)$ (fig. 2). The variation of E_a with lithium concentration shows a minimum (0.16 eV) also for [O]/[Li] around 15.

b) *Effect of the polymer chain length*: Table I presents the variation of 293 K ionic conductivity and of the activation energy as a function of the PEG chain length for chemically bonded hybrids silica-PEG with [O]/[Li] =15. The ionic conductivity σ_{amb} increases when the polymer chain length increases. It is well known that the conductivity of polymer electrolytes occurs via a liquid-like motions of cations through the segmental motion of the neighboring polymer chains [1]. Therefore, the PEG chain mobility increases as the chain length increases. Judeinstein et al [6,7] have studied by EPR the same samples without lithium and have shown that the mobility of the polymer chain near the silica clusters is much lower than the mobility of the chain at longer distance. Therefore an hybrid with short PEG chains offers

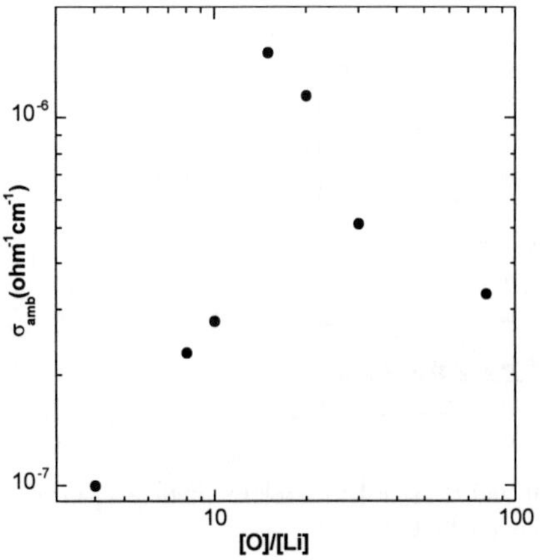

Fig 1. Variation of the room temperature conductivity, σ_{amb}, for the chemically bonded hybrid silica-PEG_{800} as a function of the [O]/[Li] ratio.

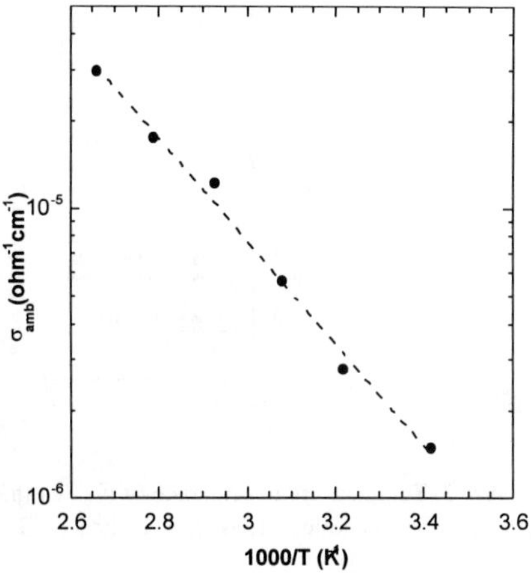

Fig 2. Temperature dependence of the ionic conductivity for the chemically bonded hybrid silica-PEG_{800} with [O]/[Li] ratio equal to 15.

only a small area of the chain with high mobility, while for longer chains the region of high mobility is much higher. This phenomenon is responsible of the ionic conductivity results. For all samples the activation energy is deduced from the temperature variation of the conductivity which follows strictly an Arrhenius law $\sigma = \sigma_0 \exp(-E_a/RT)$.

Table I: Variation of room temperature ionic conductivity and activation energy as a function of the PEG chain length for type I materials with [O]/[Li] =15.

	silica-PEG_{200}	silica-PEG_{800}	silica-PEG_{1900}
σ_{amb} (Scm^{-1})	1.3 10^{-7}	1.5 10^{-6}	1.2 10^{-5}
E_a (eV)	0.4	0.16	0.2

The effect of increasing the chain length on the mobility of the polymer in type I materials is also clearly observed in the temperature dependence of the NMR results (fig. 3). The motional narrowing of the ^7Li linewidth and the minimum of the ^7Li spin-lattice relaxation time T_1 occur at lower temperatures for longer polymer chains confirming that the polymer mobility increases as a function of the polymer chain length.

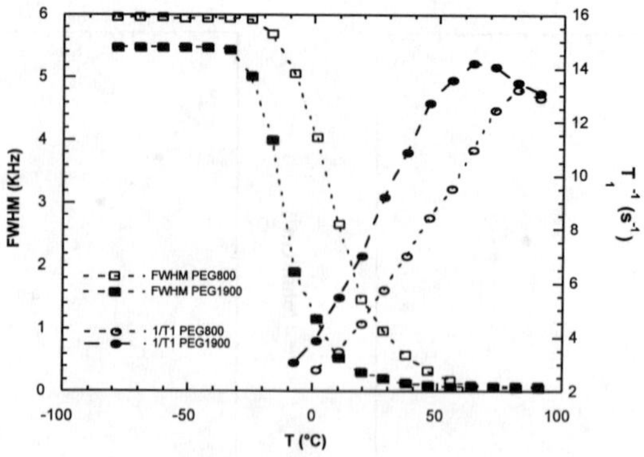

Figure 3. Temperature dependence of the ^7Li linewidth and ^7Li spin-lattice relaxation time T_1 in the type I materials: silica-PEG_{800} and silica-PEG_{1900}, both with [O]/[Li]=4.

Type II materials with weak physical bonds between the organic and inorganic network

a*) Effect of the lithium concentration*: Figure 4 presents the variation of room temperature conductivity σ_{amb} for the not chemically bonded hybrid silica-PEG_{300} as a function of the ratio [O]/[Li] (where oxygens considered are only those of the ether type). The maximum (9 10^{-4} Scm^{-1}) is observed for [O]/[Li] ≅ 8. This behavior is similar to that observed in most polymer electrolytes [8,9]. Here, only weak physical bonds (Van der Waals, hydrogen bonds) are present between the PEG and the SiO_2 network and the polymer chains have a higher mobility than in type I materials and their immobilization by interactions with the Li^+ ions occurs at higher lithium concentration.

b*) Effect of the polymer chain length*: Table II presents the variation of the room temperature ionic conductivity and activation energy as a function of the PEG chain length for not chemically bonded hybrids silica-PEG ([O]/[Li]=4). σ_{amb} and therefore the PEG chain mobility increase when the polymer chain length decreases. Because of the absence of "strong" chemical bonds between both networks the polymeric phase has a "liquid-like "behaviour [4,6]. With this configuration the interaction between the polymer chains themselves is smaller when the chains are short so that their mobility is higher than that of long PEG chains.

Again, the effect of the chain length on the mobility of the polymer in type II materials is shown by the NMR measurements. In this case, however, the motional narrowing of the ^7Li linewidth and the minimum of the ^7Li spin-lattice relaxation time T_1 which occur at higher temperatures for longer chains confirm that the chain mobility decreases with its length (fig 5)

Table II: Variation of room temperature ionic conductivity and activation energy as a function of the PEG chain length for type II materials with [O]/[Li] = 4.

	silica-PEG_{300}	silica-PEG_{600}	silica-PEG_{1000}
PEG/TEOS (weight %)	40	40	40
σ_{amb} (Scm^{-1})	9.2 10^{-5}	3.5 10^{-6}	8.0 10^{-7}
E_{a1} (eV)	0.05	0.05	0.2
E_{a2} (eV)	0.19	0.23	0.45

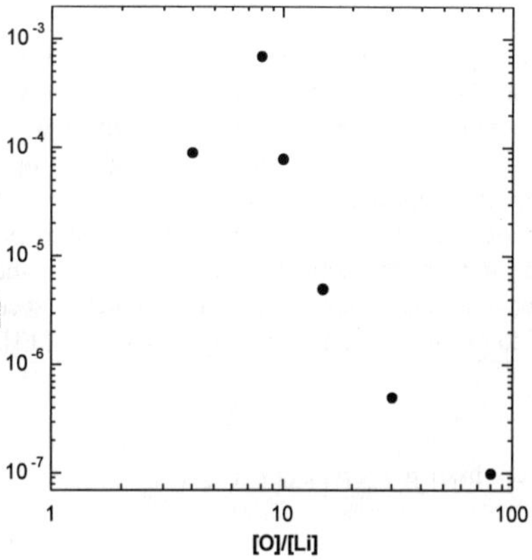

Figure 4. Variation of room temperature conductivity, σ_{amb}, for the non bonded hybrid silica-PEG_{300} as a function of the ratio [O]/[Li].

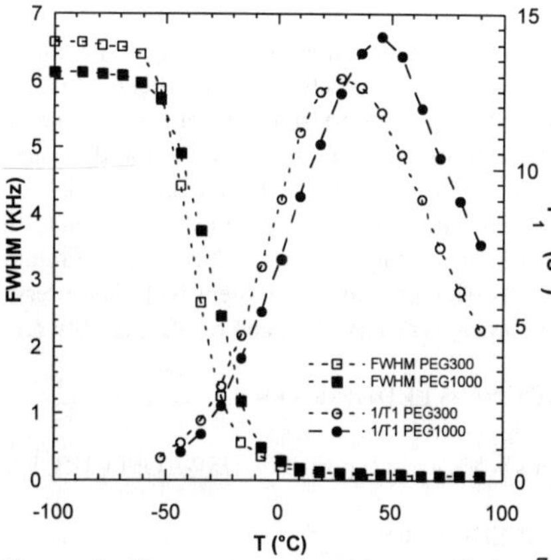

Figure 5. Temperature dependence of the ^{7}Li linewidth and ^{7}Li spin-lattice relaxation time T_1 in the type II materials: silica-PEG_{300} and silica-PEG_{1000}, both with [O]/[Li]=4

The variation of the conductivity with temperature shows discrepancies with a pure Arrhenius model. Two slopes are observed, with an intercept around 50 ^{0}C for all systems with activation energies 0.05 eV<E_{a1}<0.2 eV for T>50 ^{0}C and 0.2 eV<E_{a2}<0.5 eV for T<50 ^{0}C. Such behavior has also been observed in some polymers [11] and composite electrolytes [12] and interpreted as due to to a partial crystallization and phase separation below a critical temperature. The chemical stability of type II materials is therefore lower than that of type I.

c) Effect of the polymer concentration: Table III presents the variation of room temperature ionic conductivity and activation energy as a function of the PEG/TEOS weight ratio for not chemically bonded hybrids silica-PEG_{1000} ([O]/[Li]=4). The ionic conductivity σ_{amb} increases with the polymer concentration. In samples with high PEG concentration the influence of the silica network (Van der Waals or hydrogen interactions) is negligible for the "liquid-like" polymeric phase. When the volume of the PEG decreases, the influence of the interface between organic and inorganic domains becomes predominant and the mobility of the polymer chain is small. This influences the conductivity values.

Table III: Variation of room temperature ionic conductivity and activation energy as a function of the ratio PEG/TEOS for type II materials with [O]/[Li] = 4.

	silica-PEG_{1000}	silica-PEG_{1000}	silica-PEG_{1000}	silica-PEG_{1000}
PEG/TEOS (weight %)	20	40	80	100
σ_{amb} (Scm^{-1})	4.0 10^{-7}	8.0 10^{-7}	1.7 10^{-6}	3.3 10^{-6}
E_{a1}(eV)	0.2	0.2	0.14	0.14
E_{a2}(eV)	0.45	0.45	0.36	0.36

CONCLUSIONS

In type I materials the Li^+ ionic conductivity presents a maximum for [O]/[Li]=15, while for pure polymer electrolytes this occurs at [O]/[Li]=8. For family II the ionic conductivity increases with the polymer to silica weight ratio. The temperature dependence of the linewidth and ionic conductivity at room temperature for the various samples show that for the bonded chains the mobility increases with chain length, while the opposite happens with the unbonded chains. The linewidth transition temperatures correspond to the glass transition temperatures, determined by other techniques. Analysis of the behavior of the minimum of T_1 of the same samples confirms the above results which are also consistent with the observed conductivity values. For some samples, a second narrowing is observed around 60 ^{0}C, possibly related to the onset of an unbounded ion motion [13].

ACKNOWLEDGMENTS

This work was supported by USP-COFECUB, FAPESP, FINEP, CNPq and CAPES.

REFERENCES

1. F.M Gray, "Solid Polymer Electrolytes: Fundamental and Technological applications" (VCH publishers, New-York 1991).
2. M. Armand, Adv. Mater. 2, 278 (1990).
3. M.A. Ratner, D.F. Shriver, MRS Bull 39 (1989).
4. P. Judeinstein, J. Titman, M. Stamm and H. Schmidt, Chem. Mater 6, 127 (1994).
5. D.E. Rodrigues, A.B. Brennan, C. Betrabet, B. Wang, G.L. Wilkes, Chem. Mater. 4, 1437 (1992).
6. P. Judeinstein, M.E Brik, J.P. Bayle, J. Courtieu and J. Rault, MRS Symposium Proceedings 346 (San Francisco 1994), 937.
7. M. Brik, J. Titman, J.P. Bayle and P. Judeinstein, accepted in J.Polym. Sci. Polym. Phys. (1996).
8. M. Watanabe, S. Nagano, K. Sanui, N. Ogata, J. Power Sources 20, 327 (1987).
9. E. Tsuchida, K. Shigehara, Mol. Cryst. Liq. Cryst. 106, 361 (1984).
10. J. R. McCallum, A.C. Vincent in "Polymer Electrolytes Reviews", edited by J.R McCallum and A.C Vincent (Elsevier, London, 1987), 23
11. J.R Giles, C. Booth, R. Mobbs in "6th International Symposium on Metallurgy and Materials Science, Transport-Structure Relations in fast Ion and Mixed Conductors", edited by F.W. Hessel (1985) 329.
12. X. Huang, L. Chen, H. Huang, R. Xue, Y. Ma, S. Fang, Y. Li, Y. Jiang, Solid State Ionics 51, 69 (1992).
13. K. Dahmouche, M. Atik, N.C. Mello, T.J. Bonagamba, H. Panepucci, P. Judeinstein and M. Aegerter, accepted in J. of Sol-Gel Sci. and Technol. (1996).

MOLECULAR STATE AND MECHANICAL PROPERTIES OF EPOXY HYBRID COMPOSITE

M. HUSSAIN*, S. NISHIJIMA**, A. NAKAHIRA***, T. OKADA** and K. NIIHARA**

*Bangladesh Insulator and Sanitaryware Fact.Ltd, Bux Nagar, Mirpur, Dhaka-1216, Bangladesh
**ISIR Osaka Univ., 8-1, Mihogaoka, Ibaraki, Osaka, 567, Japan
***Dept. Inorganic Mater.,Kyoto Institute of Tech.,Matsugasaki,Sakyo-ku, Kyoto 606, Japan

ABSTRACT

A hybrid nanocomposite was prepared from the combination of an epoxy resin with a dispersion of nanometer-sized silica particles prepared by hydrolysis of silicon tetra alkoxides. A silane coupling agent was used to improve the filler-matrix interface by crosslinking into the epoxy networks. Silica dispersion caused the increase of nanometer size voids evaluated by the positron annihilation method and the decrease of Vickers hardness. The use of coupling agent also brought about the decrease of nanometer size voids and the increase in hardness.

INTRODUCTION

Inorganic oxide domains of the molecular or nanometer size dispersed in polymers have attracted much attention recently. Sol-gel processing is used to fabricate these types of nanocomposites. The key feature of this system is the introduction and crosslinking of inorganic groups from a silicon tetra alkoxide with organic groups in an epoxy matrix.

Nanocomposites derived from sol-gel processes are common in biological systems such as bones and shells[1]. Recently, nanocomposite concepts have found increasing attention for bio-ceramic materials[2]. The possibility of fabricating organic-inorganic hybrid composites for better mechanical properties at room temperature is of interest for a variety of applications[3-8]. The mechanical properties of these nanocomposites based on epoxy matrices at cryogenic temperature have been also found to be superior to those of epoxy itself.[9].

The macroscopic properties of polymers depend not only on their chemical chain structure but also on the network structure, molecular arrangements and polymer motions after polymerization. Thus, the incorporation of other groups into the polymer network strongly affects its macroscopic properties. In this work, silica was formed in the epoxy matrix to generate the novel epoxy nanocomposite by the sol-gel process. The molecular state of the nanocomposites was studied through the introduced voids into the nanocomposites.

EXPERIMENT

Materials

Tetraglycigyl-meta-xylen-diamin (TGMXDA), a tetra-functional epoxy resin was used in this experiment. The curing agent for the epoxy was 1,2-cyclohexanedicarboxylic anhydride (HHPA).The chemical structures are presented in Fig.1. To form nanometer-sized silica filler dispersion in the epoxy matrix, hydrolysis condensation of tetraethoxysilane (TEOS) was carried out with ammonium hydroxide in a solution containing the epoxy resin. After the silica was formed, the catalyst and the alcohol produced by the hydrolysis reaction were fully evaporated. The curing agent was added and the mixture was pre-cured at 363K and was post-cured at 453K for 2 hours in each case. Curing under 20 atmosphere pressure was also

Mat. Res. Soc. Symp. Proc. Vol. 435

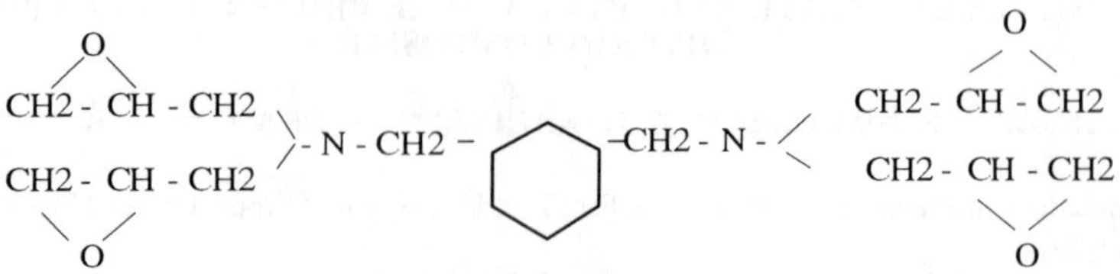

Epoxy Resin

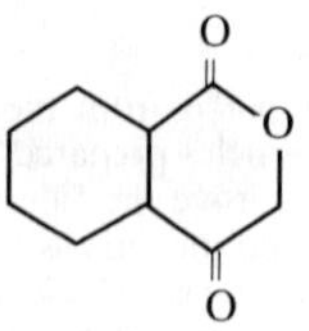

Curing Agent

Fig. 1 Chemical structure of epoxy resin and curing agent.

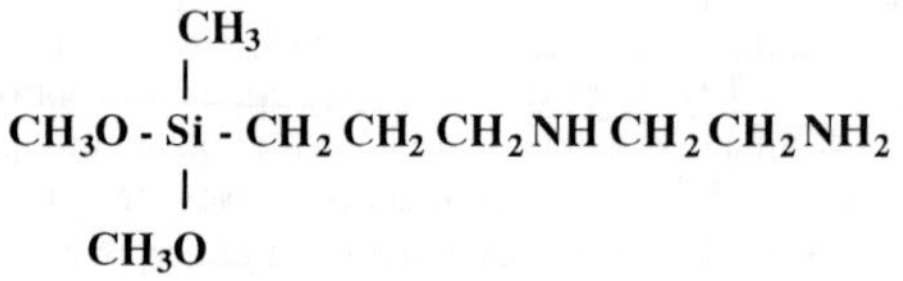

Fig.2 Chemical structure of N-β aminoethyl-γ-aminopropyl-methyl-dimethoxy-silane coupling agent

carried out to control the voids introduced by silica dispersion into the system. Amount of silica was also changed by changing the amount of TEOS in the solution and cured. The silica concentration was calculated by supposing all TEOS became silica.

In another specimen the coupling agent N-β aminoethyl-γ-aminopropyl-methl -dimethoxy - silane (Fig.2) was introduced before curing to crosslink between silica and epoxy network.

Evaluation of Free Volume and Mechanical Properties

In the hybrid materials voids will be introduced between silica and epoxy molecules during processing. These voids will be from nanometer to micrometer (or sometimes even larger) sized. Such large-sized voids are named "macrovoid" and may be determined by SEM observation. The nanometer-sized voids can be detected by the Positron Annihilation Lifetime method (PAL) which is often used to evaluate free volume in polymers [10-12]. It is important to understand that PAL cannot detect larger-sized voids (larger than several nanometers).

The positron lifetime measurement was performed by means of a fast-fast coincidence system. The positron source was 1μCi 22Na in the form of NaCl. The positron lifetime was determined as the time interval between the detection of 1.28MeV and 511 keV γ−rays. The

1.28 MeV γ−ray is emitted almost simultaneously with the emission of a positron while the 511keV γ−ray is generated on annihilation of a photon in the tested materials. In organic materials the obtained data was divided into three components where the long-lived component τ_3 corresponds to the size of free volume (or nanometer-sized void) in the materials. Thermal contraction , Vickers hardness and fracture toughness were also measure. Fracture toughness was measured by single edge noched beam method.

RESULTS AND DISCUSSION

Fig.3 shows the positron lifetime τ_3 as a function of silica concentration in the epoxy matrix. The long-lived lifetime τ_3 is related to the size of the unoccupied space (or void) by following formula,

$$1/\tau_3 = 2\{1-R/R_0+1/2\pi \sin(2\pi R/R_0)\}$$

where t_3 is the lifetime in nanoseconds, R is the void radius in angstroms and $R=R_0-\Delta R$ where $\Delta R=1.656$ Å. Consequently the increase in τ_3 means that in void. Here, the absolute value of void will not be discussed. As pointed out earlier, PAL can only sense small voids the order of several nanometers in diameter which are called "nanovoid" in this work .

It was observed that positron lifetime increased with silica concentration. Since the positron lifetime corresponds to the size of nanovoid in the polymer, such voids were found to be increased with the concentration of silica. Void size was also found to be reduced by pressurization. It was, therefore, suggested that the combination of silica dispersion and pressurization can control the nanovoids.

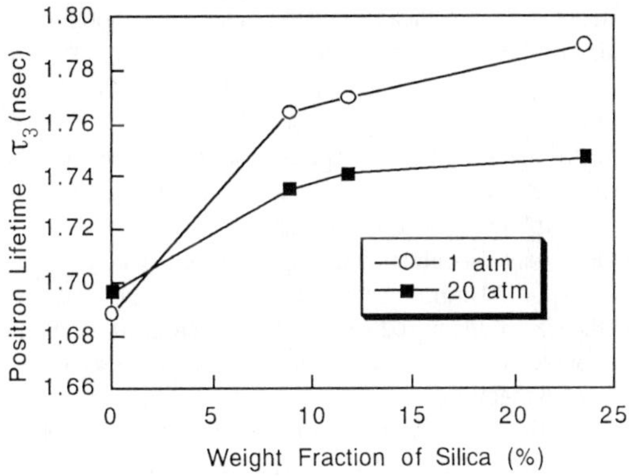

Fig.3 Positron lifetime as a function of silica content. Open and closed markings present the materials cured at ambient and 20 atm., respectively.

Table 1 Fracture toughness of hybrid materials without coupling agent cured at 1 and 20 atmospheres.

Amount of silica (% by weight)	K1c (MPa m1/2) cured at 1 atm.	K1c (MPa m1/2) cured at 20 atm.
0	0.45	0.55
3.7	0.62	0.61
4.9	0.55	0.54
9.4	0.44	0.53

In table 1 the fracture toughness of the hybrid materials are shown. First the fracture toughness of the materials increases with amount of silica and then decreases. Especially, the 3.7% silica containing material shows high fracture toughness. If a large number of macrovoids is introduced, fracture toughness will decrease. Since this is not the case, the number of macrovoids should be small in the 3.7% silica containing material. The increase of the fracture toughness of the material would be attributed to the presence of nanovoids as these will act as free volumes in polymers. Larger amounts of silica decreased the fracture toughness. This may be attributed to the formation of macrovoids.

Fig.4 shows the relationship between Vickers hardness and the positron lifetime keeping the amount of silica at 4.9% by weight. Table 1 indicates that the effect of macrovoids is small in the material with 4.9% silica. In this figure the amount of coupling agent is also shown. The epoxy material is TGMXDA. With increasing positron lifetime (size of the nanovoid) the hardness decreases. The silica dispersion in epoxy is thought to have brought about the increase in the size of the nanovoids (increase of lifetime) and the decrease of hardness. As the amount of coupling agent was increased, the size of the nanovoid decreased and at 0.49 % by weight the hardness of the hybrid material was almost the same as that of the epoxy itself. It was considered that the silica and epoxy molecules were crosslinked together by the coupling agent and hence the size of the nanovoid decreased. The pressurization was found not to be so effective in the coupling agent added materials.

The amount of coupling agent was given in the figure. Open and closed markings present the materials cured at ambient and 20 atm., respectively. There are two possible explanations of this feature depending on the size of the introduced void: the change of the Vickers hardness was induced by (i) the nanovoids or (ii) the macrovoids. In the latter case the nanovoid is thought to be accompanied with macrovoid.

The material without coupling agent in Fig 4 corresponds to the 4.9% silica added material in Table 1. This means that the 4.9% silica added material shows higher fracture toughness and lower Vickers hardness compared with epoxy. This cannot be explained only by the presence of macrovoids since macrovoids should decrease both fracture toughness and Vickers hardness. The possible explanation is that the nanovoids acts as free volume and hence the hybrid material becomes softer and tougher. This does not mean that no macrovoids are introduced. The number of the such voids is too small to decrease the fracture toughness.

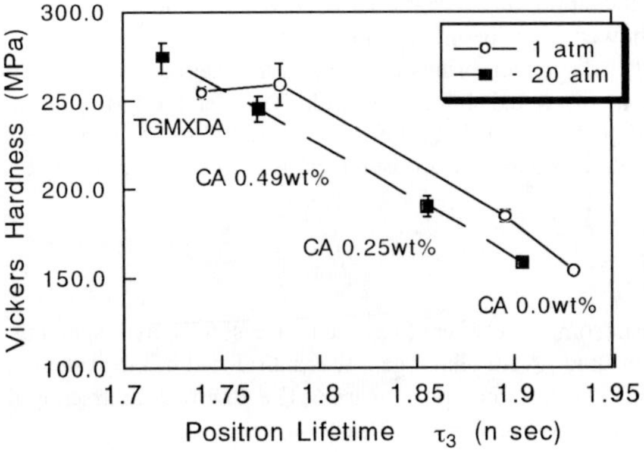

Fig.4 Relationship between hardness and positron lifetime.

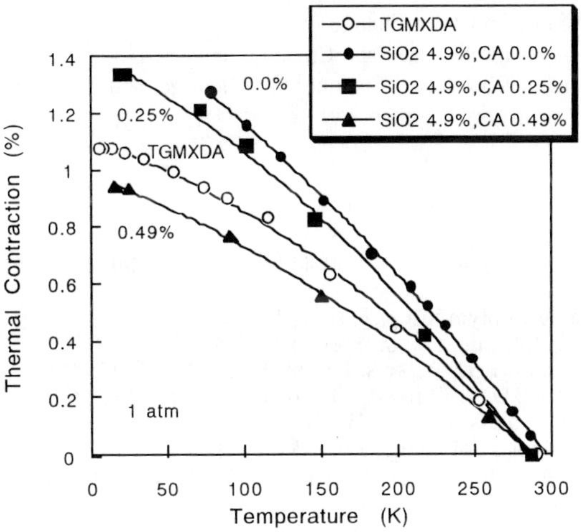

Fig.5 Thermal contraction of hybrid materials.

In Fig.5 thermal contraction down to LHeT of hybrid materials cured at 1 atm. keeping the silica concentration at 4.9 % by weight is shown. Higher thermal contraction was found in the specimen without coupling agent. Especially, the system without coupling agent (presented as closed circles) showed high thermal contraction compared with that of epoxy itself though usually inorganic filler dispersion brings decrease in thermal contraction. If the number of the macrovoids is small as discussed above, this could be understood to mean that the nanovoids act as free volume. If so Tg of this hybrid material is around room temperature. To confirm this Tg measurement should be done. This measurement is currentlyu being made.

CONCLUSIONS

Novel epoxy nanocomposites were fabricated successfully by dispersing nano-sized silica particles by sol-gel method. A coupling agent was introduced in the system to create interface bonding between the filler and the epoxy matrix. The thermal contraction down to cryogen temperature (or thermal expansion coefficient at cryogenic temperature) was decreased markedly. Possible mechanisms were proposed that is (i) nanometer size voids are introduced and they act as free volume or (ii) large size voids (macrovoids) are introduced. The macrovoids decreased the fracture toughness at higher silica concentration. The nanovoids would have important roles at lower silica concentration. The existence of the nanovoids was confirmed by positron annihilation lifetime method in this work.

ACKNOWLEDGEMENT

The authors are grateful to Dr. Y. Kobayashi in National Chemical Laboratory and Industry for his valuable discussion of positron lifetime measurement. This work was partly supported by cooperative work "Advanced Materials Creation and their Limit State Protection of Environmental Preservation" between Institute of Scientific and Industrial research and Welding Research Institute in Osaka University.

REFERENCES

1. H. J. Watzke and C. Dieschbourg, Adv. Colloid Interface Sci. **50**, 1 (1994).
2. P. Calvert, MRS Bull.Oct.37 (1992).
3. B. M. Novak, C. Davies, Polym. Prep. **32**, .512 (1991).
4. C. J. T. Landry, B. K. Coltrain, Polym. Prep. **32**, 514 (1991).
5. H. H. Hung, B. Orler and G. L. Wilkers, Polym. Bull. **14**, 557 (1985).
6. H. H. Hung, B. Orler and G. L. Wilkers, Macromolecules **20**, 1322 (1987).
7. C. Sanchez, J.Non-cryt.Solids **1**, 147 (1992).
8. A.B.Brennan, G.L.Wilkes., Polymer.,32,p.733 (1991)
9. S. Nishijima, K. Yamada, M. Hussain, A. Nakahira, Y. Honda, T. Okada and K. Niihara, Adv. Cryog. Eng. **46** (t o be published).
10. W. Brandt, Phys. Rev. **12**, 2572 (1975).
11. Q. Deng, C. S. Sunder and Y. C. Jeans, J. Phys. Chem. **96**, 492 (1988).
12. K. Okamoto, K. Tanaka, M. Katsube, O. Sueoka and Y. Itho, Radiat. Phys. Chem. **41**, 497 (1993).

ON-LINE SPECTROSCOPIC STUDIES OF GROUP IV ALKOXIDES AND THEIR INTERACTIONS WITH ORGANIC ADDITIVES DURING THE SOL-GEL PROCESS

D.WETTLING, S.TRUCHET, J.GUILMENT, O.PONCELET,
Kodak European Research Division, Laboratoire d'Analyses, Chalon/Saone, 71102 France

ABSTRACT :

The potential of vibrational spectroscopy for the study of group IV alkoxides $M(OR)_4$ has been demonstrated in several papers, but only a few of these papers have presented results from on line measurements. The monitoring of different reactions such as the stabilization of the alkoxides with organic additives, the exchange processes between different metal alkoxides (R exchange or M exchange) and the hydrolysis process can be of great importance for the development of new synthetic routes leading to materials which are easier to process.
NIR spectroscopy is a very versatile technique but lacks of specificity while IR and Raman give more interpretative results but are not always easy during processing. We used both techniques along with chemometric tools to extract relevant information on our processes. The 2D correlation allowed us to benefit from the specificity of IR and Raman to develop robust NIR methods which are able to be used on line to monitor the different steps of the sol-gel process.

INTRODUCTION :

Sol-gel processing [1] is a promising approach for the synthesis of glasses and ceramics. One of the main reasons for this interest arises from its high versatility (due to the rheological properties of sols and gels), which allows the easy fabrication of fibers, coatings and bulk materials. Sol-gel chemistry is based on inorganic polymerization reactions. Molecular precursors, mainly alkoxides [2] or carboxylates, are generally used as starting materials. A macromolecular network is then obtained via hydrolysis and condensation. However, transition metal alkoxides with a d° electronic configuration (Ti(IV), Zr(IV), Ta(V), Nb(V), ...) are very reactive towards hydrolysis, and adding water to such alkoxides quickly leads to precipitation which is not suitable for controlling the rheological properties and making films or fibers. Sols and gels must be stabilized in order to prevent precipitation. This can be done by using nucleophilic chemical additives [3] such as functional alcohols, polyols, carboxylic acids, β-diketones or alkanolamines. These additives react with alkoxides giving new molecular precursors with different structure, reactivity or solution behaviour.
Raman and Infrared spectroscopies are commonly used to study sol-gel materials, mainly during the last steps of the process [4]. We used FT-Raman and NIR spectroscopies to monitor different reactions such as the stabilization of the alkoxides with organic additives, the exchange processes between different metal alkoxides (R exchange or M exchange) and the hydrolysis process which can be of great importance for the development of new synthesis routes leading to materials easier to process. NIR spectroscopy is a very versatile technique but lacks specificity while IR and Raman give more interpretative results but are not always easy to run during processing. We used both techniques along with chemometric tools to extract relevant information on our processes. The 2D correlation allowed us to benefit from the specificity of IR and Raman to develop robust NIR methods which are able to be used on line to monitor the different steps of the sol-gel process.

Mat. Res. Soc. Symp. Proc. Vol. 435

EXPERIMENTAL :

All reactions were carried out in an atmosphere of pre purified argon at room temperature by using standard inert-atmosphere and Schlenk techniques [5].Pure alkoxides and chlorides were obtained from Aldrich or Strem Chemicals; CCl_4, iPrOH, AcacH were also obtained from Aldrich. All solvents were appropriately dried, distilled prior to use and stored under argon.
The FT-Raman spectra were recorded on a Bruker IFS66 (FRA106 accessory) using a Nd-Yag laser and a nitrogen cooled Ge detector. The NIR spectra were recorded on a Bruker IFS28/N. All the experiments were done using a 2 mm quartz circulation cell and CCl_4 as a solvent. Addition of water was done using a mixture of H_2O in iPrOH to avoid mixing problems between water and CCl_4. We used a resolution of 4 cm^{-1} for the Raman spectra and 8 cm^{-1} for the NIR spectra.
A complete description of the mathematical background of the self-modelling multivariate analysis (SIMPLISMA) as discussed here and its application to FT-IR and Raman spectra can be found elsewhere [6-7]. The principle is that Principal Componant Analysis (PCA) determines independent linear combinations of the original variables which describe the maximum variance in the data set.
The mathematical formulation is as follow :

$$D\, c{*}v = S\, c{*}f\, L'f{*}v$$

where c is the number of cases (spectra), v the number of variables (wavenumbers) and f the number of principal components (PC's). D is the original matrix which is expressed by PCA as a product of a score matrix S and a loading matrix L. The score matrix S contains the contributions of the cases in the principal components (PC's). The loading matrix L contains the contributions of the variables in the PC's. The first PC (PC1) describes the maximum possible variance in the data set, PC2 describes the maximum of the residual variance , etc... From a chemical point of view, S contains the concentrations of the pure components in the spectra and L contains the spectra of the pure components.

RESULTS AND DISCUSSION :

Chlorides and chelating agents are commonly employed to retard the hydrolysis and condensation pathway of the evolving polymer. We studied each reaction with both NIR and FT-Raman techniques and then extracted the information on the nature of the compounds and their relative concentration profile using SIMPLISMA. A 2D correlation between the two spectroscopic techniques was also carried out to try to interpret the NIR spectrum based on the FT-Raman one in the case of the complexation of $Ti(OiPr)_4$ with AcacH.

Chelating agents : acetylacetone

Addition of AcacH to $Ti(OiPr)_4$ up to 8 equivalents of AcacH per Ti leads to changes in the FT-Raman spectra in the ν(TiOC) and ν(CO) bands region indicating the interactions between the acetylacetone and the titanium alkoxide. These bands are not directly observed in the NIR spectra but perturbations can be seen on the bands in the first and second overtone. These bands are usually attributed to overtones of CH vibrations or combinations with CH vibrations perturbated by the surrounding medium.
Using SIMPLISMA, we have been able to elucidate 3 compounds appearing during the reaction with both the Raman and NIR sets of spectra. Based on the extracted spectra (figure 1), one can identify the starting material and the final material, where all the isopropoxide seems to have been exchanged, and an intermediate compound which contains iso-propanol as a constituant.

This later compound is probably a complex between Ti, AcacH and iPrOH. SIMPLISMA also allows us to calculate pseudo concentration profiles as shown on figure 2. These can become quantitative if they are corrected by the molecular absorptivity of each compounds.

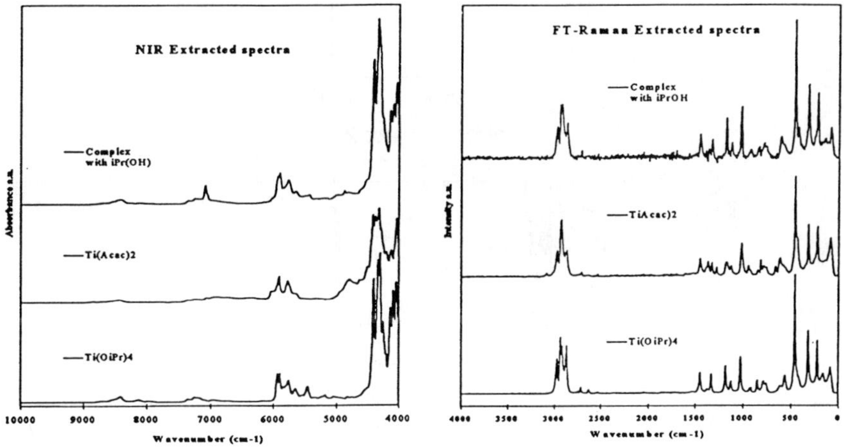

Figure 1 : NIR and FT-Raman spectra extracted using SIMPLISMA

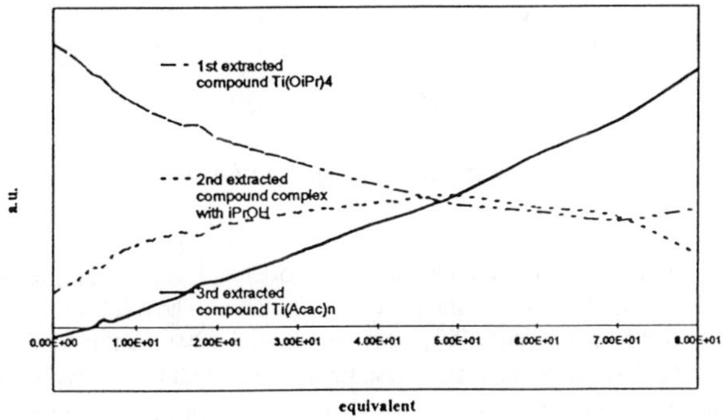

Figure 2 : SIMPLISMA concentration profiles calculated from NIR data

Finally, 2D correlation between the Raman and the NIR spectra can be used as a tool to interpret the NIR spectra or at least to identify the NIR band which gives an independant identification of the different compounds. It must be noted that most of the time, the major bands in the spectra are not pure or are shifted due to the different interactions and therefore it can be very useful to have an independant band to monitor a reaction.

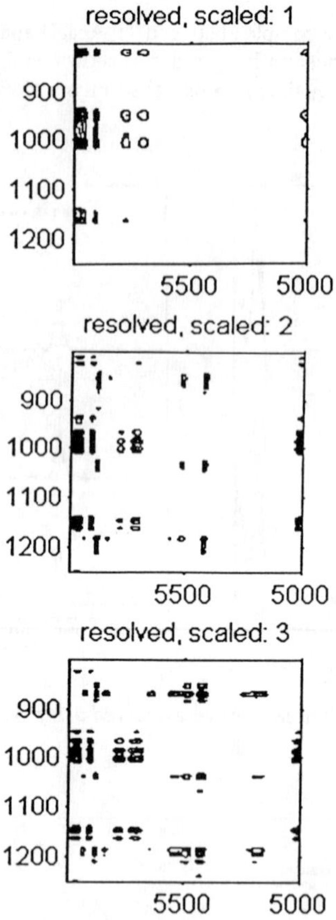

Figure 3 : Correlation between FT-Raman and NIR spectra for the 3 extracted compounds using SIMPLISMA

Figure 3 shows the 2-D correlation between the 800-1250 cm^{-1} region of the Raman set of spectra and the 5000-6000 cm^{-1} region of the NIR set of spectra. The 5452 cm^{-1} NIR band can be taken as a reference band for the $Ti(OiPr)_4$ starting material whereas the 5907 cm^{-1} band is more representative of the Ti(Acac)n species. These bands are linked respectively to 1026 cm^{-1} ν(CO) band from $Ti(OiPr)_4$ and 1173 cm^{-1} band from Ti(Acac)n in the Raman spectra. The iPrOH complex type bands are 818 cm^{-1} in Raman and 5292 cm^{-1} in the NIR spectra.

Hydrolysis :

The sol-gel process is based upon hydrolysis and condensation reactions so it is important to study the behaviour of molecular precursors during their hydrolysis. In our case, we started with a Ti(OiPr)4 alkoxide stabilized with 2 equivalents of AcacH in CCl_4. Then water was added using a mixture of isopropanol and water to allow the migration of the water molecules to the alkoxide environment.

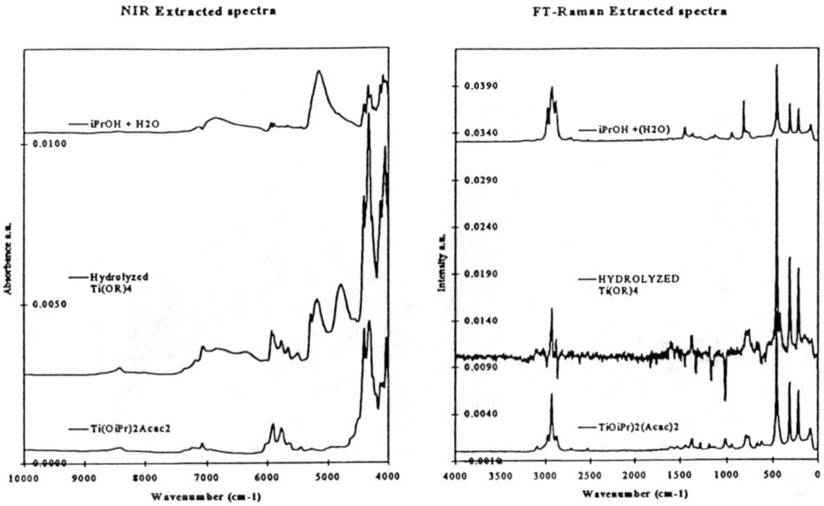

Figure 4 : NIR and FT-Raman spectra extracted using SIMPLISMA

Figure 4 presents the NIR and FT-Raman extracted spectra from the hydrolysis experiments. The complementary nature of the two techniques can be seen in this figure the complementarity of the 2 techniques as the NIR spectra are mostly based on the OH and CH vibrations while the FT-Raman spectra are not sensitive to OH vibrations but clearly show Ti-O and C-O vibrations and the perturbations associated with it. The hydrolyzed species are not clearly seen in the Raman spectrum due to the high hydroxide environment of the Ti atoms. In the NIR spectrum, these species are the main compounds.
The 2-D correlation in that case is more difficult to obtain as the main vibrations are not of the same nature.

Chloroalkoxides.

Metal chloroalkoxides can be used as starting materials in sol-gel process. Generally, they are less sensitive to hydrolysis than homoleptic alkoxides and therefore more easy to handle. These products can be prepared by the reaction of metal alkoxides with halides or hydrogen halides or by the reaction of metal chlorides with alcohols. Metal chlorides of group (IVb) react partially with alcohols to form solvated chloroalkoxides. We have chosen to synthetize these compounds by redistribution reactions between titanium alkoxides and titanium chlorides.

$$Ti(OR)_4 + x/4\ TiCl_4 \longrightarrow 2\ Ti(OR)_{4-x}Cl_x\ ,\ x = 1,2,3 \text{ ou } 4$$

The use of pure products (chloride and alkoxide) avoids the formation of solvated species and simplifys the interpretation of the spectra.

Figure 5 shows the evolution of ν(Ti-O-C) stretching with the addition of $TiCl_4$. The presence of chloride in the coordination sphere of titanium (IV) alkoxides has a strong effect on the ν(Ti-O-C) and ν(Ti-O)bands. These shifts can be due to an increase of molecular complexity (from monomeric species to dimeric or more) in the case of titanium (IV) isopropoxide.

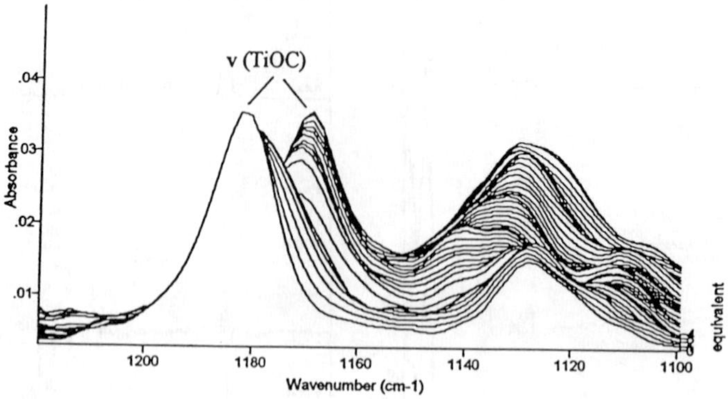

Figure 5 : Ti(OiPr)4 + TiCl4 : FT-Raman spectra

CONCLUSION

Vibrational spectroscopy is a promising analytical technique which can provide data on the first steps of sol-gel process without altering it. This method can provide information on starting materials, mainly alkoxides or modified alkoxides and confirm the significance of the choice of alkoxide in sol-gel technology. The great versatility of FT-Raman spectroscopy allows its use as an analytical tool. The results given by this technique, correlated with other analytical data, increases our knowledge of the sol-gel process. The NIR spectroscopy is complementary to FT-Raman and it is a lot easier to use in a production environment.
Using SIMPLISMA and 2-D correlation between NIR and Raman spectroscopy, one can use the interpretative power of the Raman to explain the NIR and then develop better calibrations to follow either the synthesis or hydrolysis and condensation reactions.
The results shown in this paper are preliminary results presented only to demonstrate the concept. A better interpretation of the data would be needed to implement the technique on a routine basis.

REFERENCES

[1] C.J. Brinker, G.W. Scherer, The Physics and Chemistry of Sol-Gel Processing, Academic Press, San Diego (1990).
[2] D.C. Bradley, R.C. Mehrotra and D.P. Gaur, Metal Alkoxides, Academic Press, London (1978); L.G. Hubert-Pfalzgraf, New J. Chem., 11, 663 (1987).
[3] J. Livage, M. Henry and C. Sanchez, Progress in Solid State Chemistry, 18, 259 (1988).
[4] C.C. Perry, Xianchun Li and D.N. Waters, Spectrochim. Acta, 47A (9/10), 1487 (1991); J. Chem. Soc. Faraday Trans., 87(15), 761 (1991); A. Takase and K. Miyakama, Jpn. J. Applied Phys., 30(8B), L1508 (1991); M. Aizawa, Y. Nosaka and N. Fujii, J. Non-Cryst. Solids, 128, 77 (1991).
[5] C.S. Rondestvedt, Encyclopedia of Chemical Technology, Kirk-Othmer, Wiley Interscience, New-York, 23, 176 (1983).
[6] W.Windig, J.Guilment, Analytical Chemistry, 63(14), 1425, 1991.
[7] J.Guilment, S.Markel, W.Windig, Applied Spectroscopy, 48(3), 1994.

A TG/GC/MS STUDY OF THE STRUCTURAL TRANSFORMATION OF HYBRID GELS CONTAINING Si-H AND Si-CH_3 GROUPS INTO OXYCARBIDE GLASSES

G. D. SORARÙ, R. CAMPOSTRINI, G. D'ANDREA, S. MAURINA
Dipartimento di Ingegneria dei Materiali, 38050 Trento, Italy, soraru@ing.unitn.it

ABSTRACT

The pyrolytic transformation of hybrid gels containing Si-CH_3 and Si-H groups is studied with a novel TG/GC/MS technique. Pyrolysis of a gel precursor for SiOC system shows a main decomposition step with the evolution of many different silanes and siloxane species arising from redistribution of Si-H and Si-O bonds in the polymeric network. B-containing gel, precursor for SiBOC oxycarbide glasses, displays a dramatically different pyrolysis pathway in which the redistribution reactions are completely suppressed and the Si-H and Si-CH_3 groups are partially consumed at low temperature (≈ 220°C) to form H_2 and CH_4.

INTRODUCTION

Silica-based, hybrid organic-inorganic materials, in which the inorganic network is modified by organic groups have been recently successfully synthesized by the sol-gel process [1]. They are currently studied with the aim of producing a new class of molecular-level-composites that combines the features of ceramic and organic materials.

More recently, organic-modified silica gels have been recognized as excellent precursors for novel amorphous solids in the ternary SiOC system, i.e. silicon oxycarbide glasses [2,3]. These glasses constitute an anionic modification of the silica network in which tetracoordinated C atoms partially substitute for divalent O atoms therefore leading to a more interconnected inorganic structure with improved thermal and mechanical properties [4,5].

The hybrid silica-based gels can be converted into the SiOC glasses by a pyrolysis process under inert atmosphere at elevated temperatures (≈900°C). Usually, the pyrolytic transformation does not lead to the complete incorporation of the organic C atoms present in the precursor gel into the final oxycarbide network and an appreciable amount of "free" carbon is also formed in the ceramic matrix [6]. It has already been shown that the presence of Si-H bonds in the pre-ceramic gel can strongly reduce the formation of the "free" carbon phase [6-7].

The pyrolytic transformation of hybrid silica gels into the silicon oxycarbide glasses is a very complicate process that involves several reactions with the evolution of various gases. In order to better understand the pyrolytic transformation with the ultimate aim of controlling the structure of the oxycarbide glass, several studies have been reported on the characterization of the solid residue using different spectroscopic techniques, such as MAS-NMR and XPS [8,9].

Analysis of the gaseous species evolved during pyrolysis can also greatly contribute to the understanding of the transformation mechanisms. Experimental data on the pyrolysis of hybrid silica gels to oxycarbide glasses using a conventional thermogravimetry equipment coupled with a mass spectrometric analysis (TG/MS) have been already reported in the literature [10,11]. However, detailed analysis of the mass spectra can be very complicated when many gaseous species are evolved in the same temperature interval. For this reason a gas chromatographic separation of the pyrolysis gas before the usual mass spectrometric analysis (TG/GC/MS) can greatly enhance the potentiality of the technique, allowing a better identification of the different molecular species [12]. In this TG/GC/MS study we compare the pyrolytic transformation of Si-CH_3 and Si-H containing hybrid gels precursors for SiOC and SiBOC oxycarbide glasses. The ultimate aim of this work is to elucidate the role played by the extra element (boron) in the pyrolytic synthesis of multicomponent silicon oxycarbide materials.

EXPERIMENT

Silicon and boron alkoxides were purchased from ABCR and Aldrich respectively and used as received. Gel A, precursor for SiOC glasses was prepared from a 2/1 mixture of triethoxysilane, $HSi(OEt)_3$, (T^H) and methyldiethoxysilane, $H(CH_3)Si(OEt)_2$, (D^H) using acidic water (H_2O/OEt = 1; pH=4.5) and ethanol as solvent (EtOH/Si = 2) [7]. Gel B, precursor for

Mat. Res. Soc. Symp. Proc. Vol. 435 © 1996 Materials Research Society

SiBOC glasses, was prepared from the same 2/1 T^H/D^H solution used for the preparation of Gel A and adding, after ≈10 min of hydrolysis, triethylborate, $B(OEt)_3$ (B/Si = 1). The two gelling solutions were poured into test tubes and left open for gelation at room temperature. Gels were dried by heating the samples for 2 days at 80°C.

Thermal analysis was performed with a Netzsch STA 409 equipment operating in He flow (100 ml/min) with an heating rate of 10°C/min up to 1000°C. Gas chromatographic analysis was performed on a HRGC (Carlo Erba) using He as carrier gas. Light hydrocarbons, alcohols and low-weight silanes were eluted in a PoraPLOT Q capillary column, whereas high-weight siloxanic species were separated in an OV1 capillary column. Mass spectrometric analysis was performed using a VG QMD quadrupole mass spectrometer interfaced with the gas-chromatograph. Electron impact mass spectra (70 eV) were continuously recorded (1 scan/sec) from 3 to 400 amu. Two different types of interfaces were used in this work. In the first case, a usual TG/MS coupling was obtained through an empty silica capillary column, heated at 120°C, that was placed above the sample crucible in the thermobalance and allowed the continuous direct introduction of the pyrolysis gases into the ionization chamber of the mass spectrometer. This configuration permits detection of evolved gaseous species from the trend of Total Ion Current (TIC) curve obtained from the contribution of each recorded scan. Moreover, it allows to follow the ionic current of a single m/z value, which in some cases can represent a specific molecular species, as a function of the pyrolysis temperature. TG/GC/MS coupling was obtained connecting the thermobalance with a thermostated micro sample valve Bimatic GR-8 with loops of ≈100 μl. This valve allows the injection of the gaseous species evolved at a specific temperature into the gas-chromatographic column for their separation before mass spectrometric analysis. Details of the experimental set-up are reported elsewhere [12].

RESULTS

Monolithic transparent hybrid gels were always obtained regardless the composition. The two studied gels have the same T^H/D^H = 2/1 ratio. The boron content in Gel B measured by chemical analysis (B/Si = 0.12) was lower then the nominal one (B/Si = 1). Evaporation of B-containing compounds from gelling solutions is a well known phenomenon [13]. Indeed, for Gel B we observed the formation, on the internal wall of the test tube, of a white deposit that was identified, by FT-IR analysis, as boric acid.

Thermogravimetry (TG) and its derivative (DTG) curves recorded on Gel A are reported in Figure 1. Gel A is stable up to 200°C, it shows a small weight loss (1.0%) from ≈200 to ≈400°C, a main weight loss (6.0%) from ≈400 to ≈500°C and a third one (1.6%) from ≈600 to ≈900°C. DTA analysis resulted in a featureless curve. Direct TG/MS investigation is in agreement with the TG/DTG results: the TIC curve (Figure 1) shows a main peak centred at 450°C and three smaller ones at 240, 320 and 640°C, respectively. The main products at 240 and 320°C are water and ethanol resulting from the condensation of the remaining -OH and -OEt groups.

Analysis of the main weight loss at 450°C is not so straightforward. A mass spectrum recorded at 450°C (Figure 1) revealed the presence of many ionic fragments ranging from m/z = 14 up to m/z = 182. Such a rather complicated mass spectrum, although containing clear indications of the presence of silanes and siloxane species, prevent a detailed characterization of the different molecular compounds present in the pyrolysis gas. Indeed, the TG/GC/MS analysis clearly unveiled the complexity of the gas mixture as can be seen from the gas-chromatographic elutions reported in Figure 2. Elution in the PoraPLOT Q column shows the presence of SiH_4 and CH_3SiH_3 (relative retention time, $t_R \equiv 1$ and 3.8 min respectively) and, in lower amounts CH_3CH_3 (t_R = 1.3), H_2O (t_R = 4,6) and EtOH (t_R = 31.6). The analysis of the same gas mixture in OV1 capillary column is even more impressive: it shows the presence of several peaks that have been assigned as follows [13]: the first peak ($t_R \equiv 1$), includes the simultaneous elution of SiH_4, CH_3SiH_3 and CH_3CH_3. Peaks at t_R = 1.5 and 2.5 having a relative intensity of about 20 % compared to the first one, have been attributed, after a detailed analysis of the fragmentation patterns, to $H_2Si(OH)$-OEt (t_R = 1.5) and $H_2Si(CH_3)$-OEt (t_R = 2.5) species arising from the incomplete hydrolysis and/or condensation of T^H and D^H precursors. The peak with t_R = 2.1 is due to the fragment $H_2Si(CH_3)$-O-$Si(CH_3)H_2$ arising from the co-condensation of two D^H units. Finally, other species are detected, even if their intensity can be estimated to be one order of

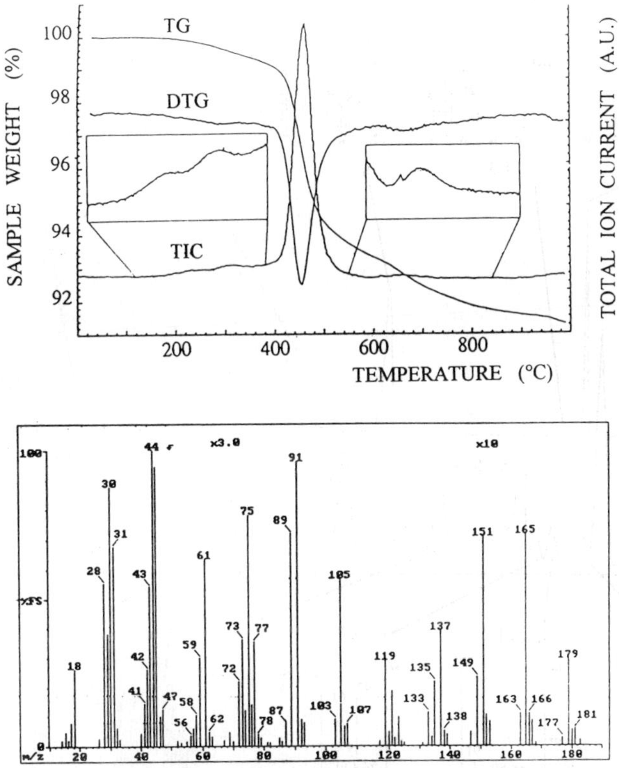

Figure 1. TG, DTG and TIC curves recorded for Gel A (above).
Mass spectrum of the gas mixture evolved at 450°C (below).

magnitude lower than the previous ones, containing 3 Si atoms such as $H_2Si(CH_3)$-O-$Si(CH_3)H$-O-SiH_3 (t_R = 4.5) and $H_2(CH_3)Si$-O-$Si(CH_3)H$-O-$Si(CH_3)H_2$ (t_R = 8.6). The main peaks at t_R = 6.4 and 10.8 are due to cyclic siloxanes (m.w. = 166 and 180 amu) analogous to the linear species eluted at t_R = 4.5 and 8.6, respectively. The formation of all the detected silanes and siloxanes arises from the redistribution of Si-O and Si-H bonds in the polymeric network [10,11]. It is worthing to note that the occurrence of these reactions produces volatile fragments of the original hybrid gel network that allow to get direct information on the polymeric gel structure. Accordingly, the identification of the fragments with t_R = 2.1 and t_R = 4.5 is an unequivocal evidence of the occurrence of co-condensation between two D^H units and between a D^H and a T^H unit respectively.

TG/GC/MS analysis of the pyrolysis gas from the third weight loss at 660°C shows that it is mainly composed by methane indicating the transformation of the polymeric network into a ceramic material. TG/MS analysis gives the evolution of CH_4 (m/z=16) and H_2 (m/z=2) as a function of the pyrolysis temperature (Figure 3). H_2 is produced mainly at ≈510°C with a smaller component at ≈640°C. CH_4, on the other hand, is formed in smaller amount at ≈510°C and has a major evolution at ≈640°C. Formation of H_2 and CH_4 can be accounted for by the homolythic cleavage of Si-H, Si-CH_3 and C-H bonds with formation of Si-C-Si bridges or C-C bonds according to the reactions [7,8]:

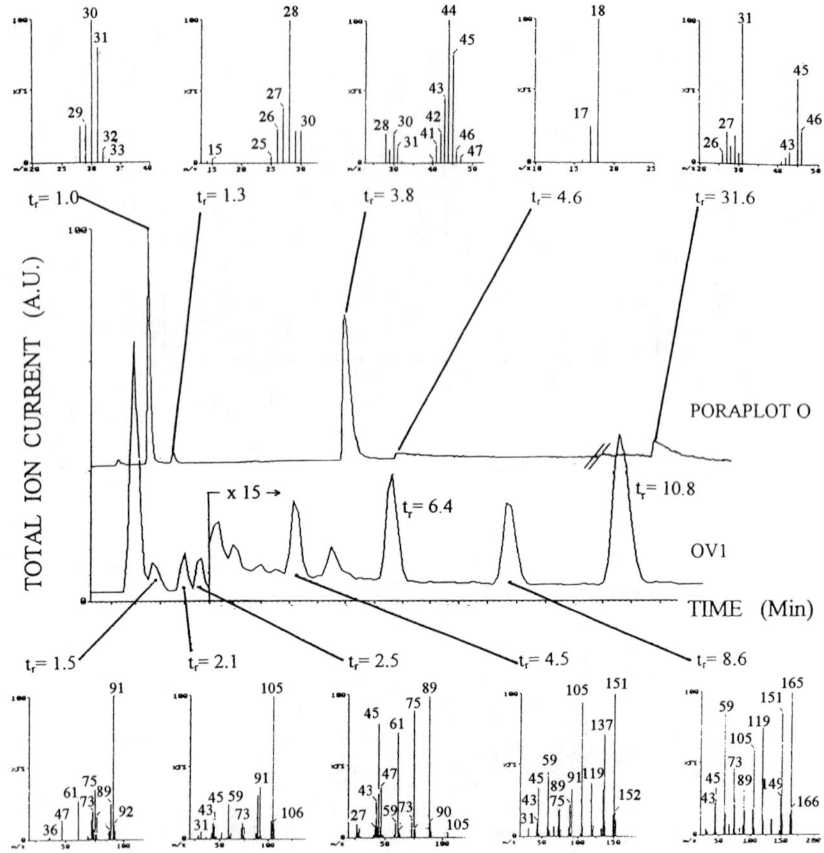

Figure 2. Gas cromathographic elutions of the gas evolved at 450°C and related mass spectra.

$$\equiv Si\text{-}H + CH_3\text{-}Si\equiv \longrightarrow \equiv Si\text{-}CH_2\text{-}Si\equiv + H_2 \qquad (1)$$
$$\equiv Si\text{-}CH_3 + CH_3\text{-}Si\equiv \longrightarrow \equiv Si\text{-}CH_2\text{-}Si\equiv + CH_4 \qquad (2)$$
$$\equiv Si\text{-}CH_3 + CH_3\text{-}Si\equiv \longrightarrow \equiv Si\text{-}CH_2\text{-}CH_2\text{-}Si\equiv + H_2 \qquad (3)$$

The prevalent evolution of H_2 at ≈510°C should be related mainly to reaction (1) that occurs through the cleavage of the more reactive Si-H bonds whereas CH_4 is mainly produced at higher temperature, via the cleavage of the stronger Si-CH_3 bonds.

The pyrolysis pathway of Gel A can then be summarized as follows: in the first 400°C only a small amount of species deriving from condensation reactions of residual -OH and -OEt groups are detected. From 400°C to 500°C a deep rearrangement of the gel network occurs due to redistribution reactions with evolution of many various silanes and siloxanes chain fragments. Ceramization of the hybrid system, with formation of bridging carbon atoms between silicon atoms starts above 500°C, as indicated by the evolution of H_2 and CH_4.

When boron is introduced in the siloxane network (Gel B), the pyrolytic transformation is dramatically modified. TG curve (Figure 3) shows two sharp weight losses: a major one (4.0%) below ≈200°C and a smaller one (1.2%) at ≈230°C. Above ≈300°C and up to ≈900°C a small and broad weight loss (1.7%) is detected. The main decomposition process detected for the Gel A at 450°C and related to the redistribution of Si-H and Si-O bonds discussed above, is absent. DTA trace shows an endothermic effect between ≈100 and ≈200°C and an exothermic effect at ≈220°C. The TIC curve, obtained from the TG/MS analysis (Figure 3) reveals more details, in agreement with the DTG pattern. Below 200°C three distinct gas evolution at 115, 145 and 165°C are present. A fourth peak is detected at ≈230°C and above this temperature the TIC curve do not indicate any major gas evolution with only a very weak and broad band in the temperature range

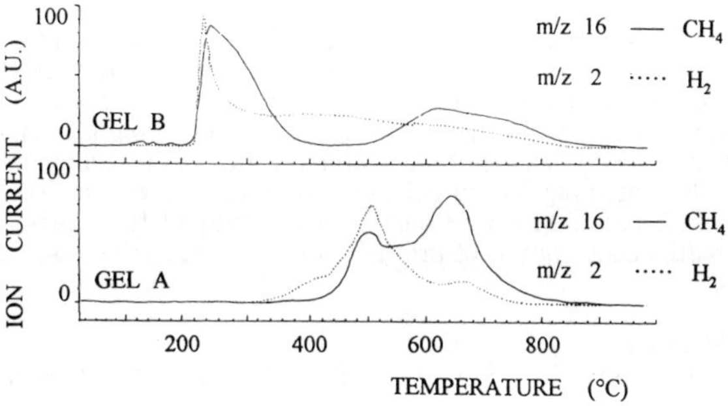

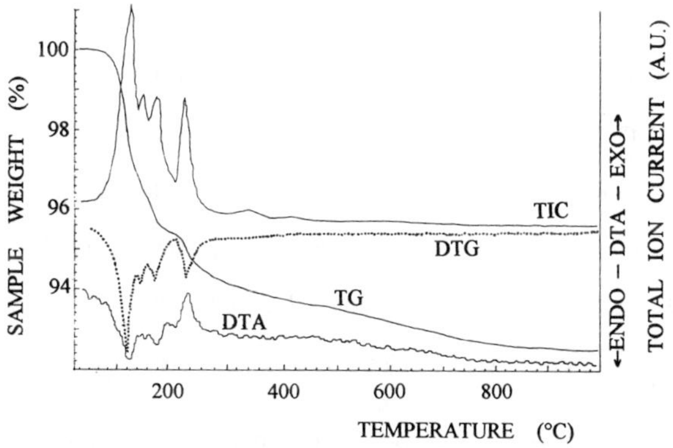

Figure 3. Comparison of H_2 and CH_4 evolution for Gel A and and Gel B (above). TG, DTG, DTA and TIC curves recorded for Gel B (below).

between 280 and 450°C. TG/GC/MS analysis of the pyrolysis gases collected at 115, 145, 165°C shows the presence of H_2O ($t_R = 4.6$) and EtOH ($t_R = 31.6$) arising from the presence in the gel of residual -OEt and -OH groups. On the other hand, at 230°C the gas mixture is composed mainly by ethane ($t_R = 1.3$), with minor amounts of H_2O ($t_R = 4.6$), EtOH ($t_R = 31.6$) and EtOEt ($t_R = 35.9$). Moreover, TG/MS analysis of the evolution of CH_4 (m/z=16) and H_2 (m/z=2) as a function of the pyrolysis temperature (Figure 3) clearly shows the formation of these species at rather low temperature (≈220°C). This facts suggest that the presence of B in the hybrid gels strongly affects the reactivity of the Si-H and Si-CH_3 functional groups. Formation of H_2 and ethane could be accounted for by the condensation reaction between the Si-H and Si-CH_3 moieties and the remaining -OH and -OEt groups of the boron alkoxide such as:

$$\equiv\text{Si-H} + \text{HO-B=} \dashrightarrow \equiv\text{Si-O-B=} + H_2 \quad (4)$$
$$\equiv\text{Si-H} + \text{EtO-B=} \dashrightarrow \equiv\text{Si-O-B=} + \text{EtH} \quad (5)$$
$$\equiv\text{Si-CH}_3 + \text{HO-B=} \dashrightarrow \equiv\text{Si-O-B=} + CH_4 \quad (6)$$

It is worthing to note that the occurrence of the above reactions leads to a partial consumption of Si-H bonds with a corresponding increase of the crosslinking degree of the gel network. This structural modification can explain the absence of the weight loss associated to the redistribution of Si-H and Si-O bonds that should occur at higher temperature (≈400°C). Moreover, consumption of Si-H bonds through reactions (4) and (5) should also reduce the efficiency of C incorporation into the silica network via reaction (1) resulting into oxycarbide glasses with a lower content of network carbon.

CONCLUSIONS

Pyrolysis of hybrid gels containing the Si-H and Si-CH_3 functionalities into SiOC and SiBOC glasses, has been studied by means of TG/GC/MS analysis. Between 400 and 500°C, Gel A undergoes a deep structural rearrangement due to redistribution reactions of the Si-H and Si-O bonds with evolution of many various silanes and siloxanes compounds. Evolution of H_2 and CH_4 suggests that the ceramization of the hybrid system, with formation of Si-C-Si bridges starts above 500°C. B-containing hybrid gel shows a dramatic modification of the pyrolysis process. Evolution of silanes and siloxane compounds is completely suppressed. This fact has been related to the partial consumption of the Si-H bonds that occurs at lower temperature with the formation of H_2.

ACKNOWLEDGEMENTS

MURST 40% and NATO (CRG 931453) are acknowledged for their financial support.

REFERENCES

1. H. Schmidt, J. Sol-Gel Sci. and Technol. 1, p. 217-231 (1994).

2. H. Zhang and C. G. Pantano, J. Am. Ceram. Soc. 73, p. 958-63 (1990).

3. R. H. Baney, M. Itoh, A. Sakakibara and T. Suzuki, Chem. Rev. 95, p. 1409-1430 (1995).

4. G. M. Renlund, S. Prochazka and R. H. Doremus, J. Mat. Res. 6, p. 2723-34 (1991).

5. G. D. Sorarù, E. Dallapiccola and G. D'Andrea, J. Am. Ceram. Soc. 79, p. 000 (1996).

6. F. Babonneau, G. D. Soraru, G. D'Andrea, S. Dirè and L. Bois, Mater. Res. Soc. Symp. Proc. 271, 789-94 (1992).

7. G. D. Sorarù, G. D'Andrea, R. Campostrini, F. Babonneau, G. Mariotto, J. Am. Ceram. Soc. **78**, 379-87 (1995).

8. L. Bois, J. Maquet, F. Babonneau, H. Mutin and D. Bahloul, Chem. Mater. **6**, 796-802 (1994).

9. R. J. P. Corriu, D. Leclerq, P. H. Mutin and A. Vioux, in Better Ceramics Through Chemistry VI, edited by A. K. Cheetham, C. J. Brinker, M. L. Meacartney, C. Sanchez, (Mater. Res. Soc. Symp. Proc. 346, Pittsburgh, PA, 1994), p. 351-356.

10. V. Belot, R. J. P. Corriu, D. Leclerq, P. H. Mutin and A. Vioux, J. Mater. Sci. Letter. **9**, 1052-1054 (1990).

11. V. Belot, R. J. P. Corriu, D. Leclerq, P. H. Mutin and A. Vioux, J. Polym. Sci., Poly. Chem. **30**, 613-623 (1992).

12. R. Campostrini, G. D'Andrea, G. Carturan, R. Ceccato and G. D. Sorarù, J. Mater. Chem. **5**, 000 (1996).

13. M. Nogami and Y. Moriya, J. Non-Cryst. Sol. **48**, 359-366 (1982).

HYDROLYSIS-CONDENSATION BEHAVIOR OF ACETYLACETONE MODIFIED TIN(IV) TETRA TERT-AMYLOXIDE

L. ARMELAO [*], F.O. RIBOT [**], C. SANCHEZ[**]
[*] Centro CNR-SSRCC, Department of Inorganic Chemistry - University of Padova, 4 Via Loredan, 35100 Padova, Italy
[**] Laboratoire de Chimie de la Matière Condensée - Université P. et M. Curie / CNRS, 4 Place Jussieu, 75252 Paris, France

ABSTRACT

The hydrolysis of $Sn(OAm^t)_2(acac)_2$ (TTA) leads to sols, gels or precipitates made of tin oxo-hydroxo polymers, surface capped with residual organic groups, which are mainly acac ligands. The residual acac/Sn ratio decreases when the hydrolysis ratio H increases. For a given H, the observed gelation-precipitation behavior and the BET surface areas of the resulting xerogels strongly depend on the nature of the solvent.

INTRODUCTION

The last years have seen an increasing interest in sol-gel technique to obtain properly tailored high-performance materials [1,2]. The soft chemical conditions provided by sol-gel chemistry allow the synthesis of new hybrid organic-inorganic materials [2-6].

These hybrids are attractive nanocomposites inside which the organic components can play a major role in determining their morphological features such as the porosity and surface area [6]. For silica based materials, much effort has already been devoted to understand the interactions and interconnections between the organic and inorganic phases [3-6]. On the contrary, very little work has been performed up to now to synthesize tin oxo based hybrid networks [7,8]. Tin based hybrids can be synthesized by using the $Sn-C(sp^3)$ bonds which are stable under hydrolytic conditions, or, as for transition metals, by using functionalized complexing ligands [9]. With this aim, it is of basic importance to understand the sol-gel chemistry of tin(IV) alkoxides, modified with complexing ligands [10,11], in terms of microstructural evolution during their transformation into the final tin oxo-hydroxo polymeric compounds.

This article describes the hydrolytic behavior of $Sn(OAm^t)_2(acac)_2$ (TTA) in different organic solvents. The nature of the solvent and the water content strongly affect the morphological properties of the growing oxo polymers. The resulting tin oxo-hydroxo based polymers, obtained after drying under vacuum, have been characterized by BET analysis, FTIR, solid state NMR (^{13}C, ^{119}Sn) and thermogravimetric analysis.

EXPERIMENT

Precursors synthesis were performed under dry argon with Shlenk-line techniques. Unless hydrolyzed with more than two water molecules per tin atom, samples were handled under dry argon (Slenk-line or gloves box).

Solution ^{119}Sn NMR experiments were performed on a Bruker AM250 spectrometer (93.27 MHz for ^{119}Sn). Solid state ^{119}Sn and ^{13}C NMR experiments were carried out on a Bruker

Mat. Res. Soc. Symp. Proc. Vol. 435 © 1996 Materials Research Society

MSL300 spectrometer (111.92 MHz for ^{119}Sn and 75.47 MHz for ^{13}C). Magic angle spinning (MAS) was used for both nuclei. ^{119}Sn NMR spectra were obtained using direct polarization, while cross-polarization (CP) with ^{1}H was used for ^{13}C NMR spectra. For ^{13}C CP-MAS NMR, samples were carefully weighted, all spectra were recorded with 1008 transients and the contact time was kept constant at 2ms. Chemical shifts are referenced (0 ppm) versus external tetramethylsilane for ^{13}C and external tetramethyltin for ^{119}Sn.

Thermogravimetric analysis were performed with a STA 409 Netzsch equipement. Samples (around 50 mg) were placed in an alumina crucible and heated under a flow of pure oxygen (5 cm^3/min), up to 900°C, with a heating rate of 10°C/min.

Surface areas were measured with Quantachrome Multipoints BET analyser. Prior to analysis, samples were heat treated at 105°C, under a flow of nitrogen, for one hour. Three points ($P/P_0 = 0.1, 0.2, 0.3$) were used in the BET analysis.

Tin(IV) tetra tert-amyloxide was synthesized according to the procedure described by Thomas [12], and purified by vacuum distillation. The obtained compound was a colorless liquid. Its purity was checked by ^{119}Sn solution NMR. The spectrum exhibited a single resonance at -370 ppm, which indicates, as expected from the bulkiness of the tert-amyloxy group, a monomeric tin(IV) tetra alkoxide [10].

$Sn(OAm^t)_2(acac)_2$ (TTA) was prepared as follow. The proper amount of pure acetylacetone (acacH/Sn=2) was added drop by drop to pure $Sn(OAm^t)_4$ under strong magnetic stirring. During addition, the system generated heat and turned to orange, indicating the exchange of some alkoxo groups by acetylacetonato ligands. The system was then dried under vacuum to remove the alcohol liberated by the reaction. The light-orange crystalline solid (m_p=70°C) so-obtained, corresponds to $Sn(OAm^t)_2(acac)_2$ [13], and was stored in a dry glove-box.

Hydrolysis experiments were carried out in different solvents, varying from strongly polar to non polar (formamide, dimethylsulfoxide, acetonitrile, tert-amyl alcohol, acetone, tetrahydrofuran and 1,4-dioxane). For all experiments, the solvent/Sn molar ratio was kept constant at 50. Half of the solvent was first added on TTA. Water was then mixed with the remaining solvent and added drop by drop to the TTA solution under strong magnetic stirring. Hydrolysis ratio ($H=H_2O/Sn$) was varied from 1 to 10. Xerogels were obtained by vacuum drying (T<50°C) of the hydrolyzed systems (sols, gels or precipitates) aged for 4 days at room temperature.

RESULTS AND DISCUSSION

Precursor

$Sn(OAm^t)_2(acac)_2$ is characterized by a single ^{119}Sn NMR resonance (-726 ppm in C_6D_6 and -722 ppm for the solid compound), about 350 ppm up-field from $Sn(OAm^t)_4$, which indicates a unique six-coordinate tin-oxygen site [10]. This environment results from two monodentate tert-amyloxy and two chelating acetylacetonato ligands. This last feature is also clearly seen by infra-red spectroscopy where the keto-enolic form of the acetylacetonato is the only one observed. Therefore, TTA is a monomeric precursor. One can point out that $Sn(OAm^t)_2(acac)_2$ is the only defined compound in the $Sn(OAm^t)_4$-acacH systems. Indeed, any acacH/$Sn(OAm^t)_4$ ratios below 2 yield a mixture of $Sn(OAm^t)_4$ and $Sn(OAm^t)_2(acac)_2$ [13].

The thermal decomposition of TTA occurs in two steps. The first one (205°C) corresponds to the combustion of the tert-amyloxy groups and the "one to one" transformation of the acetylacetonato ligands into acetato ligands ($CH_3CO_2^-$). The second step (360°C) corresponds to the combustion of the previously formed acetato ligands and yields tin oxide (around 500°C) which crystallized into SnO_2 cassiterite above 600°C. Full characterization of $Sn(OAm^t)_2(acac)_2$ will be reported elsewhere [13].

Hydrolysis of TTA

The hydrolysis of $Sn(OAm^t)_4$ always yields precipitates even for hydrolysis ratios as low as 0.2. The reactivity of many tetravalent metal alkoxides (Ti, Zr, Ce,...) towards hydrolysis has been controlled by complexation [14]. As expected, the behavior of TTA toward hydrolysis is very different from the one of un-complexed $Sn(OAm^t)_4$. Moreover, it strongly depends on the solvent in which the hydrolysis takes place ; i.e. precipitation is prevented and stable sols are obtained in very polar solvents such as formamide. Qualitative results for different solvents and hydrolysis ratio (H) are reported in Table I.

The different observed behaviors might be explained by variations in the molecular interactions between the solvent and the growing tin oxo-hydroxo polymers. As a general trend, the hydrolysis of TTA yields progressively stable sols, gels and precipitates as the solvent polarity decreases. However dielectric constant and solvent viscosity could also play an active role in the control of the growth (hydrolysis-condensation reactions may have a different degree of completion) and solvent-polymer interactions.

Table I : Hydrolysis-condensation behavior of TTA for different hydrolysis ratios ($H=H_2O/Sn$) in different solvents (S = light-orange solution, S' = light-yellow solution, S" = transparent solution, G = gel, P = precipitate).

Solvent \ H	1	2	4	10
dimethylsulfoxide	S	S	S	S
formamide	S	S	S	S
acetonitrile	S	S	G	S" + P
acetone	S + P	-	S" + G	-
tert-amyl alcohol	S' + P	S" + P	G	G
tetrahydrofurane	S' + P	S" + P	S" + P	-
1,4-dioxane	S" + P	S" + P	S" + P	S" + P

Characterization of the Xerogels

Some of the systems were processed into xerogels and characterized by N_2 adsoprtion (BET analysis), ^{13}C CP-MAS NMR, ^{119}Sn MAS NMR and thermogravimetric analysis. Xerogels are labeled TTA/solvent/H.

The BET surface results are reported in Table II. They show a clear difference between the xerogels obtained for low and high hydrolysis ratio. The surface areas for H = 1 are negligible regardless of the solvent. For higher H values the specific surface area is quite high and reflects a strong influence of the nature of the solvent. For a given H value the higher surfaces are measured for xerogels obtained from low polar solvents, those for which precipitation occurs easily.

^{119}Sn MAS NMR shows, for all xerogels with H≥4, a resonance centered on -610 ppm which exhibits a shoulder around -640 ppm. This last signal tends to disappear when H increases from 4 to 10. According to the ^{119}Sn isotropic chemical shifts of SnO_2 (-603 ppm [15]) and $K_2Sn(OH)_6$ (-570 ppm [16]), the main resonance observed for the xerogels (H≥4) reflects a six-coordinate site for tin, mainly with oxo and hydroxo ligands. The shoulder (-640 ppm) is likely related to tin atoms which still have an acetylacetonato ligand in their coordination sphere. The xerogels can therefore be pictured as tin oxo-hydroxo inorganic polymers with residual acetylacetonato ligands.

To evaluate the acac/Sn ratios of these tin oxo-hydroxo polymers, thermogravimetric analysis and ^{13}C CP-MAS NMR were performed on the xerogels. Figure 1 represents typical weight loss vs. temperature curves. X-ray diffraction experiments showed that all samples were SnO_2 cassiterite after TGA. Tin oxide contents are reported in Table II. Each thermogram can be analyzed with two weight losses. The first one, relatively smooth, takes place between 20°C and ≈250°C. It can be attributed to solvent and water removal, hydroxy condensation, combustion of the remaining tert-amyloxy groups (if any) and finally to the decomposition of each residual acetylacetonato ligand into one acetato ligand, as in TTA [13]. The second weight loss is more abrupt and occurs around 340°C. It corresponds to the combustion of the acetato ligands. From this second step, the acac/Sn ratio can be estimated for each xerogel. The results are reported in Table II.

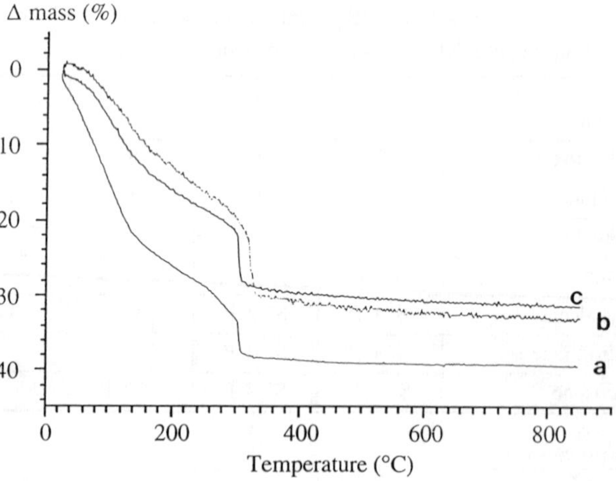

Fig. 1 : TGA of the xerogels "TTA/acetonitrile/H" ; (a) H=1, (b) H=4, (c) H=10.

Table II : BET surface area, tin oxide content (from TGA) and estimations of the acac/Sn ratio for different xerogels.

Xerogel (TTA/Solvent/H)	BET Surface (m^2/g)	SnO_2 content (%)	acac/Sn (from TGA)	acac/Sn (from ^{13}C NMR)
TTA/acetonitrile/1	6±1	49±1	1.3	1.5
TTA/acetonitrile/4	130±10	66±1	0.7	0.2
TTA/acetonitrile/10	145±10	69±1	0.5	0.1
TTA/HOAm^t/1	2±1	44±1	1.4	1.4
TTA/HOAm^t/4	185±15	64±1	0.6	0.2
TTA/HOAm^t/10	280±20	67±1	0.5	0.1
TTA/1,4-dioxane/1	2±1	41±1	1.6	1.0
TTA/1,4-dioxane/4	410±40	64±1	0.6	0.2
TTA/1,4-dioxane/10	350±30	61±1	0.5	0.1

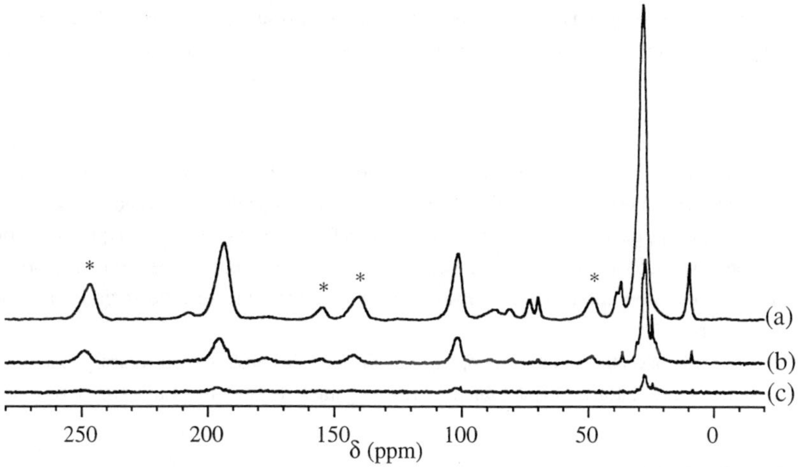

Fig. 2 : ^{13}C CP-MAS NMR spectra of the xerogels "TTA/acetonitrile/H" ; (a) H=1, (b) H=4, (c) H=10 (rotation speed : 4000 Hz, spinning side bands are indicated with asterisks).

^{13}C CP-MAS NMR gives also information on all the residual organics. ^{13}C CP-MAS NMR spectra for the xerogels "TTA/acetonitrile/H=1-4-10" are presented in figure 2. They are characteristic of the changes observed upon increasing the hydrolysis ratio. The main residual organics are the acetylacetonato ligands, as evidenced by the C=O resonance around 193 ppm and the C-H resonance around 100 ppm. Tert-amyloxy groups (quaternary carbon at 73 ppm) disappear for H>1. Solvent is also observed. Quantitative analysis of ^{13}C CP-MAS NMR spectra is usualy delicate, mainly because of variations in the cross-polarization dynamic. Yet, one can assume that the main ^1H reservoir available to cross-polarize the carbonyl or the methyne groups of the acetylacetonato ligands are the protons of these ligands. Therefore, at first order, the cross-polarization dynamic of these species can be thought to be almost independant of the total amount of residual organics. Indeed, in all samples, the intensities, measured by the integral of the isotropic resonance and associated spinning side bands, of the C=O and C-H always remained in the expected ratio for acetylacetonato ligands (2:1 ±5%). Using TTA as reference (acac/Sn=2), and scaling by the sample weight and tin oxide content, the absolute intensities of the acetylacetonato C=O and C-H ^{13}C CP-MAS NMR resonances can give a rough estimation of the acac/Sn ratios in xerogels. These estimations are reported in Table II.

The acac/Sn ratios, reported in Table II, show some discrepencies between both estimation techniques. However, both estimations show the same trends. For low hydrolysis ratios (H=1), alkoxy groups are preferencially removed as indicated by acac/Sn ratios around 1.5. Yet, when the hydrolysis ratio is increased (H>4), acetylacetonato ligand are also removed to a large extent. More strinking is the relative independence of the acac/Sn ratios to the solvent. This last feature seems to indicate that the changes observed for the specific surface areas of the xerogels (Table II) are more related to the interactions between the solvent and the oxo polymers than to changes in the hydrolysis-condensation reactions. One can think that the growing tin oxo polymers become more polar as acetylacetonato ligands are removed when the hydrolysis ratio increases. Therefore, they need strongly polar solvents to be stabilized (swelled) via polymer-solvent interactions. Note that the highly polar solvents used were also aprotic.

CONCLUSION

The hydrolysis of TTA, $Sn(OAm^t)_2(acac)_2$, leads to tin oxo-hydroxo polymers in which some "surface" sites are capped with residual acetylacetonato ligands. The amount of residual acetylacetonato groups decreases when the hydrolysis ratio increases. As a consequence, strongly polar solvents are needed to prevent precipitation. This competition between gelation and precipitation seems to be one of the key parameters that control the specific surface area of the resulting xerogels.

As far as hybrid materials are concerned, it seems difficult, because of their hydrolytic lability, to use functonalized ß-diketonate [9] to anchor any organics to the inorganic tin oxo-hydroxo polymers.

ACKNOWLEDGMENTS

One of the authors (L.A.) thanks the "Centro CNR - Stabilita e Reattivita dei Composti di Coordinazione - Padova" for financial support.

REFERENCES

1. Sol-Gel Technology for Thin Films, Fibers, Preforms, Electronics, and Especially Shapes, edited by L.C. Klein, Noyes, Park Ridge, N.J., 1988.
2. C.J. Brinker and G.W. Scherrer, Sol-Gel Science, the Physics and Chemistry of Sol-Gel Processing, Academic Press, San-Diego, CA, 1990.
3. B.M. Novak, Adv. Mater. **5**, 422 (1993).
4. C. Sanchez and F. Ribot, New J. Chem. **18**, 1007 (1994).
5. U. Schubert, N. Hüsing and A. Lorenz, Chem. Mater. **7**, 2010 (1995).

6. D.A. Loy and K.J. Shea, Chem. Rev. **95**, 1431 (1995).
7. F. Ribot, F. Banse, F. Diter and C. Sanchez, New J. Chem. **19**, 1145 (1995).
8. F. Ribot, F. Banse, C. Sanchez, M. Lahcini and B. Jousseaume, J. S. S. T. in press (1996).
9. M. In, C. Gérardin, J. Lambard and C. Sanchez, J. S. S. T. **5**, 101 (1995).
10. M.J. Hampden-Smith, T.A. Wark and C.J. Brinker, Coord. Chem. Rev. **112**, 81 (1992).
11. C. Roger, M.J. Hampden-Smith and C.J. Brinker in Better Ceramics Through Chemistry V, edited by M.J. Hampden-Smith, W.G. Klemperer and C.J. Brinker (Mater. Res. Soc. Proc. 271, Pittsburgh, PA, 1992), p. 51.
12. I. Thomas, US Patent 3,946,056 (1976).
13. L. Armelao, F. Ribot and C. Sanchez, forth comming paper.
14. C. Sanchez, F. Ribot and S. Doeuff in Inorganic and Organometallic Polymers with Special Properties, edited by R.M. Laine, NATO ASI Series, vol 206 (Kluwer, New-York, NY, 1992), p. 267.
15. C. Cossement, J. Darville, J-M. Gilles, J.B. Nagy, C. Fernandez and J-P. Amoureux, Magn. Reson. Chem. **30**, 263 (1992).
16. R.K. Harris and A. Sebald, Magn. Reson. Chem. **25**, 1058 (1987).

EFFECT OF PROCESSING PARAMETERS ON SECOND ORDER NONLINEARITIES OF AZO DYE GRAFTED HYBRID SOL-GEL COATINGS

B. Lebeau*, C. Sanchez*, S. Brasselet** and J. Zyss**.

*Laboratoire de Chimie de la Matière Condensée, URA CNRS 1466, Université Pierre et Marie Curie, Paris, FRANCE.

**Centre National d'Etudes des Télécommunications, France TELECOM, 196 Avenue Henri Ravera, BP 107, 92 225 Bagneux cedex, FRANCE.

ABSTRACT

The synthesis of transparent hybrid organic-inorganic coatings obtained from condensation between 3-isocyanatopropyltriethoxysilane containing grafted azo dyes (ICTES-Red 17) and tetramethoxysilane (TMOS) precursors has been performed. Influence of ICTES-Red 17/TMOS molar composition and thermal curing on both structure and nonlinear optical (NLO) properties of these materials have been investigated. A thermal treatment of ICTES-Red 17/TMOS coatings before poling has been found to improve nonlinear optical responses as well as its stability.

INTRODUCTION

The chemistry involved in the sol-gel process is based on hydrolysis and condensation reactions of metal alkoxides. Thess reactions lead to the formation of metal-oxo based macromolecular networks [1,2]. One paramount advantage results from the various characteristics of sol-gel process (metallo-organic precursors, organic solvents, low processing temperatures) that allow to introduce "fragile" organic molecules inside an inorganic network [3-5]. Inorganic and organic components can be mixed in virtually any ratio to obtain hybrid nanocomposites extremely versatile in their composition [5], processing and properties. These hybrids offer new approaches in the field of optics [6-8].

Azo dyes have been widely used to prepare materials for second order nonlinear optics [8-11] because of their large optical nonlinearities and their electro-optic properties. Second order nonlinearities require noncentrosymmetric alignment of the NLO chromophores. We have demonstrated that chromophore alignment in sol-gel matrices can be performed by using electrical field poling processes [12,13]. However when the dye is simply embedded in the sol-gel matrix, in the absence of electrical field, relaxation of the dipole alignment to a random configuration is generally observed [13]. The chemical bonding of the chromophore to the polymeric backbone leads to a better orientational stability of the poled NLO dye and higher concentrations of organic dye in the sol-gel film [13-15].

The optimization of the second order NLO response as well as its stability of hybrid sol-gel matrices with grafted chromophores is currently under investigation by several research groups. We have previously reported non resonant second-order optical nonlinearities as high as 78 pm/V ($\lambda = 1.34\ \mu m$) measured on siloxane-silica nanocomposites functionalized with the Red 17 (4-(amino-N,N-diethanol)-2-methyl-4'-nitroazobenzene) hybrid coatings [16]. These values are quite competitive with those reported for inorganic or polymeric materials.

In this article we report the influence of molar composition and thermal curing on the structure and the nonlinear optical (NLO) properties of these hybrid materials.

RESULTS AND DISCUSSION

1- Red 17 azo dyes in sol-gel hybrid matrices

The use of the difunctional Red 17 as an NLO chromophore allows to bond one dye molecule through two trialcoxysilyl functions. This double grafting should minimize relaxation of the dipole alignment to a random configuration. A coupling reaction between the dye and 3-

Mat. Res. Soc. Symp. Proc. Vol. 435 © 1996 Materials Research Society

isocyanatopropyltriethoxysilane (ICTES) previously reported [17] was used for grafting the Red 17. The resulting alkoxysilyl functionalized NLO precursor was called ICTES-Red 17 (figure 1).

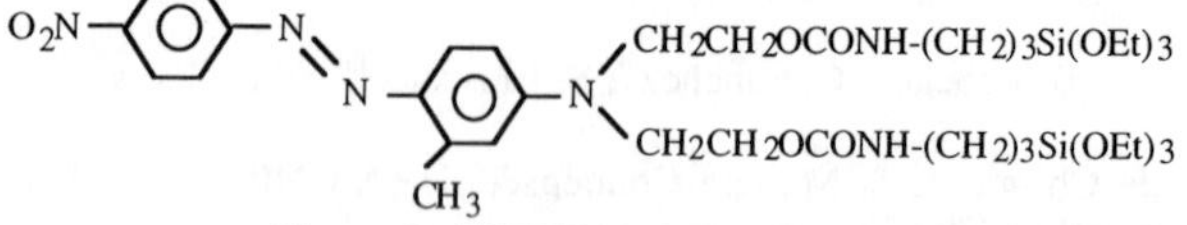

Figure 1 : ICTES-Red 17 precursor

1-a Synthesis of hybrid NLO materials : ICTES-Red 17 (T unit) was used as the siloxane network precursor carrying the NLO chromophore while TMOS (Q unit) was used as a crosslinking reagent to increase the network rigidity [15].

Sol-gel coatings with different ICTES-Red 17/TMOS molar ratios (30/70 and 50/50) were prepared as follow : precursors, ICTES-Red 17 and TMOS, were mixed in THF and co-hydrolyzed with acidic water (HCl ; pH=1). The H_2O:Si molar ratio was 1.5:1. Then, solution was stirred for 30 minutes and the resulting sols were aged for several days. From these sols, hydrophobic transparent films of several micrometers thickness, were elaborated without crack and failure. Two procedures were used to process coatings. They were prepared on ordinary soda-lime glass-sheets previously cleaned and dried by simple deposition (for NMR and FTIR experiments) or by spin-coating (for FTIR and NLO experiments). In the first process an appropriate amount of solution is poured on the glass sheet and let gelled and dried at room temperature. Thin films were also spin coated onto clean glass substrates. The spinning rate was 2500 rpm and the deposition took 30 seconds. The resulting hybrid inorganic-organic materials were first characterized by ^{29}Si MAS NMR and FTIR.

1-b Chemical characterization of hybrid NLO materials : The ^{29}Si MAS-NMR experiments have been performed on air dried deposited coatings before and after thermal curing at 150°C for 12 hours. In any case, the ^{29}Si MAS-NMR spectra of the ICTES-Red 17/TMOS materials show several components located at about -60 ppm and -100 ppm. They are respectively due to T (trifunctional) and Q (tetrafunctional) silicon-oxygen units [18,19]. From the relative intensity of each T and Q resonances, the concentration of each species can be measured. The mean degree of condensation C which corresponds to the average concentration of condensed bonds can be calculated for every T or Q set of components by taking $C(T)= \Sigma_i\, i*t_i /3$ and $C(Q)= \Sigma_j\, j*q_j /4$. t_i and q_j are the concentration measured by NMR for T_i and Q_j species (i and j the number of Si-O-Si bridging oxygen i= 0 to 3; j=0 to 4). The degree of condensation C for T and Q units versus ICTES-Red 17/TMOS molar composition before and after thermal curing are reported in table 1.

ICTES-Red 17/TMOS molar ratio	thermal curing	C(T)	C(Q)
30/70	no	68 %	76 %
	yes	79 %	80 %
50/50	no	70 %	80 %
	yes	72 %	84 %

Table 1 : degrees of condensation C for T and Q units versus ICTES-Red17/TMOS molar composition before and after thermal curing

After thermal curing, at 150°C for 12 hours, the ^{29}Si MAS-NMR spectra of 30/70 and 50/50 ICTES-Red 17/TMOS coatings are broader. The mean degrees of condensation calculated from NMR are quite higher for T and Q units. This shows better completion of the condensation reactions for cured samples.

The FTIR spectra of the room temperature air dried 30/70 and 50/50 ICTES-Red 17/TMOS spin-coated samples are presented in figure 2b and 2d, respectively. FTIR spectra exhibit the absorption bands characteristic of the vibrations of the urethane group ($\nu_{(C=O)}$=1685 cm^{-1},1715 cm^{-1}, $\nu_{(N-H)}$=3310 cm^{-1}) showing that the urethane bond between the siloxane network and the dye has been conserved [17,20]. Moreover, the absorption bands characteristic of the $\nu_{(Si-O-Si)\,asym}$ vibrations are splitted into two main components located at 1050 cm^{-1} and 1120 cm^{-1} respectively assigned to linear and cyclic siloxane species [21]. Comparison between the integrated intensities of the linear and cyclic components contributes to understanding the degree of molecular connectivity within the siloxane-silica phase. On inspecting the trend of the two main $\nu_{(Si-O-Si)}$ components it seems that the increasing TMOS content favors more linear siloxane-oxide structures [21,22].

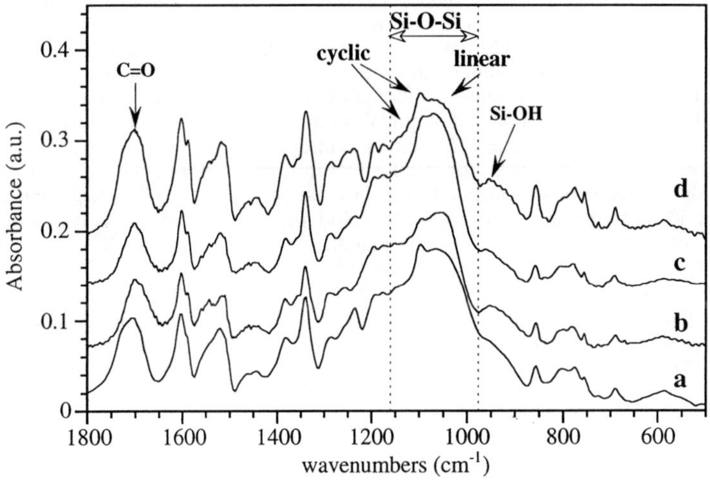

Figure 2 : FTIR spectra of the bulk 30/70 (a), not cured spin-coated 30/70 (b), cured spin-coated 30/70 (c) and the cured spin-coated 50/50 ICTES-Red 17/TMOS coatings.

Comparison of FTIR spectra of simply deposited and spin-coated 30/70 ICTES-Red 17/TMOS samples (figure 2a and 2b, respectively) show a different distribution between linear and cyclic species. For the simply deposited coating, air drying process takes more time and leads to hybrid siloxane-oxide structures more cyclic.

After thermal curing (figure 2c), at 150°C for 12 hours, the absorption bands characteristic of the vibrations of the residual Si-OH groups located at 3500 and 960 cm^{-1} are less intense. This confirm the ^{29}Si MAS-NMR results and show a better completion of the condensation reaction upon thermal curing.

2- Nonlinear optical characterization

2-a Coating composition : The nonlinear properties of two different ICTES-Red 17/TMOS Si molar composition, namely 30/70 and 50/50, were compared. After hydrolysis the sol was filtered (0.45 μm filtering) and spin-coated on microscope slides with a spinning rate of 2500 rpm. The thickness of the resulting samples were measured using a profilemeter and varied from 0.75 μm to 5 μm depending on the duration of the hydrolysis. The standard corona [23] poling technique was used to orient the chromophores in the sol-gel matrices [13], with a high voltage of 5 kV applied on a needle at 2 cm from the samples throughout the heating process. Second harmonic measurements were performed using the Maker fringe [24] method with a Y-cut quartz crystal (0.46 pm/V) as reference. We used a Nd^{3+}:Yag laser with pulses of 1 MW peak power, 10 ns duration, at 10 Hz repetition rate, and operating at 1.34 μm, which locate the measurements in the off-resonance regime (the maximum of the absorption band for such films is located near 470 nm). Previous result on ICTES-Red 17/TMOS materials [16] evidenced a d_{33} value as high as 78 pm/V, before relaxation, in the case of the 30/70 composition, the sample being spin-coated after 3 days of hydrolysis (the resulting thickness was of the order of 4.5 μm) and poled at 100°C for 3 hours and 150°C for 10 min. The temporal relaxation at room temperature was of the order of 50%. To compare 30/70 and 50/50 compositions, we realized samples of comparable thicknesses (0.75 µm for the 50/50 composition, obtained after 5 days of hydrolysis, and 1.5 µm after 2 days of hydrolysis for the 30/70 composition) and similar poling conditions, namely 1 hour at 100°C and 2h30 at 150°C. In both cases, the films were homogeneous, presenting a uniform surface with no opacity. The refractive index of these ICTES-Red 17/TMOS systems was supposed to be of the same order for the different Si molar composition, namely 1.57 at 1.34 µm [16]. The 30/70 response was less important than observed in previous studies, showing the influence of the sample preparation, namely the aging of the sol-gel before spin-coating, and of the poling conditions (choice of the amplitude and duration of the temperature steps). A longer step at low temperature (under 100°C) has been shown for example to be favorable towards a higher d_{33} value. Before relaxation, the d_{33} coefficients were respectively 38 pm/V for the 30/70 composition and 58 pm/V for the 50/50 sample, and the d_{33}/d_{31} ratio was of the order of 3.5 for both systems. The measured nonlinear coefficients are still high values in comparison with earlier reports for similar hybrid organic-inorganic systems [15,20,25-27], corresponding to theoretical electro-optic coefficients r_{33} of the order of 15 pm/V and 23 pm/V for 30/70 and 50/50 compositions respectively at 860 nm. The discrepancy between the nonlinear responses of these two systems can be essentially ascribed to the different chromophore concentrations in the films: it is indeed increased by a factor 5/3 from the 30/70 composition to the 50/50 one, which corresponds approximately to the improvement in the d_{33} coefficient, which is proportional to the number of active molecules in the film according to an oriented gas modeling of the compound [28]. Furthermore, the temporal relaxation at room temperature after poling shown in figure 3 is more favorable for the 50/50 composition: the decrease of its nonlinearity is about 35% after one month while it reaches 53% for the 30/70 composition.

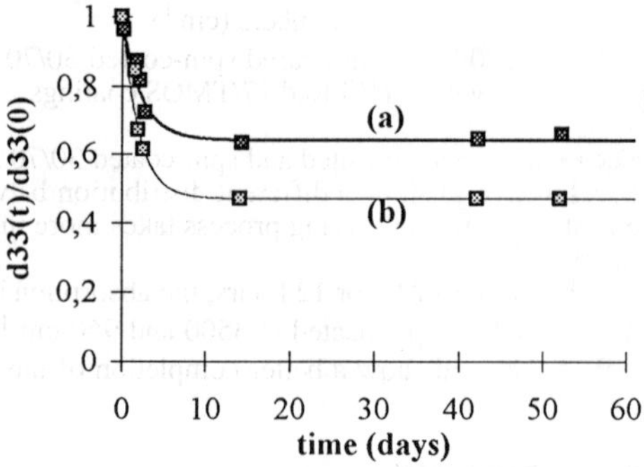

Figure 3: Evolution of the normalized d_{33} coefficient for the 50/50 (a) and 30/70 (b) Si molar compositions in the hybrid system ICTES RED 17/TMOS.

The decay curves were fitted using a simple biexponential law given by $d_{33}(t)/d_{33}(0) = d_1 e^{-t/\tau_1} + d_2 e^{-t/\tau_2}$, where the first relaxation process with time constant τ_1 is characteristic of the chromophore short term relaxation in matrices [29] and considerably faster than the second long term time relaxation constant τ_2. d_1 and d_2 are constants which measure the relative importance of the two relaxation processes and can be related respectively to the fraction of not crosslinked and crosslinked dipoles in the matrix [30]. d and τ parameters are given in table II for the two 50/50 and 30/70 compositions.

	d_1	τ_1	d_2	τ_2
50/50	35%	2.3 days	65%	∞
30/70	53%	1.9 days	47%	∞

Table II: Characteristic parameters d and τ for the room temperature relaxation of the normalized nonlinear response concerning the 50/50 and the 30/70 compositions.

For the two systems, τ_2 is found close to ∞, implying thus a good long-term stability at room temperature for the ICTES-Red 17/TMOS system. The comparable short term relaxation times show that the effect responsible for the more important orientational decay of the nonlinear response of the 30/70 system originates essentially from the limited crosslinking efficiency of the matrix and not only from the pure disorientation of the chromophores (as seen from Table II, 53% of the species are not crosslinked after the poling process). Beside these observations concerning the sol-gel composition, the shorter relaxation process could be improved by an adequate thermal treatement [25].

2-b Thermal preparation of the samples : The preliminary thermal preparation of samples is crucial towards the efficiency of the nonlinear response. We performed *in situ* poling for 1h50 for a sample which was preliminary cured at 150°C for 2 hours, and we compared its nonlinear response to that of previous poling processes. The growth of the second harmonic signal was appearered earlier [31] and its amplitude after 30 min. at 150°C was higher than that obtained without curing (see figure 4).

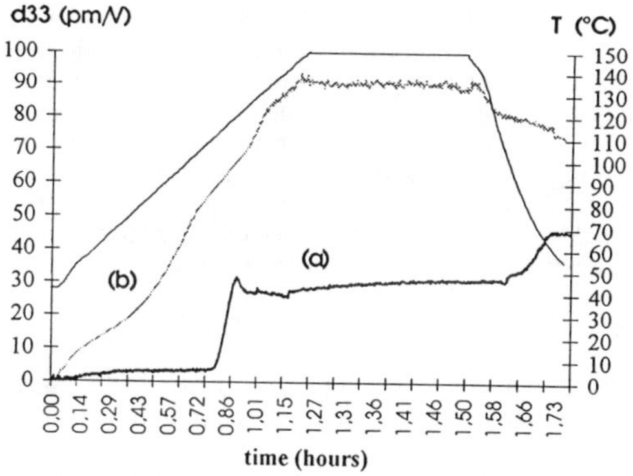

Figure 4: Influence of the thermal preparation of the 50/50 samples on d_{33} coefficient without curing (a) and after preliminary curing (b).

The d_{33} value reached 90 pm/V at 150°C for a cured sample whereas a similar poling process applied without curing resulted in a d_{33} coefficient of 30 pm/V only. Contrary to the non-cured sample for which the nonlinear signal continued to grow during the cooling, the signal of the cured sample did not undergo significative changes, showing the efficiency of a preliminary thermal treatment as to the onset of a stable structure. As the crosslinking process has been initiated beforehand, this could explain a more rapid stabilization and furthermore a more important nonlinear efficiency, further helped by the stable order created by the poling field as a result of lattice hardening. Nonlinear coefficients of 33 pm/V for the non-cured sample and 69 pm/V for the cured one were measured 10 min. after cooling and turning-off of the high voltage. After poling, films have been thermally disoriented to study the influence of the curing process on thermal stability. The temperature of disorientation shown in figure 5 was found to be around 50°C for the not pre-cured film and around 70°C for the pre-cured film.

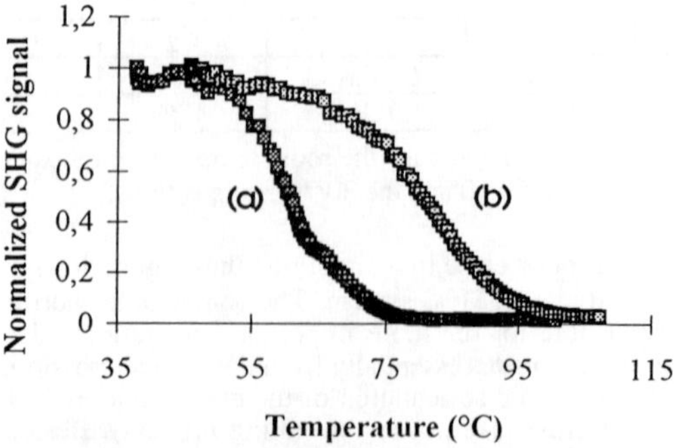

Figure 5: Effect of the thermal preparation of the 50/50 samples on the thermal stability without pre-curing (a) and with pre-curing (b).

Although differential scanning calorimetry of those sol-gels did not show any visible glass transition temperature, these behaviors are significative of the broad distribution of relaxation processes due to a very slow and uncompleted crosslinking process in the thermal preparation of the films [30]. Moreover the increase of the disorientation temperature provided by curing evidences the benefit on thermal stability of preliminary preparation of the film.

The thermal history of the coating is of paramount importance. Chemical crosslinking is not complete after room temperature air drying as shown by ^{29}Si NMR and FTIR experiments. Upon electrical field poling and thermal curing, the chemical reactions continue towards completion. The increase of the density of crosslinks modifies the thermomechanical properties of these hybrid. Consequently, as soon as hybrid sol-gel materials are processed with alcoxy functionalized chromophores the poling protocol is of major importance because these systems behave as thermosets or thermally hardened materials [15,26]. Care must be exercised so that lattice hardening is not effected before a reasonable degree of non centrosymmetric order is introduced by electric field poling. The process must be optimized with stepped increases in temperature.

CONCLUSION

The nonlinear behavior of different compositions of ICTES-Red17/TMOS films has been investigated. The thermal preparation has been shown to improve the nonlinear response as well as its stability.

Influence of thickness films on structure and nonlinear optical properties after poling is under investigation. Chemical modifications, such as the hydrolysis process, are also under study to improve the crosslinking efficiency which influences the optimization of the chromophore order during poling, and thereby improves the temporal stability of the structure.

ACKNOWLEDGMENTS:
A part of this work was sponsored by DRET. The financial support of this institute is gratefully acknowledged.

REFERENCES:
1- C.J.Brinker and G.Scherrer, Sol-Gel Science: the Physics and Chemistry of Sol-gel Processing , (Academic Press, San Diego, (1989)).
2- J. Livage, M. Henry and C. Sanchez,Progress in Solid State Chemistry, **18**, 259 (1988)
C. Sanchez, F. Ribot, S. Doeuff, Organometallic Polymers with special properties, Ed. R.M. Laine, 267 (Kluwer Academic Publisher (1992))
3- D. Avnir, D. Levy and R. Reisfeld, J. Phys. Chem., **88**, 5956 (1984)
4- H. Schmidt and B. Seiferling, Mat. Res. Soc. Symp., **73**, 10 (1986)

5- C. Sanchez and F. Ribot, New.J. Chem., **19**, 1145-1153 (1995)
R. Reisfeld, S.P.I.E.proc."Sol-gel Optics", Ed. J.D.Mackenzie and D. R. Ulrich, **1328**, 29 (1990).
6- Sol-Gel Optics I, Eds. J.D. Mackenzie and D.R. Ulrich, Proc. SPIE, **1328** (Washington, (1990))
7- Sol-Gel Optics II, Ed. J.D. Mackenzie, Proc. SPIE, **1758** (Washington (1992))
Sol-Gel Optics III, Ed. J.D. Mackenzie, Proc. SPIE, **2288** (Washington (1994))
8- Sol-gel optics, processing and applications, Ed. L.C. Klein (Kluwer Academic Publishers, Boston, (1993)) .
9- K. D. Singer,L. A. King, J. Appl. Phys., **70** (6), 3251 (1991).
10- R. J. Jeng, Y. M. Chen, A. K. Jain, J. Kumar, S. K. Tripathy, Chem. Mater., 4, 972 (1992).
11- K. Izawa, N. Okamoto, O. Sugihara, Jpn. J. Appl. Phys., **32**, 807 (1993).
12- J. Zyss, G. Pucetti, I. Ledoux, P. Griesmar, J. Livage and C. Sanchez, Eu. Patent N° 1152, (March 1991)
13- E. Toussaere, J. Zyss, P. Griesmar and C. Sanchez, Non Linear Optics , **1**, 349 (1991).
14- J. Kim, J.L. Plawsky, E. Van Wagenen and G.M. Korenowski, Chem. Mater., **5**, 1118 (1993)
15- B. Lebeau, J. Maquet, C. Sanchez, E. Toussaere, R. Hierle and J. Zyss, J. Mater. Chem., **4**, 1855 (1994)
16- B. Lebeau, C. Sanchez, S. Brasselet, J. Zyss, G. Froc and M. Dumont, New J. Chem., **20** (1), 13 (1996)
17- B. Lebeau, C. Guermeur and C. Sanchez, Better Ceramics Through Chemistry VI, Mat. Res. Soc. Symp. Proc., Eds A. Cheetham, J. Brinker, M. McCartney, C. Sanchez, **346**, 315 (1995)
18- E. Lippmaa, M. Mägi, A. Samoson, G. Engelhardt and A. R. Grimmer, J. Am. Chem. Soc., **102**, 4889 (1980)
19- R. H. Glaser, G. L. Wilkes and C. E. Bronnimann, J. Non-Cryst. Solids, **113**, 73 (1989)
20- F. Chaput, D. Riehl, Y Levy and J. P. Boilot, Chem. Mater., **5**(5), 589 (1993)
21- Q. Deng, K. A. Mauritz, R. B. Moore, Hybrid Organic-Inorganic Composites, Eds J. E. Mark, C.Y-C. Lee, P.A. Bianconi, ACS symposium Series, **585**, 66 (ACS Washington DC (1995))
22- J. L. Jr Brown, L.H. Jr Vogt, P.I. Prescott, J. Am. Chem. Soc., **86**, 1120 (1964)
23- K.D. Singer, M.G. Kusyk, W.R. Holland, J.E. Sohn, S.J. Lalama, R.B. Comizzoli, H.E. Katz and M.L. Schilling, Appl. Phys. Lett., **53**, 19 (1988)
24- P.D. Maker, R.W. Terhune, N.Nissennoff and C.M. Savage, Phys. Rev. Lett., **8**, 21 (1962)
25- S. Kalluri, Y. Shi, W. Steier, Z. Yang, C. Xu, B. Wu and L.R. Dalton, Appl. Phys. Lett. **65**, 21, 1994
26- Z. Yang, C. Xu, B. Wu, L.R. Dalton, S. Kalluri, W.H. Steier, Y. Shi, J.H. Bechtel, Chem. Mater., **6**, 1899 (1994)
27- H.W. Oviatt, K.J. Shea, S. Kalluri, Y. Shi, W. Steier and L.R. Dalton, Chem. Mater., **7**, 493 (1995)
28- K.D. Singer, M.G. Kusyk and J.E. Sohn, J. Opt. Soc. Am. B, **4**, 968 (1987)
29- A. Suzuki, Y. Matsuoka, J. Appl. Phys., **77**, 965 (1995)
30- S. Bauer, W. Ren, S. Bauer-Gogona, R. Gerhard-Multhaupt, J. Liang, J. Zyss, M. Ahlheim, M. Stähelin and B. Zysset, Proceedings 8th International Symposium on Electrets, Eds J. Lewiner, C. Alquié, D. Morisseau, IEEE, 800 (Piscataway, New Jersey, USA (1994))
31- D. Riehl, F. Chaput, Y. Levy, J.P. Boilot, F. Kajzar, P.A. Chollet, Chem. Phys. lett., **245**, 36 (1995)

SYNTHESIS OF BIOACTIVE ORMOSILS BY THE SOL-GEL METHOD

K. TSURU*, C. OHTSUKI*, A. OSAKA*, T. IWAMOTO** and J. D. MACKENZIE**
*Biomaterials Lab, Faculty of Engineering, Okayama University, Tsushima, Okayama 700 Japan
**Materials Department, School of Applied Science and Engineering, University of California, Los Angeles CA 90024, U.S.A.

ABSTRACT

Bioactive ORMOSILS (Organically Modified Silicates) were synthesized by a sol-gel method, with tetraethoxysilane (TEOS) and polydimethylsiloxane (PDMS). Ca(II) ions were incorporated into the ormosil monoliths by addition of calcium nitrate. The synthesized samples were examined on the bioactivity by the use of a simulated body fluid (the Kokubo solution). The Ca(II) containing ormosils were bioactive that deposited apatite during soaking in the Kokubo solution. The dissolution of Ca(II) from the sample favored the formation of the hydrated silica, which gave nucleation sites for apatite, while the effect of dissolved Ca(II) ions to increase the degree of supersaturation in the fluid could not be neglected.

INTRODUCTION

Some inorganic materials can directly bond to soft and hard tissues when embedded in human bodies [1-4]. They have already been used in clinical fields. However, those ceramic materials are far from the ideal bone substitute since their fracture toughness is lower than that of human cortical bone, and they thus find a limited range of use. Artificial polymers like silicone, on the other hand, that are the main stream materials employed for soft tissue substitutes are not bioactive but only biotolerant. They are surrounded by a fibrous tissue when embedded in the bodies. As a result, they may cause pain, bleeding, inflammation or other inconvenience. Thus such implant materials are in demand that not only are bioactive but have better mechanical properties.

Studies on the bone-bonding mechanism of materials indicated [5] that essential is deposition of a layer of apatite, similar to bone in composition and crystallinity, on the surface when they are in contact with the blood plasma. Ohtsuki et al. [6] concluded that Ca(II) ions dissolved from the bone-substitutes and a layer of silanol groups (Si-OH) left on their surface are the key materials for the apatite deposition: the Ca(II) ions increases the degree of supersaturation for apatite precipitation and the silanol layer provides the sites of nucleation. Therefore, it is expected that polymers having either silanol groups or Ca(II) ions in their structure should exhibit bioactivity.

In the present study, we prepared bioactive ORMOSILS (Organically Modified Silicates) starting from silanol terminated ploy(dimethylsiloxane) (PDMS) and tetraethoxysilane (TEOS) through sol-gel processing after Mackenzie et al. [7]. It is advantageous that the flexibility of these materials can be controlled by changing the mixing ratios of the organic and inorganic components and that the silanol groups (Si-OH) can be incorporated in the structure. If Ca(II), the other key species for the bioactivity, is introduced in the structure, the materials can exhibit both flexibility and bioactivity. Jones et al. prepared through the sol-gel processing [8] similar composite polymers of SiO_2-poly(methyl methacrylate), which involved Ca(II) ions and exhibited bioactivity. Mechanical properties for the polymers were unfortunately not described, and it was uncertain whether they were rubbery or not. Thus we adopted the PDMS-TEOS composite after Hu and Mackenzie [7]. We synthesized Ca(II) dispersed Ormosils and confirmed

Mat. Res. Soc. Symp. Proc. Vol. 435 © 1996 Materials Research Society

apatite formation under an in vitro experiment with a simulated body fluid (SBF, the Kokubo solution).

EXPERIMENTAL

Sol-gel preparation of Ca(II) containing ormosils

The starting materials were reagent grade calcium nitrate (Nacalai tesque), TEOS (Nacalai tesque) and PDMS (Aldrich) of a 20 centistokes solution in viscosity while reagent grade 2-propanol and tetrahydrofurane (Nacalai tesque) were used for the solvents and hydrochloric acid (Nacalai tesque) was the catalyst. TEOS (10g) and silanol terminated PDMS (5.9g) were mixed with a mixture of 2-propanol (4.8 ml) and tetrahydrofurane (3.2 ml). This solution was denoted as solution A. An appropriate amount of calcium nitrate was dissolved in distilled water, and HCl solution (35%) and 2-propanol (8.0 ml) was added to the solution in turn and stirred (solution B). Thus obtained were the precursor solutions of the mixing ratios indicated in Table 1. Solutions A and B were mixed and subsequently refluxed under stirring at 80°C for 30 min. After the reflux, the mixture was quenched to 25°C with iced water, cast into containers with a cover, and aged for gelation at 25°C. After gelation, the gels were covered with a vinylidene chloride film with a few pin holes and dried at 25°C for 1 to 2 weeks. Then the film was removed and the gel was dried at 40°C and further dried at 60°C in an oven for 48 h at each temperature. The experimental flow diagram was schematically shown in Figure 1. Rubbery ormosils (TEOS/PDMS = 60/40 in weight) free from Ca(II) ions were prepared as described elsewhere [7].

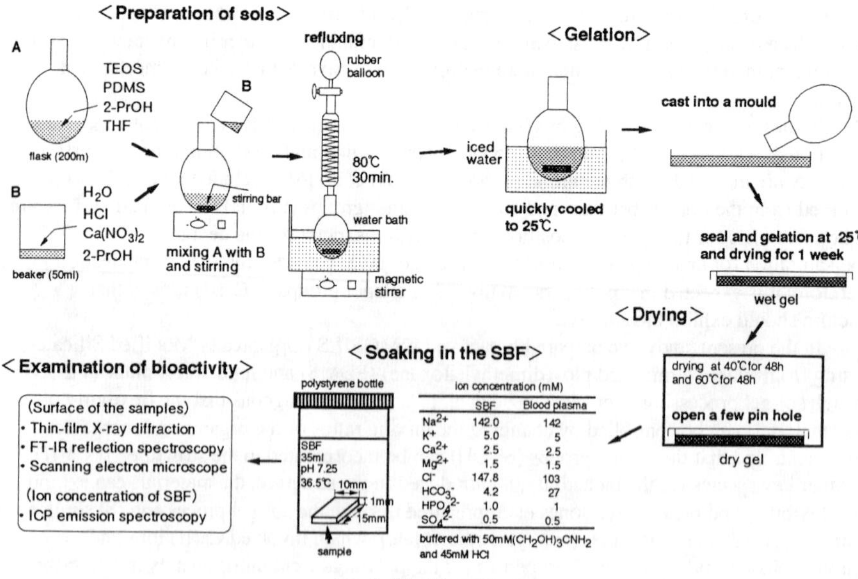

Fig. 1 The Flow diagram.

Examination of bioactivity

Table 1 Ormosils (in weight ratio).

	TEOS	PDMS	HCl	H_2O	$Ca(NO_3)_2$
Rubbery Ormosil	6	4$^{a)}$	0.3	1.6	—
A	6	3.5$^{b)}$	0.1	1.6	0.05
B	6	3.5$^{b)}$	0.1	1.6	0.25
C	6	3.5$^{b)}$	0.05	1.6	0.25

a) The average MW : 1700 (Petrach System)
b) viscosity : 20 centistokes (Aldrich Chemical)

Each ormosil was cut to 15x10x1 mm^3 in size, and the surface was polished with a #2000 Emery paper. Then it was gently rinsed with ethanol and dried. A simulated body fluid (the Kokubo solution, SBF) was prepared as described previously [6]. The SBF contains the same inorganic components as the human blood plasma in similar concentrations. It has been proved [9] that an apatite formation of implant in vivo can almost fully be reproduced in vitro experiments in the SBF. The specimen were kept in the SBF at 36.5°C up to 30 days. The concentration of the ions in the SBF was analyzed by inductively coupled plasma photometry (Seiko Electronics, SPS7700), and pH of the SBF was also monitored.

The specimen soaked in the SBF were gently rinsed with distilled water and dried. Thin film X-ray diffraction was measured with a diffractometer (Rigaku, RAD-II AX) attached with an thin-film apparatus; the incident angle was 1°. FT-IR spectra were measured with JASCO FT-IR300 spectrometer taking 75° reflection angle. Signals out of 1 μm depth could be obtained with both methods.

RESULTS

Figure 2 shows SEM photographs for the surfaces of the ormosil samples. The microstructure of the ormosils depends on the mixing ratio of the starting materials.

Figure 3 shows the thin-film X-ray diffraction (TF-XRD) patterns and FT-IR reflection spectra of the ormosil samples before and after soaking in the SBF for 30 days. The XRD and IR peaks were assigned as shown in the figures after the assignments in the literature[3,6,10,12]. The XRD patterns and IR spectra of the rubbery ormosil showed no change after the soaking. Thus the Ca free sample could not deposit the apatite layer and was not bioactive. Sample A gave no XRD peaks that could be assigned to apatite. However, the IR spectra indicated the existence of phosphate ions on the surface. The TF-XRD patterns and IR spectra indicated that sample C deposited the apatite layer within 30 days of soaking.

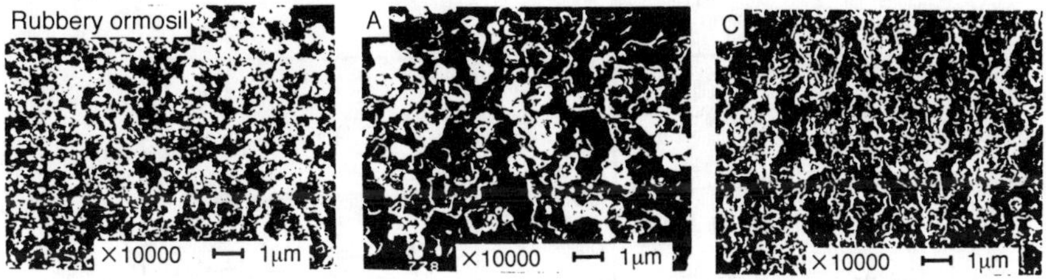

Fig. 2 SEM photographs for the surfaces of a rubbery ormosil, sample A and sample C.

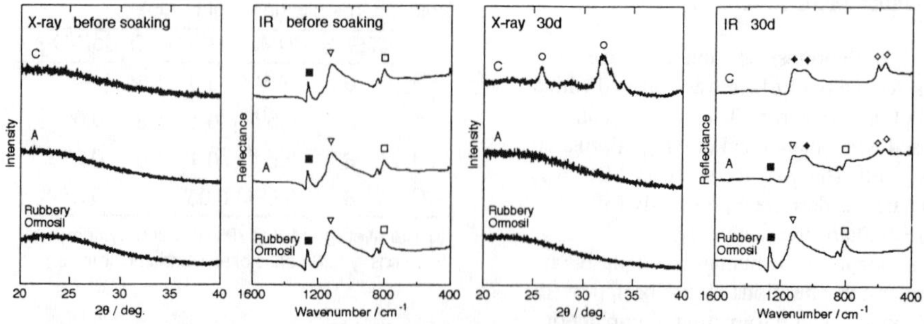

Fig. 3 Thin-film X-ray diffraction patterns and FT-IR reflection spectra for a rubbery ormosil, sample A and C before and after soaked in simulated body fluid for 30days.
○:Apatite, ▽:Si-Ostr., □:CH_3rock & Si-C str., ■:CH_3 def., ◇:P-O def., ◆:P-O str.

Figure 4 shows the SEM photographs for the surface of samples A and C after soaked in the SBF for 30 days. All these samples with Ca(II) deposited apatite. Sample C was fully covered with an apatite layer on the surface. However, Sample A deposited only a little amount of apatite. Therefore the reason for that the XRD peaks of apatite could not be detected in Fig. 3 is that sample A deposited too little an amount of apatite.

Figure 5 shows the plots of the concentrations of Ca(II), P(V) and Si(IV) ions in the SBF as a function of the soaking period for samples A and C. The values of pH of the SBF were also indicated. All the samples showed a similar trend : the concentration of the Ca(II) and

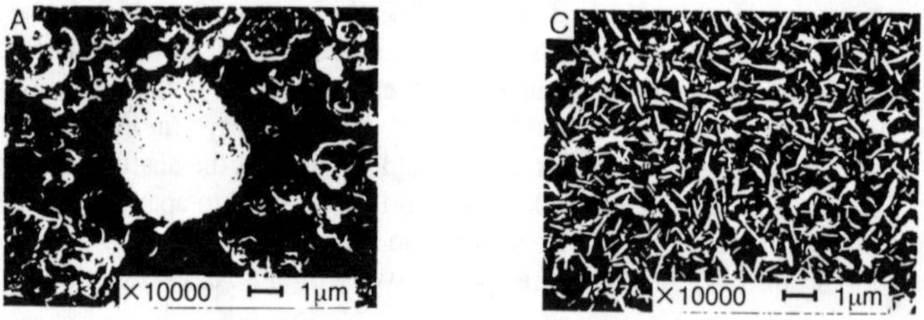

Fig. 4 SEM photographs for the surfaces of samples A and C after soaked in the SBF for 30 days. Apatite was observed for every sample but only a little on A (a spherical agglomenate at the center).

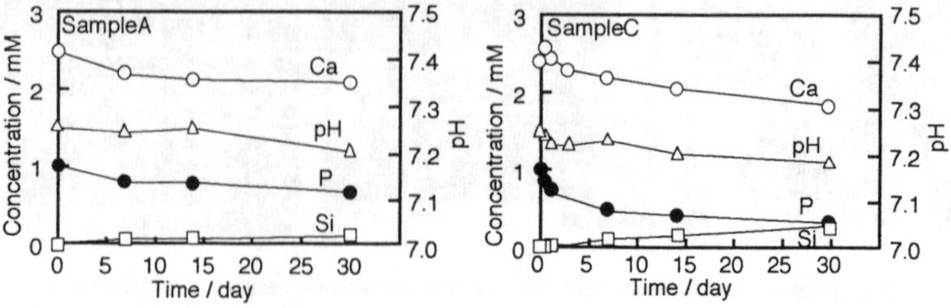

Fig. 5 The concentration of the elements and pH values of the SBF as a function of the period of soaking.

P(V) ions decreased and that of Si(IV) increased with the soaking time. The decrease in Ca(II) and P(V) corresponded to the apatite precipitation confirmed by Figs. 3 and 4. It is noted that the concentration of Ca(II) for sample C increased to a maximum in 1 day soaking and then decreased gradually.

DISCUSSION

The Ormosils consist of silica blocks, composite blocks due to copolymerization between TEOS and PDMS, and PDMS monomers left. We used calcium nitrate as the calcium source for the sample preparation. The Ca(II) ions are present in the form of $Si\text{-}O^{-}\cdots Ca^{2+}$ in the silica blocks since the PDMS blocks of the ormosils are built by covalent bonds. Then the Si-O-Ca bonds are hydrolyzed in the SBF to result in the formation of the Si-OH groups and the release of the Ca(II) and hydroxide ions [eq. (1)].

$$Si\text{-}O^{-}\cdots Ca^{2+} + H_2O \rightarrow Si\text{-}OH + Ca^{2+} + OH^{-} \qquad (1)$$

The effect of the dissolved Ca(II) ions on the apatite formation can be evaluated by the ion activity product (IAP) regarding apatite in the SBF [6]. Fig. 6 indicates the IAP for each sample as a function of the soaking time. IAP of $50CaO\bullet 50SiO_2$ glass is also shown in Fig. 6 as a reference. It increased immediately after soaking but decreased as soon as apatite formed in 1day. On the other hand, IAP of the bioactive ormosils decreased monotonously without showing a maximum. Thus the rest of the two factors is significant for the apatite deposition. That is, we conclude that bioactivity in these samples is predominantly favored by the formation of hydrated silanol groups on the surface rather than the increase of IAP due to dissolution of Ca(II).

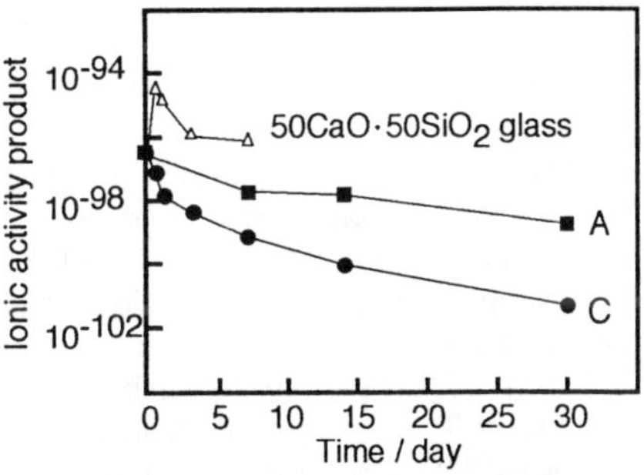

Fig. 6 Ionic activity product for samples A and C regarding apatite in the SBF.

We found in Fig. 4 that the amount of apatite deposited on sample A was smaller than that deposited on sample C. Sample C contained Ca(II) ions more than sample A and released more OH- ions (eq. (1)). Therefore, the reaction shown by equation (2) occurred frequently than in sample A.

$$Si\text{-}O\text{-}Si + OH^{-} \rightarrow 2Si\text{-}OH \qquad (2)$$

The larger concentration of Si(IV) dissolved from sample C to be SBF than that from sample A can be accounted for by eq. (2) : the terminal O-Si bonds were degraded to release Si(IV). Consequently, the absence of Ca(II) in the silica blocks caused the Ca(II) free ormosil less reactive with the SBF and it could not serve a sufficient amount of the apatite nucleation sites. This explains the fact that the Ca(II) free ormosil is not bioactive. The porous silica gels prepared by Li et al.[11] were bioactive. Their gels were more reactive with the Kokubo solution than the present ormosils. Thus the Ca(II) free ormosils may be bioactive if the reactivity with the SBF is increased.

CONCLUSION

Bioactive ormosils incorporated with Ca(II) ions were synthesized by the sol-gel process. Their bioactivity was investigated on the basis of the thin film X-ray diffraction patterns and IR reflection spectra for the specimen before and after soaking in a simulated body fluid (SBF). The Ca(II) free sample could not precipitate apatite on the surface, whereas Ca(II) containing samples deposited apatite during immersion in the SBF. The ion activity product in the SBF decreased monotonously with the soaking time. Thus the dissolution of Ca(II) ions from the ormosils effected not only increase of the degree of supersaturation in the SBF but also the ease of forming of suitable Si-OH groups for bioactivity.

ACKNOWLEDGEMENTS

One of the authors(K. T.) gratefully acknowledges Research Fellowships of the Japan Society for the Promotion of Science for Young Scientists.

REFERENCES

[1] L. L. Hench, R. J. Splinter, W. C. Allen and T. K. Greenlee, J. Biomed. Mater. Res., **2**, 117-141 (1972).

[2] J. Wilson, in Glass-Current Issues, edited by A. F. Wright and J. Dupuy, (Martinus Nijhoff Publishers, Dordrecht, 1985), pp. 662-669.

[3] L. L. Hench, J. Am. Ceram. Soc., **74**, 1487-1510 (1991).

[4] T. Yamamuro, in Introduction to bioceramics, edited by L. L. Hench and J. Wilson, (World Scientific, Singapore, 1993), pp. 89-103.

[5] T. Kokubo, J. Ceram. Soc. Japan, **99**, 965-973 (1991) .

[6] C. Ohtsuki, T. Kokubo and T. Yamamuro, J. Non-Cryst. Solids, **143**, 84-92 (1992).

[7] Y. Hu and J. D. Mackenzie, J. Mat. Sci., **27**, 4415-4420 (1992).

[8] S. M. Jones, S. E. Friberg, J. Sjoblom, J. Mat. Sci., **29**, 4075-80 (1994).

[9] T. Kokubo, H. Kushitani, S. Sakka, T. Kitsugi and T. Yamamuro, J. Biomed. Mater. Res., **24**, 721-734 (1990).

[10] R. M. Almeida, T. A. Guiton and C. G. Pantano, J. Non-Cryst. Solids, **119**, 238-241 (1990).

[11] P. Li, C. Ohtsuki, T. Kokubo, K. Nakanishi, N. Soga, T. Nakamura and T. Yamamuro, J. Am. Ceram. Soc., **75**, 2094-2097 (1992).

[12] D. R. Anderson, in Infrared, Raman and Ultraviolet Spectroscopy, edited by A. L. Smith, (JOHN WILEY & SONS, N. Y., 1974), pp. 247-286.

DIRECT DEPOSITION OF SILICA FILMS CONTAINING ORGANIC GROUPS AND DYES FROM SILICON ALKOXIDE SOLUTIONS

JUNROK OH*, HIROAKI IMAI*[†], HIROSHI HIRASHIMA*, AND KOJI TSUKUMA**

*Faculty of Science and Technology, Keio University, 3-14-1 Hiyoshi,Kohoku Yokohama 223, Japan

**Tsukuba Research Laboratory, Tosoh Corporation, 43 Miyukigaoka, Tsukuba 305, Japan

[†] to whom correspondence should be addressed

ABSTRACT

A new direct-deposition process for silica thin films was developed using silicon alkoxide solutions. Silica thin films incorporating organic groups, such as methyl (CH_3) and phenyl (C_6H_5) groups, were deposited on substrates from aqueous and ethanol solutions of organyltrialkoxysilane. Silica films without organic groups were also formed using ethanol solution of tetraalkoxysilane with aqueous ammonia. Composition of the deposited silica films were influenced by temperature and pH of the solutions and the starting alkoxides. From aqueous solutions containing organyltrialkoxysilane and organic dyes, such as Disperse Red 1 and Rhodamine B, silica films including the dyes were directly prepared on substrates. The composition of silica-organic hybrid films prepared by the deposition method was successfully controlled by selecting the deposition condition and the starting materials.

INTRODUCTION

Various types of inorganic-organic composite/hybrid materials have been prepared in recent years. The incorporation of organic groups into an inorganic film offers promising prospects for the preparation of completely new materials which are of interest commercially as well as scientifically because of their unique properties and their hybrid nature. Many methods have been demonstrated for the synthesis of hybrid materials. In particular, novel materials incorporating organic components into silicate systems through sol-gel methods have been reported [1-8]. Silica-based inorganic-organic hybrid materials were prepared by hydrosilylation [1] and co-hydrolysis/co-polymerization in solutions mixed with various kinds of alkoxides, such as tetraalkoxysilane and organyltrialkoxy- or diorganyldialkoxysilane [2-4]. Silica films containing methyl (CH_3) groups were formed on stainless steel substrates from methyltriethoxysilane (MTES) [5]. Organic dyes active for optical functions were also doped into a silica matrix, such as particles [6], gel [7] and films [8]. The alkoxides, primarily on sol-gel method, are normally chosen for their high activity and ease of reaction control. However, the sol-gel method using alkoxides requires a heat-treating process in order to obtain a dense structure and remove residual organic solvent. Thus, it is relatively difficult to obtain highly dense and durable products containing organic groups at a lower heating temperature by this method. When organic guest materials are doped into an inorganic matrix by this method, the guests are uncertainly released because of loose incorporating with host materials.

Direct deposition processes of thin films from solutions have significant advantages. Dense products can be prepared at low temperatures by immersing the substrate into a supersaturated solution. Various kinds of oxides and sulfides were produced by the deposition methods [9-10]. An aqueous solution system with fluoride ions attracts considerable interest as processing

Mat. Res. Soc. Symp. Proc. Vol. 435 © 1996 Materials Research Society

technique to prepare silica films at lower temperature [11-12]. However, this method has demerits which involves using hydrofluoric acid difficult to deal with.

Recently, silica films containing organic groups were formed to be directly deposited onto complex shapes and a wide variety of substrates from aqueous solution of alkoxides [13]. We found that the composition of the films can be controlled by the deposition condition. In this paper, we report on a new deposition method using alkoxides for preparation of silica-organic hybrid films incorporating organic groups and dyes. We discuss the effect of deposition conditions, such as pH and temperature of the solution, on the deposition process and approaches to composition control for "tailor-made" inorganic-organic hybrid films.

EXPERIMENT

We prepared three series of precursor solutions for depositing silica films: (1) aqueous solutions of organyltrialkoxysilane for films containing several organic groups, (2) ethanol solutions of organyltrialkoxysilane and/or tetraalkoxysilane with aqueous ammonia, and (3) organic-dye-dissolved aqueous solution of organyltrialkoxysilane for dye-doped silica films. (1) Methyltrimethoxysilane (MTMS) and phenyltrimethoxysilane (PTMS) were used as starting materials. Each of these chemicals was dissolved in pure water, and then vigorously stirred magnetically at 60°C for 1 hour. The obtained solution was mixed with various amounts of hydrochloric acid (HCl) or aqueous ammonia to adjust pH. The alkoxide concentration of the precursor solutions was 0.03mol/L. (2) Tetraethoxysilane (TEOS), MTES or mixture of TEOS and MTES was dissolved in ethanol. Then the alkoxide solutions were added to solution of aqueous ammonia (1N) and ethanol and magnetically stirred at 60°C for 1 hour. Typical mixing ratio was TEOS (16mL), ethanol (272mL) and aqueous ammonia (1N, 50mL). (3) Organic dye, such as Disperse Red 1 and Rhodamine B, was doped into an aqueous solution of phenyltriethoxysilane (PTES). HCl was added to accelerate hydrolysis of PTES in the mixed solution. Concentration of PTES and organic dyes were adjusted to 0.03 mol/L and between 5×10^{-4} - 2×10^{-5} mol/L, respectively. The pH of the precursor solutions was adjusted to between 0.05 - 3.

After the precursor solutions were stirred again at 60°C for 1 hour, substrates were immersed into the precursor solutions. Single-crystal silicon wafers, glass slides and nylon 6 films etc. were used as a substrate. The immersed solution was held in an oven or a refrigerator at a constant temperature between 0 and 120°C for several hours. An autoclave was used for the deposition above 100°C. In order to remove loosely bound precipitates, the substrates were washed with pure water ultrasonically. The deposited films were dried at 60°C.

The precursor solutions were characterized by ^{1}H-, ^{13}C-, and ^{29}Si-NMR spectroscopy (Nihon Denshi JNM-FX 90A and α-500). The deposited films were analyzed by a Fourier-transform infrared (FTIR) spectrometer (BIO-RAD Digilab Division FTS-65). Surface morphology of the films was observed by a scanning electron microscope (SEM)(Nihon Denshi JSM-5200). Thickness and refractive index of the films were measured by an ellipsometer (Mizojiri DHA-XA). The thickness was also estimated by the absorption intensity in the FTIR spectra and the SEM observation. Rutherford backscattering spectrometer (RBS) and elastic recoil detection analysis (ERDA) with an ion beam analyzer (Shimadzu IBA-9900) were used to determine the chemical elements. Optical absorption measurements were carried out for the organic-dye-doped silica films and the precursor solutions by an ultraviolet-visible (UV-VIS) spectrometer (Shimadzu UV-2500PC).

RESULTS AND DISCUSSION

Homogeneous films with smooth surface as shown in Fig.1 were deposited from aqueous solutions of MTMS and PTMS at pH lower than 3 or higher than 6. According to FTIR spectra, the deposited films consist of silica containing organic groups, methyl(CH_3) or phenyl(C_6H_5), which were originated from the starting materials. We did not observe the film deposition and any precipitation in the solutions at pH 3-6. In case of TEOS and fluorotrimethoxysilane (FTMS), stable aqueous solutions were not obtained because of gelation and steep formation of a large amount of precipitates, respectively. The film deposition was also influenced by temperature of the solutions. The deposition rate was accelerated with increasing deposition temperature as shown in Fig.2. The film deposition was observed on all kind of substrates used in this study. The deposition films obtained exhibited a good adhesion to the substrates, regardless of the shape and properties of the substrates such as hydrophilic and hydrophobic.

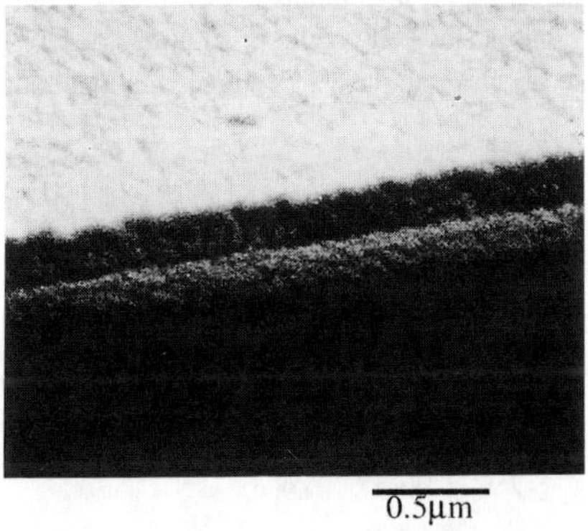

Fig.1 SEM photograph of a deposited silica film with methyl groups.

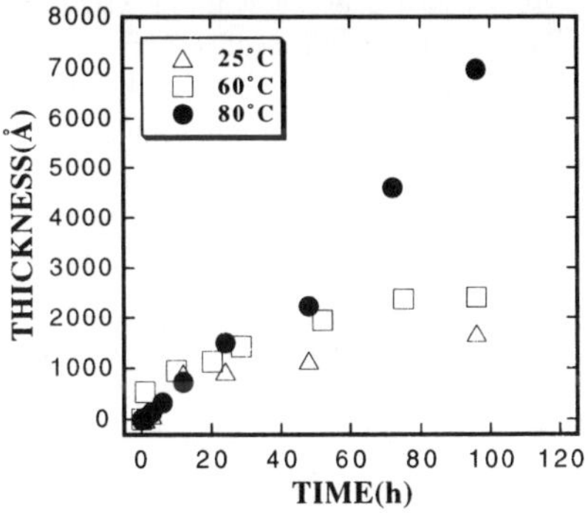

Fig.2 Film thickness as a function of deposition time at various temperatures.

^{1}H-, ^{13}C- and ^{29}Si-NMR were recorded to monitor hydrolysis reactions in the precursor solutions of MTMS. The resonances due to methanol as a hydrolysis product and methyl group coupled to a silicon were observed, while the resonances corresponding to methoxy groups (OCH_3) of the starting material were not detected. Thus, the alkoxides were suggested to be completely hydrolyzed. The ^{29}Si-NMR spectra indicate that the hydrolyzed silicon compounds exist as a monomer and oligomers in the solutions. The greater electron-providing property of CH_3 and C_6H_5 groups than alkoxy and OH groups and fluorine is suggested to be related to the formation of the stable aqueous solution with the completely hydrolyzed species which result in the film deposition. In the precursor solutions at pH lower than 3 or higher than 6, films are deduced to be deposited because the condensation reaction of the hydrolyzed species is accelerated by H^+ or OH^-.

The composition of the deposited films were also influenced by pH and temperature of the precursor solutions. Fig.3 shows the FTIR spectra of the silica films deposited at various temperatures under acidic conditions. An absorption band at 1260 cm^{-1} is assigned to the CH symmetric stretching vibrations of CH_3 group on silicon ($Si\text{-}CH_3$) and always accompanied by a

760 cm^{-1} band, which arises from CH_3 rocking and Si-C stretching vibrations [14-15]. Absorption bands at 1020 and 1120 cm^{-1} are suggested to be attributed to an asymmetric Si-O-Si stretching vibrations in siloxane compounds [16-17]. In Fig.3, the intensity of absorption peaks corresponding to the Si-C stretching vibrations normalized with the film thickness decreased with increasing deposition temperature. From the relationship between the deposition temperature and absorption coefficient of the 1260 cm^{-1} and the 760 cm^{-1} bands, which were calculated with the peak intensity and the film thickness, it is considered that the amount of CH_3 groups incorporated with Si atoms in silica films deposited at 120°C was reduced to below one-half of that in films deposited at 0°C. The element ratio of Si:C in silica-organic films deposited at 80°C is evaluated to be approximately equal to 2:1 by RBS and ERDA. In case of the higher deposition temperatures, the elimination of CH_3 groups coupled to Si is assumed to take place as well as deposition rate increasing with elevating temperature. Since the area of the 1020 cm^{-1} band in comparison with that of the 1120 cm^{-1} band increased with increasing deposition temperature, the siloxane structure of the films is suggested to be changed with the elimination of CH_3 groups. The FTIR spectra for silica films deposited from the base-catalyzed precursor solutions at 60°C were essentially the same as those of the acid-catalyzed films at above 80°C, suggesting that both films consist of the similar components.

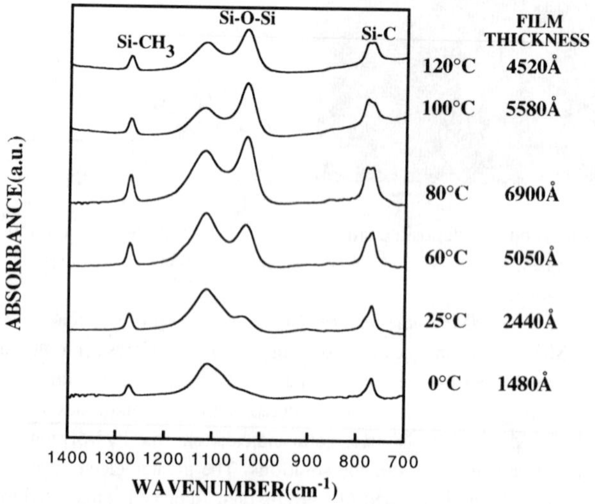

Fig.3 FTIR spectra of deposited films at various deposition temperatures, 0.03 mol/L and acidic condition(pH 0.05).

Silica films without organic groups were not obtained by the aqueous solution method because aqueous solution of TEOS was not stable as mentioned above. On the other hand, ethanol solution of TEOS with a lot of aqueous ammonia, which is known to be as a precursor solution for synthesis of large silica particles [18], was assumed to contain relatively stable hydrolyzed species. After keeping of the ethanol solution of TEOS in a oven at 60°C for 1 day, we observed the formation of homogeneous silica films with a small amount of precipitation. Silica films containing CH_3 groups were also deposited from the ethanol solutions of MTES and the mixture of TEOS and MTES. Fig.4 shows FTIR absorption spectra for silica films formed from the ethanol solutions. The TEOS- or TEOS/MTES- derived film shows no or smaller absorption peaks due to Si-CH_3 bonds than those for the films from the aqueous solution of MTMS and a clear peak at 1080 cm^{-1} due to a Si-O-Si bond as well as pure silica. The results indicate that silica films containing no or smaller amount of organic groups are also produced using the deposition processes in solutions.

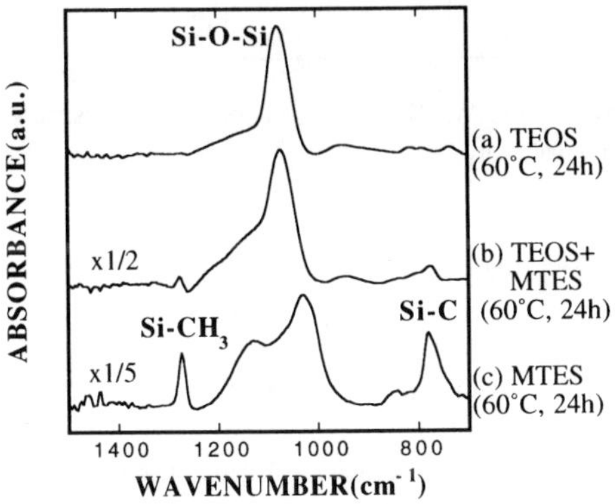

Fig.4 FTIR spectra of silica films deposited from the ethanol solutions with aqueous ammonia.

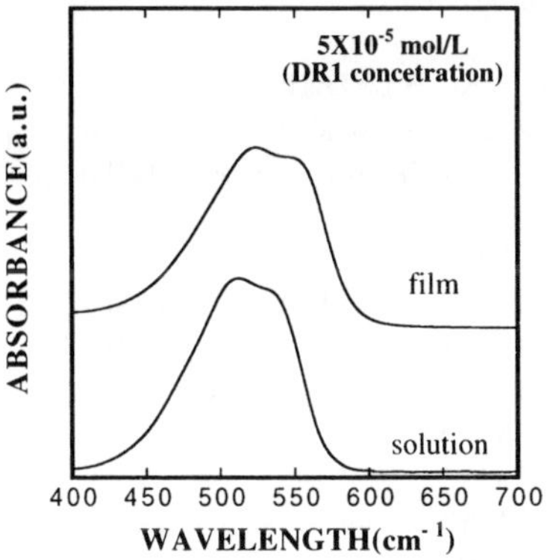

Fig.5 Absorption spectra of a DR1-doped silica film and its precursor solution.

Organic dyes, such as DR1 and Rhodamine B (Fig.6), were dissolved in aqueous solution of PTES and successfully doped into silica films deposited from the solutions. Since the hydrophilic property of the organic dyes is relatively weak, the organic molecules are assumed to be surrounded by C_6H_5 groups of PTES in the solution. Absorption spectra for the solution containing DR1 and the deposited silica films are shown in Fig.5. Since the spectra for the film and the solution are almost the same, the organic dyes doped into silica films exist with homogeneous dispersibility in a similar manner as the solution. The dyes are tentatively presumed to be trapped in a cage of phenyl groups in the films because they were not released in water. Silica films containing Rhodamine B were also formed with the same tendency as DR1.

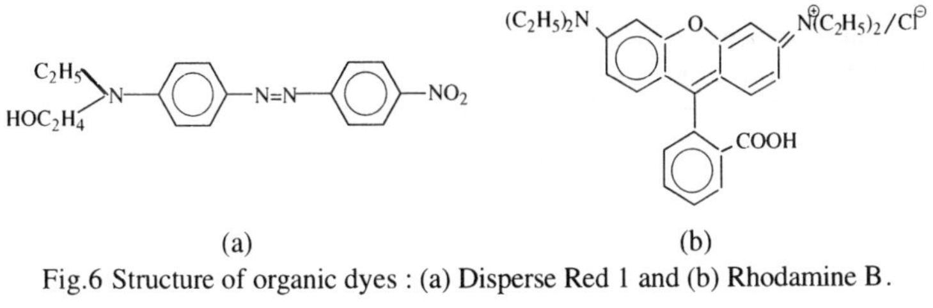

(a) (b)

Fig.6 Structure of organic dyes : (a) Disperse Red 1 and (b) Rhodamine B.

CONCLUSIONS

Direct formation using solutions of silicon alkoxides was investigated to prepare silica-organic hybrid materials. Silica films containing organic groups coupled to Si atoms and organic dyes were successfully deposited on a substrate from aqueous and ethanol solutions. When C_6H_5- and CH_3-containing alkoxides were used as a starting material, the deposited films containing C_6H_5 and CH_3 groups, respectively. Composition of the films were controlled by the starting materials and the deposition condition, such as temperature and pH values of the precursor solution. Organic dyes-doped silica films were also obtained by the deposition method with the solution containing the dyes. The direct deposition from the alkoxide solutions is expected to be a promising synthesis method for inorganic-organic hybrid films applying for various fields.

ACKNOWLEDGEMENT

The authors would like to thank Mr. T. Takayama and Mr. S. Ohya for experimental assistance of ^{29}Si-NMR spectroscopy and ellipsometry, respectively.

REFERENCES

1. P.A.Agaskar, J. Am. Chem. Soc., **111**, 6858 (1989).
2. R. H Glaser, G.L. Wilkes and C.E. Bronnimann, J. Non-Cryst. Solids, **113**, 73 (1989).
3. I. Hasegawa and S. Sakka, Bull. Chem. Soc.Jpn., **63**, 3203 (1990).
4. M.J. van Bommel, T.N.M. Bernards and A.H. Boonstra, J. Non-Cryst. Solids, **128**, 231 (1991).
5. M. Murakami, K. Izumi, T. Deguchi, A. Morita, N. Tohge and T. Minami, J. Ceram. Soc. Jpn., **97**, 91 (1989).
6. S. Shibata, T. Taniguchi, T. Yano, A. Yasumori and M. Yamane, J. Sol-Gel Sci.and Tech., **2**, 755 (1994).
7. A. Makishima, K. Morita, H. Inoue and T.Tani in Sol-Gel Optics, edited by J. D. Mackenzie and D. R. Ulrich (SPIE.Proc.**1328**, San Diego, CA, 1990) pp.264 -267.
8. Y. Zhang, P.N. Prasad and R. Burzynski, Chem. Mater., **4**, 851 (1992).
9. Raviendra D. and J. K. Sharma, J. Appl. Phys., **58**, 838 (1985).
10. R. L. Call, N. K. Jaber, K. Seshan and J. R. Whyte, Jr., Sol. Energy Mater., **2**, 373 (1980).
11. H. Nagayama, H. Honda and H. Kawahara, J. Electrochem. Soc.,**135**, 2013 (1988).
12. A. Hishinuma, T. Koda, M. Kitaoka, S. Hayashi and H. Kawahara, Applied Surface Sci., **48/49**, 405 (1991).
13. K. Tsukuma, T. Akiyama, N. Yamada and H. Imai, J. Non-Cryst. Solids (in press).
14. C. W. Young, P. C. Servais, C. C. Currie and M. J. Hunter, J. Am. Chem. Soc., **70**, 3758 (1948).
15. A. L. Smith, J. Chem. Phys., **21**, 1997 (1953).
16. N. Wright and M. J. Hunter, J. Am. Chem. Soc., **69**, 803 (1947).
17. R. C. Lord, D. W. Robinson and W. C. Schumb, J. Am. Chem. Soc., **78**, 1327 (1956).
18. W. Stober, A. Fink and E. Bohn, J. Colloid and Interface Sci., **26**, 62 (1968).

THERMAL STABILITY OF SILICON-BASED HYBRID MATERIALS CONTAINING ALUMINUM STUDIED BY ^{27}Al AND ^{29}Si SOLID STATE MAS NMR

M.P.J. Peeters*, A.P.M. Kentgens** and I.J.M. Snijkers-Hendrickx*
*Philips Research Laboratories, Prof. Holstlaan 4, 5656 AA Eindhoven, The Netherlands
**SON HF-NMR Facility, University of Nijmegen, 6525 ED Nijmegen, The Netherlands

ABSTRACT

The addition of $Al(OBu^s)_2EAA$ to silicon-based materials containing PhTES or Glymo leads to a decrease in the thermal stability of the trifunctional siloxanes. ^{29}Si MAS NMR shows that breaking of the Si-C bond and formation of an extra Si-O bond occurs after heating to 200°C of these materials. MeTES-containing materials are stable up to (at least) 200°C. The preparation of samples using $Na[AlOR_4]$ instead of $Al(OBu^s)_2EAA$ showed that the undesired decomposition reaction is caused by octahedral (non-network) aluminum (^{27}Al MAS NMR). Up to 4-5 Si-C bonds are broken per octahedral aluminum, which indicates a catalytic effect.

INTRODUCTION

For the sol-gel preparation of thick coatings with improved mechanical properties [1] $Al(OBu^s)_2EAA$ (aluminum sec-butoxide ethyl-acetoacetate [$C_6H_{10}O_3$]) is often added to organically modified silicon-based hybrid materials. The stability of the Si-C bond in materials without $Al(OBu^s)_2EAA$ is known to be good [2]. Little is known about the thermal stability of these materials in the presence of aluminum. ^{29}Si and ^{27}Al solid state MAS NMR is used to characterize the materials.

EXPERIMENT

Samples were prepared by mixing tetraethyl orthosilicate (TEOS, Merck) with 3-glycidoxypropyl trimethoxysilane (Glymo, Hüls), phenyl triethoxysilane (PhTES, Hüls) or methyl triethoxysilane (MeTES, Hüls). To this mixture was added (absolute) ethanol followed by a 1:1 (w/w) acidic water/ethanol mixture (0.005 M HCl). The last step was the addition of

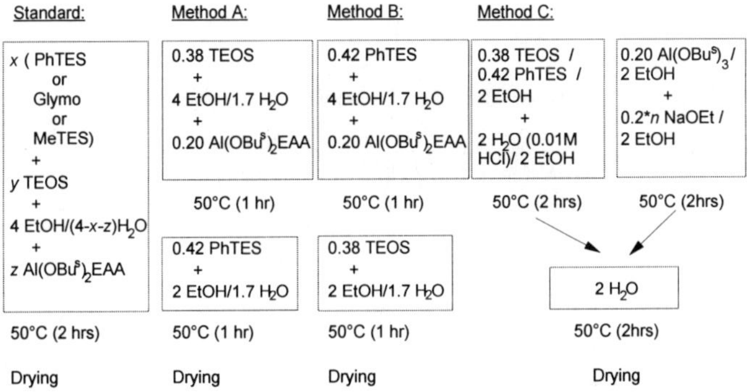

Figure 1: Schematic presentation of the different methods used for the preparation of the materials; top to bottom: mixing order.

Mat. Res. Soc. Symp. Proc. Vol. 435 © 1996 Materials Research Society

$Al(OBu^s)_2EAA$ (Gelest Inc.). All the reagents were reagent grade and were used without further purification. The general composition (in mole fractions) of the samples was: x T, y TEOS, z $Al(OBu^s)_2EAA$ (or $Al(OBu^s)_3$), 4 EtOH, $(4-x-z)$ H_2O with $x + y + z = 1$ and T is a trifunctional organically modified alkoxysilane, MeTES, PhTES or Glymo. Unless otherwise indicated, the standard procedure was used for sample preparation. Alternatively, one of the methods (*A*, *B* or *C*) described in Figure 1 was used.

Quantitative ^{29}Si MAS NMR spectra (recycle delay = 300 s) were measured on a Bruker CXP300 and a Bruker DMX300 spectrometer, operating at 59.595 MHz for silicon. Proton decoupled spectra (rf decouple strength ≈ 50 kHz) were obtained at a spinning frequency of approximately 4.0-5.0 kHz. ^{27}Al MAS NMR spectra were measured on a Bruker AM-500 spectrometer operating at 130.32 MHz (MAS speed 13.0 kHz). Short radio frequency pulses (tip angle < 15°) and a recycle delay of 2 seconds were used to obtain quantitative results (checked using α-Al_2O_3 as external standard). No proton decoupling was used.

RESULTS

A 0.20 $Al(OBu^s)_2EAA$ / 0.80 PhTES sample was prepared, which after drying at room temperature, was heated to 200 °C in air. The ^{29}Si MAS NMR spectra before and after the heat treatment are shown in Figure 2. Before the heat treatment, resonances of T_2 and T_3 sites were observed (-65 to -85 ppm). The heat treatment led to the formation of a broad resonance, centered at -100 ppm, the region ascribed to Q-atoms. The formation of resonances in this spectral region implies the breaking of the Si-C bond and the formation of an extra Si-O bond. Comparison of the intensities of the T- and Q-regions indicates that approximately 25% of the Si-C bonds was replaced by an Si-O bond. The breaking of the Si-C was confirmed by GC-IR measurements; benzene was detected in the gas produced. This decomposition reaction was observed in air as well as in a nitrogen atmosphere.

So as to be able to study the process in more detail, we measured the T/Q ratio of some samples as a function of the temperature (Figure 3) and as a function of the aluminum

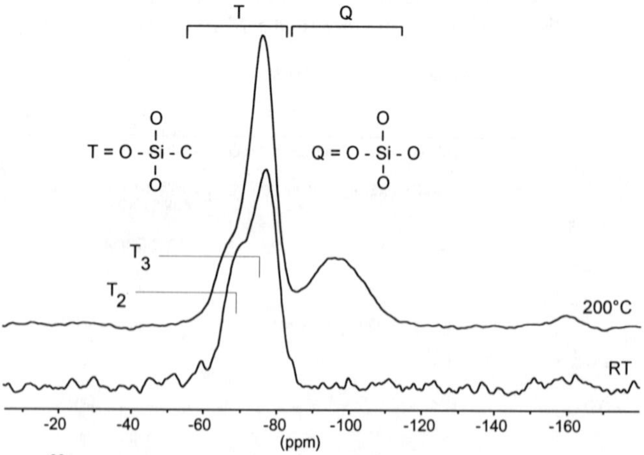

Figure 2: ^{29}Si MAS NMR spectra of a 0.80 PhTES / 0.20 $Al(OBu^s)_2EAA$ sample before and after heat treatment at 200°C.

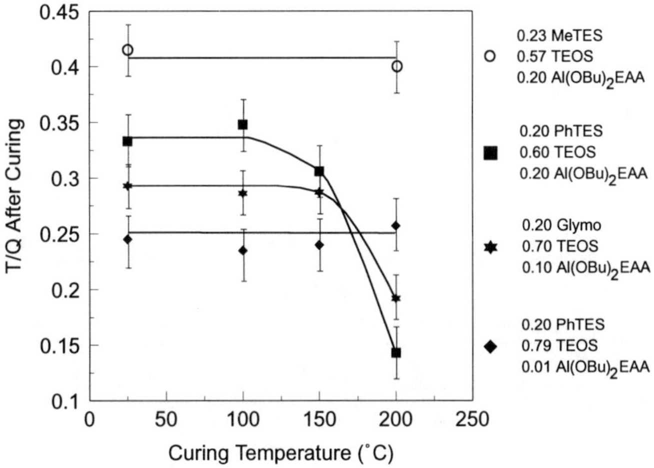

Figure 3: T/Q ratio as a function of the temperature for some samples. Heating at indicated temperature for 2 hours.

concentration after heating at 200 °C for two hours (Figure 4). From Figure 3 it can be concluded that decomposition of T-atoms in the presence of Al-atoms (breaking of the Si-C bond) took place after heating above 150 °C in the case of samples containing PhTES and Glymo with a high $Al(OBu^s)_2EAA$ content. The (undesired) decomposition reaction can only take place in the presence of aluminum. No decomposition was observed for the MeTES/TEOS/$Al(OBu^s)_2EAA$ sample after heating at 200°C.

The influence of the aluminum concentration can be more easily seen in Figure 4. The degree of decomposition of T-atoms increased with the $Al(OBu^s)_2EAA$ concentration in the Glymo/TEOS and PhTES/TEOS mixtures. The degree of decomposition observed for the

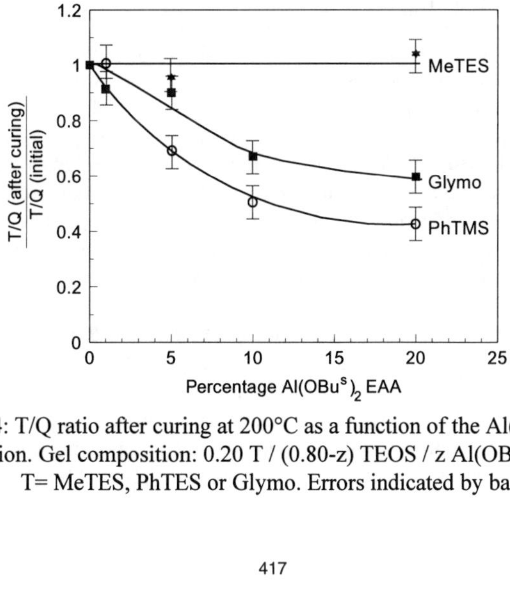

Figure 4: T/Q ratio after curing at 200°C as a function of the $Al(OBu^s)_2EAA$ concentration. Gel composition: 0.20 T / (0.80-z) TEOS / z $Al(OBu^s)_2EAA$, with T= MeTES, PhTES or Glymo. Errors indicated by bars.

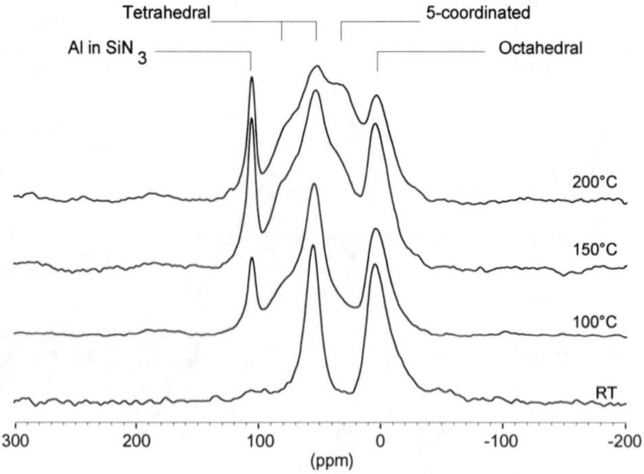

Figure 5: ^{27}Al MAS NMR (11.7 Tesla) spectra of 0.20 $Al(OBu^s)_2EAA$ / 0.80 TEOS samples after heating for 2 hours at the indicated temperature (MAS speed 13.0 kHz). The signal at 105 ppm represents an impurity in the rotor material.

PhTES/TEOS samples was slightly higher than that of the Glymo/TEOS samples. The degree of decomposition depends strongly on the $Al(OBu^s)_2EAA$ concentration. In the absence of $Al(OBu^s)_2EAA$ no decomposition of T-atoms was observed at 200°C. When the $Al(OBu^s)_2EAA$ concentration was increased the Glymo and PhTES sample were found to decompose, whereas the MeTES sample remained stable after heat treatment at 200°C in air.

^{27}Al MAS NMR was performed to see what kind of aluminum species was present in the materials. Typical spectra are depicted in Figure 5. After drying at room temperature, tetrahedral (55 ppm) and octahedral (0 ppm) coordinated aluminum species were present in approximately equal amounts. Annealing at increasing temperatures led to the formation of five coordinated aluminum (30 ppm) and a second kind of tetrahedral coordinated species (80 ppm). The amount of octahedral coordinated aluminum decreased to approximately 30-40%. The signal at 105 ppm is a background signal (rotor material). Tetrahedral coordination is normally ascribed to network aluminum whereas octahedral coordination is normally ascribed to non-network aluminum, ensuring charge compensation [3].

The addition of $Al(OBu^s)_2EAA$ to Glymo/TEOS or PhTES/TEOS mixtures leads to decomposition of the T-atoms, transforming them into Q-atoms. Furthermore, several kinds of aluminum with different coordination numbers were found to be present in the samples. A number of experiments were conducted to find out what kind of aluminum was responsible for the decomposition reaction. The composition in these experiments was fixed at 0.42 PhTES / 0.38 TEOS / 0.20 $Al(OBu^s)_2EAA$.

The mixing order of the components was varied (method *A* and *B*, Figure 1), to increase and decrease the average Al-PhTES distance. In the normal procedure PhTES, TEOS, EtOH and water were mixed, after which $Al(OBu^s)_2EAA$ was immediately added. Two extra samples were prepared where either the PhTES or the TEOS was allowed to react (1 hour, 50°C) with $Al(OBu^s)_2EAA$ before the addition of the second siloxane component. The samples obtained after heating at 200°C for two hours were analyzed with ^{29}Si MAS NMR. If the network

Table 1: Influence of the mixing order on the decomposition reaction

	Standard	Method *A*	Method *B*
T/Q (after drying)	1.0	1.1	1.1
T/Q (after annealing)	0.33	0.43	0.34
% Decomposition	50	43	52

aluminum (tetrahedral coordination) was responsible for the decomposition reaction, then the mixing order would have a pronounced influence on the degree of decomposition. From Table 1 it can be concluded that the reaction of $Al(OBu^s)_2EAA$ with TEOS prior to the addition of PhTES (Method *A*) led to a somewhat lower degree of decomposition compared to the two other methods. The differences between the different methods are however small. ^{27}Al MAS NMR showed no significant differences in the aluminum coordination between the different samples. This suggests that not the tetrahedral (network) aluminum but the octahedral (non-network) aluminum is responsible for the decomposition reaction.

To confirm the above suggestion, we prepared samples using sodium ethanolate (NaOEt) as described by Irwin et al. [3], using a slightly modified procedure (Method *C*, Figure 1). In this procedure the $Al(OBu^s)_3$ was complexed with sodium ethanolate (Na/Al = 0.25-1.0). This mixture was added (after 2 hours) to a prehydrolysed TEOS/PhTES mixture, after which an extra amount of water was added (2 mole). After drying, the samples were heated at 200°C for 2 hours. The addition of NaOEt influenced the aluminum coordination. The ^{27}Al MAS NMR spectra are shown in Figure 6. The amount of octahedral coordinated aluminum decreased with increasing NaOEt concentration since octahedral coordinated aluminum is no longer necessary for charge compensation. The amount of octahedral coordinated aluminum has a major influence on the decomposition of PhTES. The results are shown in Figure 7. It is obvious that the addition of NaOEt to the reaction mixture led to a strong suppression of the decomposition reaction; after a 1:1 complexation of $Al(OBu^s)_3$ with NaOEt the T/Q ratio remained at its initial

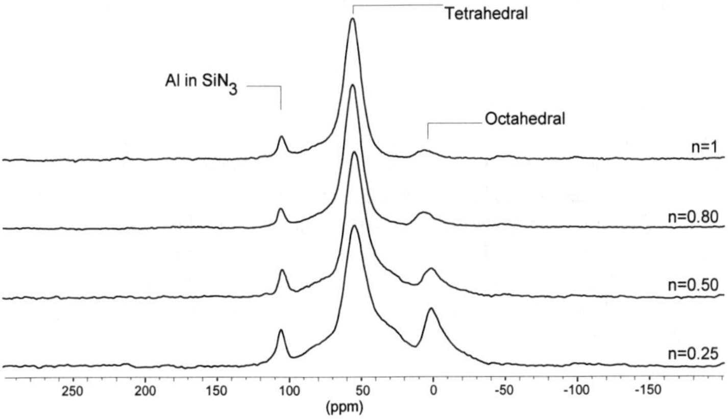

Figure 6: ^{27}Al MAS NMR (11.7 Tesla) of 0.42 PhTES / 0.38 TEOS / 0.20 $Al(OBu^s)_3$ samples after heating at 200°C (*n*= $NaOEt/Al(OBu^s)_3$ ratio). (MAS speed 13.0 kHz). The signal at 105 ppm represents an impurity in the rotor material.

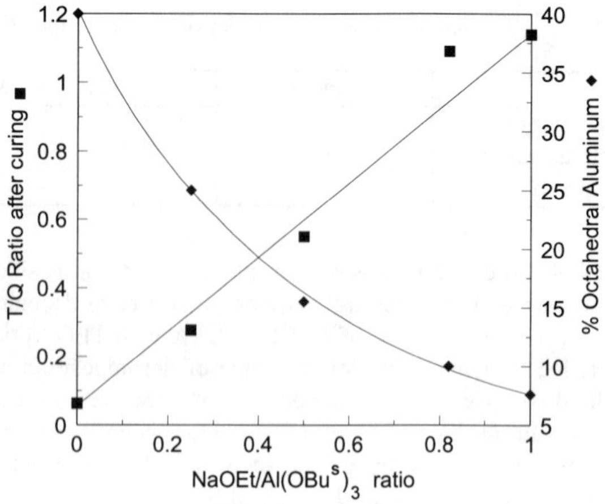

Figure 7: T/Q ratio of 0.42 PhTES / 0.38 TEOS / 0.20 $Al(OBu^s)_3$ / x NaOEt samples after heating at 200°C for 2 hours.

value of 1.1 . Using a smaller amount of NaOEt resulted in an increase of the amount of octahedral coordinated aluminum and thus in an increase of the degree of decomposition.

Also plotted in Figure 7 is the percentage of aluminum in an octahedral coordination. Complexation of $Al(OBu^s)_3$ with NaOEt leads to a decrease in the amount of octahedral coordinated aluminum (^{27}Al MAS NMR). A low concentration of octahedral coordinated aluminum was detected in the $Al(OBu^s)_3$ / NaOEt samples with a 1:1 ratio (approx. 7%). A strong correlation between the T/Q ratio after heating and the amount of octahedral coordinated aluminum was observed. Close examination of the decrease in the T/Q ratio in relation to the ^{27}Al spectra and the sol composition, revealed that up to 4-5 T-atoms were transformed into Q-atoms per octahedral aluminum present in the materials. More investigations are necessary to elucidate the underlying mechanism.

CONCLUSIONS

The addition of $Al(OBu^s)_2$EAA to materials containing PhTES and Glymo leads to a decrease in the thermal stability of the Si-C bond. Cleavage of the Si-C bond and formation of an extra Si-O bond were observed in nitrogen and atmospheric air at 200°C; in the case of PhTES benzene was released. MeTES-containing materials are stable up to (at least) 200 °C. The preparation of samples using $Al(OBu^s)_3$ /NaOEt showed that the decomposition was due to the presence of (non-network) octahedral coordinated aluminum. An increase in the amount of octahedral aluminum led to an increase in the decomposition reaction. In the absence of octahedral coordinated aluminum no decomposition was observed. A maximum of 4-5 Si-C bonds were broken per octahedral coordinated aluminum.

REFERENCES

1 M. Nogami, J.Non-Cryst. Solids **178**, p. 320-326 (1994)

2 L. Bois, J. Maquet, F. Babonneau and D. Bahloul, Chem. Mater. **7**, p. 975-981 (1995)

3 A.D. Irwin, J.S. Holmgren and J. Jonas, J. Mater. Sci. **23**, p. 2908-2912 (1988)

INVESTIGATION OF HYDROLYSIS AND CONDENSATION IN ORGANICALLY MODIFIED SOL-GEL SYSTEMS: ^{29}SI NMR AND THE INEPT SEQUENCE[†]

T. M. ALAM, R. A. ASSINK AND D. A. LOY

Properties of Organic Materials, Sandia National Laboratories, Albuquerque, NM 87185-1407

ABSTRACT

The spectral editing properties of the ^{29}Si NMR INEPT heteronuclear transfer experiment have been utilized for the identification and characterization of hydrolysis and initial condensation products in methyltrimethoxysilane (MTMS) sol-gel materials. ^{29}Si NMR assignments in MTMS are complicated by a small spectral dispersion (~0.5 ppm) and two different ^{29}Si-^{1}H J couplings. By using analytical expressions for the INEPT signal response with multiple heteronuclear J couplings, unambiguous spectral assignments can be made. For this organomethoxysilane the rate of hydrolysis was found to be very rapid and significantly faster than either the water- or alcohol-producing condensation reactions. The hydrolysis species of both the MTMS monomer and its initial T^1 condensation products follow statistical distributions that can be directly related to the extent of the hydrolysis reactions. The role of the statistical distribution of hydrolysis products on the production and synthetic control of organically modified sol-gels is discussed.

INTRODUCTION

Development of new highly crosslinked materials from organically modified alkoxysilanes continues to be an area of active research. A basic understanding of the kinetics and chemistry of the hydrolysis and condensation reactions responsible for the polymerization will allow the rational design of new and improved materials. Since hydrolysis produces the reactive silanolic species essential in subsequent condensation reactions, the relative concentrations and rate of formation of the hydrolyzed components should play an important role in the final structure of the sol-gel material. Hydrolysis and condensation chemistry in various sol-gel systems has been investigated using high resolution ^{29}Si NMR, with the majority of these investigations concentrating on tetraalkoxysilanes. The effect of organic modifications on the hydrolysis and condensation reactions is still unclear, due to the limited number of studies on organically modified alkoxysilanes[1-6].

Two difficulties encountered in these ^{29}Si NMR investigations are the *identification* and *quantification* of the different silicon environments as the reaction progresses. This is especially true in some organically modified alkoxysilanes such as MTMS where the small spectral dispersion makes even the assignment of the monomer hydrolysis products difficult. The large downfield changes in chemical shift with increasing number of hydroxyls observed in tetraalkoxysilanes is altered in organically modified alkoxysilanes resulting in smaller changes in chemical shifts or even upfield variations with increasing number of hydroxyls [7]. This can lead to chemical shift changes being different from the expected downfield progression with hydrolysis.

It is important that experimental techniques that allow for the correct assignment of ^{29}Si NMR resonances be developed. In this note the spectral editing powers of the standard INEPT

† This work supported by the US Department of Energy under Contract DE-AC04-94AL85000.

Mat. Res. Soc. Symp. Proc. Vol. 435 © 1996 Materials Research Society

sequence are used to investigate the distribution of hydrolysis species in the monomer (T^0) and initial condensation (T^1) products in methyltrimethoxysilane (MTMS).

EXPERIMENTAL

A stock solution of 2.24 M MTMS in MeOH was prepared and analyzed for hydrolysis and condensation contaminants prior to use by ^{29}Si NMR. Different H_2O/Si molar ratios (R_w) were investigated by adding chilled H_2O (273K) to the stock solution. Solutions were acidified with 1 N HCl such that the final acid concentrations were 1.58 mM, giving a nominal pH of 2.8. To reduce the ^{29}Si spin-lattice relaxation times in kinetic investigations chromium acetylacetonate ($Cr(acac)_3$) was added for a final concentration of 15.7 mM. For the INEPT experiments, no $Cr(acac)_3$ was added, as this severely degrades the performance of the pulse sequence. All ^{29}Si NMR spectra were obtained at 79.49 MHz on a Bruker AMX400 spectrometer. Kinetic investigations of the different silicon species were performed using the 10 mm broadband probe and standard inverse gate pulse sequences to reduce NOE effects. Spectra were obtained using 2-8 scan averages, an 18 s relaxation delay, and a 20 μs $\pi/2$ ^{29}Si pulse. The INEPT experiments for resonance assignments were obtained using a 5 mm probe, 8 - 16 scans, and a 2 s recycle delay. For the INEPT sequence the interpulse delay τ was 60.2 ms, while the refocussing delay Δ was varied as described in the text.

RESULTS AND DISCUSSION

The inverse gated ^{29}Si NMR spectra of the stock MTMS solution at 233 K, pH = 2.8 with 1.5 molar equivalents of water as a function of time is shown in Figure 1. At this reduced temperature only the monomer T^0, the initial condensation T^1 and associated hydrolysis products are present. The uncondensed silicon T^0 species resonate between δ = - 37 and -39 ppm, while the singly condensed T^1 species were observed between δ = -46 and -47.5 ppm. The hydrolysis products of both the monomer and condensed species show a very narrow chemical shift dispersion making resonance assignments challenging.

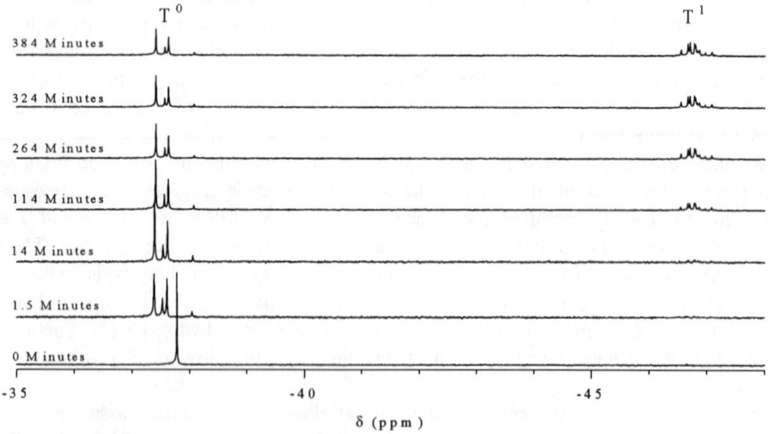

Figure 1. ^{29}Si NMR spectra of 2.24 M MTMS solution in MeOH, 233 K, pH = 2.8, with a molar water ratio of R_w = 1.5 as a function of hydrolysis time.

The response of the INEPT pulse sequence is complicated by the presence of different J(Si,H) couplings in MTMS. In order to identify the hydrolysis products of MTMS, the coupling between the methyl protons and the silicon, J(Si, H) = 8.3 Hz, is significantly different than the scalar coupling between the methoxy protons and the silicon, J(Si, H) = 3.9 Hz. The influence of heteronuclear and homonuclear couplings on the response of the INEPT pulse sequence has previously been investigated [8]. If proton-proton homonuclear couplings are zero or negligible, the intensity of the silicon signal following the INEPT pulse sequence is [7,8]

$$E_{\text{INEPT,dec}}(\tau,\Delta) = \frac{\gamma_S}{\gamma_I}\sum_p \sin[\pi\tau J(I,S^p)]\sin[\pi\Delta J(I,S^p)]\prod_{q,q\neq p}\cos[\pi\Delta J(I,S^q)] \qquad (1)$$

where the summations p and q run over all abundant nuclei (i.e. ^{1}H) J coupled to the insensitive nuclei (i.e. ^{29}Si). The variation of the INEPT sequence as function of the refocusing delay Δ for MTMS is shown in Figure 2 for both the hydrolyzed monomer T^0 species and the hydrolyzed condensed T^1 species. The signal response as a function of Δ is complex, allowing the identification and assignment of the different hydrolyzed species. For example, Figure 3 shows an expansion of the T^0 NMR spectra for different refocusing delays. In 3a (Δ = 20 ms) all the hydrolyzed species show a positive signal intensity as expected from Figure 2a. Figure 3b (Δ = 220 ms) and 3c (Δ = 90 ms) quickly allow the assignment of the fully hydrolyzed T_3^0 and singly hydrolyzed T_1^0 species. Assignment of the unhydrolyzed T_0^0 and doubly hydrolyzed T_2^0 were originally based on experiments with variation in water content [7], but are confirmed in 3d (Δ = 350 ms) where only T_2^0 and T_3^0 are expected to have positive signal intensity. Note that if the peak assignments had been based on arguments of downfield chemical shift with increasing numbers of attached hydroxyls, the assignments for the monomer MTMS hydrolyzed species would be incorrect.

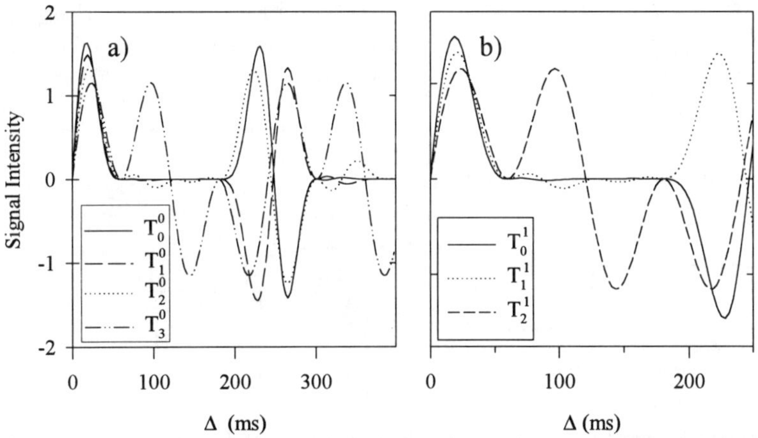

Figure 2. Theoretical signal intensity for the refocussed, proton decoupled INEPT pulse sequence as a function of the final refocusing delay Δ, for a) the monomer T^0 and b) the singly condensed T^1 hydrolyzed species. Intensities were obtained using Eqn. (1) assuming two different couplings for MTMS, J(Si,H) = 8.3 Hz for the methyl protons, and J(Si,H) = 3.9 Hz for the methoxy protons. The interpulse delay τ of 60.2 ms was optimized for the methyl coupling, $\tau_{opt} = 1/2J$(Si,H), since all species in MTMS contain this coupling.

In Figure 4 the NMR spectra of the T^1 region is shown at three different refocusing delays Δ. Spectra obtained using an inverse gate pulse sequence (not shown) allowed pairs of resonances that had the same signal intensity, plus identical variation with increasing water content, to be assigned to T^1 silicon environments in the same dimer molecule. Using this information, plus the predicted signal intensity variation (Figure 2b) for the condensed species all of the hydrolyzed condensed species were assigned. These assignments are given in Table 1. It is interesting to note that the T_1^1 environment gives rise to two distinct resonances due to the two possible diastereoisomers occurring for the chiral residues in the $T_1^1 - T_1^1$ dimer, denoted as D and D'. Differentation of diasterioisomers have been observed in polymethylhydrosiloxanes [9].

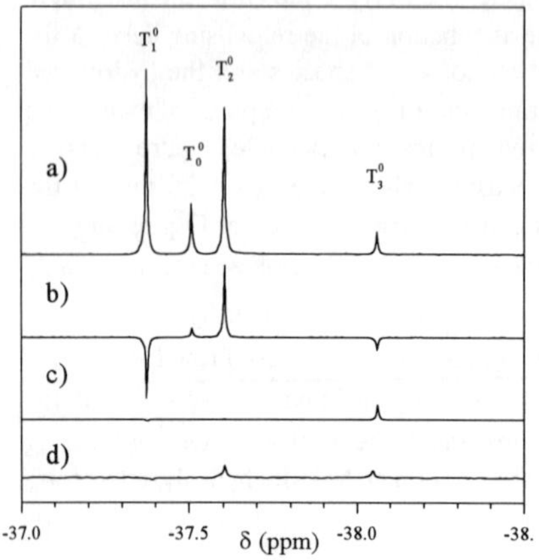

Figure 3. ^{29}Si INEPT spectra for a 2.24 M MTMS, 233 K, pH = 2.8, R_w = 1.5 solution at different refocusing delays Δ: a) 20 ms, b) 220 ms, c) 90 ms, d) 350 ms. Comparison of the observed signal intensity to Figure 2 allows for the unambiguous assignment of the hydrolyzed species.

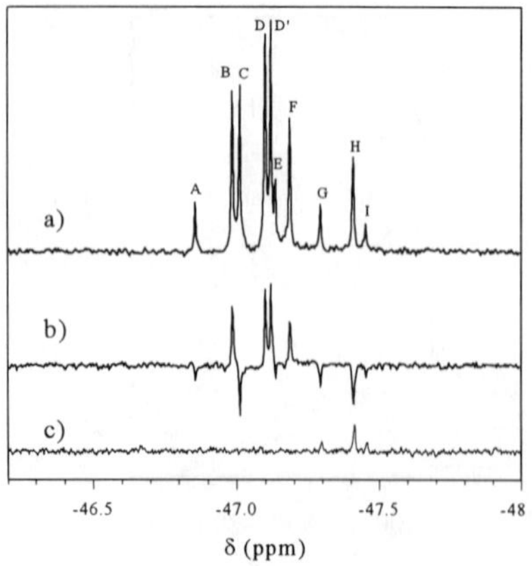

Figure 4. ^{29}Si INEPT spectra of the T^1 region for the same MTMS in Figure 3. Spectra for different refocusing delays are shown: a) Δ = 20 ms, b) Δ = 220 ms and c) Δ = 90 ms. Inspection of Figure 2 allows the identification of the hydrolyzed T^1 as detailed in Table 1.

Table 1. ^{29}Si NMR resonance assignments for T^1 hydrolysis species. Italicized correspond to the observed resonance.

	$-[T_0^1]$	$-[T_1^1]$	$-[T_2^1]$
$-[T_0^1]$	A	C	G
$-[T_1^1]$	B	D D'	H
$-[T_2^1]$	E	F	I

Inspection of Figure 1 shows that acid-catalyzed hydrolysis of the monomer MTMS species is very rapid at 233 K and is significantly faster than the condensation rate. Since the hydrolysis reaction is essentially complete within the first few minutes, monitoring the appearance of the various hydrolyzed species by NMR over time is not possible, but allows the lower limit for the hydrolysis rate to be estimated at 0.01 M^{-1} s^{-1}. However, the ability to assign the individual ^{29}Si NMR resonances for both the monomer and singly condensed species does allow the population distributions to be evaluated. These population distributions have been shown to be very sensitive to the rates of hydrolysis, allowing information about the hydrolysis kinetics to be determined.

It has been shown that for hydrolysis in alkoxysilanes, both *reversible* or *irreversible* models predict relative populations of hydrolyzed species that follow a binomial distribution, if the relative ratio of the hydrolysis rates for subsequent reactions are governed purely by statistical processes [6]. This implies that the rates for hydrolysis or esterification (reverse hydrolysis) are directly proportional to the number of reactive groups within a molecule. For example, the rate for the first hydrolysis reaction (MTMS to singly hydrolyzed) will be 1.5 and 3 times faster than the subsequent 2nd and 3rd hydrolysis reactions for the MTMS monomer. A similar trend is observed for the esterfication reaction.

If the observed number of SiOH bonds surrounding a silicon is governed entirely by statistics (not influenced by the chemical identity) then the concentration of the various hydrolysis products can be defined by a binomial distribution. For hydroxyl groups distributed randomly over n Si-O bonds the population or probability of silicons having ν hydroxyl groups, $P(\nu,n)$ is given by [6]

$$P(\nu,n) = C_{\nu}^{n} p^{\nu} q^{n-\nu} \tag{2}$$

where p gives the probability of hydrolysis for a single bond, $q = 1 - p$ defines the probability for an alkoxy SiOR bonds, and C_{ν}^{n} is the binomial coefficient. The probability p is equivalent to the extent of reaction for hydrolysis (ε), and defines the probability of a Si-O bond belonging to a hydroxyl. For the monomeric T^0 hydrolysis products in MTMS, the extent of reaction is given by

$$\varepsilon = \frac{[\mathrm{SiOH}]}{[\mathrm{Si-O}]} = \frac{[\mathrm{Si-OH}]}{[\mathrm{Si-OCH_3}]_0} = \frac{[\mathrm{T}_1^0]+2[\mathrm{T}_2^0]+3[\mathrm{T}_3^0]}{3[\mathrm{T}_0^0]_0} \tag{3}$$

where $[\mathrm{T}_1^0],[\mathrm{T}_2^0]$, and $[\mathrm{T}_3^0]$ are the observed concentrations of the various hydrolysis species and $[\mathrm{T}_0^0]_0$ is the initial concentration of the unhydrolyzed monomer. For MTMS there are three possible sites for hydrolysis (n = 3). With Eqn. 2 the distribution for each of the hydrolyzed monomer species can be easily evaluated. The extent of reaction for the T^1 condensed species is defined as

$$\varepsilon = \frac{[\mathrm{T}_1^1]+2[\mathrm{T}_2^1]}{2\sum_{i=0,2}[\mathrm{T}_i^1]} \tag{4}$$

where $[\mathrm{T}_0^1]$, $[\mathrm{T}_1^1]$ and $[\mathrm{T}_2^1]$ are the observed concentrations of the hydrolyzed singly condensed species, allowing Eqn. 2 to predict the distribution in the T^1 hydrolyzed species.

Figure 5 shows the distribution of the hydrolyzed species from MTMS and its singly condensed dimeric derivatives for different extents of reaction. The hydrolyzed T^0 species are well described by a simple binomial distribution for the entire range of extent of reaction ε. The condensed T^1 species also follow the binomial distribution for the range of ε observed. The ε for the T^0 and T^1 species were nearly identical regardless of the water concentrations studied.

Conclusions

Hydrolysis of MTMS is extremely rapid (< 0.01 M^{-1} s^{-1}), and is over a hundred times more rapid than either water- or alcohol-producing condensations. From the analysis of the concentration profiles, information about the relative hydrolysis rates in the MTMS monomer and single condensed dimer species can be determined. It has been shown that the hydrolysis reaction is reversible in MTMS [6]. The observation of binomial distributions for both the T^0 and T^1 hydrolyzed products suggest that the ratio of rates for consecutive hydrolysis reactions and consecutive esterfication reactions are governed by simple statistical arguments. The similarity of ε between T^0 and T^1 species suggest equilibrium partitioning of H_2O between monomer and condensed species.

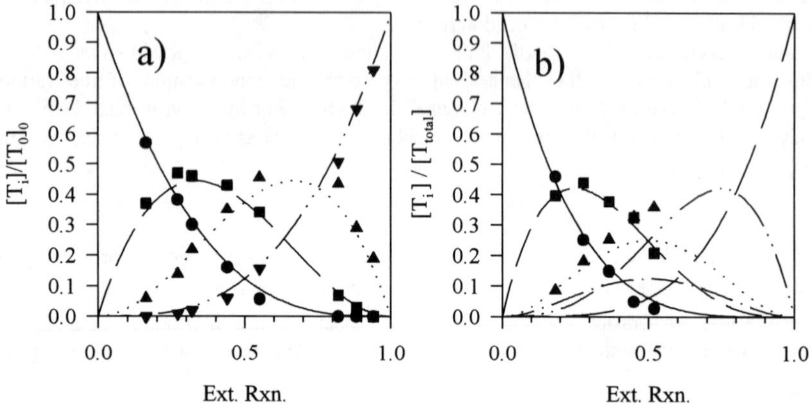

Figure 5. Experimental and theoretical concentration ratios for a 2.24 M MTMS (pH = 2.8) for different extents of reactions. Theoretical lines were obtained assuming a binomial distribution. a) Concentration ratios for the hydrolyzed monomer $[T_i^0]/[T_0^0]_0$, theoretical (——) i = 0, (— —) i = 1, (••••) i = 2, (— •• —) i = 3, experimental (●) i = 0, (■) i = 1, (▲) i = 2, (▼) i = 3. b) Concentration ratios for selected species in the hydrolyzed dimer $[T_i^1 - T_j^1]/[T^1 - T^1]_{total}$, theoretical (——) $i = j = 0$, (— — —) $i = 0, j = 1$, (••••) $i = j = 1$, (—•• —) $i = 0, j = 2$, (—••—) $j = 2,\ j = 1$, (— —) $i = j = 2$, experimental (●) $i = j = 0$, (■) $i = 0, j = 1$, (▲) $i = j = 1$.

REFERENCES

1. K. A. Smith, Macromolecules **20**, 2514 (1987).
2. R. C. Chambers, W. E. Jones, Y. Haruvy, S. E. Webber, M. A. Fox, Chem. Mater. **5**, 1481 (1993).
3. L. Delattre, F. Babonneau in *Better Ceramics Through Chemistry VI*, edited by A. K. Cheetham, C. J. Brinker, M. L. Mecartney, C. Sanchez (Mater Res. Soc. Proc. 346, Pittsburgh, PA 1994), p. 365-370.
4. Y. Sugahara, S. Okada, S. Sato, K. Kuroda, C. Kato, J. Non Cryst. Solids **167**, 21 (1994).
5. S. Suda, M. Iwaida, K. Yamashita, T. Umegaki, J. Non Cryst. Solids **176**, 26 (1994).
6. T. M. Alam, R. A. Assink, D. A. Loy (Submitted for Publication).
7. T. M. Alam, R. Assink, S. Prabakar, D. A. Loy, Magn. Reson. Chem. (In Press)
8. K. V. Schenker, W. von Phillipsborn, J. Magn. Reson. **61**, 294 (1985).
9. Y.-M. Pai, W. P. Weber, K. L. Servis, J. Organometallic Chem. **288**, 269 (1985).

SCHIFF BASE MEDIATED SOL-GEL POLYMERIZATION

D.A. LINDQUIST, C.M. HARRISON, B. WILLIAMS, R.D. MORRIS
Department of Chemistry, University of Arkansas at Little Rock, Little Rock, AR 72204

ABSTRACT

Formation of a Schiff base imine by reacting a primary amine with either an aldehyde or ketone was initiated by an aluminum compound acting as a Lewis acid catalyst. The water byproduct of the reaction then was used as an *in situ* reagent for subsequent hydrolysis and sol-gel condensation of the aluminum species. These reactions yielded a gel network containing the entrained Schiff base. Two examples of this synthetic approach are described with two different aluminum catalyst/reagents: a diethylaluminum diethylphosphate ester $[(CH_3CH_2)_2Al\text{-}O\text{-}P(O)(OCH_2CH_3)_2]$ and triethyl aluminum $[Al(CH_3CH_2)_3]$. Anhydrous ammonia and acetone were used as the Schiff base precursors.

INTRODUCTION

To date, the vast majority of sol-gel studies have employed the hydrolysis and condensation of metal alkoxides, and to a lesser extent, organo substituted metal alkoxides and metal halides [1] [2]. Few studies have been made using hydrolyzable metal alkyl precursors. This is due to the extreme hydrolytic sensitivity of metal-carbon bonds, with the exception of the silicon-carbon bond.

The reaction series described here allow for controlled *in situ* formation of water followed by hydrolysis of the sensitive aluminum-carbon bond to yield a gel. The first reaction is the Schiff base condensation of a primary amine with either a ketone or aldehyde (1).

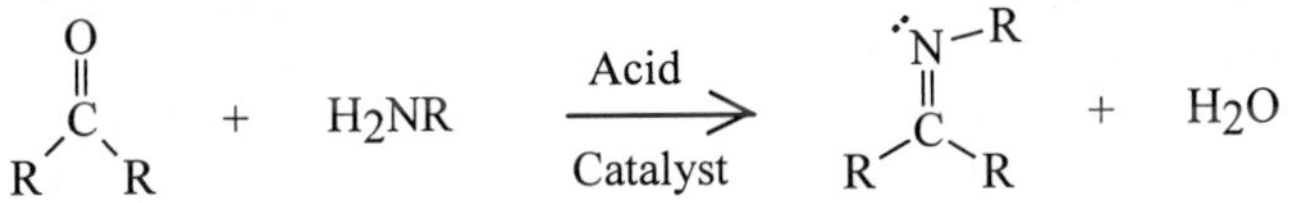

(1)

Reaction (1) is catalyzed by an aluminum alkyl derivative.

The water produced by (1) then reacts to oxidize and cross-link the aluminum alkyl functions with evolution of the alkane as shown in simplified form by reaction (2).

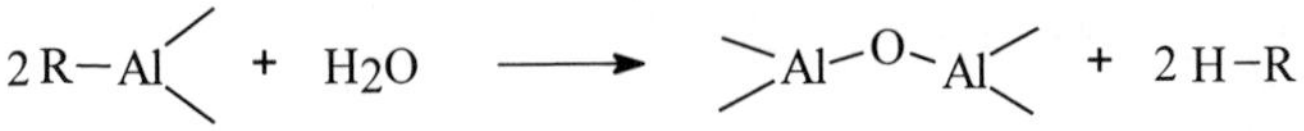

(2)

The cross-linking chemistry is certainly more complicated than as shown by (2). For example, the average coordination number of aluminum is increased to six as oxidation proceeds.

The net result of the overall process is a sol-gel matrix containing an intimately associated imine. The composition of the product is governed by various factors including: the specific imine from reaction (1), the rate and quantity of water produced in the Schiff base reaction, and the specific aluminum catalyst/reagent.

Mat. Res. Soc. Symp. Proc. Vol. 435

EXPERIMENTAL STUDIES

Ethyl Aluminum Ethyl Phosphate Ester

In a previous MRS proceedings [3] we described the synthesis of an aluminum phosphate gel by bubbling ammonia gas through a dry acetone solution of an ester of aluminum and phosphorus $[(CH_3CH_2)_2Al\text{-}O\text{-}P(O)(OCH_2CH_3)_2]$. High surface area aluminum phosphate (BET surface area = 530 m^2/g) was obtained and characterized, but we did not know the mechanism of the gelation at the time. It now is clear that an important step in the reaction sequence involves formation of an imine or Schiff base. The mechanism of the process is described here; for specific synthetic details the reader is referred to the original report [3].

When anhydrous ammonia is bubbled through the acetone solution of the diethylaluminumdiethylphosphate ester, there is a latent period of about one minute where no visible changes occur, followed by a prolonged gas evolution from the solution. The delay of the onset of gas evolution is due to the rate limiting formation of the ammonia-acetone adduct intermediate on aluminum illustrated in Figure 1.

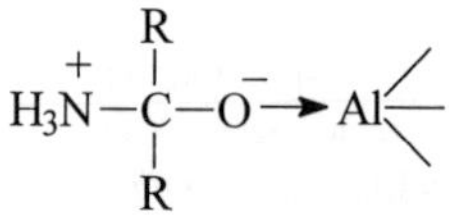

Figure 1. Aluminum activated imine intermediate

Formation of the intermediate in Figure 1 is facilitated by aluminum which activates the carbonyl function of acetone for nucleophilic attack by ammonia. The intermediate eventually decomposes to form acetoimine and water according to reaction (1) above. As water is produced, subsequent gas loss corresponds to production of ethane as alkyl aluminum functions are hydrolyzed according to reaction (2). Ethane evolution was verified by gas phase infrared spectroscopy of the reaction volatiles. Formation of a colorless gel occurs in a few minutes to hours depending on the rate at which ammonia is added. The Schiff base mechanism just described is supported by the fact that no reaction is observed when ammonia is added to the neat ester in the absence of acetone nor when hydrocarbon solvents such as hexane or benzene are used in place of acetone. It is interesting to note that if the colorless gel is aged in a sealed container it gradually develops a dark red color over a few weeks time. The color is likely due to condensation reactions involving the unstable acetoimine Schiff base to form electrically conjugated species.

Aluminum Amide

One may note that acetone does not react directly with the ester just described above. This is due to the fact that the electrophilic character of aluminum has been tempered by the oxygen ester bond. In general, it appears that heteroatom substituted aluminum alkyls are more suitable reagents than pure aluminum alkyls because undesirable side products from direct reactions between the carbonyl compound and aluminum are avoided [4]. This reasoning has been applied in the reaction of ammonia with triethyl aluminum prior to addition of the carbonyl compound acetone. Adapting a published alkyl aluminum amide synthesis [5], triethyl aluminum was reacted with anhydrous ammonia in dry benzene to form cyclic alkyl aluminum amides $\{(CH_3CH_2)_2AlNH_2]_x\}$. Two equivalents of dry acetone then were slowly added to the solution causing ethane gas evolution. Anhydrous ammonia then was bubbled through the stirring solution for a few minutes to form a gel. The gel was aged for one week at room temperature and then heated in air at 500 °C for 3 hours. The resulting granular solid had a BET surface area of 450 m^2/g.

CONCLUSION

In the syntheses described here, the composite nature of these materials has not been addressed. This is because our original interest was in preparing high surface area materials with the organic fraction removed by calcining. Now that we have some understanding of the process we are beginning to focus attention on the organic portion of the gels. The nature of the deep red color which develops in the gels is currently being investigated.

There are possibilities to make interesting composites by this approach since any of a number of primary amines and ketones or aldehydes can be used as reagents. For example, some useful Schiff bases which warrant study include: the blue and green macrocyclic phthalocyanine dyes, cage compounds such as hexamethylene tetraamine, and Schiff base aluminum complexes for polymerization catalysts [6].

REFERENCES

1. C. Jeffrey Brinker and George W. Scherer, Sol-Gel Science, Academic Press, San Diego, 1990.

2. L. Mascia, Trends in Polymer Science, **3**, p. 61 (1995).

3. K.B. Babb, D.A. Lindquist, S.S. Rooke, W.E. Young, M.G. Kleve. in Advances in Porous Materials, edited by S. Komarnani, D.M. Smith, and J.S. Beck, (Mater. Res. Soc. Proc. 371, Pittsburgh, PA 1995), p. 211-215.

4. J.J. Eisch in Comprehensive Organometallic Chemistry Volume 1, edited by G. Wilkinson, F.G.A. Stone, and E.W. Abel; Pergamon; New York, 1982; pp. 555-682.

5. F.C. Sauls, W.J. Hurley, L.V. Interrante, P.S. Marchetti, G.E. Maciel, Chem. Mater., **7**, p. 1,361 (1995).

6. H. Sugimoto, C. Kawamura, M. Kuroki, T. Aida, S. Inoue, Macromolecules **27**, p., 2,013 (1994).

PRECERAMIC POLYMER APPLICATIONS - PROCESSING AND MODIFICATIONS BY CHEMICAL MEANS

H. J. WU*, Y. D. BLUM*, S. M. JOHNSON*, C. KANAZAWA, J. R. PORTER**, AND D. M. WILSON***

*SRI International, Menlo Park, CA 94025
**Rockwell Science Center, Thousand Oaks, CA
***3M Corporation, St. Paul, MN

ABSTRACT

The use of preceramic polymers in diverse ceramic applications requires control of a complex set of chemical characteristics to ease processing and develop suitable ceramic properties. The desirable characteristics of a good precursor system are controllable viscosity (preferably without the use of solvents), inhibition of curing during fabrication followed by rapid curing, minimal release of volatiles during curing, high ceramic yield, and controllable final composition and stoichiometry.

This paper describes the use of chemical concepts to synthesize, modify, and manipulate polymeric precursors to silicates and the importance of curing (crosslinking) mechanisms and processing conditions as critical elements in developing ceramic matrix composites.

INTRODUCTION

Chemical synthesis approaches for processing advanced materials have attracted much attention over the past two decades. Preceramic polymer technology has undergone considerable development and now provides a way to prepare products that either cannot be produced using traditional means or require expensive processes that limit economic feasibility. Preceramic polymers generally consist of inorganic skeletons with organic pendant groups. They can be processed similarly to organic polymers and are converted to ceramics by a pyrolytic process followed by further heat treatment.

The two applications for preceramic polymers that have attracted most interest are fibers and matrices for composites. In both cases the processes are derived from well-established techniques developed for similar organic and carbon (graphite) products. In ceramic matrix composites, the preceramic polymer can be infiltrated into fiber preforms and then pyrolyzed and heated at higher temperatures. Typically, multiple infiltration-pyrolysis cycles are required to obtain acceptable density, primarily because the polymer shrinks by 40% to 80% during each pyrolysis/heat treatment cycle.

Resin transfer molding (RTM) is a promising low-cost approach to fabricating ceramic fiber-reinforced composites using preceramic polymers to form the matrix. RTM has been widely used for the manufacture of reinforced plastics and is well-suited to rapid mass production of complex shapes [1]. RTM can significantly reduce the cost of fabrication and provide excellent dimensional control [2]. However, the need for reinfiltration and repeated pyrolysis when preceramic polymers are used to form the matrix minimizes the economic advantage of RTM.

Most of the ceramic matrix composites under development are based on nonoxide fibers and matrices. Clearly, oxidation and possible degradation during service in oxidizing environments is a concern.

One approach [#] to overcoming these disadvantages is described in this paper. The objective is to develop a net-shape fabrication process for oxide-oxide composites for use in

Mat. Res. Soc. Symp. Proc. Vol. 435 © 1996 Materials Research Society

structural applications up to 1200°C. A high strength aluminosilicate fiber developed by 3M (Nextel 720®) is being coated with lanthanum phosphate. The usefulness of lanthanum phosphate as an interfacial coating was discovered recently at Rockwell [3]. The matrix, being developed at SRI, is based on low-cost, silicon-based preceramic polymers that are liquids with very low viscosity and can be cured easily and rapidly by a transition metal catalytic approach. Suitable filler powders are added to compensate for the polymer shrinkage and adjust the final composition of the matrix. In this paper we discuss some important aspects in the development of the matrix precursor for forming the composite by RTM.

EXPERIMENTAL

The basic polymer and filler powders used in this study are commercially available. Polyhydridomethylsiloxane ($[CH_3SiHO]_x$, PHMS) with a viscosity of ~20 centipoise is the initial polymeric material. The polymer and powders were dried before mixing by ball milling. No solvent is used. The slurries were mixed with the catalyst system and rapidly ball-milled for approximately 10 minutes just before the RTM process. The matrix was cured slightly in the mold to solidify the slurry, then removed from the mold and given further curing, pyrolysis, and heat treatments.

A Brookfield digital viscometer, model RVTDV-II with spindle 21, was used to measure the viscosity of polymers and slurries at room temperature. Viscosity and shear stress values were read directly from the viscometer. A Perkin-Elmer thermogravimetric analyzer equipped with a TAC 7 instrument controller was used for thermogravimetric analyses (TGA).

RESULTS AND DISCUSSION

Polymer Requirements

PHMS is one of the cheapest inorganic polymers. It is derived from a major by-product of the silicone (polysiloxane) industry and is available in large quantities. Cyclomers are produced by hydrolysis/condensation of dichloromethylsilane. The polymer with high molecular weight is formed by ring-opening polymerization [4]. Because it is a linear polymer, by itself PHMS has a negligible ceramic yield on pyrolysis, even when a relatively high molecular weight polymer is used. As shown in Figure 1, TGA shows that untreated PHMS is thermally stable up to ~350° and then undergoes a rapid weight loss between 350°C and 450°C, yielding a very small residuum of ceramic. The large weight loss is due to thermally induced depolymerization by a metathesis reaction, resulting in the formation of volatile siloxane products.

Over the past ten years SRI has developed a dehydrocoupling reaction catalyzed by transition metal catalysts [5]. This approach improves the ceramic yield of PHMS dramatically, making it a very useful preceramic polymer. As shown in Figure 1, the ceramic yield of PHMS, increases from 15 wt% to 95 wt% in air and 90 wt% in nitrogen after adding 100 ppm of $Ru_3(CO)_{12}$, the dehydrocoupling catalyst. This dramatic increase is attributed to interaction of the catalyst with the Si-H bonds of PHMS, allowing them to dehydrocouple with moisture, thus forming a highly crosslinked structure before it is heated to the temperature where depolymerization begins. The crosslinking suppresses the thermally induced depolymerization by eliminating the mobility of the polymer chains and introducing significant steric hindrance at the crosslinking junction points.

Filler Requirements

Considerable shrinkage occurs during the polymer-to-ceramic conversion as the polymer densifies and some volatile species are lost. Shrinkage and the formation of pores can be

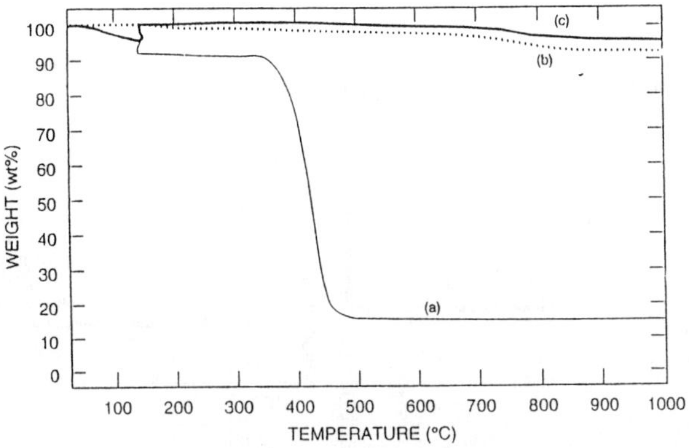

Figure 1. TGA of polyhydridomethylsiloxane.
(a) Without catalytic treatment, pyrolyzed in N_2.
(b) With catalytic treatment, pyrolyzed in N_2.
(c) With catalytic treatment, pyrolyzed in air.

diminished significantly by incorporating fillers in the polymer. However, the presence of fillers is a challenge in RTM because fillers can change the rheological properties of the polymer dramatically [6]. Another major issue is infiltration of the filler between the fibers in a woven tow and between tows [7]. These are significant problems for RTM when low viscosity and solventless slurries are needed. The general requirements for a suitable slurry prepared by mixing a polymer and filler are that

- The filler must be able to form a homogeneous mixture with the polymer without formation of agglomerates.
- A low viscosity slurry (<1000 cps, and preferably <500 cps) is needed to avoid the use of high pressure in RTM.
- The filler should infiltrate the fiber tow and not be filtered out.
- The slurry should not contain solvent.
- The slurry should be stable both chemically and rheologically before molding.
- The slurry should cure rapidly inside the mold after infiltration.
- No volatile gases should be formed inside the mold.

A variety of powders could be mixed homogeneously with the polymer without forming agglomerates. The shape, size, and surface properties of the powder affect the homogeneity and viscosity of the slurry. A powder that can be mixed easily and which has the smallest effect on viscosity was selected for use.

The most critical requirement for RTM operation is a low viscosity (preferably less than 1000 cps) to avoid high pressure which can deform fiber preforms and be a safety hazard. The viscosity of the polymer used in this study was about 20 cps. When fillers are incorporated into the polymer, the rheological properties change significantly. Because the slurry infiltrates the fiber preforms in a closed mold, the use of solvents is not desirable. For this reason we studied the effect of different fillers on viscosity. For example, one filler (Filler A) can infiltrate fiber tows well but dramatically increases the viscosity even with a very low filler:polymer ratio. As shown in Figure 2, a rapid increase in viscosity occurs as the amount of Filler A increases above the ratio of 0.5:1.0 (filler: polymer), resulting in a viscosity too high for RTM. A modified version of this filler (with the same material composition, referred to as Filler B) that also infiltrates the fiber tows shows a dramatic improvement in rheological properties. As shown in Figure 3, even with the loading ratio of 3:1, the viscosity is still acceptable for RTM.

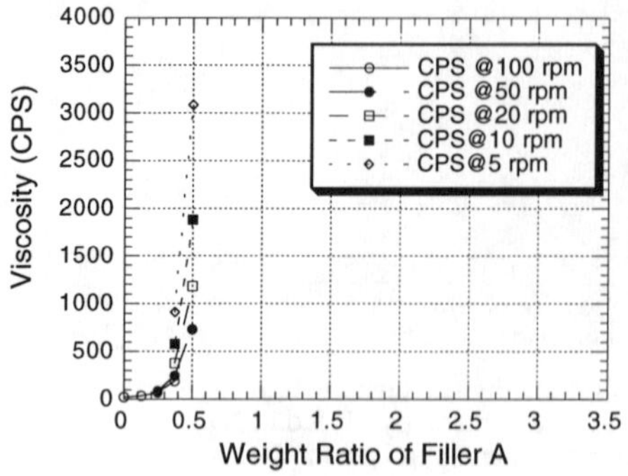

Figure 2. Slurry viscosity as a function of Filler A (Filler A:1.0 polymer).

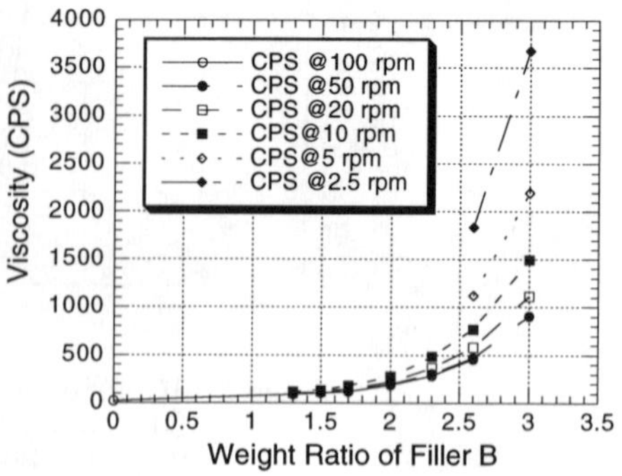

Figure 3. Slurry viscosity as a function of Filler B (Filler B:1.0 polymer).

The fillers can penetrate the fiber tow very well, as shown in Figure 4, although the filler used is not predominantly submicron. This result is very significant compared to other infiltration techniques where submicron fillers must be used to infiltrate fiber preforms [8]. This improvement is attributed to the lubricating effect of the low viscosity polymer.

Figure 4. Infiltration of fiber tows by polymer-powder slurry.

Curing of the Slurry

Once the fiber preforms are infiltrated, the slurry must be cured in a controlled manner. As mentioned earlier, SRI's catalytic approach, used to improve ceramic yield, can also be used to control curing of the slurry. Curing occurs while the slurry is still in a closed mold, so the use of external crosslinking reagents such as gaseous moisture or ammonia, commonly used by us in other applications, is not feasible.

We have developed an internal curing agent approach in which the curing catalyst system is mixed with the slurry shortly before molding. The curing rate is controlled so that the slurry cures after it has completed infiltration of the fiber preform. During crosslinking, the viscosity increases rapidly after a short induction period and the slurry solidifies in about an hour. Control of the curing kinetics in the process is essential to satisfy all the processing requirements.

As shown in Figure 5, with a different concentration of catalyst, the curing time can change from 620 minutes to less than 3 minutes at 25°C. By changing the curing temperature, we can also change the curing time from 240 minutes (at 25°C) to 9 minutes (at 90°C) with 5 ppm of catalyst (calculated on the basis of the polymer). With this degree of control over the crosslinking reaction rates, we have developed slurries that can be mixed and infiltrated in a convenient time and cured in the mold in a short time and with high efficiency. The product cured in the mold does not contain cracks or bubbles.

Further curing and heat treatments are required after the material is removed from the mold. The polymer still has some latent functionality after the initial curing, and it is important to control the curing/heat treatment to ensure full and even curing, maximize ceramic yield, and dimensional stability. When the current process is used, the fully cured matrix material shrinks less than 5% after heating at 1300°C for 48 hours. Prolonged exposure at 1200° to 1300°C in air does not result in degradation of the matrix; if anything, the microstructure improves. Further high temperature exposure tests are in progress.

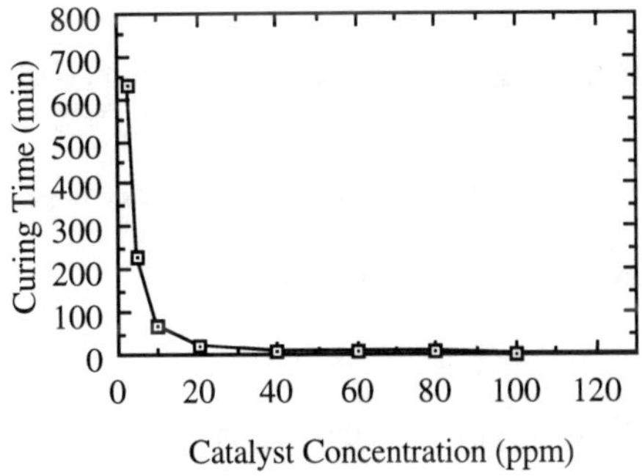

(a) Catalyzed at 25°C.

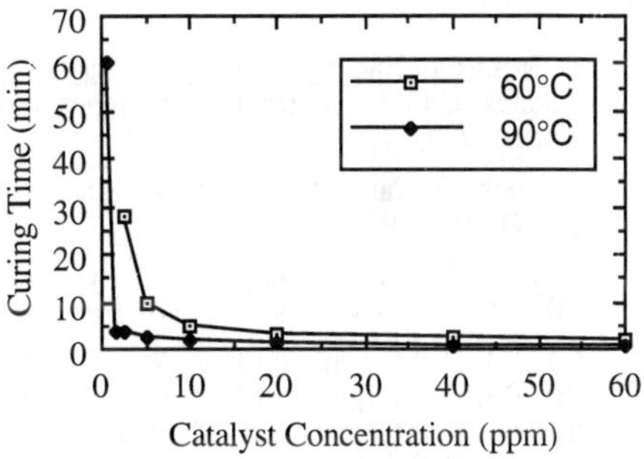

(b) Catalyzed at 60°C and 90°C.

Figure 5. Curing time of slurry.

SUMMARY AND CONCLUSIONS

A slurry suitable for making a low-cost oxide-oxide composite by resin transfer molding has been developed. Progress in the matrix development is summarized as follows:

- A low-cost slurry based on a cheap preceramic polymer and filler powders can be made.
- A low-viscosity slurry can be made by choosing appropriate filler powders and a very low-viscosity liquid polymer.
- The viscosity of the slurry can be modified as needed by altering the amount and characteristics of the fillers.
- The slurry can infiltrate a stack of woven Nextel fibers, and some infiltration of powder particles between the fibers within the tows is observed.
- Transition metal catalysts are efficient for curing the molded materials and improving the ceramic yield of the polymer from negligible to >90 wt%.
- The curing rate of the slurry is controlled and allows sufficient processing time followed by rapid curing after infiltration.
- After curing in the mold, the matrix contains no cracks or bubbles.
- A two-stage curing process is being developed to maximize ceramic yield and dimensional stability.

The investigation of this matrix in terms of microstructure, high temperature behavior, and mechanical properties is still in progress. In this study, we have demonstrated that careful understanding and control of chemical reactions and chemical properties at the different steps in processing a preceramic polymer is crucial in achieving proper rheological properties for processing, good control of curing, high ceramic yield, and dimensional stability.

REFERENCE

1. D. Rosato in Encyclopedia of Polymer Science and Engineering edited by H. F. Mark, N.M. Bikales, C G. Overberger, G. Menges, and J.I. Kroschwitz, Vol. 14, (John Wiley, New York, 1988), pp. 377-391.

2. R. Leek, G. Carpenter, J. Madsen, and T.M. Donnellan, Ceramic Eng. & Sci. Proceedings **16** (4), 191-199 (1995).

\# The Mullite Matrix Composites (M^2C) Consortium consists of 3M, Rockwell, and SRI International. It is funded on a cost-sharing basis by ARPA.

3. (a) P.E.D. Morgan and D.B. Marshall, Mater. Sci. Eng. **A162**, 15-25 (1993).
(b) P.E.D. Morgan and D.B. Marshall, J. Amer. Ceram. Soc. **78**, 1553-63 (1995).
(c) P.E.D. Morgan, D.B. Marshall, and R.M. Housley, Mater. Sci. Eng. **A195**, 215-222 (1995).

4. M.S. White in Siloxane Polymers, edited by S.J.C. Larson and J.A. Semlyen (PTR Prentice Hall, Englewood Cliffs, 1993), pp. 245-308.

5. (a) Y.D. Blum, US Patent 5,246,738 (June 7, 1994).
(b) Y.D. Blum, US Patent 5,319,121 (September 21, 1993).

6. J. D. Kiser and M. Singh, Ceramic Eng. & Sci. Proceedings **16** (5), 1107-1114 (1995).

7. M. Erdal, B. Friedrichs, and S. Guceri, Ceramic Eng. & Sci. Proceedings **16**(5), 1097-1100 (1995).

8. J. Jamet, J.R. Spann, R.W. Rice, D. Lewis and W.S. Coblenz, Ceramic Eng. & Sci. Proceedings **5** (7-8), 677-689 (1984).

NMR CHARACTERIZATION OF HYBRID SYSTEMS BASED ON FUNCTIONALIZED SILSESQUIOXANES

C. BONHOMME*, F. BABONNEAU*, J. MAQUET*, C. ZHANG**, R. BARANWAL** & R. M. LAINE**
*Université Pierre et Marie Curie / CNRS, Chimie de la Matière Condensée, Paris, France.
** University of Michigan, Departments of Chemistry and Material Science and Engineering, Ann Arbor, MI, USA.

ABSTRACT

In this paper, we present recent NMR investigations on several silsesquioxanes $(RSiO_{1.5})_8$ with R = H, CH_3, $CH{=}CH_2$, $OSi(CH_3)_2R'$ (R' = H, CH_3, $CH{=}CH_2$). The octameric polyhedral "cubane like" derivatives were analyzed by means of high resolution ^{13}C and ^{29}Si solid state NMR including CP (Cross-Polarization) and MAS (Magic Angle Spinning) techniques. The CP sequence including variable contact time was used in order to extract quantitative data. The 1D IRCP sequence (Inversion Recovery Cross Polarization), based on the standard CP scheme, allowed us to investigate the CP dynamics of the involved sites and to propose a complete spectral editing of the spectra. Furthermore, local molecular motions were determined through the careful analysis of CP dynamics. Finally, the NMR results related to crystalline cubane derivatives were extended to hybrid systems obtained by cross-coupling of monomeric entities (via hydrosilylation).

INTRODUCTION

During the last 15 years, the development of material with tailored properties led to the growing interest in silsesquioxanes and their functionalized derivatives. Several review articles were published [1-3]. Much attention was paid to new synthetic schemes [4] to prepare novel functionalized silsesquioxane monomers that upon subsequent polymerization, were transformed in polysilsesquioxane-based networks with improved microscopic and macroscopic properties. But besides their unique topology that makes them very attractive building blocks for hybrid polymeric networks, these polyhedral monomers $(RSiO_{1.5})_8$ represent valuable models for spectroscopic investigations : X-ray diffraction [5], force-field calculations, vibrational spectroscopies [5a], motions in crystals [6], liquid state ^{13}C, ^{29}Si and ^{1}H NMR, including elegant 2D correlation techniques [7] etc... The "famous" "Q_8M_8" silsesquioxane (R = $OSi(CH_3)_3$) appeared in the pioneering works of Lippmaa and Grimmer [8] concerning the application of ^{29}Si CP MAS NMR to the study of silicates and aluminosilicates. More recently, Hoebbel et al. investigated a large number of silsesquioxanes including octameric and decameric derivatives [9-10]. However, these studies were mainly based on the determination of isotropic chemical shifts. In this paper, we show that the CP and **IRCP** (**I**nversion **R**ecovery **C**ross **P**olarization)

Mat. Res. Soc. Symp. Proc. Vol. 435

techniques are valuable tools for a better understanding of magnetization transfer processes and can provide:

(i) isotropic chemical shifts <u>and</u> chemical shift anisotropies (or CSA), which can be interpreted through local symmetry of the sites and molecular motions if present;

(ii) a complete spectral editing, i.e. it becomes possible to distinguish between CH, CH_2 and quaternary C groups as well as between Si-H and non-protonated Si groups. Such spectral editing can be easily extended to amorphous or ill-crystallized compounds, for which strong overlap of the different NMR lines generally occurs;

(iii) quantitative measurements of the unreacted Si-CH=CH_2 and Si-H groups, as well as the corresponding Si-CH_2-CH_2-Si groups obtained by hydrosilylation between two different octameric entities.

EXPERIMENTAL SECTION

Basic T_8 cubes were obtained upon hydrolysis of R-$SiCl_3$ including Fe catalyst [1]. Q_8M_8 cubes were obtained by reaction between R'$(CH_3)_2$SiCl and $[OSiO_{1.5}]_8^{8-}$ [11]. Six silsesquioxanes were studied by ^{13}C and ^{29}Si NMR. Schematic representations and abbreviated names are given below.

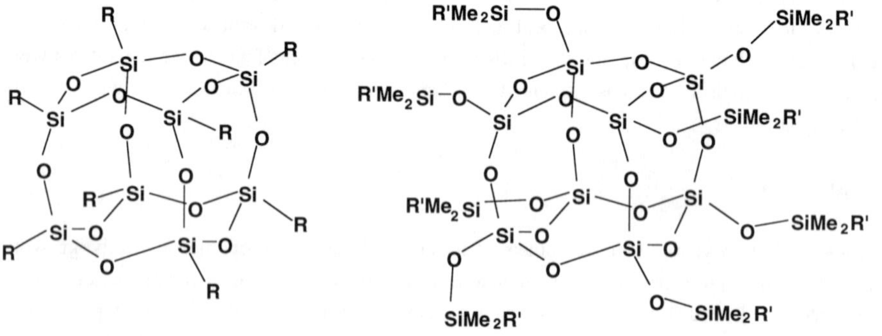

R = H : T_8^H; R = CH_3 : T_8; R = CH=CH_2 : T_8^v; R' = H : $Q_8M_8^H$; R' = CH_3 : Q_8M_8 and R' = CH=CH_2 : $Q_8M_8^v$.

The "T" cubanes involve T^3 silicon units. The "Q" cubanes involve Q^4 and M^1 units. Two polymeric systems obtained by hydrosilylation and involving the following silsesquioxanes, T_8^H/T_8^v and $Q_8M_8^H/Q_8M_8^v$ were also investigated.

Solid state ^{13}C and ^{29}Si CP MAS NMR spectra were obtained at 75.46 MHz and 59.62 MHz using a Bruker MSL 300 spectrometer and commercial MAS probe (rotation speed = 4000 Hz). Standard CP experiment using a single Hartmann-Hahn (H-H) contact was used, including ^1H high power decoupling. The 90° pulse width for ^1H was 5.5 µs. Recycle delays were set between 6 and 15 s and carefully checked for each sample. The H-H match condition was set experimentally using either adamantane, glycine (^{13}C) or T_8^v cube (^{29}Si). Chemical shifts are given with respect to TMS using external samples of adamantane and T_8^v cube as secondary references. The IRCP sequence [12-13], in its 1D version, is a simple derivation of the standard CP sequence. After a H-H contact during t_{CP}, a 180° phase shift is operated on the ^{13}C channel, during the ***inversion time*** $\mathbf{t_i}$. This leads to an inversion of spin temperature. The M^0 magnetization, first created after t_{CP}, decreases, passes through zero and becomes negative. It has been proven both theoretically and experimentally that the decrease in magnetization could be analyzed though coherent transfer of magnetization (between ^{13}C or ^{29}Si and directly bonded protons) and incoherent transfer (or spin diffusion) [13]. In other words, strongly coupled sites, such as rigid ^{13}CH, $^{13}CH_2$, ^{29}SiH will experience magnetization inversions characterized by two regimes : the first and rapid one corresponds to coherent transfer ; the second and much slower one corresponds to spin diffusion. As these regimes are characterized by very different time constants (one to two orders of magnitude), they are well separated by a sharp turning point located at $[(1-n)/(1+n)]M^0$: n corresponds to the number of directly bonded protons at a given site. On the other hand, weakly coupled sites (such as quaternary ^{13}C, non protonated ^{29}Si) are characterized by a single exponential decrease of the magnetization during IRCP experiments. CH_3 groups have an intermediate behavior as the dipolar coupling is strongly reduced by rapid reorientation. The IRCP sequence appears as a valuable tool for spectra editing, as in the case of the well-known liquid-state INEPT and DEPT sequences, which are based on J-coupling constants. At this stage, it should be noted that the setting of the H-H condition is crucial for IRCP experiments. Indeed, deviations from the H-H condition may lead to important distortions in the IRCP curves [12]. Therefore, glycine was used as a standard compound for the IRCP sequence : under minimal mismatch of the H-H condition, the $^{13}CH_2$ magnetization is already negative for t_i = 25 μs. This simple test was done several times during the IRCP experiments.

RESULTS

Points **(i)**, **(ii)** and **(iii)** (see above) will be discussed for two particular derivatives i.e. the crystalline T_8^v cube and the polymeric T_8^H/T_8^v system. The same approach has been applied for all other compounds and polymers.

Crystalline compounds : T_8^v cube

(i) *^{29}Si CP MAS* : the T_8^v spectrum shows two closely positioned lines at δ = -79.9 ppm and δ = -80.4 ppm (ΔW = 7-10 Hz), in full agreement with crystallographic data [14]. CSA is about +35 ppm, in agreement with previously published data [15].

^{13}C CP MAS : the T_8^v spectrum shows lines located at 128.7 ppm and (137.9-138.4) ppm. CSA are unusually small when compared with those observed for "rigid" vinyl groups [16]. Such an observation could be related to rapid local motions, leading to reduced shift anisotropies. All chemical shifts (^{13}C and ^{29}Si) relative to the various cubanes are given in Table I.

Compound	^{13}C (ppm)	^{29}Si (ppm)
T_8	-3.9	-65.8/-66.4
T_8^H		-83.9/-85.8/-86.5
T_8^v	137.9/138.4 (CH_2v) 128.7 (CHv)	-79.9/-80.4
Q_8M_8	2.6/2.4/2.1	11.9/11.7 (M units) -108.2/-108.5/-109.2/-109.5 (Q units)
$Q_8M_8^H$	1.2	-2.1/-2.9 (M units) -109.0/-109.2 (Q units)
$Q_8M_8^v$	137.9/137.6 (CHv) 133.6/133.2 (CH_2v) 0.27 (CH_3)	-0.5 (M units) -108.7/-109.3 (Q units)

Table 1 : Chemical shift values relative to the various cubanes.

(ii) *^{13}C IRCP MAS* : the curves relative to T_8^V are presented in Figure 1 :

The inversion of the magnetization M^0 (created after t_{CP} = 5 ms) is plotted versus the inversion time t_i. Clearly, two regimes are observed, in agreement with strongly coupled ^{13}CH and $^{13}CH_2$ vinyl groups. The turning points that are roughly -1/3 (δ = 137.9-138.4 ppm) and 0 (δ = 128.7 ppm) allow us to definitely assign the different NMR lines to CH_2 (δ = 137.9-138.4 ppm) and CH (δ = 128.7 ppm) groups. However, the CH_2 and CH magnetizations decrease rather slowly during the first tens of μs, when compared with rigid CH_2 and CH groups [13,17]. As for CSA values, local reorientations of the vinyl groups may be invoked [14]. Large amplitude reorientations are also present in the case of the

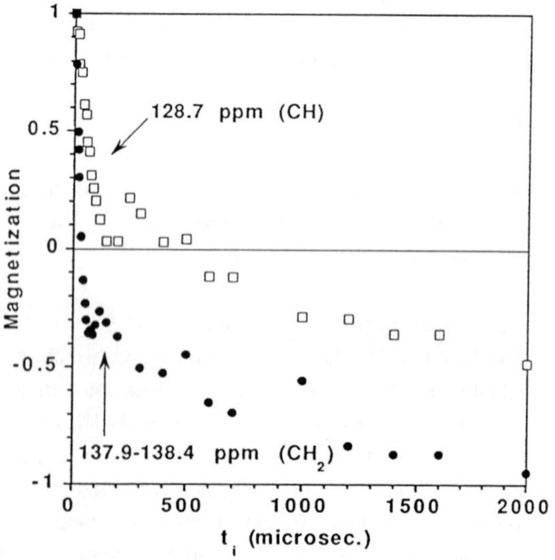

Figure 1 : Evolution versus inversion time of the ^{13}C IRCP MAS-NMR signal intensities for T_8^v.

$Q_8M_8^v$ cubane, leading to even much more reduced dipolar coupling and therefore to unusually slow inversions of the CH_2 and CH magnetizations.

Amorphous compounds : polymeric T_8^H/T_8^v system

Hydrosilylation reaction between T_8^H and T_8^v cubanes should lead to aliphatic C-C bonds, following two possible schemes :

Si-H + H_2C=CH-Si → Si-CH_2-CH_2-Si **α-hydrosilylation**

Si-H + H_2C=CH-Si → Si-CH(CH_3)-Si **β- hydrosilylation**

(ii) *^{29}Si IRCP MAS* : to our knowledge, very few data concerning the ^{29}Si IRCP sequence has been published in the literature [18]. Recently, this technique appeared however as a valuable tool for the study of polysilanes and poly-carbosilanes [19]. Spectra obtained for variable inversion time are presented in Figure 2. Straightforward assignments are possible : δ = - 81 ppm → residual Si-H (rapid inversion due to strong dipolar coupling) ; δ = -64 ppm → Si-CH_2 (intermediate inversion) ; δ = -78 ppm → residual Si-CH(vinyl) (slow inversion, as the dipolar coupling between Si and H is reduced by the vinyl group reorientation - see above). The IRCP technique may be applied to samples where the overlap of lines is much more severe [19].

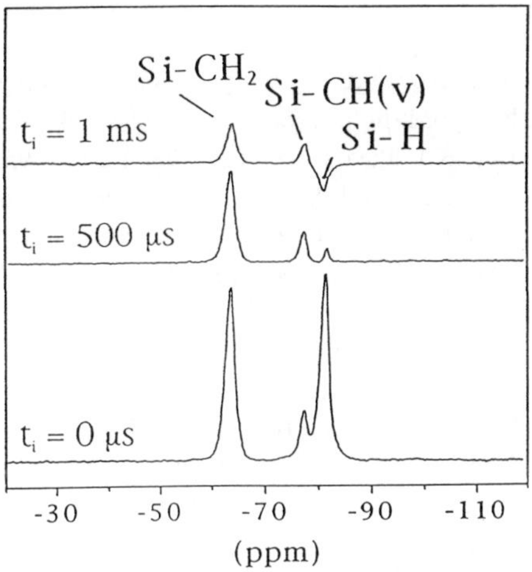

Figure 2 : ^{29}Si IRCP MAS-NMR spectra of sample obtained from hydrosilylation reaction between T_8^H and T_8^V.

(iii) *Quantitative measurements* : it is now well known that CP experiments are generally not quantitative, as the transfer of magnetization depends on the effective dipolar coupling at a given site and on the relaxation of ^{1}H in the rotating frame. However, quantitative data can be extracted when dealing with complete CP curves (with variable contact time) [20].

By measuring the relative intensities of Si-H, Si-CH(v) and Si-CH_2 lines, the hydrosilylation rate was estimated to be 45 % in the case of the T_8^H/T_8^v system. This rate increased up to 80 % when considering the $Q_8M_8^H/Q_8M_8^v$ system. Presence of M units based spacers clearly enhanced the efficiency of the polymerization reactions, by bringing some mobility to the network. The ^{13}C IRCP sequence showed that the obtained aliphatic sites were mainly CH_2 groups. Such an observation tends to prove that mainly **α–hydrosilylation** occurred.

In this paper, we showed that the combination of CP/IRCP MAS techniques allowed us :
* to assign all NMR spectra even in the case of poorly crystallized derivatives.
* to detect molecular motions which can lead to substantial reduction of dipolar coupling.
* to quantify the hydrosilylation yields for several systems.
Work is in progress in the laboratory to develop 2D solid state NMR techniques [21], which could connect chemical shifts and dipolar couplings, as well as sequences which could globally identify the topology of the cubane "cages".

REFERENCES

1. M.G. Voronkov, V.I. Lavrent'yev, Top. Curr. Chem., **102**, 199 (1982).
2. R.H. Baney, M. Itoh, A. Sakakibara, T. Suzuki, Chem. Rev., **95**, 1409 (1995).
3. D.A. Loy, K.J. Shea, Chem. Rev., **95**, 1431 (1995).
4. see for instance : (a) F.J. Feher, K.J. Weller, Organometallics, **9**, 2638 (1990) (b) M. Moran, M. Casado, I. Cuadrado, J. Losada, Organometallics, **12**, 4327 (1993) (c) J.D. Lichtenhan, A.O. Yoshiko, M.J. Carr, Macromol., **28**, 8435 (1995) (d) S.E. Yuchs, K.A. Carrado, Inorg. Chem., **35**, 261 (1996).
5. (a) G. Calzaferri, R. Imhof, K.W. Törnroos, J. Chem. Soc., Dalton Trans., 3123 (1994) (b) K.W. Törnroos, H.B. Bürgi, G. Calzaferri, Acta Cryst., **B51**, 155 (1995).
6. H.B. Bürgi, Acta Cryst., **B51**, 571 (1995).
7. B.J. Hendan, H.C. Marsmann, J. Organomet. Chem., **33**, 483 (1994).
8. E. Lippmaa, M. Mägi, A. Samoson, G. Engelhardt, A.R. Grimmer, J. Am. Chem. Soc., **102**, 4889 (1980).
9. D. Hoebbel, I. Pitsch, D. Heidemann, H. Jancke, W. Hiller, Z. Anorg. Allg. Chem., **583**, 133 (1990).
10. I. Pitsch, D. Hoebbel, H. Jancke, W. Hiller, Z. Anorg. Allg. Chem., **596**, 63 (1991).
11. I. Hasegawa, S. Sakka, Zeolite Synthesis, ACS symposium Ser. 398, 8 (1989).
12. M.T. Melchior, Poster B29, 22nd exp. NMR conference, Asilomar (1981).
13. X. Wu, K.W. Zilm, J. Magn. Res., **A102,** 205 (1993).
14. C. Bonhomme, L. Bonhomme-Coury, J. Livage, P. Boch, to be published.
15. R.K. Harris, T.N. Pritchard, E.G. Smith, J. Chem. Soc., Faraday Trans I, **85**, 1853 (1989).
16. W.S. Veeman, Prog. NMR Spectr., **16**, 193 (1984).
17. C. Bonhomme, J. Maquet, J. Livage, G. Mariotto, Inorg. Chim. Acta, **230**, 85 (1995).
18. N. Zumbulyadis, J. Chem. Phys., **86**, 1162 (1987).
19. F. Babonneau, J. Maquet, C. Bonhomme, R. Richter, G. Roewer, D. Bahloul, Chem. Mater., in press.
20. M. Mehring, Principles of High Resolution NMR in Solids, Springer (1983).
21. P. Palmas, P. Tekely, D. Canet, J. Magn. Res., **A104**, 26 (1993).

ORGANICALLY-MODIFIED Eu^{3+}-DOPED SILICA GELS

V.C. COSTA *, B.T. STONE **, K.L. BRAY **
*Dep. de Eng. Metalurgica- UFMG- R. Espirito Santo 35, 2° andar, MG, 30160-030, Brasil
**Department of Chemical Engineering, University of Wisconsin, Madison, WI, 53706-1691

ABSTRACT

We present results of fluorescence line narrowing (FLN) and lifetime studies of Eu^{3+}-doped ormosils prepared from $Si(OCH_3)_4$ and $CH_3Si(OCH_3)_3$, $(CH_3)_2Si(OCH_3)_2$, $(C_2H_5)_2Si(OCH_3)_2$, and $(n\text{-}C_3H_7)Si(OCH_3)_3$ in various proportions. Similar results are also presented for Eu^{3+}-doped gels derived from $Si(OCH_3)_4$ and fluorinated Eu^{3+} precursors ($Eu(fod)_3$, $(CF_3SO_3)_3Eu$, and $(CF_3CO_2)_3Eu \cdot 3H_2O$). The FLN studies indicated that significant Eu^{3+} clustering occurs in densified samples of both the organically modified and fluorinated compositions. Lifetime studies of the organically modified compositions showed longer Eu^{3+} lifetimes at low heat treatment temperatures relative to an unmodified sample. The difference in lifetime between modified and unmodified compositions decreased at high heat treatment temperatures as the organic substituents were driven from the matrix. The longest Eu^{3+} lifetimes were observed when the fluorinated Eu^{3+} precursors were used and persisted at high heat treatment temperatures. The lifetime studies indicate that the fluorinated precursors, and to a lesser extent organically modified precursors, are effective at reducing the water content in densified gels.

INTRODUCTION

Sol-gel synthesis has been used to produce solid state inorganic materials with a variety of compositions. An important application has been the preparation of optical materials. The properties of sol-gel optical materials have been studied extensively and a variety of applications have been demonstrated or proposed including solid state lasers, non-linear optics, chemical sensors, ferroelectrics and photochromic glasses [1-3].

Rare earth doped sol-gel materials are a promising class of materials. Rare earth ions have luminescence properties that are potentially useful in laser, waveguide and optical amplifier applications. Rare earth ions are also useful optical probes of the sol-gel process and the structural evolution of sol-gel materials during heat treatment [4,5].

The luminescence efficiency of rare earth ions in sol-gel host materials is currently limited by the tendency of rare earth ions to form clusters and by the presence of residual water and hydroxyl groups. Clustering results in concentration quenching of fluorescence due to nonradiative energy transfer between rare earth ions within the clusters. This effect is undesirable because it leads to poor luminescence efficiency. Previous work [6] indicates that clustering occurs in Eu^{3+}-doped silica gel-glasses even at low concentration.

Recent work has shown that codoping with Al^{3+} [6-8] or other metal ions [9] inhibits rare earth ion clustering and promotes a more uniform distribution of rare earth ions in silica gels and silicate glass matrices. Dopant ion encapsulation by complex-forming ligands has also been successfully used to protect individual rare earth ions and inhibit clustering [5,10].

In order to minimize the hydroxyl content of sol-gel matrices, the use of organically modified silicon alkoxide sol-gel precursors [11], and the addition of fluorine as a dehydroxylating agent [12] have been demonstrated to reduce hydroxyl content in silica gel-

Mat. Res. Soc. Symp. Proc. Vol. 435 © 1996 Materials Research Society

glasses. Organically modified precursors contain hydrophobic alkyl groups that act to inhibit water and hydroxyl retention. Encapsulation shields dopant ions from water and hydroxyl groups and fluorination treatments eliminate residual water and hydroxyl groups in silica gels.

In this study, Eu^{3+} fluorescence is used to probe the effect of organic modification and fluorinated Eu^{3+} precursors on clustering and quenching due to water and hydroxyl groups. Fluorescence line narrowing spectroscopy will be used to characterize Eu^{3+} clustering and Eu^{3+} lifetime measurements will be used to assess hydroxyl quenching.

EXPERIMENTAL PROCEDURE

The gels were derived from combinations of $Si(OCH_3)_4$ (TMOS) separately with four modified silicon alkoxides: $CH_3Si(OCH_3)_3$ (MTMS), $(CH_3)_2Si(OCH_3)_2$ (DMDMS), $(CH_3)_2Si(OC_2H_5)_2$ (DMDES) and $C_3H_7Si(OCH_3)_3$ (PTMS). Organically modified silica gels were prepared by mixing TMOS, the modified silicon alkoxide and an aqueous solution in which $Eu(NO_3)_3{\cdot}6H_2O$ was previously dissolved. The molar ratio of water to the combined silicon precursor content (TMOS + modified silicon precursor) was 16. The molar ratios of TMOS to modified alkoxide were 19/1, 9/1, and 3/1. Reaction and gelation occurred at room temperature. After gelation the gels were aged and dried at 60°C for two days and at 90°C for two days. The samples were ultimately heated to 800°C. All samples contained 1.0 wt% Eu_2O_3. The europium concentration in a sample is expressed in terms of the weight percent of Eu_2O_3 that would be present in a fully densified SiO_2 glass in which water and organics are absent.

A series of samples prepared from fluorinated Eu^{3+} precursors and TMOS were also studied. The Eu^{3+} precursors used were: Tris (6,6,7,7,8,8,8-heptafluoro-2,2-dimethyl-3,5-octanedionate)Eu(III) ($Eu(fod)_3$), Eu(III) trifluoromethanesulfonate ($(CF_3SO_3)_3Eu$), and Eu(III) trifluoroacetate ($(CF_3CO_2)_3Eu{\cdot}3H_2O$). The Eu^{3+} precursors were dissolved in water ($(CF_3SO_3)_3Eu$, $(CF_3CO_2)_3Eu{\cdot}3H_2O$) or DMF-methanol ($Eu(fod)_3$) and combined with an aqueous solution of TMOS (water:TMOS = 16). HNO_3 was added to lower the pH to 1.5. Samples containing 1.0 and 5.0 wt% Eu_2O_3 were prepared. The heat treatment procedure for these samples was the same as that used for the organically-modified samples.

Room temperature broadband (non-selective) emission spectra were obtained by exciting samples with the 514.5 nm line of an argon ion laser. Luminescence was collected with a 1-meter monochromator and detected by a photomultiplier tube. Fluorescence line narrowing measurements were completed with a tunable dye laser pumped by a Q-switched Nd:YAG laser. The dye laser provided the tunable emission from 571-580 nm needed to selectively excite the $^7F_0 \rightarrow {}^5D_0$ transition of Eu^{3+}. Spectra were normalized to the $^5D_0 \rightarrow {}^7F_1$ peak intensity and were collected at 77 K. Luminescence decay measurements were recorded using a digital storage oscilloscope. Since the decay curves were slightly non-exponential, average lifetimes were obtained by calculating the area under normalized decay curves [13].

RESULTS

Organically Modified Compositions

Figure 1 presents broadband fluorescence spectra for several 1.0 wt% Eu_2O_3 doped ormosils densified at 800°C. The spectra for the different compositions are similar. Enlargement of the non-degenerate $^5D_0 \rightarrow {}^7F_0$ transition near 578 nm revealed differences in the linewidth of the transition for the different compositions. Larger linewidths were observed in the organically

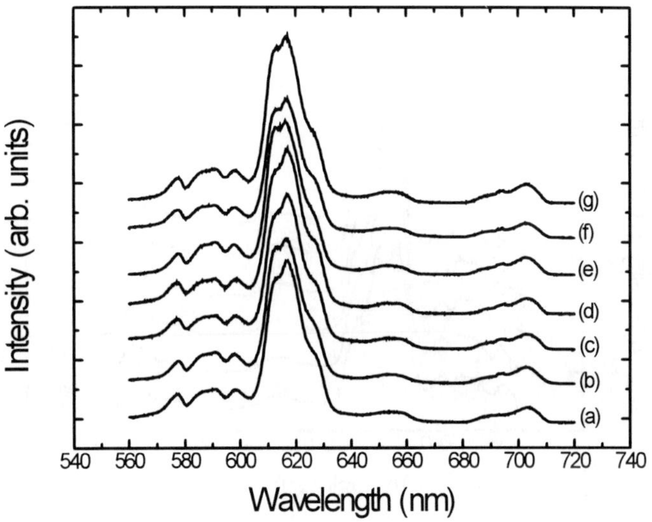

Figure 1. Room temperature fluorescence spectra of 1.0 wt% Eu_2O_3 doped ormosils (λ_{ex}=514.5 nm). (a) Eu^{3+}-doped sol-gel silica; (b) TMOS/DMDMS=9 ; (c) TMOS/MTMS=9; (d) TMOS/DMDES=9; (e) TMOS/DMDES=3; (f) TMOS/MTMS=3; (g) TMOS/MTMS=19.

modified compositions relative to the unmodified composition, a result that indicates a wider range of bonding sites for Eu^{3+} in the organically modified systems. The linewidth broadening occurred on the high energy side of the transition and became more pronounced as the amount of organically modified precursor used in the preparation was increased. For a given TMOS:modified alkoxide ratio, the linewidth decreased as DMDES > MTMS > DMDMS.

Figure 2 shows 77 K FLN spectra for a 1.0 wt% Eu_2O_3 doped sample made from a 3:1 molar ratio of TMOS:DMDES and densified at 800°C. The conventional broadband luminescence spectrum is shown at the top of the figure for comparison. The spectra obtained upon selective excitation were similar to the non-selective broadband spectrum. The lack of spectral sensitivity to excitation wavelength reflects an inability to obtain fluorescence from selectively excited subsets of Eu^{3+} ions within the inhomogeneous distribution of bonding sites. We attribute the finding to efficient energy transfer among Eu^{3+} ions. Since efficient energy transfer occurs only when Eu^{3+} ions are spatially close to each other, we are led to conclude that Eu^{3+} ions aggregate or cluster in the matrix. This indicates that the distribution of Eu^{3+} ions in the material is non-uniform. Similar results were obtained for the other organically modified compositions considered in this study and were also obtained previously for unmodified materials derived from TMOS and TEOS [6].

The lifetime of Eu^{3+} is highly sensitive to hydroxyl concentration in sol-gel materials. Hydroxyl groups have high phonon energies and introduce an efficient non-radiative decay pathway for rare earth ions. We measured lifetimes of several samples at different heat treatment temperatures in order to examine the effect of organic modification on the hydroxyl content. Representative results are shown in Table I. At the wet gel stage, the organically modified systems show slightly longer lifetimes than the unmodified control sample. The difference in

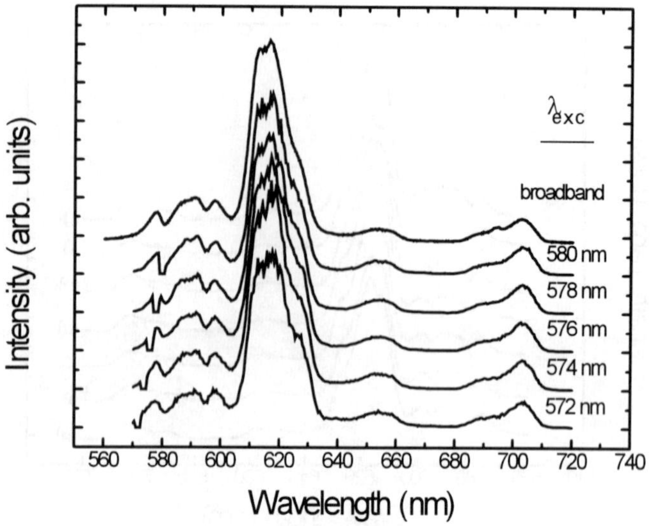

Figure 2. FLN spectra at 77 K of 1.0 wt% Eu_2O_3 doped ormosil (molar ratio TMOS/ DMDES = 3) heated to 800 °C. Intensities were normalized to the $^5D_0 \rightarrow {}^7F_1$ emission at each pump wavelength. Upper spectrum, fluorescence spectrum at 300 K (λ_{ex} = 514.5 nm).

Table I. Selected 77 K Eu^{3+} lifetimes (μsec) in organically modified gels at various heat treatment temperatures. Excitation and detection wavelengths were 579 ± 0.5 nm and 616 ± 1 nm, respectively. The small changes in these wavelengths were needed to account for the evolution of the spectrum upon heat treatment.

Modified Precursor	TMOS: Modified Precursor	wet gel	90 °C	200 °C	800 °C
none	---	107	152	298	1065
MTMS	9	133	195	342	
DMDMS	9	125	182	327	1048
DMDES	3		190	376	1102
DMDES	9	119	195	340	1001
PTMS	9		178	370	856

lifetime increases with heating temperature up to 200°C, but is no longer evident in samples heated to 800°C. These results indicate that the presence of organically modified precursors affects the hydroxyl content of the gels. The longer lifetimes at low heating temperatures are evidence of lower hydroxyl content and are consistent with the expected hydrophobic character of the alkyl groups incorporated in the precursors. The organic substituents appear to facilitate removal of water from the matrix at low heating temperatures. At high heating temperatures, the alkyl groups decompose and are driven from the matrix. The lifetime results indicate that the moisture inhibiting effects of the alkyl groups are lost during the decomposition process.

Fluorinated Eu(III) Precursors

Room temperature broadband fluorescence spectra of 1.0 wt% Eu_2O_3 doped samples prepared from three fluorinated Eu^{3+} precursors and densified at 800 °C were measured and compared to the spectrum of a sample prepared from $Eu(NO_3)_3 \cdot 6H_2O$. The spectrum of the sample prepared from $(CF_3SO_3)_3Eu$ was similar to that of the non-fluorinated sample prepared from $Eu(NO_3)_3 \cdot 6H_2O$. The other two samples had spectra that differed in two important ways from the spectrum of the non-fluorinated sample. First, the $^5D_0 \rightarrow {^7F_0}$ transition was broader and second, the $^5D_0 \rightarrow {^7F_2}$ transition intensity was much stronger. The first observation suggests that a wider range of Eu^{3+} bonding sites is present in the samples prepared from $Eu(fod)_3$ and $(CF_3SO_3)_3Eu$ and the second observation implies that more highly distorted/less symmetrical Eu^{3+} bonding environments are present. Selectively excited fluorescence spectra measured at 77 K for the samples prepared from the fluorinated Eu^{3+} precursors revealed no line narrowing effect. As in the case of the organically modified systems, this result indicates that fluorinated Eu^{3+} precursors do not inhibit Eu^{3+} clustering.

The fluorinated precursors had a pronounced effect on the Eu^{3+} lifetime. Table II summarizes 77 K lifetime results of samples containing 5.0 wt% Eu_2O_3 heated to 800°C for two excitation (λ_{exc}) and two detection (λ_{em}) wavelengths. A clear increase in lifetime was observed when fluorinated precursors were used. Since the fluorescence line narrowing spectra indicate that appreciable clustering is present, the longer lifetimes indicate that the use of fluorinated Eu^{3+} precursors leads to lower hydroxyl content in the densified gels.

Table II. Lifetime values (ms) measured at 77 K for 5.0 wt% Eu_2O_3 doped silica gels prepared from $Eu(NO_3)_3 \cdot 6H_2O$ and three fluorinated Eu^{3+} precursors.

	λ_{exc} = 577 nm		λ_{exc} = 579 nm	
λ_{em} (nm)→	610	625	610	625
$Eu(NO_3)_3 \cdot 6H_2O$	1.17	1.14	1.16	1.17
$(CF_3CO_2)_3Eu \cdot 3H_2O$	1.63	1.44	1.69	1.61
$(CF_3SO_3)_3Eu$	1.63	1.63	1.74	1.67
$Eu(fod)_3$	1.60	1.63	1.67	1.64

CONCLUSIONS

We have presented the results of fluorescence line narrowing and lifetime studies of two series of densified gels: Eu^{3+}-doped organically modified gels (based on TMOS and each of $CH_3Si(OCH_3)_3$, $(CH_3)_2Si(OCH_3)_2$, $(C_2H_5)_2Si(OCH_3)_2$, and $(n\text{-}C_3H_7)Si(OCH_3)_3$) and unmodified Eu^{3+}-doped TMOS gels prepared using fluorinated Eu^{3+} precursors ($(CF_3CO_2)_3Eu \cdot 3H_2O$, $(CF_3SO_3)_3Eu$, and $Eu(fod)_3$). The FLN studies indicate that significant Eu^{3+} clustering occurs in both series of gels. Lifetime studies of the organically modified gels suggest that at low heat treatment temperatures, the hydrophobic alkyl substituents lead to lower hydroxyl content in the gels relative to an unmodified sample prepared with the same Eu^{3+} precursor($Eu(NO_3)_3 \cdot 6H_2O$). At high heat treatment temperatures, however, the alkyl groups are driven from the gels and only a small difference in hydroxyl content is inferred from lifetime measurements. Much longer lifetimes, relative to samples prepared from $Eu(NO_3)_3 \cdot 6H_2O$, were observed in gels heated to 800°C that were prepared from the fluorinated Eu^{3+} precursors. This result suggests that the presence of fluorine in the pores of the gel significantly facilitates the removal of water during heat treatment and that the use of fluorinated precursors may lead to rare earth doped sol-gel materials with higher luminescence efficiency.

ACKNOWLEDGMENT

The authors acknowledge financial support from National Science Foundation-Division of Earth Sciences, the UW Graduate School and the Brazilian National Council of Research, CNPq.

REFERENCES

1. Sol-Gel Optics III, edited by J.D. Mackenzie (SPIE Proc. **2288**, Bellingham, WA, 1994).
2. R. Reisfeld and C.K. Jørgensen, in Structure and Bonding 77, 207 (1992).
3. Better CeramicsThrough Chemistry VI, edited by A.K. Cheetham, C.J. Brinker, M.L. Mecartney and C. Sanchez (Mater. Res. Soc. Proc. **346**, Pittsburgh, PA, 1994).
4. M. Ferrari, R. Campostrini, G. Carturan and M. Montagna, Phil. Mag. B **65**, 251 (1992).
5. L.R. Matthews, X. Wang and E.T. Knobbe, J. Non-Cryst. Solids **178**, 44 (1994).
6. M.J. Lochhead and K.L. Bray, Chem. Mat. 7, 572 (1995).
7. S. Sen and J.F. Stebbins, J. Non-Cryst. Solids **188**, 54 (1995).
8. I.M. Thomas, S.A. Payne and G.D. Wilke, J. Non-Cryst. Solids **151**, 183 (1992).
9. V.C. Costa, M.J. Lochhead and K.L. Bray, Chem. Mat. **8**, 783 (1996).
10. N. Sabbatini, and M. Guardigli, Mat. Chem. and Phys. **31**, 13 (1992).
11. C.A. Capozzi and A.B. Seddon, in Sol-Gel Optics III, edited by J.D. Mackenzie (SPIE Proc. **2288**, Bellingham, WA, 1994) p. 340.
12. E.J.A. Pope and J.D. Mackenzie, J. Am. Cer. Soc. **76**, 1325 (1993).
13. G. Armagan, B. DiBartolo, and A.M. Buoncristiani, J. Lumin. **44**, 129 (1989).

Detection of *Cryptosporidium parvum* in antibody-doped gels

E. HONG*, E. BESCHER*, L. GARCIA**, and J. D. MACKENZIE*
*UCLA Department of Materials Science and Engineering, Los Angeles, CA 90095
**UCLA Department of Microbiology, Los Angeles, CA 90095

ABSTRACT

Doping porous oxides with organic functionalities has become increasingly useful for the detection of many types of compounds and molecules. So far the detection of microorganisms by this process has not been firmly established. There is, however, a need to develop easy to use, sensitive devices for the detection of microorganisms in potable water supplies. For example, the need for a sensor that can detect a pathogenic parasite called *Cryptosporidium parvum* has emerged in the United States. In this report, we describe the production of such a device via the sol-gel method. The sol-gel sensor can detect trace amounts of *Cryptosporidium parvum*. The advantages of the porous oxide matrix lead to high sensitivity. The sensor is easy to use and inexpensive. The presence of *Cryptosporidium parvum* is indicated by a color change in an otherwise colorless gel. Absorbance intensities for the enzyme-linked antibody used in the study correlated appropriately to oocyst concentration in the solution.

INTRODUCTION

Over the past three years, the presence of *Cryptosporidium parvum* in the water supply has become the most prominent water safety concern in the United States. *Cryptosporidium parvum* is a small (2-6μm) protozoan parasite which causes chronic illnesses in otherwise healthy individuals and is fatal to immuno-suppressed patients [1]. AIDS and HIV-impacted persons are particularly at risk for cryptosporidiosis, and their support organizations are calling for appropriate monitoring of the US water supply. There has been a large scale outbreak of this parasite in Milwaukee in 1993 [2] and a smaller outbreak in Las Vegas. *Cryptosporidium parvum* is present in most water systems in the United States and is hard to filter and even harder to kill. Studies have shown that *C. parvum* is present in 87% of tested raw water supplies in the United States [3]. Filtration techniques currently being applied do little to reduce the presence of *C. parvum*: a reported 27% of tested filtered drinking water supplies are contaminated or show occurrence of *C. parvum* oocysts [4]. Ozonation is the most effective destruction technique but even then the oocysts are not all made benign. There have been attempts to detect *C. parvum* in water systems but the current methods are time consuming and costly. There is currently no legislation enforcing *C. parvum* monitoring in water systems but pending legislation will require shorter and more cost-effective detection times. Enzyme-Linked Immunosorbent Assay (ELISA) testing is a widely known and accepted method of determining and identifying microorganisms. Incorporating this established technique, we produced a viable sensor by encapsulating antibodies in a porous oxide gel.

Mat. Res. Soc. Symp. Proc. Vol. 435 © 1996 Materials Research Society

Oxides have long been produced by the sol-gel technique, one of their most prominent features is their high porosity. Organic-inorganic hybrids add functionality to a purely inorganic network. Among the most interesting of these new materials, protein or enzyme-doped gels show promise as new sensing materials [5,6,7]. For example, Livage et al. have encapsulated microorganisms in gels for antibody detection employing the same ELISA principles [8].

EXPERIMENTAL

Using the ELISA technique as a blueprint for our method, we have encapsulated antibodies into our porous oxide matrix and kept the antibody active. Maintaining antibody activity was achieved using an alcohol-free technique in producing the gel. The only way to produce gels this way is through sonication. A TEOS, water and HCl mixture was sonicated for 15 minutes. Sodium hydroxide, a base, was added at the end of the process before the addition of the antibodies to the sol-gel solution. The overall process seen in Fig. 1 shows the method used to produce the gels .

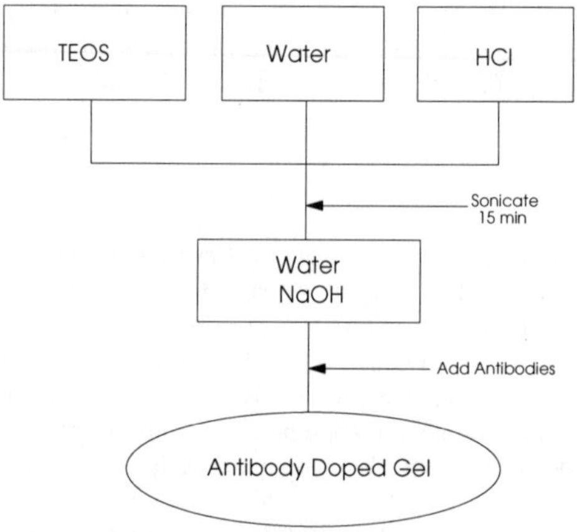

Figure 1- Sol-gel processing for active antibody encapsulation

The gel was dried in air at room temperature for a week before testing for sensing capabilities. *Cryptosporidium parvum* solutions were prepared by diluting a stock solution of concentration ca. 120,000 oocysts/mL with distilled water. The stock solution was obtained from the Department of Microbiology at UCLA. The gels were tested as follows: the gels are exposed to the positive or negative control solutions by immersion in a prepared solution. The gels were first exposed to the solution for one hour and washed with a buffer solution. Next the gels were exposed to enzyme-linked antibody and allowed to interact for one hour, were subsequently washed with the buffer solution.

Finally the chromogen was added and a color change occured if *C. parvum* was present in the solution. No color change occured if the antigen was not present in the solution.

RESULTS AND DISCUSSION

A diagram of the test procedure is presented in Fig. 2, and can be summarized as follows. First, the antibody in the gel is introduced to a solution containing *Cryptosporidium parvum* (a). Then the *Cryptosporidium parvum* specific antigen which is found on the outer cell wall attaches itself to the antibody reception site (b). Since the antibodies are made specifically for *Cryptosporidium parvum,* there is no chance of cross-contamination or cross-reactions. Antibodies are matched with the antigen by protein and DNA recognition sites. Next the device is washed to insure that there are no stray antigens. Then an antibody to which an enzyme, Horse Radish Peroxidase (HRP) has been attached is introduced (c). The antibody attaches itself to the antigen if present. We wash again to remove all non-binding enzyme-linked antibody, thus preventing a false positive. Finally Tetra Methyl Benzidine (TMB), a chromogen, is added and changes color due to the presence of the enzyme or remains colorless if the enzyme is not present (d). Fig. 3 is a photograph of the sensors showing a color change for the positive detection on the left and a colorless gel for the negative control on the right.

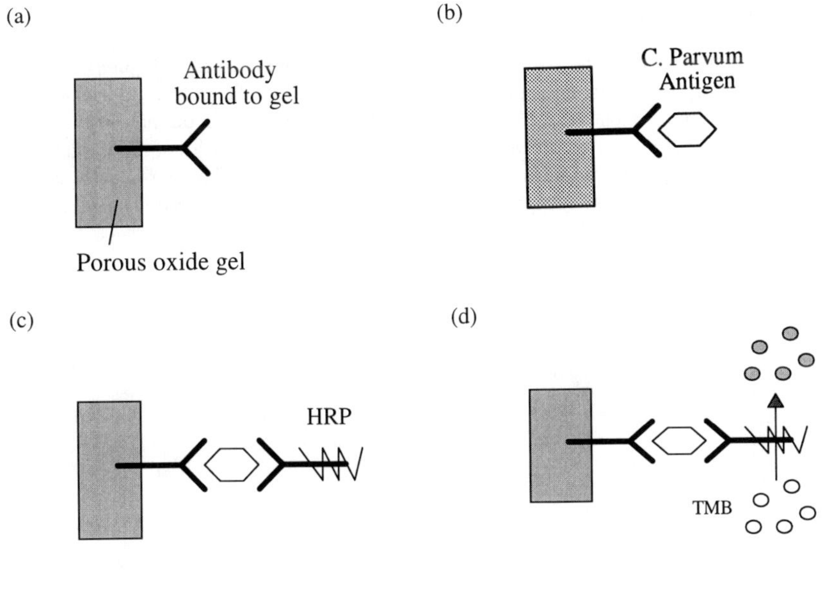

Figure 2- Outline of testing procedure

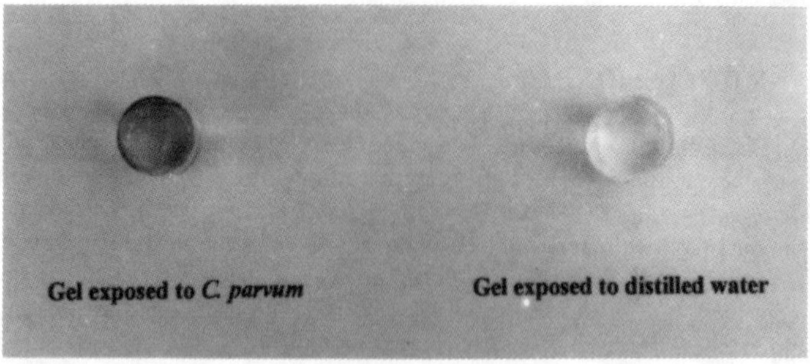

Figure 3- Sol-gel sensors tested with positive (6 oocysts/mL) and negative (0 oocysts/mL) solutions
The gel on the left is the positive (colored) and the gel on the right is negative (colorless)

The most important step of the testing process is the washing to make sure that there are not any false positives. Because the test is dependent on the second antibody attaching itself to the antigen, washing is required to eliminate all stray enzymes. The test is indirect since the color change is dependent not on the antigen itself but on the enzyme linked-antibody.

The absorbance spectrum of the positive control sensing device is shown in Figure 4. It exhibits two absorption peaks, one at 460 nm which accounts for the yellow coloration, and another at 270 nm which is due to the enzyme linked-antibody. A good indication of the oocyst concentration may be obtained from this peak at 270 nm, as it is proportional to the amount of antigen which have bound themselves to the gel. The color change is less qualitative since the chromogens or color substrates (TMB) do not attach themselves directly onto the gel.

(a)

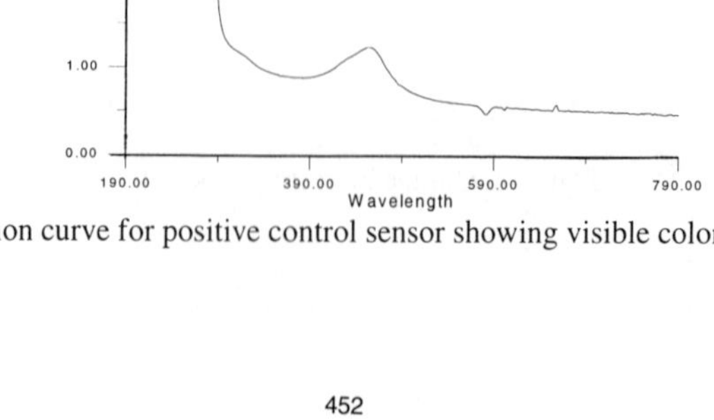

Figure 4 (a) Absorption curve for positive control sensor showing visible color change at ~460nm

(b)

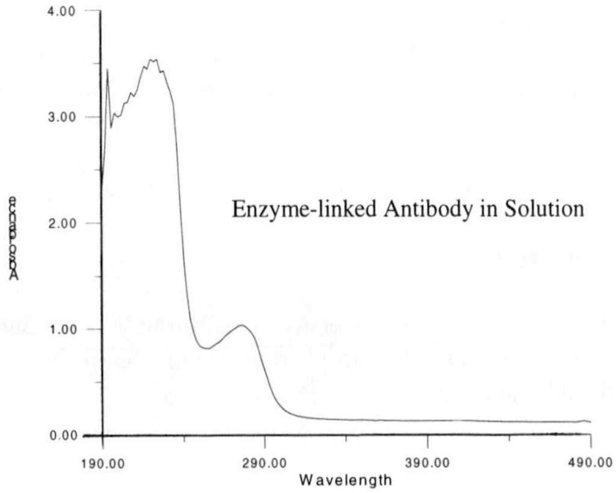

Figure 4- (b) Absorption curve for enzyme-linked antibody solution

Experiments show that the enzyme linked-antibody attachment is directly proportional to the *C. parvum* concentration in the tested solution. It is confirmed in Figure 5 that the amount of antibody bound to the gel exposed to 100 oocysts/mL is far higher than for the sensor exposed to 6 oocysts/mL. The current detection levels of 100 oocysts/mL in water [9] have been somewhat improved by the sol-gel sensor.

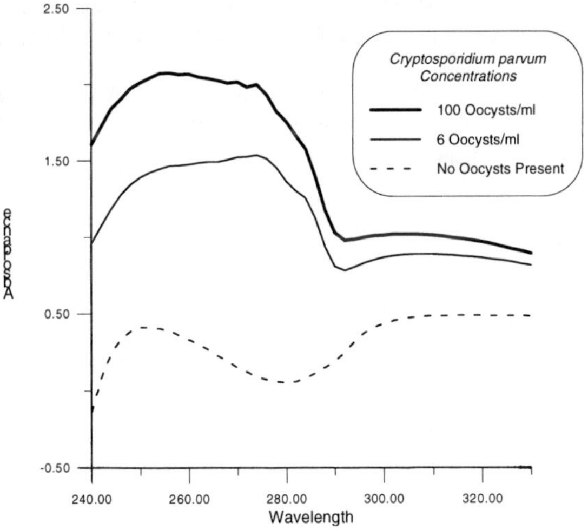

Figure 5- Absorption peaks for sensors exposed to different levels of *C. parvum*

CONCLUSIONS

A device for the detection of *Cryptosporidium parvum* in water has been fabricated using the sol-gel technique. The sensor successfully determines the presence of this microorganism and shows quantitative properties. Concentrations in *C. parvum* as low as 6 oocysts/mL in water have led to a color change. The presence of the microorganism may be detected either through a color change in the visible part of the spectrum or in the UV where the enzyme-linked antibody absorption occurs.

ACKNOWLEDGMENTS

This research is supported by the University of California's Toxic Substances Research & Teaching Program (TSR&TP). Spectrophotometric measurements were carried out in Professor Staffsudd's laboratory.

REFERENCES

1. Timothy P. Flanigan and Rosemary Soave, Progress in Clinical Parasitology Volume III, (Springer-Verlag, NY 1993) chapter 1.
2. Brooke A. Masters, The Washington Post, A39 (Friday December 10, 1993).
3. Mark W. Lechevallier, William D. Norton, and Ramon G. Lee, Applied and Environmental Microbiology, **57**, pp. 2610-2616 (September 1991).
4. Mark W. Lechevallier, William D. Norton, and Ramon G. Lee, Applied and Environmental Microbiology, **57**, pp. 2617-2621 (September 1991).
5. David Avnir et al., United States Patent #5300564, (April 5, 1994).
6. Shuguang Wu, Lisa M. Ellerby, J.S. Cohan, Bruce Dunn, M. A. El-Sayed, J Selverstone Valentine, and Jeffrey I. Zink, Chemistry of Materials, **5**, pp. 115-120 (1993).
7. Bruce S. Dunn et al., United States Patent #5200334, (April 6, 1993).
8. J. Livage et al., SPIE Sol-Gel Optics III, **2288**, pp. 493-503 (1994).
9. Private Communication (November, 1995).

INORGANIC ORGANIC COMPOSITE MATERIALS AS ABSORBERS FOR ORGANIC SOLVENTS

V. Gerhard, H. Schmidt and U. Dreier
Composite Department, Institut für Neue Materialien GmbH, Im Stadtwald, Geb. 43, D-66123 Saarbrücken, Germany

ABSTRACT

Inorganic organic composite materials have been developed in the form of non-porous films on various substrates and as porous materials with specific surface areas of about 700 m^2/g. Both types of materials have been tested for absorption and adsorption of solvents from the gas phase (butyric acetate, benzene, toluene and xylene). The results show that the non-porous films could be loaded with solvents up to 30 wt.% and unloaded at 130 °C within a few minutes. The porous materials could be loaded up to 38 wt.%. The kinetics of desorption are slowed down to values of some hours, being however still fast compared to activated carbon. Adsorption in wet atmospheres do not affect adsorption capacity but leads to a decrease of adsorption kinetics due to a slowing down of the diffusion or a replacement of H_2O against solvent within the pores.

INTRODUCTION

Porous materials are used as adsorbers and catalysing materials, for scientific examination and technical applications. Common adsorption materials are activated carbon, activated aluminum oxide, silica gel or zeolithes. These materials normally show high adsorption capacities but also very long desorption times (e. g. activated carbon: >>1 day). The active surface of adsorbers can be poisoned by compounds, which can not be desorbed any more. In contrast, diffusion controlled absorption processes do not show these disadvantages. Since the material properties of inorganic organic hybrid materials can be tailored in wide ranges by variation of their composition and the process parameters during their synthesis, these composites are predestined as absorber materials. Composites, which are adapted to special separation problems can be applied as coatings on different surfaces by common coating techniques, e. g. dip or spray coating, or can be used as bulk materials for absorption columns [1, 2].

The aim of the investigations was to compare the sorption behaviour of inorganic organic composites and commonly used absorbers for different organic solvents.

EXPERIMENTAL

The inorganic organic composite coating materials were prepared by adding dropwise the stoichiometric amount of 0.1 n hydrochloric acid to the organically modified alkoxysilane or mixture of different silanes. Two phases are perceptible. The mixture is stirred at room temperature for 3 h, until the phase separation has disappeared. The material is applied by dip or spray coating and is then thermally cured at 130 °C. Inorganic organic composites were prepared as bulk materials from solutions of tetramethylorthosilicate and methyltriethoxysilane by adding dropwise the stoichiometric amount of 6 n hydrochloric acid at room temperature. Spontaneous gelation of the mixture occurs immediately after the complete addition of the acid.

Mat. Res. Soc. Symp. Proc. Vol. 435 © 1996 Materials Research Society

After drying at 130 °C, the sorption characteristics of the obtained solids were measured in an absorber column with butyric acetate, benzene, toluene and xylene as model solvents (column length: 300 mm, diameter: 20 mm). The flow rate of the air stream was 2 l/min. The air speed calculated for the empty column was 10.6 cm/s. The volume of the composite in the column was 45.2 cm^3, which gives an effective air speed of 20.4 cm/s. The concentration of the solvent in the air stream acting on the sensors leads to a change in the potential, which is detected. The surface areas of the solids were determined by BET analysis of nitrogen adsorption data, collected at 77 K, measured with an ASAP 2400 from Micromeritics.

RESULTS

Both inorganic oranic composite coatings and bulk materials were synthesized via the sol gel process, but different paths were used. Figure 1 shows the reaction scheme for the synthesis of bulk materials and coatings.

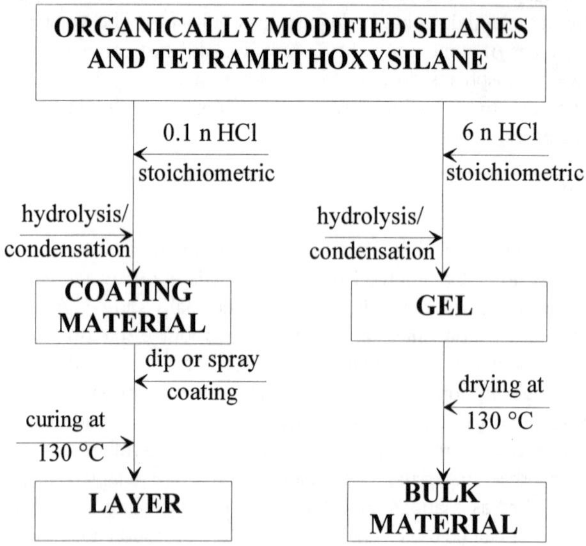

Figure 1: reaction scheme for the synthesis of inorganic organic composite coatings and bulk materials for absorption.

<u>Inorganic organic coating materials</u>

For the determination of the absorption characteristics, the takeup of the layers in a saturated solvent atmosphere was measured as a function of time. The butyric acetate takeup of the synthesized materials depends on the composition of the composites and is shown in figure 2. In particular, the content of different organic groups like methyl, propyl, phenyl or vinyl groups plays an important role on the capacity of such materials.

The high relative takeup observed with butyric acetate is remarkable since the coatings are non-porous and the specific surface area is correspondent to the film surface area. It can be

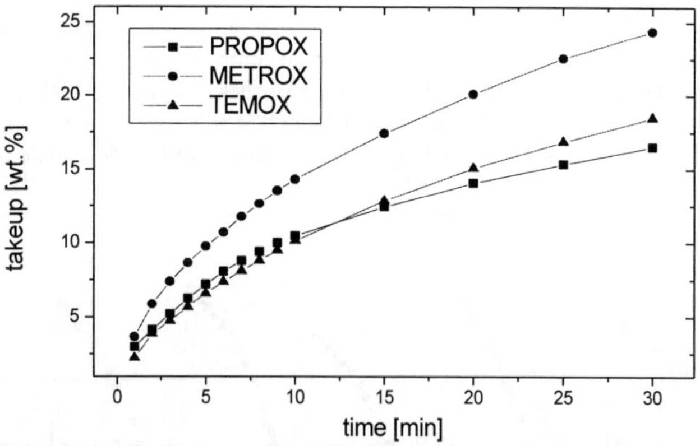

Figure 2: normalized takeup of butyric acetate of different inorganic organic composite layers (coating thickness: 5 µm), PROPOX: propyltrimethoxysilane; METROX: 25 mole % methyl-trimethoxysilane, 75 mole % diphenyldihydroxysilane; TEMOX: 66 mole % tetraethylortho-silicate, 17 mole % diphenyldihydroxysilane, 17 mole-% vinyltriethoxysilane

shown that the kinetics of solvent takeup are controlled by a diffusion limited process. The takeup can be described by equation (1) which can be derived from Fick's laws for diffusion [3].

$$\frac{M_t}{M_\infty} = \frac{4}{b} \bullet \left(\frac{D}{\pi}\right)^n \qquad (1)$$

M_t is the relative weight gain at the time t, M_∞ is the relative weight gain in the equilibrium state, D is the coefficient of diffusion, b is the thickness of the layer and n is a constant. For Fickian diffusion n is equal to 0.5 and the diffusion coefficient is independent of time. The plot of the takeup of butyric acetate versus $\sqrt{t}$ shows that each of the coatings show a linear dependence which proves that the solvent takeup is a diffusion contolled process and takes place by absorption and not by adsorption (figure 3).

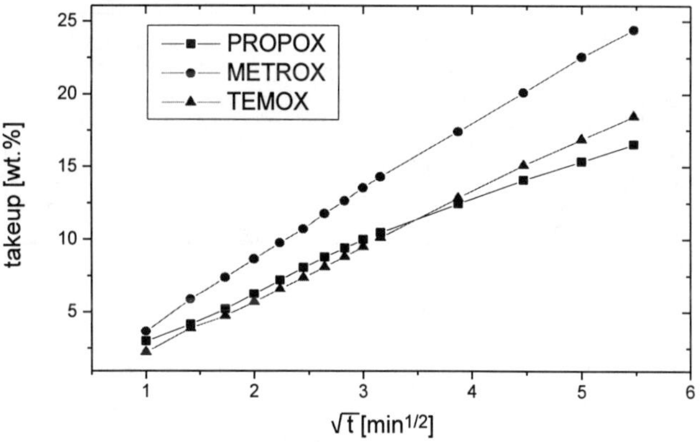

Figure 3: development of the takeup of butyric acetate of different coating systems as a function of $\sqrt{t}$

In contrast to highly porous adsorbers such as activated carbon, where in many cases the complete desorption takes very long times (>> 1 day) at elevated temperatures (in the range of 130 to 150 °C), the absorbed solvents can be desorbed completely from composite layers within a few minutes. Therefore the regenerative capacity of these coatings, which is of interest for practical application, is equal to the maximum takeup of a solvent after a given time of absorption. Figure 4 shows the absorption and desorption behaviour of butyric acetate on a composite layer.

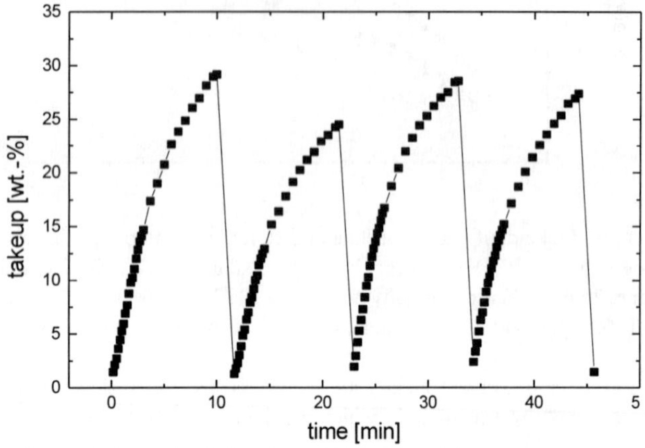

Figure 4: absorption and desorption behaviour of butyric acetate on a composite layer, layer-thickness: 1 μm, absorption at room temperature, desorption at 130 °C

The desorption time is a very important economic aspect. As can be seen in figure 4 the desorption time of composite layers is very short. By means of comparison, activated carbon has been investigated, which can adsorb up to 50 % of solvent, needs more than 30 h for a complete desorption at the same temperature. Figure 5 shows as an example the desorption behaviour of activated carbon loaded with toluene at 130 °C compared to that of an inorganic organic composite layer.

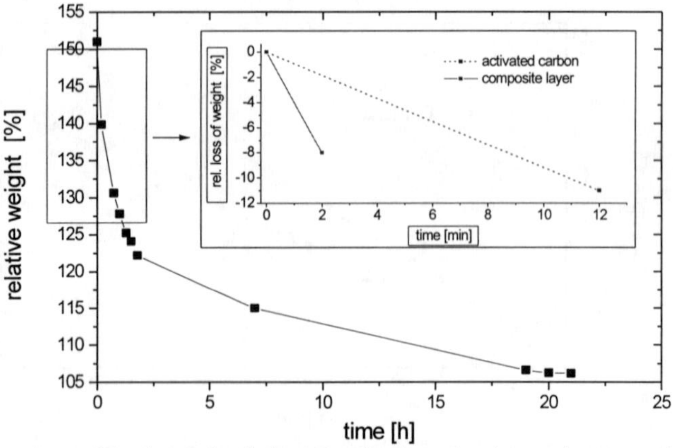

Figure 5: comparison of the desorption behaviour of activated carbon and a composite layer

Since the solubility of a given solvent depends on the composition of the absorber material, it is possible to influence the selectivity of the composites for different solvents by tailoring the composition. The coating materials synthesized can be applied to different surfaces by common coating techniques, e. g. dip coating or spray coating, and can be cured thermally. To gain large capacities in combination with fast kinetics it is necessary to limit the thickness of the layers and to coat large surface areas, e. g. porous materials or structures similiar to an heat exchanger.

Porous inorganic organic composite materials

The aim of the investigations was the development of solid absorber materials with high regenerative capacities for organic solvents in comparison to activated carbon. The investigation of absorbing composite coatings has shown that a maximum solvent takeup of 30 wt.% can be achieved within 30 minutes. Based on these results, inorganic organic composite bulk materials were developed, which can be used as absorption materials for organic solvents. Solid phases derived from methyl substituted silanes and tetramethylorthosilane, which have BET-surfaces of more than 700 m^2/g, show the best results in solvent takeup. As with the coating materials, butyric acetate, benzene, toluene and xylene were chosen as solvents for the investigations. A maximum takeup of solvent of nearly 38 wt.% within 2 - 3 hours can be observed depending on the composition of the materials. For the determination of the absorber characteristics, the materials were placed in a column, which is flooded with an air stream loaded with the appropriate organic solvent. The solvents are detected at the entrance and the top of the column by sensors as shown in figure 6.

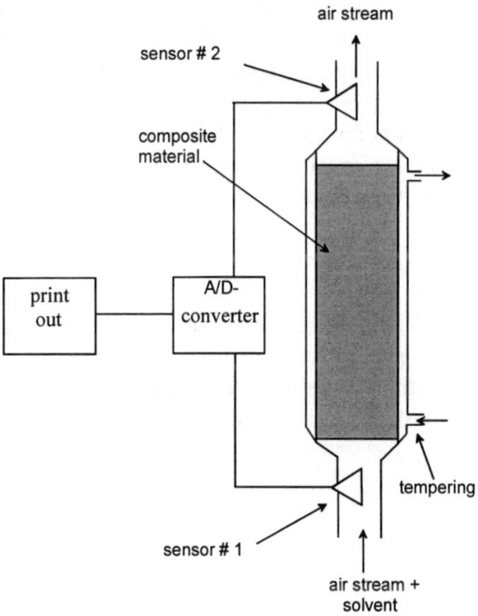

Figure 6: scheme of the apparatus for absorption measurements on bulk materials

As can be seen in figure 7, the solvent is completely held up by absorption for over 2.5 hours. The capacity at the point of breakthrough is 30 %.

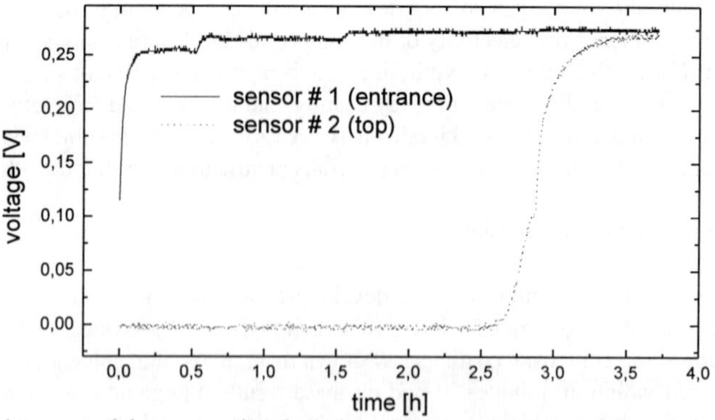

Figure 7: development of the sensor signals at the entrance and the top of the column as function of time, flow rate: 2 l/min, vapour pressure of the solvent (butyric acetate): 52 mbar

The desorption takes place at a temperature of 130 °C, whereby 90 % of the absorbed solvent can be removed in the first hour. The complete desorption takes 4 - 5 hours. In comparison, activated carbon with a specific surface of 1300 m²/g, loaded with 50 % of solvent, requires more than 30 hours for a complete desorption at the same temperature. Absorption experiments with wet air show that the water takeup of the bulk material developed is in the same order of magnitude as the takeup of organic solvents. By comparison, composite coatings show a water takeup of only 2 - 3 wt.%. The large specific surface area of the bulk materials, which is caused by the high porosity, could be the reason for this behaviour. If the surface of the material is covered with water due to adsorption and capillary condensation in the pores only a small part of the surface is able to absorb non polar organic solvents. This leads to very long diffusion paths and to slow absorption kinetics, which results in a rapid break through of solvents in the column. To avoid this the specific surface area, the pore structure and the polarity of the surface of the materials have to be tailored.

CONCLUSION

The investigations have shown that both inorganic organic composite coatings and bulk materials are able to absorb organic solvents from the gas phase. Compared to classic adsorbers the maximum capacities of 30 wt.% with coatings and 38 wt.% with bulk materials are not as high as that of e g. activated carbon but the necessary desorption time is much shorter than that of activated carbon.

REFERENCES

1. H. Schmidt, A. Kaiser, M. Seiferling; presented at the Enviceram '91, 2. Int. Symposium, Saarbrücken, Germany, 1991 (unpublished).

2. V. Gerhard, U. Dreier, H. Schirra, H. Schmidt; presented at the workshop "Teschnologie und Anwendung von Sol-Gel-Schichten, Frankfurt a. M., Germany, 1996 (unpublished).

3. Gad Marom; Polymer Permeability, Belfast University Press, Belfast, 341 (1985).

NMR AND IR SPECTROSCOPIC EXAMINATION OF THE HYDROLYTIC STABILITY OF ORGANIC LIGANDS IN METAL ALKOXIDE COMPLEXES AND OF OXYGEN BRIDGED HETEROMETAL BONDS

D. HOEBBEL, T. REINERT, H. SCHMIDT
Institut für Neue Materialien, Im Stadtwald, Geb. 43, D-66123 Saarbrücken, Germany

ABSTRACT

IR and ^{13}C NMR investigations of the hydrolytic stabilities of the saturated and unsaturated ß-keto ligands acetylacetone (ACAC), ethylacetoacetate (EAA), allylacetoacetate (AAA), methacryloxyethyl-acetoacetate (MEAA) of the Al-, Ti- and Zr-butoxide complexes show a strong dependence on the type of the metal alkoxide and the structure of the organic ligands. The hydrolytic stabilities of the ligands decrease in the order Al->Zr->Ti-alkoxide and ACAC>AAA>EAA≥MEAA. Sol-gel reactions of complexes having a weak ligand stability leads to a larger water consumption and to larger particle sizes in sols than those with stable ACAC ligands. Heterometal bonds, i.e. Si-O-Al, Si-O-Ti and Si-O-Zr, in the system diphenyl-siloxanediol/metal alkoxide (complex) proved by ^{29}Si and ^{17}O NMR are hydrolysed to a different extent depending on the water amount, the type of the Si-O-M bond and the structure of the heterometal species. The degradation of the heterometal bonds leads to a separation of M-O-M and Si-O-Si bonds which can entail a decreased homogeneity of the materials at a molecular level.

INTRODUCTION

Metal alkoxides, their organic derivatives and oxygen bridged heterometal bonds play an important role in the synthesis of advanced inorganic-organic hybrid materials and heterometal polymers via the sol-gel process. These polymeric materials are used for e.g. optical and electronic devices, functional coatings and the preparation of ceramics and glasses [1-4].

A demand on advanced heterometal or hybrid materials is the high homogeneity in the distribution of the material components at a molecular scale, i.e. in the distribution of heterometal atoms and in the links between the organic and inorganic network structures [4,5]. A homogeneous heterometal structure of the materials is influenced by the condensation rate of the different metal alkoxides [6] and by the hydrolytic stability of the heterometal bond in the course of the sol-gel process [7]. Furthermore, the hydrolytic stability of the functional organic ligands of metal alkoxide complexes is a prerequirement for a continuous link of the inorganic and organic network via metaloxo groups [4,8]. So far, only insufficient knowledge exists about the hydrolytic stability of the organic ligands in metal alkoxide complexes (L-M bonds in Fig.1) and of the heterometal bonds (M-O-M*) during the sol-gel process.

The objective of this work is the examination of the hydrolytic stabilities of reactive and non-reactive ß-keto ligands in Al-, Ti- and Zr-alkoxide complexes and of the heterometal bonds, i.e. Si-O-Al, Si-O-Ti and Si-O-Zr, in the system diphenylsilanediol/metal alkoxide (complex) by means of IR, ^{13}C, ^{17}O and ^{29}Si NMR spectroscopy with regard to a more controlled sol-gel reaction leading to materials with defined structure and properties.

Mat. Res. Soc. Symp. Proc. Vol. 435 © 1996 Materials Research Society

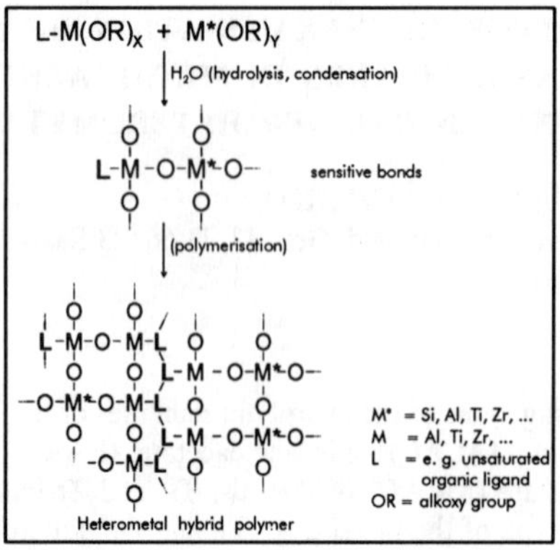

Figure 1. Scheme of the synthesis of a heterometal hybrid polymer by the sol-gel process

EXPERIMENT

Titanium-n-butoxide $Ti(OBu^n)_4$, aluminium-sec-butoxide $Al(OBu^s)_3$ and zirconium-n-butoxide $Zr(OBu^n)_4$ (84 % in n-butanol) were used for the complexation with acetylacetone (ACAC), ethylacetoacetate (EAA) and the unsaturated ß-ketoesters allylacetoacetate (AAA) and methacryloxyethylacetoacetate (MEAA) at a molar ratio metal(M):ligand(L)= 1:1. The metal alkoxide concentration in the solutions was kept at 1 M.

The hydrolysis of the $Ti(OBu^n)_3L$, $Zr(OBu^n)_3L$ and $Al(OBu^s)_2L$ complexes were carried out with a water/butanol solution at a hydrolysis ratio h (molar ratio water/OR) =1.

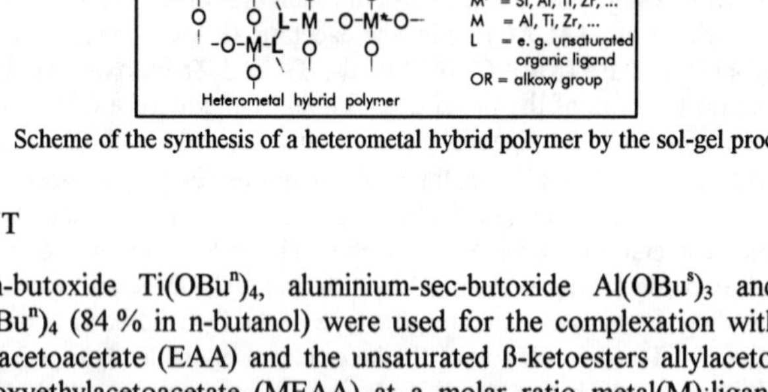

Figure 2. ^{13}C NMR and IR spectra of the $Zr(OBun)_3AAA$ complex before and after hydrolysis at h= 1

Fig.2 describes the changes in the IR and ^{13}C NMR spectra before and after the hydrolysis of the $Zr(OBu^n)_3AAA$ complex. The appearance of IR absorptions at 1740 and 1715 cm^{-1} (a*,b*) with increased hydrolysis time indicates that AAA ligands are partially released from the complex by the addition of water. The ^{13}C NMR spectra of the $Zr(OBu^n)_3AAA$ hydrolysate show sharp signals at δ= 201(b*) and 167 ppm (a*) for the keto form of AAA which confirm the presence of released AAA. The semi-quantitative determination of the released ligands was made by integration of the respective absorptions bands in the IR spectra.

The heterometal solutions were synthesised by mixing of diphenylsilanediol (DPSD) with $Al(OBu^s)_3$ or the metal alkoxide complexes $Ti(OBu^n)_3ACAC$ and $Zr(OBu^n)_3ACAC$ at a molar ratio Si:M= 1:0.75 in tetrahydrofuran solution and 0.7 molar concentration in DPSD. The solutions were hydrolysed with a water/THF solution at a molar ratio H (H_2O/metal)= 1, 6 and 9. The spirocyclic titanosiloxane $Ti[O_3Si_2(C_6H_5)_4)]_2$ (A) was prepared by the reaction of tetraphenyldisiloxanediol with $Ti(OPr^i)_4$ at a molar ratio 4:1 in dioxan solution. The spirocyclic titanosiloxane $Ti[O_5Si_4(C_6H_5)_8)]_2$ (B) was synthesised according to the Zeitler method [9].

The ^{29}Si and ^{17}O NMR spectra were obtained using liquid state (Bruker AC200) spectrometer operating at 4.7 T. *^{29}Si NMR*: external reference: tetramethylsilane, internal standard: phenyltrimethylsilane, repetition time: 40 s, pulse angle: 63°, number of scans: 20-100. *^{17}O NMR:* single pulse, repetition time: 600 ms, pulse angle: 90°, reference: H_2O (1 % ^{17}O), number of scans: 4000-10000.

RESULTS

Hydrolytic Stability of Organic Ligands in Metal Alkoxide Complexes

Metal alkoxide complexes with hydrolytically unstable ligands can release a part of themselves during the hydrolysis/condensation reactions in the course of the sol-gel process according to the scheme in Fig.3.

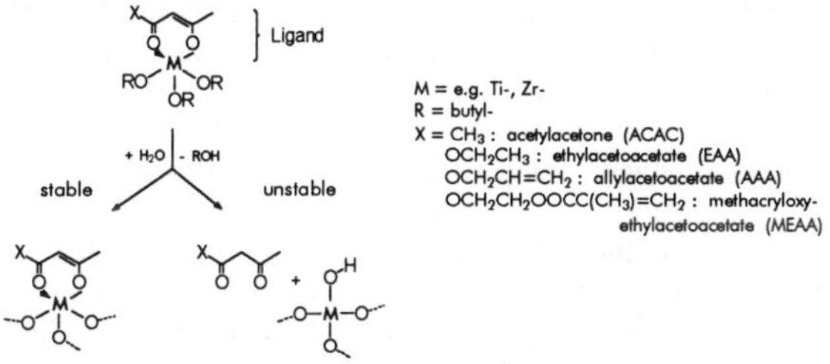

Figure 3. Scheme of the hydrolysis of ß-keto ligands from metal alkoxide complexes

The release of polymerisable ligands entails an increasing number of OH/OR groups at the metal atom. Fig.4 summarises the IR spectroscopic results on the hydrolytic stability of different ß-keto ligands (see Fig.3) attached to Al-, Ti- and Zr-butoxides at a molar ratio 1:1 and at h= 1.

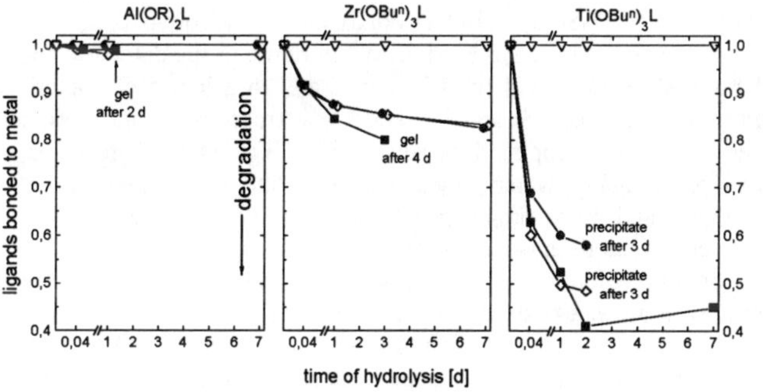

Figure 4. Hydrolytic stability of ß-keto ligands L of $Al(OR)_2L$, $Zr(Bu^n)_3L$ and $Ti(Bu^n)_3L$ complexes at h= 1. L: ACAC (∇), EAA (◊), AAA (●), MEAA (■).

The Al-sec-butoxide complexes with AAA, EAA and MEAA and the Al-butoxyethoxide complex with ACAC as ligands show a high hydrolytic stability between 95 and 100 % within a reaction time of 7 days. The ACAC ligands of Zr- and Ti-butoxide complexes show a likewise high hydrolytic stability in accordance with the literature [10]. The other ligands of the Ti- and Zr-butoxide complexes (AAA, MEAA and EAA) show a decreased hydrolytic stability. About 15 % of the AAA and MEAA-ligands are released in hydrolysates of Zr-butoxide complexes after 1 day. The AAA-, EAA- and MEAA-ligands at Ti-butoxide complexes are released up to 50 % after the same time.

The different hydrolytic stability of the ligands at one and the same metal alkoxide is visible at the example of the Ti-butoxide complexes (Fig.4). The ligand stability decreases in the order ACAC>AAA>EAA≥MEAA. Furthermore, the hydrolytic stability of one and the same ligand at different metal alkoxides decreases in the order $Al(OBu^s)_3 > Zr(OBu^n)_4 > Ti(OBu^n)_4$. The results make sure that the hydrolytic stability of the complex ligands is significantly influenced by the structure of the ligands and the type of the metal alkoxide.

A more detailed investigation of the hydrolysates of the metal alkoxide complexes [11] showed that the hydrolytically stable ligands of Al-butoxide complexes and the ACAC ligand of Ti- and Zr-butoxide complexes consume 0.3-0.4 mol water/OR at h= 1 and complexes with unstable ligands (i.e. MEAA) 0.4-0.7 mol within 1 day. The additional amount of water is required for the splitting off and a partial chemical degradation of the unstable ligands. As consequence, additional OH/OR groups appear at the metal atoms which contribute to more extended condensation reactions of the complexes and to larger particles in the sols. Complexes with the stable ACAC ligand form in their hydrolysates particles with nearly time-independent sizes of about 3-4 nm within 7 d whereas the hydrolysates of unstable Zr/MEAA and Ti/MEAA complexes contain particles between 6 and 10 nm after 1 day hydrolysis.

The results show that only the complexes $Al(OBu^s)_2AAA$, $Zr(OBu^n)_3AAA$ and $Zr(OBu^n)_3MEAA$ have a sufficient hydrolytic stability of their unsaturated ligands. Therefore, these complexes should be suitable precursors for the preparation of sols with a defined particle size and for the synthesis of hybrid polymers with a high homogeneity in the interconnections between the inorganic and organic networks.

Formation and Characterisation of Heterometal Bonds

So far, it has not been sufficiently proved whether or not an optimal homogeneous distribution of the heterometal bonds at a molecular scale exists in the sol-gel derived products [12,13]. Recent NMR investigations [14] have indicated a remarkable degradation of heterometal Si-O-Ti bonds to Si-O-Si and Ti-O-Ti bonds during hydrolysis/condensation reactions. The objective of the second part of this work is to contribute to the detection and identification of heterometal Si-O-Al, Si-O-Ti and Si-O-Zr bonds and to a better characterisation of their hydrolytic stability by means of ^{29}Si and ^{17}O NMR spectroscopy. Diphenylsilanediol (DPSD) has been chosen as model compound because of its high stability towards condensation reactions and its difunctionality, which only allows chain-like or cyclic structures of the homocondensation products and more simple structures in the hetero-condensation products with metal alkoxides as compared to those of multifunctional alkoxysilanes.

In general, a broad distribution of different siloxanes and heterometal species is detected in mixtures of DPSD with metal alkoxides or their complexes [15]. In the course of our examination of the reactions of DPSD with $Al(OBu^s)_3$, $Ti(OBu^n)_3ACAC$ and $Zr(OBu^n)_3ACAC$ we found special compositions of solutions in which mainly one structural type of Si atoms in the heterometal compounds prevails.

Fig.5-IV,V,VI shows the ^{29}Si and ^{17}O NMR spectra of the chosen systems with a molar

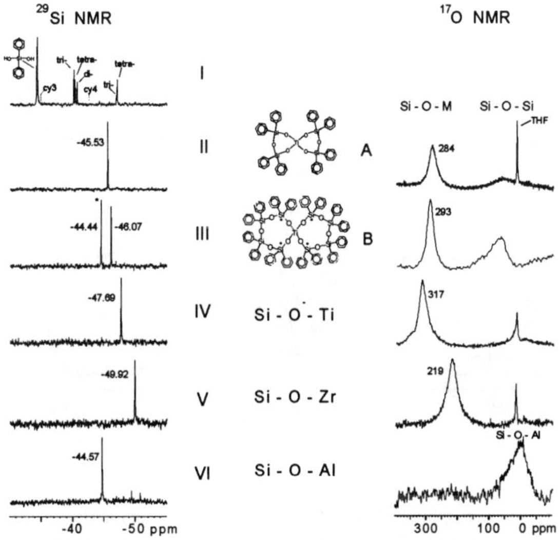

Figure 5. ^{29}Si and ^{17}O NMR spectra of a condensed DPSD solution (I), the spirocyclic titanosiloxanes A and B (II,III), and the reaction products of DPSD with $Ti(OBu^n)_3ACAC$ (IV), $Zr(OBu^n)_3ACAC$ (V) and $Al(OBu^s)_3$ (VI) at a molar ratio 1:0.75.

ratio DPSD/0.75 metal alkoxide (complex) in THF solution. Two heterometal compounds of known structure, the spirocyclic titanosiloxanes $Ti[O_3Si_2(C_6H_5)_4)]_2$ (A) and $Ti[O_5Si_4(C_6H_5)_8)]_2$ (B), are included in the series. The structures of the monomolecular compounds in the solutions IV-VI are still unknown. For comparison in Fig.5-I the ^{29}Si spectrum of a condensed DPSD solution is shown. Practically, all expected signals of the condensation products (chain-like di-, tri- and tetrameric phenylsiloxanediol) are visible in the spectrum [15]. Additionally, the positions of the chemical shifts of the cyclic tri- (cy3) and cyclic tetrameric phenylsiloxanes (cy4) are marked in the spectrum. The ^{29}Si NMR spectra IV-VI show mainly a single signal with chemical shifts between δ= -44.4 and -49.9 ppm. The chemical shift of the signals does not correspond with those of the condensed siloxanes in Fig.5-I. From this it can be concluded that the silicon atoms are bonded in monomolecular, probably, heterometal species.

Additionally, ^{17}O NMR spectra were recorded from the solutions II-VI to confirm the existence of heterometal bonds (Fig.5). The spectra of the ^{17}O labelled DPSD/$Ti(OBu^n)_3ACAC$ and DPSD/$Zr(OBu^n)_3ACAC$ solutions show intensive signals at δ= 317 and 219 ppm, resp., which confirm the presence of Si-O-Ti and Si-O-Zr bonds [14] (Fig.5-IV,V). The chemical shift of the oxygen atoms in Si-O-Al bonds is expected in the region δ= 30 to 80 ppm [16]. The detected broad signal in the spectrum of the DPSD/$Al(OBu^s)_3$ solution between δ= 0-70 ppm overlaps with the region in which the signals of the oxygen atoms in Si-O-Si and Si-OH bonds appear. Therefore, an exact identification of the Si-O-Al bond cannot be made.

Hydrolytic Stability of the Heterometal Bonds

Fig.6 shows as an example the ^{29}Si and ^{17}O NMR spectra of the ^{17}O labelled spirocyclic compound A before and after hydrolysis in THF solution at H= 9.

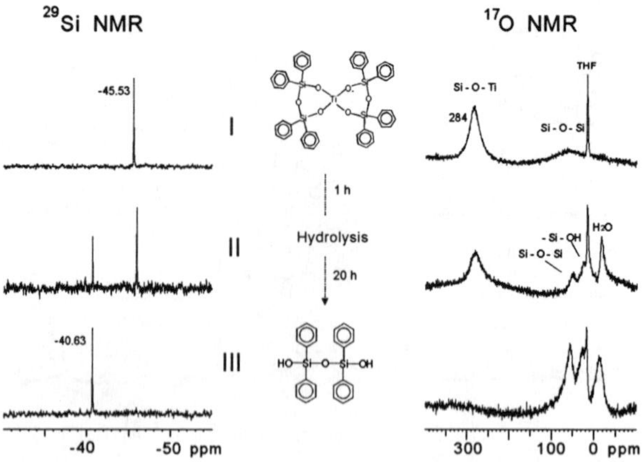

Figure 6. ^{29}Si and ^{17}O NMR spectra of the spirocyclic titanosiloxane A before and after hydrolysis at H= 9

It follows from the ^{29}Si spectra that the spirocyclic structure is degraded to the tetraphenyldisiloxanediol within about 20 h. The typical signal of Si-O-Ti bonds at δ= 285 ppm disappears in the ^{17}O spectrum after 24 h in favour of signals caused by oxygen atoms in Si-O-Si and SiOH bonds of the disiloxanediol.

The results of the examination of the hydrolytic stability of the Si-O-Al, Si-O-Ti and Si-O-Zr bonds at molar ratio H= 1 and 6 by means of ^{29}Si and ^{17}O NMR are summarised in Fig.7.

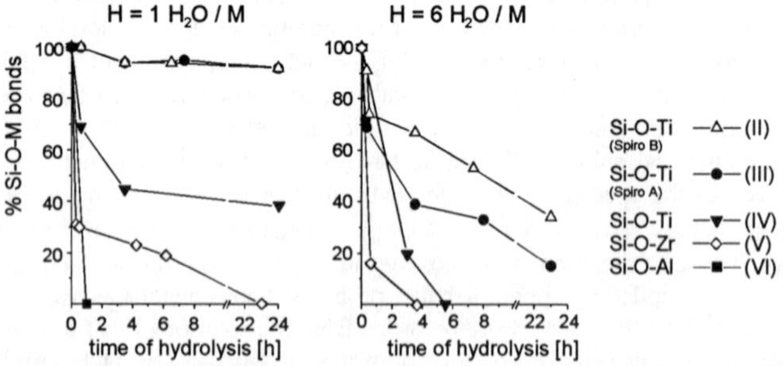

Figure 7. Hydrolytic stability of the Si-O-M bonds in the spirocyclic titanosiloxanes (A,B) and in solutions of DPSD/0.75{$Ti(OBu^n)_3$ACAC (IV) or $Zr(OBu^n)_3$ACAC (V) or $Al(OBu^s)_3$ (VI)} at H= 1 and 6

The results show that already at H= 1 a complete hydrolysis of the Si-O-Al bonds occurs within 1 h. At the same time of hydrolysis about 70 % of the Si atoms in Si-O-Zr bonds and 30 % in Si-O-Ti bonds are hydrolysed. The spirocyclic titanosiloxanes A and B show in contrast to the Si-O-Al, Si-O-Ti and Si-O-Zr species in the solutions IV-VI a high hydrolytic stability of its Si-O-Ti bonds. A larger water amount (H= 6) leads to significantly stronger degradation reactions. The Si-O-Ti and Si-O-Zr bonds in the solutions IV and V are completely hydrolysed within 5 h. The Si-O-Ti bonds of the spirocyclic structures A and B are hydrolysed to 60 and 40 %, resp., after 5 h. A complete hydrolysis of the Si-O-Ti bonds in the compound (A) only occur at H= 9 after 20 h (see Fig.6). The results show a decreased stability of the Si-O-M bonds in the solutions IV-VI in the order Si-O-Ti>Si-O-Zr>Si-O-Al. Furthermore, the results show that the heterometal Si-O-Ti bonds in the spirocyclic structures are considerably more stable towards hydrolysis than those of the solution IV.

CONCLUSIONS

1. The significant differences in the hydrolytic stability of the organic ligands (ß-diketone and ß-ketoester) in Al-, Ti- and Zr-butoxide complexes are to be taken into consideration for more controlled syntheses of sols with a defined particle size, of organically modified polymers and of inorganic-organic hybrid polymers. A highly hydrolytically unstable ligand is connected with additional condensation reactions of the MOR/OH groups and leads to larger particle sizes in the sols. Furthermore, the instability of unsaturated ligands can lead to incomplete connections between inorganic and organic networks in the hybrid polymers.

2. The heterometal Si-O-Al, Si-O-Ti and Si-O-Zr bonds in co-condensation products of diphenylsilanediol with Al-butoxide and Ti- and Zr-butoxide complexes are already degraded to a great extent at low water amounts (H= 1). The Si-O-Ti bonds in spirocyclic titanosiloxanes show a considerably higher hydrolytic stability at identical hydrolysis conditions. The results lead to the conclusion that the hydrolytic stability of the Si-O-Ti bonds is strongly influened by the structure of the heterometal compounds. The instability of heterometal bonds can lead to a separation of Si-O-Si and M-O-M bonds rich regions, which lowers the homogeneity of the materials at a molecular scale.

ACKNOWLEDGMENTS

The authors gratefully acknowledge the Volkswagen-Stiftung and the Fonds für die Chemische Industrie for the financial support and thank Prof. E. Arpac, Akdeniz-University Antalya, Turkey, for helpful discussions, Ms S. Carstensen for preparative work and Dipl.-Chem. M. Nacken for NMR measurements.

REFERENCES

1. R.C. Mehrotra, Chemistry, Spectroscopy and Applications of Sol-Gel Glasses, edited by R. Reisfeld, C.K. Jorgensen (Springer-Verlag, Berlin,Heidelberg, 1992).
2. C.J. Brinker, D.E. Clark, D.R. Ulrich, Better Ceramics Through Chemistry, (Elsevier Science Publishers, New York, 1984).
3. H. Schmidt, J. Non-Cryst. Solids **100**, 51-64 (1988).
4. U. Schubert, N. Hüsing, A. Lorenz, Chem. Mater. **7**, 2010-2027 (1995).
5. C.J. Brinker and C.W. Scherer, Sol-Gel Science, (Academic Press, New York, 1990).
6. P.J. Dirken, M.E. Smith, H.J. Whitfield, J. Phys. Chem. **99**, 395 (1995).
7. J. Jonas, A.D. Irwin and J.S. Holmgren, in Ultrastructure Processing of Advanced Materials, edited by D.R. Uhlmann and D.R. Ulrich (J. Wiley, 1992) pp. 303-314.
8. C.Sanchez, F. Ribot, New J. Chem. **18**, 1007-1047 (1994).
9. V.A. Zeitler and C.A. Brown, J. Am. Chem. Soc. **79**, 4618-4621 (1957).
10. P. Toledano, M. In, C. Sanchez, C.R. Acad. Sci. Paris **313**, 1247 (1991).
11. D. Hoebbel, T. Reinert, H. Schmidt, E. Arpac, to be published (1996).
12. H. Schmidt and B. Seiferling, Mat. Res. Soc. Symp. Proc. **73,** 739-750 (1986).
13. C.L. Schutte, J.R. Fox, R.D. Boyer and D.R. Uhlmann, in Ultrastructure Processing of Advanced Materials, edited by D.R. Uhlmann and D.R. Ulrich (J. Wiley, 1992) pp. 95-102.
14. F. Babonneau, Mat. Res. Soc. Symp. Proc. **346**, 949-960 (1994).
15. D. Hoebbel, T. Reinert, H. Schmidt, J. Sol-Gel Sci. Technol. in press (1996).
16. G.A. Pozarnsky, A.V. McCormick, J. Non-Cryst. Solids **190**, 212-225 (1995).

results show that the formation of N–O–Ti bonds in these [illegible] are [illegible] more stable toward [illegible] the [illegible].

CONCLUSIONS

1. The important differences in the hydrogen stability of the [illegible] and [illegible] ... [illegible] ... with [illegible] and particle size ... [illegible] ... [illegible] ... [illegible] ... the [illegible] ... tion between organic and inorganic phases in the hybrid polymers.

2. The [illegible] ... [illegible] ... [illegible] ... [illegible] ... [illegible] ... [illegible] ... [illegible] ... [illegible] ...

ACKNOWLEDGMENTS

[illegible]

REFERENCES

1. [illegible] ... Sol-Gel Science [illegible] ... [illegible] ... [illegible]
2. [illegible] ... [illegible] ... [illegible]
3. [illegible]
4. [illegible] ... Chem. [illegible] ... (1985).
5. [illegible] ... (1988).
6. [illegible]
7. [illegible] ... (1992) ... [illegible]
8. [illegible] ... (1990).
9. [illegible] ... (1975).
10. [illegible]
11. [illegible] ... (1990).
12. [illegible]
13. [illegible]
14. [illegible]
15. [illegible] ... (1994).
16. [illegible]

THE *IN-SITU* GENERATION OF SILICA REINFORCEMENT IN MODIFIED POLYDIMETHYLSILOXANE ELASTOMERS+

S. PRABAKAR*, S. E. BATES**, E. P. BLACK**, T. A. ULIBARRI**,
D. W. SCHAEFER**, G. BEAUCAGE** AND R. A. ASSINK*,**
*Advanced Materials Laboratory, University of New Mexico, Albuquerque, NM 87106,
**Sandia National Laboratories, Albuquerque, NM 87185-1407.

ABSTRACT

The structure and properties of a series of modified polydimethylsiloxane (PDMS) elastomers reinforced by *in situ* generated silica particles were investigated. The PDMS elastomer was modified by systematically varying the molecular weight between reactive groups incorporated into the backbone. Tetraethoxysilane (TEOS) and partial hydrolyzate of TEOS were used to generate silica particles. The chemistry and phase structure of the materials were investigated by ^{29}Si magic angle spinning nuclear magnetic resonance (NMR) spectroscopy and swelling experiments.

INTRODUCTION

The *in-situ* generation of silica particles by the sol-gel method provides an alternative route to prepare reinforced polydimethylsiloxane polymers [1,2]. Although the chemistry of the individual components have been the subject of numerous investigations, the interaction of the two phases has only recently been the subject of investigation [3]. We have begun a systematic study of the *in situ* growth of the silica reinforcement phase within the PDMS elastomer. Reactive functional groups were incorporated into the elastomer backbone in order to modulate the degree of interaction between the matrix and filler phases:

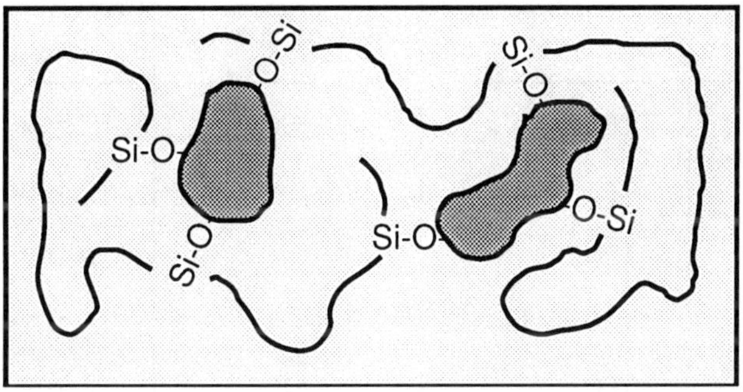

The local chemical structures and extents of reaction of the materials were measured by solid state ^{29}Si NMR spectroscopy. Solubility and swelling experiments were use to determine the extent to which the two phases were chemically coupled, while ^{29}Si NMR relaxation times were used to determine the extent to which the two phases were physically coupled.

EXPERIMENTAL

Preparation of $(Me_2SiO)_x(MeOMeSiO)_y$ Copolymers

A mixture of dimethylsiloxanes (DC 2-0409), linear methylhydrogen siloxanes (DC-1107) and hexamethyldisiloxane were used to prepare $Me_2SiO/MeHSiO$ copolymers capped with

+ This work supported by the United States Department of Energy under contract DE-AC04-94AL8500.

trimethylsilyl. Trifluoromethanesulfonic acid (0.1 weight %) was added to the three component system and the solution was heated to 70 °C for 4 hours. The resulting reaction mixture was neutralized by the addition of a ten-fold excess of sodium bicarbonate. The copolymers were then filtered and used without further purification. To methoxylate the silicon hydride moiety, the $(Me_2SiO)_x(MeHSiO)_y$ copolymer, anhydrous methanol, and N,N-diethylhydroxylamine (1 weight %) were combined and stirred in a sealed vessel until complete substitution was achieved. Completion of the reaction was indicated by the absence of the Si-H band at 2150 cm^{-1} in the FT-IR. The resulting copolymers were used without any further purification.

Preparation of Tetraethoxysilane Partial Hydrolyzate (TEOS-PH)

Tetraethoxysilane partial hydrolyzate was prepared by the addition of substoichiometric amounts of deionized water in absolute ethanol to an acidified solution of tetraethoxysilane. After stirring, the acid was neutralized with sodium bicarbonate and volatiles were removed by distillation. The solution was then filtered to remove salts and used without further purification.

Preparation of Elastomers

The silica-filled elastomeric materials were prepared by mixing the matrix polymer, in the presence of dibutyltin dilaurate catalyst (0.2 weight %), with TEOS or TEOS-PH, in amounts sufficient to both crosslink and fill the matrix to 10 weight % silica. Samples 2 to 3 mm thick were allowed to cure in a 60% relative humidity (R_H) environment for 1 week. The final elastomer compositions are shown in Table 1 where the x/y ratios correspond to the relative values of x and y in the $(Me_2SiO)_x(MeHSiO)_y$ copoolymer.

Table 1: The elastomeric structures, the silica precursors and the degree of polymerization for the filled elastomers.

Copolymer x/y	Silica Precursor	Degree of Polymerization (dp)	Copolymer Formula
10/1	TEOS	506	$M[D_{10}T^{OMe}]_{46}M$
10/1	TEOS-PH	506	$M[D_{10}T^{OMe}]_{46}M$
20/1	TEOS	504	$M[D_{20}T^{OMe}]_{24}M$
20/1	TEOS-PH	504	$M[D_{20}T^{OMe}]_{24}M$
50/1	TEOS	510	$M[D_{50}T^{OMe}]_{16}M$
50/1	TEOS-PH	510	$M[D_{50}T^{OMe}]_{16}M$

Characterization

Solid state ^{29}Si NMR spectra were recorded at 39.6 MHz on a Chemagnetics console interfaced to a General Electric 1280 data station described previously [4]. The samples were spun about the magic angle in a 7 mm ceramic rotor at 2.5 kHz. A crosspolarization sequence was used with various crosspolarization times and a pulse delay of 4s.

Equilibrium swelling measurements were conducted by swelling pre-weighed samples in heptane for 4 days. The excess heptane was removed and the samples were weighed in the swollen state. The samples were then allowed to dry and their final weights measured. Using the theory of the Flory [5], the molecular weights between crosslinks and the crosslink density were calculated using the following equations:

$$M_c = V_s r_P(C^{1/3} - C/2)/[\ln(1-C) + C + CC^2] \quad (1)$$

$$r_c = r_P \text{ (Avogadro's \#)}/M_c \quad (2)$$

$$C = W_0/(r_P V_\infty) \quad (3)$$

$$V_\infty = W_0 r_P + (W_\infty - W_0)/r_S \quad (4)$$

where M_c = Average molecular weight of network chains (g/mole); V_s = Molar volume of the solvent (146.6 cm^3/g-mole), C = Flory-Huggins coefficient (Chi parameter, 0.39); r_P = Polymer density (1.0 g/cm^3); r_s = Solvent density (0.684 g/cm^3; r_c = Crosslink density; C = Relative concentration (dimensionless); V_∞ = Final polymer volume (cm^3); W_0 = Initial polymer weight (Since the soluble fraction of many of these materials was high, the weight used as the initial polymer weight was the extracted or final dry weight); W_∞ = swollen polymer weight. The soluble fraction was calculated using the following equation:

$$\text{Soluble Fraction} = 100(W_0 - W_f)/W_0 \quad (5)$$

where W_f = final dry weight after swelling.

RESULTS AND DISCUSSION

The solid state ^{29}Si NMR spectrum of the 10/1 TEOS material is shown in Fig 1. The resonance at -21.4 ppm corresponds to the D backbone silicons, the resonances at -56.9 and -65.3 ppm correspond to the T backbone silicons which are functionalized and the resonances at -100.7 and -108.6 ppm correspond to the Q silicons of the silica filler formed by TEOS. The D backbone silicons are not functionalized so a single resonance is observed. The T silicons begin as T^2 silicons, bonded to two neighboring silicons in the chain, and can only undergo a single condensation reaction to form a T^3 silicon. The Q silicons enter the reaction as Q^0 species and undergo reactions to form predominately Q^3 and Q^4 species.

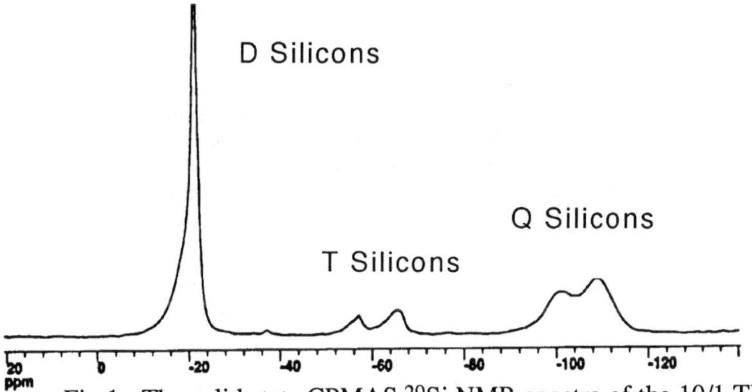

Fig 1. The solid state CPMAS ^{29}Si NMR spectra of the 10/1 TEOS material showing the D, T and Q silicon species.

The spectra for each material were recorded as a function of crosspolarization time so that the relative product distributions for each silicon species could be determined. The signal intensities as a function of crosspolarization times are shown in Fig 2. The experiments were only carried out to 15 ms so the longer $T_{1\rho H}$ relaxation times were not measured. The magnetization buildup was fit by a single exponential function where the time constant was set equal to τ_{cp}, the crosspolarization relaxation time. The shapes of the resonances corresponding to the T and Q silicons were independent of time for cp times ranging from 7 to 15 ms. The long plateau region exhibited by the signal intensities, coupled with a constant spectral shape for each silicon species gave us confidence that the spectrum components for the T and Q silicon species were semiquantitative.

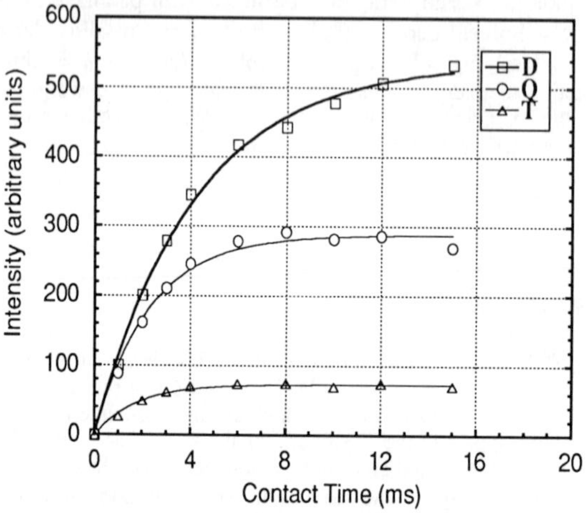

Fig 2. The magnetization buildup of the various silicon species as a function of crosspolarization time.

Deconvolution of the Q resonances showed that the extents of reaction of these silicons ranged from 86 to 92 %. These values are somewhat higher than observed for acid-catalyzed TEOS sol-gels. We attribute the greater extent of reaction of TEOS to its mobility when dispersed in the elastomeric matrix. The T silicons begin the reaction as T^2 silicons and condense to form T^3 silicons. The extents of reaction of the T silicons are approximately 65 % for the two 10/1 materials. The lower extent of reaction of the T silicons is probably due to the networking which they experience before the reaction formally begins.

Equilibrium swelling measurements of the elastomers in heptane were used to determine the soluble fraction and to measure the average molecular weight between crosslinks. The results are shown in Table 2. The soluble fraction for elastomers prepared using TEOS was generally higher than the soluble fraction for TEOS-PH elastomers. This is likely due to the fact, that for the hydrolyzate, more of the tetrafunctional silane molecules are locked into the structure with a single condensation reaction. An increase in the soluble fraction was also found for elastomers prepared with decreasing amounts of functionality on the backbone. Fewer reactive sites on the backbone result in fewer attachment sites for nucleation and growth of silica, thereby increasing the number of unattached TEOS derivatives. The molecular weight between crosslinks (M_c)

follows the expected trend with the distance between crosslinks increasing as the distance between functionality on the matrix polymer backbone increases.

Table 2. Soluble fraction, cross-link density, and average molecular weight between cross-links (M_C).

Elastomer	Silica Precursor	Soluble Fraction	M_C	Cross-link Density
$MDT^{OMe}M$ 10/1	TEOS	12.5	1746	3.45E+20
$MDT^{OMe}M$ 10/1	TEOS-PH	11.5	1893	3.18E+20
$MDT^{OMe}M$ 20/1	TEOS	14.5	3187	1.89E+20
$MDT^{OMe}M$ 20/1	TEOS-PH	15.8	4246	1.42E+20
$MDT^{OMe}M$ 50/1	TEOS	27.4	12841	4.69E+19
$MDT^{OMe}M$ 50/1	TEOS-PH	23.8	11000	5.47E+19

The extent to which the various phases are physically coupled can be probed by examining the crosspolarization times of each type of silicon. These times are shown in Fig 3. The crosspolarization times of the T and Q silicons are similar to each other indicating that these two phases have similar mobilities. The crosspolarization times of the D backbone silicons are considerable longer than those of either the T or Q silicons demonstrating that on the average, these silicons possess greater mobilities. F. Babonneau has reported that the resonance of the D silicons in a material prepared from PDMS and TEOS consists of both a broad and a narrow component [3]. The broad component was associated with the silicons closely coupled to the silica phase. We observe similar behavior for these materials. For short cross-polarization times, the broad component is accentuated relative to the narrow component indicating that these backbone silicons have lower mobilities than the silicons associated with the narrow component.

CONCLUSIONS

Approximately 65 % of the reactive groups incorporated into the backbone of the PDMS chain were found to have undergone reaction. The concentration of T and Q silicons in the 10/1 formulation are comparable, while the number of T silicons is substantially less than the number of Q silicons for the 20/1 and 50/1 formulations. Thus, on a statistical basis we expect that a significant number of chemical bonds are formed between the T silicons in the PDMS chain and the Q silicons in the silica particle. These expectations were confirmed by a decrease in the solubility fraction and an increase in the crosslink density for materials prepared with a high fraction of reactive silicons in the PDMS chain. Cross-polarization relaxation times show that the T silicons in the PDMS chain are also physically coupled to the silica filler particles. We expect that the intimate chemical and physical bonding of the elastomeric and filler phases of these materials will be reflected in their mechanical properties.

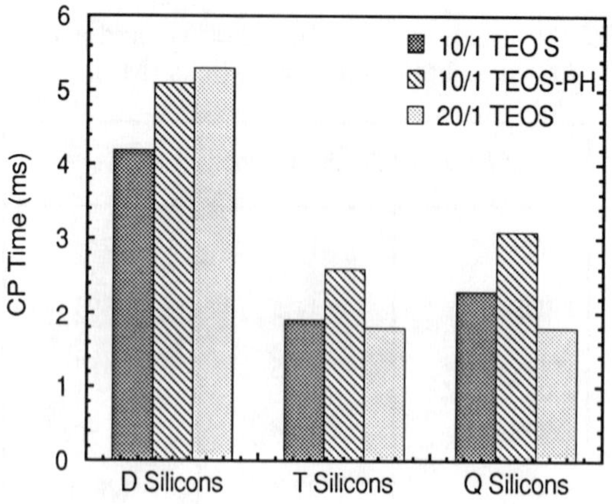

Fig 3. Comparison of cross-polarization relaxation times for the D, T and Q siliconspecies for selected materials.

REFERENCES

1. For a review of hybrid nanocomposite technologies, see *Hybrid Organic-Inorganic Composites*. J. E. Mark, C. Y-C Lee, P. A. Bianconi, Eds., ACS Symposium Series **585**, American Chemical Society: Washington, D. C. (1995).

2. a. B. M. Novak, Adv. Mater. **5**, 422 (1993). b. H. Schmidt in *Chemical Processing of Advanced Materials*; L. L. Hench, J. K. West, eds., John Wiley & Sons: New York, NY, p. 727 (1992). c. J. E. Mark, Chemtech **19**, 230 (1989). d. G. L. Wilkes, "Ceramers: Hybrid Inorganic/Organic Networks." in *The Polymeric Materials Encyclopedia: Synthesis, Properties and Applications*, in press.

3. F. Babonneau, New J. Chem. **18**, 1065 (1994).

4. S. Prabakar, R. A. Assink, N. K. Raman and C. J. Brinker, Mat. Res. Symp. Proc. **346**, 979 (1994).

5. P. J. Flory, Macromolecules **12**, 119 (1979).

ANISOTROPY IN HYBRID MATERIALS : AN ALTERNATIVE TOOL FOR CHARACTERIZATION

V. Dessolle[1], E. Lafontaine[2], J.P. Bayle[1] and P. Judeinstein[1]

[1] Laboratoire de Chimie Structurale Organique (URA 1384),
Université Paris Sud, Bât. 410, 91405 Orsay - France

[2] DGA/CREA
16, Bis Avenue Prieur de la Côte d'Or
91414 Arcueil Cedex - France.

ABSTRACT

Hybrid materials presenting mechanical rubbery properties are prepared from silicon alkoxides modified by organic polymers [i.e. poly(propyleneoxide), poly(ethyleneoxide), poly(butadiene) of moderate molecular weight ($M_W \approx$ 3000-4000)]. These materials were characterized by ^{29}Si NMR. They can be swollen by deuterated probes which can be used with ^{2}H NMR to follow the network structure. Under unidirectionnal mechanical strain, uniaxial anisotropy of the samples is obtained and the macroscopic ordering is evidenced by quadrupolar deuterium NMR of swelling probe.

INTRODUCTION

Applications of hybrid organic-inorganic nanocomposite materials in the fields of optics, coatings, electronics, biomaterials and specialized mechanical properties are under active development (1). Improvement of the material properties is strongly related to the knowledge of their intricate architecture. Advancements arise mainly from the development of characterizing procedures, techniques and spectroscopies (2). Generally, these materials obtained by mixing and reacting together organic and inorganic components are at least biphasic (3). Furthermore, the properties of the final materials are related to the chemical nature of the starting materials and the synthesis pathway (4).Their description should then involve the structure and the topology of the network, the dynamics and motional modes at the molecular and macromolecular levels.

This work is mainly concerned with materials presenting mechanical rubbery properties at room temperature. These properties are obtained when long and flexible chains are cross-linked to form a three-dimensional network. Generally, polymers based on poly(butadiene), poly(siloxane) or poly(ether) chains of high molecular weight (M_W >> 5000) produce rubbery materials with high yield ratio (5). The use of alkoxysilane end-capped polymers [$(OEt)_3$Si-polymer-Si$(OEt)_3$] is a new approach to get rubbery materials. The crosslinking of these precursor molecules is easily performed by silica hydrolysis-condensation reactions and leads to the formation of silica nodes. These materials can be swollen by different solvents or oligomers which are compatible with the polymeric phase. Deuterated probes are used, and the materials are strained in the magnetic field of an NMR spectrometer. Then ^{2}H quadripolar NMR probes the uniaxial ordering inside the structure. This technique is very sensitive to the induced order parameter and reflects the interactions between the polymeric structure and the entangled inner liquid phase in the gels.

Mat. Res. Soc. Symp. Proc. Vol. 435 © 1996 Materials Research Society

EXPERIMENTAL

Synthesis

All chemical reagents are commercially available and were used as purchased : 3-isocyanatotriethoxyslilane (IsoTreOS), O,O'-bis(2-aminopropyl)-poly(propylene)-glycol (PPO ; $M_W = 4000$) (or poly(ethylene) ; PEO ; $M_W = 200$), poly(butadiene) hydroxyl terminated (PBu ; $M_W = 2800$, 0.5 meq/g OH), ammonium fluoride, formic acid and tributyltin methoxide. THF was dried over molecular sieves and deionized water was used in all experiments.
The synthesis procedure and the conditions to obtain the different materials are reported in Scheme 1.

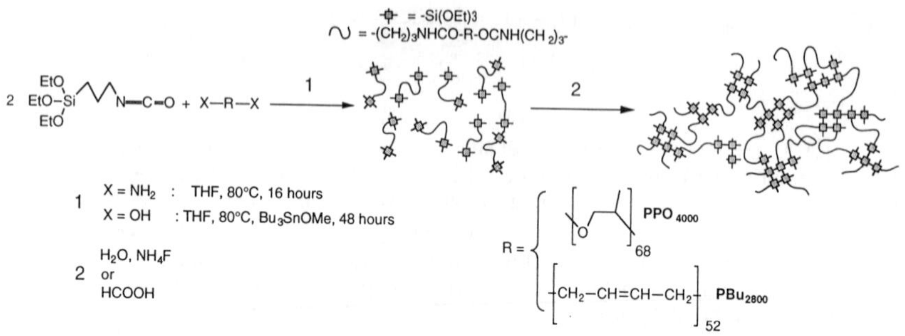

Scheme 1 synthesis of the hybrid materials

In a first step, the silicon end-capped polymer $(OEt)_3Si\cup M_n\cup Si(OEt)_3$ is obtained. When the diamino polymer is available, the precursor is easily synthesized in mixing stoichiometric quantities of IsoTreOS and NH_2-M_n-NH_2. The mixture is stirred in refluxing THF for 6 hours. THF is evaporated and pure $(OEt)_3Si\cup M_n\cup Si(OEt)_3$ is obtained (6). In the case of the polybutadiene derivative, only the dihydroxyl terminated precursor is available. The reactivity of OH groups towards isocyanate function is much lower than NH_2 groups and the reaction needs to be catalyzed (7). To a stoichiometric mixture of HO-pBu-OH (10g, 3.6 10^{-3} mol) and IsoTreOS (1.24g, 5 10^{-3} mol) in THF is added 0.1g (3.1 10^{-4} mol) of tributyltin methoxide. The mixture is kept under reflux for 48 hours and then THF is evaporated. The reaction was monitored by ^{13}C NMR ($\delta_{(N=C=O)} = 122$ ppm; $\delta_{(NH\text{-}COO)} = 157$ ppm) and the reflux was stopped when the isocyanate peak disappeared.

The condensation of the silicon atoms leads to the formation of the rubbery material. Compatible chemicals and solvents should be mixed to get a homogeneous solution. The synthesis of polyether based materials by hydrolysis in presence of NH_4F as a catalyst ($[H_2O]/[Si] = 5$, $[NH_4F]/[Si] = 10^{-2}$) has been already described (6). In the case of PBu based materials, the problem is more tricky because of the demixing of $(OEt)_3Si\cup PBu_n\cup Si(OEt)_3$-water mixtures. A non-hydrolytic condensation path was followed (8). 0.05g (10^{-4} mol) of formic acid were mixed with 3.3g (10^{-4} mol) of the precursor. A transparent rubbery homogeneous material is then obtained in a few minutes. Liquid solutions are cast in a Teflon mold in order to get homogeneous monoliths ($30*5*2$ mm^3).

RESULTS AND DISCUSSION

Si-29 Solid-State NMR Characterization

^{29}Si CP-MAS NMR spectra were obtained with a Bruker MSL200 spectrometer using standard conditions (9). Figure 1 presents these spectra for materials based on PPO_{4000} and PBu_{2800} polymers and they are compared with those of similar materials obtained with shorter polymer chains (PEO_{200}). All the spectra presents wide signals centered at -59 and -66 ppm characteristic of T^2 ($RSi(OSi)_2(OH)$) and T^3 ($RSi(OSi)_3$) sites. They show the condensation of the silicon alkoxides and the formation of silica nodes which act as crosslinking units of the polymeric network.

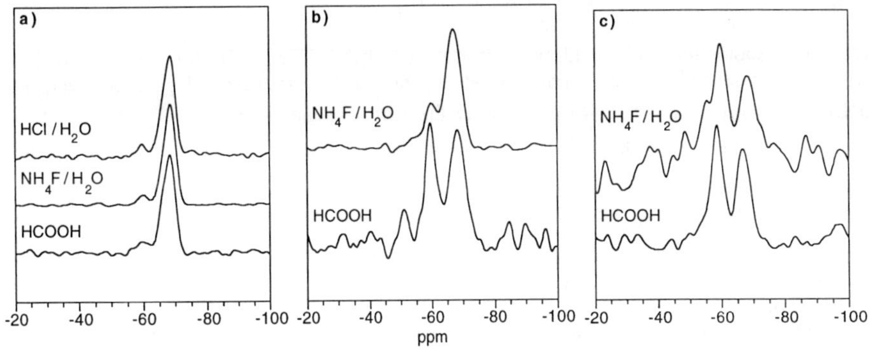

Figure 1 : ^{29}Si CP-MAS NMR spectra of SiO_2/polymer obtained under different condensation processes a) PEO_{200} b) PPO_{4000} c) PBu_{2800}

However, strong differences in the respective signal intensities are observed. For materials with shorter chains (PEO_{200}), signals from T^3 sites are predominant whatever is the catalyst (HCl, NH_4F) and show the high condensation degree of the silica nodes. Moreover, the same spectrum is again obtained when condensation is initiated with non-hydrolytic process by reaction of formic acid and silicon alkoxide in stoechiometric quantities. For materials with longer polymer chains, the intensity of signals from T^2 sites increases, demonstrating a lower degree of condensation. The effect of the condensation initiation process is different for the two materials. This effect is very pronounced for PPO_{4000} based materials, in which the degree of condensation increases greatly for hydrolytic process, while no noticeable difference are measured for PBu_{2800} based materials.

Obviously, these measurements demonstrate the effect of the organic component on the hybrid materials condensation. They are certainly related to the probability for silicon atoms to react together, which depends on the viscosity of the initial solution, the steric factor, the probability for two atoms to react but also the polarity of the silicon surroundings.

After drying (60°C - 48 hours), the materials could be swollen by compatible solvent for the polymer phase, i.e. hydrocarbons and aromatics in PPO and PBu, water and aromatics in PEO. The use of deuterated solvents permits the quadripolar NMR study. The swelling ratio s_r(s_r = $100.(W-W_0)/W_0$) is measured by weighting the dry (W_0) and swollen samples (W). Relative changes of swelling ratios are less than 10% during NMR measurements, indicating a slow solvent evaporation.

Quadripolar NMR in oriented media

When a polymer chain is stretched, the solvent molecules undergo an anisotropic reorientational motion around the straining direction as a consequence of orientated chain-solvent interactions. Deuterium NMR is a powerful tool for studying local order in oriented medium : liquid crystals, micelles, pure and swollen strained elastomers (10). In an isotropic liquid, the spectrum of a deuterium consists of a single line. When fast motions are anisotropic, the molecules are partially oriented, and the average of the quadripolar coupling becomes nonzero, which splits the line into two components. The separation of each doublet defines the quadrupolar splitting $\Delta\nu_Q$. In the high field approximation and assuming a negligible asymetry parameter, this splitting, in frequency units, can be expressed as (11) :

$$\Delta\nu = \frac{3}{2}\nu_Q \left| P_2(\cos\Omega) \right| \left| \overline{P_2(\cos\theta(t))} \right|$$

where ν_Q denotes the static quadrupolar coupling constant (≈ 220 kHz for C-D bonds). The angles in the second Legendre polynomial depend on the experimental geometry and the molecular dynamics : Ω is the tilt angle between the constraint and the spectrometer magnetic field and $\theta(t)$ is the instantaneous angle between the constraint and the electric field gradient at the deuterium which is along the molecular C-D axis ; $P_2\cos(\theta(t))$ is the Legendre polynomial $(3\cos^2\theta-1)/2$. The overbar denotes a time average over the motions which are faster than the characteristic ^{2}H NMR time ν_Q^{-1}. $\overline{P_2(\cos\theta(t))}$ is the mean degree of orientational order, S, of the C-D bond with respect to the strain director.

^{2}H NMR spectra were measured at 38.37 MHz on a Bruker AM250 spectrometer. The samples are stretched with a device similar to that described in reference (12) and the stretching direction is coincident with the axis of the NMR sample ($\Omega = 0°$). The extension ratio $\lambda = L/L_0$ was estimated from the stretched L_0 and unstretched L lengths measured and determined within ±0.2 mm. Because of the low molecular weight of the polymer chains, limited extension ratios ($\lambda<1.2$) are attainable before the fracture of the sample.

The ^{2}H spectrum of pure toluene-d_8 presents two sets of peaks representative of the methyl group CD_3 (small chemical shift, high field) and the aromatic part C_6D_5 (large chemical shift, low field) with the relative expected intensities (3:5) and chemical shift differences ($\delta_{C6D5} - \delta_{CD3}$ = 3.5 ppm) ; the linewidth is 0.5 Hz.

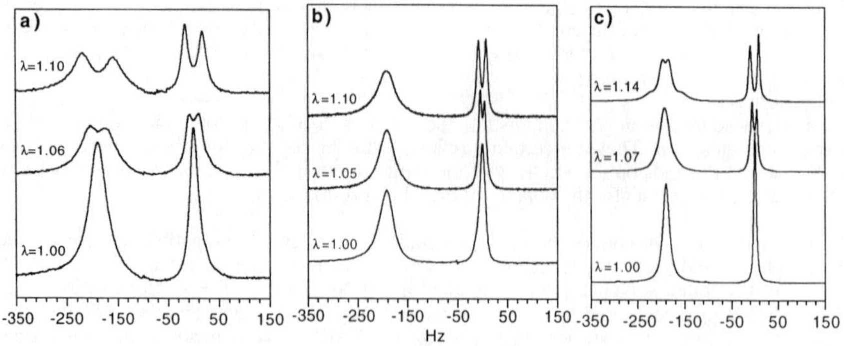

Figure 2 : ^{2}H NMR spectra of hybrid materials swollen by toluene ($CD_3C_6D_5$)
a) PBu $_{2800}$ (s_r= 12%) b) PPO $_{4000}$ (s_r=9%) c) PPO $_{4000}$ (s_r=25%)

Figure 2 presents the ^2H NMR spectra of deuterated toluene-d_8 ($C_6D_5CD_3$) probe dissolved in PPO and PBu based hybrid materials for different λ values.
A moderate broadening of the two lines (less than 10 and 20 Hz respectively) is measured when the toluene is entrapped inside the PBu_{2800} materials. This is related to the slight hindrance of molecular mobility of molecules dissolved in a polymer above its glass transition temperature T_g (13) and the inhomogeneities of the material. When the materials is stretched, the shape of the spectrum is modified, and each peak tends to split into a quadrupolar doublet. The finite values of $\Delta\nu$ are indicative of the anisotropic reorientational diffusion for the toluene-d_8 molecules inside the uniaxial field generated by the stretched network chains.

Each doublet was deconvoluted then splitting and linewidth values were measured for each signal. An increase of the splitting values is observed with the extension ratio, while the linewidth remains constant with the exception of PPO_{4000} and s_r<12%. Figure 3 presents the variations of $\Delta\nu$ vs. $\lambda^2 - \lambda^{-1}$ for methyl and aromatic ^2H signals.
Kuhn and Grün stated the proportionality between the time averaged Legendre polynomial describing the polymer chains and the classical term of elasticity $[\lambda^2 - \lambda^{-1}]$ for affine deformation of the chains and uncorrelated reorientation of the segments of the polymer chain (5). In our compounds, if this proportionality is still valid for the solvent molecules, the mean orientation of the C-D segments of the toluene-d_8 in strained materials is proportional to the mean orientation of the chains.

Strong couplings between solvent molecules and elastomeric chains are then measured and indicate the strong physical interactions between the gel matrix (polymer chains) and the molecules of the solvent. The slope is different for the two families of deuterons evidencing a different order parameter for each sites. A slight increase of both slopes is measured in decreasing the swelling ratio as it was already mentioned in rubber and PDMS based polymers (14).

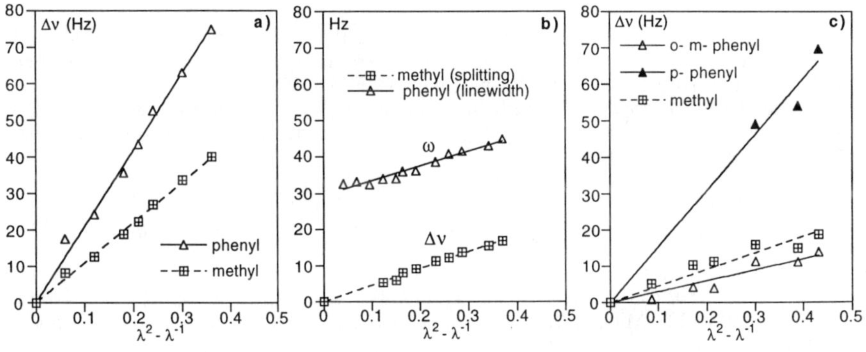

Figure 3 : $\Delta\nu$ vs. $\lambda^2 - \lambda^{-1}$
a) PBu_{2800} (s_r= 12%) b) PPO_{4000} (s_r=9%) c) PPO_{4000} (s_r=25%)

For PPO based materials two regimes are observed depending the swelling ratio (Fig. 2b and 2c.). When s_r is smaller than 15%, the methyl group splits while the aromatic region do not. However, a sensitive variation of the linewidth of the aromatic part is observed. The evolution of both the quadripolar coupling of the CD_3 group and the linewidth of the C_6D_5 part follow the classical elasticity theory (Fig. 3b). It means that in fact both parts of the molecules follow the uniaxial constraint of the polymer chains but certainly, a strong orientational disorder explains the unsplitting of the aromatic peak. When the swelling ratio is larger than 15%, the shape of the aromatic region is more complex and could be deconvoluted as two doublets with the same chemical shift but two different splittings (Fig. 3c ; λ = 1.14). From the signal intensities, the larger doublet corresponds to the *para* deuterium, while the narrower corresponds to the *ortho* and *meta* deuterium positions. These differences reflect the order parameter for each deuterium due to the particular mean orientation of the probe molecule inside the strained network. The main influence of the swelling ratio is certainly related to the molecular mobility which decreases the linewidth of the signals with increasing the toluene content, as it was observed also in other gels and nanocomposites (15).

Unequivocally, the order parameters of the different deuterons of the toluene-d_8 do not follow the same trends in the two matrices (PPO and PBu). The molecular weight of these two polymers and the crosslink functionality are rather similar, but for equivalent stretching and swelling ratios, the chain conformations are certainly different because of the nature of the polymer chain. Then, local order parameters of the polymer chain itself should be different. Moreover, the nature of the inter-molecular interactions between the polymer and the toluene should involve different association, in explaining different molecule/polymer positioning. Further studies are in progress to study the precise molecule ordering inside these polymers.

CONCLUSION

In these different gels (nanocomposite + solvent systems), the deuterated solvent probes the orientational order of the flexible polymer chain under uniaxial mechanical strain. However, differences of polymer conformation and/or solvent-polymer couplings are observed from the solvent order parameter. Further work is in progress to measure the influence of polymer chain length, the solvent, the nature of the polymer and of the silica cross-link node geometry. These data are primordial for the knowledge of supramolecular forces between solvent molecules and functional materials and gels.

BIBLIOGRAPHY

(1) a) in *Better Ceramics Through Chemistry VI*, Mat. Res. Symp. Proc., **346**, (1994) and references therein. b) in *Hybrid Organic-Inorganic Composites*, Eds. J. E. Mark, C. Y.-C. Lee and P.A. Bianconi, ACS Symp. Ser., Washington, DC, **585**, (1995).

(2) in *Hybrid Organic Inorganic Materials*, special issue of *New J. Chem.*, **18**, (1994).

(3) a) B. M. Novak, *Adv. Mater.*, **5**, 422, (1993). b) U. Schubert, N. Hüsing, A. Lorenz, *Chem. Mater.*, **7** , 2010, (1995).

(4) P. Judeinstein, C. Sanchez, *J. Mater. Chem.*, **6**, (1996), in press.

(5) L. B. Treolar, *The Physics of Rubber Elasticity*, Clarendon Press, Oxford, (1975).

(6) P. Judeinstein, M. E. Brik, J. P. Bayle, J. Courtieu, J. Rault, *Mat. Res. Symp. Proc.*, **346**, 937, (1994).

(7) D. P. N. Satchell, R. S. Satchell, Chem. Soc. Rev., **4**, 231, (1975).

(8) K. G. Sharp, *J. Sol-Gel Sci. & Tech.*, **2**, 35, (1994).

(9) G. Engelhardt and D. Michel, *High Resolution Solid State NMR of Silicates and Zeolites*, J. Wiley and Sons Ltd., New York, (1987).

(10) E. T. Samulski, *Polymer*, **26**, 177, (1985).

(11) J. W. Emsley, J. C. Lindon, *NMR spectroscopy Using Liquid Crystal Solvents*, Pergamon Press, Oxford, pp 221-257, (1975).

(12) B. Deloche, M. Beltzung, J. Herz, *J. Physique - lettres*, **43**, L-763, (1982).

(13) M. E. Brik, J. J. Titman, J. P. Bayle, P. Judeinstein, *J. Polym. Sci., Polym. Phys.*, in press.

(14) P. Sotta, B. Deloche, J. Herz, *Polymer*, **29**, 1171, (1988).

(15) P. Judeinstein, P. W. Oliveira, H. Krug, H. Schmidt, *Chem. Phys. Lett.*, **220**, 35, (1994).

SYNTHESIS OF INORGANIC-ORGANIC HYBRIDS FROM METAL ALKOXIDES AND ETHYL CELLULOSE

Ikuko YOSHINAGA, Noriko YAMADA and Shingo KATAYAMA
Nippon Steel Corporation, Advanced Technology Research Laboratories,
1618 Ida, Nakahara-ku, Kawasaki 211, JAPAN, ikko@lab1.nsc.co.jp.

ABSTRACT

Inorganic-organic hybrids were synthesized by the reaction of inorganic species of $Ti(OC_2H_5)_4$, $Nb(OC_2H_5)_5$, $Ta(OC_2H_5)_5$ and $Fe(OC_2H_5)_3$ with an organic polymer of ethyl cellulose. The studies of FT-IR spectroscopy revealed that M-O-cellulose bonds were formed by the reaction of ethoxy groups of metal ethoxides with hydroxy groups of ethyl cellulose. The dielectric constants of the resulting Ti-O-, Nb-O- and Ta-O-cellulose hybrids were higher than that of ethyl cellulose. The molar magnetic susceptibility of the Fe-O-cellulose hybrid was independent of iron content and was lower than that of $Fe(OC_2H_5)_3$.

INTRODUCTION

Inorganic-organic hybrids are of commercial and scientific interest because of their potential for providing unique combinations of inorganic and organic properties. There have been some attempts to prepare inorganic-organic hybrids such as organically modified silicates (ORMOSILs) [1] and ceramics (ORMOCERs) [2] by copolymerization of metal alkoxides as inorganic monomers and organic-substituted silicon alkoxides as organic monomers.

As an alternative route, the synthesis of inorganic-organic hybrids from inorganic monomers such as metal alkoxides and organic polymers can be considered [3,4]. Cellulose is a candidate for the organic polymer because its hydroxy groups can react with the metal alkoxides. However, there are only a few reports on the modification of cellulose derivatives with metal alkoxides in order to improve mechanical and/or thermal properties of cellulose materials [5-8]. The authors have previously investigated the reaction of $Ti(OC_2H_5)_4$ with ethyl cellulose as an new route for the synthesis of inorganic-organic hybrids [9].

In this work, inorganic-organic hybrids were synthesized from various transition-metal alkoxides and ethyl cellulose. The reaction was investigated by FT-IR spectroscopy. Dielectric and magnetic properties of the inorganic-organic hybrids were measured.

EXPERIMENTAL

$Ti(OC_2H_5)_4$ (WAKO PURE CHEMICAL INDUSTRIES, LTD.), $Nb(OC_2H_5)_5$, $Ta(OC_2H_5)_5$, $Fe(OC_2H_5)_3$ (KOUJUNDO CHEMICAL LABORATORY CO., LTD.) and ethyl cellulose (TOKYO CHEMICAL INDUSTRY CO., LTD., 49% cp10) were used as starting materials. Ethyl cellulose, which is the ethyl ether of cellulose, had the ethyl-substitution of about 50%. Metal ethoxide and ethyl cellulose were dissolved in toluene in equal weights and heated in an oil bath at 130℃ for 2 hours under a nitrogen atmosphere.

Mat. Res. Soc. Symp. Proc. Vol. 435 © 1996 Materials Research Society

The solid products were obtained after removal of the solvent by distillation.

The samples were examined with a transmission electron microscope (TEM; HITACHI H800) equipped with an energy dispersive spectrometer (EDS; HORIBA EMAX-3000). Differential scanning calorimetric (DSC) analyses of Ti-O-cellulose hybrid and ethyl cellulose were conducted on a differential scanning calorimeter (Mettler DSC 30) at a heating rate of 10℃/min under nitrogen. Infrared spectra of starting materials and refluxed solutions were taken on a FT-IR spectrometer (JEOL JIR 5500). KBr disk-method was employed for solid type samples and KBr-plates were used for liquid type samples. The relative dielectric constants of the Ti-O-, Nb-O- and Ta-O-cellulose hybrids were measured with a impedance analyzer (YHP4194A) in the frequency range from 1kHz to 10MHz. The relative dielectric constants of ethyl cellulose and TiO_2-dispersed cellulose were also measured. The sample of ethyl cellulose was prepared by evaporation of the toluene solution. The TiO_2-dispersed cellulose was prepared by evaporation of a toluene solution of ethyl cellulose in which amorphous-TiO_2 powder was dispersed. The amorphous-TiO_2 powder was prepared by hydrolysis of $Ti(OC_2H_5)_4$. The ratio of Ti : ethyl cellulose for TiO_2-dispersed cellulose was the same as that for Ti-O-cellulose hybrid. Magnetic properties of Fe-O-cellulose hybrids at room temperature were measured with a vibration sample magnetometer (VSM; TOEI INDUSTRY CO., LTD.). The weight ratio of $Fe(OC_2H_5)_3$: ethyl cellulose was 1:2, 1:1, 2:1 and 5:1.

RESULTS AND DISCUSSION

Figure 1 shows the photograph of inorganic-organic hybrids synthesized by incorporation of $Ti(OC_2H_5)_4$, $Nb(OC_2H_5)_5$ and $Ta(OC_2H_5)_5$ into ethyl cellulose. These inorganic-organic hybrids were transparent without any precipitates. They did not re-dissolve, but swelled slightly in toluene used as a solvent for their synthesis. Though the metal ethoxides react readily with water to form precipitates, the inorganic-organic hybrids can keep their transparency even in water. This suggests that the metal ethoxides in the inorganic-organic hybrids are stabilized.

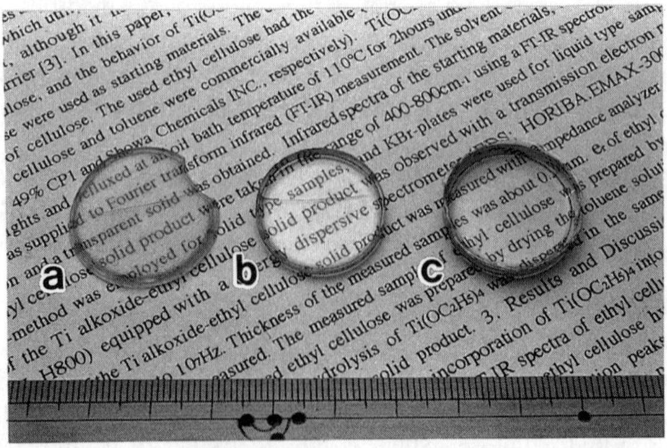

Fig. 1 Appearance of a) Ti-O-, b) Nb-O- and c) Ta-O-cellulose hybrids.

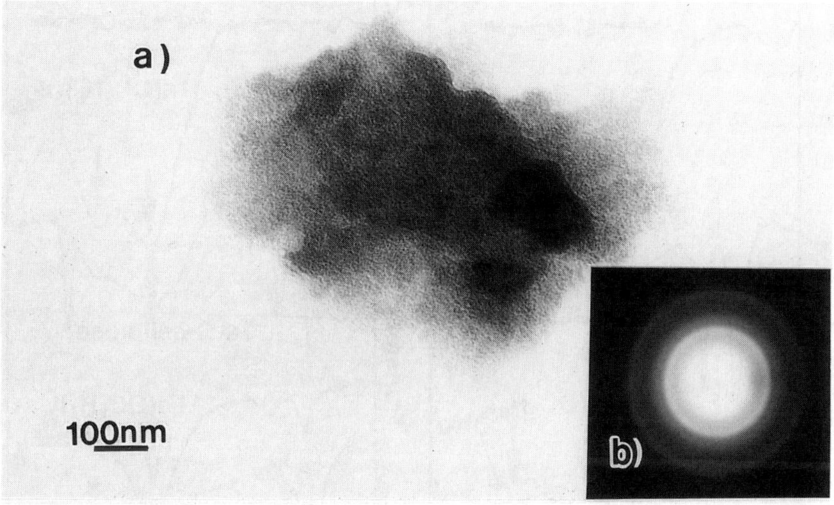

Fig. 2 TEM photograph (a) and electron diffraction pattern (b) of Ti-O-cellulose hybrid.

The TEM photograph of the Ti-O-cellulose hybrid is shown in Fig. 2(a). Inorganic precipitates such as large domains of titanium oxides were not observed by TEM. The halo pattern of the electron diffraction in Fig. 2(b) also supports that crystalline precipitates were not formed. Titanium was detected everywhere in the sample by EDS. Thus, the inorganic specie of titanium is homogeneously dispersed in ethyl cellulose.

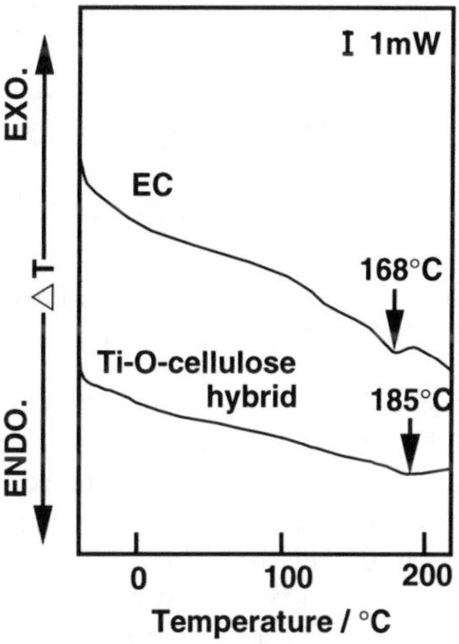

Fig. 3 DSC curves of ethyl cellulose and Ti-O-cellulose hybrid.

The DSC curves of ethyl cellulose and Ti-O-cellulose hybrid are shown in Fig. 3. The DSC curve of ethyl cellulose exhibited a endothermal peak at 168 ℃ attributed to melting. In the Ti-O-cellulose hybrid, the corresponding peak was observed at a higher temperature of 185℃. This suggests that ethyl cellulose is cross-linked by the reaction of $Ti(OC_2H_5)_4$ and ethyl cellulose. The formation of cross-links is also supported by the fact that the hybrids did not dissolve in toluene used as a solvent.

In addition, the removed solvent contained ethanol which is the by-product of the esterification reaction between the metal ethoxides and ethyl cellulose. Therefore, it is

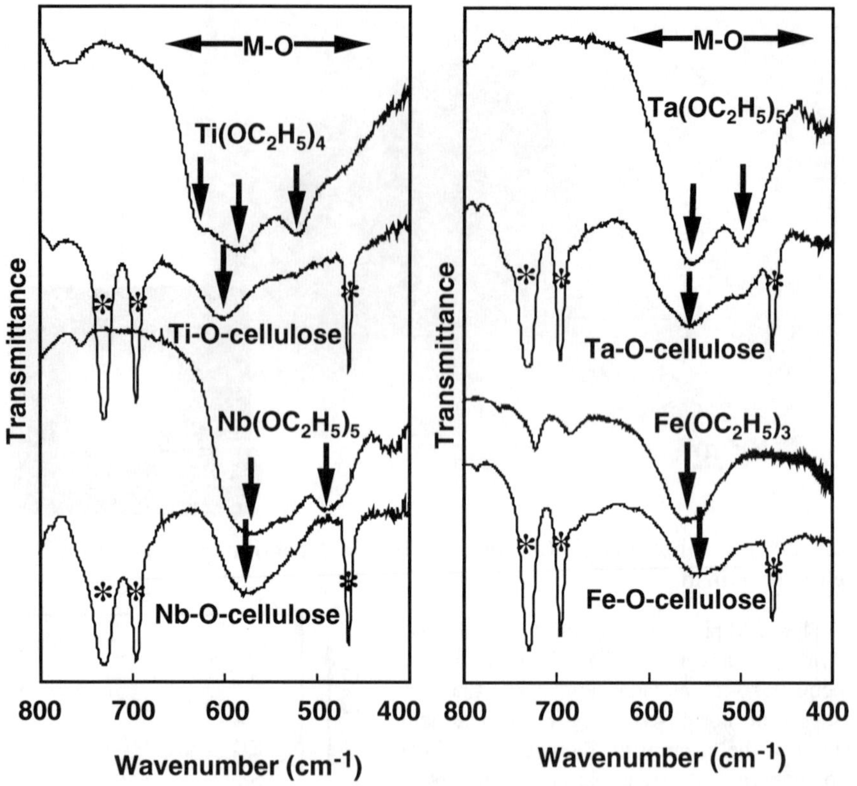

Fig. 4 FT-IR spectra of metal alkoxides and refluxed toluene solutions of metal alkoxide and ethyl cellulose. * : toluene

assumed that ethoxy groups of metal ethoxides react with hydroxy groups of ethyl cellulose, leading to form such as cellulose-O-M-O-cellulose cross-linking bonds.

The reaction of the metal alkoxides with ethyl cellulose was further investigated by FT-IR spectroscopy. Figure 4 compares the spectra of the metal alkoxides and refluxed solutions of metal alkoxides and ethyl cellulose. $Ti(OC_2H_5)_4$ has three absorption peaks at 523, 583 and 625 cm^{-1} attributed to the Ti-O vibrations of terminal and bridging ethoxy groups [10]. However, the refluxed solution had only one absorption peak at 612 cm^{-1} which was different from those of $Ti(OC_2H_5)_4$. The shift of peaks corresponding to the Ti-O vibrations means that the strength and/or length of M-O bonds were changed by refluxing a solution of titanium tetraethoxide and ethyl cellulose. The change of the absorption peaks corresponding to the M-O vibrations was also observed in the Nb, Ta and Fe systems. The refluxed solution of $Nb(OC_2H_5)_5$ and ethyl cellulose had only one peak at 580 cm^{-1} attributed to the Nb-O vibration, although $Nb(OC_2H_5)_5$ had peaks at 571 and 484 cm^{-1} attributed to the Nb-O vibrations of terminated and bridging ethoxy groups,

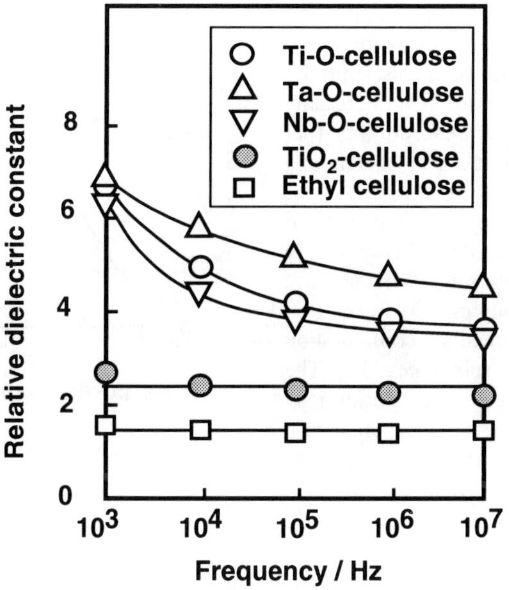

Fig. 5 Frequency dependence of relative dielectric constants for M-O-cellulose hybrids, TiO_2-dispersed cellulose and ethyl cellulose.

respectively. Although $Ta(OC_2H_5)_5$ had two peaks at 550 and 500 cm^{-1} attributed to the Ta-O vibrations of terminal and bridging ethoxy groups, respectively, the refluxed solution of $Ta(OC_2H_5)_5$ and ethyl cellulose had a peak at 557 cm^{-1}. The refluxed solution of $Fe(OC_2H_5)_3$ and ethyl cellulose had a peak at 540 cm^{-1}, although $Fe(OC_2H_5)_3$ had a peak at 560 cm^{-1} attributed to the Fe-O vibration. The results in the Nb, Ta and Fe systems are similar to those observed in the Ti system in that M-O-cellulose bonds were formed between metal and ethyl cellulose.

Figure 5 shows the frequency dependence of relative dielectric constants for M-O-cellulose hybrids, ethyl cellulose and TiO_2-dispersed cellulose. The dielectric constant of Ti-O-, Nb-O- and Ta-O-cellulose hybrids were higher than that of ethyl cellulose in the range of measured frequencies. It is considered that the high dielectric constant does not only result from titanium, niobium and tantalum species with high electronic polarizabilities, but also from the formation of a dipole moment. This is supported by the following two facts. First, the Ti-O-cellulose hybrid exhibited a higher range of dielectric constant than the TiO_2-dispersed cellulose. Second, the dielectric constant of Ti-O-cellulose hybrid decreased with increasing frequency, although dielectric constant of ethyl cellulose and TiO_2-dispersed cellulose was steady in the range of measured frequencies. The formation of a dipole moment in the M-O-cellulose hybrid appears to arise from a partial charge formed by the incorporation of inorganic species which are different in electronegativity from ethyl cellulose.

Figure 6 shows the magnetic susceptibility versus iron content for Fe-O-cellulose hybrids. The mass susceptibility (per gram-hybrid) increased with increasing iron content. The molar susceptibility (per mol-Fe) was independent of iron content and was lower than that of $Fe(OC_2H_5)_3$. It is reported that the magnetic moment of $Fe(OC_2H_5)_3$ was lower than the spin only value expected for high spin Fe (III) because of the spin-spin coupling between iron atoms and/or the intermediate configuration between high and low spin states [11]. The lower value of Fe-O-cellulose hybrids may also result from the higher spin-spin coupling and/or the intermediate configuration close to a low spin state.

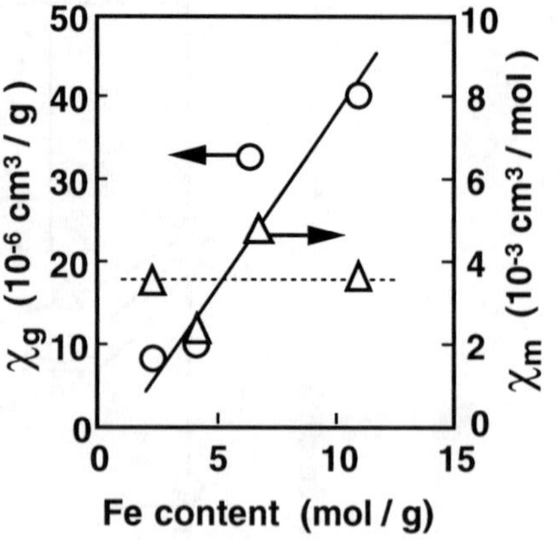

Fig. 6 Magnetic susceptibilities of Fe-O-cellulose hybrids.

CONCLUSIONS

Inorganic-organic hybrids containing various inorganic species have been synthesized by incorporation of $Ti(OC_2H_5)_4$, $Nb(OC_2H_5)_5$, $Ta(OC_2H_5)_5$ and $Fe(OC_2H_5)_3$ into an organic polymer of ethyl cellulose. Ethoxy groups of metal alkoxides were found to react with hydroxy groups of ethyl cellulose to form chemical bonds of M-O-cellulose. The dielectric constants for Ti-O-, Nb-O- and Ta-O-cellulose hybrids were higher than that of ethyl cellulose or oxide-dispersed cellulose. The molar magnetic susceptibility of Fe-O-cellulose hybrid was independent of iron content and was lower than that of $Fe(OC_2H_5)_3$.

REFERENCES

1. G. Philipp and H. K. Schmidt, J. Non-Cryst. Solids **63**, 283 (1984).
2. H. K. Schmidt, Mater. Res. Soc. Symp. Proc. **180**, 9361 (1990).
3. Z. Ahmad, S. Wang, J. E. Mark, J. P. Chen and F. E. Arnold, Poly. Mater. Sci. Eng. **70**, 303 (1994)
4. C. S. Betrabet and G. L. Wikles, Chem Mater. 7, 535 (1995)
5. H. H. Beacham, U. S. Patent **2686133** (Aug. 10, 1954)
6. H. H. Beacham, Adv. Chem. Ser. **23**, 282 (1959)
7. R. J. Speer and D. R. Carmody, Ind. Eng. Chem. **42**, 251 (1950)
8. H. C. Gulledge and G. R. Seidel, Ind. Eng. Chem. **42**, 440 (1950)
9. I. Yoshinaga and S. Katayama, J. Sol-Gel Sci. Tech. **6**, (1996), (in press).
10. D. C. Bradley, R. C. Mehrotra and D. P. Gaur, Metal Alkoxides, (Academic Press, New York, 1978), p.120.
11. R. W. Adams, R. L. Martin and G. Winter, Aust. J. Chem. **19**, 363 (1966)

SYNTHESIS AND CHARACTERIZATION OF TITANIUM OXO-ALKOXIDES THROUGH SOLVATOTHERMAL PROCESS

N. STEUNOU[1], Y.DROMZEE[2], F. ROBERT[2] and C.SANCHEZ[1]*

[1]Chimie de la Matière Condensée (URA CNRS 1466) and [2]Chimie des Métaux de Transition (URA CNRS 419). Université Pierre et Marie Curie. 4, place Jussieu. 75252 PARIS. FRANCE.

ABSTRACT :

The non hydrolytic synthesis of several titanium oxo-alkoxo clusters $Ti_3O(OPr^i)_8(OR)_2$, $Ti_{11}O_{13}(OPr^i)_{18}$, $Ti_7O_4(OEt)_{20}$ and $Ti_{16}O_{16}(OEt)_{32}$ have been performed via thermolysis of different titanium alkoxides ($Ti(OEt)_4$, $Ti(OPr^i)_4$). The structures of these clusters were mainly determined by X-ray diffraction, ^{17}O NMR and ^{13}C NMR.

INTRODUCTION :

Sol-gel chemistry allows, via the hydrolysis of metal alkoxides precursors, the room temperature synthesis of metal-oxo based networks[1,2]. The hydrolysis of organically functionnalized precursors $MZ(OR)_{n-1}$ (Z being an organic group carrying any specific function) leads to the formation of hybrid organic-inorganic materials[3-7]. The characterization of these materials is often made difficult by the amorphous and polydisperse nature of these systems. Crystalline transition metal oxo clusters can sometimes be obtained upon controlled hydrolytic conditions. Metal-oxo based clusters with well defined structures are particularly interesting nanobuilding blocks that can be used as precursors for oxide based ceramics[8], for the synthesis of supramolecular assemblies [9] or for the design of new hybrid organic-inorganic networks [10,11]. However, hydrolytic conditions are not always favorable to their storage. Another approach is to synthesize these metal-oxo clusters via non hydrolytic processes [12]. Since a few years non hydrolytic processes have gained interest [12-16], because they allow the formation of anhydrous metal-oxo based networks in numerous organic solvents and even in water immiscible organic solvents (toluene , CCl_4 etc..).

Three main routes can be used to form metal-oxygen bridges through non hydrolytic processes:

i) The reaction between transition metal halides MX_n and transition metal alkoxides $M(OR)_n$ with RX alkylhalide elimination (X= Cl, Br; M= Ti, Zr ..) [13,14]

M-Cl + M-OR.-----> M-O-M + RCl

This reaction occurs with a good yield at 110°C.

ii) The reaction at room temperature between transition metal alkoxides and transition metal carboxylates with ester elimination RCOOR'[16,17].

M-OCOR' + M-OR.-----> M-O-M + RCOOR'

This reaction occurs at room temperature but it can be improved by heating the reaction bath at about 100°C

iii) The reaction between transition metal alkoxides with ether elimination R_2O

$$\text{M-OR + M-OR.-----> M-O-M + } R_2O$$

This reaction may occur at room temperature with some transition metal alkoxides such as $Nb(OR)_5$ [18], $M(OR)_6$ or $MO(OR)_4$ (M=Mo, W) [19] but it can be improved by heating the reaction bath at about 100°C-150°C.

As far as thermolysis processes are used care must be taken because at high temperatures (220°C-250°C) alcohol dehydration may also occur (especially with secondary or tertiary alkoxide groups) leading to the formation of alkenes, water and thus hydrolysis as a secondary reaction [20].

This paper addresses the non hydrolytic synthesis of titanium-oxo alkoxo clusters made through the thermolysis of different titanium alkoxides ($Ti(OEt)_4$, $Ti(OPr^i)_4$). Several titanium-oxo alkoxo clusters are characterized by X-ray diffraction, ^{17}O NMR and ^{13}C NMR and ^{13}C CP MAS NMR.

EXPERIMENTAL:

Different titanium alkoxides precursors ($Ti(OEt)_4$ or $Ti(OPr^i)_4$) neat or in presence of a small quantity of dry acetylacetone (acacH), were heated at 150°C in a steel bomb during 5 days or were refluxed (T= 120°C) during 2 days. Both procedures (bomb or reflux) were separately investigated for all precursors. After slow cooling at room temperature, orange or yellow solutions were obtained for all preparations. An equivalent volume of dried organic solvent (ethanol or acetone) was then added to this solution. After a few days, four different kinds of crystals labelled A, B, C, D were grown from the reaction mixtures. The presence of a small quantity of acacH led to the formation of crystals A and a very small amount of crystals B. The same synthesis performed without acacH produced only crystals B alone. The crystals D and C exhibit differents shapes and and were big enough to be sorted. The synthesis conditions and the precursors are summarized in Table I.

precursor / molar ratio	heating mode	crystallization solvent	crystalline compounds
$Ti(OEt)_4$ / acacH, 1/ 0,3	bomb	EtOH	A+ ε B
$Ti(OEt)_4$	reflux/bomb	EtOH	B
$Ti(OPr^i)_4$	reflux/bomb	acetone	C
$Ti(OPr^i)_4$ / acacH, 1/ 0,1	reflux/bomb	acetone	D + C

Table I : Synthesis conditions and precursors for the preparation of the crystalline compounds A - D

The four crystalline compounds A-D have been characterized by ^{17}O, ^{13}C - CP MAS NMR spectroscopy, X-ray powder diffraction or single crystal X-ray diffraction .

Solid State NMR : The ^{13}C -CP NMR spectra were recorded at 100.6 MHz on a model MSL 400 Bruker spectrometer by using MAS rotors filled with freshly ground crystals within a dry glove box.

Liquid state NMR : Fresh crystals were dissolved in organic solvent. The ^{13}C liquid state NMR spectra were recorded at 75.46 MHz on a model AC 300 Bruker spectrometer and the liquid ^{17}O NMR spectra at 54.2 MHz on a model MSL 400 Bruker spectrometer. The whole ^{13}C NMR data will be published in a full article[21]. In this proceeding the ^{13}C NMR data for compound D will be summarized.

Powder X-ray diffraction : The powder X-ray diffraction patterns were recorded on a Philips diffractometer using θ-2θ Bragg-Brentano reflection geometry and Cu Kα radiation.

Single crystal X-ray diffraction : The single-crystals were inserted into Lindeman glass capillary tubes. All measurements were carried out at room temperature using a CAD4-Enraf-Nonius diffractometer

RESULTS and DISCUSSION :

Many titanium oxo-alkoxide clusters have already been characterized by single crystal X-ray diffraction [22-26]. Single crystal X-ray data from litterature were used to calculate the corresponding powder X-ray diffraction diagrams by using the Full Prof Program. Good agreement was obtained between the experimental diffraction powder pattern and the calculated diffractogram for compound A, which obviously corresponds to the $Ti_{16}O_{16}(OEt)_{32}$ titanium oxo-ethoxide cluster already reported by Mosset et al [25].

New experimental powder diffraction patterns which do not fit any calculated diagrams from litterature data, were recorded for compounds A, B, and C.

The structures of B and C compounds were determined using single crystal X-ray diffraction. The crystalline compound B has a titanium oxo-ethoxide core which corresponds to $Ti_7O_4(OEt)_{20}$. This metal oxo core was already found in two titanium oxo-ethoxide structures reported in the literature [25,26]. However, in the present work $Ti_7O_4(OEt)_{20}$ crystallized in the space group $P2_1/n$, which implies a different molecular arrangement of the cluster in the cell.

	$Ti_7O_4(OEt)_{20}$ (B)	$Ti_7O_4(OEt)_{20}$ [25]
space group	$P2_1/n$	P-1
crystal system	monoclinic	triclinic
volume (Å^3)	12948	3124
a (Å)	23.213	13.91
b(Å)	22.381	20.212
c(Å)	27.007	12.162
α(°)	90	90.49
β(°)	112.66	108.20
γ(°)	90	74.65

Table II : Crystallographic data for $Ti_7O_4(OEt)_{20}$ (B) and $Ti_7O_4(OEt)_{20}$ [25]

There are two isomers in the cell and the comparison of both clusters shows that the main differences concern the relative orientations of ethoxy groups. A summary of the crystallographic data for the compound B and the triclinic form previously obtained [25] is given in Table II.
It is noteworthy that the cell volume of the monoclinic form is much larger than the triclinic one. This is related to the fact that the monoclinic cell contains eight titanium oxo cores whereas the triclinic one contains only two.

The structure determination using single crystal X-ray diffraction reveals that crystals C contain discrete $[Ti_3(\mu_3\text{-O})(\mu_2\text{-OPr}^i)_3(\text{OPr}^i)_5(\mu_3\text{-OR})(\text{OR})]$ units. Each Ti(IV) is octahedrally coordinated by one μ_3- oxo, two trans μ_2-OPr^i, one μ_3 - OR, and two cis terminal alkoxy ligands OPr^i or OR. However the presence of disorder in the cell prevents an accurate determination of the nature of the two alkoxide groups OR. The two OR groups are tertiary alkoxy groups created upon heating, as shown by ^{13}C CP MAS NMR. The ^{13}C NMR spectrum and the INEPT ^{13}C NMR spectra of compound D were recorded in solution in C_6D_6. ^{13}C NMR spectra of compound D in solution exhibited ten ^{13}C resonances in the 82-75 ppm range. These ^{13}C resonances are located in the range corresponding to the OCH_x (x=0,1,2) carbons of alkoxy groups bonded to titanium (IV)[27, 28]. From INEPT data eight resonances located at high field (75.4, 75.9, 76, 76.7, 76.9, 78, 78.4, 79.7 ppm) were assigned to the methine carbon resonances OCH of isopropoxy groups bonded to titanium atoms.

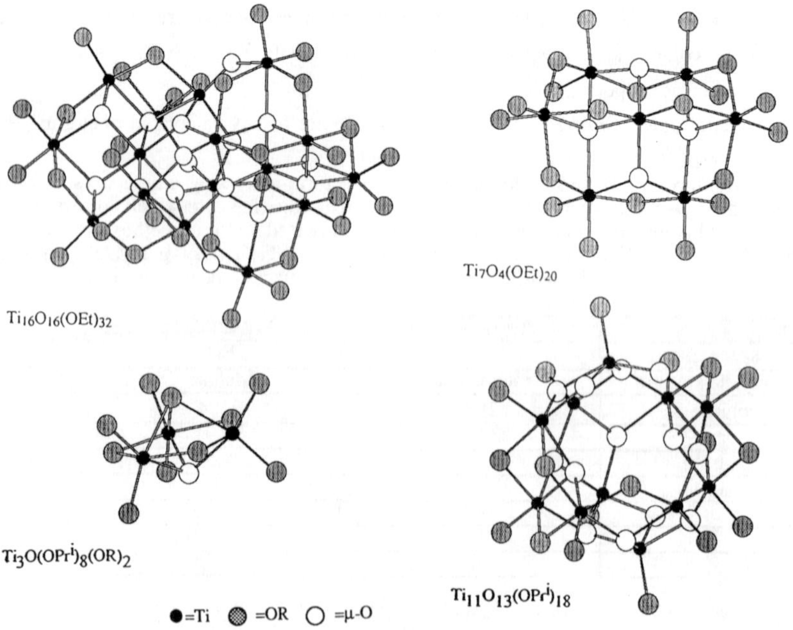

Figure1: Ortep view of the different titanium-oxo-alkoxo clusters

The two resonances located at low field (81.1 and 79.9 ppm) were assigned to the quaternary carbon resonance of a tertiary alkoxy group bonded to titanium R_3-C*-O-Ti.
These resonances are also observed in the solid state ^{13}C-CP MAS NMR. The high steric hindrance provided by the bulky μ_3 - OR group located at the bottom of the molecule should probably be responsible for the disorder existing in this structure. Compound C was fully characterized by using ^{17}O experiments (vide infra).

The structure of compound D has not yet been resolved by X-ray diffraction. However the molecular structure of the corresponding titanium oxo-alkoxo core of the D compound was determined by ^{17}O and ^{13}C NMR experiments [23] (vide infra). Figure 1 shows the ORTEP drawings of the four clusters. The different titanium oxo-alkoxo cores correspond to $Ti_{16}O_{16}(OEt)_{32}$ (crystal A) $Ti_7O_4(OEt)_{20}$ (crystal B), $Ti_3O(OPr^i)_8(OR)_2$ (crystal C) and $Ti_{11}O_{13}(OPr^i)_{18}$ (crystal D)

A ^{17}O NMR spectroscopic study of these four clusters was also performed in solution in CCl_4 and C_6D_6. The ^{17}O NMR resonances recorded for the different solutions and their assignment are summarized in Table III. The ^{17}O NMR resonances of the alkoxy groups bonded to titanium located at about 260-300 ppm [29], are not reported in table III.

clusters	chemical shift (ppm)	assignment
$Ti_{16}O_{16}(OEt)_{32}$ (A)	750 561 553 381	μ_2 - O μ_3 - O μ_3 - O μ_4 - O
$Ti_7O_4(OEt)_{20}$ (B)	537 364	μ_3 - O μ_4 - O
$Ti_3O(OPr^i)_8(OR)_2$ (C)	556	μ_3 - O
$Ti_{11}O_{13}(OPr^i)_{18}$ (D)	770 729 711 537 529 515	μ_2 - O μ_2 - O μ_2 - O μ_3 - O μ_3 - O μ_3 - O

Table III : ^{17}O NMR data for the four compounds A-D in solution in CCl_4 and C_6D_6

^{17}O NMR data are in very good agreement with those reported by W. Klemperer et al [8, 23, 24], who assigned the ^{17}O NMR chemical shifts of many titanium oxo alkoxide clusters synthesized via hydrolysis of titanium alkoxides with ^{17}O enriched water.
The ^{17}O NMR chemical shifts of compounds A and B in solution are in good agreement with those already observed for $Ti_7O_4(OEt)_{20}$ and $Ti_{16}O_{16}(OEt)_{32}$ entities[8] confirming the structures resolved by X-ray diffraction.

For compound C, the NMR spectrum displays only one sharp resonance at 556 ppm which corresponds to a μ_3-O bridging oxygen. This ^{17}O NMR data is in agreement with the fact that the $Ti_3O(OPr^i)_8(OR)_2$ cluster contains only one μ_3-O bridge.

Other $Ti_3O(OPr^i)_8(OMe)_2$ and $Ti_3O(OPr^i)_{10}$ clusters are known to exhibit similar ^{17}O NMR resonances respectively located at 553 and 554 ppm [24].

The ^{17}O NMR spectrum of compound D exhibits one broad resonance located at 300 ppm corresponding to the oxygen of Ti-OPr^i groups and the same six sharp resonances (similar intensity and position) as those reported for the $Ti_{11}O_{13}(OR)_{18}$ oxide core [23].Moreover the characteristic vibration of acetylacetonato ligands bonded to titanium were not detected by FTIR(ν_{CO} and $\nu_{C=C}$: 1605, 1525, 1425 cm^{-1}). As a consequence compound D corresponds to a $Ti_{11}O_{13}(OPr^i)_{18}$ titanium oxo cluster.

CONCLUSION :

Several titanium oxo alkoxo clusters $Ti_3O(OPr^i)_8(OR)_2$, $Ti_{11}O_{13}(OPr^i)_{18}$, $Ti_7O_4(OEt)_{20}$ and $Ti_{16}O_{16}(OEt)_{32}$ have been obtained via thermolysis in dry conditions. Their titanium-oxo alkoxo cores correspond to those of clusters synthesized through controlled hydrolysis of titanium alkoxides. This result emphasizes that transition metal alkoxides could contain oxo forms. These oxo bridges are probably obtained during aging or thermally driven purification steps[30]. For a given set of experimental conditions the structure of thermally synthesized clusters depends strongly on the nature of alkoxy group. The reaction pathways are under investigation by ^{17}O NMR and chemical analysis. These clusters are stable when they are dissolved in non polar organic solvents. Therefore they could be used as nanobuilding blocks for the synthesis of hybrid organic-inorganic compounds.

REFERENCES

[1] C.J. Brinker and G. Scherrer, Sol-Gel Science, *the Physics and Chemistry of Sol-gel Processing,* Academic press, San-Diego, (1989)

[2] J. Livage, M. Henry and C. Sanchez, *Progress in Solid State Chemistry* , **18**, 259, (1988)

[3] G.L. Wilkes ,B.Orler and H.H.Huang, *Polymer Prep.,* **26(2)**, 300, (1985)

[4] G-S Sur and J.E. Mark, *Eur. Polym. J.*, 1985, **21(12)**, 1051.

[5] H. Schmidt and B. Seiferling, *Mat. Res. Soc. Symp. Proc.*, **73**, 739, (1986)

[6] B.M.Novak , *Adv. Mater.*, **5**, 422, (1993)

[7] C. Sanchez and F. Ribot, *New Journal of Chemistry,* **1**, 1007, (1994)

[8]Y.W.Chen, W.G.Klemperer and C.W.Park, *Mat. Res. Soc. Symp. Proc*, **271**, 57, (1992)

[9] F. Ribot, F.Banse, F. Diter and C. Sanchez, *New J. Chem*, **19**, 1145, (1995)

[11] D. Hoebbel, K. Endres, T. Reinert, H. Schmidt, *Mat. Res. Soc. Symp. Proc*, **346**, (1994)

[10] F. Ribot, F.Banse, C. Sanchez, M. Lahcini and B. Jousseaume, *J.Sol Gel Science and Technology*, (1996) (in print)

[12] A. Mosset and J.Galy, *C.R. Acad. Sci.Paris Ser. II,* **307**, 1747, AS(1988)

[13] R.J.P. Corriu, D.Leclecq, P.Lefèvre, P.H. Mutinand A.Vioux, *J. Non-Cryst. Solids*, **146**, 301, (1982)

[14] R.J.P. Corriu, D.Leclecq, P.Lefèvre, P.H. Mutinand A.Vioux, *Chem. Mater.*, **4**, 961, (1992)

[15] P.E.D.Morgan, H.A. Bump, E.A. Pugar nad J.J.Ratto,

[15] P.E.D.Morgan, H.A. Bump, E.A. Pugar nad J.J.Ratto, *Science of Ceramic Chemical Processing*, Eds Wiley, (1986)
[16] M.Jansen and E.Guenther, *Chem.Mater*, **7**, 2110, (1995)
[17] R.C. Mehrotra, *J.Indian Chem.Soc.*, **38**, 509, (1961)
[18] V.G. Kessler, N.Ya. Turova, A.I Yanovskii, A.I Belokon, Y.T Struchkov, *Russ. J. Inorg. Chem*, **36(7)**, 938, (1991)
[19] N.Ya. Turova, V.G. Kessler, and S.I.Kucheiko, *Polyhedron*, **10, 22**, 2617.(1991)
[20] D.C.Bradley and M.M. Faktor, *Trans. Faraday Soc.*, **55**, 2117, (1959)
[21] N. Steunou, C. Sanchez, Forth coming paper
[22] V.W. Day, T.A. Eberspacher, W.G. Klemperer, C.W. Park, F.S. Rosenberg, *J.Am.Chem.Soc.*, **113**, 8190-8192, (1991)
[23] V.W. Day, T.A. Eberspacher, W.G. Klemperer, C.W. Park, *J.Am.Chem.Soc.*, **115**, 8469-8470, (1993), and supplementary material,
[24] V.W. Day, T.A. Eberspacher, Y. Chen, J. Hao, W.G. Klemperer, *Inorg. Chim. Acta*, **229,** 391-405, (1995)
[25] R. Schmidt, A. Mosset, J. Galy, *J.Chem.Soc.Dalton Trans.*, 1999, (1991)
[26] K. Watenpaugh, C. Caughlan, *Chem.Comm.*, (1967)
[27]E. Albizzati, L. Abis, E. Pettenati, *Inorg. Chim. Acta,* **120**, 197-203, (1986)
[28] ^{13}C chemical shift of δ_C(Ti-O-C*) = 82.76 ppm for $Ti(OAm^t)_4$ - S. Barboux - Doeuff, C. Sanchez - Private communication
[29] J. Blanchard, S. Barboux - Doeuff, J. Maquet, C. Sanchez, *New J.Chem*, **19**, (1995)
[30] N.Ya Turova, N.I.Koslova, E.P. Turevskaya, T.V. Rogova,and V.G.Kessler, *Mat. Res. Soc. Symp. Proc.*, **346**, 261, (1994)

DISPLACEMENT OF POLY(ETHYLENE OXIDE) FROM LAYERED NANOCOMPOSITES

C. O. ORIAKHI, M. M. LERNER
Department of Chemistry and Center for Advanced Materials Research,
Oregon State University, Corvallis, Oregon 97331, lernerm@ccmail.orst.edu

ABSTRACT

The reaction of tetraethylammonium salt with the nanocomposites $K_{0.2}(C_2H_4O)_{2.3}M_{0.9}PS_3$ (M = Mn, Cd) under ambient conditions results in the rapid and quantitative displacement of the polymer to form the alkylammonium intercalate. The displacement reactions go to completion with no polymer degradation after initial formation of the nanocomposite. Pseudo-reaction rates (0.0 to 7.8 min^{-1}) are obtained using temporal XRD studies for different cations, concentrations, and polymer molecular weights. A strong rate dependence is observed for alkylammonium concentration, and displacement occurs very slowly, or only to a limited extent, for tetramethylammonium, tetrapropylammonium, and tetrabutylammonium salts. Rapid displacement is also reported for other layered nanocomposites containing poly(ethylene oxide), including those with Li_xMoS_2, Na-montmorillonite, and Li_xMoO_3.

INTRODUCTION

Several reports have appeared on the formation of layered nanocomposites containing PEO or other poly- or oligo-ethers. The inorganic structures studied include montmorillonite [1], MoS_2 [2], TiS_2 [3], MoO_3 [4], and $MnPS_3$ [5]. In each of these examples, the PEO layers at maximum polymer incorporation have dimensions of 8 - 9 Å along the stacking direction, which is consistent with a bilayer of the polymer adsorbed onto the encasing inorganic surfaces. The polymer layers also contain alkali metal cations, which are solvated by PEO and compensate the negative charge of the inorganic layers.

Profound differences exist between nanocomposited and bulk crystalline PEO: for example, no melting transition is observed in the former case. Studies have included IR, NMR, electrical, and compositional analyses [2,4,6], but many questions remain difficult to answer. Besides conformational analyses, fundamental polymer properties such as the average molecular weight can be difficult to assess within nanocomposites. Although the primary PEO structure, maintained by covalent linkages, should not be affected by nanocomposition with montmorillonite, but this is not a safe assumption when chemically-active hosts are involved. A more complete understanding the nature of the PEO / inorganic adsorption interaction and how this affects polymer conformation and viscoelasticity would also be useful, because these molecular properties underlie macroscopic properties such as ionic conductivity and rheology.

In this study, we report for the first time a rapid, quantitative, and general method for extracting PEO from layered hosts. The polymer is displaced using aqueous tetraethylammonium salt under ambient conditions. The reaction with $K_x(C_2H_4O)_yM_{1-x/2}PS_3$ (M = Mn, Cd) is analyzed in detail and kinetic effects described.

EXPERIMENTAL

Lithiated MoS_2 was prepared by reaction of MoS_2 (Aldrich, 99%) with butyl lithium (Aldrich, 1.6 M in hexane) following a standard procedure [7]. Li_xMoO_3 was prepared by

Mat. Res. Soc. Symp. Proc. Vol. 435 © 1996 Materials Research Society

reaction of MoO_3 (Aldrich, 99.5%) with $Na_2S_2O_4$ (Mallinckrodt, reagent) to afford Na_xMoO_3, and subsequent ion-exchange [8].

MPS_3 (M = Mn, Cd) was prepared by heating high-purity Mn (99.9%) or CdS (> 99.9%) with a stoichiometric amount of red phosphorus (Aldrich, 99.99%) and sulfur (Aldrich, 99.999%) *in vacuo* in a silica ampule at 650 or 680 °C for 1 week. The potassium hydrate intercalation compound, $K_xMn_{1-x/2}PS_3 \cdot \delta H_2O$, was subsequently formed via reaction of $MnPS_3$ with 3M KCl. The product was repeatedly centrifuged and washed with water, then dried. For $CdPS_3$, this reaction was carried out in the presence of EDTA and a buffer. A composition of $K_{0.3}Mn_{0.85}PS_3 \cdot 1.0H_2O$ was determined using TGA and K analysis.

The PEO / $K_xM_{1-x/2}PS_3$ (M = Mn, Cd) nanocomposites were prepared using the procedure described previously [5]. PEO / MoS_2 and PEO / MoO_3 nanocomposites were prepared by exfoliation of the corresponding lithiated hosts (1.00 g / 200 mL H_2O) using ultrasound, followed by the slow addition of excess aqueous PEO (0.50 g / 50 mL H_2O). The mixtures were stirred for 2 h. until a solid flocculated, and the products isolated as above. The PEO / montmorillonite nanocomposite was prepared by the addition of aqueous PEO (0.50 g / 50 mL) to a dispersion of sodium montmorillonite in water (1.00 g / 200 mL).

Aqueous dispersions of the nanocomposites were added to aqueous solutions containing at least 50-fold excess of alkylammonium salt. The reactants were stirred for a proscribed period, and a solid product isolated by centrifugation, washing with water, and drying *in vacuo* at 80 °C. In kinetic studies, reaction times were counted from the initial combination of the alkylammonium salt solution (0.05 M - 0.20 M, 10 mL) and aqueous nanocomposite dispersion (containing 0.035 g solids) until the product had been centrifuged (2-3 min) and the filtrate decanted. All aqueous phase reactions were conducted under ambient conditions.

XRD powder patterns were obtained from sample films cast onto a Pyrex disc. Infrared spectra were recorded on samples pressed into KBr disks. Thermal analyses of powder samples (10 - 20 mg) were carried out at 10 °C / min in flowing air (50 mL / min). Molecular weight profiles and polymer concentrations were obtained by elution of aqueous polymer solution (20 μL) through a series of 2 ultrahydrogel linear columns (Waters) and a Waters R401 differential refractometer at 0.80 mL / min. Elemental analyses for CHN were performed by Desert Analytics Laboratory, Tucson, AZ, K content was determined by flame spectrophotometry using a standard procedure.

RESULTS

The interaction of $K_{0.3}M_{0.85}PS_3$ (M = Mn, Cd) with an aqueous solution of high-molecular-weight PEO results in the uptake of PEO and an increase in the stacking dimension. A phase expanded by 4.8 Å appears at lower polymer stoichiometries, when this is increased a phase appears with stacking repeat distance = 15.6 Å (Δd = 9.1 Å). The unreacted $K_{0.3}M_{0.85}PS_3 \cdot \delta H_2O$ is no longer evident when the reaction stoichiometry exceeds 0.20 g/g .

These data are consistent with the accommodation of PEO as a bilayer between the MPS_3 sheets. Observations of both monolayer and bilayer phases for PEO incorporation, with Δd of ~ 4 - 6 Å and ~ 9 Å, respectively, have been noted in montmorillonite and MoO_3.

Thermogravimetry of the bilayer phase, obtained using a reaction stoichiometry of 0.30 g PEO / g $K_xM_{1-x/2}PS_3$, shows mass losses of 7% and 1% for M = Mn and Cd, respectively, below 200 °C, which is ascribed to dehydration. A mass loss of 39% (M = Mn) and 32% (M = Cd) is centered at 320 °C, within the expected decomposition range for PEO. The nanocomposite stoichiometry was thus found to be $K_{0.2}(C_2H_4O)_{2.3}Mn_{0.9}PS_3 \cdot 1.2H_2O$ [calc. C=17.9, H=3.7, N=0.0, K=2.5; found C=18.0, H=3.1, N=0.0, K=2.4 %], which agrees with previous results [5].

Displacement Reactions using aqueous $(C_2H_5)_4NBr$

The reaction of $K_{0.2}(C_2H_4O)_{2.3}Mn_{0.9}PS_3$ with an aqueous solution of tetraethylammonium bromide results in the rapid, quantitative and irreversible displacement of both PEO and alkali metal cations to form the alkylammonium intercalate. An XRD pattern of the product after reaction for 1 - 2 h. shows the stacking repeat distance to decrease from 15.4 to 11.2 Å, with no other phases evident. Subsequent reactions of the alkylammonium intercalate with PEO do not alter the XRD pattern or otherwise demonstrate polymer uptake.

Polymer Characterization by GPC

A gel permeation chromatograph of the solution containing PEO displaced from $K_{0.2}(C_2H_4O)_{2.3}Mn_{0.9}PS_3$ shows that M_p is unchanged from the original polymer standard (M_p = 1.7 x 10^5). The peak becomes asymmetric, however, with increased intensity on the longer-elution side, indicating some scission products in the displaced polymer. GPC traces for the displaced polymer do not evolve with increased stirring time with alkylammonium. Similarly, the traces obtained for the displaced polymer are independent of the nanocomposite age. These observations suggest that polymer scission occurs during the initial preparation.

Some hydrogen sulfide can be detected when $K_{0.3}Mn_{0.9}PS_3$ reacts with aqueous PEO (though not during the displacement reaction), and scission therefore may be related to nucleophilic attack of HS^- at the etheric carbon during PEO incorporation.

A chromatograph of the solution after the initial reaction with $K_{0.3}Mn_{0.85}PS_3$ shows that unincorporated PEO exhibits scission to a far greater extent than the displaced PEO. The trace obtained for the excess PEO solution also resembles the difference of traces obtained for the displaced and standard PEO. These observations lead to the conclusion that polymer scission occurs in forming the nanocomposite. A correlation of GPC peak area vs. PEO concentration for standard solutions and the displaced polymer indicates that the polymer displacement is quantitative.

Displacement Reactions with other layered hosts

Rapid and quantitative displacement of PEO can also be effected with other layered hosts. Figure 1 illustrates the XRD patterns obtained following reaction between aqueous $(C_2H_5)_4NBr$ and PEO-containing nanocomposites with Li_xMoS_2, $K_{0.3}Cd_{0.85}PS_3$, Na-montmorillonite, or Li_xMoO_3. In each case, the basal repeat distance decreases by ~ 3 - 4 Å, consistent with the displacement of PEO bilayer and associated alkali metal cations by tetraethylammonium.

Displacement Kinetics

A series of XRD traces corresponding to the product of 0.15 M $(C_2H_5)_4NBr$ and $K_{0.2}(C_2H_4O)_{2.3}Mn_{0.9}PS_3$ (PEO M_w = 5 x 10^6 D) illustrates the rapid decrease in peak intensities for the nanocomposite (d = 15.4 Å) and simultaneous increases in peaks associated with the alkylammonium intercalate (d = 11.2 Å). (Figure 2) The unconverted solid fraction is determined using the following relation:

Figure 1. XRD powder patterns for PEO-containing nanocomposites and products following reaction with aqueous $(C_2H_5)_4NBr$; (a) MoS_2, (b) $CdPS_3$, (c) montmorillonite, and (d) MoO_3.

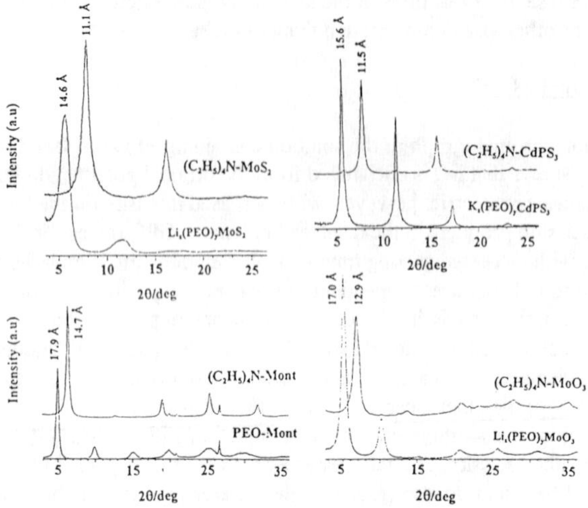

Figure 2. XRD powder patterns for products obtained from 0.15 M $(C_2H_5)_4NBr$ and $K_{0.2}(C_2H_4O)_{2.3}Mn_{0.9}PS_3$ (using M_w (PEO) = 5 x 10^6 D) after reaction times indicated.

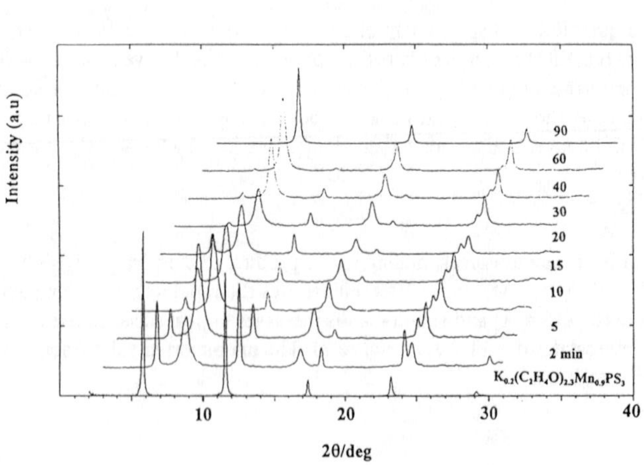

$$\chi(t) = \frac{I_{001,\ reactant}}{I_{001,\ reactant} + I_{001,\ product}} \qquad (1)$$

where I_{001}'s correspond to *(001)* reflection peak intensities for the reactant, $K_{0.2}(C_2H_4O)_{2.3}Mn_{0.9}PS_3$, or product, $[(C_2H_5)_4N]_{0.3}Mn_{0.85}PS_3$, and $\chi(t)$ provides the fraction of unconverted solid in each XRD pattern. The extent of reaction, $1 - \chi$, will therefore range between 0 and 1.

When the calculated values for χ are plotted against time, the displacement reactions at $[(C_2H_5)_4N^+] = 0.15$ are seen to go to completion in 15 min. - 2 h. The plots generated under a variety of reaction conditions are linearized by taking $1 / \chi$ vs. time (Figure 3), and line slopes indicate pseudo-reaction rates. (Table 1) A strong correlation between $[(C_2H_5)_4N^+]$ and rate is evident, with a reaction order between 2 and 3, indicating a significant mechanistic role for the dissolved ion.

Figure 3. $1 / \chi$ vs. time for reactions of aqueous R_4NX with $K_{0.2}(C_2H_4O)_{2.3}Mn_{0.9}PS_3$ with various alkylammonium ions and concentrations, and polymer molecular weights.

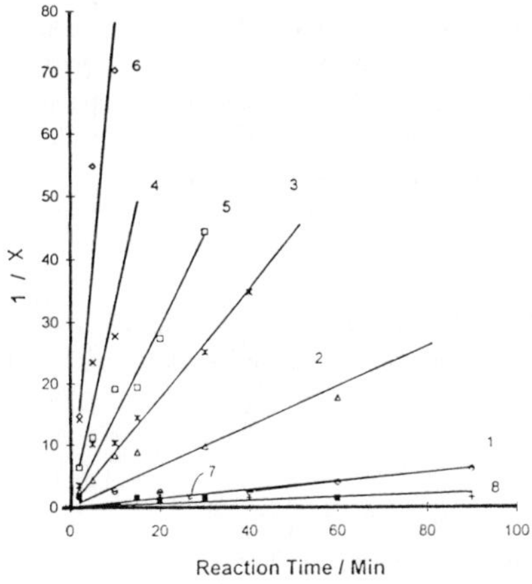

Table 1. Effects of alkylammonium size and concentration and polymer M_w on k.

#	k *min^{-1}*	R	$[R_4NX]$ *M*	polymer *M_w / D*
1	0.07	C_2H_5	0.05	5×10^6
2	0.33	"	0.10	"
3	0.89	"	0.15	"
4	3.3	"	0.20	"
5	1.5	"	0.15	9×10^5
6	7.8	"	0.15	1×10^5
7	0.03	CH_3	0.20	5×10^6
8	0.03	C_3H_7	"	"
9	n/a	C_4H_9	"	"

The kinetic effect of the polymer molecular weight is also examined. Nanocomposites prepared from 5 x 10^6, 9 x 10^5 , and 1 x 10^5 D PEO react at rates of 0.9, 1.5, and 7.8 min^{-1}, respectively. The increase in reaction rates by factors of 1.2 and 8.7 for the lower-molecular-weight polymers are similar to the relative polymer diffusivities of 2.4 and 7.1 estimated from $M_p^{-1/2}$. This correlation suggests that polymer diffusion may play a role in the displacement mechanism.

An unexpected result is obtained when other symmetric tetraalkylammonium cations, R_4NX (R = methyl, propyl, and butyl), are reacted under the same conditions as $(C_2H_5)_4NBr$. The displacement reactions are slower by at least 30 - 100 x, in fact, reactant and product concentrations for the tetramethyl and tetrapropyl ions do not appear to change significantly after a small peak due to the alkylammonium intercalate appears within an hour. The initial product may arise from an impurity of amorphous material contained in the nanocomposite, rather than displacement of the ordered nanocomposite. No displacement reaction can be observed with tetrabutylammonium salt.

The tetraethylammonium ion has a diameter similar to the thickness of a PEO monolayer adsorbed onto an anionic sheet - this is reflected in the similar basal-repeat distances of the monolayer nanocomposite and tetraethylammonium intercalate (11.2 and 11.4 Å, respectively). The PEO bilayers are far less densely packed than bulk PEO[1,2], suggesting that the polymer galleries form with a porous arrangement. A coincidence of cation and pore dimensions should lead to the more favorable chemisorption of $(CH_3)_4N^+$ at the nanocomposite surfaces. Significant cation occupancy within these pores would lead to coulombic repulsion and particle delamination. A mechanism of this type might explain both the strong rate dependence on tetraethylammonium ion concentration and also the anomolous behavior of this ion.

ACKNOWLEDGMENTS

The authors gratefully acknowledge supporting grant DMR-9322071 from the National Science Foundation. K analyses were performed by Kartik Ramachandran at O.S.U.

REFERENCES

1. J. Wu, M. Lerner, Chem. Mater. **5**, 835 (1993).
2. J. Lemmon, M. Lerner, Chem. Mater. **6**, 207 (1994).
3. J. Lemmon, M. Lerner Solid State Commun. **94**, 533 (1995).
4. L. Nazar, H. Wu, W. Power, J. Mater. Chem. **5**, 1985 (1995).
5. I. Lagadic, A. Leaustic, R. Clement, J. Chem. Soc., Chem. Commun. 1396 (1992).
6. P. Aranda, E. Ruiz-Hitzky, Acta Polymer. **45**, 59, (1994).
7. M. Dines, Mater. Res. Bull. **10**, 287 (1975).
8. D. Thomas, E. McCarron, Mater. Res. Bull. **21**, 945 (1986).

FUNCTIONAL PHOSPHATE ALKOXYSILANES FOR FACILITATED TRANSPORT MEMBRANE MATERIALS

N. HOVNANIAN*, M. SMAIHI*, A. CARDENAS*$, C. GUIZARD*
*LMPM, UMR 9987 CNRS, UMII, ENSCM, 8 rue de l'Ecole Normale, 34053 Montpellier, France smaihi@crit1.univ-montp2.fr
$Laboratorio FIRP, Facultad de Ingenieria,Universidad de Los Andes, Mérida, Venezuela

ABSTRACT

Sol-gel processing of heteropolysiloxanes containing phosphate groups has been investigated for potential applications in the synthesis of facilitated transport solid membrane. The co-hydrolysis-condensation of SiP (diethylphosphatoethyltriethoxysilane) with each of the three following alkoxysilanes: tetraethoxysilane (TEOS), methyltrimethoxysilane (MTMOS) and $C_6H_4[C(O)NH(CH_2)_3Si(CH_3)(OEt)_2]_2$-1,4 (abbreviated P1) have been studied. The gels produced have been characterized by ^{29}Si and ^{31}P MAS NMR spectroscopy to determine the chemical structure of these new materials. The connectivity and the hydrophilic properties of the materials are controlled by the chemical nature of the alkoxysilanes and the SiP concentration. Dense flexible membranes have been obtained by tape casting and have been used for the facilitated transport of Ni^{++} ions.

INTRODUCTION

The selective removal of metal ions from a mixture is a challenge for many industrial applications. One way for the recovery of high cost and/or toxic metals is to use fixed site carrier membranes. The latter have the advantage of facilitated transport liquid membranes (i.e., high selectivity) without one of their drawbacks (loss of carrier)[1]. The complexing agent (carrier) must be covalently bonded to the polymer backbone. Since heteropolysiloxanes are organic-inorganic polymers that exhibit a silica backbone modified by the incorporation of specific organofunctional groups, they are suitable materials for the preparation of fixed site carrier membranes. These hybrid materials are prepared at temperatures below 100°C compatible with the presence of organic carrier, and yield dense materials which are required for facilitated transport.

Only few examples of separation carried out in liquid media, by complexing agents grafted onto polymeric membranes for facilitated transport applications, are described in the literature[2]. We first demonstrated that facilitated transport of K^+ versus Li^+ ions takes place through fixed-site crown-ether carrier membranes derived from HPS materials[3]. Now we are studing the possibility of facilitated transport of divalent ions through a HPS matrix containing phosphate groups obtained by the sol-gel process.

EXPERIMENTS

Materials

TEOS (Fluka), SiP (Roth-Sochiel) and MTMOS (Roth-Sochiel) are used as received. P1 is synthesized from 3-aminopropylmethyldiethoxysilane and terephthaloyl chloride[4]. HF (Fluka), NaF (Merck) or HCl (Fluka) are used as catalysts and deionized water (18 MΩ) is used for the hydrolysis.

Mat. Res. Soc. Symp. Proc. Vol. 435

Contact angle measurements

These measurements give a qualitative information on the hydrophilicity of the membrane materials. The contact angle was measured at 20°C, using the sessile drop method. The equipment consisted of a zoom lens connected to a computer, a support for the flat membrane and a light source. The contact angle was measured from photos of a drop of distilled water, carefully placed on the top of the membrane, using an image treating program.

Transport experiments

Metal ions transport measurements of these membranes were performed using a cell described by Lacan et al[3]. The compositions of the source and the receiving phases are $Ni(NO_3)_2$ (0.1M) and water respectively. Ion flows were followed by conductivity measurements. The cation amounts in the receiving phase were measured by a Varian Spectra AA 400 flame atomic absorption spectrometer.

RESULTS

Precursor materials

SiP has been chosen since this heteropolysiloxane precursor contains a phosphate group, often encountered in liquid membrane carriers[6]; they are efficient complexants for many metal ions. This compound overcomes the problem of the instability of the P-O-Si bonds in water when a mixture of phosphorus and silicon alkoxides is hydrolysed[7]; no cleavage of the Si-C and P-C bonds is observed under hydrolytic and acidic conditions. Since the hydrolysis-condensation of SiP yields cyclic chains which prevent gelation[8], the presence of a cross-linking agent is required to build a three dimensional network. Each of the three network formers (TEOS, MTMOS, P1) is hydrolysed with SiP, yielding three hybrid materials A, B and C respectively.

Influence of the nature of the network former and the molar concentration of SiP:

a) On the structure of the network

The molecular structure of the gels has been studied by ^{31}P and ^{29}Si MAS NMR spectroscopy. Figure 2 displays the relative proportions of the D^i, T^i and Q^i species present in the gels of these three systems as a function of SiP molar concentration. Although the hydrolysis conditions are not the same for each system (table I), we note that whatever the molar ratio of SiP, the most reticulated species (Q^3-Q^4-T^3, T^3, T^3-D^2) are predominant. This shows that a three dimensional network is always obtained. Moreover the absence of T^1 and/or Q^1 resonances in the TEOS/SiP and MTMOS/SiP networks proves that there are no dangling ends. For the TEOS/SiP system, at [SiP]=0.25 mol/L, there is a small Q^2 resonance which is due to a few linear segments in this gel. As the molar concentration of SiP increases, the relative proportion of Q^4 and Q^3 species decreases whereas T^3 increases progressively. When [SiP]=0.56 mol/l, T^3 species are more numerous than Q^3 ones. These results show that the SiP connectivity is important and increases with its concentration in the mixture. For the MTMOS/SiP system, the relative proportion of T^2 and T^3 units does not vary significantly with the SiP concentration. The presence of SiP in the network does not modify the connectivity. For the P1/SiP system, there are little D^1 and T^2 resonances that reveal the presence of residual Si-OY (Y=R or H) groups. At [SiP]=0.1 mol/L, the mean units are D^2 responsible of the formation of small oligomers. They decrease as the SiP concentration increases, whereas T^3 units increase due to branched species.

Sample preparation

Reactions were studied in a silane-alcohol-aqueous solution system, and transparent gels were obtained under the conditions described in Table I. The amount of SiP was varied. For the preparation of the sols, an aqueous solution of catalyst was slowly poured into the mixture of precursors and alcohol. The homogeneity of the solution was checked after stirring for a few minutes at room temperature. For tests of gelation, the aging of the solutions was performed in closed glass containers at room temperature.

Table I. Composition of the gels yielding materials A, B and C

material	system	[alcohol]/[alkoxide]	hydrolysis ratio h	[catalyst] (M)
A	TEOS/SiP/EtOH/HF	2.34	3.33	0.37
B	MTMOS/SiP/MeOH/NaF	1.55	2.80	0.05
C	P1/SiP/iPrOH/HCl	24.0	2.00	1

^{29}Si and ^{31}P NMR

Molecular structures of the solid gels were determined from 79.49 MHz ^{29}Si and from 161.98 MHz ^{31}P MAS NMR spectra recorded on a Bruker ASX 400 spectrometer. The sample was placed in an alumina rotor which was rotated at 5 KHz. The spectra were obtained with a pulse width of 7 and 3 μs respectively and a repetition time of 60 s (for ^{29}Si) and 10s (for ^{31}P). The spectra width is 15kHz and 16 K of datapoints are used. The chemical shifts are given with reference to tetramethylsilane or phosphoric acid. The quantitative analysis of line intensities was carried out off-line with the program WIN-NMR[5]. Depending on the signal-to-noise ratio, error margins of the integration are estimated to be ± 5%. The ^{29}Si species are called using the conventional notation Q^n, T^n and D^n where n is the number of bridging oxygen atoms. Two parameters were calculated on the basis of experimental spectra :
(i) the relative proportion r^i of the X^i (X = Q, T, D) species , $r^i = S^i/\Sigma S^i$ where i = number of siloxane bridges and S^i = resonance peak area and
(ii) the condensation ratio t which enables the quantification of the kinetics of the condensation reactions, $t = i.r^i/f$ where f = organoalkoxide functionality.

Membrane preparation

Membranes were obtained by a tape-casting technique, starting from precursors of materials A, B and C. The membrane was obtained by depositing the sol onto a flat α–alumina support after drying at 80°C. They were crack-free, except those obtained from material A, and around 4 μm thick (Figure 1). Gas permeation measurements with helium at 3 bar on these membranes revealed that they are dense and pinhole-free.

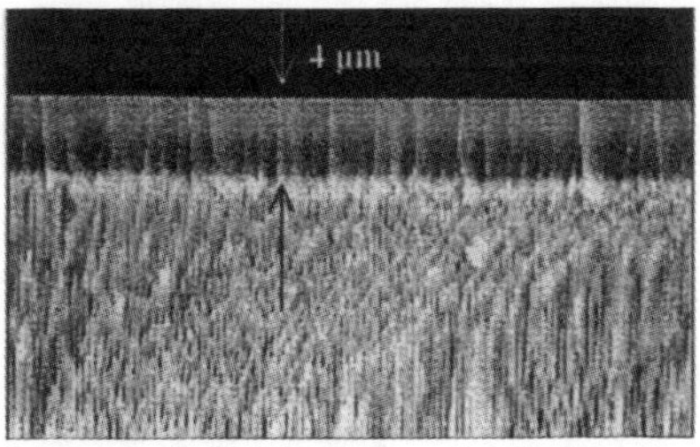

Figure 1. A SEM cross-section image of a membrane obtained from material B

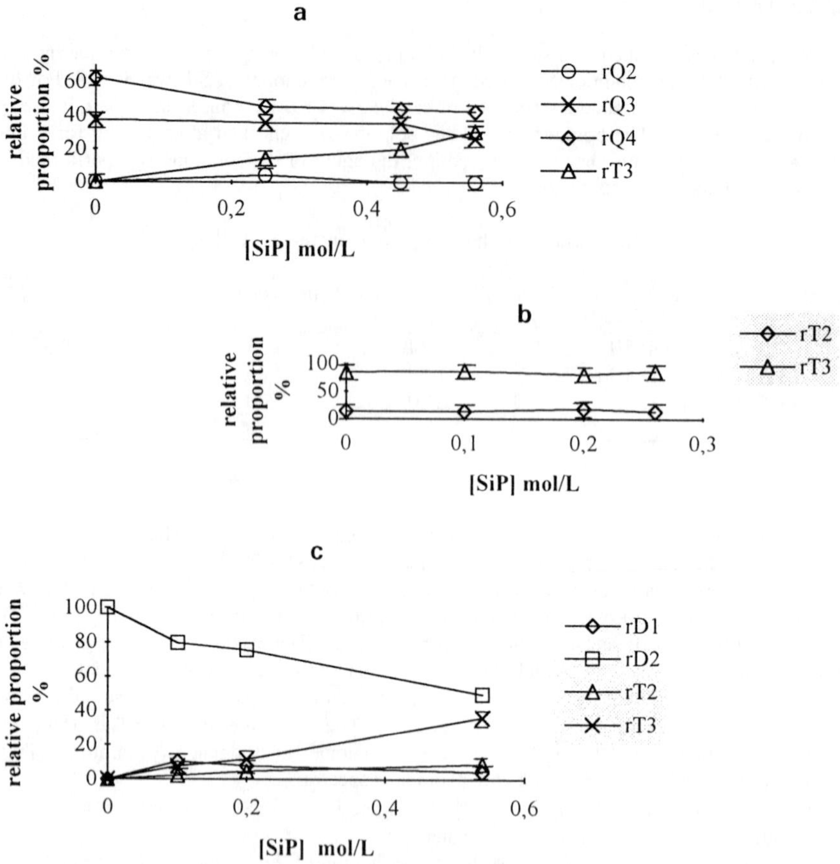

Figure 2. Relative proportions of the Q^i, T^i and D^i species as a function of the SiP molar concentration in a) TEOS/SiP system b) MTMOS/SiP system c) P1/SiP system.

Another way to evaluate the network connectivity is to examine the condensation ratio of D, T and Q species. Figure 3 presents the evolution of the condensation degree of the D, T and Q units for the various systems as a function of SiP molar concentration. Although the hydrolysis conditions are quite different for each system, the degree of condensation of the T^i species increases slowly with the SiP molar concentration for TEOS/SiP and P1/SiP systems whereas the tD^i and tQ^i condensation rates decrease respectively. For the same SiP concentrations, the degree of condensation of the T^i species is higher for the system containing P1 than that containing TEOS. It does not change significantly in the MTMOS/SiP networks.

The ^{31}P MAS NMR spectra of the gels of each system show a single peak around 33 ppm with a weak half-line width[8] in accordance with the resonance of the pure SiP alkoxide (32.9 ppm)[9]. So in each gel, there is no change of the phosphorus environment.

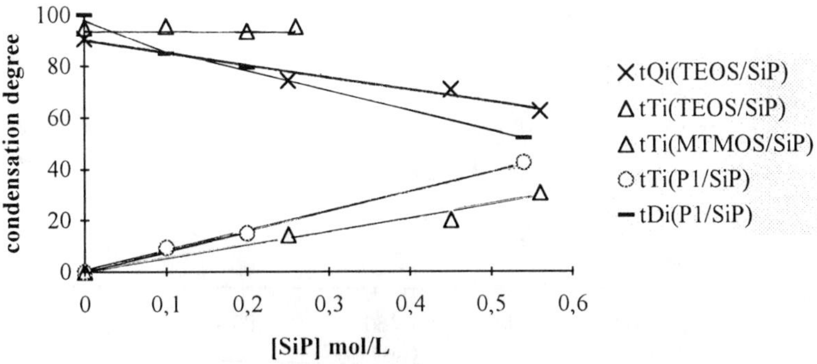

Figure 3. Evolution of the condensation degrees of the D^i, T^i and Q^i units as a function of the SiP molar concentration in the following systems: TEOS/SiP, MTMOS/SiP and P1/SiP.

b) On the hydrophilicity of the membrane materials

By selecting the chemical nature of the cross-linking agent, it is possible to tailor the hydrophilicity of the membrane materials. Table II shows that the contact angle decreases from material C to material A. Therefore, TEOS/SiP systems yield the most hydrophilic materials. Whatever the cross-linking agent, the hydrophilicity increases with the SiP concentration in the material. Since the hydrophilic character of the HPS material promotes the solubility of cations through the membrane, material A seems to be the most suitable for facilitated transport. However it is found to be very brittle, although the brittleness decreases as the SiP proportion increases; its dense structure does not allow enough flexibility to obtain a membrane without any cracks.

Table II. Contact angle measurements of the membrane materials

θ(°)	90.0	67.55	66.63	65.42	65.37	65.33	59.16	52.54
HPS	C	C	C	C	B	C	B	A
%SiP	0	10	20	40	0	54	26	25

c) On transport properties

Since the presence of specific carrier groups in the membrane enhances selective transport, we have tested membranes obtained from materials B and C that contain a high site carrier density (26% SiP with MTMOS and 54% SiP with P1). The concentration gradient is the driving force which controls mass transport through the membrane. First results are presented in Figure 4. A facilitated transport of Ni^{++} is observed since no ions go through the free carrier membranes. The complexing character of the phosphate groups is conserved even when trapped in a rigid matrix. Nevertheless, low diffusion rates are found independent of the network former. Modifying the complexant by $PO(OH)_2$ for proton driven ion transport, in order to enhance the transport rates, is in progress.

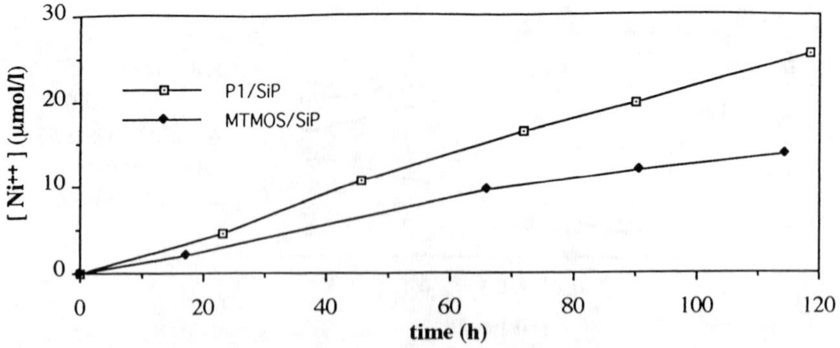

Figure 4. Nickel stripping phase concentration as a function of time.

CONCLUSIONS

Heteropolysiloxanes used for Ni facilitated transport membrane materials, can be obtained using a sol-gel process by mixing SiP with a network former (TEOS, MTMOS, P1). Depending on the nature of the network former and the SiP concentration, the material is more or less connected and/or hydrophilic.With TEOS, dense and brittle materials are formed whereas the MTMOS and P1 precursors yield membrane materials that are adapted for transport applications.

REFERENCES

1. A.M. Neplenbroek, D. Bargeman and C.A. Smolders, Desalination, **79**, p.303 (1990).

2. M. Yoshikawa, S. Shudo, K. Sanui and N. Ogata, J. Membrane Sci., **26**, p 51 (1986). T. Fyles, Polymer membrane for proton driven ion transport, US Pat. 4,906,376, (1990). L.M. Dulyea, T.M. Fyles and G.D. Roberton, J. Membrane Sci., **34**, p. 87 (1987).

3. P. Lacan, C. Guizard, P. Le Gall, D. Wettling L. Cot, J. Membrane Sci., **100**, p. 99 (1995).

4. C. Guizard and P. Lacan, New J. Chem., **18**, pp. 1097-1107 (1994).

5. WIN-NMR, Brüker Analytische Messtechnik GmbH, Wikingerstr. 13, W-75000 Karlruhe, Germany.

6. C.F. Colman and J.W. Moddy, Solvent Extraction Revs., **1**, pp. 63-91 (1971).

7. J.C. Schrotter, A. Cardenas, M. Smaihi and N. Hovnanian, J. Sol-Gel Sci. and Tech., **4**, pp. 195-204 (1995).

8. A. Cardenas, N. Hovnanian and M. Smaihi, J. Applied Polym. Sci., in press.

9. L. Ernst, Org. Magn. Reson., **9**, pp. 35-43 (1977).

POLYSTYRENE-POLY(VINYLPHENOL) COPOLYMERS AS COMPATIBILIZERS FOR ORGANIC-INORGANIC COMPOSITES

Christine J. T. Landry, Bradley K. Coltrain and David M. Teegarden
Eastman Kodak Company, Rochester, New York 14650-2116.

ABSTRACT

Random, graft, and block copolymers of polystyrene (PS) and poly(4-vinylphenol) (PVPh), and PVPh homopolymer are shown to act as compatibilizers for incompatible organic-inorganic composite materials. The VPh component reacts, or interacts strongly with the polymerizing inorganic (titanium or zirconium) alkoxide. The organic components studied were PS, poly(vinyl methyl ether), and poly(styrene-co-acrylonitrile). The use of such compatibilizers provides a means of combining *in situ* polymerized inorganic oxides and hydrophobic polymers. This is seen as a reduction in the size of the dispersed inorganic phase and results in improved optical and mechanical properties.

INTRODUCTION

Morphology and phase separation control are critical in the generation of organic-inorganic composite (OIC) materials via *in situ* polymerization of metal alkoxides. It has been shown[1-10] that highly transparent, homogeneous OIC materials can be produced using differing means to control morphology. High degrees of homogeneity have been achieved using organic polymers functionalized with trialkoxysilane groups that can co-react with added inorganic monomers thereby retarding phase separation.[2-4] Homogeneity has also been obtained without the use of trialkoxysilane groups by selecting polymers with appropriate functionalities such as ester, ether, ethylene oxide, or amide groups, that can interact with the growing inorganic oxide network.[5-10]

There are limitations to the methods described above in controlling phase separation in OIC materials. Composites prepared with trialkoxysilane modified polymers are hydrolytically unstable. Additionally, the use of trialkoxysilane functionality does not insure homogeneity in the composite. Not all unfunctionalized polymers interact sufficiently with the inorganic oxide networks to retard phase separation, particularly hydrophobic polymers. Thus, it is desirable to develop alternate means to homogenize organic polymers and inorganic oxide networks in order to optimize composite properties.

Many organic polymers are inherently incompatible and require a third constituent, or compatibilizer (such as a random, graft, or block copolymer) to improve the blend properties. The compatibilizer diffuses to the interface between the two immiscible polymers and reduces the interfacial tension, thereby reducing the dispersed phase particle size. Additionally, the compatibilizer can increase the adhesion between the two phases.

This study investigates the effectiveness of block, graft and random copolymers of styrene (S) and 4-vinylphenol (VPh) as compatibilizers for OIC materials. PVPh interacts strongly with silica, suggesting that copolymers with this monomer might be effective compatibilizers. Polystyrene (PS), poly(vinyl methyl ether) (PVME), and poly(styrene-co-acrylonitrile) (SAN), found to be incompatible with the growing inorganic oxide network and were chosen as matrices.

EXPERIMENTAL

Tetrahydrofuran (THF) and 4-ethylphenol (Kodak), titanium isopropoxide (Fluka), and zirconium butoxide (Alfa) were used as received. PS (Scientific Polymer Products (SPP) M_w =

Mat. Res. Soc. Symp. Proc. Vol. 435 © 1996 Materials Research Society

275 000 (by size exclusion chromatography (SEC)). SAN was Tyrill 880 (Dow Chemical) and contains *ca.* 23 wt % acrylonitrile (AN). PVME was obtained as a toluene solution from SPP. The toluene was removed by rotary evaporation prior to redissolution of the polymer in THF.

PVPh (M_w = 67 500 (PS equivalents)) and P(S-co-VPh) random copolymer (with 55 wt % VPh, M_w = 108, 500 (absolute)) were synthesized as described in reference 11. The diblock copolymers [P(S-b-VPh)] (peak molecular weight (MW) of 141,000 and 50-60 wt % VPh) was prepared by living, anionic polymerization and sequential addition techniques. The graft copolymer [P(VPh-g-S)] (VPh content of < 20 wt %) was prepared by conventional free radical copolymerization of a methacrylate terminated polystyrene oligomer (Sartomer) and p-(tert-butyloxycarbonyloxy)styrene (Kodak). It had a peak MW value of 113,000 and ca. 7 PS arms of MW = 13,000 each. Further details of the synthesis of the block and graft copolymers is provided in reference 11.

The titanium or zirconium alkoxides were prehydrolyzed in a manner similar to a procedure reported by Wilkes and co-workers.[12] Alternatively, hydrolysis and condensation of the alkoxides could be effected *in situ* by allowing atmospheric moisture to diffuse into the composites. No other source of water was added in these cases.

Composite samples were prepared by dissolving the polymer and the P(S-VPh) copolymers (or PVPh homopolymer) separately in THF. The inorganic component was introduced by adding a stock solution from prehydrolyzed alkoxide, or by adding the inorganic alkoxide directly. Variations on the order of addition, the drying temperature, as well as prehydrolysis of the titanium, were tested. The resulting solutions were generally yellow-orange and were immediately cast and dried slowly, allowing contact with atmospheric moisture. Samples for mechanical testing were cut to the desired shape and cured at 100 °C *in vacuo* for 3 h.

The relative weight ratios of the components in the starting (initial feed) solutions are used to identify the composites. Thus, a composite made from 2 g PS, 1 g titanium isopropoxide, and 0.05 g of block copolymer would be designated PS/Ti/block, 2/1/0.05. The Ti content represents the weight percentage of titanium alkoxide, not TiO_2.

Dynamic mechanical thermal analysis (DMA) was performed using a Rheometrics Solids Analyzer RSA II at 10 Hz with a heating rate of 2-3 °C/min. Mechanical properties were measured on a Sintech 20 testing machine. Miniature dogbone samples, ASTM D638M-III format, were tested at a constant crosshead speed of 2.5 mm/min (strain rate = 0.1 min^{-1}). PS and SAN samples for transmission electron microscopy (TEM) were prepared by first sectioning at ambient temperature, producing sections about 95 nm in thickness. These sections were examined using a JEOL 100CXII microscope at 100 kV. The PVME samples were dry sectioned, dry transferred, and examined under liquid nitrogen conditions. The intrinsic contrast between the titania-rich phase and the organic polymers was sufficient and no additional staining was required.

RESULTS AND DISCUSSION

The addition of titanium alkoxides to solutions of PVPh results in immediate color change and gelation. However, prehydrolyzing (ph) titanium isopropoxide prior to addition to PVPh results in a stable solution, and a transparent orange film is obtained upon casting and drying. Gelation of the PVPh/Ti solutions is presumably due to alcohol exchange on the titanium resulting in crosslinking via Ti-phenol bonds. The addition of tetraethoxysilane (TEOS) (and 4 equivalents of 0.15 N HCl) to PVPh produces a transparent film, whereas the addition of prehydrolyzed zirconium butoxide to PVPh results in a clear gel. These results are indicative of strong interactions between the inorganics and PVPh.

Organic polymers such as PS, PVME, and SAN formed opaque, white composites when titanium was added, in the absence of compatibilizer.

Table I summarizes physical appearance, the average particle size of the dispersed inorganic phase, and the weight percentage of compatibilizer and metal oxide for selected samples. The percentage of metal oxide is calculated from the known amount of Ti in the feed, assuming full conversion to MO_2, and confirmed by neutron activation analysis.

Increasing Ti or compatibilizer content increased shrinkage and crack formation. Composites for mechanical property determination prepared with PS were therefore limited to less than 10 wt % titania.

Prehydrolysis of the alkoxide was effective in minimizing gelation. The morphologies of the resulting films did not seem to be dramatically affected by the mode of inorganic addition. Gelation was more pronounced as the VPh content of the copolymer increased. However, a minimum concentration of VPh is required for efficient compatibilization.

Morphology. Scanning electron microscopy of the fracture surface of PS/Ti, 2/1 shows a large degree of phase separation. The fracture surface of this same PS/Ti composition with 8.1 wt % block copolymer revealed a dramatically improved sample homogeneity and a different

Table I. Samples

Components	Feed ratio	Wt % TiO_2	Wt % coplymer	Appearance	Particle size (by TEM)
PS/Ti	2/0.5 and 2/1	6.6, 12.3	-	opaque	0.5-1 μm
PS/Ti/block	2/0.5/.025	6.5	1.2	opaque	-
"	2/0.5/.050	6.4	2.3	clear	10-20 nm
"	2/0.5/0.10	6.3	4.5	clear	<10 nm
"	2/0.5/0.15	6.1	6.5	clear	<10 nm
PS/Ti/random	2/0.5/0.025	6.5	1.2	translucent	0.1-0.2 μm
"	2/0.5/0.05	6.4	2.3	clear	"
"	2/0.5/0.20	6.0	8.5	clear	50-100 nm
PS/Ti/PVPh	2/0.5/0.025	6.5	1.2	opaque	0.1-0.2 μm
"	2/0.5/0.10	6.3	4.5	clear	50-100 nm
PS/Ti/EPH	2/1/0.20	11.3	up to 8.1	opaque	0.5-1 μm
SAN/Ti	2/1	12.3	-	opaque	0.5-1.0 μm
SAN/Ti/block	2/1/0.20	11.3	8.1	translucent	<10 nm
PVME/Ti	2/1	12.3	-	opaque	1 μm
PVME/Ti/block	2/1/0.20	11.3	8.1	translucent	<10 nm
PS/Zr	2/1		-	opaque	1-5 μm
PS/Zr/graft	2/1/0.1 and 0.2			translucent	-

fracture mechanism. TEMs of the same composite with and without block copolymer compatibilizer are shown in Figure 1a and 1b. It is evident that the inorganic phase size has been significantly reduced, from 0.5 - 1 mm in PS /Ti, to < 10 nm upon addition of the block copolymer. Results for the SAN/Ti composites are similar.

Slightly different morphologies were observed for PVME/Ti, 2/1 (Figures 1c (with no compatibilizer) and 1d (with 4.2 wt % block copolymer)). Unlike the PS and SAN examples, the inorganic domains (1c) appear as dense spheres approximately 1 mm in diameter. With block copolymer the inorganic domains were substantially smaller (20-50 nm) and more homogeneously dispersed.

The morphologies of the PS/Ti, 2/0.5 series are similar. When the diblock was added at 2.3 wt %, the size of the titania particles decreased to about 10 nm. These small titania particles aggregate into clusters ranging from 50 to 250 nm in size. The clusters are more loosely formed than in the composites with more titania and the individual titania particles are more visible in the micrograph.

The morphology for the PS/Ti, 2/0.5 series with added random copolymer was slightly different from that with the diblock. The size of the titania particles remained approximately 0.05-0.2 microns in diameter, regardless of the amount of added copolymer (within the range 1.2 to 8.5%). However, as the copolymer concentration was increased, the inorganic phase became more dispersed, forming what appeared to be a network structure. Also, the titania phase seems to become less dense with increasing copolymer concentration. Addition of the block copolymer results in much smaller and well-dispersed titania phases than does the addition of similar quantities of random copolymer.

One question is whether the phenolic copolymers are actually serving as compatibilizers or whether the phenolic moieties simply modify the polymerization mechanism of the inorganic monomers. To test this, the small molecule 4-ethylphenol (EPH) was substituted for the copolymer. TEMs for PS/Ti 2 /1 with 2.1 and 8.1 wt % EPH reveal that the inorganic domain size remained large and was virtually independent of the amount of added EPH, suggesting that the phenolic moieties are not simply affecting the titanium alkoxide reaction kinetics.

PVPh homopolymer is highly immiscible with PS, so it would be anticipated to be ineffective as a compatibilizer. However, the addition of increasing amounts of PVPh improved the composite homogeneity, similar to the results with the random copolymer. It is probable that the VPh could effectively shield the surface of the growing titania particles, resulting in the exterior of the particles appearing very similar to PS. Such shielding could also produce an emulsification effect in which the polymer prevents titania aggregation and gross phase separation.

<u>Mechanical properties.</u> No rubbery plateau in the elastic modulus (E') above the T_g of PS is observed for PS/Ti composites. However, addition of a minimum of 2.4 wt % copolymer to the PS/Ti composites resulted in a clearly defined plateau in E' (above T_g) extending to 300 °C. The plateau modulus increased with increasing copolymer (E' = 5E06 Pa with 6.5 wt % copolymer) with little corresponding increase in T_g. The E' plots are virtually indistinguishable for added random, block, or homopolymer compatibilizers, even though there are morphological differences between the samples.

The magnitude of E' in the plateau region is lower for the compatibilized PS/Ti composites (ca. 1E06 Pa) than reported for the PVAc/TEOS composites[5] (ca. > 1E07 Pa). This is consistent with crosslinking, as opposed to the presence of a load-bearing, continuous, inorganic network that was proposed for the PVAc system.[5]

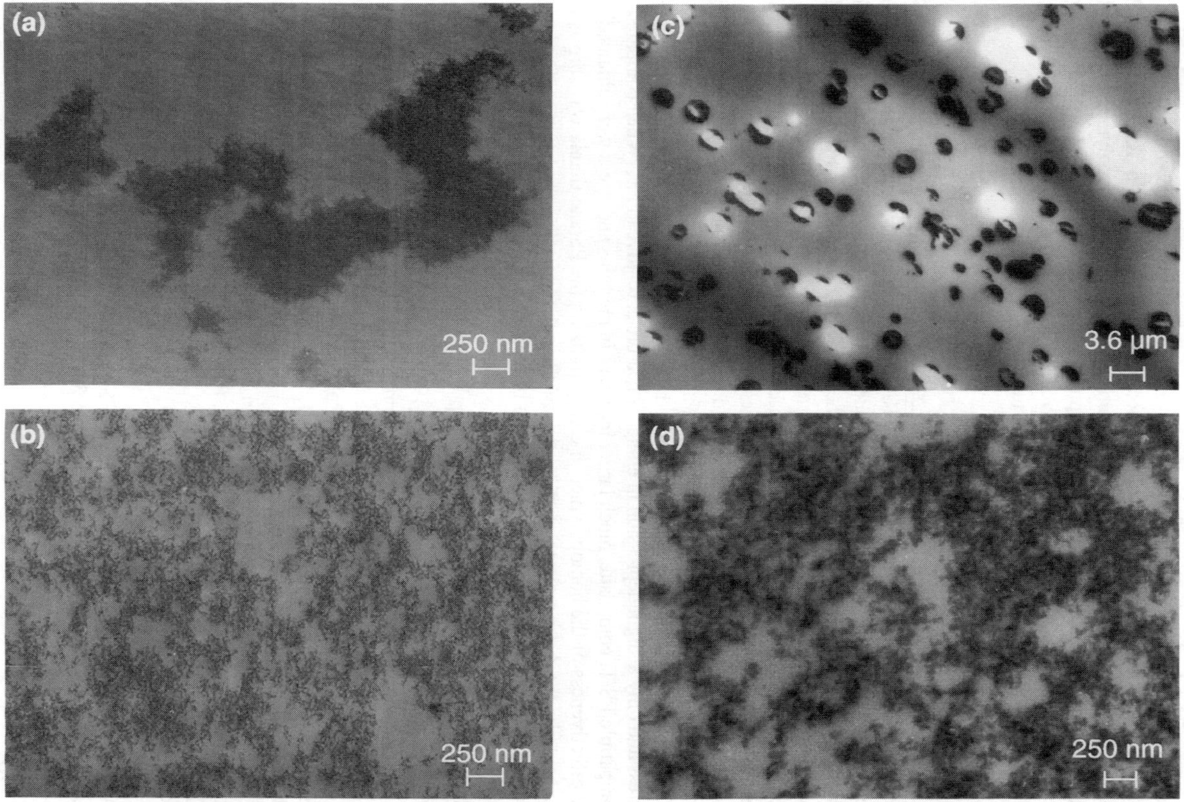

Transmission electron micrographs of PS/Ti 2/1 composite with (a) 0 wt %, (b) 4.2 wt % block copolymer, and PVME/Ti 2/1 with (c) 0 wt %, (d) 4.2 wt % block copolymer.

The addition of titanium alkoxides (6.5 - 8.5 wt % titania (PS/Ti, 2/0.5)) is found to substantially reduce the ultimate mechanical properties (modulus, stress-to-break, and strain-to-break) of PS. These properties are shown to increase with the addition of block, graft or random copolymer, or PVPh homopolymer compatibilizer, relative to the uncompatibilized PS/Ti composite. A maximum is observed in the mechanical properties as a function of wt % compatibilizer, which is typical of the use of compatibilizers in organic polymer blends. In general, the optimum amount of compatibilizer appears to be between 2 and 3 wt %. Further detail can be found in reference 11. The improvement in properties is likely a result of the reduced size and better dispersion of the titania particles.

CONCLUSIONS

The results with these test cases demonstrate the viability of using PS-co-PVPh compatibilizers for improved properties in OIC materials. The compatibilizer allowed homogeneous introduction of titania to organic polymers with which it is inherently incompatible. Microscopy results indicated that the size of the dispersed inorganic phase could be substantially reduced by addition of the compatibilizer. Possible mechanisms include a reduction in interfacial tension between the phases or steric stabilization of titania primary particles by VPh. DMA results showed that a plateau in the tensile modulus above the T_g of PS could be obtained with added compatibilizer. Ultimate mechanical properties showed that the effect of the compatibilizer on the mechanical properties was generally optimized at about 2-3 wt % copolymer. Modulus values could be improved by over 200 MPa with virtually no change in break stress and only a slight depression in elongation at break relative to PS. When compared to the incompatible PS/Ti composite, much larger increases in modulus and stress were obtained.

The effectiveness of the homopolymer in compatibilizing the PS and titania was surprising, and is apparently not due simply to modification of the polymerization mechanism of titanium isopropoxide, as the small molecule EPH proved to be ineffective.

REFERENCES

1. C. -C. Sun and J. E. Mark, Polymer **30**, p. 104 (1989).
2. H. H. Huang, B. Orler and G. L. Wilkes, Macromolecules **20**, p. 1322 (1987).
3. B. K. Coltrain, J. M. O'Reilly, S. R. Turner, J. S. Sedita, V. K. Smith, G. A. Rakes and M. R. Landry, in Proc.5th Annu. Int. Conf. Crosslinked Polym., Switzerland (1991), 11.
4. D. E. Rodrigues, A. B. Brennan, C. Betrabet, B. Wang, and G. L. Wilkes, Chem. Mater. **4**, p. 1437 (1992).
5. J. J. Fitzgerald, C. J. T. Landry and J. M. Pochan, Macromolecules, **25** p. 3715 (1992).
6. C. J. T. Landry, B. K. Coltrain and B. K. Brady, Polymer **33**, p. 1486 (1992).
7. M. Toki, T. Y. Chow, T. Ohnaka, H. Samura and T. Saegusa, Polym. Bull. **29**, p. 653 (1992).
8. B. K. Coltrain, W. T. Ferrar, C. J. T. Landry, T. R. Molaire and N. Zumbulyadis, Chem. Mater., **4,** p. 358 (1992).
9. K. A. Mauritz, R. F. Storey and C. K. Jones, ACS Symp. Ser. **395**, p. 401 (1989).
10. I. A. David and G. W. Scherer, Polym. Prepr. **32**, p. 530 (1991).
11. C. J. T. Landry, B. K. Coltrain, D. M. Teegarden, T. E. Long, V. K. Long, accepted for publication in Macromolecules.
12. B. Wang, A. Gungor, A. B. Brennan, D. E. Rodrigues, J. E. McGrath, and G. L. Wilkes, Polym. Prepr. **32**, p. 521 (1991).

SYNTHESIS OF AN INORGANIC/ORGANIC NETWORK POLYMER BY THE HYDROLYSIS/CONDENSATION OF POLY(DIETHOXYSILYLENEMETHYLENE) AND ITS PYROLYTIC CONVERSION TO SILICON OXYCARBIDE

Q. LIU, T. APPLE, Z. ZHENG and L. V. INTERRANTE
Chemistry Department, Rensselaer Polytechnic Institute, Troy, NY 12180-3590

ABSTRACT

A linear, ethoxy-substituted polycarbosilane has been prepared by ring-opening polymerization of 1,1,3,3-tetraethoxy-1,3-disilacyclobutane. After hydrolysis, condensation and drying, the obtained gel was pyrolyzed under nitrogen to give a silicon oxycarbide glass. The gel and its pyrolysis chemistry were characterized by elemental analysis, thermogravimetric analysis, ^{29}Si and ^{13}C MAS NMR, FTIR and mass spectrometric analysis of the gaseous pyrolysis products. The conversion of the $[SiOCH_2]_n$ network, which has a nearly pure SiC_2O_2 micro environment at the outset, into a silicon oxycarbide that contains a full distribution of the five possible $SiC_{4-x}O_x$ environments occurs between 600 and 1000^{0}C. This suggests the occurrence of redistribution reactions involving the exchange of Si-O and Si-C bonds during the pyrolysis.

INTRODUCTION

Due to its excellent mechanical strength and toughness, its high temperature stability (up to ca. 1500°C)[1,2], its resistance to oxidation and corrosion, and its amorphous character, silicon oxycarbide, a ceramic material having the general formula SiO_xC_y, would be expected to be a good material for many applications[3,4] including protective coatings[5,6]. The increased chemical and thermomechanical stability of these glasses compared to silica is apparently due to the incorporation of C in place of oxygen in the silica structure, thereby leading to a "tightening" of the network structure[7]. However, due to its intrinsically low solubility, the incorporation of carbon by direct reaction with silica at high temperatures does not provide a very effective route to this ceramic material.

In the last two decades, there has been much interest in the synthesis of ceramics from polymeric precursors. This new approach offers many potential advantages over the conventional synthesis and fabrication methods, such as low processing temperature, control of the ceramic composition and microstructure, high purity, easy of production of complex shapes, etc. In the case of SiO_xC_y, crosslinked organosilicon compounds or polymers have been typically used as the precursors, resulting in substantial incorporation of carbidic carbon, invariably, along with a considerable amount of free carbon[8,9]. This route to silicon oxycarbide typically involves the hydrolysis/condensation (sol-gel processing) of organically modified alkoxy silanes, having the general formula of $R'_nSi(OR)_{4-n}$, to form crosslinked organic/inorganic hybrid polymers and subsequent pyrolysis of the obtained polymers. This route is attractive as an approach to silicon oxycarbide glasses since it provides a relatively easy way of introducing carbon, as Si-C bonded organic substituents, into the gel structure and relatively low temperatures are required to convert these gels into SiO_xC_y glasses; moreover, the use of the sol gel method offers distinct processability advantages over the ceramic approach.

There have been a number of studies on the synthesis of silicon oxycarbide from polymeric precursors[6,8,9] resulting in an improved understanding of the precursor-to-ceramic conversion process[10]. However, the complexity of the structure of the starting gels that have been investigated thus far have made it rather difficult to follow this conversion process in detail.

In the present study, the sol-gel conversion of an linear, alkoxy-substituted polycarbosilane, poly(diethoxysilylenemethylene) (L-EPCS), to $[Si(O)CH_2]_n$, and its pyrolytic conversion to SiO_xC_y has been examined by solid state NMR and other methods. Unlike most of the other precursor systems that have been studied as SiO_xC_y precursors, this one is

Mat. Res. Soc. Symp. Proc. Vol. 435

remarkably simple in terms of the Si and C microenvironments, thereby permitting a detailed study of the structural changes occurring during pyrolysis to silicon oxycarbide.

EXPERIMENTAL SECTION

The linear poly(diethoxysilylenemethylene) (L-EPCS) was synthesized by ring opening polymerization of 1,1,3,3-tetraethoxy-1,3-disilacyclobutane (CBS) with Karsted's catalyst in a manner similar to that described previously for other poly(silylenemethylenes)[11,12]. The preparation of 1,1,3,3-tetraethoxy-1,3-disilacyclobutane has been described in detail elsewhere[6,11]. The L-EPCS was converted to an inorganic/organic network polymer by the usual sol-gel method, with an acid(HCl) catalyst[6]. The (charcoal) decolorized L-EPCS was dissolved in ethanol (molar ratio of 4 to Si) at ca. 50^0C. Then H_2O (molar ratio of 4 to Si, in the form of 1M HCl aqueous solution) was introduced. The gelled solution was aged overnight at 50~60^0C. Following that, the solvents were stripped off and the gel was dried under vacuum at ca. 60^0C for at least 6 hours. A white solid was obtained.

The pyrolysis was conducted in a CM High Temperature Furnace under flowing N_2 at 5^0C/min. Samples were removed from the furnace at various temperatures for elemental analysis, solid-state NMR and IR studies.

Elemental analysis were performed by either Galbraith Laboratories Inc. or Leco Corporation. TGA studies were carried out on a Perkin-Elmer TGA 7 thermogravimetric analyzer with a heating rate of 5^0C/min. under N_2. IR spectra were obtained by using a Perkin-Elmer 1800 Fourier Transform Infrared Spectrometer. The solid samples were first ground to a fine powder using a mortar and pestle, followed by mixing with KBr, and then pressed into pellets for IR transmission studies. Solid state NMR were recorded on a Chemagnetics CMX-360 SSNMR spectrometer. The cross polarization-magic angle spinning technique (CP-MAS) was used to record ^{13}C NMR spectra with a contact time of 3 ms, pulse delay of 3 s, spinning speed of 4 kHz, and as well as ^{29}Si NMR spectra with a contact time of 3 ms, pulse delay of 5 s, spinning speed of 5 kHz. Single pulse with H decoupling experiments (1pda) were used for ^{29}Si NMR spectra with pulse widths of 2 us (Θ=30^0), pulse delays of 60 s and 1000 acquisitions. Peaks will be labeled with the usual C, M_n, D_n, T_n and Q_n notation.[9] C, M, D, T and Q respectively refer to $SiC_{4-x}O_x$ units with x=0, 1, 2, 3, or 4, n is the number of bridging O atoms surrounding Si. A Hiden analytical quadrupole mass spectrometer (200 amu range) with an electron multiplier detector was used to detect the gaseous species produced during the pyrolysis. The sample was placed into a quartz tube which was connected to the mass spectrometer. A needle valve was used to control the amount of gaseous species introduced into the probe of the mass spectrometer. After pumping and baking out the system for several days, a tube furnace was used to heat the sample from room temperature to 1000^0C at 5^0C/min. The heating process was controlled by CN8600 Series Process Controller (OMEGA Engineering, INC.). Scans were collected at certain temperature throughout the pyrolysis process.

RESULTS AND DISCUSSION

The reactions involved in the synthesis of this hybrid carbosilane/siloxane gel and its pyrolytic conversion to SiO_xC_y are as follows:

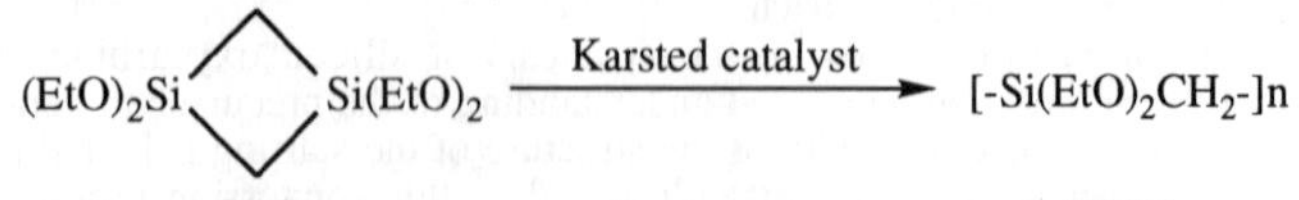

(1)

CBS L-EPCS

$$[\text{-Si(EtO)}_2\text{CH}_2\text{-}]\text{n} \xrightarrow[\text{hydrolysis}]{H_2O\ (H^+)} \xrightarrow[\text{condensation}]{\text{-}H_2O/EtOH} [\text{-Si(O)CH}_2\text{-}]\text{n} \qquad (2)$$

$$[\text{-Si(O)CH}_2\text{-}]n \xrightarrow[\text{pyrolysis}]{} \text{SiOxCy} \quad (3)$$

dried gel ceramic

<u>Characterization of Dried Gel</u>

The ^{29}Si NMR spectrum of the dried gel (Fig.1) shows three peaks at -59.9, -17.4, and 8.2ppm which can be assigned to $SiCO_3$ units (T), SiC_2O_2 units (D), and SiC_3O units (M)[9]. The respective chemical shift values for the T and D units are intermediate between the expected values for T_2(-58ppm) and T_3(-66ppm) units, and between the values for D_1(-13ppm) and D_2 (-22ppm) units. This could arise from the presence of terminal oxygen ligands, such as -OH, bonded to Si. The slightly enhancement of intensity and the relatively narrowing of D unit peak in the ^{29}Si CP spectrum (Fig.1b) is consistent with a more protonated environment for the Si. These uncondensed species are also evidenced by the OH ($\nu_{OH} \approx$ 3400cm^{-1}) and Si-OH ($\nu_{as} \approx$ 900cm^{-1}) bands observed in the IR spectrum of the gel (Fig.2). The ^{13}C CP-MAS NMR spectrum of the gel (Fig.3) exhibits a broad peak at 5.9 ppm due to the Si-CH_2-Si groups, and two sharp peaks at 17.1 and 58.0 ppm, due to O-CH_2CH_3 groups. The latter peaks can be assigned to residual ethoxy groups, that have not been hydrolyzed, or to trapped ethanol.

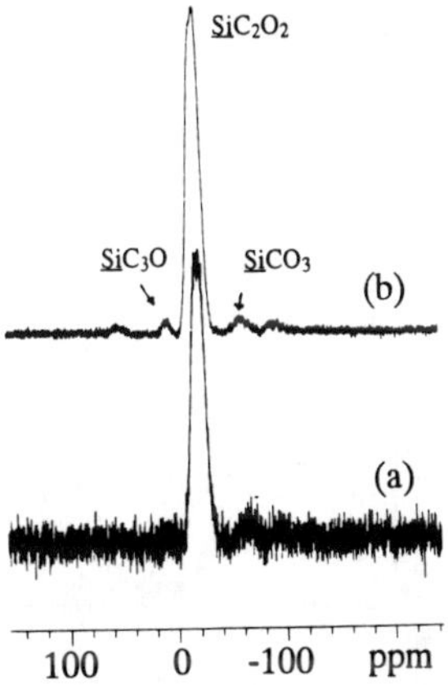

Fig.1 ^{29}Si SSNMR spectra of the dried gel (a) 1pda-MAS; (b) CP-MAS

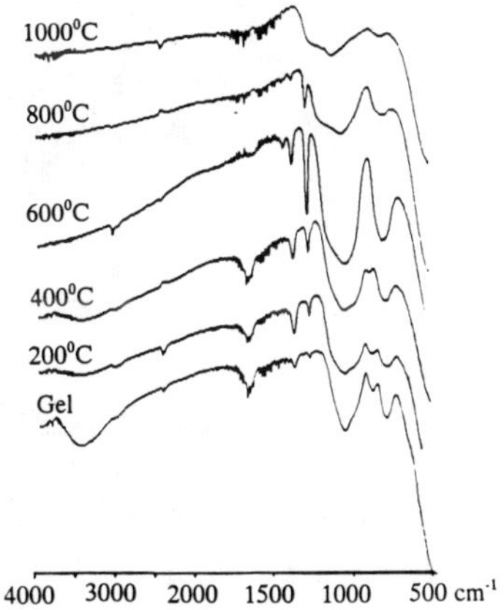

Fig.2 FT-IR spectra of the gel and the products after being pyrolyzed to different temperatures

The integration and deconvolution of the ^{29}Si NMR spectrum shows that the major component is SiC_2O_2 (D, ca. 94.77%). Also, $SiCO_3$ (T, ca. 4.48%) and SiC_3O (M, ca. 0.75%) are present to a much lesser extent. Assuming that all oxygens bond to Si, and that the carbons bonded to Si are in bridging CH_2 groups, we find that, O/Si=1.02, CH_2/Si=0.98, which are consistent with the ratios in the theoretical [-Si(O)CH_2-] structure.

Elemental analyses for the gel [Si, 45.77; C, 18.90; H, 5.13; O, 30.20 (by difference); $SiO_{1.15}C_{0.96}H_{3.14}$] are consistent with those from the theoretical calculation, as well as those from NMR data. The slightly higher contents of O and H presumably arise from H_2O and/or uncondensed -OH groups

<u>Conversion of the Dried Gel into Silicon Oxycarbide</u>

Studies by TGA (Fig.4) and mass spectrometry show three regions of weight loss, from room temperature to 200°C (ca. 3%), from 200 to 600°C (ca. 6%) and from 600 to 1000°C (ca. 6%). The first weight loss region below 200°C is attributed to the loss of solvent and water. The other two stages are attributed to the completion of the condensation process and the decomposition of the polymer, respectively. The polymer gave a ca. 84% ceramic yield. This is lower than the ceramic yield of the gel obtained by direct hydrolysis/condensation of the ethoxy-substituted disilacyclobutane (CBS) used in the ROP process (ca. 91%)[6], but higher than that of the highly branched ethoxypolycarbosilane produced by Mg coupling of $ClCHSiCl_3$, followed by ethanolysis, $[Si(OEt)_2CH_2]$ (B-EPCS) (ca. 72%)[9]. The difference between the CBS case and the linear polymer studied here may be due to the fact that CBS is a small molecule and would be expected to be more completely hydrolyzed and more extensively crosslinked than the polymer, which can be seen from the relatively small $-OCH_2CH_3$ peaks in the ^{13}C NMR of the CBS gel[6]. On the other hand, B-EPCS is highly branched which should lead to a further increase in the steric hindrance toward hydrolysis and condensation relative to the linear polymer, thereby giving a higher initial weight and a higher overall weight loss during pyrolysis. This should be directly provable by a comparison of the formulas obtained from elemental analyses of these gels: B-EPCS: $SiO_{1.75}C_{1.46}H_{3.81}$; L-EPCS: $SiO_{1.15}C_{0.96}H_{3.14}$; CBS: $SiO_{1.07}C_{0.99}H_{1.95}$. From these formulas we can see that B-EPCS has the highest relative amounts of O, C and H, indicating a significant degree of incomplete hydrolysis and condensation. On the contrary, the composition of the CBS gel is close to that expected for the theoretical $[-Si(O)CH_2-]$ structure. The L-EPCS gel lies in between these two. These results are also consistent with their weight loss during pyrolysis.

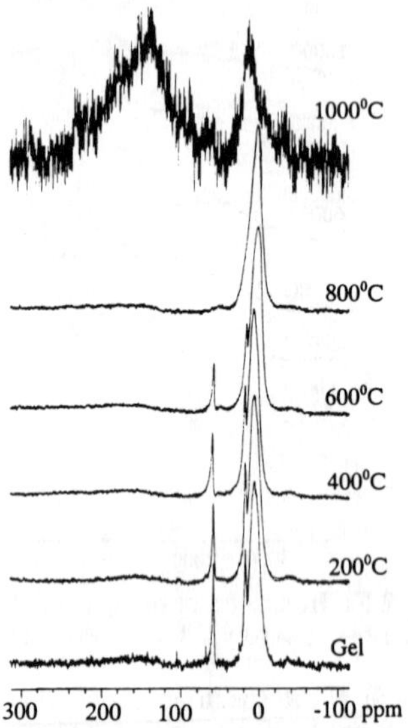

Fig.3 ^{13}C SSNMR(CP-MAS) spectra of the gel and the products after pyrolysis to different temperatures

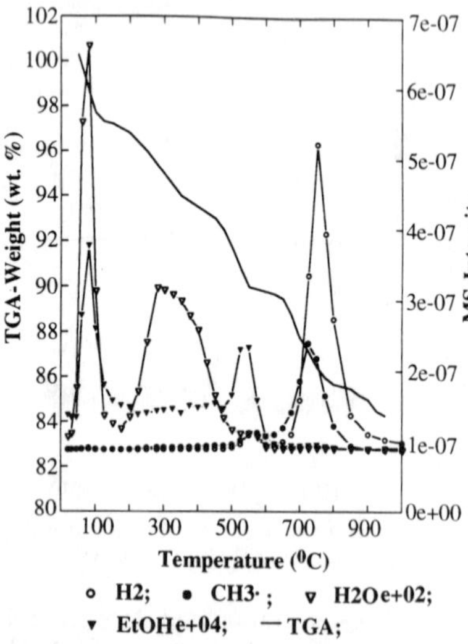

Fig.4 TGA-MS curves of the gel during pyrolysis (under N_2)

<u>Structural Evolution During the Conversion of the Gel into Silicon Oxycarbide</u>

Solid state NMR (^{13}C and ^{29}Si) as well as IR spectra were obtained at various stages in the pyrolysis process and they all evidence the onset of significant structural changes at ca. 600°C. The weak IR peak due to the uncondensed Si-OH structure at 900 cm^{-1} decreased as the temperature increased and finally disappeared at 600°C (Fig.2), which is consistent with the completion of the condensation process in this temperature range. There is a significant increase in the relative intensity of the band corresponding to the CH_3 structure at 1260 cm^{-1} (absent in the original poly(silylenemethylene)), as well as the Si-O structure at 1050 cm^{-1} and the Si-C structure at 800 cm^{-1}. The ^{29}Si solid state NMR spectra (Fig.5) show a decrease in the SiC_2O_2

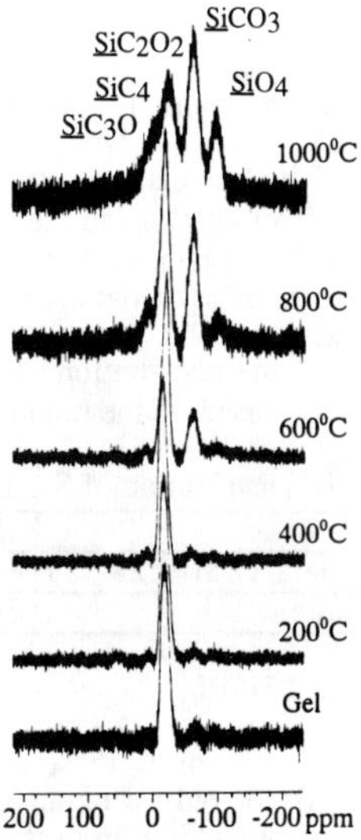

Fig.5 ^{29}Si SSNMR(1pda-MAS) spectra of the gel and the products after being pyrolyzed to different temperatures

(D) unit content above 600°C, a significant increase in the $SiCO_3$ (T) unit content, and the emergence and increase in the SiO_4 (Q) unit content which is absent entirely in the initial gel structure. ^{13}C CP-MAS NMR spectra of the pyrolyzed samples (Fig.3) show that the peaks due to OCH_2CH_3 groups have totally disappeared by 800°C and a single peak remains due to Si<u>C</u>H_2Si groups with a slight shoulder on the low field side. This could be caused by a small amount of CH_3 groups generated by redistribution reactions, as is suggested by the IR spectral data. Above 800°C, the quality of the spectrum changes indicating a strong modification in the carbon environment and a poor efficiency of the cross-polarization technique due to the low proton content in the sample. The peak due to aliphatic carbon sites broadens and is shifted to a lower field (δ=10.9 ppm), presumably due to the presence of a distribution of sites which are less and less protonated. A new peak appears at 139.8 ppm. This signal is attributed to aromatic carbon atoms and is presumably related to the formation of a free carbon phase.

All of these observations suggest that the environment of the silicon atoms during the heating of this material from ca. 600°C to 1000°C has changed from that of a nearly pure SiC_2O_2 structure to a full distribution of the five possible $SiC_{4-x}O_x$ environments. The reformation of the environments of the silicon atoms may be accounted for by the occurrence of redistribution reactions involving the Si-O and Si-C bonds, which starts at ~500°C and reaches a metastable equilibrium state at ca. 900°C[13].

The comparison of the structure and composition of the ceramics obtained at 1000°C from different gels is listed in Table I.

Table I. Chemical Analysis and ^{29}Si NMR Data for Ceramics From Different Gels

		B-EPCS[9]	CBS[6]	L-EPCS
chemical analysis	Si,%	45.41	51.9	49.42
	C,%	17.31	17.4	15.25
	formula	$SiO_{1.44}C_{0.89}$	$SiO_{1.04}C_{0.78}$	$SiO_{1.25}C_{0.72}$
	free carbon,%	68.5	38.5	47.9
NMR data	% Q units	23.4	29.1	14.36
	% T units	32.2	38.3	39.65
	% D units	31.6	23.0	36.63
	% M units	2.8	2.0	5.88
	% C units	10.1	7.6	3.49
	O/Si ratio	1.28	1.39	1.28
	Si-C bonds/Si	1.44	1.21	1.44

The ratio of Si-C bonds/Si atoms reflects the amount of carbidic carbon content in the ceramic. For the CBS gel, during pyrolysis, ring opening reactions, presumably initiated by nearby Si-OH groups, were evidenced, leading to the formation of dangling CH_3 groups and an average "$O_{3/2}SiCH_2Si(O)CH_3$" formula[6]. This is presumably the origin of the higher O/Si ratio and lower Si-C bonds/Si ratio observed in this case relative to that of the B-EPCS polymer. The build up of Si-CH_3 and consequent reduction of Si-OH observed by IR spectroscopy between 200 and 600 °C in the case of the L-EPCS gel (vide infra), along with the somewhat higher O/Si ratio and lower Si-C bonds/Si ratio observed for the corresponding 1000 °C ceramics compare to what is expected from the formula, suggests that this Si-OH + Si-CH_2-Si --> Si-O-Si + Si-CH_3 reaction also proceeds to some extent in the case of the gels derived from B-EPCS and L-EPCS.

The free carbon content of our final SiO_xC_y glass, as calculated from elemental analysis results, is quite close to the anticipated 50%, assuming that, for the $(CH_2)SiO$ system, the O/Si ratio does not change during pyrolysis and the usual valences of C, Si, and O are maintained Therefore, 50% of the carbon should be lost or converted into free carbon[9]. Moreover, the O/Si ratios obtained from the elemental analysis and from quantitative solid state NMR, and the results of the distribution of $SiC_{4-x}O_x$ sites from experiment and from a statistical calculation (Table II) based on the random distribution structural model[12] are also quite consistent.

Table II. Distribution of $SiC_{4-x}O_x$ Sites

	% Q	% T	% D	% M	% C
NMR data, O/Si=1.28	14.36	39.65	36.63	5.88	3.49
calculation	16.78	37.75	31.85	11.94	1.68

CONCLUSION

In summary, these detailed studies of the linear poly(silylenemethylene) polymer has provided not only a better understanding of the structural evolution during the polymer-to-ceramic conversion process, but has also given a unique perspective on the role of oxygen and OH groups attached to Si in the initial polymer in determining the pyrolysis chemistry and thereby the composition of the ceramic end product. It seems quite clear from these results that the thermodynamic stability of the Si-O bond, coupled with the availability of relatively facile Si-O/Si-C redistribution reactions, leads to the effective retention of essentially all of the initial siloxy oxygen functionality and elimination of any excess carbon present from the final $SiO_{2-x}C_{x/2}$ network as free carbon. Moreover, this process appears to be enchanced in the presence of uncondensed Si-OH groups which apparently facilitate the conversion of bridging methylene to terminal methyl groups while adding bridging oxygen to the oxycarbide network structure.

ACKNOWLEDGMENT

This work was supported by the National Science Foundation (CHE-950930) and the NSF Summer Research Program in Solid State Chemistry for Undergraduates (for Mr. Z. Zheng).

REFERENCES

[1]. H. Zhang and C. G. Pantano, Mater. Res. Soc. Symp. Proc., **271**, 783 (1992).
[2]. G. M. Renlund, French Patent 2,647,777, 1990.
[3]. G. T. Burns, R. B. Taylor, Y. Xu, A. Zangvil and G. A. Zank, Chem. Mater., **4**, 1313(1992).
[4]. F. I. Hurwitz, S. C. Farmer and F. M. Terepka, J. Mater. Sci., **26**, 1247(1991).
[5]. R. Kasemann, H. Schmidt and E. Wintrich, Mater. Res. Soc. Symp. Proc., **346**, 915(1994).
[6]. W. Shi, Ph.D Thesis, Rensselaer Polytechnic Institute, 1995.
[7]. G. M. Renlund, S. Prochazka and R. H. Doremus, J. Mater. Res. **6**, 2716(1991); **6**, 2723(1991).
[8]. a) D. G. White, S. M. Oleff, R. D.Boyer, P. A. Budinger and J. R. Fox, Adv. Cerm. Mater. **2(1)**, 45(1987); **2(1)**, 53(1987); b) F. Babonneau, L. Bois and J. Livage, J. Non-Cryst. Solids 1992, **147&148**, 280; c) E. Breval, M. Hammond and C. G. Pantano, J. Am. Ceram. Soc. **77(11)**, 3012(1994); d) F. Babonneau, K. Thorne and J. D. Mackenzie, Chem. Mater. **1**, 554(1989)
[9]. F. Babonneau, L. Bois, C-Y.Yang and L. V. Interrante, Chem. Mater. **6**, 51(1994)

[10]. F. Babonneau, in Applications of Organometallic Chemistry in the Preparation and Processing of Advanced Materials, edited by J. F. Harrod and R. M. Laine (Kluwer Academic Publishers, The Netherlands, 1995) p. 103.
[11]. Q. Shen and L. V. Interrante, Preprints of 35th IUPAC international Symposium on Macromolecules, Akron, USA, 345(1994); Polym. prepr. **36(2)**, 378(1995)
[12]. H-J. Wu and L. V. Interrante, Chem. Mater. **1**, 564(1989); Macromolecules, **25**, 1840(1992)
[13]. R. J. P. Corriu, D. Leclercq, P. H. Mutin and A. Vioux, J. Mater. Sci. **30**, 2313 (1995)

MICRON SCALE PATTERNING OF SOLUTION-DERIVED CERAMIC THIN FILMS DIRECTED BY SELF-ASSEMBLED MONOLAYERS

P. G. CLEM, N. L. JEON, R. G. NUZZO, AND D. A. PAYNE
Department of Materials Science and Engineering, School of Chemical Sciences, Seitz Materials Research Laboratory, and Beckman Institute for Advanced Science and Technology,
University of Illinois at Urbana-Champaign, Urbana, IL 61801

ABSTRACT

Self assembled monolayers patterned by microcontact printing have been used in conjunction with sol-gel processing to selectively deposit oxide thin films with micron-scale lateral resolution. This simple, three step process allows ambient, lithography-free patterning of oxide thin films for integrated microelectronics, optoelectronics, and sensor applications. A variety of patterned structures, such as capacitors and waveguides, have been fabricated from $LiNbO_3$, Ta_2O_5, $PbTiO_3$, and $BaTiO_3$ on technologically important substrates, including Si, Al, Pt, sapphire, and TiN. The technique involves functionalization of substrate surfaces by microcontact printing of octadecyltrichlorosilane (OTS) self-assembled monolayers (SAMs). Sol-gel precursors are then spin-coated on the SAMs-patterned surfaces and heat treated to deposit 20 nm-300 nm amorphous oxide layers. Oxide on derivatized regions is removed with mild polishing, yielding patterned films with features as small as 5 μm. For example, data are reported for 80-200nm Ta_2O_5 films to demonstrate the potential applications and mechanisms involved. The effects of sol-gel precursor chemistry, heat treatment, and other processing variables are reported. These results suggest unique potential for microfabrication of ceramic thin films using molecular self assembly and low temperature processing of solution-derived thin films.

Mat. Res. Soc. Symp. Proc. Vol. 435

INTRODUCTION

Self-assembled monolayers (SAMs) have been extensively studied in recent years.[1] So far, most of the studies focused on utilizing the SAMs (of organic thin films) alone as the thin films for applications. Microcontact printing (μCP)[2-4] of SAMs has been shown to be an effective technique for modifying the surface properties of a variety of substrate surfaces. Microcontact printing of alkylsiloxane monolayers and their application in selective deposition of metal by CVD has been reported previously.[5,6] The ability of μCP to form patterned SAMs of high quality on metal and non-metal (SiO_2/Si, sapphire, ITO and glass) surfaces is well established. We report here the use of patterned SAMs to direct the deposition of solution-derived oxide thin films.

Integrated functional ceramic thin films show promise for a variety of applications as capacitors, actuators, sensors, electrochromics, and memory cells. For all such applications, the ability to pattern films into useful device architectures is a key consideration. The inherent chemical stability and refractory nature of ceramics often presents difficulty in this regard, resulting in requirements for severe chemical etching or ion milling that increase the level of processing complication. Current technology of patterning oxides normally involves blanket oxide deposition, followed by post-deposition chemical or reactive ion etching. Such processing is often complicated by nonuniform etch rates, difficulties with fine geometries, and requirements for controlled atmosphere processing.

An alternative approach is presented here: oxide film patterning enabled by initial patterned functionalization of substrates with SAMs.[7] It was expected that patterning of substrates to create hydrophobic and hydrophilic regions would enable preferential wetting of sol-gel precursors. Sol-gel processing routes for oxide thin films are well known to enable integration of oxide devices such optical waveguides and integrated capacitors at low processing temperatures (400-600°C).

EXPERIMENTAL

Microcontact printing of octadecyltrichlorosilane (OTS) was used to selectively functionalize surfaces before sol-gel deposition. The PDMS stamps used for microcontact printing were fabricated according to a published procedure.[2-4] Substrates were rinsed with DI water, acetone, and isopropanol, and dried with a stream of Ar. Substrates were modified with UV/ozone treatment for 5-10 minutes prior to stamping to promote formation of the SAM. A solution of OTS in hexane (10 mM) was used as the "ink". This solution was applied to the PDMS stamp using a photoresist spinner (3000 rpm for 30s). The stamp was carefully brought into contact with the substrate for ~30s to deposit patterns of hydrophobic, CH_3-terminated (OTS) monolayer.

Sol-gel precursors for Ta_2O_5 were prepared from tantalum ethoxide solutions in ethanol. The concentration was varied from 0.03 to 1 M to optimize film properties. Additionally, levels of hydrolysis water were adjusted for tantalum ethoxide solutions to promote formation of dense films. The best results were obtained for 0.3 M solutions with the molar ratio of water to tantalum, $R_w = 1$. These solutions were spin coated on selectively functionalized substrates of Si/SiO_2, Al/Al_2O_2, TiN, Pt, and indium tin oxide (ITO). Figure 1 outlines the procedure employed.

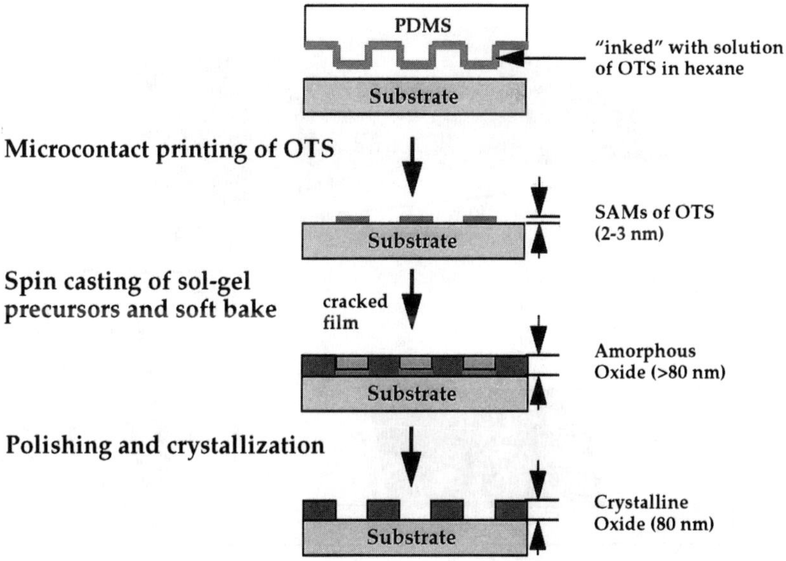

Figure 1. Schematic outline of the procedures used for the patterning of oxide thin layers.

As-spun thin films were heat treated to an intermediate pyrolysis temperature of 150-300°C to remove residual organic species and drive film densification. Preferential film adhesion was observed on unfunctionalized (hydrophilic) regions of substrate surfaces. These films were then polished on a non-abrasive felt pad wet with isopropanol. The resulting patterned, amorphous films were subsequently fired at higher temperatures to achieve crystallization.

RESULTS

Atomic force microscopy (AFM) was used to image microcontact printed OTS monolayers in topography and lateral force modes. Figure 2a shows a 3 μm wide patterned monolayer of OTS, revealed in topography mode AFM. Ellipsometry indicates 25 Å monolayers of this siloxane are deposited by printing of a 10 mM OTS/hexane solution. The idealized structure of this monolayer is shown in figures 2b and 2c. The chemically and thermally robust nature of these monolayers was instrumental in the pattern definition of sol-gel thin films. Thermal cycling of these monolayers indicated stability up to at least 300°C.

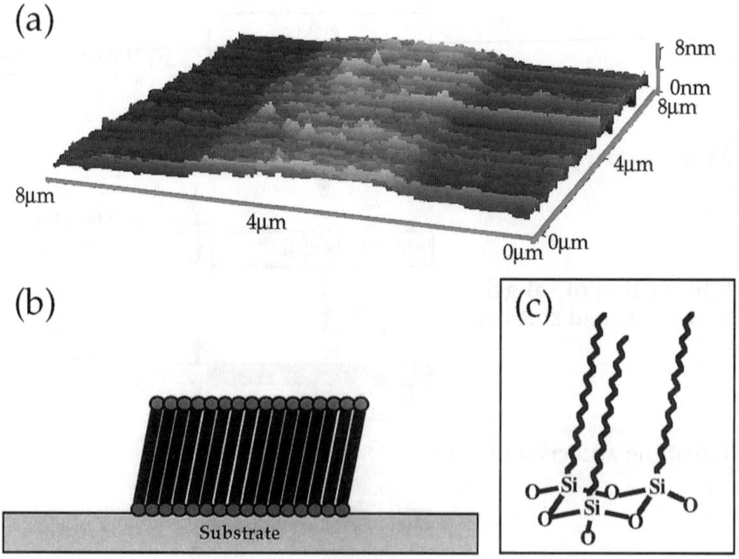

Figure 2 (a) AFM image of 3μm wide patterned OTS monolayer on SiO_2, (b) cartoon of OTS self-assembled monolayer, (c) idealized OTS monolayer structure.

Sol-gel thin films were deposited by spin coating at 3000 rpm for 30s. While preferential wetting of the solution precursors was observed, differences in adhesion after pyrolysis of as-spun films resulted in pattern definition. After pyrolysis, regions of amorphous layers above the OTS functionalized regions were severely cracked and showed poor adhesion. In contrast, well-bound, continuous layers were formed above underivatized regions. Figure 3a shows a typical thin film microstructure immediately after pyrolysis. The loosely adhered layers were selectively removed by mild mechanical abrasion with cotton felt wetted with isopropanol, yielding patterned microstructures as shown in figure 3b. Edge resolution for layers was as small as ±75 nm. Heat treatment to 700°C was sufficient for crystallization of these amorphous layers. Thin films of amorphous and crystalline Ta_2O_5 were electrically insulating with dielectric constants K' ≈ 30 and tan δ = 0.007-0.02 measured for 80 nm layers.

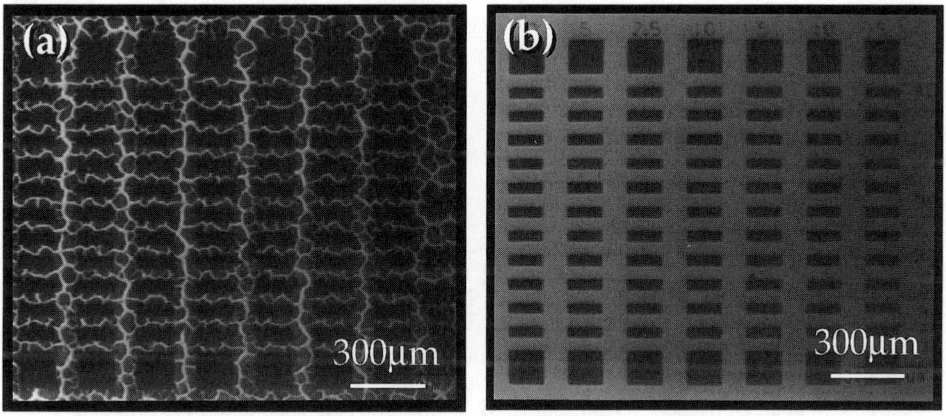

Figure 3. The post-pyrolysis difference in adhesion of films on SAM modified surface versus unmodified surface results in pattern definition. The (a) de-adhered layer on SAMs derivatized regions is removed to yield (b) patterned 200 nm oxide structures on unmodified substrate regions.

CONCLUSIONS

Microcontact printing of alkylsiloxane monolayers combined with sol-gel processing was found to be effective in producing patterned oxide thin layers on a variety of substrates. Continuous oxide thin layers were selectively deposited on areas not modified by μCP of SAMs. Insulating dielectric structures were prepared on a variety of technologically important materials such as Si, Al, Pt, TiN and ITO. The capability of this technique to deposit patterned dielectric layers at reduced processing steps and temperatures offers distinct advantages over conventional ceramic thin film processing. This technique complements the selective deposition of metals by μCP and CVD to allow photolithography-free fabrication of micron-scale metal-oxide-silicon devices. Combining the flexibility of the μCP process and the ease of oxide thin layer preparation by sol-gel deposition suggests a wide range of potential technological applications for integrated oxide devices such as capacitors and sensors.

ACKNOWLEDGMENTS

We are grateful for the support of the National Science Foundation (CHE 9500995) and the Department of Energy (DEFG02-91ER45439). We acknowledge the indispensable contributions made by the central facilities of the Frederick Seitz Materials Research Laboratory operated with the support of the DOE. PGC acknowledges the support of a Beckman Institute Research Assistantship for interdisciplinary study.

REFERENCES

1. A. Ulman, Ultrathin Organic Thin Films, (Academic Press, San Diego, 1991).
2. A. Kumar and G. M. Whitesides, Appl. Phys. Lett. **63**, 2002 (1993).
3. A. Kumar, H. Biebuyck, and G. M. Whitesides, Langmuir **10**, 1498 (1994).
4. J. Wilbur, A. Kumar, E. Kim, and G. M. Whitesides, Adv. Mat. **6**, 600 (1994).
5. Y. Xia, M. Mrksich, E. Kim, and G. M. Whitesides, J. Am. Chem. Soc. **117**, 9576 (1995).
6. N. L. Jeon, R. G. Nuzzo, Y. Xia, M. Mrksich, and G. M. Whitesides, Langmuir **11**, 3024 (1995).
7. N. L. Jeon, P. G. Clem, D. A. Payne, and R .G. Nuzzo, J. Mater. Res. **10**(12), 2996 (1995).

VANADIUM-OXO BASED HYBRID ORGANIC-INORGANIC COPOLYMERS

A. CAMPERO*, A.M. SOTO*, J. MAQUET**, C. SANCHEZ**
*Departamento de Química, Universidad Autónoma Metropolitana (Iztapalapa), México DF 09340, México, acc35@xanum.uam.mx
**Laboratoire de Chimie de la Matière Condensée-URA 1466, Université Pierre et Marie Curie, 4, place Jussieu, Paris, France

ABSTRACT

We describe the synthesis and characterization of hybrid organic-inorganic copolymer materials formed by the reaction of the transition metal alkoxide $VO(OAm^t)_3$ with the chelating monomer ligand acetoacetoxyethylmethacrylate (AAEM). By the simultaneous induction of the organic and inorganic polymerization reactions, covalent bonds are formed between both types of interpenetrating components. The organic chelating moiety of AAEM is linked through its ß-diketo function to the vanadium-oxo species.

INTRODUCTION

Many well established procedures for the synthesis of advanced materials of technological importance are available through the use of the sol-gel method. Until recently, efforts were centered almost exclusively on inorganic oxides [1-3]. However, presently a strong interest is emerging to study the materials formed by the incorporation of organic components into inorganic metal-oxo networks [4,5].

A specially important feature characteristic of the sol-gel process is its use of metallo-organic precursors, organic solvents, and ambient or low processing temperatures [6-8], thus avoiding the destruction of organic moiety, which disperses through the inorganic gel network, establishing physical or chemical links with it [8].

We describe here the molecular design and synthesis, as well as a preliminary spectroscopic characterization, of a nanocomposite hybrid organic-inorganic material exhibiting electronic properties. It yields sols which can be easily deposited as coatings. Vanadium-oxo-AAEM copolymers have been synthesized from $VO(OAm^t)_3$ modified at the molecular level by AAEM, a complexing ß-diketone chelating ligand. This ligand has been previously used in the synthesis of oxo-zirconium materials [9]. It contains both a strongly chelating β-diketonate and the highly reactive methacrylate groups, and is quite stable towards hydrolysis. Its covalent bonding to an alkoxide of a transition metal with several oxidation states, for example the vanadium alkoxide described in this work, leads to the synthesis of important new materials with possible conductive and redox properties.

Mat. Res. Soc. Symp. Proc. Vol. 435 © 1996 Materials Research Society

EXPERIMENTAL

Synthesis

0.2 M solutions of $VO(OAm^t)_3$ in tert-amyl alcohol Am^tOH were made to react with the AAEM monomer (Eastman Kodak, 98%), in the presence of the radical initiator azobisisobutyronitrile AIBN (0.2%). Water for hydrolysis was added in appropriate amounts. It was then refluxed for 24 hours which initiates a simultaneous development of both the networks. We prepared two sets of samples: a).- at a constant hydrolysis ratio $h=[H_2O/V]=2$, the complexation ratio $r=[AAEM/V]$ was increased in steps ($r=0.2$) from 0.2 to 0.8, samples were labeled A1 to A4, respectively; b.- at a constant $r=[AAEM/V]=0.2$, the value of $h=[H_2O/V]$ was varied from 2 to 8 in steps of 2: samples were labeled B1 to B4, respectively. We use xerogel B2 as example of a general behavior.

RESULTS

Infrared Spectroscopy

The infrared spectra of $VO(OAm^t)_3$ and AAEM, the two reaction precursors, are shown in Figure (1A) and Figure (1B), respectively.

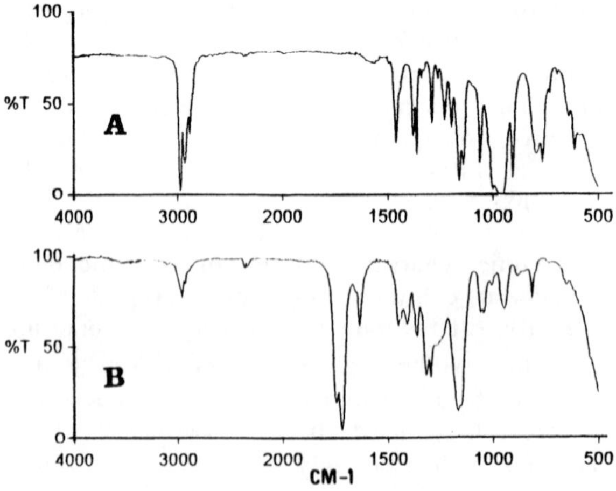

Fig. 1. Infrared spectra of A) $VO(OAm^t)_3$, B) AAEM.

In Fig. (1B) the spectrum of AAEM is shown. Its most characteristic bands for our discussion are situated at 1749 and 1720 cm^{-1}. They correspond to ν(C=O) stretching vibrations of the carbonyl functions belonging a).-to the ketonic form of the ß-diketone group, and b).- to the carbonyl of the ester of the methacrylate group, respectively. The band at 1637 cm^{-1} is known to correspond to the C=C double bond of the latter group.

Fig. (2) depicts the spectrum of the AAEM modified gel B2 and we notice several prominent changes with respect to its precursors $VO(OAm^t)_3$ and AAEM.

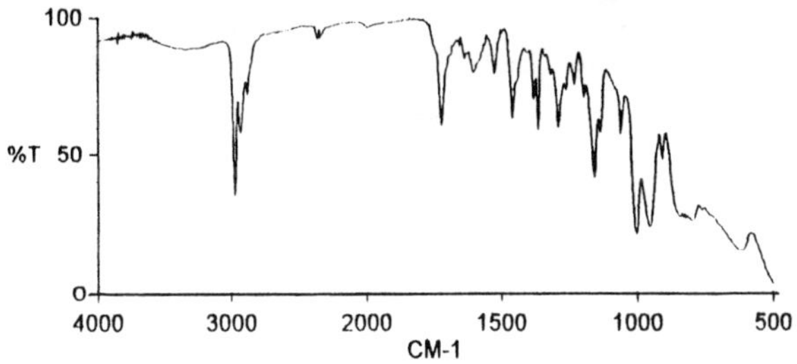

Fig. 2. Infrared spectrum of hybrid copolymer B2 (r=0.2, h=4).

Firstly, notice that two new bands have appeared, at 1605 and 1525 cm^{-1}, corresponding to stretching vibrations of the enol double bond and of the C-O bond in AAEM, respectively.Secondly, that the strong band at 1749 has disappeared.. These two observations show, of course, that no free AAEM exists in the gel, and that it has bonded covalently to the vanadium of the inorganic polymer. The C=C band at 1637 cm^{-1} has practically disappeared, an indication of AAEM polymerization. Finally, a new broad band at 626 cm^{-1} appears, which most probably corresponds to V-O-V bonds [10] , showing that the vanadium inorganic network has been formed.

NMR Spectroscopy

The solid state ^{13}C MAS NMR spectrum (Fig. 3) of the olive green hybrid organic-inorganic xerogel, sample B2 (r=0.2, h=4), has been obtained in a Bruker 300 ASX spectrometer. It gives several broad peaks which are probably due to a slight paramagnetism due to some incipient V(IV) entities. In order to simplify the analysis of the spectrum, we label the resonance peaks. The identification of these peaks with the carbon sites in the polymer molecule is shown in Fig. 3.

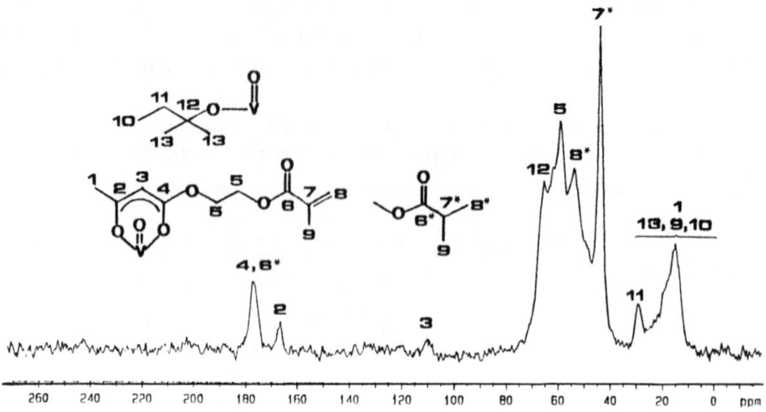

Fig. 3. ^{13}C MAS NMR spectrum of hybrid copolymer B2 (r=0.2, h=4)

Most of these are assigned on the basis of the spectra of free AAEM and VO(OAm^{t}). However, the important chemical information is supplied by the following features of the spectrum.

The peak at 43.9 ppm is assigned to the quaternary carbon 7* of the polymerized AAEM. This peak appears neither in free AAEM nor in xerogels prepared in the absence of the organic initiator AIBN. This is a clear evidence that organic polymerization has taken place. Also, the peak at 53.5 ppm is asigned to carbon 8*, the $\underline{C}H_2$ of the polymerized methacrylate function. This is further confirmed by the disappearance of the resonance line around 125 ppm of the carbon 8 in the non-polymerized methacrylate group. Other important peaks are those of the carbons 3 and 4, at 110.4 and 177.1 ppm, respectively; this proves unambiguosly that AAEM in its enol form is chemically bonded to the vanadium atom.

Fig. 4 shows the solid state ^{51}V MAS NMR spectrum of sample B2 (r=0.2, h=4). It consists of an extremely broad band almost 100 ppm wide, centered around -650 ppm. At least three shoulders are found superposed on the broad band, namely at -670.5, -654.0, and -643.6 ppm, whose partial assignment can be based on the earlier literature reports [11, 12]. The first two peaks most probably belong to various types of alkoxide hydrolysis products [12]. We can assign tentatively the shoulder at -643.6 ppm to vanadium oxo-alkoxide oligomers [11]. Moreover, our NMR results do suggest the diamagnetic V(V) oxidation state for vanadium species in the hybrid xerogel although the green color of the polymers indicates some

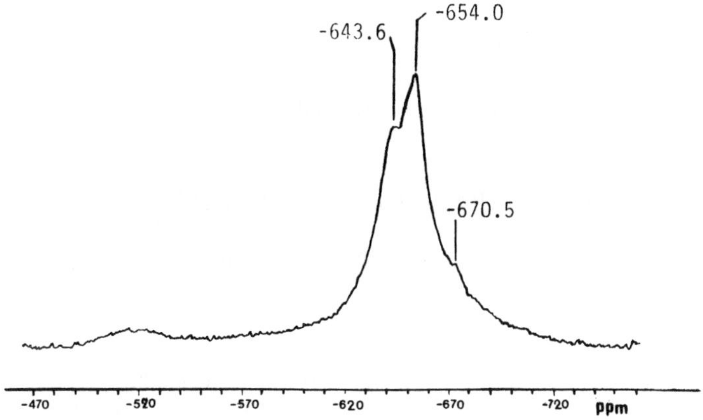

Fig. 4. ^{51}V MAS NMR spectrum of gel B2 (r=0.2, h=4).

weak paramagnetism. Probably polymerization of this alkoxide leads to oxo polymers in various arrangements, with a predominant coordination number 5.

CONCLUSIONS

Vanadium-oxo based hybrid organic-inorganic copolymer materials with electronic properties have been synthesized by means of the reaction between the vanadium alkoxide $VO(OAm^t)_3$ and the organically functionalized organic ligand AAEM. The organic and the inorganic components are covalently bonded, and these links exist both in the sols and in the gel materials. The complexation ratio [AAEM/V] and the hydrolysis ratio [H_2O/V] are the key parameters controlling their structure.

ACKNOWLEDGMENTS

The present study was financially supported in part by Consejo Nacional de Ciencia y Tecnología (México). The authors would also like to thank Atilano Gutierrez for NMR spectra.

REFERENCES

1. J. Livage, M. Henry and C. Sanchez, Progress in Solid State Chemistry, **18**, 259 (1988).

2. C. J. Brinker, and G. Scherrer, Sol-Gel Chemistry, the Physics and Chemistry of Sol-Gel Processing, Academic Press, San Diego, 1989.

3. C.D. Chandler, C. Roger, and M. Hampden-Smith, Chem. Rev. **93,** 1205 (1993).

4. D. Avnir, D. Levy and R. Reisfeld, J. Phys. Chem. **88**, 5956 (1984).

5. H. Schmidt and B. Seiferling, (Mat. Res. Soc. Proc. 73,1986), p.739.

6. B.M. Novak, Adv. Mater. **5**, 422 (1993).

7. B.K. Coltrain, C.J.T. Landry,, J.M. O'Reilly, A.M. Chamberlain, G.A. Rakes, J.S.S. Sedita, L.W. Kelts, M.R. Landry and V.K. Long, Chem. Mater. **5,** 1445 (1993).

8. C. Sanchez and F. Ribot, New Journal Chemistry **18,** 1007 (1994).

9. C. Sanchez and M. In, J. Non-Cryst. Solids **147&148**, 1 (1992).

10. K. Nakamoto, Infrared and Raman Spectra of Inorganic and Coordination Compounds, J. Wiley, New York, 1978.

11. C. Sanchez, B. Alonso, F. Chapusot, F. Ribot and P. Audebert, J. Sol-Gel Sci. & Techn. **2,** 161 (1994).

12. D.C. Crans, H. Chen, R.A. Felty, J. Am. Chem. Soc. **114**, 4543 (1992).

Part VII

Electrical and Optical Properties of Organic/Inorganic Hybrid Materials

NANOSTRUCTURED MATERIALS FOR PHOTONICS

N. D. KUMAR, G. RULAND, M. YOSHIDA, M. LAL, J. BHAWALKAR, G. S. HE AND P. N. PRASAD
Photonics Research Laboratory, Department of Chemistry,
State University of New York at Buffalo, NY 14260-3000.

ABSTRACT

Nanocomposite materials for application in photonics were developed by sol-gel processing and reverse micellar microemulsion techniques. The capability of incorporating many materials with different functional properties in sol-gel processed glass matrices has been explored in making these materials. The large pore volume fraction and the enormous surface area of the sol-gel glasses enables one to introduce many materials in a phase separated fashion, where the phase separation is in the nanometer range. It is possible to introduce an active material on to the pore surface by solution infiltration and subsequent removal of the solvent, then filling the pores with a monomer containing another active material, and polymerizing inside the pores. Using this approach we have developed composite materials for optical power limiting applications at different wavelengths and a tunable solid state dye lasing medium

Optically transparent polyimide:TiO_2 composite waveguide materials were prepared by the dispersion of nano-sized TiO_2 particles into a polyimide matrix. The particles were produced through reverse micelles using the sol-gel method, and were incorporated into the fluorinated polyimide solution. A polyimide:TiO_2 (4 wt %) composite waveguide was produced from the solution. Since the particle size is so small, no noticeable scattering loss was observed. The measured optical propagation loss at 633 nm was 1.4 dB/cm, which is equivalent to that of the pure polyimide. The refractive index was increased from 1.550 to 1.560 by the incorporation of TiO_2.

INTRODUCTION

Much of the current interest in photonics arises due to the applicability of photonics to current and future information and image processing technologies. The realization of this advanced technology rests on the development of multifunctional materials which simultaneously satisfy many functional requirements.[1,2] Progress during the last decade in the design of organic systems allows one to prepare new materials with promising lasing and nonlinear optical properties.[3,4] A major effort has been centered on introducing the optically active organic materials into a photostable medium such as a glass matrix, allowing their use as building blocks for photonic devices. Although a commonly used solid matrix for hosting organic molecules are organic polymers,[5] they suffer from lack of photostability and optical quality. This provides motivation to successfully introduce nonlinearlly active organic molecules such as polymers and dyes into more photostable matrices such as inorganic glasses. In the early 80's, when the sol-gel process became a popular method for preparing glasses at room temperature, it was demonstrated that an

Mat. Res. Soc. Symp. Proc. Vol. 435 © 1996 Materials Research Society

organic dye can be successfully doped within an inorganic glass.[6] Sol-gel processed materials have been used in areas ranging from novel solid-state lasers to platforms for chemical and biosensors.[7-11] Also, the sol-gel technique has been used to fabricate nonlinear optically active composites for their use in optical communications.[1]

Composite glasses prepared via the sol-gel technique are of high optical quality[12] and one can make large size monolithic bulk forms for various photonic functions such as: lasing,[13] optical power limiting,[14,15] and nonlinear optical response,[16] etc. An advanced approach is to form multiphasic nanostructured composites.[17-19] By using a multi-step impregnation method, one can dope two (or more) different optically responsive materials, each of which resides in different phases of the matrix (the silica phase, the organic polymer phase and the interfacial phase). For example; in the silica phase, one can dope any species that can survive the thermal treatment. Examples are inorganic species, such as rare-earth ions, which exhibit amplifying performance (Er^{+3}, Eu^{+3}) or semiconductors (ZnS, CdS) that due to the nanoscale size of the pores can be quantum dots with promising nonlinear properties.. In the interfacial phase, one can dope inorganic or organic species. Examples are C_{60}, which exhibits power limiting effect (predominantly due to reverse saturable absorption) and materials with high third order optical succeptibility ($\chi^{(3)}$).[17-19] Ionic dyes which exhibit strong two-photon absorption can also be incorporated in the interfacial phase for power limiting.[20] The only limitation for doping in the interfacial phase is that the dopant is not soluble in the monomer liquid (such as methylmethacrylate) that will be polymerized in-situ. In the polymer phase, one can dope laser dyes and $\chi^{(3)}$ or other active organic chromophores.[21]

Composite materials consisting of sol-gel processed inorganic materials and polymers are useful for fabrication of optical waveguides, since they have improved thermal and mechanical properties over equivalent pure polymer. Furthermore they provide freedom of manipulation of refractive index.[22-24] The improved mechanical stability is an especially favorable characteristic for the fabrication of channel waveguide, since hard materials provide a better edge polishing capability, resulting in lower coupling losses.

Polyimide is one of the most intriguing materials for optical waveguide applications since it has outstanding thermal stability, high mechanical strength, and low thermal expansion coefficient due to its rigid molecular structure. Therefore, polyimide is an ideal polymer for composite materials. The polyimide composite materials with inorganic components such as silicon dioxide and/or titanium dioxide also have been studied by many researchers for improving the mechanical strength of the polymer by forming three dimensional inorganic networks.[25-27] These methods generally involve the mixing of an alkoxide precursor for the inorganic phase and a polyimide precursor, which may lead to phase separation in macro level. Here we suggest a new method for the fabrication of titania:polyimide composite materials for optical waveguides using the reverse micelle technique.

Ultra fine particles have received a great deal of attention because of the novel properties they exhibit which greatly differ from the bulk properties.[28] These include quantum size effect on photochemistry,[29] nonlinear optical properties in semiconductors[30] and metallic nanoclusters.[31] Various chemical methods have been proposed for the synthesis of these particles. Since the particles are prone to aggregate into larger particles, it is necessary to prevent uncontrolled aggregation in the course of preparation. To be able to make these nanostructures, reverse micelles are good candidates.[32]

Reverse micelles are spheroidal aggregates formed by the dispersion of a surfactant in an organic solvent.[33] They can be formed both in the presence and absence of water. The presence of water is, however, necessary to form large surfactant aggregates. Water can be readily dissolved in the polar core, forming a "water pool". The size of the "water pool" is determined by W_O, the water-surfactant molar ratio (W_O=[H2O]/[S]). The aggregates containing a small amount of water (below Wo=15) are usually called reverse micelles whereas droplets containing a large amount of water (above W_O=15) are called microemulsions. When using isooctane as a continuous medium, the water pool radius, Rw is found to be linearly dependent on the water content (Rw =1.5W_O)[34] which in turn is related to the core size of the reverse micelle. As a result, particles of various sizes can be synthesized by varying the size of the water pool which serves as a micro-reactor for carrying out various reactions. Confinement in these small compartments inherently limits particle growth thereby resulting in controlled size, monodispersed particles.

SAMPLE PREPARATION

Preparation of Composite glass

Hydrolysis and condensation of silicon alkoxide under carefully controlled conditions followed by partial densification produces porous silica monoliths of nanoscale pore size. These pores can be successfully used as a platform for loading organic molecules to form inorganic-organic composite glass.[35-37] The silica skeleton of the composite glass is prepared by a two-step hydrolysis of tetraethyl orthosilicate. First, an acid catalyst was used for hydrolysis, which lead to a high degree of cross-linkage, followed by a base catalysis to accelerate the gelation. The resultant product was a highly porous monolith with small pores. These monoliths yielded high quality porous glass after a moderate heat treatment. The porous glass was then impregnated with an organic monomer (such as methylmethacrylate), which diffuses into the pores, and polymerized in-situ to form a composite glass. The final composite glass consisted of 32% by volume silica and 68% by volume PMMA.[21] Nitrogen adsorption measurements indicated that the composite glass are pore free and hence mechanically stable.[38] The observed density (1.447 g/cm^3) and refractive index (1.472) are in good agreement with the expected values from the average macroproperties of the separate phases. These composite glasses exhibit excellent transmission over most of the visible range since the domains are of nanometer size.

A composite glass containing C_{60} and an organic dye, bisbenzothiazole 3,4-didecyloxy thiophene (BBTDOT) was prepared for optical power limiting application at different wavelengths. At first C_{60} was deposited on the pore surface of the silica glass by impregnating a saturated solution of C_{60} in toluene into the pores and subsequent removal of the solvent by evaporation. The pores were then filled with a solution of the dye in methylmethacrylate and polymerized in-situ. The final composite glass thus contains C_{60} deposited on to the pore surface and the dye in the polymethylmethacylate phase (PMMA). The composite glass is then subjected to fine polishing to improve the optical quality.

Using the above approach, two different organic dyes were incorporated into two different phases of the composite glass to achieve tunable solid state lasing from the composite glass. Organic dyes Rhodamine-6G and a new laser dye *trans*-4-[p-(N-ethyl-N-(hydroxyethyl)amino)phenylstyryl]-N-(hydroxyethyl) pyridinium iodide (ASPI) were introduced in to the composite glass by the same approach, where ASPI was deposited onto the walls of the porous glass and Rhodamine-6G in the PMMA phase.

Fabrication of Titania dispersed Polyimide composite waveguide

In the present work, the synthesis of TiO_2 particles was done by carrying out the hydrolysis of titanium isopropoxide (TIP) in a AOT/isooctane reverse micellar system. TIP was dissolved in the bulk oil continuous medium (isooctane) and water required for the hydrolysis was provided by the aqueous pools of reverse micelles. When the solution containing TIP was mixed to the reverse micellar solution containing water in the reverse micelles, TIP slowly diffused into the reverse micelles and hydrolysis and condensation took place in reverse micelles, resulting in the formation of TiO_2 particles in the reverse micellar micro-reactors.

The reverse micellar solution was prepared by dissolving sodium bis(2-ethylhexyl)sulfosuccinate (AOT) in isooctane (2,2,4-trimethylpentane) to form a 0.1M solution. The resulting solution was filtered using a 0.2 micron membrane filter. This was followed by the addition of the required amount of filtered distilled water. The Wo was varied in the range 1-10. Titanium isopropoxide (TIP) was diluted with 2-propanol to be 13 wt %, and the hydrolysis of TIP was carried out in a conical flask (100 ml) at 25°C. The reaction was initiated by the injection of the TIP solution (200 ml) into the reverse micellar solution (30 ml) with mild stirring on a magnetic stirrer. The TiO_2 particles prepared with $W_0=2$ were then extracted from the reverse micellar solution by the addition of a known volume of N-methylpyrrolidinone (NMP) (4 ml). Immediate phase separation occurred and the particles were extracted in the lower NMP phase. To this NMP solution containing TiO_2 particles, the polyimide solution was added and a uniform, clear and viscous solution was obtained.

To fabricate slab waveguides, the TiO_2 particle-containing-polyimide solution was cast on a glass slide. The film was immediately baked on a 100°C hot plate for 10 min, otherwise the film became opaque due to moisture intake from the air. The film was then baked at 200°C for 30 min, followed by a heat treatment of 300°C for 30 min. in a nitrogen atmosphere. The concentration of TiO_2 in the resultant film was calculated to be 4 wt % from the dose of TIP.

RESULTS AND DISCUSSION

Multiphasic Nanostructured Composites for Power Limiting

Optical power limiters are devices which limit the optical output to a certain value. There are several physical mechanisms of optical power limiting action, for example: nonlinear absorption, reverse saturable absorption, nonlinear refraction and thermal effects. Each may be suitable for a specific condition. The multiphasic nanostructured composites may be specifically suitable for this purpose because one can incorporate different power limiting functions in different phases. A good example of a multifunctional power limiting material is a monolith which contains C_{60} at the walls of the pores and bisbenzothiazole 3,4-didecyloxy thiophene (BBTDOT) in the polymer, PMMA, phase. C_{60} has shown optical power limiting behavior, which has been suggested to be due to a multitude of mechanisms such as reverse saturable absorption, nonlinear refraction and thermal effect [39] BBTDOT shows a strong two-photon nonlinear absorption.

The power limiting action of this multiphasic composite was studied at two wavelengths: (i) at 532 nm, where C_{60} is active due to reverse saturable absorption,[40] and (ii) at 800 nm where the dye BBTDOT is active due to two-photon nonlinear absorption.[17] The source at 532 nm was a frequency doubled, Q-switched Nd:YAG laser which delivered 8 ns pulses at a repetition rate of 10 Hz. The source at 800 nm was a dye laser with an IR dye (LDS 820) pumped at 532 nm which generated ~5 ns pulses at 800 nm. At each wavelength the transmitted intensity was recorded as a function of the incident intensity and the data was normalized to a transmissivity of unity at a low intensity incidence beam. An incident beam with pulse flux power up to 300 MW/cm^2 was used in both cases. It is important to point out that the composites can stand up to 400 MW/cm^2 pulse flux power.

Figure 1 presents the normalized output fluence as a function of the input fluence for the multiphasic composite glass containing both C_{60} and BBTDOT as well as for the two composite glasses containing each molecule separately. In figure 1, panel A represents the results of optical power limiting measurements on the composite glasses at a wavelength of 532 nm and panel B that at 800 nm. From figure 1A, it is clear that C_{60} is more active as an optical limiter at 532 nm than at 800 nm, as expected from a one photon induced reverse saturable absorption (RSA) process. BBTDOT has a linear absorption maximum at approximately 400 nm and, therefore, has almost no absorption at either 532 or 800 nm light at low intensity. This was supported by the measured transmissivity (at a low intensity incidence beam) of the pale yellow BBTDOT doped sample, which was 0.98 at 800 nm and 0.70 at 532 nm.

At higher intensities (>50 MW/cm^2) of the 800 nm beam, a strong two-photon absorption (TPA) induced blue fluorescence was clearly visible in the sample. As expected, this glass shows a significantly higher nonlinearity at 800 nm than at 532 nm. On the other hand, the glass doped with C_{60} + BBTDOT had a light brown color and showed transmissivity (at a low intensity incidence beam) of 0.92 at 800 nm and 0.30 at 532 nm. The lower transmissivity at 532 nm is due to linear absorption of the C_{60} molecules. It can be clearly seen from figure 1 that this sample shows excellent optical power limiting behavior at both 532 nm and 800 nm. The nonlinear behavior at 532 nm appears to be slightly enhanced compared to the sample doped with only C_{60}. This suggests that there may be a favorable contribution from BBTDOT. On the other hand, at 800 nm, the nonlinearity is slightly less than that due to BBTDOT alone. This is probably due to the fact that the local intensity in the sample decreases due to the presence of C_{60} (some linear absorption). Therefore, the nonlinear absorption in BBTDOT, which varies as the square of the incident intensity, is reduced.

The multiphasic composite glass doped with both C_{60} and BBTDOT exhibited effective optical power limiting at both wavelengths. The fact that we observed significant optical limiting using nanosecond pulses in the nanostructured composites is further evidence that in the multiphasic nanostructured composites the fullerene mimics its properties in solution, but is still phase-separated from the host material and other dopants. Therefore, this result represents the fascinating possibilities embodied in the multiphasic composite glasses for fabricating multifunctional devices for photonics.

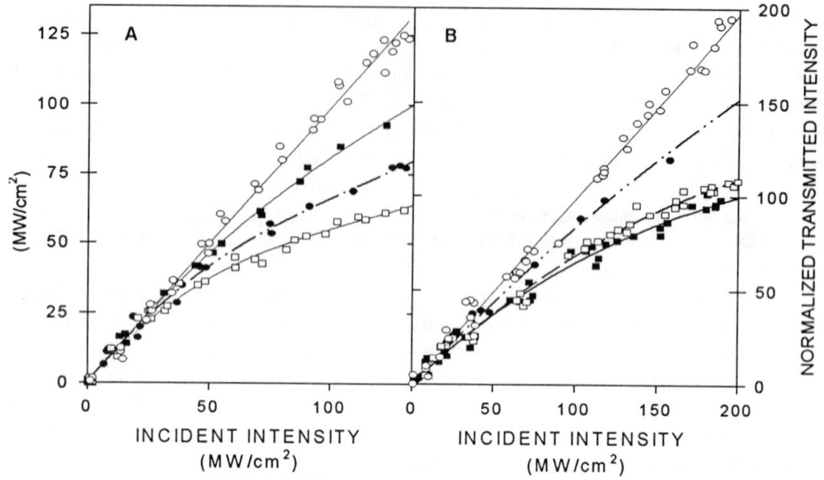

Figure 1. The intensity-dependent normalized transmission curves measured at two different wavelength: 532 nm (panel A) and 800 nm (panel B). In both cases: C_{60} plus BBTDOT doped composite (empty squares), C_{60}-doped composite glass (filled circles), BBTDOT-doped composite glass (filled squares), and pure composite glass (empty circles).

Multi-dye Tunable Solid State Laser

The multidye solid state tunable laser consisted of ASPI, which resides in the interfacial phase and Rhodamine-6G, which resides in the polymer phase. Figure 2 presents the fluorescence emission of a multiphasic composite glass containing both dyes. For comparison the fluorescence emission of two composite glasses containing each dye separately are presented as well. Emission is observed from both dyes in the multiphasic composite glass with no significant quenching of emission from either dye. However, in the solution state a mixture of the two dyes, at the same concentration as that of the glass, exhibits complete quenching of the Rhodamine-6G emission.

Tunable lasing was also observed across the region of the individual dyes. For comparison the lasing tunability of the dyes in the solution state were also measured. Tunable narrow band laser outputs were observed in a cavity consisting of a grating as the back reflector and a ~ 70 % reflecting output coupler. Figure 3 presents the lasing tunability of the dyes in the solution state. Three curves are presented: one of a Rhodamine-6G solution, one of an ASPI solution, and one of a Rhodamine-6G/ASPI solution, each with ethanol as the solvent. It is clear from this figure that in the Rhodamine-6G/ASPI solution the lasing emission from Rhodamine-6G is completely quenched.

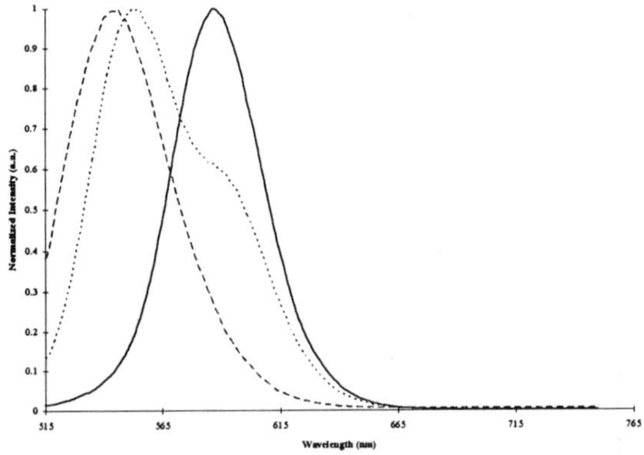

Figure 2. Fluorescence emission of Rhodamine-6G composite glass (dashed curve), ASPI composite glass (solid curve), and the composite glass containing both dyes (dotted curve).

Figure 4 presents the lasing tunability of a multiphasic composite glass containing both dyes. For reference, figure 4 also presents the lasing tunability of two composite glasses containing each dye separately. The FWHM of the tunability spectra for the ASPI composite glass is ~ 21 nm and that for Rhodamine-6G ~ 12 nm. For the multiphasic composite glass containing both dyes it is ~ 37 nm. From this data it is evident that the glass containing both dyes is tunable across the range of both dyes, (560 - 610 nm), where as in the solution state the Rhodamine-6G emission is quenched.[17] We believe that the quenching in the solution state is a result of Förster energy transfer. Therefore, due to the extremely large ratio between the specific surface area and the pore volume, which is ~ 8.5 x 10^6, it is possible to separate the two dye molecules, (which reside in different phases) to a distance where the Förster energy transfer is not effective.

The rate of energy transfer between the donor molecule (Rhodamine-6G) and the acceptor molecule (ASPI) in the Förster energy transfer mechanism can be described by the following equation.[41]

$$k_T = \frac{9(\ln 10)\kappa^2 Q_d J}{128\pi^5 n^4 N_a \tau_d R^6} \qquad (1)$$

where κ^2 is the orientational factor for the dipole - dipole interaction, Q_d is the fluorescence quantum yield of the donor molecule without the acceptor molecule, n is the refractive index of the medium, N_a is Avogadro's number, τ_d is the fluorescence lifetime of the donor molecule without the presence of the acceptor, R is the distance between the centers of the donor and the acceptor molecules, and J is the normalized spectral overlap integral. From equation (1) the rate of energy transfer is directly proportional to the orientational factor κ to the second power and inversely proportional to the sixth power of the distance between the center of the molecules. It is these two factors that we believe are significantly changed in the multiphasic composite glass. By increasing the distance between the molecules by a factor of 1.6 we have decreased the rate of energy transfer by more than one order of magnitude.[19] Also, in the solution state, the molecules are free to rotate and sample most if not all of the orientational possibilities during the excited state lifetime of Rhodamine-6G. In the solid state the molecules are effectively frozen in place

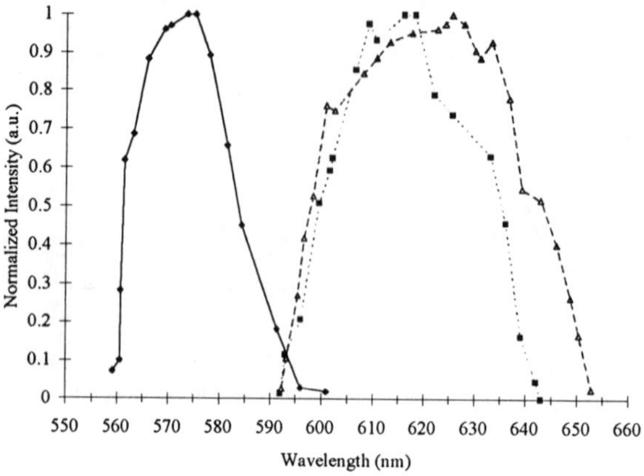

Figure 3. Lasing Tunability of Rhodamine-6G (diamonds), ASPI (squares), and a mixture of Rhodamine-6G and ASPI (diamonds) all in ethanol.

with little or no rotation allowed leading to a further decrease in the energy transfer.[19] Due to the increase in the distance between the molecules and the decrease of the orientational factor, Förster energy transfer (and other potential quenching mechanisms) has been prevented in the multiphasic glass. This result represents another fascinating possibility embodied in the multiphasic composite glasses for fabricating multifunctional devices for photonics.

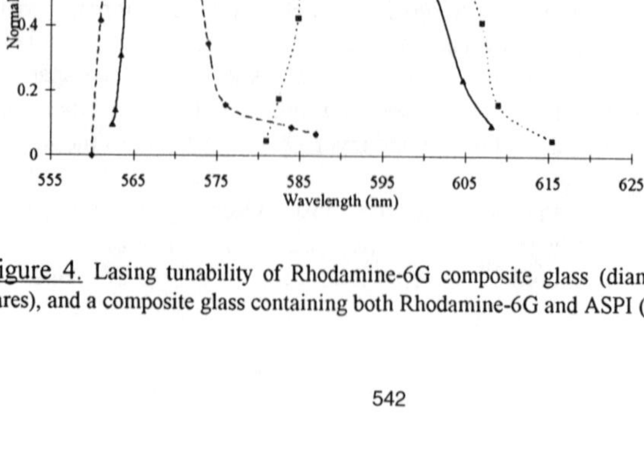

Figure 4. Lasing tunability of Rhodamine-6G composite glass (diamonds), ASPI composite glass (squares), and a composite glass containing both Rhodamine-6G and ASPI (triangles).

TiO2:Polyimide Composite wave guide

The formation of TiO_2 particles within the reverse micelle was characterized through the UV-Visible absorption spectra recorded on a Shimadzu UV-visible spectrophotometer (Model UV 3101 PC). The results are shown in figure 5. In all the cases, peaks around 300 nm suggesting the formation of TiO_2 were observed; however, we found $W_O \leq 2$ gave the best result in terms of the optical transparency. The TiO_2 particles made with W_O=3 or higher were so large that the solutions became opaque. Figure 5 shows the noticeable increases of background absorption due to scattering, for W_O=3 or higher.

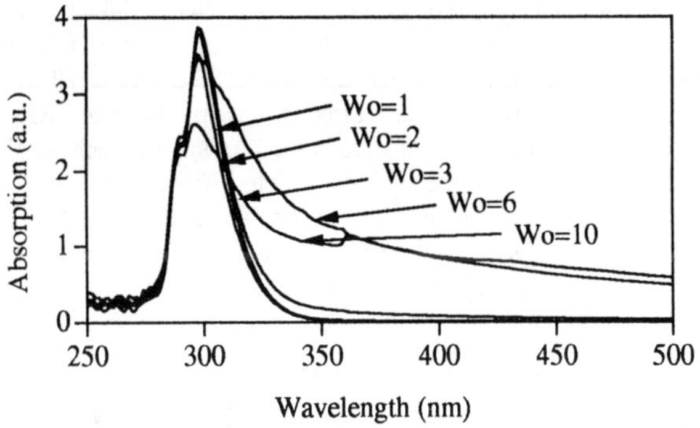

Figure 5. UV-visible spectra of TiO_2 particles in a reverse micellar solution for different water to surfactant ratio (W_O).

Refractive index and film thickness were calculated from coupling angles measured by the prism coupling technique using a TiO_2 rutile prism and a He-Ne laser (Melles Griot, wavelength; 632.8 nm, output power; 3 mW, TE mode). Optical propagation loss was measured by observing the optical intensity decay of the streak line with a CCD camera (Electrim Corp., Computer camera EDC-1000). The intensity decay in each pixel of the CCD camera was digitized and plotted in dB unit with respect to the propagation distance in cm unit. The slope is the optical propagation loss in dB/cm unit.

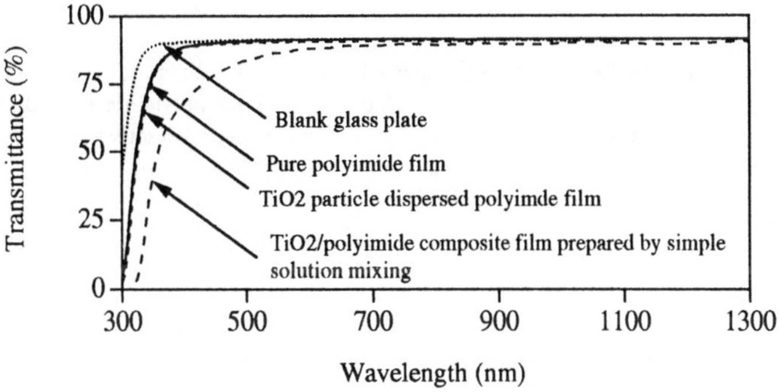

Figure 6. Optical transmission spectra of various polyimide-processed film.

The UV-visible transmission spectra of the TiO_2 particle dispersed polyimide film, along with a pure polyimide film and a blank glass substrate, are shown in figure 6. These results suggest that the impregnation of TiO_2 particles does not produce any noticeable absorption loss due to discoloration of the material. Prior to this work, we simply mixed TIP/isopropanol solution to the polyimide solution and prepared a film. However, this method produces a yellow film showing a strong absorption at shorter wavelengths, resulting in high optical loss at 633 nm. This strong absorption must be due to the formation of a titanium complex from the reaction of TIP and polyimide or the impurity of polyimide. The spectrum of this film is also shown in figure 6 for comparison.

The TiO_2 particle dispersed polyimide film was evaluated as an optical waveguide. The measured optical intensity decay along the streak line is shown in figure 7 and the measured optical propagation loss was 1.41 dB/cm. This number is the same as that of pure polyimide. This suggests that particle size of TiO_2 is small enough to avoid scattering loss, and no absorption loss is created by the dispersion of TiO_2 particles. This is because TiO_2 is dispersed in the solid state in the present approach, so the formation of various kinds of absorbent species are avoided, which is prone to happen for solution state impregnation. The optical propagation loss of the sample prepared by simple mixing of the TIP solution and the polyimide solution was also evaluated; however, no visible streak line was observed because of too high absorption loss.

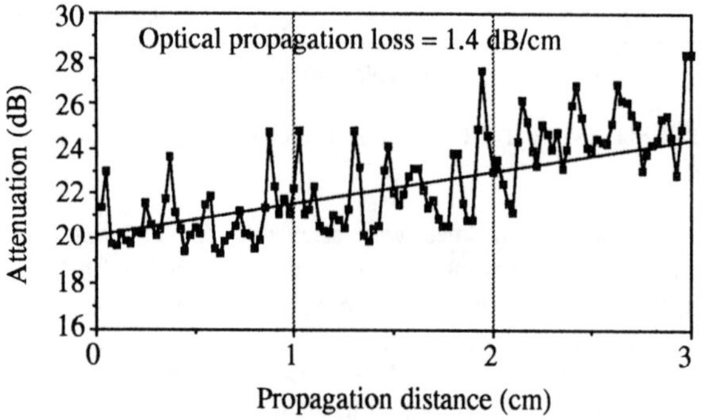

Figure 7. Optical intensity decay along the streak line in TiO_2 nano particles dispersed polyimide waveguide.

The refractive index of the TiO_2 particle dispersed polyimide film was also calculated to be 1.560, which is 0.010 higher than that of pure polyimide (1.550). The refractive index of amorphous TiO_2 is around 2.2, so the expected refractive index of 4 wt % TiO_2 in polyimide should be 1.576. The lower measured value may be due to a portion of TiO_2 particles being lost during the process, which would lead to the real concentration lower than 4 wt %.

CONCLUSIONS

Multifunctional nanostructured composites for photonics applications have been accomplished through sol-gel processing and reverse micelle approach. We were able to incorporate different optically active materials in sol-gel glass in different phases, with phase separation in nanometer scale. A composite glass doped with C_{60} and an organic dye exhibited efficient optical power limiting at different wavelengths. Two different dyes doped in two different phases of a composite glass showed broad band lasing in the gain region of both the dyes. Two-photon pumped, solid state cavity lasing was also achived in a sol-gel processed nanocomposite glass. Ultra fine TiO_2 particles synthesized by the reverse micelle technique were dispersed in polyimide thin films for waveguide applications. This approach for making inorganic:organic composite thin film could be useful in manipulating the refractive index of a thin film and achieving better optical quality and machanical stability.

ACKNOWLEDGMENTS

This work was supported by the Air Force Office of Scientific Research and the Polymer Branch of Wright Laboratory.

REFERENCES

1. R. Burzynski and P. N. Prasad, *Photonics and Nonlinear Optics with Sol-Gel Processed Inorganic Glass:Organic Polymer Composite*, Ed. by L. C. Klein, Kluwer Academic, Boston, 1994, Chapter 19.
2. P. N. Prasad, F. V. Bright, U. Narang, R. Wang, R. A. Dunbar, J. D. Jordan and R. Gvishi, *Hybrid Organic-Inorganic Composites*, Ed. by J. E. Mark, C.Y-C. Lee and P. A. Bianoni, ACS symposium series 585, Washington, 1995,Chapter 25
3. P. N. Prasad and B. A. Rhinhardt, *Chem. Mat.* **2**, 660 (1990).
4. D. M. Burland, R. D. Miler, and C. A. Walsh, *Chem. Rev.* (1993).
5. H. L. Wang and L. Gample, *Optics Comm.* **18,** 4 (1976).
6. D. Avnir, D. Levy and R. Reisfeld, *J. Phys. Chem.* **88,** 5956 (1984).
7. F. Salin, G. Le Saux, P. Georges, A. Brun, C. Bagnall and J. Zarzycki, *Optics Lett.* **14,** 785 (1989).
8. J. M. Mckiernan, S. A. Yamanaka, B. Dunn and J. I. Zink, *J. Phys. Chem.* **94,** 5652 (1990).
9. S. Braun, S. Rappoport, R. Zusman, D. Avnir and M. Ottolenghi, *Mater. Lett.* **10,** 1 (1990).
10. U. Narang, F. V. Bright and P. N. Prasad, *Appl. Spectrosc.* **47**, 229 (1993).
11. R. Gvishi and R. Reisfeld, *J. of Non-Crystalline Solids* **128**, 69 (1991).
12. X. Li, T. A. King and F. Pallikari-Viras, *J. Non-Cryst. Solids* **170**, 243 (1994).
13. R. Gvishi, R. Reisfeld and Z. Burshtein, *J. Sol-Gel Sci. Tech.* **4**, 49 (1995).
14. G. S. He, R. Gvishi, P. N. Prasad, B. A. Reinhardt, J. C. Bhatt and A. G. Dillard, *Optics Comm.,* In press, (1995).
15. G. S. He, R. Gvishi, P. N. Prasad, B. A. Reinhardt, J. C. Bhatt and A. G. Dillard, *CLEO 95 Conference,* (1995).
16. R. Gvishi, P. N. Prasad, B. A. Reinhardt and J. C Bhatt, *JSST special issue on "Sol-Gel preparation of Nonlinear Optical Material",* Submitted for publication (May 1995).
17. R. Gvishi, J. Bhalwakar, N. D. Kumar, G. Ruland, U. Narang and P. N. Prasad, *Chem. of Mat.* 7, 2199 (1995).

18. P. N. Prasad, R. Gvishi, G. Ruland, D. N. Kumar, J. Bhalwakar and U. Narang, *SPIE Proc.* **2530**, 128 (1995).
19. G. Ruland, R. Gvishi, and P. N. Prasad, *J. Am. Chem. Soc.*, In Press, (1996).
20. C. F. Zhao, G. S. He, J. D. Bhawalkar, C. K. Park and P. N. Prasad, *Chem. Mat.* In press (August 1995).
21. Gvishi, R. *Ph.D. Thesis, The Hebrew University of Jerusalem, Jrusalem, Israel*,(1993).
22. G. Phillip, H. Schmidt, J. Non-Cryst. Solids **283,** 63 (1984).
23. C. Li, J. Y. Tseng, K. Morita, C. Lecher, Y. Hu, J. Mackenzie, Proc. SPIE **1758,**410 (1992).
24. M. Yoshida, P. N. Prasad, Chem. Mater. (1995).
25. Y. Imai, J. Macromol. Sci. **A28,** 1115 (1991).
26. M. Nandi, J. A. Conklin, S. Lawrence, A. Sen, Chem. Mater. **3,** 201 (1991).
27. A. Kioul, L. J. Mascia, Non-Cryst. Solids **175,** 169 (1994).
28. V. Swayambunathan, D. Hayes, K. H. Schmidt, Y. X. Liao, D. Meisel, J. Am. Chem. Soc. **112,** 3831 (1990).
29. Y. Wang, A. Suna, J. McHugh, J. Chem. Phys. **92,** 6927 (1990).
30. Y. Wang, W. Mahler, Optics Comm. **61,** 233 (1987).
31. I. Lisiecki, M. Bjorling, L. Motte, B.Ninham, M. P. Pileni, Langmuir **11,** 2385 (1995).
32. A. R. Kortan, R. Hull, R. L. Opila, M. G. Bawendi, M. L. Steigerwald, P. J. Carroll, L. E. Brus, J. Am. Chem. Soc. **112,** 1327 (1990).
33. G. J. M. Koper, W. F. C. Sager, J. Smeets, D. Bedeaux, J. Phys. Chem. **99,** 13291 (1995).
34. C. Petit, P. Lixon, M. P. Pileni, J. Phys. Chem. **94,** 1598 (1990).
35. E. J. A. Pope, M. Asami and J. D. Mackenzie, *J. Mater. Res.* **4**, 1018 (1989).
36. L. L. Hench and J. L. Nogues, *Sol-Gel Optics: Processing and Applications*, Ed. by L. C. Klein, Kluwer Academic, Boston, 1993, Chapter 3.
37. L. C. Klein, *Sol-Gel Optics: Processing and Applications,* Ed. by L. C. Klein Kluwer Academic, Boston, 1993, Chapter 10.
38. R. Gvishi, G. Ruland and P. N. Prasad, *Opt. Comm.*, In Press, (1996).
39. L.W.Tutt and T.F.Bogges, Prog. Quant.Electr., **17**, 299 (1993).
40. J. Catalán and J Elguero, *J. Am. Chem. Soc.* **115**, 9249 (1993).
41. M. R. Eftink, *Topics in Fluorescence Spectroscopy Vol. 2*, Ed. by J. R. Lakowicz, Plenum Press, NewYork, 1991 Ch. 2.
42. K. H. Drexhage, *Dye Lasers*, Ed. by F. P. Schäfer Springer-Verlag, Berlin, 1990, Chapter 5.

PERFLUOROARYL SUBSTITUTED INORGANIC-ORGANIC HYBRID MATERIALS

C. ROSCHER and M. POPALL#
Fraunhofer-Institut für Silicatforschung, Neunerplatz 2, D-97082 Würzburg, Germany

ABSTRACT

A new class of perfluoroaryl substituted inorganic-organic copolymers (ORMOCER*s) for integrated optics have been synthesized via hydrolysis and condensation (sol-gel processing) of novel perfluoroarylalkoxysilanes followed by organic crosslinking reactions. Curing leads to transparent coatings with low NIR optical loss factors. Patterned layers for optical interconnection were fabricated by photolithographic procedures.

INTRODUCTION

For optical applications two types of materials have gained importance: glasses and optical transparent polymers. Although inorganic glasses remain unsurpassed as optical transmission media, they are not universally suitable for the fabrication of optoelectronic devices due to their low flexibility, high brittleness and weight [1]. Polymer based optical materials like PMMA or polystyrene (PS) also have several significant disadvantages such as low heat-resistance (PMMA 80°C, PS 70°C) and relatively poor adhesion between both components and the substrate material. Several applications in optical data communication and processing demand optical materials with high flexibility, heat-resistance and low processing temperatures.

A solution for these contrasting problems may be provided by silicate-based inorganic-organic copolymers (ORMOCERs) [2]. In contrast to the well known silicones, ORMOCERs allow better tailoring towards the intended purpose of the materials and the demands of the application technology due to their flexible design [2].

ORMOCERs are hybrid inorganic-organic materials [3] composed of inorganic oxidic structures cross-linked or substituted by organic groups. They are prepared from organosilane precursors by sol-gel-processing [4] in combination with organic crosslinking of polymerizable organic functions. As a result of these functionalities the properties of the ORMOCERs can be adjusted to particular applications [2,5]. Systematic variation of composition combined with adaption to micro system technology allows great flexibility in processing. In addition to the properties resulting from the network, special functional organic groups (network-modifiers) effect additional material properties.

For applications where ORMOCERs are to be employed as optical wave guides (buffer, core or cladding) within specific data communication devices such as optical switches, distributors and amplifiers it is necessary to develop materials with a minimum of light transmission loss in the NIR-region (1300-1550 nm), where high output power optical sources like laser diodes are available. Considering the fact that CH vibrational absorption is a major loss factor in this region [1], this entails minimizing the number of carbon-hydrogen bonds per cubic centimeter. One possibility to achieve this goal is to replace the carbon-hydrogen bonds by carbon-fluorine bonds, a practice employed in conventional polymer optical materials [1]. The most efficient method with regards to the present study is to use fluorinated organosilane

* registered trademark of Fraunhofer-Gesellschaft in Germany

Author to whom correspondence should be adressed

Mat. Res. Soc. Symp. Proc. Vol. 435 © 1996 Materials Research Society

precursors in the initial step of the ORMOCER-synthesis. All commercially available fluorinated silanes to date are long-chained polyfluorinated alkylsilanes which are unsuitable in the ORMOCER-synthesis because of their very low polarity. Moreover, the low refractive index of polyfluoroalkyl-derivatives presents a problem as the majority of applications require an index greater than 1.45. However, perfluoroarylsilanes are expected to have appropriate refractive indices. In this paper the preparation of novel fluorinated alkoxysilanes and the results concerning the optical loss of fluorinated ORMOCERs based on these silanes are described.

GENERAL ASPECTS

A structural scheme of the novel perfluoroaryl silane type is shown in Fig. 1. The alkoxy silyl groups allow the formation of an inorganic Si-O-Si-network by hydrolysis and condensation reactions. The attached perfluoroaryl group provides the means by which network modification or thermally or photochemically induced polymerization to build up an additional organic network may be achieved, depending on the functional group in para position of the phenyl ring.

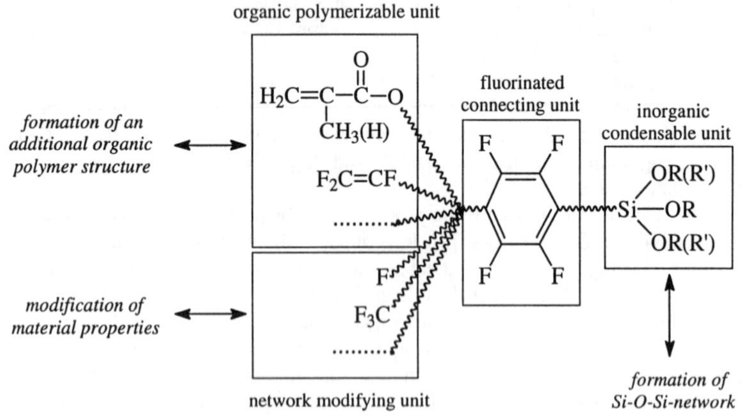

Fig. 1. General formula of the novel perfluoroaryl silanes

EXPERIMENTAL

^1H-, ^{19}F- and ^{13}C-NMR spectra were obtained in $CDCl_3$ as solvent and internal standard using a Bruker WM 400 spectrometer. Vibrational spectra were recorded using a FTIR-spectrometer (Bio-Rad FTS-25) and an UV-VIS-NIR-spectrometer (Shimadzu UV 3100).

Synthesis of the fluorinated precursors was carried out by modification of a method developed by Wittingham and Jarvie [6]:

24.7 g (100 mmol) pentafluorobromobenzene, 2.65 g (110 mmol) magnesium turnings, 41.7 g (200 mmol) freshly distilled tetraethoxysilane and a few crystals of iodine were mixed together at room temperature and diethyl ether added dropwise to the vigorously stirred mixture until an exothermic reaction was observed (approx. 30 ml). After stirring and refluxing for 16 h the mixture was cooled to room temperature. An excess of n-heptane was added to precipitate the magnesium salts. Filtration gave a clear pale yellow solution. Solvents were removed and the residue fractionally distilled under reduced pressure to yield 11.6 g (35 %) pentafluorophenyltriethoxysilane, b.p. 67-70°C/0.2 mbar and 4.5 g (20%) bispentafluorophenyldiethoxysilane, b.p. 99-101°C/0.2 mbar.

3.92 g (10 mmol) p-bromoperfluorostyrene [8], 0.53 g (22 mmol) magnesium turnings, 8.33 g (40 mmol) freshly distilled tetraethoxysilane, 1.87 g (10 mmol) 1,2-Dibromoethane and 10 ml diethyl ether were mixed together and brought to reflux conditions until an exothermic reaction was observed (approx. 15 min). The stirring mixture maintained under those conditions for 16 h and subsequently cooled to room temperature. An excess of n-heptane was added to precipitate the magnesium salts. Filtration gave a clear pale yellow solution. Solvents were removed and the residue fractionally distilled under reduced pressure to yield 11.6 g (40 %) p-triethoxysilyl-perfluorostyrene, b.p. 74-75°C/0.2 mbar.

As a starting point for developing the low loss system, an established and well characterized four component system [9] was chosen with the single modification of a perfluorophenylsilane in exchange for the phenylalkoxysilane constituent. The four components were γ-glycidyloxypropyltrimethoxysilane (GLYMO), γ-methacryloxypropyltrimethoxysilane (MEMO), bispentafluorophenyldiethoxysilane (P_F2) and tetraethoxysilane (TEOS). This composition was found to be suitable for thermal and/or photochemical curing in order to form the organic network.

The preparation of pentafluorophenyl substituted ORMOCER-lacquer was carried out by sol-gel processing: In the initial step the four components were mixed together without any requisite solvent; subsequent addition of the stoichometric amount of water (1.5 mol relative to Si [7]) facilitated the construction of an inorganic network by hydrolysis/condensation reactions.

For final curing a photoinitiator (Irgacure 184, Ciba-Geigy) was added and the materials were applied onto substrates (glass or silicon wafer) by spin coating. Photo curing was carried out by UV-radiation (1000 W high pressure mercury lamp). Thermal curing took place at T > 120°C.

RESULTS AND DISCUSSION

Precursors

The perfluoroaryl substituted alkoxysilanes can be synthesized by a Barbier-Grignard reaction of bromoperfluoroaryl-derivatives with tetraethoxysilane. A general reaction scheme is shown in Fig. 2.

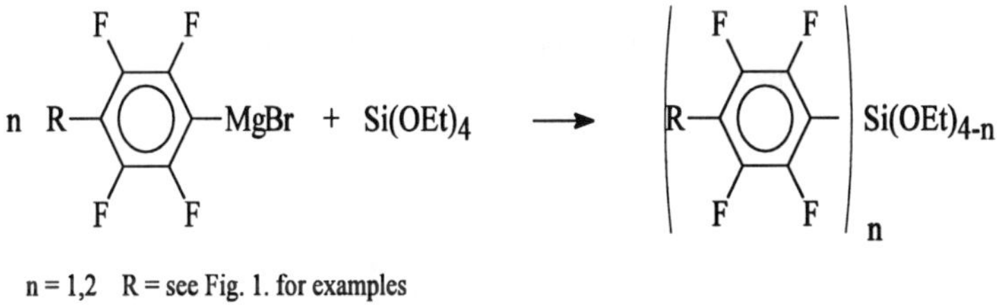

Fig. 2. Reaction scheme for the precursor synthesis

For initial studies the most convenient compounds, pentafluorophenyltriethoxysilane and bispentafluorophenyldiethoxysilane, were chosen. p-Triethoxysilyl-perfluorostyrene was synthesized as a potential network-former. NMR- and IR-data are given in Table I.

Table I. IR- and NMR-data of fluorinated silanes

Pentafluorophenyltriethoxysilane	
IR-data:	ν = 2981, 1644, 1518, 1468 1387, 1292, 1169, 1092, 973, 792, 688 cm^{-1}
^{1}H-NMR-data:	δ 3.94 (q, 6H, J = 7.0 Hz, $OC\underline{H_2}CH_3$), 1.27 (t, 9H, J = 7.0 Hz, $OCH_2C\underline{H_3}$)
^{13}C-NMR-data:	δ 149.7, 143.1, 137.6, 105.0, 59.5, 18.1
^{19}F-NMR-data:	δ -127.2, -151.0, -161.9
Bispentafluorophenyldiethoxysilane	
IR-data:	ν = 2983, 1644, 1519, 1469, 1387, 1293, 1167, 1094, 974, 794 cm^{-1}
^{1}H-NMR-data:	δ 4.01 (q, 4H, J = 7.0 Hz, $OC\underline{H_2}CH_3$), 1.30 (t, 6H, J = 7.0 Hz, $OCH_2C\underline{H_3}$)
^{13}C-NMR-data:	δ 149.5, 143.6, 137.6, 105.8, 60.5, 17.9
^{19}F-NMR-data:	δ -128.1, -149.0, -161.0
p-Triethoxysilyl-perfluorostyrene	
IR-data:	ν = 2981, 1785, 1454, 1394, 1308, 1208, 1169, 1107, 1087, 986, 901, 794, 687 cm^{-1}
^{1}H-NMR-data:	δ 3.96 (q, 6H, J = 7.0 Hz, $OC\underline{H_2}CH_3$), 1.28 (t, 9H, J = 7.0 Hz, $OCH_2C\underline{H_3}$)
^{13}C-NMR-data:	δ 154.1, 149.3, 144.1, 118.0, 114.3, 109.3, 59.6, 18.1
^{19}F-NMR-data:	δ -96.3, -111.9, -127.0, -138.1, -171.9

Fluorinated ORMOCERs [10]

In order to synthesize the inorganic-organic backbone, the alkoxysilanes were hydrolyzed and polycondensed by addition of water. Water consumption and therefore the progress of the reaction step was monitored by Karl-Fischer-titration. During the course of the reaction, the ethoxy IR-absorption at 2983 cm^{-1} decreased. IR-spectroscopy also provided evidence that the sensitive Si-C-bond was not cleaved significantly under the sol-gel conditions employed. There was no decrease in intensity of the pentafluorophenyl absorption at 1644 cm^{-1}, which would be expected under conditions where bond cleavage occurs. The reactivity of the pentafluorophenyl-ethoxysilanes within the sol gel process is insignificantly greater than that exhibited by their unfluorinated analogons, yet lower than the unfluorinated phenylmethoxy-silanes. This can be explained by the fact that the electron-withdrawing effect of the pentafluorophenyl group is rather low [11].

The obtained lacquers are homogeneous and can be used as coating materials with good wetting and adhesion properties e.g. on PMMA, glass or silicon wafer. The final organic polymerization can be induced by a photochemical or thermal process after addition of an UV- or thermal initiator.

The resulting layers are transparent and optically homogeneous. The DTA trace shows no endothermic events up to the range of initial decomposition (>200°C), indicating that there is no phase transition. This behaviour is consistent with a material of amorphous nature, which is also supported by X-ray diffraction analysis.

Optical properties

The NIR spectra of a novel perfluoroaryl substituted ORMOCER-system as lacquer (containing ethanol and methanol from sol gel processing, in 1 mm cuvette), resin (after evaporation of solvents, in 1 mm cuvette) and finally cured coating (thickness 1 mm) are shown in Fig. 3. As indicated earlier, a prerequisite for low optical losses in the NIR-range is a reduction of CH-bonds, which generally effects absorptions at 1310 nm, in addition to reduction of remaining SiOH- and SiOR-bonds, which are concerned with the absorption at 1550 nm. The dramatic decrease in absorption intensity upon reduction of the above mentioned groups is shown in Fig. 3.

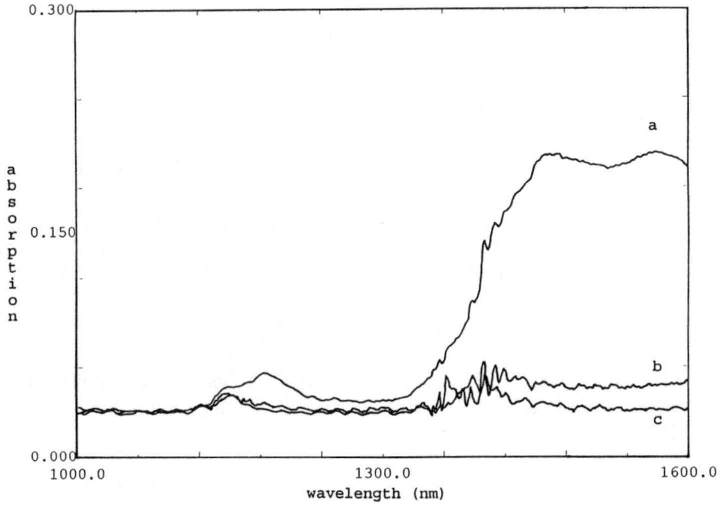

Fig.3. NIR spectra of an ORMOCER-system, containing 50 mol% P_F2 as lacquer (a), resin (b) and finally cured layer (c)

In optical loss and refractive index measurements on layers and waveguides losses of <0.4 dB/cm at 1310 nm, <0.8 dB/cm at 1550 nm and indices of >1.49 at 633 nm were found.

CONCLUSIONS

Synthesis of novel perfluoroarylalkoxysilanes has enabled the development of amorphous hybrid inorganic-organic materials with low optical loss in the NIR-range. The network exhibits better thermal stability in comparison to the majority of pure organic polymers. The material is highly adaptable and ideal for efficient device and component production, due to the possibility of either photo-initiated or thermal crosslinking.

ACKNOWLEDGEMENTS

We wish to thank the Deutsche Forschungsgemeinschaft for its financial supports. We also like to acknowledge the work of our technical staff members and thank P. Dannberg, Fraunhofer-Einrichtung für angewandte Optik und Feinmechanik, Jena, Germany, for optical measurements.

REFERENCES

[1] T. Kaino in Polymers for Lightwaves and Integrated Optics, edited by L. A. Hornak (Marcel Dekker, Inc., New York, 1992), pp. 1-38.

[2] M. Popall, J. Kappel, J. Schulz, H. Wolter in Micro System Technologies '94, edited by H. Reichl, A. Heuberger (VDE-Verlag, Berlin, 1994), pp. 271-280.

[3] Review articles: (a) U. Schubert, N. Hüsing, A. Lorenz, Chem. Mater. **7**, 2010 (1995). (b) C. Sanchez, F. Ribot, New J. Chem. **18**, 1007 (1994). (c) B. M. Novak, Adv. Mater. **5**, 422 (1993).

[4] (a) L. L. Hench, J. K. West, Chem. Rev. **90**, 33 (1990). (b) C.J. Brinker, G. Scherer, Sol-Gel-Science, (Academic Press, London, 1990).

[5] (a) M. Popall, X.-M. Du, Electrochim. Acta **40**, 2305 (1995). (b) H. Wolter, W. Storch, H. Ott, in Better Ceramics Through Chemistry VI, edited by A. K. Chestham, C. J. Brinker, M. L. Mecartney, C. Sanchez (Mater. Res. Soc. Proc. **346**, Pittsburgh, PA, 1994) pp. 143-149. (c) M. Popall, J. Kappel, M. Pilz, J. Schulz, G. Feyder, J. Sol-Gel Sci. Technol. **2**, 157 (1994). (d) J. Kron, S. Amberg-Schwab, G. Schottner, J. Sol-Gel Sci.Technol. **2**, 189 (1994).

[6] A. Wittingham, A. W. Jarvie, J. Organometal. Chem. **13**, 125 (1968).

[7] S. Amberg-Schwab, E. Arpac, W. Glaubitt, K. Rose, G. Schottner, U. Schubert, in High Performance Ceramic Films and Coatings, edited by P. Vincencini (Elsevier Sci., Amsterdam, 1991), pp. 203-210.

[8] E. J. Soloski, W. E. Ward, C. Tamborski, J. Fluorine Chem. **2**,361 (1972/73).

[9] M. Popall, H. Meyer, H. Schmidt, J. Schulz, in Better Ceramics Through Chemistry IV, edited by C. J. Brinker, D. E. Clark, D. R. Ulrich, B. J. J. Zelinsky (Mater. Res. Soc. Proc. **180**, Pittsburgh, PA, 1990) pp. 995-1001.

[10] C. Roscher, M. Popall, to be published.

[11] W. A. Sheppard, J. Am. Chem. Soc. **92**, 5419 (1970).

Fabrication of GRIN-Materials by Photopolymerization of Diffusion-Controlled Organic-Inorganic Nanocomposite Materials

P. W. Oliveira, H. Krug, P. Müller and H. Schmidt
Institut für Neue Materialien, Composite Technology Group, Im Stadtwald 43, 66123 Saarbrücken, Germany.

ABSTRACT

New photopolymers have contributed significantly to the recent growth of holographic and lithographic applications. Photopolymerizable organic-inorganic hybrid materials, based on methacrylate functionalized silane and zirconia particles as holographic recording material, are presented. Thick films of this composite system were prepared and volume diffractive gratings were fabricated by a two laser beam interference technique. The formation of the gratings is based on the diffusion of high refractive index components (ZrO_2-nanoparticles) to areas with high irradiation intensity with subsequent immobilization by full irradiation of the film. The influence of the zirconia particles as the main component for obtaining highly efficient gratings is presented and the correlation between particle concentration and refractive index profile is shown.

INTRODUCTION

The sol-gel process allows the production of ceramic colloidal particles in the presence of organoalkoxysilanes, bearing various functions. The synthesis of multifunctional transparent inorganic-organic composites is possible, if the particle size can be kept in the lower nanometer range. Composites with a high refractive index generally require the incorporation of inorganic components such as ZrO_2 or TiO_2 as inorganic network formers [1]. A material system based on 3-methacryloxypropyl trimethoxysilane (MPTS) and zirconium propylate (Zr), chelated by methacrylic acid (MA) has shown to be suitable for index matching, by varying the Zr/MA-concentration [2]. As reported earlier, these materials have been developed for optical applications like gratings, waveguid [3,4] and microlenses [5]. The organic polymers can be photopolymerized and micropatterned using mask techniques. Those patterns can then be fixed using a subsequent development step. Without this development step, the refractive index difference between the polymerized and unpolymerized zones reduces during a full irradiation of the micropatterned area with a UV lamp, as already known from the study of organic polymers.

The question arises how far double bonds attached to inorganic high refractive index nanoparticles in a low viscosity polymerizable matrix are able to migrate into the irradiated area. In this case, after full area irradiation, differences in the refractive index n between preirradiated areas and the areas being exposed only to the full area treatment should exist. This would lead to a permanent gradient information in the films (holographic process).

In this paper, the possibility of fabricating high efficient gratings using this process was investigated. First evidence for the existence of such a diffusive process in case of the MPTS/Zr/MA system is described in [5]. A fundamental understanding of diffusion behaviour is necessary for the prediction of index profiles, in order to produce micro optical components with well defined optical properties [6,7].

Mat. Res. Soc. Symp. Proc. Vol. 435

EXPERIMENTAL

The inorganic-organic nanocomposite is based on 3-methacryloxypropyl trimethoxysilane (MPTS), and zircononium-n-propylate complexed with methacrylic acid (Zr/MA). The synthesis process is described elsewhere [1]. The compositions of the materials employed for holographic experiments were varied according to their mole ratios (MPTS/ZrMA) as follows 10:0:0, 10:1:1, 10:2:2, 10:4:4 and 10:10:10. Irgacure 369 was employed as a photoinitiator (0.4 mole %/C=C). Butanol was taken as a solvent to adjust the viscosity of the sols to the same value for all compositions. Films of 10 μm thickness were prepared between two glass plates as substrates [6]. The distance between the substrates was set using glass fibers of 10 μm thickness, and the sol was filled between the substrates by capillary forces.

The two-wave mixing experiment was carried out with a laser beam of wavelength 351 nm (from an Ar^+-laser), divided into two beams of exactly the same power by a beam splitter and directional mirrors. These two writing beams interact producing an interference wavefront with a periodic change of the irradiated intensity within the sample. A detailed description of the experimental set up is given in [3]. For the given angle of 2 degrees between the two writing beams, the energy density of 0.8 W/cm^2 is sinusoidal spatially modulated with a period of 10 μm. The width of the interference region is 1.5 mm. Concentration profiles of Zr/MA were detected by micro-Raman spectroscopy and energy dispersive X-ray analysis (EDX) in a scanning electron microscope (SEM).

RESULTS AND DISCUSSION

The illumination of the sample with an interference pattern produces a periodical variation of the optical thickness (refractive index -**n** multiplied to the film thickness -**d**). This leads to a diffraction effect, when the sample is illuminated with coherent light from a probe laser (He-Ne-laser 632,8 nm), and the variation of the optical thickness can be monitored in real time by following the diffraction efficiency as a function of illumination time. To complete the hologram-storage-process, the patterned region was illuminated with an UV light [5]. The preparation of the film between the two glass substrates (see experimental section) eliminates changes in thickness due to shrinkage and only changes in the index of refraction can be detected.

The grating formation was followed in real time with the probe laser. In figure 1 a the measured diffraction efficiency in dependence of time and Zr/MA concentration during the writing process is shown. For all concentrations, the illumination by the interference pattern was completed after 20 minutes and the behaviour of the diffraction efficiency was followed for an additionally 15 minutes, using only the reading beam of the experimental set-up during the fixing process (fig.1 b)

From figure 1a it is obvious that a maximum saturation of the diffraction efficiency exists for all MPTS/Zr/MA concentrations, which decreases with increasing Zr/MA concentration. For the pure prehydrolysed MPTS system, which contains no nanoparticles, the saturation diffraction is lower in comparison to the particulate systems. It is also obvious, that for this pure MPTS system with fastest kinetics of grating formation, the curve shows no irregularities and represents a one step process. A different behaviour can be observed in the presence of nanoparticles.

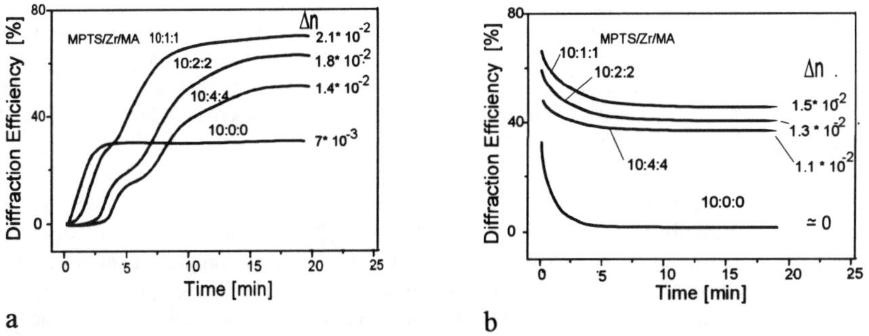

Fig 1 Real time diffraction efficiency measurement and Δn values of different nancomposite systems, a) Writing process b) fixing process. The Δn values are calculated by coupled wave theory (see text).

A two step process can be observed which is evidenced by a point of inflection in the diffraction efficiency curves. The position of each point of inflection is shifted to longer illumination times with increasing Zr/MA concentration. Absorption effects can be neglegted, as there is no interaction between the reading beam at 633 nm and the polymerizing systems.

The illumination of the sample with an interference pattern induces a polymerization of the regions corresponding to the bright fringes. This photochemical reaction produces a conversion of the monomer to polymer and additional monomer molecules containing zirconia diffuse to these regions from non illuminated areas. Due to the progressive formation of the polymer network, large molecules are prevented from migrating into the recording layer, hence a one-way diffusion process occurs, which only affects the unreacted monomers. This effect has as a consequence the increasing of the optical thickness of the polymerizing regions. This goes along with an increase in concentration of the monomer-derived polymer and a decrease in binder concentration in the irradiated regions

The hypothesis of a two step process, characterized by two different kinetics, was checked by calculating the photosensitivity ε. The kinetics of grating formation are directly correlated to the sensitivity, ε, of a photosensitive material. ε is defined as the slope of the linear region of the diffraction efficiency curve plotted against the incident light energy. The calculated sensitivities for the different compositions, shown in figure 2 are summarized in table I.The results indicate that the addition of Zr/MA reduces the sensitivity and hence the grating formation kinetics of the material. But there are differences in the sensitivity before and after the inclination points. Therefore a two step process can be postulated and is attributed in a first hypothesis to the MPTS and Zr/MA species. The sencitivity reduction process due the which has to be confirmed in future.

Table I Calculated holographic sensitivity ε in dependence of Zr/MA concentration, ε_1 before inclination point and ε_2 after the inclination point in curves shown in fig. 1a.

MPTS/ZR/MA	10:0:0	10:1:1	10:2:2	10:4:4
ε_1 [J/cm^2]	89.5	11.,8	7.5	1.5
ε_2 [J/cm^2]		10.22	7.12	1.43

The reduction of the sensitivity ε_1 and ε_2 during the grating writing process can be explained through the grating formation kinetics processes, photopolymerization and diffusion. In the first case, the increasing of the Zr/MA concentration in the MPTS/Zr/MA increases the light absorption of the sol and less UV-light energy will be used to induce a photochemical process. Whereas in the Second case, the increasing of Zr/MA concentration in the MPTS/Zr/MA leads to a increase of the Zr-particle radius, which decreases the mobility and consequently the diffusion of the Zr-particles.

The Δn values indicated in figure 1 for each particle concentration were obtained by the so called coupled wave theory. The refractive index modulation Δn is calculated by the diffraction theory described by Kogelnik [6]. As predicted by the coupled wave theory, the diffraction efficiency of gratings produced by two wave mixing techniques, exhibits a $\sin^2$ behaviour as a function of the incident angle of the reading beam. The holographic information can be reconstructed only under a narrow viewing angle on both sides of the Bragg incidence. The diffraction efficiency of such holographic elements depends on the following relationship:

$$\eta = \sin\left(\left(v^2 + \xi^2\right)^{\frac{1}{2}}\right)^2 \left(1 + \frac{\xi^2}{v^2}\right)^{-1} \text{ with } v = \frac{\pi \Delta n d}{\lambda \cos\psi} \qquad \text{and } \xi = \frac{\Delta\psi \pi d}{\Lambda}$$

(η: the diffraction efficiency, d: the film thickness, Δn: the variation of refractive index, ψ: the Bragg angle, $\Delta\psi$: difference between the Bragg incidence and the playback angle in the recording medium, Λ: grating periodicity and λ: wavelength of the probe laser). A typical angular response of the diffraction efficiency measurement of the 1st diffracted order of a grating stored in MPTS/Zr/MA is shown in figure 2

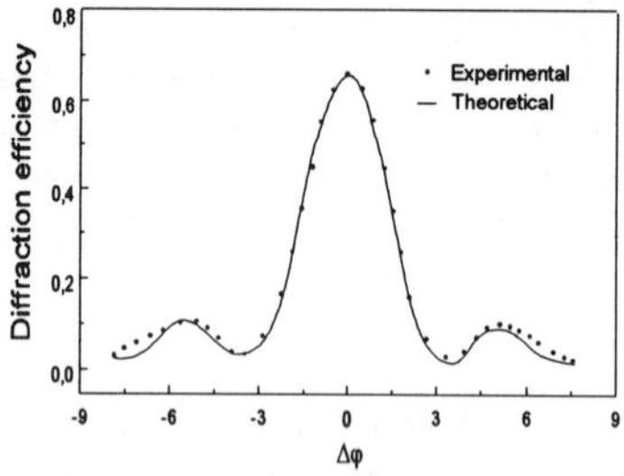

Fig.2. Diffraction efficiency in dependence of $\Delta\psi$, Λ=10 μm, d=10μm and λ=0.633 μm.

The calculated and measured curves exhibit a very good agreement. This is also the case for the calculated thickness of the film, which differs only 3% from the measured one. Unpolymerized C=C double bonds in the regions of lower intensities in the interference pattern are polymerized during the fixing process (figure. 1b). Diffraction efficiency drops to zero for the pure MPTS system, and the grating formed during the two wave mixing experiment is destroyed as a result of the homogeneous index of refraction over the whole illuminated area. For the particle-containing MPTS system, the diffraction efficiency does not fall to zero, which can only be explained by a fixed modulation of the index of refraction. The reduction of the index of refraction after the writing process up to the end of the fixing process for all Zr/MA concentrations can be explained by the conversion of C=C double bonds to C-C single bonds which reduces the mole refraction in the less illuminated areas during the writing process which are now postpolymerized during the fixing process. By this effect, the phase shift is lowered and the diffraction efficiency is decrease but does not drop to zero which can be only contributed by a concentration profile.

Raman spectra were recorded for different positions on the film, in order to elucidate the mechanism, which leads to a permanent modulation of the refractive index. In figure 3, the Raman spectra for the composite 10:2:2 is shown.

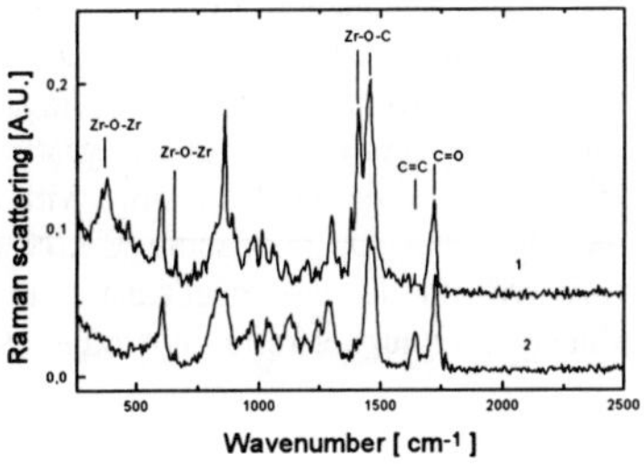

Fig.3 Raman spectra taken from irradiated (1) and unirradiated (2)

The Raman spectrum taken from irradiated (1) and unirradiated areas (2) shows that the signal intensity of the Zr-O-Zr bands (363 and 665 cm^{-1}) and Zr-O-C bands (1452 and 1576 cm^{-1}) increases in region 1. It can be conducted that the irradiated area (during the hologram storage) contains more zirconia than the non-illuminated areas. Furthermore, the C=C signal of region 2, shows that the conversion is not complete even after the fixing-process. To verify the results obtained by Raman-spectroscopy, films (10:2:2) and substrate were end-face polished and Zr-concentrations in the film plane was detected by energy dispersive X-ray (EDX) in a scanning electron microscope.

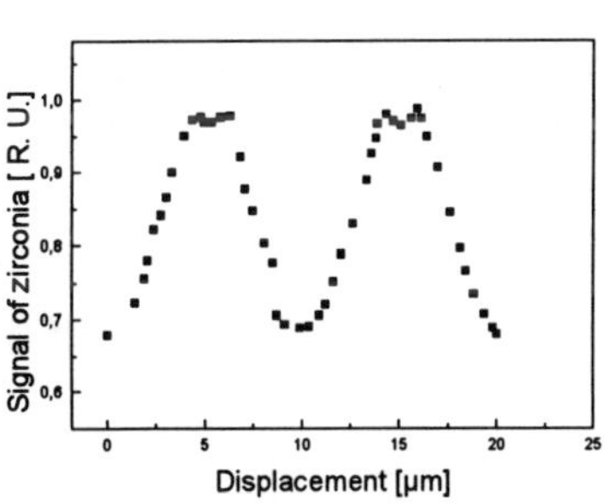

Fig.4: EDX signal of zirconia in a holographic micropatterned film of MPTS/Zr/MA (10:2:2)

From figure 4, it is evident that the maxima of the Zr-profile, as detected by EDX, coincides exactly with the grating period of 10 μm, which was set in the two wave mixing experiment. It follows that during the storage process an irreversible transport of the zirconia particles from the unirradiated to irradiated region is produced, which corresponds exactly with the refractive index modulation.

CONCLUSION

It could be conclusively demonstrated that, using organic-inorganic composite containing nanoscaled ceramic particles, a permanent gradient can be produced by applying a two laser beam interference technique. The gradient is caused by a modulation of the index of refraction, which is is derived from a concentration gradient of the complexed zirconia particles, as detected by micro-Raman-spectroscopy and EDX. The concentration gradient arises from diffusion of Zr/MA-nanoparticles into the (during the writing process) laser-illuminated sections of the film. The creation of phase gratings with a refractive index modulation up to $1.5*10^{-2}$ is obtainable by fixing this profile with a full area illumination step. This process has an enormous potential for producing GRIN (gradient refractive index)-materials. Future work will involve kinetic and mechanistic investigations of the diffusive process and the scope for producing even higher refractive index modulations.

REFERENCES

[1] R. Nass, H.Schmidt and E. Arpac Sol-Gel Optics I-SPIE, Vol 1328 ,258 (1990)

[2] H. Krug, F. Tiefensee, P.W. Oliveira and H. Schmidt, Sol-Gel optics II-SPIE, vol 1758, 448 (1992)

[3] H. Krug and H. Schmidt, New J. Chemistry, vol 18, no. 10, 1125 (1994)

[4] H.Schmidt, H. Krug, R. Kasemann, F. Tiefensee, SPIE vol 1590, 36 (1991)

[5] P.W. Oliveira, H. Krug, H. Künstle and H. Schmidt, Sol-Gel optics III-SPIE vol 2288, 554, (1994)

[6] J.P. Foussier , Radiation Curing in Polymer Science and Technology, edited by:
J.P. Foussier and J.F. Rabek (Elsevier Science Publishers, New York, 1993) pp 1-62.

[7] M. Yamane,"Graded Refractive index Materials via Sol-Gel Processing", in Chemical Processing of Advanced Materials. edited by Larry L. Hench and Jon K. West, pp 863-874 (1992)

DIELECTRIC PROPERTIES OF ORGANIC-INORGANIC HYBRIDS: PDMS-BASED SYSTEMS

G.Teowee*, K.C.McCarthy*, C.D.Baertlein*, J.M.Boulton*, S.Motakef**, T.J.Bukowski**, T.P.Alexander** and D.R.Uhlmann**
*Donnelly Corporation, 4545 East Fort Lowell, Tucson, AZ 85712
**Department of Materials Science and Engineering, University of Arizona, Tucson, AZ 85721.

ABSTRACT

Organic-inorganic hybrids, with tailorable properties via control of their chemistries, offer great potential for many optical, electrical and mechanical applications. PDMS-based materials have been fabricated, having low optical losses of < 0.15 dB/cm but the dielectric properties of these hybrids have rarely been explored or reported. In the present study, the dielectric properties of PDMS:SiO_2:TiO_2 films are explored as a function of composition and curing temperature using an impedance analyzer. Dielectric spectroscopy was also performed to investigate the dielectric relaxation and dispersion behaviors. Results indicate that ε_r at 1 MHz ranges from 3 to 5. Residual hydroxyl and alkoxy species in the films contribute to the overall polarizabilities especially at low frequencies (< 100 kHz).

INTRODUCTION

The class of materials known as inorganic-organic hybrids (or organically modified inorganic composites) yields an interesting range of multifunctional properties. A notable advantage is the possibility of obtaining thick crack-free coatings and bulk monoliths compared to polymeric alkoxide-derived gels, which suffer from cracking. These materials offer tailorable optical and electrical properties depending on the inorganic and organic functionalities chosen, which can be utilized in a wide range of applications. An important consideration impacting the material properties is the degree of homogeneity (or controlled heterogeneity) obtained in the gel network during processing.

Silanol-terminated poly(dimethylsiloxane) (PDMS) is an attractive candidate for the synthesis of hybrids due to the terminal Si-OH groups which can co-condense with alkoxides resulting in a covalently bonded network. The mechanical behavior of PDMS materials have been explored in detail[1-7]. The network structure of the PDMS-based hybrids depends on the chemistry employed, e.g. pH of reaction, water content and the molecular weight of PDMS[1-5]. Small angle X-ray scattering (SAXS) revealed that homogeneity is enhanced at high tetraethoxysilane (TEOS) contents and low pH. The presence of titanium alkoxides enhances the rates of condensation and network cross-linking[6,8-10].

The optical properties of PDMS-based hybrids have been studied, namely their refractive index and waveguiding loss[11]. The inorganic matrix consisted of SiO_2-TiO_2 or SiO_2-GeO_2. Optical losses as low as 0.15dB/cm can be obtained, illustrating the high degree of homogeneity possible in these systems. The major contribution to optical losses in the PDMS-based hybrids arise from surface roughness of the films and not from volume inhomogeneities[11].

Organic polymers are attractive in electronic packaging applications due to their low values of dielectric constant; however their mechanical durability, thermal properties and

Mat. Res. Soc. Symp. Proc. Vol. 435 © 1996 Materials Research Society

adhesion to selected substrates are inferior to those obtained with inorganic materials. The electrical behaviors for hybrids is not as well explored compared to their optical properties. Materials requirements of hybrids in electronic packaging applications include low dielectric constants, low dielectric losses (or dissipation factors), good thermal conductivities or diffusivities and low dielectric relaxation or dispersion. Schmidt et. al.[12,13] have reported dielectric constants in the range of 2 - 3 and a surface conductivity of $10^{16}\ \Omega$ for organically-modified coatings based on vinylmethyl-diphenylsilane[14].

EXPERIMENTAL PROCEDURES

Hybrids were synthesized from $Si\ (OEt)_4$ and $Ti\ (O^iPr)_4$ and silanol-terminated PDMS with a molecular weight of 400- 700. PDMS occupied 60, 75 and 80% by volume of the total network in the films. The alkoxides were added to the solution at Si to Ti molar ratios of 1 : 0.5, 1 : 1, 1 : 1.5 and 1 : 2. In synthesizing the solutions, TEOS and water (acidified to 0.15 M HCl) were refluxed in ethanol at a H_2O:TEOS molar ratio of 2:1 for 30 min. to hydrolyze partially the TEOS. The solution was then cooled, the polymer and titanium alkoxide were added, and stirring was carried out for 15 min., followed by concentration using rotary evaporation.

The substrates consisted of ITO coated glass substrates The precursor solutions were filtered using a syringe filter (0.1 μm) in a clean room to minimize particulate contamination. Tape was positioned on one corner of the substrates to provide a baseline for thickness measurement. The green films were dried under vacuum at 125°C - 225°C to consolidate the films. Pt top electrodes were sputtered via a shadow mask placed on top of the coatings to produce dots with diameters of 1.0 mm (area of $7.84 \times 10^{-3}\ cm^2$). Monolithic Pt-PDMS hybrid-ITO capacitors were completed by scraping portions of the films with a blade to yield the contacts.

The dielectric properties (dielectric constant and dissipation factor) were obtained using a Hewlett Packard 4192A Low Frequency Impedance Analyzer inside a shielded probe station connected via GPIB to a PC computer. The signal amplitude was 1V rms at 1MHz while for dispersion studies, the probing frequency ranged from 500Hz to 1 MHz.

RESULTS AND DISCUSSIONS

The dielectric constant of PDMS is typically low, ranging from 2.20 - 2.75[15], due to the low dipole moment of the monomer unit and absence of conjugated bonds. The dielectric constant of TEOS films cured at 100C under vacuum for 48 hr was reported previously as 13 [16]. Sol-gel derived TiO_2 films in this experiment were also found to exhibit dielectric constants of 13. At 100 - 175°C, both TiO_2 and SiO_2 are still not fully densified and may contain residual organics which lead to elevated dielectric constant. Thus the dielectric constant of the hybrid films is expected to lie in the range 2.2 to 13 based on the rule of a simple linear mixture, relatively insensitive to SiO_2/TiO_2 content at least for materials which contain significant amounts of residual OH and thereby exhibit increased dielectric constants. The dielectric constant and dissipation factor of the PDMS-based hybrids cured at 175°C for 2 days as a function of SiO_2/TiO_2 molar ratio including neat SiO_2 and TiO_2 are shown in Fig. 1. The value of dielectric constant was about 5.5 to 6.5 and the dissipation factor was fairly low, about 0.01, irrespective of the SiO_2/TiO_2 molar ratio.

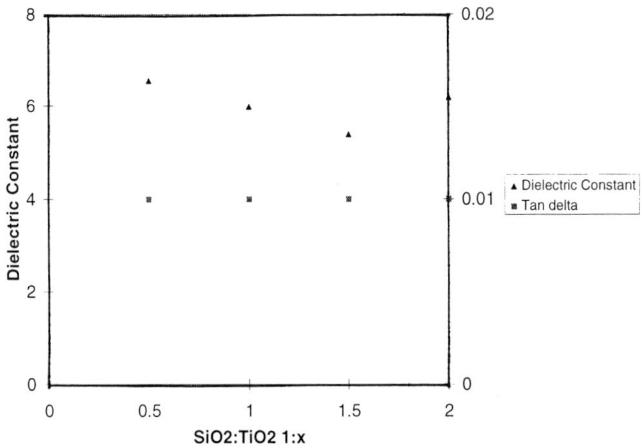

Fig.1 Dielectric constant and dissipation factor of PDMS-based films (80 vol. % PDMS) cured at 175°C for 3 days as a function of SiO_2/TiO_2 molar ratio

The effect of curing temperature on dielectric constant of PDMS-based hybrid coatings with a SiO_2/TiO_2 molar ratio of 1:1 is shown in Fig. 2. The dielectric constant decreases at a curing temperature of about 175°C, likely reflecting a decrease in M-OH and M-OR bonds by that temperature. Here M denotes a metal, e.g., Si-O-Ti.

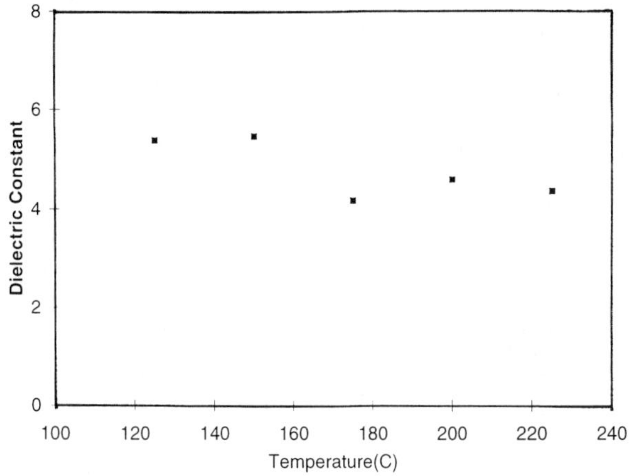

Fig. 2 Dielectric constant of PDMS-based hybrid coatings with 80 vol.% PDMS and SiO_2/TiO_2 molar ratio of 1:1 as a function of curing temperature for 48 hr.

To ascertain the effect of curing on the films, curing was performed at 175°C for various time periods ranging from 1 - 48 hours. The results are shown in Fig. 3.

The dielectric constant is < 2 for short curing times and increased with long curing times. This increase is likely indicative of enhanced densification and elimination of porosity. The low values of dielectric constant at short earlier curing times attests to the relatively porous network(air having dielectric constant of 1) in the incompletely condensed gel network. The presence of such pores resulted in the low dielectric constants.

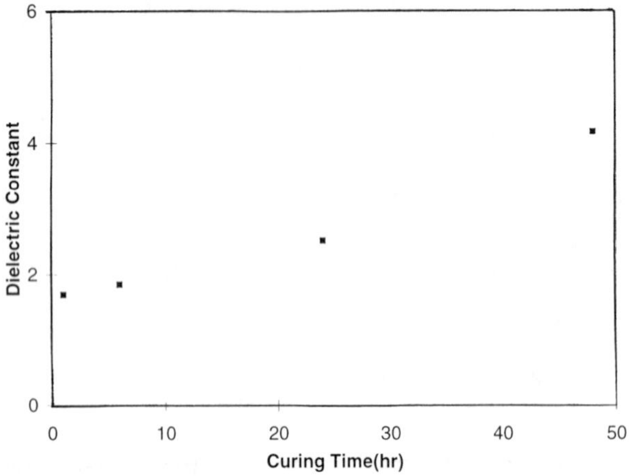

Fig. 3. Dielectric constant of PDMS-SiO_2-TiO_2 (with 80 vol. % PDMS and SiO_2:TiO_2 ratio of 1:1) films as a function of curing time at 175°C.

Table 1 shows the measured and theoretical values of dielectric constant of PDMS-based coatings(SiO_2:TiO_2 ratio of 1:0.5) as a function of polymer content where the calculated values are based on a linear rule of mixture. The dielectric constant of TEOS-based films fired to 100°C has been measured as 13[18] while that of PDMS is 2.20. The measured dielectric constants are in the range of, but lower than the calculated values. Since PDMS has a lower dielectric constant than SiO_2-TiO_2, incorporation of a higher amount of PDMS should yield films with lower dielectric constants. The presence of porosity in the films contributed to the lowering of the measured dielectric constants. Based on a linear mixture of the two end members, the lowest dielectric constant achievable in **fully dense** PDMS-based films is no more than 2.20.

Table 1. Measured dielectric constant and theoretical values in PDMS-based hybrids (SiO_2/TiO_2 of 1:0.5 at 200°C for 4 hr).

Film	Measured	Expected
60% PDMS	5.63	6.52
75% PDMS	4.90	4.99
80% PDMS	3.95	4.36

To explore the nature of dielectric relaxation in the hybrid coatings, the capacitance and dissipation factor were measured across a wide range of frequencies. Fig. 4 shows the effect of curing a hybrid PDMS-SiO_2-TiO_2 film with 80% v/v PDMS and SiO_2/TiO_2 ratio of 1:0.5.

There is a strong dispersion of dissipation factor in the lower frequencies(< 100kHz) of the film cured for 1 h, indicative of dc conduction. This is probably due to remnant hydroxide groups or alkoxide groups. The film cured for 24 h exhibited very little relaxation. This observation is consistent with electronic-type DC conduction in the former film[15]. In both coatings, there is a monotonic decrease in capacitance with frequency denoting a gradual relaxation of polarizability in the films.

a)

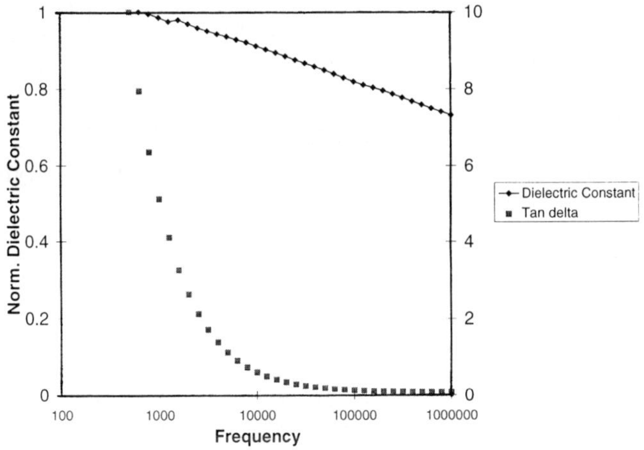

b)

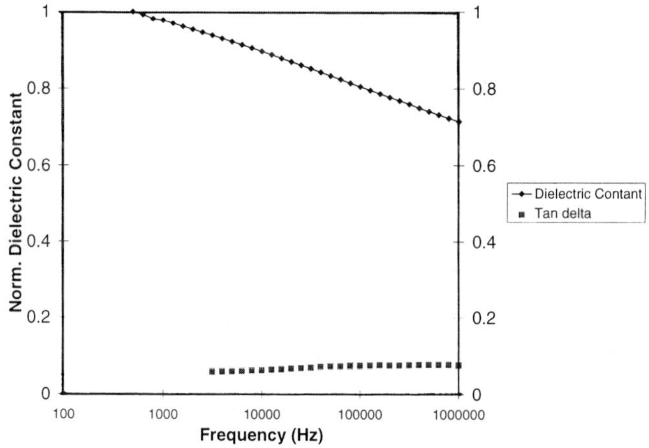

Fig. 4. Normalized dielectric constant versus frequency of PDMS-SiO_2-TiO_2 (80 vol.% PDMS and SiO_2:TiO_2 mole ratio of 1:0.5) films cured at 175°C for (a)1 h and (b) 24 h.

CONCLUSIONS

Various PDMS-based hybrid coatings were prepared on conductive glass substrates. The effect of chemistries (polymer content and SiO_2/TiO_2 ratio) and processing conditions (curing time and temperature) were investigated. Dielectric constants of 4 - 6 can be obtained in these films. Lower values of dielectric constant (< 2) can be obtained only in porous hybrid films. The dielectric constant decreased with increasing polymer content but increased with longer curing times. Sufficient curing time and temperature (i.e. 175°C for 48 h) is required to remove remnant alkoxide or OH groups in the films which lead to conductivity related dispersion at low frequencies.

ACKNOWLEDGEMENTS

The financial support of the Donnelly Corporation and the Air Force Office of Scientific Research is greatly appreciated.

REFERENCES

1. H.H.Huang, B.Orler and G.K.Wilkes, Polym. Bull., **14**, 557(1985).
2. H.H.Huang, B.Orler and G.K.Wilkes, Polym. Preprints, **26**, 300(1985).
3. H.H.Huang, B.Orler and G.K.Wilkes, Macromol., **20**, 1322(1987).
4. H.H.Huang, B.Orler and G.K.Wilkes, Polym. Preprints, **28**, 434(1987).
5. R.H.Glaser and G.L.Wilkes, Polym. Preprints, **28**, 236(1987).
6. C.S.Parkhurst, W.F.Doyle, L.A.Silverman, S.Singh, M.P.Anderson, D.McClurg, G.E.Wnek and D.R.Uhlmann, MRS Symp. Proc., **73**, 769(1986).
7. R.H.Glaser and G.L.Wilkes, Polym. Bull., **19**, 51(1988).
8. C.Sanchez and M.In, J.Non-Cryst. Solids, **147-148**, 1(1992).
9. I.Gauthier-Muneau, A.Mosset and J.Daly, J.Mater. Sci., **25**, 3739(1990).
10. D.E.Rodrigues, A.B.Brennan, C.Betrabet, B.Wang and G.L.Wilkes, Chem. Mater., **4**, 1437(1992).
11. S.Motakef, T.Suratwala, R.L.Roncone, J.M.Boulton, G.Teowee and D.R.Uhlmann, J.Non-Cryst. Solids, **178**, 37 (1994).
12. H.Schmidt in Sol-gel Science & Technology (ed. M.A.Aegerter, M.Jafelicci, Jr., D.F.Souza and E.D.Zanotto, World Scientific, Singapore, 1989), 432
13. H.Schmidt, DVS-Berichte, **110**, 54(1988).
14. H.Schmidt and H.Wolter, J.Non-Cryst. Solids, **121**, 428(1990).
15. Silicon Compounds Register and Review, United Chemical Technogies, Inc., 1991.
16. G.Teowee, J.M.Boulton, H.H.Fox, A.Koussa, T.Gudgel and D.R.Uhlmann, MRS Symp. Proc., **180**, 407(1990).
17. A.K.Jonscher, Dielectric Relaxation, Chelsea Dielectric Press, London, 1983.

PHOTOCHEMICAL STUDIES USING ORGANIC-INORGANIC SOL-GEL MATERIALS

B.C. DAVE,*,** F. AKBARIAN,** B. DUNN,* J.I. ZINK**
*Department of Materials Science and Engineering, **Department of Chemistry and Biochemistry
University of California, Los Angeles, CA 90095.

ABSTRACT

The flexible solution chemistry of the sol-gel process has been used to encapsulate a wide variety of organic, organometallic, and biomolecules in inorganic solids. This paper reviews different types of photochemical reactions which we have used to produce specific products or to generate oxygen within sol-gel matrices. By controlling synthesis conditions, the molecules can be made to exhibit desired reactivities when trapped in the sol-gel matrix.

Three different examples of photochemical reactions are presented in this paper. An organometallic gold precursor compound, dimethyl (hexafluoroacetylacetonato)gold, dissolved in a silicate sol is used to produce gold nanoparticles of desired sizes. The second example is based on sol-gel encapsulated photochromic (and thermochromic) spiropyran, that converts to a colored form using thermal energy or UV radiation. The synthesis strategies for selectively isolating the colored or the colorless form in sol-gel materials are presented. Materials of these types may be useful in write-once-read-many (WORM) optical data storage applications. The third example involves sol-gel matrices doped with a biosystem, the green plant photosystem II (PS II). The resulting aged gels and xerogels are photoactive and are capable of photooxidizing water. Oxygen illumination was measured under white light and there is an indication that PS II particles are stabilized by the encapsulation process.

INTRODUCTION

Sol-gel synthesis of SiO_2 glasses offers a convenient low-temperature route for microencapsulation of different molecules in an inorganic matrix [1]. The synthesis of sol-gel glasses containing suitable dopant molecules is well established as a way to induce new optical properties in the inorganic glass or to study the effects of encapsulation within the oxide glasses [2]. Using the sol-gel process, a wide variety of molecules of organic, organometallic and biological origins can be encapsulated and their reaction chemistry can be established [3]. The glasses formed by hydrolysis of alkoxy silicate precursors are porous and optically transparent in the visible region [4]. The sol-gel glasses are an especially attractive matrix to study photochemical reactions due to the fact that the porous structure provides a solvent-rich reaction matrix while the optical transparency is conducive to induce and to study photon-mediated reactions.

Photoinduced reactions are strongly affected by the local environment [5]. The usually low energies of such reactions are highly sensitive to medium dependent structural changes. In these reactions the energy of activation associated with the reaction is quite small and consequently other sources of free energy make a significantly larger contribution to the overall free energy of activation for the reaction. The molecular organization in sol-gel glasses places the photoactive molecule within an inorganic enclosure with *sticky* pore walls that are capable of showing H-bonding interactions [6]. The dimension (or dimensions) along which spatial and translational freedom of the trapped molecules is restrained or altered determines the ability of the sol-gel pore to influence reactivity. The effective reaction cavity of the trapped molecule is limited to the dimensions of the pore. This may place severe limitations on the reaction kinetics. As a result only those reaction pathways, where the reactants and products that conform to the geometry of the pore, remain accessible. In this way, the porous gel may influence the stability, reactivity, and the structure of the reactants and final products.

The present paper describes the use of inorganic silica sol-gel glasses as a transparent host matrix to carry out different photoinduced and photochemical reactions within the pores of the amorphous material. Three specific examples are described. The first one focuses upon the use of the finite pore dimensions of the sol-gel to influence the structure of the product. This strategy is used to

Mat. Res. Soc. Symp. Proc. Vol. 435 © 1996 Materials Research Society

photochemically fabricate gold nanoparticles with controlled particles sizes. The second example is based upon use of the sol-gel matrix to preferentially stabilize the colored or the colorless form of the photochromic spiropyran 6-nitro-8-methoxybenzospiropyran (NM-BIPS). The methodology adopted here is to influence the photochromic reaction by changes in pore sizes or by altering the nature of interactions provided by the pore walls via a selective functional modification of the parent silica framework. The third example is that of stabilization of a biological system facilitated by the protective network provided by the structurally rigid silica sol-gel inorganic matrix. The reactivity of the biological photosystem II component of the plant photosynthesis apparatus stabilized in the silica sol-gel glasses is presented.

PHOTOGENERATION OF CONTROLLED-SIZE GOLD NANOPARTICLES WITHIN SOL-GEL GLASSES.

One of the new directions in sol-gel chemistry has been its novel application in the area of metal particle entrapment and surface related chemistry [7]. The pores of the silica gel matrix serve as an appropriate host for formation of metal particle centers [8]. Most of the research on metal particle encapsulation has focused on generation of metal nuclei from precursor molecules mixed in with the sol-gel reaction mixture. The metal particles can be generated chemically, or alternatively, via photolysis of precursor molecules. An interesting application of metal particle entrapment was the use of photogenerated gold particles for surface adsorption of organic molecules for surface enhanced Raman spectroscopy (SERS) [9]. For this purpose, the precursor molecules employed were dimethyl(trifluoroacetylacetonato)gold, $(CH_3)_2Au(tfac)$, and dimethyl(hexafluoroacetylacetonato)-gold, $(CH_3)_2Au(hfac)$. Typical preparation of gold particles involved mixing the precursors with the sol-gel reaction mixture to generate gels containing the precursor molecules. The gold particles can then be directly generated within the pores of the gels by irradiation of these gels with 351-nm laser light or light from an unfiltered 100 W mercury vapor lamp.

The size of the particles is governed by the physical state and the porosity of the sol-gel, and can be monitored by absorption spectroscopy. The band maximum in the absorption spectrum of the colloidal gold particles provides a good estimate of the particle sizes; the wavelength of the maximum is directly related to the gold particle diameters. In this way, by a control of pore morphology the desired particle sizes could be obtained. It was found that the gold particle sizes were different depending on the physical state of the sol-gel. Bigger particle sizes resulted when fresh gels were irradiated. On the contrary, irradiation of dried xerogel resulted in photogeneration of smaller particles. The absorption spectra of the gold particles generated in aged gels and xerogels are shown in figure 1. As can be seen, the absorption due to colloidal gold photogenerated in aged gels shows a broad band with a maximum centered around ~650 nm. On the other hand, the absorption maximum of the particles in xerogels shows a narrowed peak at 535 nm. The presence of smaller particle sizes in xerogels can be easily calculated using equations derived by Mie [10]. The gold nanoparticles in aged gels have an average radius of 750 Å, while in the xerogels the radius is 250 Å. More importantly, the influence of the gel pore sizes on the relative distribution of the photogenerated gold particles can also be inferred. The wide range of pore radii in the aged gels results in formation of nanoparticles that show a distribution in terms of size as evidenced by the broad absorption peak in the optical spectrum. The narrow band width of the absorption peak due to gold particles generated in the xerogels suggests more uniformly distributed particle diameters. This is a clear indication of the influence of the finite pore sizes of the sol-gel matrix to influence a given reaction and alter its dimensional characteristics. Since the diameters of the gold particles are limited by the pore sizes of the matrix, the sol-gel materials can be used to generate stabilized gold particles with controlled diameters. The feasibility to fabricate controlled metal particles may find suitable application in catalysis and surface enhanced resonance Raman spectroscopy.

The characteristics of these gold particles were investigated by X-ray diffraction, and transmission electron microscopy (TEM) which revealed the presence of gold particles of varying morphologies. X-ray diffraction was used to confirm the presence of elemental gold. The particle sizes calculated form XRD measurements are in reasonable agreement with those calculated from absorption spectroscopy. TEM measurements on the sol-gel trapped nanoparticles showed irregularly shaped particles. The morphological features were found to be a function of both the physical state of the sol-gel as well as the irradiation treatment.

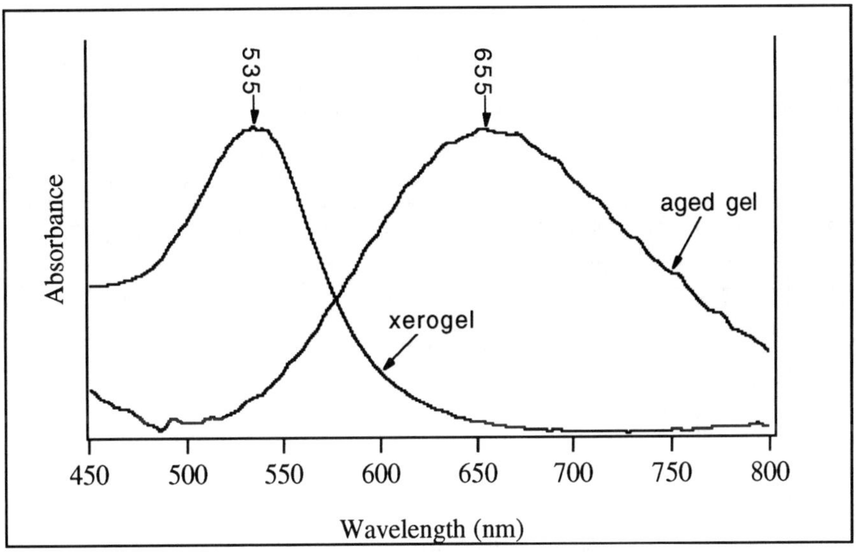

Figure 1. Absorption spectra of photogenerated gold nanoparticles in aged and dried silica sol-gels.

ISOLATION OF COLORLESS OR COLORED SPIROPYRAN IN SOL-GEL MATRICES

The spiropyrans are a class of organic compounds known to undergo photoinduced structural and optical changes [11]. The reversible changes in optical spectra caused by light or thermal energy confer the properties of photochromism and thermochromism to these compounds. In general spiropyrans contain a 2H pyran ring in which the 2-C atom of the ring is involved in a spiro linkage. The 1,2-single bond cleavage induced by photons or by thermal energies leads to formation of a zwitterionic structure which is colored. The 6-nitro derivatives of the parent 1',3',3'-trimethylspiro-[2H-1-benzopyran-2,2'-indoline] (BIPS) are usually photochromic in solution. These spiropyran derivatives undergo a structural change to the colored form upon irradiation with UV light. The colored solution then thermally fades to the original colorless form [11]. The thermochromic and photochromic behaviors of the spiropyrans have been investigated both in aluminosilicate and silicate sol-gels [13]. In general, the reaction rates of both the ground and the excited states are altered inside the porous sol-gel matrices.

The generalized reaction for the colored and colorless forms of the spiropyrans can be represented as follows:

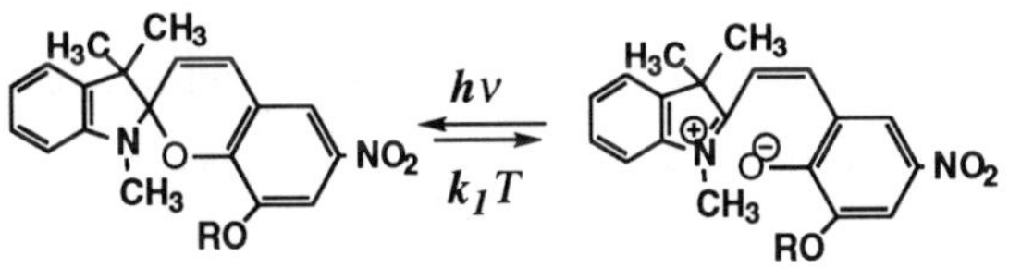

Colorless (I) **Colored (II)**

In a solution, under a given set of conditions, an equilibrium exists between the colorless (I) and colored (II) forms as shown in the scheme above. Parameters such as solvent, temperature, ionic strength, and pH of the solution can shift the equilibrium changing the relative concentrations of the two forms. In general, the behavior of a solution containing the spiropyran can be easily investigated by absorption spectroscopy. An absorption band in the visible indicates a presence of the open form (structure II) of the spiropyran, while the closed form (structure I) is characterized by an absorption band in the UV only. The changes in the equilibrium caused by encapsulation within the sol-gel matrix can therefore be studied by monitoring the absorption band in the visible. Moreover, irradiation with UV light generates the colored form, which reverts back to the clear form in dark. The optical spectrum of NM-BIPS in toluene demonstrates the changes caused by UV irradiation. The original light yellow colored solution of NM-BIPS in toluene shows no bands in the visible region (Fig. 2). The spiropyran in toluene is photochromic and exposure to UV light (365-nm; Hg vapor lamp) causes a photocoloration of the solution to purple (λ_{max} = 615 nm).

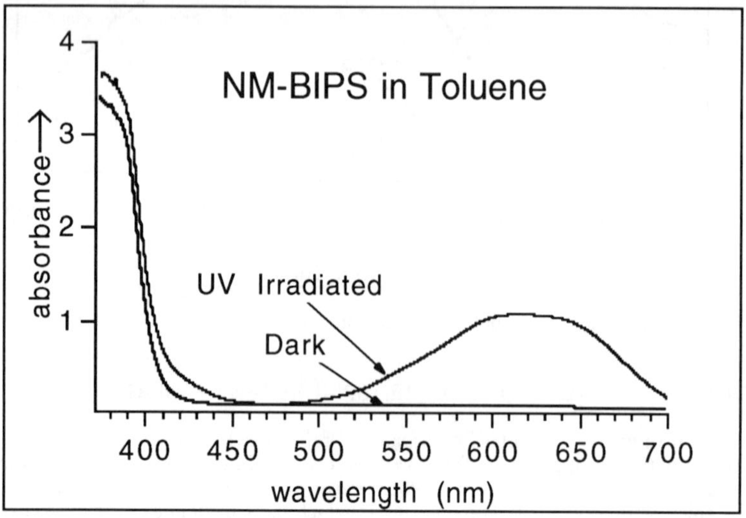

Figure 2. Optical spectra of NM-BIPS in toluene at room temperature under dark and photo-illuminated conditions.

In order to determine the behavior of NM-BIPS in porous ceramic matrices, it was encapsulated within silica sol-gels that were prepared by using the tetramethoxysilane (TMOS) precursor [12]. The initially formed gels were light yellow in color and subsequently became a red-purple color as the gels dried. The drying of the gels was performed under dark conditions as the dried, light red-purple xerogels show reverse photochromism and irradiation with UV light leads to photobleaching of the color.

The photochromic activity of the NM-BIPS in sol-gel derived matrices has been elaborated previously [13] and in this paper we focus only on the changes in the ground state properties of the encapsulated NM-BIPS. The equilibrium between the two forms of NM-BIPS is also affected by the physical state of the gel and, depending upon whether the sol-gels are aged or dried, one form of the spiropyran can be selectively isolated. The freshly formed sol-gels contain the colorless form of NM-BIPS and as the gels are dried ambiently, the colored isomer is preferentially stabilized. The ground state equilibrium of the NM-BIPS was investigated using optical absorption spectroscopy. The results show that color formation accompanies the shrinkage of the gels (Fig. 3). The ground state equilibrium between the two forms of the spiropyran appears to change as the gels are dried and the pore structure of the gel collapses. The formation of colored species, as evidenced by absorption spectroscopy, suggests that the sol-gel matrix preferentially stabilizes the colored form as the materials are dried. In other words, drying causes the equilibrium to shift towards the open zwitterionic from (structure II). The change in color is more likely due to pore collapse and not due to evaporation of the organic solvent phase as the changes continue to take place even after the gels have lost more than half their original weight. As shown in inset of figure 3, the increase in intensity of the visible band at ca. 550-nm correlates linearly with the % weight loss (error limits ±3 %) of the sol-gels as they continue to shrink due to drying. Spiropyrans are known to form the colored isomer upon surface adsorption on particulate silica [14]. The linear relationship between increase in intensity and weight loss are not consistent with adsorption effects which would exhibit a nonlinear change according to adsorption isotherms. The linear changes observed in the spectral properties of TMOS sol-gel encapsulated NM-BIPS suggest that the formation of the colored form is mainly due to shrinkage of the pore structure which facilitates the formation of the open form of the spiropyran. It is tempting to ascribe the isolation of the colored form to internal stresses that occur during drying of the gels that lower the activation energy associated with breaking of the C-O bond, thereby facilitating a shift of the equilibrium towards the open colored form. The open colored form would then be stabilized via hydrogen bonding interactions of the phenoxy and quaternary ammonium groups with the pore walls of the silica sol-gels.

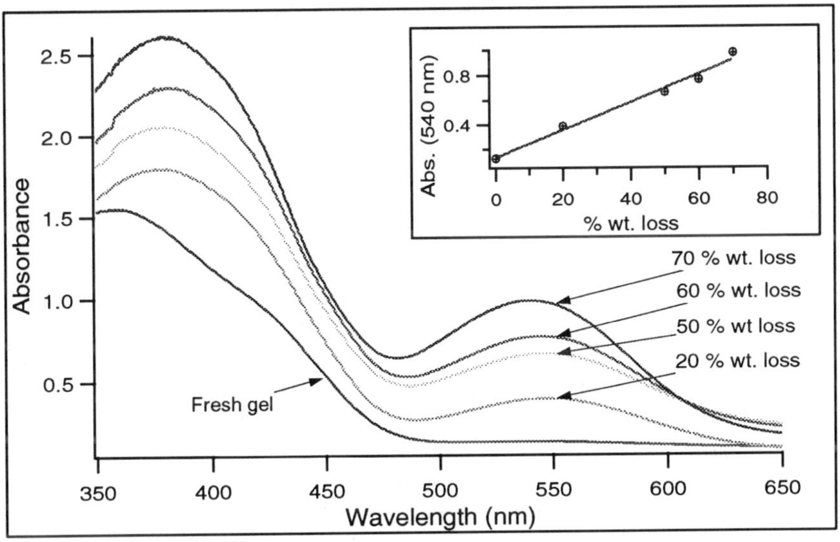

Figure 3. Absorption spectra of NM-BIPS in silica sol-gel showing the stabilization of the colored form as the gels are dried in the ambient. Inset correlates the changes in intensity of the 540-nm band with drying.

The relative concentration of the colored and clear forms is also governed by medium dependent factors such as dielectric constant and hydrogen bonding with the solvent. Thus, a change in the composition of the sol-gel is expected to result in differential stabilization of one form over the other. When organically modified hydrophobic sol-gels are used, the colored form is predominantly stabilized in the ground state [15]. The matrix dependence of the photochromic spiropyrans is especially important from the perspective of fabricating materials whose properties can be photoregulated [16].

The effects of *sticky* matrices capable of noncovalent interactions with the photochromic NM-BIPS were also investigated in this study. This work considered the effect of matrices capable of hydrogen bonding on the relative stabilities of the spiropyran derivatives. A structural modification of the silica framework with strongly hydrogen bonding functionalities such as $-NH_2$ groups is likely to alter the spiropyran equilibrium between the colored and the colorless states. With this in mind, transparent gel glasses were prepared by hydrolysis of 3-aminopropyl trimethoxysilane (ATMOS). The formation of sol-gels was accomplished by adding 5 mL of water to 10 mL of the precursor. Addition of water caused a spontaneous reaction and the solution turned warm. The clear homogeneous reaction mixture was then sonicated for about ten minutes to yield clear sol. The spiropyran was added directly to the sol to give light yellow colored solutions. The gels were then cast in polystyrene cuvettes. Gelation occurred in about 2-3 days with formation of transparent light yellow colored sol-gel. The color of these samples was found to be independent of both the physical state of the gel as well as irradiation; drying and UV light caused no coloration. Thus, the photochromic and thermochromic spiropyran NM-BIPS, when encapsulated in the ATMOS-derived sol-gels, showed no changes in the optical properties from either pore shrinkage or irradiation with white light. The thermochromic behavior, however, was still retained, and heating to ~180 °C caused the materials to turn blue-purple. The color faded slowly as the gels were cooled to room temperature.

The isolation of the colorless form in ATMOS is counter-intuitive as one would expect a shift in equilibrium towards the zwitterionic form stabilized by hydrogen bonding interactions with the $-NH_2$ groups from pore walls. Our results indicate an extensive stabilization of the closed form by the $-NH_2$ groups from the sol-gel pore walls. One of the critical requirements for formation of the colored form is that of stabilization of the phenoxy group, most likely by an adventitious proton. The high basicity of the amino groups in the ATMOS material ensures that the concentration of free protons in the material is minimal. Moreover, if the spiro oxygen (involved in C-O bond) is stabilized extensively by H-bonding interactions with the pore walls it would experience an increased energy of activation to convert to the phenoxy anion. A combination of factors that stabilize the colorless form and raise the activation energy to generate the colored form are likely to be factors that are responsible for selective isolation of the colorless form in ATMOS-derived sol-gels .

To summarize, in this study we have outlined approaches to use sol-gel based matrices to alter the properties of spiropyrans. Ordinary TMOS-derived silica gels shift the ground state equilibrium towards the colored isomer as the gels are dried. Moreover, the sol-gel matrix also affects the photochemical properties and the encapsulated spiropyran demonstrates a reverse photochromism in the solvent-free dried xerogels. A structural modification of the matrix to aminopropyl substituted sol-gels stabilizes the clear form of NM-BIPS. Additionally, the organically modified sol-gel matrix also renders the trapped spiropyran non-photochromic. The resulting material is insensitive to light and remains clear, although the thermochromic properties of the spiropyran are still retained and thermal heating causes coloration. Thus, in TMOS based materials drying and thermal heating causes coloration while in ATMOS based matrices only thermal heating leads to coloration. The materials with suppressed photochromism while still retaining thermochromic behavior may find suitable applications as write-once-read-many (WORM) type optical data-storage media with high intensity infra-red laser beams as recording agents.

STABILIZATION AND PHOTOREACTIVITY OF PHOTOSYSTEM II IN POROUS SILICA SOL-GELS.

The TMOS-derived silica gels have been successfully employed to encapsulate a variety of biological molecules [17]. One especially attractive feature underlying the bio-gel methodology is that given the range of functional processes mediated by biological molecules in nature, a variety of potentially useful biomaterials can be prepared. Herein, we report on the encapsulation of green plant photosystem II isolated from spinach leaves.

One of the most important biological processes responsible for life on earth is photosynthesis, the biological mechanism that derives chemical free energy via conversion and storage of solar energy [18]. The photosynthetic process involves absorption of light by extensively folded chlorophyll with the subsequent migration of electrons across precisely arranged intrinsic transmembrane proteins. This process generates sufficient redox potential to oxidize water and the released electrons ultimately form reducing compounds that fix carbon dioxide. The entire photosystem can be divided into two parts, termed photosystems I and II. It has been recognized that the oxidation of water is localized on photosystem II (PS II). It is now believed that PS II is composed of at least six subunits. Four of the largest subunits have molecular weights ranging from 30-50 kDa. The reactions of PS II take place in a highly organized matrix where individual components work in a cooperative fashion fulfilling complementary roles like architectural support, light absorption, primary charge separation followed by energy transfer and electron transfer. The PS II particles liberate dioxygen upon irradiation. The overall reaction can be given as

$$2\,H_2O + 4\,h\nu \xrightarrow{\text{PS II}} O_2 + 4\,H^+ + 4\,e^-$$

Analogous to the natural system, artificial systems in which light-induced electron transfer events lead to splitting of water into oxygen and hydrogen are believed to be some of the most promising alternate energy sources and a convenient means of harnessing solar energy. Various strategies have been devised to achieve charge separated systems that could be utilized for this purpose, and moderate success has been attained. Although several chemical systems have been found to be effective in generating hydrogen from water, relatively few promising approaches have been identified for dioxygen liberation. The conversion of water to dioxygen by the PS II component of green plant photosystem remains the most efficient way to photooxidize water, and, in spite of intense research efforts, a synthetic functional model complex of the water oxidation enzyme still remains elusive.

An alternative approach to the water photosplitting problem is the possibility that the natural PS II may itself be used in conjunction with an artificial hydrogen generating system to achieve photosplitting of water. The natural abundance of the plant PS II in the ecosystem and the relative ease with which it may be isolated from green leaves ensures an ample supply of it. However, a direct use of PS II encounters problems associated with reusability and long term storage of the protein due to bacterial degradation and denaturation caused by various external stimuli such as heat, acid, etc. A prolonged, effective shelf-life of the biosystem is a necessary requirement for its practical applications.

In the course of our studies on sol-gel glasses doped with biomolecules, we have found that encapsulation of biomolecules in the inorganic matrix stabilizes the dopant with respect to microbial attack. Also, the physical entrapment of the dopant in a porous cage renders the biomolecules less susceptible to unfolding and thermal denaturation. The sol-gel glasses show a distribution of pore sizes and biosystems of different molecular dimensions can be effectively encapsulated. Provided the dopant concentration is millimolar or lower, only a small fraction of the pores (< 1%) is of the large size required for encapsulating the protein. The size of a vast majority of pores is determined by sol-gel synthesis conditions (pH, aging conditions etc.) and is in the range of 50-100 Å for our typical preparations. As the gel network grows around the dopant, the size of the protein or enzyme may determine the geometry of the pore in which it resides. As such the dopant size does not prevent the entrapped protein from exhibiting the same chemical function and biological response as observed in solution. Although proteins and enzymes have been shown to retain their functional integrity in the sol-gel matrix, there are still questions of whether the structural changes during aging and drying of the gels affect the internal organization of the trapped biosystems. The PS II represents an excellent probe wherein such an effect can be observed. The PS II function depends on the precise architectural organization of various subunits in the thylakoid membrane. Thus, a structural perturbation brought about by pore collapse during the aging or drying processes should be reflected by a change in the structure and reactivity of the system.

The encapsulation of an entire system as opposed to a distinct protein provides an opportunity to study effects of changes in the structure and reactivity of the system upon immobilization. The optically transparent nature and the ability to prepare bulk samples are vital for carrying out the optical assay of the activities in aged gels and rehydrated xerogels. The photoreactivity of PS II particles shows that they retain their functional organization within the porous gels. The trapped photosystem effectively carries out photooxidation of water to dioxygen, and silica gels doped with particles of PS II liberate dioxygen on exposure to light.

Both direct as well as indirect means can be used to characterize the reactivity of PS II in the sol-gel matrix. A direct measurement of evolved oxygen can be carried by using an oxygen-sensitive electrode. Alternatively, the oxygen evolution can be coupled to photoreduction of the redox-active organic dye 2,6-dichlorophenol indophenol (DPIP), such as

$$H_2O + DPIP_{ox} \xrightarrow[h\nu]{PS\ II} 1/2\ O_2 + DPIP_{red}$$

The absorption changes accompanying this reaction (Hill reaction) serve as a useful optical assay method for PS II reactivity.

The PS II particles were isolated using a modification of the protocol developed by Berthold et al [19]. The particles were immobilized using the buffered sol-gel route. The initial sol was prepared using TMOS as the precursor. The reactivity of PS II particles was monitored by two different methods. Direct oxygen evolution was monitored using a Clark-type electrode incorporated in YSI model 53 oxygen meter (Yellow Springs Instrument Co., Inc.; Yellow Springs, Ohio). The gels were crushed to enhance the surface area for dioxygen evolution experiments and suspended in deoxygented 50 mM phosphate buffer (pH 7.5) made 1 mM in DPIP. A calibration baseline was obtained in the dark. The samples were then irradiated with a 75 W tungsten filament lamp. A 630-nm cut-off filter was used to avoid thermal and high-energy excitation of the sample. Hill photoreduction of the electron acceptor DPIP was monitored using a tungsten filament lamp. The crushed gel samples were suspended in a solution of DPIP (~ 1 mM) and irradiated in the absence of stirring. At appropriate time intervals, the supernatant DPIP solution was tested for changes in the absorption intensities at 600 nm. The absorption changes in the blank solution were negligible and no correction was applied to the absorbance data.

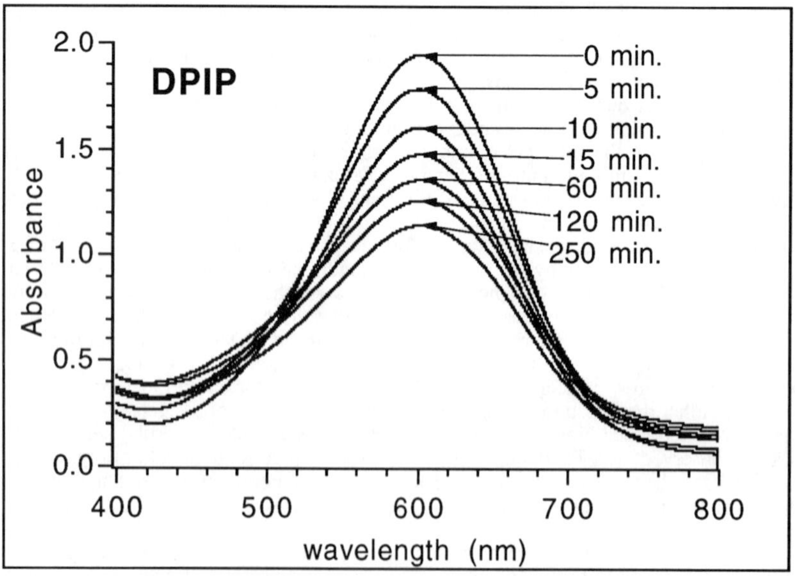

Figure 4. Optical absorption spectra showing the photoreduction of DPIP by sol-gel encapsulated PS II particles.

The sol-gel glasses containing PS II are photoactive towards water splitting in both the aged and dried gel forms. The photogeneration of dioxygen from water constitutes a series of electron transfer events initiated by photon capture and subsequent charge migration. The quantum harvesting is carried by chlorophyll pigments with high absorption coefficients in the visible region to produce the chlorophyll cation. The transfer of energy from the antenna complexes to the main reaction center is believed to take place via resonance energy transfer. The reaction center actively involved in dioxygen formation is called the oxygen evolving center (OEC). Under the conditions of continuous illumination in water, a saturating oxygen evolution is observed. If the Hill acceptor DPIP is used instead, the absorbance changes can be monitored optically due to reduction of the oxidized form (λ_{max} = 600 nm).

The photochemical reactivity of the PS II containing gels is evidently retained as the absorption spectrum of DPIP solution containing crushed gels shows a decrease in absorbance with respect to illumination indicative of photoreduction (Fig. 4). The initial reduction of the electron acceptor is fast but continuous illumination results in saturation kinetics. Whereas the PS II solution data showed saturation in about 10 min, the aged gels and the xerogels required about 10 times longer to achieve saturation. Although the same concentration of PS II was employed for all the measurements, the fact that one sample was a solution while the others were crushed gels with finite exposed surface area may give rise to differences in reactivity and kinetics. For the latter, one must consider that the photoreduction of DPIP in the external solution is dependent upon diffusion of the oxidized form into the pores of the gel, and rediffusion of the reduced form back into the external solution. Thus it is not surprising that saturation in the aged gels and the xerogels takes longer to occur than in solution. Nonetheless, the results in figure 4 demonstrate that the aged gel samples are photoactive.

A direct measurement of evolved oxygen was undertaken in order to confirm that the encapsulated PS II particles were photosynthetically active. Oxygen evolution was observed upon irradiation of the PS II encapsulated in aged gel and xerogel samples. Upon irradiation, an increase in the oxygen concentration in the solution was observed, which slowly decayed to background levels when the light was switched off. Thus, changes in dioxygen concentration could be directly correlated with presence or absence of light. Undoped aged gel and xerogel samples showed no changes in the background dioxygen concentration during illumination. As the effective light intensity and the exposed surface of different experiments varied, no attempt was made to quantify the rate of oxygen generation. The photoreactivity was found to be reversible for both the aged gels and rehydrated xerogels and dark-adapted gels could be utilized again for oxygen generation. Additionally, the reactivity of the photoactive xerogels was checked in the presence and absence of the PS II inhibitor, hydroxylamine, NH_2OH. The inhibitor was added to the aqueous medium in which the xerogel was immersed. Subsequent experiments showed substantial inhibition as a relative decrease of ~75 % in oxygen current was observed. It is rather interesting to note that the dried xerogel samples retain an appreciable photoactivity and are able to perform photosplitting of water. Since the ambiently dried xerogel samples still retain an aqueous environment, the structural organization of the encapsulated particles is maintained.

The stability of the encapsulated PS II was compared with solution samples. Oxygen evolution measurements were used to characterize reactivities of sol-gel immobilized PS II particles (stored at 4 °C) with those stored in solution at 77 K. The sol-gel encapsulated particles showed about 20 % oxygen evolution after 3 weeks as compared to particles stored in buffer at 77 K. On the other hand, the particles stored in solution buffer at room temperature lose their photochemical ability to evolve oxygen. The physical encapsulation of the biosystem in porous matrices is believed to stabilize the biosystem and prevent denaturation pathways that are usually accessible to the biological system in solution media.

SUMMARY

In this paper, we have described three different systems that are highly disparate in their structural and functional characteristics. We have outlined approaches to suitably modify their properties by using the method of simple encapsulation within a porous matrix. The photoactive systems in porous sol-gel based materials still retain their characteristic solution chemistries, however, the porous matrix alters the structure, reactivity, and stability equilibria. It is shown that a combination of both physical and chemical interactions of the encapsulated systems with plain or functionally modified silica can be used as a means to alter the reaction chemistries of photoinduced processes.

ACKNOWLEDGMENTS

This research was supported by a grant from National Science Foundation (DMR 9408780) for which we are most appreciative.

REFERENCES

1. J. I. Zink, J. S. Valentine, and B. Dunn, New J. Chem. **18**, 1109 (1994).
2. B. Dunn, and J. I. Zink, J. Mater. Chem. **1**, 903 (1991).
3. D. Avnir, S. Braun, O. Lev, and M. Ottolenghi, Chem. Mater. **6,** 1605 (1994).
4. L. L. Hench, and J. K. West, Chem. Rev. **90,** 33 (1990).
5. M. A. Fox, Photochem. Photobiol. **52**, 617 (1990).
6. V. Ramamurthy, R. G. Weiss, and G. S. Hammond Adv. Photochem. **18**, 67, (1993).
7. M. A. Cauqui, and J. M. Rodriguez-Izquierdo, J. Non-Cryst. Solids, **147-8**, 724 (1992).
8. T. Lopez, L. Herrera, J. Mendez-Vivar, R. Gomez, and R. D. Gonzalez, J. Non-Cryst. Solids **147-8**, 753 (1992).
9. F. Akbarian, B. Dunn, and J. I. Zink, J. Phys. Chem. **99,** 3982 (1995).
10. G. Mie, Ann. Phys **25**, 377, (1908).
11. R. C. Bertelson in Techniques of Chemistry, Vol III: "Photochromism", G. H. Brown, Ed., Wiley-Interscience, New York, (1971), p 100.
12. E. H. Lan, B. C. Dave, B. Dunn, J. S. Valentine, and J. I. Zink, MRS Symp Proc. (1995).
13. D. Preston, T. Novinson, W. C. Kaska, B. Dunn, and J. I. Zink, J. Phys. Chem. **94**, 4167 (1990).
14. T. R. Evans, A. F. Toth, and P. A. Leermakers, J. Am. Chem. Soc. **89**, 5060 (1967)
15. D. Levy, and D. Avnir, J. Phys. chem. **92**, 4737 (1988).
16. M. Ueda, H.-B. Kim, and K. Ichimura, J. Mater. Chem. **4**, 883 (1994).
17. B. C. Dave, B. Dunn, J. S. Valentine, and J. I. Zink, Anal. Chem. **66**, 1120A (1994).
18. G. Renger, Angew. Chem., Int. Ed. Engl. **26**, 643 (1987).
19. D. A. Berthold, G. T. Babcock, and C. F. Yocum, FEBS Lett, **134**, 231 (1981).

THE FORMATION OF LASER ACTIVE COMPOSITE FILMS FROM SILICATE CERAMICS

L.L. BEECROFT*, R.T. LEIDNER*, C.K. OBER*, D.B. BARBER‡, AND C.R. POLLOCK‡
Departments of Materials Science and Engineering* and Electrical Engineering‡, Cornell University, Ithaca, NY 14853

ABSTRACT

Composite films containing solid state laser nanoparticles have been created which show optical amplification in the technologically important near IR range. The solid state laser materials studied, Cr-forsterite (Cr-Mg_2SiO_4) and Cr-diopside (Cr-$CaMgSi_2O_6$), were prepared using a dispersion polymerized polymer precursor. The nanoparticles produced during the polymer synthesis acted as size templates, creating fine crystalline powders of the Cr-forsterite and Cr-diopside upon calcination. These fine powders were dispersed in a refractive index matched polymer matrix and cast as 5-10 µm thick films. The resulting composite containing Cr-forsterite showed optical amplification of 300 dB/m at 1.24 µm.

INTRODUCTION

Composite materials have long been useful because of their ability to take advantage of the desirable properties of dissimilar materials. Solid state laser materials have useful optical properties in many areas, including optical communications and medicine, but must be laboriously grown into large single crystals to be of use. These large crystals are difficult to incorporate in the increasingly integrated applications for which they are required. Outside of transparency, polymers typically do not have interesting optical properties, however they are often easy to process in film and fiber forms. By combining the interesting optical qualities of solid state laser materials and the processability of polymers, improved composite materials can be constructed.

Optically homogeneous composite films have been created by embedding small solid state laser particles in a polymer matrix with the same refractive index (RI) as the particles. If the particles and the RI difference between the particles and the matrix are small enough, significant scattering can be avoided. Calculations based on Rayleigh scattering show that for a 0.015 difference in refractive index, 80 nm is the upper limit on particle size.[1] Taking advantage of the film forming ability of the polymer matrix, composite films were cast and exhibited the desired laser activity of the particles.

The optical composite system our group has focused on concerns chromium doped forsterite (Cr-Mg_2SiO_4), a tunable laser in the technologically attractive near IR regime (1167-1345 nm).[2] 1312 nm is a dispersion minimum for silica waveguide materials and is currently used in the majority of installed optical fibers.[3] Using the composite concept, laser amplifying films containing Cr-forsterite embedded in a specially tailored matrix material have been created.[1] The measured amplification of 300 dB/m at 1240 nm exceeds the gain observed in single crystal forsterite as well as the gain of 3 dB/m found in widely used Er-silica amplifiers at 1.55 µm. If the amplification could be optimized at 1312 nm, these composites might be very important as optical amplifiers for long haul fiber optic communications.

A second system, Cr-diopside (Cr-$CaMgSi_2O_6$), may also be very interesting. Cr-diopside powder displays strong fluorescence in the near IR between 700-1200 nm, centered about 1000 nm. Although this is a good indication of possible laser activity in the material, a single crystal cannot be grown due to incongruent melting. The optical composite strategy we have developed may allow the first measurements of laser activity in this material. Using the same principles, Cr-diopside has been synthesized and can be cast into composite films using the matrix material created for the Cr-forsterite composites.

Mat. Res. Soc. Symp. Proc. Vol. 435 © 1996 Materials Research Society

Three polymers were developed in order to make these composites. The first was a dispersion polymerized precursor to Cr-forsterite which resulted in small (100-500 nm) particles.[4] These particles acted as size templates as the polymer was heated to 1000 °C to remove the organic material and crystallize the Cr-forsterite. The second polymer was a close relative to the Cr-forsterite precursor, with the correct composition to result in the formation of Cr-diopside. The third polymer acted as the RI tailored matrix material to the forsterite and diopside particles. By copolymerizing napthyl methacrylate (NM, 1.63) and tribromostyrene (TBS,1.66) a polymer with the refractive index of forsterite (1.652) was prepared.[5] This polymer was transparent in the near IR and formed suitable films for the amplification experiments.

The principles developed in this work can be extended beyond solid state laser systems to include non linear optical materials and photochromic materials among others. It may be very interesting in the study of ceramic materials which display fluorescence at an attractive wavelength, but for which the single crystal cannot be grown to test possible lasing.

EXPERIMENT

Cr-forsterite and Cr-diopside nanoparticle synthesis[4]

Magnesium methacrylate (MgMA), calcium methacrylate (CaMA), (both Monomer-Polymer and Dajac Laboratories, Inc.), chromium trioxide (CrO_3), hydroxypropyl cellulose (HPC), (all Aldrich) and methanol (Fisher) were used without further purification. 3-methacryloxypropyltrimethoxy silane (MOSiMA, Aldrich) was distilled under vacuum. Azobis(isobutyronitrile) (AIBN, Kodak) was recrystallized from methanol/acetone. A reaction vessel was charged with MgMA, HPC (10-20% by monomer weight), CrO_3 (Cr/Si 0.004-0.05), methanol, and MOSiMA under positive N_2 pressure. Diopside precursors also contained CaMA. This solution was heated to 60 °C and stirred vigorously until the HPC dissolved. Then distilled water was added (0-15 vol %) and the solution was degassed with N_2 for 30 minutes. The AIBN (1-5% by monomer weight) was dissolved in a small amount of methanol and added to the reaction mixture. The solution was left to reflux under N_2 overnight. In a typical reaction to form forsterite 1.104g MgMA, 0.5 ml MOSiMA, 160 mg HPC, 30 ml methanol, 2.1 mg CrO_3 and 48 mg AIBN were used. In a typical reaction to form diopside 0.2044g MgMA, 0.2209g CaMA, 0.5 ml MOSiMA, 95 mg HPC, 18 ml methanol, 2 ml H_2O, 2.1 mg CrO_3, and 28 mg AIBN were used. The resulting particles were centrifuged, supernatant methanol was decanted, then they were redispersed in fresh methanol. This was repeated several times to remove unreacted monomer and excess HPC. The particles were dried then fired (Lindberg Model 55322 tube furnace) at 5°C/min to 1000 °C for 8 hours under flowing humid air. SEM samples were prepared by dispersing the particles in methanol, ultrasonicating ca.10 minutes, then dropping onto hot Si substrates. They were coated with ca. 10 nm of Au/Pd.

Synthesis of the tailored RI copolymers[5,9]

NM was prepared using a Schotten-Baumann type synthesis modifying the procedure of Saric et al.[6] TBS was graciously provided by Jerry Garrison at AmeriBrom and used without further purification. Methanol (Fisher) and anhydrous toluene (sure-seal, Aldrich) were used as received. For the copolymer discussed in this paper (55/45 TBS/NM) a round bottom flask was charged with 1.995g NM, 3.9242g TBS, and 4.9 mg AIBN then degassed with N_2. Toluene (20 ml) was added and the solution was degassed with N_2 for 15-20 minutes. It was then lowered into a 60°C bath and left to polymerize overnight. The polymer solution was precipitated in methanol. The resulting polymer was filtered, dissolved into toluene again (required heat), reprecipitated, and dried overnight under vacuum. It had a weight average molecular weight of 780,000 g/mol with polydispersity of 10 (Waters GPC system, 490 UV detector, 410 RI detector, compared to polystyrene standards).

Composite synthesis[1,5]

Cyclohexanone and isobutyltrimethoxy silane were used as received from Aldrich. A 5 wt% solution of the TBS/NM copolymer (RI=1.66, Becke Line Technique) in cyclohexanone was prepared and the particles and isobutyltrimethoxysilane were added. The isobutyltrimethoxysilane acted as a compatibilizer binding at one end with the -OH groups expected on the surface of the forsterite and diopside particles and mixing with the organic polymer at the other. The solution was ultrasonicated until the particles were dispersed, filtered to 1 μm using a syringe tip filter, and puddle cast on a level glass substrate. After drying slowly overnight, the film was annealed at 200 °C for two hours to remove excess solvent (T_b cyclohexanone = 155 °C) and to relieve stresses in the film. This method resulted in films from 5-10 μm thick and 2-3 cm long. Finally the ends were cleaved to make smooth surfaces for end-coupling in the amplification experiment.

Instrumental

A Leica 440 Scanning Electron Microscope was used to examine particles visually. X-ray analysis was performed using a Scintag PAD X Automated Powder diffractometer using Cu K_α (λ=1.54Å) radiation. Near IR measurements of the matrix were made on a Cary UV-Vis-NIR spectrophotometer. For the amplification experiment collinear beams of 1.24 μm light generated by a continuous wave Cr-forsterite laser (reference signal) and 1.06 μm light from a Nd:Yittrium Aluminum Garnet (YAG, pump signal) laser were focused into the cleaved end of the film. Before reaching the InGaAs detector, the light was passed through a monochromator at 1.24 μm to filter out the pump beam as well as any broad band light from spontaneous emission that might be occurring in the film. A shutter allowed the pump beam to be blocked independently of the reference.

RESULTS AND DISCUSSION

Nanoparticle Synthesis

We have developed a novel technique to form fine ceramic powders using a polymeric precursor as a size template.[4] Specifically we have synthesized Cr-forsterite and Cr-diopside nanoparticles using this technique. The technique involves methacrylate type monomers containing Mg, Si, and Ca which are randomly dispersion copolymerized. The dispersion polymerization results in polymer precursor particles from 100-500 nm in size. After heating to 1000 °C to remove the organic components and crystallize the laser material, the particles have retained their shape and a large number are near the targeted 100 nm size as can be seen in Figures 1a and 1b. As seen in these micrographs, Cr-diopside sinters considerably more than Cr-forsterite resulting in a low yield of particles of the desired size.

The conversion of both polymeric precursors has been studied by high temperature X-ray diffraction. These three dimensional experiments are represented by the contour plots shown in Figure 2. The Cr-forsterite precursor yields phase pure material (2Θ = 36.2, 35.4) which begins to crystallize at 750 °C without the formation of a second phase as can be seen in Figure 2a. The Cr-diopside precursor (Figure 2b) exhibits an unidentified transitional phase seen in the two theta peaks at 32.8, 34.2, and 34.8, which begins to crystallize at 700 °C. This phase begins to disappear at 925 °C leaving phase pure Cr-diopside (2Θ = 29.9, 30.3, 30.9, 35.5) by 1000 °C. The pure diopside phase begins to crystallize at about 825 °C. The transitional phase may contribute to the enhanced sintering seen in this material as compared to Cr-forsterite. This enhanced sintering may also be a result of the relative melting temperatures of diopside (T_m=1391 °C) and forsterite (T_m=1910 °C).[7] The intimate mixing of the metals in the polymer precursors result in the remarkably low crystallization temperatures seen in these ceramics.

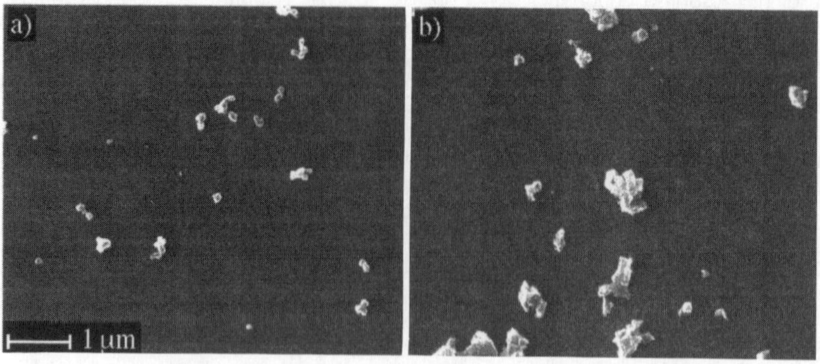

Figure 1. SEM micrographs of a) Cr-forsterite and b) Cr-diopside nanoparticles. Both use the same scale bar.

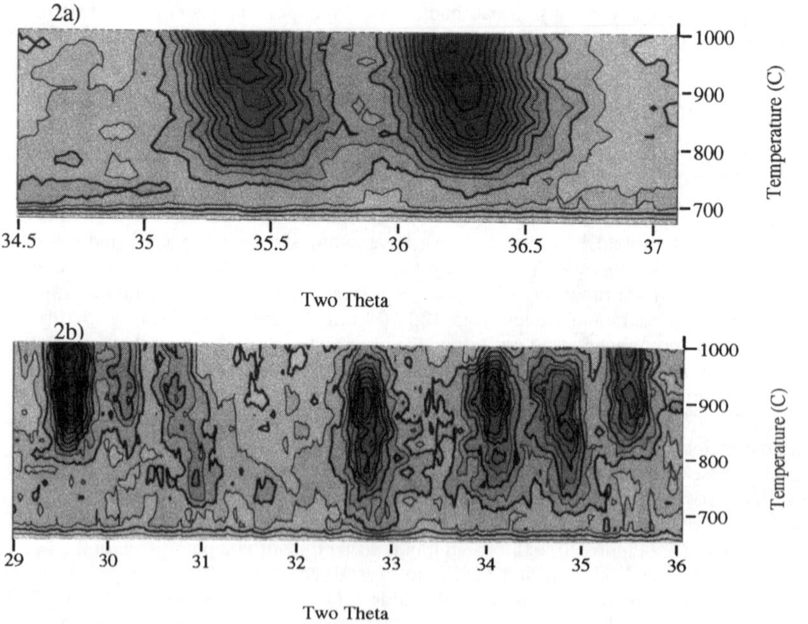

Figure 2. High temperature X-ray diffraction contour plots for a) Cr-forsterite and b) Cr-diopside

Both powder types show strong fluorescence in the near IR indicating the likelihood of laser activity.[4] In Cr-forsterite a large Cr^{4+}/Cr^{3+} ratio can be identified which is desirable because Cr^{4+} is the ion responsible for lasing.[8] Cr-diopside fluoresces from 700-1200 nm, however the role of Cr^{4+} in this material has not been studied.

This technique for the preparation of fine ceramic powders using dispersion polymerized precursors can be extended to other systems as long as monomers which can undergo dispersion polymerization can be identified. Alternatively, emulsion polymerization may be a viable precipitation technique for some systems.

RI tailored matrix synthesis

The refractive index of forsterite is 1.652,[7] relatively high for a polymer, but attainable using brominated and aromatic monomers with RIs as high as 1.7. The strategy used to tune the refractive index of the matrix was to copolymerize monomers with RIs bracketing the desired RI. In doing so, several copolymer systems were created which can obtain the desired RI of 1.65.[9] The TBS/NM copolymer (RI=1.66) was chosen for the composite work due to its desirable film forming ability. Although the refractive index of diopside is slightly higher than that of forsterite (1.677),[7] the same matrix material was used. The average refractive index mismatch for the Cr-forsterite composites is 0.015 and for the diopside composites is 0.017 with this matrix.

The matrix material must be transparent at the wavelengths used in the optical experiment, 1064 nm (YAG pump) and 1200-1300 nm (amplification). Figure 3 shows the absorption of a TBS/NM copolymer in different concentrations of THF over 1 cm (20% polymer over 1 cm is comparable to 2 mm solid material). Both regions of interest show no absorption, indicating the polymer will not interfere with the optical measurements. The small absorption from 1100-1150 nm has been identified as an artifact caused by the THF. These wavelengths are not probed in the amplification experiment. Composites containing this matrix material were repeatedly exposed to the pump energy for 5 minute intervals without visible damage.

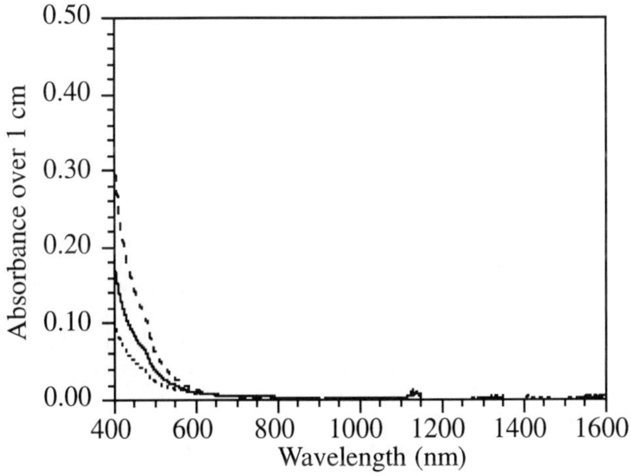

Figure 3. Near IR absorbance of TBS/NM copolymer . Three concentrations, 20% (dashed), 10% (solid), and 5% (dotted) were measured in THF.

Composite results

Composites prepared with Cr-forsterite nanoparticles in the TBS/NM matrix were studied in an amplification experiment, shown schematically in Figure 4. The addition of the pump energy caused excitation of the Cr^{4+} ions. During pumping, the reference signal stimulated emission in the excited Cr^{4+} ions causing an increase in detected signal indicating amplification. When the pump signal was blocked, the reference signal passed through the film and was unaffected by the Cr-forsterite because the Cr^{4+} ions were in their ground state. By comparing the detected signal with and without the pump the amplification factor for the film was obtained. This amplification factor was converted into gain/m by the simple formula: gain (dB) = 20 $\log_{10}$(amplification factor) divided by the film length.

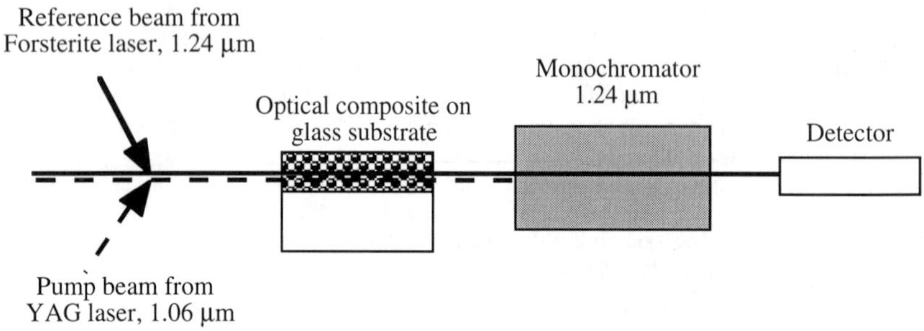

Figure 4. Schematic of amplification experiment.

Figure 5 shows the gain as a function of pump power for a Cr-forsterite containing film about 2 cm in length. An amplification factor greater than 2 was obtained at the highest pump power (using low signal power). This result is shown in Figure 5 as a gain of about 300 dB/m, which can be compared to widely used Er-silica fibers which show a gain of 3 dB/m. This composite gain also exceeds the saturated gain of single crystalline Cr-forsterite. The high amplification is in part due to the high Cr concentration which can be obtained in the nanoparticles as compared to single crystalline forsterite. Samples used in this experiment had 0.02 Cr/Si while single crystals typically contain 0.00007-0.0008 Cr/Si concentrations. In the single crystal growth process the Cr evaporates at the high processing temperature and disrupts crystal growth above a certain concentration.[10]

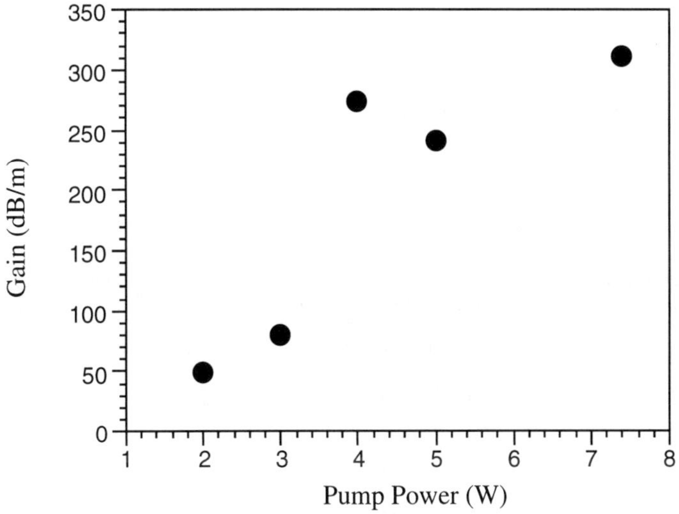

Figure 5. Amplification in Cr-forsterite optical composites as a function of pump power.

CONCLUSION

Silicate ceramics, Cr-forsterite and Cr-diopside, were synthesized using dispersion polymerized prepolymers as size templates to form nanoparticles. The Cr-forsterite nanoparticles were in the targeted 100 nm size regime, while the Cr-diopside particles were larger primarily due to sintering effects. These ceramic precursors crystallize at relatively low temperatures resulting in phase pure materials. The Cr-diopside precursor exhibits a transitional phase which is not present in the final material. This new method of fine particle preparation can be extended to any system which can undergo dispersion polymerization. In some cases, emulsion polymerization may be a viable alternative in the preparation of the prepolymer.

A tailored RI copolymer, with the unusually high refractive index of 1.66, was synthesized as a matrix material for the laser active nanoparticles. This matrix is transparent at the wavelengths of operation and forms high quality films.

Using a novel optical composite concept, laser active films have been synthesized based on Cr-forsterite. By combining the matrix and nanoparticles, optical composites showing amplification of 300 dB/m at 1240 nm have been created. This is the first time to the authors' knowledge that such a structure has been used to prepare solid state laser films. The general principles outlined here may be useful for non linear optical materials and photochromic materials in addition to solid state lasers. We are looking forward to amplification results with Cr-diopside composites showing the first laser activity in this material.

ACKNOWLEDGMENTS

We would like to thank the National Science Foundation, Department of Education, and Cornell Materials Science Center for support of this project.

REFERENCES

1. L.L. Beecroft, D.B. Barber, C.K. Ober, and C.R. Pollock, submitted to Advanced Materials, 1996.

2. V. Petricevic, S.K. Geyen, R.R. Alfano, K. Yamagishi, H. Anzai, and Y. Yamaguchi, Appl. Phys. Lett., **52**, 1040 (1987).

3. C.R. Pollock, Fundamentals of Optoelectronics, (Irwin, Boston, 1995).

4. L.L. Beecroft and C.K. Ober, Advanced Materials, **7**, 1009 (1995).

5. L.L. Beecroft, D.B. Barber, J.L. Mass, J.M. Burlitch, C.R. Pollock, and C.K. Ober, Proc. ACS Div.: Polym. Mat.: Sci. and Eng., **73**, 162 (1995).

6. K. Saric, Z. Janovic, and O. Vogl, Croatica Chemica Acta, **58,** 57 (1985).

7. D.R. Lide, ed., CRC Handbook of Chemistry and Physics, (CRC Press, Inc., 1994), 4-72, 4-48, 4-137.

8. V. Petricevic, S.K. Gayen, and R.B. Alfano, Appl. Phys. Lett., **53**, 2590 (1988).

9. L.L. Beecroft and C.K. Ober, manuscript in preparation.

10. J.L. Mass, J.M. Burlitch, S.A. Markgraf, M. Higuchi, R. Dieckmann, D.B. Barber, and C.R. Pollock, accepted in J. Crystal Growth, (1996).

PHOTOREFRACTIVE SOL-GEL MATERIALS

F. CHAPUT *, B. DARRACQ **, J.P. BOILOT *, D. RIEHL **, T. GACOIN *, M. CANVA **, Y. LEVY **, A. BRUN **
* Groupe de Chimie du Solide, Laboratoire de Physique de la Matière Condensée, URA CNRS 1254 D, Ecole Polytechnique, 91128 Palaiseau (France), fc@pmcsun1.polytechnique.fr
** Groupe d'Optique Non Linéaire, Institut d'Optique Théorique et Appliquée, URA CNRS 14, Bâtiment 503, B.P. 147, 91403 Orsay Cedex (France).

ABSTRACT

We report the synthesis and characterization of photorefractive sol-gel materials that possess covalently attached push-pull azobenzene and carbazole moieties. Molecular structural characterization of the modified silane monomers was achieved by ^1H NMR and infra red spectroscopy. The second-order nonlinear optical properties of the organic-inorganic hybrid films prepared from modified silane monomers were evaluated by second-harmonic generation. The stabilized value of the second harmonic coefficient, d_{33}, of films poled by corona discharge, at 1064 nm fundamental wavelenght was found to be 107 pm/V. Photorefractivity was clearly displayed from a two beam coupling experiment.

INTRODUCTION

Since the accidental discovery of the photorefractivity in 1966 in $LiNbO_3$ and $LiTaO_3$ numerous inorganic materials like ferroelectric crystals ($BaTiO_3$, $KNbO_3$, $(PbLa)(ZrTi)O_3$), sillenites ($Bi_{12}SiO_{20}$) and semiconductors (GaAs, InP) have been shown to exhibit the photorefractive effect [1]. They were studied and they are still studied for their potential application to optical information processing and holographic image storage. The photorefractive effect is observed in materials which exhibit both photosensitivity, photoconductivity and the electro-optic effect. This effect arises when charge carriers, photogenerated in the lighted regions of a nonuniform light intensity pattern, migrate and become trapped in the dark regions to produce an internal electric field. The space charge field changes the refractive index via the electro-optic effect. In the case of two light beam interferences, the photorefractive grating is 90° shifted to the light intensity distribution leading to asymmetric exchange of energy between the two beams.

In the past five years organic materials have emerged as an important new class of photorefractive media. Photorefractivity in organic crystals [2] liquid crystals [3] and composite polymers [4] has been reported. There is considerable interest in the development of photorefractive polymers owing to their large nonlinearities, low dielectric constant, structural flexibility, low cost and because they are easier to produce than inorganic and organic crystals. They are easily prepared into thin films as required for several applications.

From the above it is clear that to prepare sol-gel material exhibiting photorefractivity, different specific molecules or monomers which provide the indispensable properties have to be present inside the inorganic network. They can be attached as a pendant side group, or doped into the silica-based backbone. Charge transport is generally supported by the polymeric matrix itself. High volume fraction of charge transporting moiety is thus obtained. Charge generation and transport in organic polymers are intensively studied due to their xerographic application [5]. The most commonly used charge transporting group is carbazole group. A small amount of 2,4,7-trinitro-9-fluorenone (TNF) is added to increase the photosensitivity in the visible. It is well known that TNF forms a charge transfer complex with carbazole moities that absorbs photons to create mobile charges.

Second order optical nonlinear activity in the sol-gel materials is obtained by dipole orientation of nonlinear active chromophores previously dispersed in the sol. High volume fraction is generally attained by using nonlinear optical molecules grafted on silicon alkoxide [6]. In the past decade second order optical nonlinearity in organic polymer [7] and later in sol-gel material [6, 8] has also been intensively investigated for their potential application to frequency doubling and electro-optic devices. High electric field poling is required to generate dipole

Mat. Res. Soc. Symp. Proc. Vol. 435 © 1996 Materials Research Society

orientation in centrosymmetric matrices. Electric field poling consists in raising the temperature of the material, applying a strong electric field and then cooling with the field still applied. Concerning device applications, the major problem of organic polymers is the poor thermal and orientational stability of the poled materials. Sol-gel technology provides an attractive route to the preparation at low temperature of a rigid amorphous inorganic matrix which could slow down the molecular orientational motion which leads to the randomization.

We report in this paper the design, synthesis and optical characterizations of photorefractive sol-gel material, which contains a second-order chromophore (disperse red one=DR1) and a charge transporting molecule (carbazole groups) covalently attached to the silica based backbone. A small amount of TNF is added as doping molecule. We show that gel films exhibit large and stable quadratic nonlinearities as deduced from second harmonic measurements. Moreover, photorefractivity is confirmed by two-beam coupling measurements.

EXPERIMENTAL

Synthesis and characterization of side-group units

Carbazole attached alkoxysilane was synthesized by reacting carbazole-9-carbonyl chloride (CCC) with 3-aminopropyltriethoxysilane (APTES) in benzene in the presence of pyridine as HCl acceptor at room temperature for 1h (scheme 1).

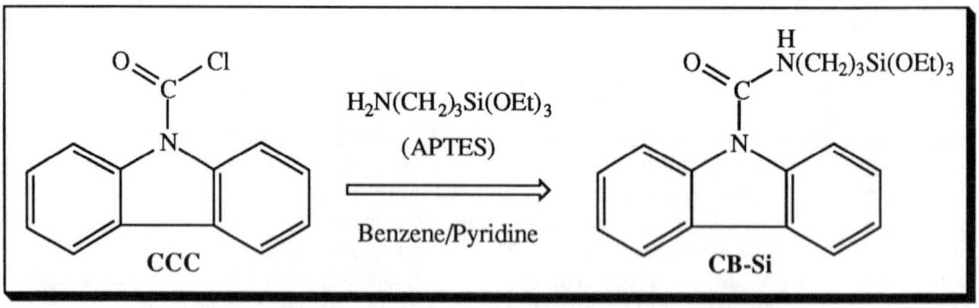

Scheme 1 : Synthesis of carbazole attached alkoxysilane.

Under N_2 atmosphere the APTES solution in benzene was added dropwise to the mixture of benzene, CCC and pyridine while stirring. The resulting salt was separated by filtration and the solvent was evaporated under reduced pressure at room temperature to give a brown product (CB-Si). Yield : 100%. ^1H, spectra were recorded on a AC200 Bruker spectrometer. IR spectra were recorded on a Perkin Elmer spectrometer.

^1H NMR ($CDCl_3$) : δ(ppm) = 0.8 (t, 2H, -**CH**$_2$-Si-), 1.2 (t, 9H, -O-CH$_2$-**CH**$_3$), 1.9 (q, 2H, -**CH**$_2$-CH$_2$-Si-), 3.6 (q, 2H, -CO-NH-**CH**$_2$-), 3.8 (q, 6H, Si-O-**CH**$_2$-CH$_3$), 6.10 (t, 1H, -CO-N**H**-CH$_2$-), 7.35 (td, 2H, aromatic protons), 7.50 (td, 2H, aromatic protons), 8.05 (d, 4H, aromatic protons).

IR (KRS5) : 3280 (s; NH), 1660 cm^{-1} (vs; C=O).

The NLO chromophore functionalized monomer was prepared according to published procedure [9]. The synthetic route was as follows : under slightly elevated temperatures (50-60°C), the hydroxy group of the azobenzene dye (Disperse red one : DR1) reacted with isocyanate group of the silicon precursor (3-isocyanatopropyltriethoxysilane : ICPTEOS) in pyridine. The dye functionalized monomer (DR1-Si) was isolated by vacuum distillation of the pyridine, washed with toluene and dried under vacuum.

^1H NMR ($CDCl_3$) : δ(ppm) = 0.65 (t, 2H, -**CH**$_2$-Si-), 1.25 (t, 9H, -O-CH$_2$-**CH**$_3$), 1.6 (m, 2H, -**CH**$_2$-CH$_2$-Si-), 3.20 (q, 2H, -CO-NH-**CH**$_2$-), 3.6 (q, 2H, =N-**CH**$_2$-CH$_3$), 3.7 (t, 2H, =N-**CH**$_2$-CH$_2$-O-), 3.9 (q, 6H, Si-O-**CH**$_2$-CH$_3$), 4.35 (t, 2H, =N-CH$_2$-**CH**$_2$-O-), 5.05 (t, 1H, -CO-N**H**-CH$_2$-), 6.90 (d, 2H, aromatic protons), 8.00 (dd, 4H, aromatic protons), 8.35 (d,

2H, aromatic protons).

IR (KRS5) : 3320 (s; NH), 1680 cm^{-1} (vs; C=O).

Film preparation

Coating solutions were obtained from copolymerization of modified silane monomers with tetraethoxysilane (TEOS) (Scheme 2). The hydrolysis was performed under acidic conditions with tetrahydrofuran (THF) as a common solvent.

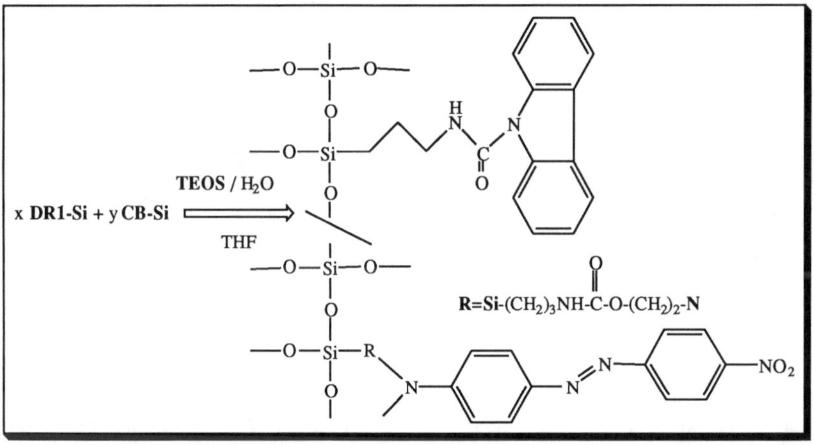

Scheme 2 : Synthesis of the silica-based backbone containing required functionalities for the photorefractive effect as pendant side groups.

The alkoxysilane : water : THF initial molar ratios were 1: 4 : 10. The atomic molar silicon concentration in alkoxysilane was defined as :

$Si_{alkoxysilane}=Si_{TEOS}+Si_{azobenzene}+Si_{carbazole}$ with $Si_{azobenzene}$: Si_{TEOS} : $Si_{carbazole}$ = 1 : 0.5 : 2.

A small amount of TNF was also dissolved in the sol (TNF : Carbazole molar ratio was 1 : 0.0137).

After hydrolysis for one hour at room temperature, the as-prepared sols were filtered (0.45 µm) to remove particle impurities before deposition. Thin films were made by spin-coating on glass, or indium tin oxide or gold covered substrates depending on the requirements of the intended characterization technique. Thicker films were also prepared from a multi-coating procedure. The angular speed of the spinner varied from 200 to 10000 rpm (figure 1). The samples thus obtained were finally dried at 120°C for 5 minutes until the surface of the film became tack-free. The spin-coated xerogel films are smooth and uniform with a bright red color.

Orientation of the nonlinear optical side groups in the sol-gel films was achieved by single-point corona poling technique using a sharp metallic needle as electrode suspended 12 mm above the film surface to which a voltage of 5.6 kV was applied. The sample was poled at 160°C for 15 minutes and cooled down (3 °C/min) to room temperature with the electric field maintained.Figure 2 shows the electronic absorption spectra of a gel film (spin coating rotation speed = 4000 rpm) before poling, shortly after poling and 22 days after poling, showing the relaxation of the poling-induced orientation.

NONLINEAR OPTICAL PROPERTIES

Second harmonic coefficients d_{13} and d_{33} were deduced from Maker fringe second-harmonic generation experiments. Measurements were performed on poled samples, at 1.0642 µm with a

Q-switched Nd:YAG laser. Maker fringes were recorded in sp- and pp-polarisation configurations, and calibrated by using 1 mm-thick α-quartz sample (d_{11}=0.46pm/V) as a reference. The value of d_{13} was extrapoled from the sp-curve and used to determine the d_{33} value in the fitting procedure of the pp-curve. We found 118 pm/V for d_{33} and 34 pm/V for d_{13}, immediately after poling. The d_{33} value stabilized to 107 pm/V after 2 months. Electro-optic measurements were performed at 633 nm on a 3.5-μm-thick gel film using an ellipsometric method. We found 18 pm/V 1 day after corona poling and 15 pm/V after 3 weeks. A more detailed study has been previously published [10].

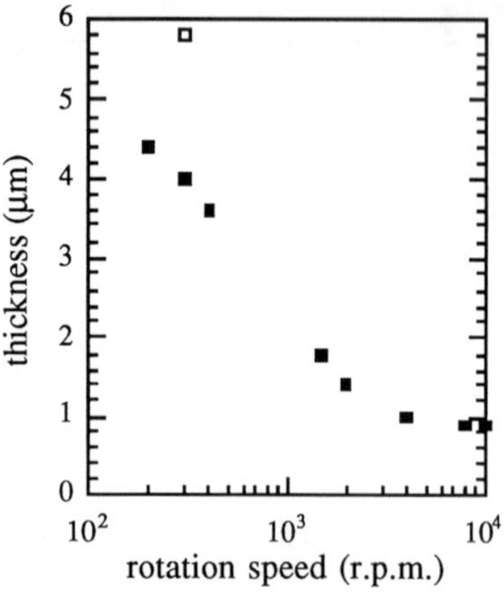

Figure 1 : Thicknesses of the film versus angular velocity of the spinner.
The open square indicates the thickness of a three-layer sample, spin-coated at 300 rpm.

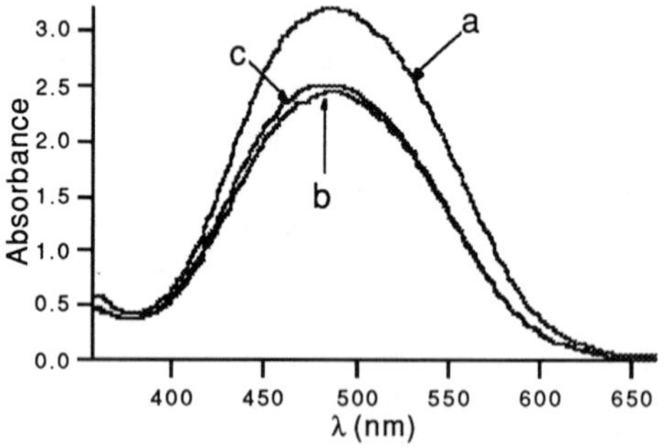

Figure 2 : Absorption spectra of a 4000 rpm DR1/carbazole gel film,
a) before poling,
b) shortly after poling,
c) 22 days after poling.

TWO BEAM COUPLING EXPERIMENTS

In order to unambiguously display photorefractivity in our samples, two beam coupling (2BC) experiments were performed on a 3.4-μm-thick poled sol-gel films without applying external electric field. A laser beam issued from a continuous wave He-Ne laser (632.8 nm) was divided into two s-polarized beams of equal intensity (2.5 mW). They were superposed on the same spot of 1 mm diameter onto the sample. The angle between the two beams was 46° and the light interference pattern created had a grating spacing of 0.81 μm. The sample was rotated by +17° from the bisectrix direction of the beams. Figure 3a shows the normalized intensity I/I(t=0) of both writing beams. Results obtained from two separate experiments are shown in this figure :

-a) In the first experiment, the normalized transmission of one beam was recorded without the presence of the second one. The fact that the normalized transmission was not constant is probably due to the photoisomerization of DR1 molecules under the linearly polarized light [11, 12]. Alignment of the molecules was induced perpendicularly to the polarisation of the writing beams and consequently the transmission of the s-polarized light increased.

-b) In the second experiment, the transmission of beam 1 and beam 2 were recorded for t=300 s. Then every 10 s, the two beams were alternatively switched off for 1 s to measure their transmission (figure 3b). An asymmetric energy exchange was observed and if the sample was rotated by 180° around its rotation axis or by -17° from the bisectrix direction the phenomenon was reversed.

These results are the clear signature of the photorefractive effect and this effect was not observed on unpoled samples. In order to observe and measure energy transfer only due to photorefractive grating, values of I/I(t=0) were divided by the corresponding normalized transmission values for each beam. Figure 4a shows the corrected curves which allowed us to measure the asymmetric change of intensity of the writing beam due to the 90° phase shifted photorefractive grating and to determine the BC gain coefficient by the following formula :

$$\Gamma = [\ln (m\gamma) - \ln (1 + m - \gamma)] / L$$

where L is the optical path (for beam with gain), m the ratio of the two writing beam energies before crossing the sample (m=1 in our case) and γ the beam coupling ratio. The saturation corrected ratio $I/I(t=0)=\gamma=1.08$ of beam 1 leads to a value of Γ of about 445 cm^{-1}. This value must be compared to the absorption coefficient α of 562 cm^{-1} at the same wavelength. So as to achieve a net internal gain, necessary for most applications, Γ-α must be positive, which is not yet the case. However, such a value of Γ has been obtained without applying external electric field during the 2BC experiment and without optimizing the sample structure, more especially its spectral response. Erasure time of photorefractive grating was also recorded under uniform illumination by one of the beam while the other is switched off (at t=320 s) then switched on every 60 s for 1 s to performe the measurement (figure 4b). Decay time at 90% of the saturation value is about 350 s.

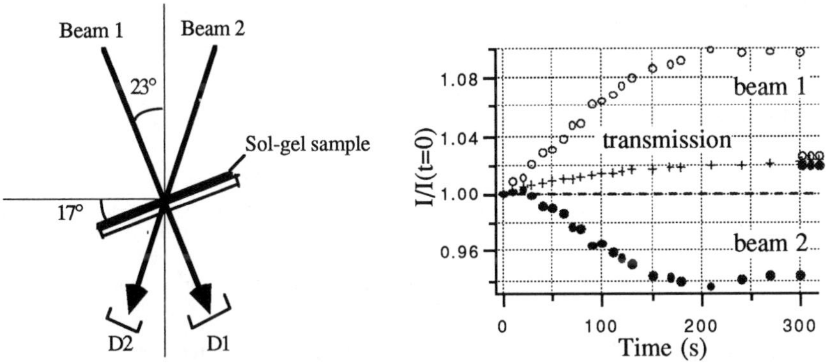

Figure 3 : (a) Two beam coupling arrangement (D1 and D2 are photodiode detectors), (b) 2BC experimal results at 633 nm for poled sol-gel films (see text for further information). Crosses correspond to the evolution of the normalized transmission for one beam while the other is switched off.

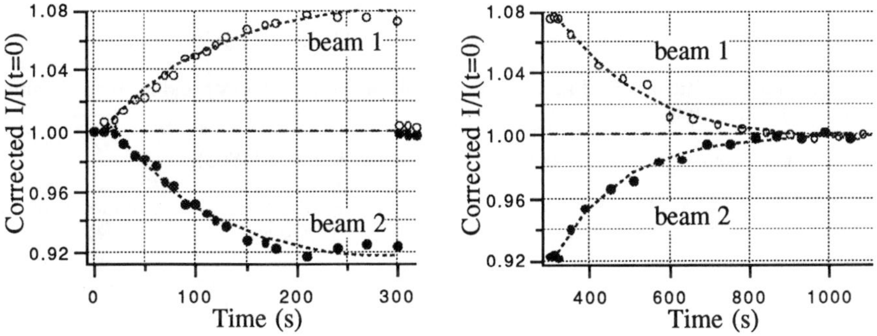

Figure 4 : (a) Corrected experimental values for 2BC experiments taking into account the change of the transmission for each beam, (b) Optical erasure of the photorefractive grating. Monoexponential fits (dashed lines) are guides for the eyes. Rise and decay times for beam 1 at 90% of the saturation are respectively 170 and 350 s.

Several reasons can explain long response times for photorefractive grating formation and optical erasure. Among them, two probably dominate : the mobility of charge carriers is only due to drift and the low dielectric constant ($\varepsilon \approx 4$) can favour geminate recombination of photogenerated charges [4] and thus limit photoconductivity. This analysis is under investigation.

CONCLUSION

We have prepared organic-inorganic hybrid films from modified silane monomers which exhibit second-order nonlinear optical properties. The value of the second harmonic coefficient, d_{33}, deduced from second-harmonic generation experiments near the resonance, of films poled by corona discharge, was found to be close to 100 pm/V.

Besides, photorefractivity was clearly displayed from a two beam coupling experiment. Even if a net internal gain is not yet achieved, the value of the Γ coefficient is very promising. Further experiments are now in progress : photoconductivity measurements, 2BC experiments with external electric field and degenerate four wave mixing. Better insight of the photorefractive effect in sol-gel materials will be thus evaluated.

REFERENCES

1. P. Günter and J.P. Huignard, eds. "Photorefractive Materials and their Applications I and II", Topics in Applied Physics vol. 61 and 62, Springer- Verlag (1988).

2. K. Sutter, K. and P. Grünter, J. Opt. Soc. Am. B7, **12**, p.2274 (1990).

3. I.C. Khoo, H. Li and Y. Liang, Opt. Lett., **19**, p.1723 (1994).

4. W.E. Moerner and S.M. Silence, Chem. Rev., **94**, p.127 (1994). G.G. Malliaras, PHD dissertation, University of Groningen (1995).

5. P.M. Borsenberger and D.S. Weiss, eds. "Organic Photoreceptors for Imaging Systems", Optical Engineering vol. 39, Marcel Dekker, Inc. (1993).

6. D. Riehl, F. Chaput, Y. Lévy, J.P. Boilot, F. Kajzar and P.A. Chollet, Chem. Phys. Lett. 245, p.36 (1995).

7. D.M. Burland, R.D. Miller and C.A. Walsh, Chem. Rev., **94**, p.31 (1994).

8. G. Puccetti, E. Toussaere, I. Ledoux, J. Zyss, P. Griesmar and C. Sanchez, Polymer Pepr., **32**, p.61 (1991).

9. F. Chaput, D. Riehl, Y. Lévy and J.P. Boilot, Chem. Mater., **5**, p.589 (1993).

10. F. Chaput, D. Riehl, J.P. Boilot , K. Cargnelli, M. Canva, Y. Levy and A. Brun, Chem. Mater., **8**, p.312 (1996).

11. S. Xie, A. Natansohn and P. Rochon, Chem. Mater., **5**, p.403 (1993).

12. F. Chaput, J.P. Boilot, D. Riehl and Y. Levy, SPIE's 1994 International Symposium on Optics, Imaging and Instrumentation, SPIE vol. 2288 Sol-Gel Optics III, p.286 (1994).

SILICONE-POLYOXOMETALATE (SiPOM) HYBRID COMPOUNDS

Dimitris E. Katsoulis and John R. Keryk
Dow Corning Corporation, Midland MI 48686, usdcccp5@ibmmail.com

ABSTRACT

Oligomeric and polymeric SiPOM materials have been prepared by reacting trichloro endblocked PDMS [degree of polymerization, ~ 14 **(I)** and 60 **(II)**] with α-$K_8SiW_{11}O_{39}$. NMR and IR spectroscopy show that the POM units are bonded to the siloxane via Si-O-W bonds. The reaction of the short PDMS segment **(I)** with the POM produces powdery materials, (isolated as tetrabutylammonium salts) while the longer PDMS **(II)** forms elastomeric materials. For the latter the POM anions act as reinforcing fillers. Other important properties of the hybrid materials are their UV absorbency, their electrochromicity, their electrical properties (volume resistivity, $\sim 10^{11}$ ohm·cm; low loss tangent, memory capability) and their film formation.

INTRODUCTION

The class of polyoxometalates[1] (POMs) of Mo, W and V has received considerable attention over the years and many uses have been identified for these compounds in areas of technology such as catalysis, photochemistry, biochemistry/medicine, chemical analysis, etc.[2] More recently POM research activity has been expanded to include sol-gel chemistry,[3] doping of polymers,[4] silica chemistry,[5] and silane chemistry.[6]

The class of silicone polymers also offers properties that make them very attractive for a large number of applications.[7] Due to the wide range of functionalities that can be easily grafted on the siloxane backbone, they lend themselves as starting oligomers and polymers for the synthesis of new families of materials. We decided to take advantage of the rich chemistries of siloxanes and POMs and create hybrid compounds (SiPOMs) that could combine the features of both classes of materials and have the potential to exhibit additional properties, possibly superior to the properties of their constituents.

We describe herein some initial results on the synthesis and characterization of two types of SiPOM hybrids from the reaction of α-$K_8SiW_{11}O_{39}$ with two trichloro endblocked PDMS oligomers that differ only on degree of polymerization [$Cl_3SiO(Me_2SiO)_zSiCl_3$, z≈14 **(I)** and ~60 **(II)**].

For the preparation of the SiPOM hybrid compounds we expanded the synthetic rationale first developed by Knoth[6] and later followed by Judeinstein.[7] These authors have shown that the lacunary (deficient) $SiW_{11}O_{39}^{8-}$ anion (abbreviated SiW_{11}, Fig. 1a) reacts with trifunctional silanes, $RSiX_3$ (X=Cl or OR; R = C_2H_5, C_6H_5, $NC(CH_2)_3C_3H_5$, $CH=CH_2$, $CH_2CH=CH_2$, etc.) to form organofunctional $(RSi)_2W_{11}SiO_{40}^{4-}$ anions (Fig. 1b).[6] Spectroscopic evidence indicate that the two silanes are covalently bonded to the surface of the $SiW_{11}O_{39}^{8-}$ anion via Si-O-W bonds between the silicon atoms of the silanes and the four oxygens that define the "hole" of the deficient anion. A new siloxane bridge [Si-O-Si] formed by the condensation of the third functional group of the organic precursors joins the two silanes right above the vicinity of the POM "hole". In a simplified view the POM acts as a molecular site for assembling siloxane molecules.

The hybrid materials described herein exhibit properties characteristic of both precursors. They absorb UV light, they are photoreducible and electrochemically active. The POM molecules act as fillers for the elastomer produced from the reaction with PDMS **(II)**. In addition, moderate conducting properties are measured for these hybrids, attributable to the presence of POMs.

Mat. Res. Soc. Symp. Proc. Vol. 435 © 1996 Materials Research Society

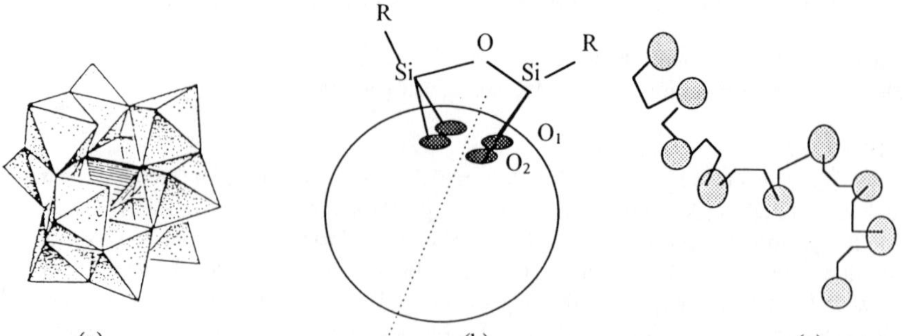

(a) (b) (c)

Fig. 1. (a) Octahedron representation of α-$SiW_{11}O_{39}^{8-}$ anion. (b) Schematic representation of the $(RSi)_2SiW_{11}O_{40}^{4-}$ anion. The four shaded circles depict the POM oxygens that are involved in the bonding with the organosilane groups. The dotted line lies on the C_s plane of symmetry that dissects the POM and results to two equivalent pairs of oxygens. (c) Idealized schematic representation of the **SiPOM-(II)** polymer.

EXPERIMENTAL

Materials and Methods: ^{29}Si NMR of SiPOM adducts were recorded using a pulse width of 12.4 μs and a relaxation delay time of 60 s in order to obtain complete relaxation of the silicon nuclei.

Synthesis of SiPOM-(I) adduct: Fifteen grams (~$5x10^{-3}$ moles) of dried (overnight at 115-117°C) $K_8SiW_{11}O_{39}$ were stirred for 2 hours in a mixed solvent of $CH_3CN/C_6H_5CH_3$ (404 g each). To the white slurry 5.31 g ($4x10^{-3}$ moles) of oligosiloxane **(I)** dissolved in 20 g toluene were slowly added over a period of ~0.5 hour. The mixture was stirred for 72 hours at ambient conditions. Any insoluble material was filtered off, 14.18 g ($4.4x10^{-2}$ moles) $(n\text{-}C_4H_9)_4NBr$ were added, and the mixture was allowed to stir for 1.5 hours prior to another filtration. The clear filtrate that contained the final product was stripped in a rotary evaporator (P<5mmHg and T≈55-57°C) and after ~45 min. 29.7 g of a creamy opaque, stiff and tacky material [contained excess of $(n\text{-}C_4H_9)_4NBr$] was obtained. The product was re-dissolved in CH_3CN and reprecipitated by the addition of water as a finely divided white powder.

Synthesis of SiPOM(II) adduct: An amount of 1.38 g ($3x10^{-4}$ moles) oligosiloxane **(II)** was dissolved in a mixed solvent of 20 g CH_3CN and 27 g $C_6H_5CH_3$. Then, 1.5 g ($5x10^{-4}$ moles) dried $K_8SiW_{11}O_{39}$ were added and the mixture was allowed to stir overnight. Next day the solids were separated by centrifuging and decanting the mother liqueur and 0.048 g ($1.3x10^{-4}$ moles) $(n\text{-}C_4H_9)_4NBr$ were added. The slightly hazy solution was centrifuged, the clear layer was decanted and its volume was reduced by rotary evaporation until it obtained an opaque, milky appearance. The product produced an elastomer when it was allowed to completely dry on a substrate such as a glass plate.

RESULTS AND DISCUSSION

Heterogeneous reactions were carried out in CH_3CN / $C_6H_5CH_3$ mixed solvent between solid $K_8SiW_{11}O_{39}$ and the oligosiloxanes **(I)** or **(II)**. The POM was kept at a slight excess (1.1:1 ≤ mole ratio, POM : siloxane ≤ 2.5:1). The SiPOM adducts produced from **(I)** {referred herein as **SiPOM-(I)**} were isolated as off-white powders by precipitation with $(n\text{-}C_4H_9)_4NBr$.

The adducts were soluble in DMSO, DMF, CH_3CN. They were insoluble in EtOH, IPA, CH_3COCH_3, THF and PDMS. Their infrared spectra suggest that the compounds retain a modified (substituted) Keggin structure that includes the principle siloxane stretches (Table I).

Table I. Selective infrared stretching frequencies (cm^{-1}) for the reaction product of $K_8SiW_{11}O_{39}$ with oligosiloxane **(I)** **{SiPOM-(I)}** and oligosiloxane **(II)** **{SiPOM-(II)}**; precipitated as $(C_4H_9)_4N^+$ salt (cation frequencies omitted).

SiPOM-(I)	SiPOM-(II) (film)	Assignment
1260	1262	$\nu(CH_3)_{sym}$
1097, 1028	1093, 1026	$\nu(Si\text{-}O\text{-}Si)_{asym}$
1050 (br)		$\nu(SiOW)_{ext}$
996 (sh)		$\nu(SiO)_{central}$
964	977	$\nu(W{=}O)_{ter}$
923, 906, 886 (sh), 873 (sh)	924, 881	$\nu(WOW)_{asym}$
846		
803	795	$\nu(SiC)_{asym}$
762 (sh), 738(sh), 717(sh)		$\nu(W\text{-}O_c)$
615		
549 (sh)		$\nu(WOW)_{sym}$
534	536	

The band at 1050 cm^{-1} is assigned to Si_{ext}-O-W bonds in accordance to assignments by Knoth[6] and Judeinstein[7] for model compounds of the general formula $SiW_{11}O_{39}O(SiR)_2^{4-}$.

^{29}Si and ^{183}W NMR spectra (Table II) support the argument that the POM anions are covalently bound to the siloxane segments via Si-O-W bonds. The ^{29}Si NMR spectrum of **SiPOM-(I)**, besides the prominent $-Me_2SiO-$ resonance at ~ -22 ppm, contains a set of three resonances at - 84.7, -85.1 and -85.4 ppm and a third set at -92 ppm. This last resonance is assigned to the external silicon nuclei that link the oligosiloxane **(I)** to the POM via the two non-equivalent oxygens of the "hole" $[-OSiO(O_1W_1)(O_2W_2)]$. The three major resonances at around -85 ppm indicate that the SiPOM is not of a single composition but a mixture of siloxane substituted POMs. The resonance at -85.1 ppm is assigned to the central $SiO_{4/2}$ of the complete $SiW_{12}O_{40}^{4-}$ Keggin anion. (It is well known that under acidic conditions the lacunary $SiW_{11}O_{39}^{8-}$ reverts to the $SiW_{12}O_{40}^{4-}$ Keggin anion).[1] The resonance at -84.7 ppm is assigned to the central $SiO_{4/2}$ tetrahedra of monosubstituted POMs (only one siloxane oligomer per POM). Two of the four oxygens that define the "hole" of the lacunary structure are utilized as linkages between the siloxane and the POM. The other two remain free. The resonance at -85.4 ppm is assigned to the central $SiO_{4/2}$ tetrahedra of disubstituted POMs (two siloxane oligomers per POM). Each oligomer is anchored to two non-equivalent oxygens of those that define the "hole" of the deficient POM structure via the trifuntional terminal silicon group . The third functionality is used to link the two siloxane oligomers through a siloxane bond over the area of the "hole". This latter arrangement retains the C_s symmetry of the POM, while the former one reduces it to C_1. The approximate ratio of monosubstituted siloxane -POMs to the disubstituted siloxane-POMs is estimated from the ^{29}Si spectra to approximately 1/2. The average degree of polymerization of the coordinated siloxane was also estimated from the relative intensities to be ~40, consistent with the above assignments.

Table II. NMR data of the reaction product of $K_8SiW_{11}O_{39}$ with oligosiloxane **(I)** {**SiPOM-(I)**}. Solvent, DMSO-d_6. Chemical shifts for ^{29}Si referenced to TMS; for ^{183}W referenced to 2M Na_2WO_4.

Nucleus	Chemical shift, δ(ppm); (no. of W)
^{29}Si	-19.3, -22 [-$OSiMe_2$]-
	-84.7
	-85.1 [$SiW_{11}O_{40}^{4-}$]
	-85.4
	-92 [Si_{ext}-O-W]
^{183}W	-105(2W), - 107.1(2W), -115.3(1W), -129(2W), -178(2W), -256(2W) [disubstituted POM]
	-68.2, -86.1, 100.2, -102.8, -108[a], -124.4, -128[a], -149.4, -169.7, 174.4, -215 [monosubstituted POM]
^{183}W for $SiW_{11}O_{39}^{8-}$[b]	-98.5(2W), -114.4(2W), -120.5(1W), 127.5(2W), 141.1(2W), -173.5(2W)

[a] shoulder on resonance of the disubstituted POM. [b] solvent, D_2O; temperature, 90°C.

The ^{183}W NMR spectrum of **SiPOM-(I)** adduct compliments the ^{29}Si NMR spectrum. (Table II). A peak at -93 ppm is attributed to the twelve equivalent tungsten nuclei of $SiW_{12}O_{40}^{4-}$ anion. The remaining of the complex pattern can be analyzed in two sets. A six line pattern (relative intensity, 2:2:1:2:2:2) attributed to the disubstituted POM component of the mixture (C_s symmetry)[9] and an 11 line pattern (all of equal intensity) which is due to the monosubstituted POM component (C_1 symmetry). The large chemical shifts (up to 100 ppm) observed for some of the tungsten nuclei as compared to those of the $SiW_{11}O_{39}^{8-}$ precursor, indicate important shielding effects induced by the anchoring of the siloxane oligomers to those tungsten atoms.

Low angle laser light scattering experiments in (DMF) indicated that the formula weight of the **SiPOM-(I)** adduct was < 10000 g/mol. This information (although of unknown accuracy and precision for these type of compounds) suggests that the adduct is most probably composed of monomeric (and/ or possibly dimeric) siloxane-POM complexes rather than large copolymers. The exact arrangement and ratios of the siloxane and POM in these complexes still remains to be determined.

Properties both of the POM and the siloxane components were observed in **SiPOM-(I)** adduct. For example the material absorbed in the UV (λ_{max} = 261 nm) and was electrochemically active. The cyclic voltammogram of **SiPOM-(I)** in CH_3CN/0.1 M $(C_4H_9)_4NClO_4$ shows four well defined redox couples observed at potentials slightly more negative to those of the parent Keggin anions (Table III). The TGA analysis of the material in Air and in He indicated that the siloxane portion of the adduct remained intact up to 200°C (5 wt% reduction at ~300°C). This data compares well with the usual stability of PDMS oligomers.

In contrast to **SiPOM-(I)** which resembled an inorganic salt, the reaction product of $K_8SiW_{11}O_{39}$ with the longer PDMS chain **(II)** was an opaque, fairly tough, elastomeric material {**SiPOM-(II)**}. It was obtained upon complete evaporation of the solvent ($CH_3CN/C_6H_5CH_3$) from a reaction mixture POM : siloxane (mole ratio 1.5-2:1) to which $(C_4H_9)_4N^+$ cations had been added. The rubbery material had good solvent resistance in common polar and nonpolar solvents.

TEM micrographs (Fig. 2) revealed the presence of at least two phases. High electron density irregular particles (20-300 nm) that contained Si, W and K (phase separated SiW_{11} and SiW_{12} salts) were embedded into a high Si content amorphous region (PDMS rubber). In this material the POMs were acting as reinforcing fillers.

Table III. Oxidation / Reduction potentials ($E_{1/2}$) for **SiPOM - (I)** and **(II)** adducts. Solvent, CH_3CN/0.1 M $(C_4H_9)_4NClO_4$.

Material	$E^{red}_{1/2}$ (V)	$E^{ox}_{1/2}$ (V)
SiPOM-(I)	-0.83, -1.35, -1.58, -2.00	-0.70, -1.24, -1.47, -1.83
SiPOM-(II)[a]	-0.83, -1.33, -1.80, -2.27	-0.72, -1.11, -1.56, -1.91
$[(C_4H_9)_4]_4SiW_{12}O_{40}$	-0.71, -1.22, -1.92	-0.61, -1.13, -1.82

[a] CV recorded from a casted film on carbon glass electrode.

Fig. 2. TEM of the **SiPOM-(II)** elastomer, from the reaction of α-$K_8SiW_{11}O_{39}$ and $Cl_3SiO(Me_2SiO)_zSiCl_3$, z≈60. Bar, 500 nm.

Evidence from ^{29}Si NMR experiments indicated the existence of covalent bonding between the siloxane oligomers and the POMs (resonances at ~ -91 ppm). Additional resonances at -99 ppm were also observed which were assigned to silicon nuclei of "T" structure due to condensation reactions between siloxane oligomers. These "T" structures cured even further upon evaporation of the solvent to produce the silicone-POM hybrid elastomer. These data suggest that a portion of the POM has been incorporated into the PDMS network, anchored via Si-O-W bonds {similarly to the **SiPOM-(I)** adduct}, and a larger portion has been trapped as a second phase into the network. Under the acidic conditions of the chlorosilane **(II)**, it is almost certain that a significant portion of the lacunary $SiW_{11}O_{39}^{8-}$ has been converted to the complete Keggin $SiW_{12}O_{40}^{4-}$ structure (which has no coordinative sites). An idealized structure of the **SiPOM-(II)** polymer is shown in Fig. 1c.

DSC analysis showed that the material had two glass transition temperatures at -117°C (assigned to the Tg of the siloxane component) and at +27°C (assigned to the Tg of the polymer region in the vicinity of POMs). A thin film of the polymer formed on a quartz substrate absorbed UV light below 320 nm.

The hybrid nature of the elastomer was demonstrated by measuring its electrical properties. Its volume resistivity was $7x10^{10}$ ohm·cm, indicating a moderately insulating material. Its conductivity did not decrease with time indicating that there were free electrons in the material. The low energy loss measured (0.620, at 100 Hz, ambient temperature) suggested that the POMs were held quite well within the PDMS elastomers. The dielectric constant of the elastomer decreased greatly from 53 at 100 Hz to 6.5 at 100 kHz reflecting the large size of the POMs and their sluggish response to higher frequencies. However, the loss tangent remained in the same order of magnitude (0.620 at 100 Hz vs. 0.380 at 100 kHz) indicating that there was some electronic nature to the loss; in most materials, the loss tangent also will decrease substantially from 100 Hz to 100 kHz. When the electrical measurements were repeated for a second time larger dielectric constant values were obtained suggesting that some memory capabilities might exist in this hybrid material.

Polymer film castings from **SiPOM-(II)** reaction mixture on glassy carbon electrodes showed electrochemical activity. Four pairs of redox waves were obtained in the cyclic volatmmogram (Table III). They are assigned to the covalently bound POM on the PDMS polymer.

CONCLUSIONS

Hybrid silicone-polyoxometalate materials were easily prepared by the reaction of $SiW_{11}O_{39}^{8-}$ anion and trichloro endblocked PDMS oligomers. Spectroscopic data indicate that the siloxane and POM are linked together via Si-O-W bonds. The hybrid materials exhibit properties of both components. They can be elastomeric, photoreducible, electrochemically active, ionic conductors, UV absorbers, etc.

ACKNOWLEDGMENTS

We thank Fred Dall for the electrical measurements and Rebecca Durall for the electron microscopy studies. We are grateful to Dow Corning Corporation for supporting this project.

REFERENCES

1. M.T. Pope in Heteropoly and Isopoly Oxometalates (Springer Verlag, New York, 1983).
2. Polyoxometalates: From platonic solids to anti-retroviral activity, edited by M.T. Pope and Achim Müller (Kluwer Academic Publishers, Dordrecht, 1993).
3. M. Tatsumisago, K. Kishiba, T. Minami Solid State Ionics **59,** 171, (1993)
4. B. Fabre, G. Bidan, Adv. Mater. **5,** 646, (1993).
5. N. Azuma, R. Ohtsuka, Y. Morioka, H. Kosugi, J. Kobayashi, J. Mater. Chem. **1**, 989, (1991).
6. W.H. Knoth, J. Am. Chem. Soc. **101,** 759, (1979)
7. E.G. Rochow in Silicon and Silicones (Springer Verlag, New York, 1987); F.O. Stark, J.R. Falender, A.P. Wright in Comprehensive Organometallic Chemistry, edited by G. Wilkinson (Pergamon Press, New York, 1982), p. 305.
8. P. Judeinstein, C. Deprun, L. Nadjo, J. Chem. Soc. Dalton Trans. 1991, 1991; P. Judeinsten, Chem. Mater. **4**, 4, (1992).
9. R. Acerete, C.F. Hammer, L.C.W. Baker, J.Am. Chem. Soc. **104,** 5384, 1983; C. Brevard, R. Schimp, G. Tourne, C.M. Tourne, ibid. **105,** 7059, 1983.

NOVEL SOL-GEL PROCESSED PHOTOREFRACTIVE MATERIALS

RYSZARD BURZYNSKI[†], SASWATI GHOSAL[†], MARTIN K. CASSTEVENS[†], YUE ZHANG[‡]
[†]Laser Photonics Technology, Inc., 1576 Sweet Home Rd., Amherst, NY 14228
[‡]ROI Technology, Optical Materials Division, 2000 Cornwall Road, Monmouth Junction, NJ 08852

ABSTRACT

We report the development and characterization of a new photorefractive multifunctional ormosil consisting of a second-order nonlinear optical chromophore and a charge transporting group covalently bound to a silicon atom. The sol-gel technique is used to process this ormosil into a homogeneous, single-phase material which exhibits electrooptic and charge transporting properties. When doped with a photocharge generation sensitizer, the material shows photorefractivity as evidenced by the electric field dependence of the four-wave mixing diffraction efficiency and that of the two-beam coupling gain.

INTRODUCTION

Photorefractive materials can be used in many applications such as rewritable holographic information storage, image reconstruction, phase conjugation, and optical interconnects.[1] Polymeric photorefractive materials have attracted great attention[2-6] in the past few years because of their potential advantages over inorganic crystals which have been studied for over three decades. These advantages include larger nonlinearities, lower dielectric constants, greater processability, and lower cost. Composite photorefractive materials based on side-chain second-order nonlinear optical polymers,[3] photoconductive polymers,[4] and even inert polymer binders[5] have been processed and characterized. Several multifunctional polymers which combine all the required functional groups attached to a polymer backbone have also been reported.[6]

Although high diffraction efficiencies and beam coupling gains have been obtained in photorefractive polymer composites, their use in practical applications can be limited because of potential phase separation in these materials, especially at high doping levels, which results in the loss of optical quality. Sol-gel processed optically transparent glass and ceramic materials provide a new generation of molecular composites extremely useful in the design of optical devices.[7] Low temperature sol-gel processing permits countless varieties of molecules with exceptional optical and chemical properties to be incorporated into oxide glasses. More recently, sol-gel processing has been applied to organically modified silanes (ormosils)[8] to produce high optical quality materials possessing the properties of both oxide glass and organic compounds.

We recently reported the development of the first sol-gel processed ormosil exhibiting electrooptic and photoconductive properties.[9] When doped with a photocharge generation sensitizer, 2,4,7-trinitrofluorenone (TNF)-carbazole charge-transfer complex, the material showed photorefractivity. Low temperature sol-gel processing techniques have been utilized to obtain optical quality films in a wide range of thicknesses from several to more than 200 μm. This paper describes the detailed properties of another ormosil where a thiapyrylium dye is used as a charge generation sensitizer. The electrooptic coefficient, photoconductivity, four-wave mixing diffraction efficiency, and two-beam coupling gain have been measured. The electric field dependence of these parameters prove the photorefractivity.

Mat. Res. Soc. Symp. Proc. Vol. 435 © 1996 Materials Research Society

EXPERIMENTS

The molecular structures of ormosils used in sol-gel processed composites are presented in Figure 1. The ormosils were synthesized by reacting trimethoxysilylpropylisocyanate (TMSC) with the second-order NLO chromophore 4-(N,N-bis(β-dihydroxyethyl)amino)-4'-nitrostilbene (DHD) or the charge transporting agent N-(β-hydroxyethyl)carbazole (HCAR) under reflux and nitrogen gas using a catalyst and chlorobenzene as a solvent. Upon completion, chlorobenzene was distilled off under vacuum. Butanol solutions of the ormosils were mixed in 3:1

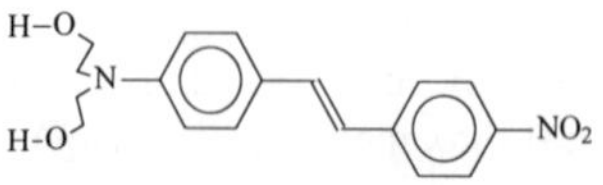

(1)

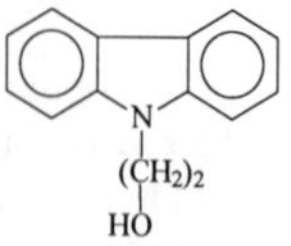

(2)

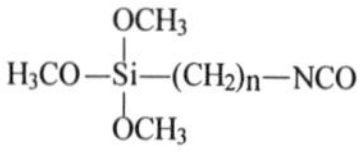

(3)

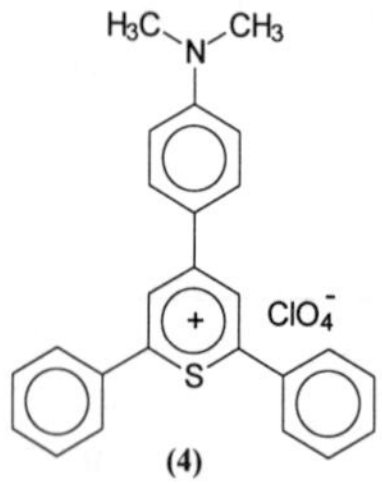

(4)

Figure 1 Molecular components of the sol-gel composite: (1) second-order NLO chromophore, DHD; (2) charge transporting molecule, HCAR; (3) trimethoxysilylpropylisocyanate, TMSC, and (4) photocharge generation agent, TPY.

(HCAR:DHD) molar ratio and hydrolyzed (50 °C ultrasonic bath, 30 min.) by adding water (3:1 H_2O:Si molar ratio) and formic acid as a catalyst. A solution of thiapyrylium dye (TPY, a well known photocharge generating dye[10]) was added to the hydrolyzed ormosils at room temperature at a concentration of 0.2 wt.% or 0.4 wt.% to introduce a charge generation property. Films 2-3 μm thick were deposited on indium-tin-oxide (ITO) coated glass substrates by spin coating filtered ormosil solutions (0.2 μm pore size Teflon® membrane). Thicker films, 80-150 μm thick, were prepared by casting approximately 0.7 mL of the filtered ormosil solution on 2.5×5 cm microscope slides coated with a 1 cm wide strip ITO electrode centered along its length. The films were dried at room temperature for 3 days and, then, in a vacuum oven at 80 °C for 24 hours. A thin layer of semitransparent gold electrode was deposited on the top of thick films used for four-wave and two-wave mixing experiments and a set of thicker, circular gold electrodes was deposited on thin films used for the measurements of photoconductive and electrooptic properties.

The characterization of ormosil composites included measurements of the electrooptic coefficient, photoconductivity, four-wave mixing diffraction efficiency, and two-beam coupling gain. The EO coefficient was measured by an ellipsometric technique using a setup similar to that described by Teng and Mann.[11] The photoconductivity was measured using a simple photocurrent technique in which a He-Ne laser beam illuminates the sample through the glass substrate.[12] The photocurrent I_{ph}, calculated from the measured voltage drop on a resistor connected in series to the film, was used to compute the photoconductivity σ and the photoconductivity sensitivity S of the material according to the following equations:[13]

$$\sigma = \frac{L}{AV/I_{ph}} \; ; \quad S = \frac{\sigma}{I_0} \tag{1}$$

Here, L is the sample thickness, A the illuminated area (the area of the metal electrode), V the applied dc voltage, and I_0 the fluence of the laser beam.

The photorefractive properties of the material were studied by four-wave mixing and two-beam coupling experiments using an experimental setup described earlier.[13] Holographic gratings were written by two laser beams from a He-Ne laser operating at 633 nm. Two writing beams with an equal intensity of 250 mW/cm^2 intersected in the sample at incidence angles of θ= 45° and θ= 56° (in air), respectively, creating an intensity grating with a periodicity of approximately Λ=3.4 μm. The grating wavevector is oriented at an angle of θ_G = 30.5° with respect to the film surface. The reading beam from another He-Ne laser at 633 nm with an intensity of 10 mW/cm^2 propagated in the direction opposite to one of the writing beams (Beam 2). The diffracted signal propagated in the direction opposite to that of Beam 1, reflected off a beam splitter, and then detected by a photodiode. In a two-beam coupling experiment, the energy transfer between the writing beams was observed by monitoring the intensity of each of the writing beams when an external electric field was applied. The same energy transfer was also observed by the increase (or decrease) of the beam intensity while the other writing beam was open and blocked.

RESULTS AND DISCUSSION

Figure 2 (a) displays the electric field dependence of the photoconductivity sensitivity for the two composites containing different amounts of TPY photosensitizer. The filled circles and open squares correspond to materials with 0.2 wt.% and 0.4 wt.% of TPY dye, respectively, in a DHD/HCAR ormosil composite. Films with the higher concentration of TPY exhibit about an order of magnitude higher photoconductivity sensitivity at all fields. It is clear that the efficiency of photocharge generation is strongly dependent on the concentration of the charge generating dye. However, a compromise between the material's transparency and the strength of the photoresponse has to be reached in order for the materials to be practical, i.e., too large an optical density may render a material of limited use in applications requiring visible wavelengths regardless of its good photocharge generation and transport properties.

The photoconductivity of the same composites was also measured as a function of the illumination intensity (see Figure 2 (b)). The observed data can be fit to a sublinear dependence, $\sigma(I_0) \sim I_0^x$ with a best fit parameter x of 1.05 and 0.79 for TPY concentrations of 0.4 and 0.2 wt.%, respectively. Assuming that the charge mobility is independent of the illumination intensity, the photocharge generation quantum efficiency can then be expressed as

$$\phi(I_0) = d\, I_0^x \; , \tag{2}$$

where d is a constant. This type of response has also been observed in inorganic photorefractive crystals such as barium titanate ($BaTiO_3$)[14] as well as in a number of polymeric composites, with x values ranging from 0.72[13] to 0.97.[15] The origin of a sublinear dependence of the material photoconductivity on the light intensity in photorefractive polymeric composites has not yet been determined and, in most cases, has been assigned to the presence of shallow traps (or acceptors) as in inorganic crystals. We propose a similar mechanism in ormosil composites; the almost linear dependence for the composite with a doubled amount of TPY dye may indicate that the concentration of photocharges exceeds that of shallow traps.

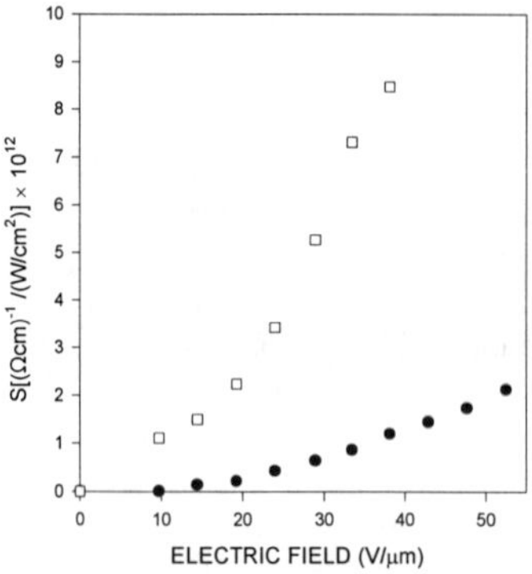

(a)

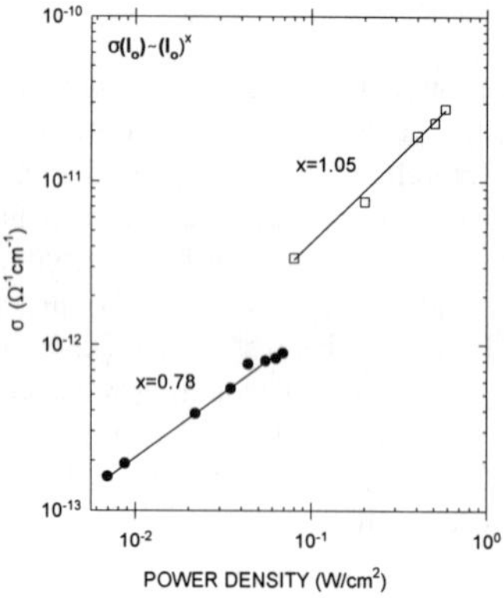

(b)

Figure 2 Charge transporting properties of ormosils. (a) Electric field dependence of photoconductivity sensitivity for composites containing 0.2 (filled circles) and 0.4 (open squares) wt.% of TPY. (b) Photoconductivity as a function of illumination ligth intensity for the two composites

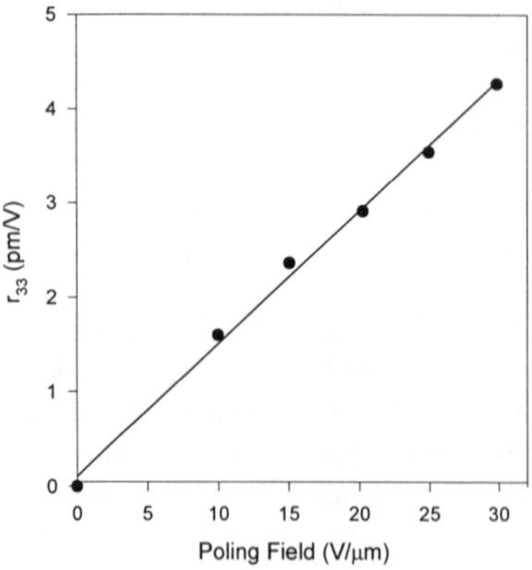

Figure 3 Electric field dependence of the electrooptic coefficient obtained from the ormosil containing 0.2 wt. % of TPY. The EO coefficient of a composite containing 0.4 wt.% TPY is virtually identical.

The EO coefficient was measured on 2.5 μm thick films as a function of the poling electric field and a linear dependence was obtained up to 30 V/μm. As expected, the EO coefficient was found to be independent of the TPY concentration within experimental error. Figure 3 shows the field dependence of r_{33} for a composite containing 0.2 wt.% of TPY; similar results were obtained on films containing 0.4 wt.% of TPY. At a field of 95 V/μm, the maximum electric field applied in the four-wave mixing experiments, the electrooptic coefficient (r_{33}) was measured to be 9 pm/V.

The photorefractive properties of the material were evaluated using a four-wave mixing arrangement. The diffraction efficiency was measured as a function of the external electric field (see Figure 4 (a)). The diffraction efficiency increases rapidly with the applied electric field and reaches a value close 0.065 % at an external field of 95 V/μm. It should be pointed out that the field dependence of the diffraction efficiency also includes the field dependence of the composite's electrooptic coefficient. The electrooptic coefficient at low field can be expressed as[16]

$$r_{33} = 3r_{13} = -\frac{8\pi}{n_0^4}\frac{fN\mu\beta E_0}{5kT} \tag{3}$$

where f is the appropriate local field factor, N the number density of the NLO chromophores, μ the molecular dipole moment, β the first hyperpolarizability, E_0 the applied electric field, n_0 the refractive index of the composite, k the Boltzman constant and T is the temperature. The holographic diffraction efficiency with a p-polarized reading beam is given by[2]

$$\eta_p = \sin^2\left(\frac{\pi\, n_0^3\, r_{eff} E_{sc} d \cos(2\theta_0)}{2\,\lambda\,(\cos\theta_1 \cos\theta_2)^{1/2}}\right) \tag{4}$$

where θ_1 is the incidence angle of beam 1, θ_2 the incidence angle of beam 2 and θ_0 the full angle between the writing beams, all inside the material. The space charge field, E_{sc}, is given by:[1, 2]

$$E_{sc}^2 = \frac{E_d^2 + E_{0g}^2}{(1+b)^2 + b^2(E_{0g}/E_d)^2} \tag{5}$$

where $E_d = K_g kT/e = 0.105$ V/μm, K_g is the grating wavevector, E_{0g} is the component of E_0 along the grating wavevector and b is a constant of the material. The experimentally measured diffraction efficiency, η_p, was fitted to Eqs.(3)-(5) and are shown as solid lines in Figure 4 (a).

The holographic grating formation dynamics in the material have been studied by measuring the grating writing and erasure processes and their electric field dependence. The inset of Figure 4 (b) shows a typical four-wave mixing signal with the writing process fitted to $I_s \propto E_{SC}^2[1 - \exp(-t/\tau)]^2$ and the erasure fitted to $I_s \propto E_{SC}^2[\exp(-t/\tau)]^2$ where E_{SC} is the saturation space charge field and τ is the writing or erasure time constant. The measured field dependence of the writing rate is presented in Figure 4 (b) where the writing rate is plotted against the square root of the electric field on a semi-log scale. The linear dependence is consistent with the field dependence of the charge mobilities as the writing time constant is related with the charge mobility by $\tau \propto 1/\mu$ and $\tau \propto \exp(E^{1/2})$.

To prove the photorefractive nature of the observed holographic gratings, two-beam coupling experiments were performed using two p-polarized writing beams with approximately equal intensities. The intensity of each writing beam was monitored while the other beam was switched on and off in the presence of the applied electric field. Clear asymmetric energy transfer has been observed as the intensity of beam 1 increases when beam 2 is switched on, while that of beam 2 decreases by an almost equal amount when beam 1 is on. This asymmetric beam intensity change results from a finite phase shift between the refractive index grating induced by the space charge field through a linear electrooptic effect and the interference pattern of the writing beams. The measured two-beam coupling gain as a function of the applied electric field is shown in Figure 4 (a) (open triangles).

(a)

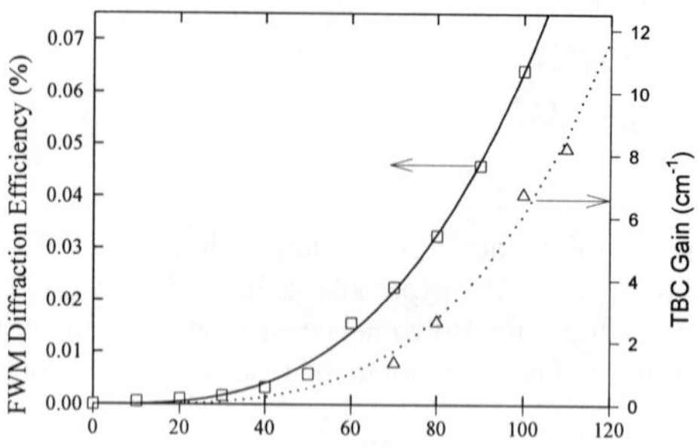

(b)

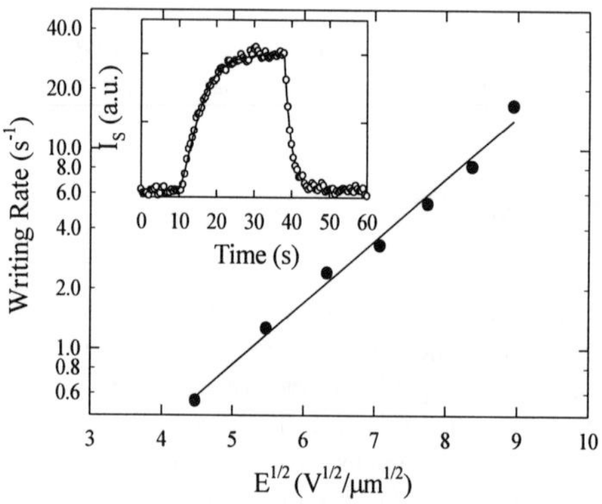

Figure 4 (a) Field dependence of diffraction efficiency and two beam coupling gain in TPY (0.2 wt.%) sensitized photorefractive ormosil composite. Open squares: diffraction efficiency; open triangles: TBC gain. Solid lines represent computer fit curves to experimental data. (b) Dynamics of holographic grating formation in TPY sensitized composite. Inset shows a typical buildup and erasure of a holographic grating.

Since these materials can potentially be used in information storage applications, it is important to determine the longevity of the holographic gratings. Figure 5 shows the dark decay of the gratings in the presence of the reading beam, and when the reading beam and the electric field are both turned off. In the presence of the reading beam, which partially erases the gratings, the readout signal quickly decays to zero in about 150 seconds with a time constant of approximately 59 seconds. This decay results from the reading beam erasure in the presence of the electric field. When the field is turned off after the grating was written, the grating decay process becomes slower, as shown by the filled circles in Figure 5. The true dark decay process was determined by turning off the electric field and the reading beam. At intervals of about 50 seconds, the reading beam and the electric field are turned on for 1 second and the readout monitored. Open triangles in Figure 5 show the measured dark decay curve of the holographic gratings with a time constant of 310 seconds.

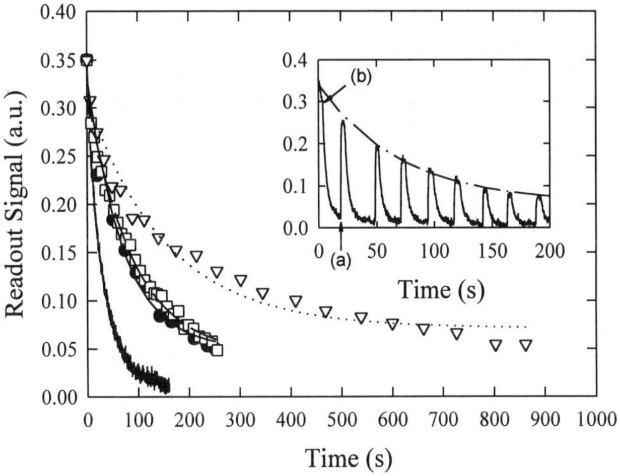

Figure 5 Dark decay of the holographic gratings under various conditions. Solid line: grating decay with the reading beam and electric field on; Open squares: decay with field applied and the reading beam switched on each 10 seconds; Filled circles: decay with reading beam on and field turned on for short intervals; Open triangles: decay with both field and reading beam on for 1 second at 50 second intervals. Inset: grating decay with reading beam on and switching on (at point a) and off (at points b) the applied electric field.

CONCLUSIONS

In summary, we have demonstrated that sol-gel processed ormosils with second-order NLO and charge transporting functionalities show great potential as photorefractive materials when doped with photocharge generating molecules. The greatest advantage of ormosil-based photorefractive materials is their excellent optical quality (single phase homogeneity), and chemical/environmental stability. Another benefit of using ormosils is the potential of inducing temporally stable noncentrosymmetric alignment of NLO chromophores by poling the films at elevated temperatures. As an example, in a separate experiment we have performed corona discharge poling (5 kV, 1.3 cm between a discharge wire and the film surface) of a 2.4 μm thick film at 160 ^{0}C for 30 minutes and achieved an electrooptic coefficient of 20 pm/V at 633 nm. Curing at this temperature causes partial densification of the ormosil network, permitting us to achieve EO coefficients that are stable at room temperature for over 1500 hours.

The relatively low values of diffraction efficiency and two beam coupling gain obtained in current studies may be attributed to the fact that all measurements were performed on unpoled films, i.e., films without permanently induced non-centrosymmetry.[17] It has been observed that photorefractive polymer composites capable of retaining electric field induced orientation of NLO chromophores exhibit photorefractive responses (i.e., asymmetric energy transfer) even without the presence of external electric field.[18]

It is clear that there are applications for amorphous photorefractive materials. LPT's research group has developed a wide number of amorphous photorefractive composites and makes them available to photonics community. Although many photorefractive polymers developed in last five years have exhibited good diffraction efficiencies in the visible, these low T_g materials have often demonstrated inadequate optical quality or other performance parameters. Ormosil photorefractive composites have unique molecular structures which, with continued work, are expected to yield materials more suitable for device applications.

ACKNOWLEDGEMENTS

The authors gratefully acknowledge the financial support of this research by the Air Force Office of Scientific Research through Contract No. F49620-92-C-0061.

REFERENCES:

1. Photorefractive Materials and Their Applications, I & II, edited by P. Gunter and J. P. Huignard, Topics in Applied Physics, Vols. 61 and 62 (Springer-Verlag, New York, 1988); P. C. Yeh, Introduction to Photorefractive Nonlinear Optics (Wiley, New York, 1993).

2. For a complete review, see Y. Zhang, R. Burzynski, S. Ghosal and M. K. Casstevens, Adv. Mater., **8**, 111 (1996); W. E. Moerner and S. M. Silence, Chem. Rev., **94**, 127 (1994).

3. S. Ducharme, J. C. Scott, R. J. Twieg, and W. E. Moerner, Phys. Rev. Lett., **66**, 1864 (1991); Y. P. Cui, Y. Zhang, P. N. Prasad, J. S. Schildkraut, and D. J. Williams, Appl. Phys. Lett., **61**, 2132 (1992).

4. Y. Zhang, Y. P. Cui, and P. N. Prasad, Phys. Rev. B, **46**, 9900 (1992); Y. Zhang, C. A. Spencer, S. Ghosal, M. K. Casstevens and R. Burzynski, Appl. Phys. Lett., **64**, 1908 (1994); K. Meergolz, B. L. Volodin, Sandalphon, B. Kippelen and N. Peyghambarian, Nature, **371** (1994); M. E. Orczyk, B. Swedek, J. Zieba and P. N. Prasad, J. Appl. Phys., **76**, 4995 (1994).

5. K. Yokoyama, K. Arishima, T. Shimada, K. Sukegawa, Jpn. J. Appl. Phys., **33**, 1029 (1994); R. Burzynski, Y. Zhang, S. Ghosal, M. K. Casstevens, J. Appl. Phys., **78**, 6903 (1995); Y. Zhang, S. Ghosal, M. K. Casstevens and R. Burzynski, Appl. Phys. Lett., **66**, 256 (1995); S. M. Silence, J. C. Scott, J. Stankus, W. E. Moerner, C. R. Moylan, G. C. Bjorklund, and R. J. Twieg, J. Phys. Chem., **99**, 4096 (1995).

6. B. Kippelen, K. Tamura, N. Peyghambarian, A. B. Padias, H. K. Hall, Jr., J. Appl. Phys., **74,** 3617 (1993); L. P. Yu, K. M. Chen, W. K. Chan, Z. H. Peng, Appl. Phys. Lett., **64**, 2489 (1994).

7. Sol Gel Technology for Thin Films, Fibers, Preforms, Electronics and Speciality Shapes, edited by L. C. Klein (Noyes Publications, Park Ridge, 1988).

8. R. Burzynski, P. N. Prasad in Sol-Gel Optics. Processing and Applications, edited by L. C. Klein, (Kulver Academic Publisher, Boston, 1994) Chapter 19, pp. 417-450.

9. P. N. Prasad, M. E. Orczyk, J. Zieba, B. Swedek, C. F. Zhao, H. K. Park, R. Burzynski, Y. Zhang, S. Ghosal, and M. K. Casstevens, Proc. SPIE, **2527**, 231 (1995).

10. J. H. Perlstein in Electrical Properties of Polymers, edited by D.A. Seanor (Academic Press, New York, 1982).

11. C. C. Teng and H. T. Man, Appl. Phys. Lett., **56**, 1734 (1990).

12. J. S. Schildkraut, Appl. Phys. Lett., **58**, 340 (1991).

13. Y. Zhang, C. A. Spencer, S. Ghosal, M. K. Casstevens, and R. Burzynski, J. App. Phys., **76**, 671 (1994).

14. D. Mahgerefeth, J. Feinberg, Phys. Rev. Lett., **64**, 2195 (1990).

15. J. C. Scott, L. Th. Pautmeier, and W. E. Moerner, J. Opt. Soc. Am. B, **9**, 2059 (1992).

16. P. N. Prasad and D. J. Williams, Introduction to Nonlinear Optical Effects in Polymers and Molecules (Wiley, New York, 1991).

17. The rationale in performing experiments in this manner was to obtain electrooptic data that would be close for both thick and thin films. A non-centrosymmetric order and, thus, second-order optical nonlinearity induced by the electric field poling in 100 μm or more thick films can be quite different from that obtained in thin films because of difficulties in applying external fields of the same magnitude. As an example, 1 μm thick film can easily be poled at fields as high as 100 V/μm or 1MV/cm; to obtain the same field across a 100 μm thick field would require application of a 10 kV DC bias.

18. L. Yu, Y. M. Chem, and W. K. Chan, J. Phys. Chem., **99**, 2797 (1995).

DIPOLAR ORGANIC/ FERROELECTRIC OXIDE HYBRIDS

Eric Pascal Bescher, Edward Hong, Yu-Huan Xu and John D. Mackenzie
Department of Materials Science and Engineering
University of California Los Angeles
Los Angeles CA 90095-1595

ABSTRACT

Organic-inorganic hybrids in which both constituents exhibit electrical dipolar properties have been fabricated. The inorganic constituent was either $LiNbO_3$ or $BaTiO_3$, and the organic constituent was triethoxysilyl 2,4-dinitrophenylaminosilane (TDP) cross-linked with SiO_2. Transparent films were made in a 1:1 organic:inorganic ratio. The films were characterized optically. Preliminary indication of interactions between organic and inorganic constituents, particularly in the case of $BaTiO_3$-TDP-SiO_2, has been observed.

INTRODUCTION

Ferroelectric materials have been fabricated by the sol-gel technique [1] for quite some time now. Many oxides such as $LiNbO_3$ and $BaTiO_3$ have been successfully fabricated. They have many applications as thin films in integrated circuitry. Many of these oxides exhibit ferroelectric-like properties at or near room temperature. This behavior shows that electric dipoles are present in the material at extremely low temperatures, before crystallization temperatures.

Concurrently, there has been increased interest lately in the properties of organic-inorganic hybrids. For example chromophores bound to an inorganic matrix are preferred to a pure polymer because they provide high density of non linear groups and better system stability. These materials combine the features of inorganic and organic constituents within the same matrix. Indeed, many new hybrids have been fabricated which exhibit novel characteristics: mechanical, chemical or physical properties [2]. Some examples of organic-inorganic interactions in optical hybrids have been demonstrated [3]. Hybrids containing an inorganic ferroelectric and an organic molecule with a dipole moment have not been studied yet, but might exhibit interesting synergystic properties.

In this study, triethoxysilyl 2,4-dinitrophenylaminosilane (TDP) has been chosen as the organic component. If poled, the silane has shown second harmonic generation properties [4,5,6,7]. $BaTiO_3$ and $LiNbO_3$ have been chosen for their high dielectric constants and high non-linear coefficients.

EXPERIMENTAL

Triethoxysilyl 2,4-dinitrophenylaminosilane (TDP) was obtained from Gelest. Lithium niobium and barium-titanium double alkoxides were obtained from Gelest. 45$BaTiO_3$-55SiO_2 and 45 $LiNbO_3$-55SiO_2 films were fabricated with mixing the alkoxide with acetone in a 1:20 ratio. For 45TDP-45$BaTiO_3$-10SiO_2 and 45TDP-45$LiNbO_3$-10SiO_2 films; TDP, acetone and TEOS (1:20:0.1) were stirred in air for 4 hours before addition of the appropriate amount of double alkoxide. The solution was filtered before spinning onto fused quartz substrates at 4000 rpm. Films were then heat-treated at 200°C for two hours. Film thickness was measured using a Dektak profilemeter, and the values obtained confirmed by ellipsometry (Ellipsometer Rudolf) which was also used for refractive index measurement. Corona poling was carried out at 120°C using a field of $10kV/cm^{-1}$.

Mat. Res. Soc. Symp. Proc. Vol. 435 © 1996 Materials Research Society

RESULTS AND DISCUSSION

The chromophore represented in Figure 1 is a benzene derivative having the desired donor-acceptor structure for high hyperpolarizability. TDP-SiO_2 films have been fabricated [8] and shown to exhibit a non-linear response whether or not the chromophore is cross-linked with TEOS [9,10] The nitro groups in ortho and meta positions are responsible for two absorption peaks at 350 nm and 415 nm [11]. All films made in this study were transparent. The TEOS-TDP films, about 1200 Å thick for one coating, exhibited the characteristic yellow color of the chromophore. $BaTiO_3$-SiO_2 and $LiNbO_3$- SiO_2 films were colorless with a UV cut-off around 300 nm (Figure 2 left). Absorption of all films is presented in Figure 2. A photograph of all films on quartz substrate is presented in Figure 3

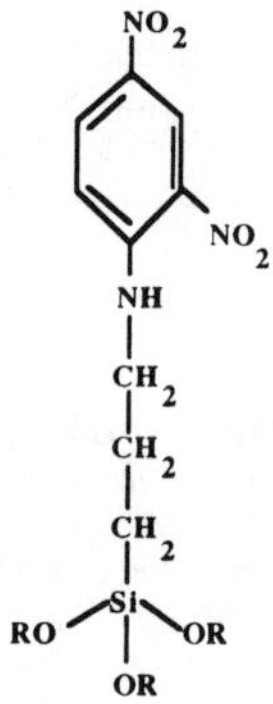

Figure 1- [Triethoxysilylpropyl] 2,4 dinitrophenylamine

45TDP-45$LiNbO_3$-10SiO_2 also exhibits two absorption maxima due to both nitro groups, and the spectrum is not radically diferent from that of 45TDP-55SiO_2. However, a significant change occurs for 45$BaTiO_3$-45TDP-10SiO_2 films. As seen in Figure 2 (right), a new peak appears at 510 nm, and the samples are dark red instead of bright yellow. This peak appears to be a shift of the longer wavelength absorption of TDP, which is due to the nitro group in meta position.
The refractive indices of binary systems are 1.70 to 1.80, whereas that of the ternary systems are1.90-2.00.

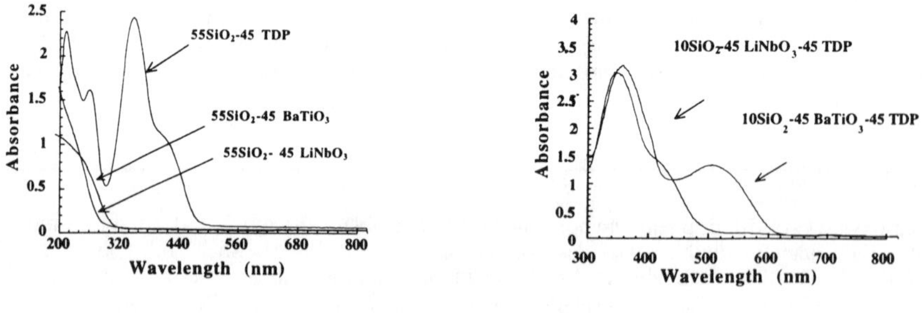

Figure 2- Visible absorption spectra of 45$BaTiO_3$-55SiO_2, 45$LiNbO_3$-55SiO_2, 45TDP-55SiO_2, (left) and 45$LiNbO_3$-55TDP-10SiO_2, 45$BaTiO_3$-45TDP-10SiO_2 films (right)

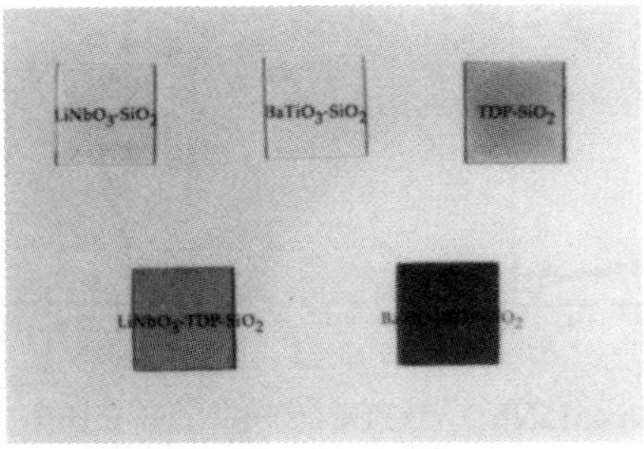

Figure 3- Photograph of $45BaTiO_3$-$55SiO_2$, $45LiNbO_3$-$55SiO_2$, 45TDP-$55SiO_2$, $45LiNbO_3$-55TDP-$10SiO_2$ and $45BaTiO_3$-45TDP-$10SiO_2$ films, showing the effect of the Ba-Ti double alkoxide on the absorption of TDP (lower right sample).

Thermogravimetric analyses showed that TDP in SiO_2 starts to decompose around 230 °C. Therefore, it is not expected that poling for two hours at 120°C altered the integrity of TDP. Figure 4 shows that, after poling, the dipoles aligned in the direction of the field. The order parameter ϕ is defined as

$$\phi = 1 - A_p/A_o$$

where A_p and A_o are, respectively, the absorbance at the maximum of absorption after and before poling[8]. Measurements are made with light polarized in a direction perpendicular to that of the electric field. The value is an indication of how the chromophores have aligned. For TDP-SiO_2 films, 38% of the organic chromophores aligned in the direction of the field, a value close to that obtained by Kim et al (35%).

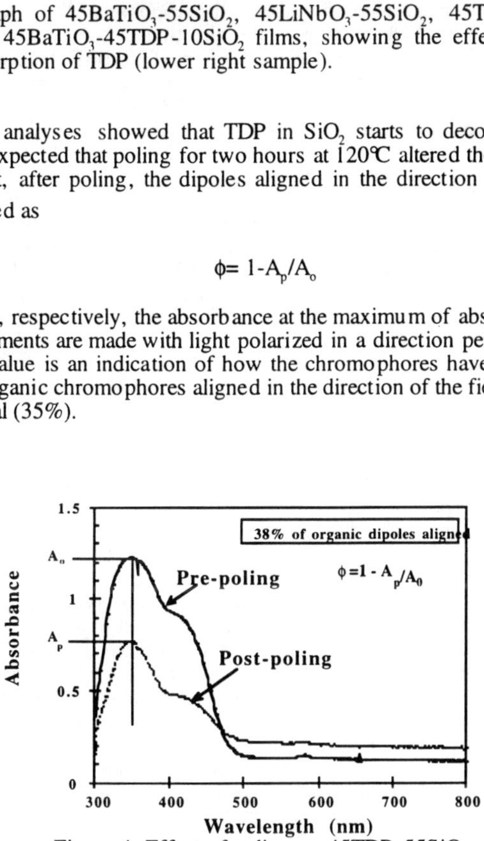

Figure 4- Effect of poling on 45TDP-$55SiO_2$

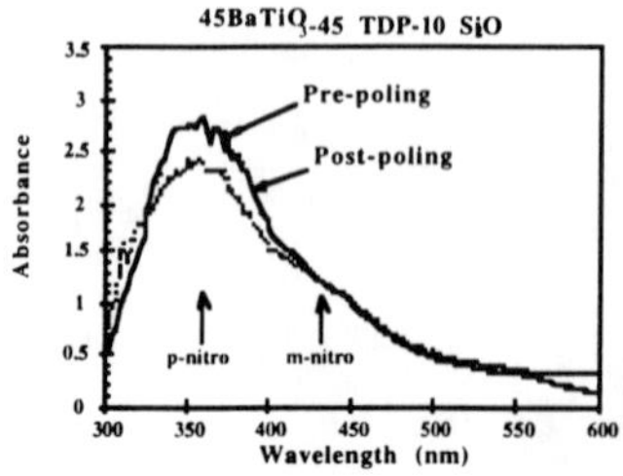

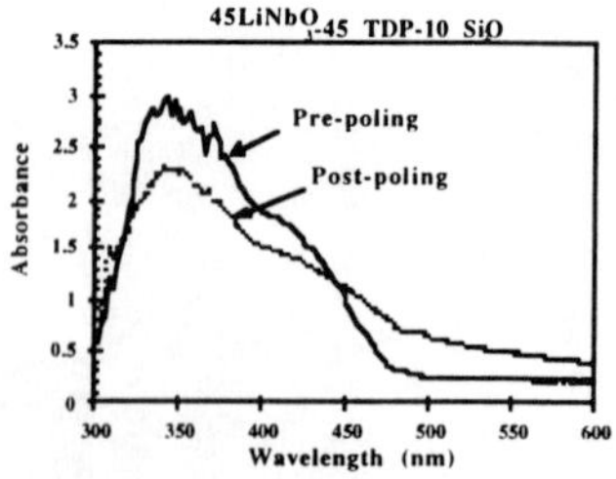

Figure 5- Effect of poling on 45LiNbO_3-55TDP-10SiO_2 and 45BaTiO_3-45TDP-10SiO_2 films

The presence of the double metal alkoxides altered the response of the chromophore to poling. In particular, the ability of the TDP to align in the direction of the field was diminished. For TDP-LiNbO_3 films, 25% of the dipoles aligned. For TDP- BaTiO_3-SiO_2, only 15% of the dipoles aligned. In addition, it appeared that the longer wavelength component of the absorption of TDP was not as affected by the poling as the short wavelength component. This is somewhat consistent with the observed color change from yellow to red. If the color change is due to the reactivity of the nitro in meta position (see Figure 2), it is conceivable that the molecule would be pined on its side at the reaction site, in addition to one or two anchoring points on the alkoxy groups. The molecule might still be able to align somewhat, as the decrease in the absorption due to the ortho nitro shows. In any case, it seems that the interaction between the inorganic and the organic might have interesting implications such as a stabilization of the relaxation of the chromophore. The reason for the difference between TDP-LiNbO_3 (no color change) and TDP-BaTiO_3 is not clear yet but might be linked to the structure of the Ba-Ti alkoxide.

Microstructure

The microstructure of the films was observed by transmission electron microscopy. 55SiO_2-45BaTiO_3 and 55SiO_2-45LiNbO_3 fired at temperatures above 400℃ showed clearly defined crystallinity within an amorphous matrix, with crystal size above 20 nm. Samples fired at 200℃ exhibited an essentially amorphous microstructure with areas of local ordering. Such areas are two small to lead to an X-ray diffraction signal. The presence of TDP did not affect this microstructure drastically. 45TDP-45BaTiO_3-10SiO_2 and45TDP-45LiNbO_3-10SiO_2 samples both exhibited similar features: an essentially amorphous structure with some areas of local ordering (ferrons). Such ferrons have previously been observed in sol-gel derived pure ferroelectrics, but this is the first time they have been observed in an organic-inorganic hybrid. In 45TDP-45BaTiO_3-10SiO_2, the size of such ferrons in on the order of 5 nm, whereas for 45TDP-45LiNbO_3-10SiO_2 ferrons on the order of 15 nm have been seen (Figure 6). This difference is ascribed to the difference in crystallization temperature of both oxides, as well as to the structure of the double alkoxides. The structure of barium-titanium double alkoxide has been studied [9] and has been shown to be Ba_4Ti_{13} clusters with 42 coordinated oxygen atoms. The central Ti in the $Ti_{13}O_{42}$ core in which the central Ti is coordinated to six oxygen atoms. If one considers this cluster to be the building block, a mere 10 such units are sufficient to lead to an observable "ferron" such as the one shown in Figure 6. It is believed that such entities play a significant role in the ferroelecrric-like properties of the materials.

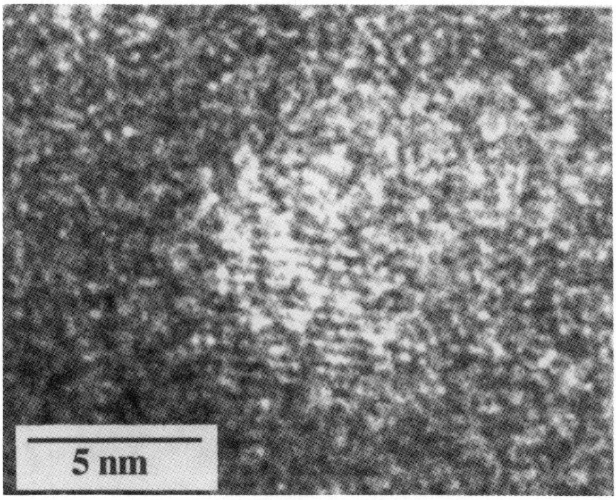

Figure 6- Ferron in $45LiNbO_3$-45TDP-$10SiO_2$ film fired at 200°C for 2 hours

Ferroelectric loops have been observed in the $45BaTiO_3$-$55SiO_2$, $45LiNbO_3$-$55SiO_2$ systems. They have also been observed in the TDP-ferroelectric hybrid systems and will be reported in an upcoming publication. Since both the organic and inorganic constituents in these materials exhibit a NLO response, experiments are currently under way to determine the SHG response of the organic-inorganic combination.

CONCLUSIONS

Organic-inorganic hybrids in which both constituents exhibit dipolar characteristics have been fabricated. A high density of a chromophores (TDP) has been placed in a matrix containing either $LiNbO_3$ or $BaTiO_3$. The chromophore absorption spectra have shown that the donor-acceptor characteristics are affected by the presence of the double metal alkoxide of the ferroelectric, in the case of $BaTiO_3$-TDP. The inorganic also affected the poling response of the chromophore. The microstructure of the materials was essentially amorphous, with islands of partially ordered inorganic phase, which size varied with the nature of the inorganic. Further work is under way to assess the NLO charactristics of these new hybrids.

REFERENCES

[1] Yuhuan Xu and John D. Mackenzie, "Ferroelectric Thin Films Prepared by Sol-Gel Processing", Integrated feroelectrics, 1992 Vol.1, pp.17-42

[2] Proceedings of the First European Workshop on Hybrid Organic-Inorganic Materials (Synthesis, Properties, Applications), C. Sanchez and F. Ribot Editors, November 8-10 1993, France

[3] Eric Bescher, J.D.Mackenzie, T. Ohtsuki and N. Peyghambarian, "Rare earth/Organic dye nanocomposites by the sol-gel method", Mat. Res. Soc. Symp. Proc. Vol. 351, 1994, 135-139

[4] J. Kim and J. Plawsky, "Second Harmonic Generation in Organically Modified Sol-gel Film", Chem. Mat. 1992,4, 249-252

[5] E. Toussaere, J, Zyss, P. Griesmar and C. Sanchez, Non Linear Optics, 1991, 3

[6] P. Griesmar, C. Sanchez, G. Puccetti, I. Ledoux and J. Zyss, Molecular Engineering, 1991, 1, 205

[7] B. Lebeau, J. Maquet, C. Sanchez, E. Toussaere, R. Hierle, J. Zyss, J. Mater. Chem., 4 (12) 1994, 1855.

[8] Jongsung Kim and Joel Plawsky, "Non-Linear Optical Material Fabrication via Sol-Gel Processing", Mat. Res. Soc. Symp. Proc. Vol 247, 1992, pp 135-140.

[9] Jongsung Kim, Joel L. Plawsky, Elizabeth Van Wagenen and Gerald Korenowski, "Effect of Processing Parameters and Polymerization on the Non-Linear Optical Response of Sol-Gel Materials", Chem. Mater. , **5**, 1993, pp.1118-1125

[10] B. Lebeau, PhD Thesis, Universite Pierre et Marie Curie, Paris, France, 1996.

[11] F.H. Hartman, "Light Absorption of Organic Colorants, Theoretical treatment and Empirical rules", Reactivity and Structure Concepts in Organic Chemistry",Volume 12; K.HafnerSpringer-Verlag Publ., 1980, pp.83-36.

[9] "Synthesis of Bimetallic Barium Titanium Alkoxides as Precursors for Electrical Ceramics", Inorg. Chem. 1991, **30**, 3244-3245

ACKNOWLEDGMENTS

This research is supported by the Air Force Office of Scientific Research. Spectrophotometric measurements were carried out in Professor Staffsudd's laboratory. TEM was carried out at Arizona State University with the help of John Wheatley.

VANADIUM OXIDE/POLYPYRROLE AEROGEL NANOCOMPOSITES

B.C. DAVE**, B. S. DUNN** , F. LEROUX*, L. F. NAZAR*, H.P. WONG**
*Department of Chemistry, University of Waterloo, Waterloo, Ontario Canada, N2L 3G1; lfnazar@chemistry.uwaterloo.ca
**Department of Materials Science and Engineering, UCLA, Los Angeles, CA, 90024

ABSTRACT

Vanadium pentoxide/polypyrrole aerogel (ARG) nanocomposites were prepared by hydrolysis of $VO(OC_3H_7)_3$ using pyrrole/water/acetone mixtures. Monolithic green-black gels with polypyrrole/V ratios ranging from 0.15 to 1.0 resulted from simultaneous polymerization of the pyrrole and vanadium alkoxide precursors. Supercritical drying yielded high surface (150-200 m^2/g) aerogels, of sufficient mechanical integrity to allow them to be cut without fracturing. TEM studies of the aerogels show that they are comprised of fibers similar to that of V_2O_5 ARG's, but with a much shorter chain length. Evidence from IR that the inorganic and organic components strongly interact leads us to propose that this impedes the vanadium condensation process. The result is ARG's that exhibit decreased electronic conductivity with increasing polymer content. Despite the unexpected deleterious effect of the conductive polymer on the bulk conductivity, at low polymer content, the nanocomposite materials show enhanced electrochemical properties for Li insertion compared to the pristine aerogel.

INTRODUCTION

In contrast to the well-known aqueous sol-gel route to V_2O_5 xerogels, hydrolysis of vanadium alkoxides results in rapid reaction to form monolithic V_2O_5 gels that can be supercritically dried to form aerogels. As shown by previous studies, these materials display extremely high surface areas and controllable porosity.[1] These characteristics make them extremely attractive as cathodes in rechargeable lithium batteries where kinetic limitations due to Li access and transport within the crystalline lattice pose difficulties in attaining theoretical capacities. Improved capacity for Li insertion, for example, has been observed for aerogel[2] and xerogel V_2O_5[3], over that of highly crystalline V_2O_5 prepared by high temperature (HT) synthesis.

An intriguing idea is that of combining the conductive and redox properties of conjugated polymers with those of the V_2O_5 framework in its aerogel form, to maximize surface area, redox capacity and possibly introduce capacitance effects. Numerous xerogel (but not aerogel) V_2O_5-polymer nanocomposites have been reported, including those of polyaniline[4] and polypyrrole[5] in which the polymer chains are interleaved between the oxide sheets. At high polymer content, these show improved conductivity compared to pristine V_2O_5 xerogels. Furthermore, our recent electrochemical studies on polyaniline-nanocomposites of $HTaWO_6$[6] and MoO_3 have shown enhancement in lithium ion mobility.[7] These findings have prompted us to develop methods for the synthesis of polymer/oxide aerogels, and examine their electrochemical properties.

EXPERIMENTAL

Monolithic polymer/vanadium aerogels were prepared by hydrolyzing a vanadium alkoxide solution in acetone with water/pyrrole/acetone mixtures at 0 °C. Two methods of preparation were employed. In the first method, the reaction mixture with a typical molar composition 1.0 $VO(OC_3H_7)_3$: 40 H_2O: 17 $(CH_3)_2CO$: y C_4H_4NH was rapidly mixed, and immediately transferred

Mat. Res. Soc. Symp. Proc. Vol. 435 © 1996 Materials Research Society

to a test-tube. Gelation ocurred within 10-30 seconds for most samples, resulting in deep green-black gels. The ratio of pyrrole was varied from 0.15 to 1.0 moles /$VO(OR)_3$, and the water-acetone molar ratio was also varied extensively over a wide composition range. The second method involved the pre-oxidation of the pyrrole monomer with $(NH_4)_2S_2O_8$ (0.075 mole/V) prior to its reaction with the alkoxide. Upon addition of the oxidant, the pyrrole/water/acetone mixture changed from translucent to opaque dark green; after 30 minutes, polypyrrole precipitated. The solution was allowed to polymerize to different extents before it was cooled to ~0°C and added to the alkoxide (~3°C). Gelation occurred within 10 seconds.

The wet gels were aged for 4 days in capped tubes that had a 1mm hole in the cap, to allow for slow evaporation of the solvent within the wet gel. This resulted in a small amount of shrinkage (15%) and densification that strengthened the gels. After aging, the gels were removed, and immersed in acetone for a day to promote exchange of the water in the gel pores for acetone. This procedure was repeated 4 times to ensure complete exchange. The gels were transferred to the pressure vessel (Polaron E3000 Critical Point Dryer) for supercritical extraction by CO_2.[1] The resultant dark-green cyclinder-shaped monoliths had dimensions of approximately 1x3 cm.

The density of the aerogels was determined using a pycnometer vial filled with Hg. Surface area measurements (N_2 absorption) were conducted on a Quantachrome Autosorb-1 system and analyzed by the BET method. The electrical conductivity of the aerogels was measured between 25 and 180 °C by the complex impedance method, similar to that reported previously.[1] Two probe measurements were performed in flowing argon on disc-shaped samples that were heat-treated at 140 °C for 24 hours, using a HP Precision LCR meter (20Hz to 1MHz). TEM was performed on a Phillips CM30. FTIR data were obtained as KBr pellets on a Nicolet FT-IR spectrometer operating between 400 cm^{-1} to 4,000 cm^{-1}.

PPY was intercalated into the xerogel by a variation of the method reported by Kanatzidis *et al.*[5] Powder XRD patterns of the precipitate matched those reported, with an interlayer spacing of 13.8Å. The mole fraction of polypyrrole per V_2O_5 was determined to be 0.40 by TGA.

For electrochemical studies, the aerogel was mixed with 30-85 wt % Ketjen carbon black, and 2 wt % ethylene propylene diene monomer (EPDM). The resulting composite was pressed onto a fine stainless steel grid and heated at 100 °C for 2 hours to form the electrode with a surface area of 1 cm^2. A 1.0 M solution of $LiClO_4$ in propylene carbonate served as the electrolyte. Swagelock-type cells were assembled in an argon atmosphere, and studied under galvanostatic conditions between constant voltage limits using a MAC-PILE.™ The discharge cut-off voltage was chosen based on the stability of the electrolyte previously studied with V_2O_5 (HT).[8]

RESULTS

Surface Area/morphology. The synthesis of vanadium oxide aerogels by an alkoxide based sol-gel route has been described previously. In this study, vanadium alkoxide was hydrolyzed and co-polymerized with pyrrole during sol formation to form polypyrrole/vanadium oxide aerogel nanocomposites. Incorporation of the polymer resulted in fragile aerogels when the same conditions were used to age and dry the samples as those for V_2O_5 aerogels. We found that this could be compensated by partial drying of the wet gel in air under ambient conditions, which resulted in strengthening (and shrinking) of the nanocomposites. The degree of shrinkage varied between samples, resulting in densification of the gels compared to those that were not subjected to partial drying. For pure V_2O_5 aerogels, the latter exhibited average densities of 0.1 g /cc; gels subjected to a partial drying regime yielded densities in the range of 0.2 g/cc. The addition of pyrrole has little effect on aerogel density, as the [PPY]V_2O_5 ARGs (all partially dried) exhibited slightly increased densities of about 0.2 g/cc (Table 1). The nanocomposites also have surface areas similar to those of the V_2O_5 aerogel (about 150-200 m^2/g). These values are slightly lower than the 300-400 m^2/g range reported previously prepared using different synthesis conditions.

Table 1: Properties of $[PPY]_{2y}V_2O_5$ ARG Nanocomposites

Pyrrole/V ratio (y)	*ρ (g/cc)*	*σ (S/cm) T=25°C*	*surface area(m^2/g)*	*Sample*
0	0.1	$2x10^{-4}$	150	*1
0.15	0.25	a	a	*2
0.3	0.19	$3x10^{-5}$	a	*3
0.4	0.18	$1x10^{-6}$	a	*4
0.5	0.22	$2x10^{-6}$	a	*5
0.7	0.20	$4x10^{-7}$	a	*6
0.9	0.25	$2x10^{-8}$	a	*7
1.0	0.15	a	160	*8
0.5	0.12	$1x10^{-6}$		#9
1.0	0.12	$2x10^{-6}$	200	#10

[a] not measured

* samples 1-8 prepared *without* oxidizing agent

samples 9-10 prepared *with* oxidizing agent

XRD studies showed that both V_2O_5 and $[PPY]V_2O_5$ ARG's exhibit a very low degree of crystallization compared to their respective 2D xerogel forms. TEM studies also showed a distinct difference in their fibrous morphology. Whereas the fibers for the V_2O_5 aerogel displayed a long, ribbon-like structure similar to that previously reported for vanadate xerogels,[1] those of the polymer-aerogel were much shorter in length, and were aggregated in a random fashion.

FTIR. Figure 1 shows the FTIR spectrum of the aerogel nanocomposite between 2000 and 400 cm^{-1}. The bands in the region 1600-900 cm^{-1} are diagnostic of polypyrrole, and their position and intensities show that the conductive form of the polymer is produced (Table 2). The bands exhibit

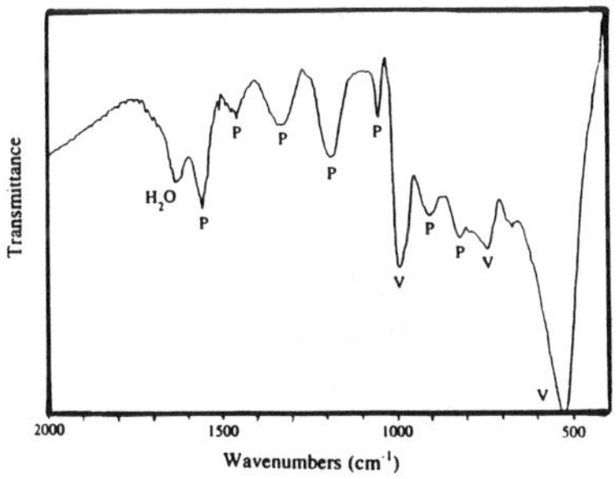

Fig.1 IR spectrum of $[PPY]_{0.5}V_2O_5$

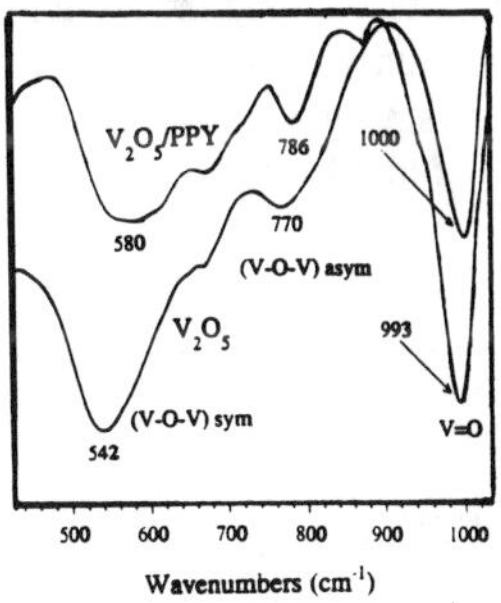

Fig. 2 IR of V_2O_5 and $[PPY]_{2.0}V_2O_5$ in the region 1000-400 cm^{-1}

shifts from those of bulk, p-doped PPY,[9] suggesting that a substantial interaction with the V_2O_5 framework occurs. The chain length and conductivity of the PPY can be estimated from the ratio of the A'/B' bands at 1580 and 1460 cm^{-1} respectively. Studies of bulk PPY show a linear relationship between this ratio and the log of the conductivity of the polymer.[9] Using this data, together with the A'/B' value in the nanocomposite, (≅ 5.0), interpolation indicates that the PPY component appears to be relatively conductive (1 S/cm).

Table 2: FTIR Peak Positions for Polypyrrole and PPY/V_2O_5 Nanocomposites

Material	*A' cm^{-1}*	*B' cm^{-1}*	*C' cm^{-1}*	*D' cm^{-1}*	*E' cm^{-1}*	*F' cm^{-1}*
$PPY_{.5}V_2O_5$ ARG	1561	1461	1331	1184	1053	905
[9]conductive PPY	1560	1480	1320	1178	1047	901
[9]insulating PPY	1523	1440	1207	1076	1038	919
[10]$PPY_{.4}MoO_3$	1551	1465	1321	1189	1049	906

The bands corresponding to the inorganic component are also assigned in Figure 1. Due to the presence of V-O-V moieties in the structure, two vibrational modes are expected in the 400-800 cm^{-1} region, corresponding to the symmetric (ν_{sym}) and asymmetric (ν_{asym}) stretch. These were assigned to the bands at 542 cm^{-1} and 772 cm^{-1} respectively, based on our studies using ^{18}O-labelled H_2O that will be reported elsewhere. We found that both modes shift to higher wavenumber with increasing PPY content (Figure 2). For example, in samples containing a 1:1 molar ratio of PPY:V, the ν_{sym} mode shifts up by 38 cm^{-1} to 580 cm^{-1}, and the asymmetric mode shifts to 786 cm^{-1}. In contrast, the vanadyl stretch ($\nu_{V=O}$) which occurs at 1000 cm^{-1} in the V_2O_5 ARG, shifts down to 993 cm^{-1} on incorporation of the polymer, implying either increased V^{4+} content in the lattice or a weakening of the V=O bond. The V-O-V band "up-shift" is inconsistent with increased V^{4+} content: however, it could be explained by a reduced V-O-V bond length caused by weak binding of the nitrogen atom of the pyrrole ring to the sixth apical coordination site on the square-pyramidal V site. Binding of nitrogen-containing ligands with SP-coordinated V in vanadates has been observed for pyridine, for example.[11] This would also partially account for the down-shift of the V=O band, as the vanadyl bond would lengthen as a result of such an interaction. The major factor in the shift of the V=O band, however, is the contribution of H-bonding interactions. Spectra obtained on PPY-V_2O_5 aerogels prepared in D_2O showed a sensitivity to deuterium substitution (typically, the V=O mode shifted from 993 to 987 cm^{-1}) in contrast to the pure V_2O_5 ARG's which showed no change. These results further confirm the presence of highly interacting inorganic-organic components in these hybrid materials.

Conductivity. The samples were heat treated prior to AC impedance measurements to partially remove the water within the interlamellar gap of the V_2O_5. TGA of the aerogels indicated 10% mass loss at 180°C, corresponding to a composition $[PPY]V_2O_5{\cdot}0.6\,H_2O$. The conductivity of pure V_2O_5 aerogel varies linearly with respect to density. The conductivity increases by a factor of 5 when the density is increased from 0.1 to 0.25; however, the conductivity of the $[PPY]V_2O_5$ samples *decreases* with increasing polypyrrole content, despite the increase in density. Figure 3 shows the results of the AC

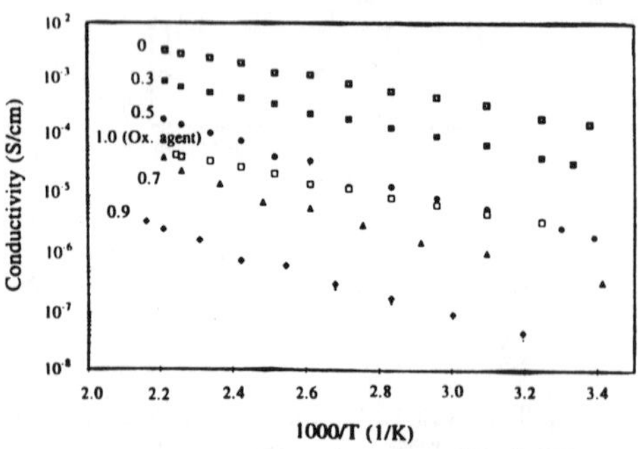

Fig. 3 Log σ *vs* 1000/T for $[PPY]_{2y}V_2O_5$ at varying ratios y of PPY/V (denoted on the fig.)

highest RT conductivity of $2x10^{-4}$ S/cm. The solid data points represent $[PPY]V_2O_5$ synthesized using Method 1. There is a dramatic reduction in conductivity as the ratio of pyrrole monomer to vanadyl alkoxide reaches 0.9. We found, however, that this could be compensated by using an external oxidizing agent. The open data points in Figure 3 represent $[PPY]_{2.0}V_2O_5$ samples prepared using method 2. The conductivity in this sample is increased by two orders of magnitude, although the value is still less than that of pure V_2O_5 aerogel. The reason for the decrease in conductivity with increasing polymer content is not evident at this stage, but is clearly related to the significant changes in the IR spectra of these materials, and the strong interaction of the polymer with the inorganic framework. Future studies will focus on samples with higher ratios of oxidizing agent and lower amounts of polypyrrole.

Electrochemistry. By insertion of a conductive polymer in V_2O_5, we can expect active participation of the organic component in the redox process of the cathode. Reaction of pyrrole, during the sol-gel hydrolysis step, reduces a portion of the V^{5+} to V^{4+}, according to the reaction:

$$yC_4H_4NH + V_2O_5 \text{ ----> } [\text{-}(C_4H_2NH)\text{-}]_yV_2O_5^{(2y+0.3)-} + 2H^+$$

This takes into account the $2e^-$/monomer oxidative polymerization of the pyrrole; and a further 0.3 e^- oxidation to obtain the doped polymer.[12] The reaction implies that the reducing power arises solely from the V_2O_5, although it is very likely that atmospheric oxygen plays a major role in the oxidation process, as well as oxidizing highly reduced vanadium sites. Nonetheless, incorporation of more conductive polymer in the nanocomposite results in overall less V^{5+} sites that are available for reduction.

Carbon black was added to the nanocomposite material to increase the electronic conductivity, which is standard practice for the formation of the positive electrode in lithium batteries. For 30% carbon black, the faradaic capacity decreases from 3.1 to 1.0 as the polypyrrole content (y) increases from 0-> 0.5 -> 1.5 moles. This is accompanied by an equivalent decrease in the open-circuit voltage from 3.52V-->3.35V, consistent with increasing V^{4+} content. The experimental capacities are still, however, much less than the theoretical value, suggesting that there is either an electronic and/or ionic limitation. The first could be offset by increasing the carbon black fraction in the composite electrode. Electrodes prepared from $[PPY]_{0.5}V_2O_5$ with 85%C results in a substantial increase in capacity, showing that there is indeed an electronic limitation. The carbon black also has an intrinsic faradaic capacity characteristic of behavior that results from interfacial adsorption/desorption phenomena.[13] Carbon aerogels and several high-surface area carbons can display this effect, which typically gives rise to capacitance on the order of 20-100 F/g. In our case, the capacitance of about 65 F/g (determined from cyclic voltammetry) accounts for approximately 40% of the response of the composite electrode.

The effect of an initial charge on subsequent electrochemical processes has different effects on the pristine *vs* the $[PPY]V_2O_5$ aerogel. During a charge, both V^{4+} and PPY can be oxidized; the respective electrochemical processes corresponding to the deintercalation of Li and adsorption of ClO_4^-. For $[PPY]V_2O_5$, a charge to 3.8V appears to be reversible in terms of faradaic capacity, and is equivalent to a charge after beginning the cycle with a discharge (Figure 4b). Conversely, for the PPY-free V_2O_5 aerogel, an initial charge *increases* the capacity of the subsequent discharge (compared to a second discharge after beginning the cycle in discharge first) and hence changes the reduction process compared to the polymer aerogel (Figure 4a). In this case, the dx/dV (electronic density/mole) curves show a shift of the main redox process to 2.6V (vs Li) from that at 2.2V observed for the V_2O_5 ARG in charge first. This indicates that the first processes are not equivalent; with respect to either the capacity, or the Li potential sites that are occupied. Moreover, polymer/V_2O_5/C electrodes exhibit a lower degree of polarization compared to V_2O_5/C in the same voltage range, at the same carbon content (85%). This effect, in addition to the other factors, gives rise to an increase in the faradaic capacity from 4.0 to 5.0 F/mole.

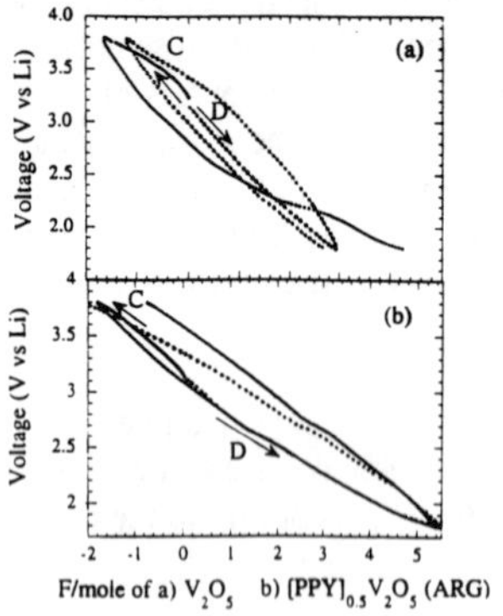

Fig. 4 Effect of a charge/discharge on the capacity of the ARG's with and without PPY in the voltage range 3.8-1.8V *vs* Li.

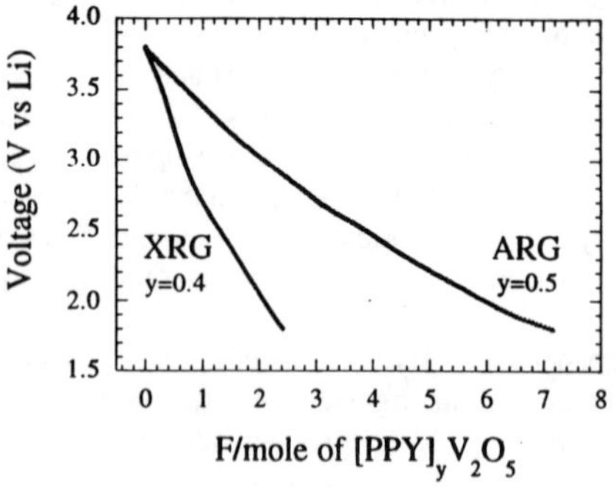

Fig. 5 Comparison of the capacity (on first discharge) of $[PPY]_yV_2O_5$ in aerogel and xerogel forms.

Furthermore, comparison of both the aero and xerogel nanocomposites (containing the same fraction of PPY, and the same percentage of carbon black) shows that both curves possess two distinct slopes (Figure 5). The aerogel electrode, however, exhibits a capacity almost 3 times greater than that of the corresponding xerogel.

ACKNOWLEDGMENTS

LFN and BSD gratefully acknowledge NSERC (Canada) and ONR (U.S.A), respectively, for funding this research.

REFERENCES

1. F. Chaput, B. Dunn, P. Fuqua, and K. Salloux, *J. Non-Cryst. Solids*, **188**, 11 (1995).
2. D.B. Le, S. Passerini, A.L. Tipton, B.B. Owens, and W.H. Smyrl, *J. Electrochem. Soc.*, **142,** L102 (1995).
3. H.-K. Park and W.H. Smyrl, *J. Electrochem. Soc.*, **141,** L25 (1994); 1215 (1993); R. Baddour, J.P. Pereira-Ramos, R. Messina and J. Perichon, *J. Electroanal. Chem.*, **314**, 81 (1991); K. West, B. Zachau-Christiansen, T. Jacobsen, and S. Skaarup, *Electrochimica Acta*, **38**, 1215 (1993).
4. M. Kanatzidis, C-G Wu, H.O. Marcy, and C.R. Kannewurf, *J. Am. Chem. Soc.*, **111**, 4139 (1989).
5. C-G. Wu, M. Kanatizidis, H.O. Marcy, and C.R. Kannewurf, *Polym. Mater. Sci. Eng*, **61**, 969 (1989).
6. B.E. Koene, and L.F. Nazar, *Solid State Ionics*, in press.
7. L.F. Nazar, H. Wu, and W.P. Power, *J. Mater. Chem.*, **5**, 1985 (1995); T.A. Kerr, H. Wu, L.F. Nazar, *Chem. Mater.*, in press.
8. C. Delmas, and H. Cognac-Auradou, *J. Power Sources*, **54**, 406 (1995).
9. J. Lei, Z. Cai and C. R. Martin, *Synth. Met.*, **46**, 53(1992); Synth. Met., **48**, 301 (1992).
10. H. Wu, and L. F. Nazar, unpublished data.
11. J. W. Johnson, A.J. Jacobson, J.F. Brody and S.M. Rich, *Inorg. Chem.*, **21**, 3820 (1982).
12. B.R. Saunders, R.J. Fleming, and K.S. Murray, *Chem. Mater.*, **7**, 1082 (1995).
13. B.E. Conway, *J. Electrochem. Soc.*, **138**, 1539 (1991); S.T. Mayer, R.W. Pekala, and J.L. Kaschmitter, *ibid*, **140**, 446 (1993).

Er^{3+}-DOPED SILICA AND HYBRID ORGANIC/INORGANIC SILICA GELS

B.T. STONE, K.L. BRAY
Department of Chemical Engineering, University of Wisconsin-Madison, Madison, WI, 53706

ABSTRACT

The fluorescence properties of Er^{3+} in densified sol-gel silica and alkyl-modified silicates are presented. In sol-gel silica, strong infrared ($^4I_{13/2} \rightarrow {}^4I_{15/2}$) emission was observed over a wide range of Er^{3+} concentrations. The effect of metal ion co-dopants, which are known to inhibit clustering of Eu^{3+} in sol-gel silica, on Er^{3+} fluorescence are also considered. The co-dopants La^{3+}, Y^{3+}, Yb^{3+}, and Al^{3+} are increasingly more effective at inhibiting Er^{3+} clustering and promoting a more uniform spatial distribution of Er^{3+} ions. Lifetime studies were also conducted to assess the extent of hydroxyl quenching.

Organic/inorganic hybrid silica gels are expected to contain a lower amount of water than simple sol-gel silica gels and should be better hosts for rare earth ions. Er^{3+}-doped hybrid organic/inorganic gels were prepared using $Si(OC_2H_5)_4$ and $CH_3Si(OC_2H_5)_3$, $(CH_3)_2Si(OC_2H_5)_2$, or $C_2H_5Si(OC_2H_5)_3$. Er^{3+} fluorescence spectra and lifetime studies of these gels are presented and compared to simple Er^{3+}-doped sol-gel silica. The effect of organic modification on Er^{3+} clustering and hydroxyl retention are discussed. Er^{3+} upconversion properties are also discussed.

INTRODUCTION

Rare earth ion-doped glass devices are of great interest for a wide range of applications [1,2]. Er^{3+}-doped glasses in particular are being considered for use as laser materials [3] and as amplifiers for optical communications [4]. The $^4I_{13/2} \rightarrow {}^4I_{15/2}$ electronic transition in Er^{3+} produces light with a wavelength of approximately 1.54 μm [5]. This wavelength is outside the range of wavelengths of light focused on the retina of the human eye [6], and is also close to the wavelength of lowest signal attenuation in a low-loss silica glass fiber [7]. Additionally, Er^{3+}-doped glasses co-doped with Yb^{3+} are being investigated as upconversion laser materials providing laser wavelengths in the blue and green regions of the visible spectrum [8].

The sol-gel method provides a convenient alternative to the traditional melt preparation of rare earth ion-doped glasses for optical studies. Some of the advantages of the sol-gel process over melt preparation of glass include lower temperature processing, higher sample homogeneity and purity, and the opportunity to prepare new non-crystalline solids [9]. The high level of hydroxyl impurities that remain in sol-gel glasses are a disadvantage of the sol-gel method [9], especially in the case of rare earth ion-doped glasses. Hydroxyl impurities decrease the fluorescence efficiencies and shorten luminescent level lifetimes of dopant ions in glasses [10], adversely affecting optical device performance. Techniques for lowering the hydroxyl content of sol-gel glasses include high temperature treatment with carbon tetrachloride [11] or the inclusion of hydrofluoric acid in the initial reaction mixture [12].

Another problem affecting rare earth ion-doped sol-gel glasses is ion clustering [13]. Clustered rare earth ions are more susceptible to concentration quenching through cross-relaxation or energy transfer processes which lead to a decrease of fluorescence efficiencies and shorter lifetimes [14,15]. The dispersion of the rare earth ion Eu^{3+} in sol-gel silica glasses containing Al^{3+} was recently demonstrated in a fluorescence line narrowing study [16].

Mat. Res. Soc. Symp. Proc. Vol. 435 © 1996 Materials Research Society

The use of organically-modified silicon alkoxides allows the preparation of hybrid organic/inorganic materials using the sol-gel process. Zhang et al. recently reported the preparation of methyl-modified silica gel [17]. These authors found that the gels became hydrophobic with the introduction of methyl groups [17]. Hybrid silica gels may provide rare earth ions with lower hydroxyl environments than purely inorganic silica gels.

The goals of this study are to investigate the influence of the co-dopants Al^{3+} and Yb^{3+} on the optical properties of Er^{3+} in sol-gel silica glasses and to determine how hybrid organic/inorganic silica gels affect those properties. Initial attempts to measure upconversion fluorescence are also discussed.

EXPERIMENT

Silica sol-gel samples were prepared using $Si(OC_2H_5)_4$) (TEOS). Deionized water was used for hydrolysis of the TEOS, and ethanol was added as a solvent. The molar ratio of TEOS/water/ethanol in the initial sols was 1/4/4. $Er(NO_3)_3{\bullet}5H_2O$, $Al(NO_3)_3{\bullet}9H_2O$ and $Yb(NO_3)_3{\bullet}5H_2O$ were used as the sources of Er^{3+}, Al^{3+}, and Yb^{3+}. The initial sols were mixed until homogeneous, cast in plastic vials which were then sealed, and left to react at room temperature. Hybrid organic/inorganic silica sol-gel samples were prepared using TEOS and $CH_3Si(OC_2H_5)_3$ (MTES), $(CH_3)_2Si(OC_2H_5)_2$ (DMDES), or $C_2H_5Si(OC_2H_5)_3$ (ETES). Solutions were made with 10: 1 and 5: 1 ratios of TEOS: modified alkoxide. Deionized water and ethanol were added to achieve a molar ratio of (total Si)/water/ethanol equal to 1/4/4. Er^{3+} was added as above in the form of the nitrate salt, and a small amount of concentrated nitric acid was also added as a catalyst to the initial sol. The hybrid sols were also mixed until homogeneous, cast in plastic vials and sealed and left to react at 35°C.

Gel times ranged from one to five weeks depending on the reaction conditions, alkoxides used, and the amount of salt in the initial sol. After gelation, the gels were aged for two days at 60°C, followed by two days at 90°C. Samples were then heated in air to a final densification temperature of 800°C, with 24 h dwell times at 200° intervals. The concentration of erbium in a sample is expressed in terms of the weight percent of Er_2O_3 that would be present in a fully densified SiO_2 glass in which water and organics are absent. Most of the organic groups in the hybrid gels have decomposed after being heated to 800°C [18], but to distinguish these samples from those prepared with TEOS only, they are still referred to as hybrid gels. All of the fluorescence spectra and fluorescence lifetimes presented in this paper were measured with samples heated to 800°C.

Fluorescence spectra and lifetimes were measured using the 488 nm line of an argon ion laser. Laser power densities ranging from 0.5 to 2.5 W/cm^2 were used for the fluorescence measurements. Sample fluorescence was focused on the entrance slit of a 1-meter monochromator. Infrared fluorescence was detected with a germanium detector and measured using standard lock-in amplifier techniques. The spectra are uncorrected. Fluorescence decay traces of the 1.54 μm $^4I_{13/2} \rightarrow {}^4I_{15/2}$ emission were recorded on a digital oscilloscope from the output of a fast (~ 200 μs rise time) germanium detector. The 488 nm radiation of an argon ion laser was chopped at a frequency of 10 Hz to simulate pulsed excitation. Low temperature measurements were made using a variable temperature cryostat. Attempts to measure upconversion fluorescence were made using 650 nm laser excitation from a tunable dye laser pumped by an argon ion laser.

RESULTS AND DISCUSSION

The visible and infrared fluorescence spectra measured for the Er^{3+}-doped silica gels co-doped with Al^{3+} or Yb^{3+} and the Er^{3+}-doped hybrid gels were all similar in general appearance to those measured previously for Er^{3+}-doped silica gels [19]. Er^{3+} $^4I_{13/2} \rightarrow {}^4I_{15/2}$ fluorescence in 1.0 wt % Er_2O_3 silica gel samples containing only erbium, and containing Al/Er and Yb/Er ratios of 1 are shown in Figure 1. The individual traces are normalized to a peak intensity of 1. The addition of both Al^{3+} and Yb^{3+} caused a broadening of the $^4I_{13/2} \rightarrow {}^4I_{15/2}$ emission, with the influence of Al^{3+} being more pronounced than that of Yb^{3+}. The broadening of Er^{3+} fluorescence suggests a wider distribution of Er^{3+} bonding sites and that both ions act to disperse Er^{3+} in the glass, with Al^{3+} being more effective than Yb^{3+}. Evidence that Al^{3+} directly influences the Er^{3+} bonding environment is found in the small shift of the more intense peak to shorter wavelengths. This shift suggests that Er-O bonding is more ionic in the presence of Al^{3+}.

Weak fluorescence at 980 nm, shown in Figure 2, corresponding to the $^4I_{9/2} \rightarrow {}^4I_{15/2}$ transition of Er^{3+}, was measured in samples containing Yb^{3+}. This fluorescence had not been detected previously in samples containing only Er^{3+} or containing Er^{3+} and Al^{3+}. The fluorescence peak centered at about 1030 nm corresponds to the $^2F_{5/2} \rightarrow {}^2F_{7/2}$ electronic transition of Yb^{3+}. This peak was found to increase in overall intensity as the Yb/Er ratio was increased, as shown in Figure 2. The traces are not normalized, but are on the same scale. Since Yb^{3+} is not directly excited with 488 nm excitation, the Yb^{3+} fluorescence is due solely to nearly resonant transfer from the $^4I_{11/2}$ level of Er^{3+} to the $^2F_{5/2}$ level of Yb^{3+}. The presence of $Er^{3+} \rightarrow Yb^{3+}$ energy transfer is direct evidence for close association of some of the Er^{3+} and Yb^{3+} ions in the matrix.

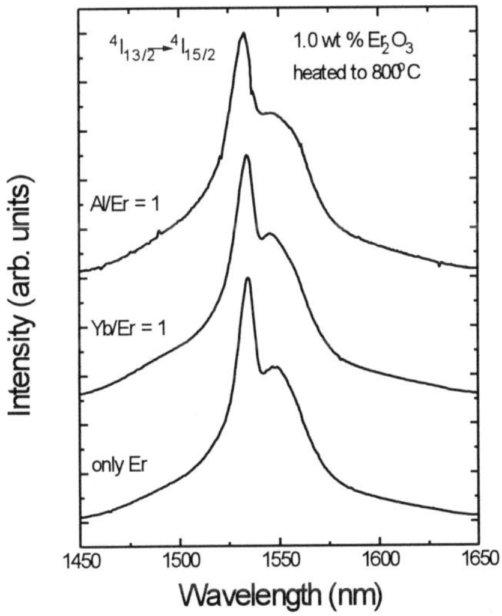

Figure 1. Er^{3+} $^4I_{13/2} \rightarrow {}^4I_{15/2}$ fluorescence in purely inorganic silica gels.Co-dopants and amounts are indicated on the figure.

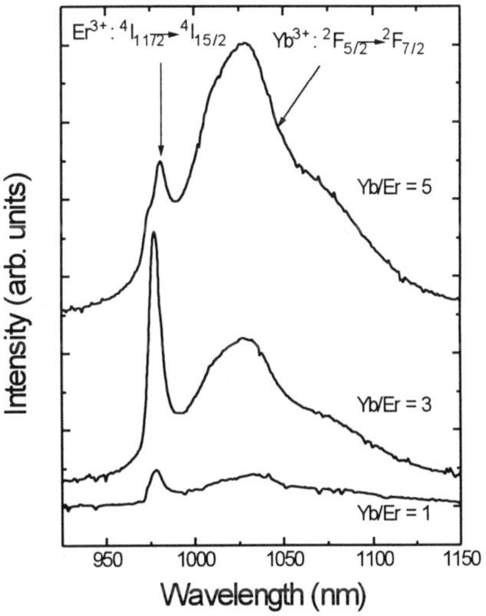

Figure 2. Er^{3+} $^4I_{11/2} \rightarrow {}^4I_{15/2}$ fluorescence and Yb^{3+} $^2F_{5/2} \rightarrow {}^2F_{7/2}$ fluorescence in purely inorganic silica gels.

Figure 3 depicts the infrared fluorescence of a 1.0 wt % Er_2O_3 silica gel sample and 1.0 wt % Er_2O_3 hybrid samples. The individual traces are normalized to a peak intensity of 1. The width of the fluorescence band was similar in all of these spectra. The similarity in the widths of the spectra obtained for 1.0 wt % Er_2O_3 doped silica and hybrid silica samples indicates that the use of organically-modified precursors does not improve the dispersion of Er^{3+} in the glass.

Fluorescence lifetimes of the Er^{3+} $^4I_{13/2}$ level in various doped silica and hybrid organic/inorganic samples are listed in Table I. The lifetimes in Table I were calculated from the area under the normalized decay curves recorded on the oscilloscope. The decay curves were only slightly non-exponential and were also fit using single exponential functions. The values obtained from the areas and from the exponential fits were in good agreement.

The $^4I_{13/2}$ level lifetime in Er^{3+}-doped silica gels decreased with increasing Er_2O_3 concentration, and decreased with increasing temperature. The $^4I_{13/2}$ level lifetime is decreased by the effects of hydroxyl quenching [10] and concentration quenching via energy transfer [14,15]. The effect of energy transfer is expected to be enhanced by clustering of Er^{3+} ions in a glass. The decrease in the lifetime of Er^{3+}-doped silica samples with increasing Er^{3+} content is due to the increasing importance of concentration quenching. The decrease in the lifetime with increasing temperature is attributed to the increasing importance of non-radiative decay due to hydroxyl groups. At room temperature, both hydroxyl and concentration quenching are important in Er^{3+}-doped silica samples.

The addition of Al^{3+} or Yb^{3+} caused a decrease in the room temperature lifetimes measured for all Er_2O_3 concentrations, although the decrease for 1.0 wt % Er_2O_3-doped samples did not occur until a high co-dopant/Er^{3+} ratio was reached. The addition of Al^{3+} and Yb^{3+} to Er^{3+}-doped silica reduces the extent of clustering, and should also reduce the importance of

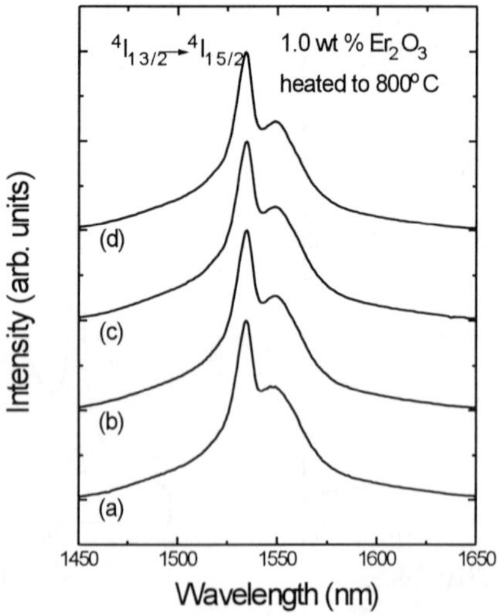

Figure 3. Er^{3+} $^4I_{13/2} \rightarrow$ $^4I_{15/2}$ fluorescence in hybrid organic/inorganic silica gels. The letter under each trace indicates the silicon alkoxides used to prepare each sample: (a) - TEOS only; (b) - TEOS: MTES = 10: 1; (c) - TEOS: DMDES = 10:1; (d)- TEOS: ETES = 10: 1.

Table I. Er^{3+} $^4I_{13/2}$ level lifetimes (ms) in densified silica and alkyl-modified silica gels.

Er^{3+}-Doped Silica Samples

Er_2O_3 concentration	T = 12 K	T = 100 K	T = 200 K	T = 295 K
1.0 wt %	8.49	8.06	7.17	7.19
3.0 wt %	7.24	6.78	6.29	6.21
5.0 wt %	5.63	5.42	6.21	4.86
10.0 wt %	-	-	-	4.10

Er^{3+},Al^{3+}-Doped Silica Samples (room temperature)

Er_2O_3 concentration	Al/Er = 1	Al/Er = 3	Al/Er = 5	Al/Er = 10
1.0 wt %	6.95	7.04	6.96	5.10
3.0 wt %	4.82	3.11	-	-

Er^{3+},Yb^{3+}-Doped Silica Samples (room temperature)

Er_2O_3 concentration	Yb/Er = 1	Yb/Er = 3	Yb/Er = 5	Yb/Er = 10
1.0 wt %	6.97	7.14	5.16	5.03
3.0 wt %	4.70	3.11	-	-

Er^{3+}-Doped Hybrid Silica Samples (concentration = 1.0 wt % Er_2O_3)

TEOS/MTES	T = 12 K	T = 100 K	T = 200 K	T = 295 K
10	5.30	5.05	4.69	5.22
5	5.95	5.58	5.28	5.77
TEOS/DMDES	T = 12 K	T = 100 K	T = 200 K	T = 295 K
10	5.14	5.16	5.18	5.64
5	7.13	6.93	6.74	-

concentration quenching. The addition of Al^{3+} also leads to increased hydroxylation of rare earth ion dopants [16], and increased hydroxyl quenching. The high ionic charge/ionic radius ratio of Al^{3+} allows it form stronger complexes with a larger number of hydroxyl groups than Er^{3+}. The decrease in the Er^{3+} $^4I_{13/2}$ level lifetime with the addition of Al^{3+} is attributed to increased hydroxyl quenching. The addition of Yb^{3+} is not expected to increase the hydroxyl concentration since it is similar in size and charge to Er^{3+}. The decrease in the Er^{3+} lifetime caused by Yb^{3+} may be the result of energy transfer between the two types of ions.

The lifetimes obtained from 1.0 wt % Er_2O_3-doped hybrid gels are shorter than the corresponding lifetime from a purely inorganic gel for all modified alkoxides used in this study and for ratios of TEOS: modified alkoxide of 10: 1 and 5: 1. In each hybrid system, the lifetime increases with decreasing TEOS: modified alkoxide ratio.The hydroxyl content of the hybrid silica sol-gel samples is expected to be lower, so the decrease in the lifetime relative to the silica samples is probably due to an increase in concentration quenching. This suggests that Er^{3+} ion clustering is more extensive in the hybrid samples than in the purely inorganic samples, and that using organically-modified precursors for lower hydroxyl gels has an overall negative effect on the optical properties of Er^{3+} in silica. This result suggests that rare earth ions may be incompatible with the organic part of the hybrid matrix during gelation and may preferentially segregate away from the organic groups to produce a non-uniform, high local concentration in the inorganic part of the matrix.

The possibility of upconversion fluorescence at 545 nm ($^4S_{3/2} \rightarrow {}^4I_{15/2}$) and 410 nm ($^2H_{9/2} \rightarrow {}^4I_{15/2}$) upon 650 nm excitation ($^4F_{9/2} \leftarrow {}^4I_{15/2}$) was checked in samples containing only

Er^{3+}, and samples co-doped with Al^{3+} and Yb^{3+}. Measurements were made at 12K and at room temperature, but no upconversion fluorescence was detected from these samples. The absence of upconversion fluorescence in Er^{3+}-doped samples with and without Al^{3+} or Yb^{3+} is attributed to hydroxyl quenching of the weakly-populated excited states that produce upconversion.

CONCLUSIONS

The addition of Al^{3+}, and to a lesser extent, Yb^{3+}, leads to a broader distribution of Er^{3+} sites in sol-gel silica, but also leads to an increase in local concentration of hydroxyl groups near the Er^{3+} ions. Alkyl-modified gels, while expected to contain a lower hydroxyl content than purely inorganic silica gels, exhibit a greater degree of rare earth ion clustering, causing an overall decrease in the radiative lifetime.

ACKNOWLEDGMENTS

We acknowledge partial financial support from the National Science Foundation and the University of Wisconsin Graduate School. This work was supported in part by a National Science Foundation Graduate Fellowship.

REFERENCES

1. D.C. Hanna, in Solid State Lasers: New Developments and Applications, edited by M. Inguscio and R. Wallenstein (Plenum Press, New York, 1993), p. 231.
2. D.J. DiGiovanni, in Optical Waveguide Materials, edited by M.M. Broer, G.J. Sigel, Jr., R.T. Kersten and H. Kawazoe (Mater. Res. Soc. Proc. **244**, Pittsburgh, PA, 1992), pp. 135-142.
3. V.P. Gapontsev, S.M. Matitsin, A.A. Isineev and V.B. Kravchenko, Optics and Laser Tech. **14**, 189 (1982).
4. B.J. Ainslie, J. Lightwave Tech. **9**, 220 (1991).
5. E. Snitzer and R. Woodcock, Appl. Phys. Lett. **6**, 45 (1965).
6. D.C. Winburn, Practical Laser Safety, (Marcel Dekker, New York, 1985), p. 14.
7. T. Miya, Y. Terunuma, T. Hosaka and T. Miyashita, Electron. Lett. **15**, 106 (1979).
8. J.L. Jackel, A. Yi-Yan, E.M. Vogel, A. Von Lehmen, J.J. Johnson and E. Snitzer, Appl. Opt. **31**, 3390 (1992).
9. J.D. Mackenzie, J. Non-Cryst. Solids **48**, 1 (1982).
10. A.J. Berry and T.A. King, J. Phys. D: Appl. Phys. **22**, 1419 (1989).
11. J. Phalippou, T. Woignier and J. Zarzycki, in Ultrastructure Processing of Ceramics, Glasses and Composites, edited by L.L. Hench and D.R. Ulrich (Wiley, New York, 1984), p. 70.
12. E.J.A. Pope and J.D. Mackenzie, J. Am. Ceram. Soc. **76**, 1325 (1993).
13. T. Fujiyama, M. Hori and M. Sasaki, J. Non-Cryst. Solids **121**, 273 (1990).
14. R.S. Quimby, W.J. Miniscalco and B. Thompson, J. Appl. Phys. **76**, 4472 (1994).
15. J. Nilsson, P. Blixt, B. Jaskorzynska and J. Babonas, J. Lightwave Tech. **13**, 341 (1995).
16. M.J. Lochhead and K.L. Bray, Chem. Mat. **7**, 572 (1995).
17. Z. Zhang, Y. Tanigami, R. Terai and H. Wakabayashi, J. Non-Cryst. Solids **189**, 212 (1995).
18. F. Babonneau, L. Bois and J. Livage, J. Non-Cryst. Solids **147&148**, 280 (1992).
19. B.T. Stone and K.L. Bray, J. Non-Cryst. Solids, in press.

Part VIII

Particulates and Layered Films

METAL/CERAMIC NANOCOMPOSITES BY SOL-GEL PROCESSING OF TETHERED METAL IONS: OPTIMIZATION OF THE PARTICLE-FORMING STEP

CLAUS GÖRSMANN[a], ULRICH SCHUBERT[a,b+], JÜRGEN LEYRER[c], AND EGBERT LOX[c]
[a]Institut für Anorganische Chemie der Universität, Am Hubland, D-97074 Würzburg, Germany
[b]Institut für Anorganische Chemie der Technischen Universität Wien, Getreidemarkt 9, A-1060 Wien, Austria, uschuber@fbch.tuwien.ac.at
[c]Degussa AG, P.O.Box 1345, D-63403 Hanau 1, Germany

ABSTRACT

The nanocomposite Pt · 62 SiO_2 was prepared by sol-gel processing of $Pt(acac)_2$, two equivalents of $H_2NCH_2CH_2NH(CH_2)_3Si(OEt)_3$ and 60 equivalents of $Si(OEt)_4$, followed by heating of the obtained metal complex-containing gel in air. When the temperature treatment is performed at T ≥ 550°C, elemental Pt is the only crystalline phase. The influence of the calcination conditions (maximum temperature, period of heating) on the size and size distribution of the resulting Pt particles and on the morphology of the SiO_2 matrix (surface area, pore volume) was investigated. When the temperature does not exceed 550°C, small Pt particles (3-5 nm) are formed which have a very narrow size distribution. Their size and size distribution are not affected by the period of heating. Gels heated to 550°C still contain about 0.1 - 0.25 wt% carbon. Carbon-free particles are obtained at 950°C. However, at temperatures between 750°C and 950°C the average size of the particle increases considerably, and the size distribution gets very broad, particularly for extended periods of heating. The specific surface area increases from about 50 m^2/g in the metal complex-containing gel to about 230 m^2/g during the oxidation step. At temperatures above 500°C considerable sintering of the matrix occurs. The pore volumes show the same trend. Heating the gels to 500°C results in the formation of nanopores with radii below 2 nm, which disappear again on heating to higher temperatures.

INTRODUCTION

One of the advantages of preparing oxide materials by the sol-gel method is the possibility to control their microstructure and homogeneity. For most applications very homogeneous materials are desired. However, R. Roy already pointed out in the early eighties that the advantages of sol-gel processing can also be exploited for the preparation of di- or multiphase materials [1].

Our approach for getting a nanometer-sized metal phase in an oxide host phase is a molecular dispersion of the metal precursor while the oxide phase is formed during sol-gel processing [2]. This is achieved by complexation of metal ions and tethering of the metal complexes to the oxide network by organofunctionally substituted alkoxides [3]. The metal nanophase is then obtained by controlled thermal treatment and reduction in the later steps of the preparation procedure. The inherent advantages of the sol-gel process can additionally be exploited, such as the tailoring of the microstructure of the oxide matrix, or the preparation of coatings. The used organofunctional alkoxides are of the type $(RO)_3Si$-X-A, where A is a functional group capable of coordinating metal ions

+ To whom correspondence should be addressed at the Technische Universität Wien

Mat. Res. Soc. Symp. Proc. Vol. 435 © 1996 Materials Research Society

(NH_2, $NHCH_2CH_2NH_2$, CN, $[CH_3(O)C]_2CH$, etc.), and X is a chemically inert spacer, mostly a $(CH_2)_n$ chain.

The dispersed metals are prepared by a three-step procedure, as previously reported [4,5]. In the first step a solution of a metal salt, the silane $(RO)_3$Si-X-A and $Si(OR)_4$ is processed by the sol-gel method. Metal complexes of the type $L_nM[A\text{-}X\text{-}Si(OR)_3]_n$ are formed in situ by reaction of the metal ions with $(RO)_3$Si-X-A. By the M:Si ratio of the starting compounds the metal loading of the composite is determined. In the second step the metal complex-containing gels are dried and then heated in air to oxidize all organic moieties. Due to the high dispersion of the metal ions in the first step, small metal oxide particles are formed, which are then reduced to metal particles having diameters of a few nm. In passing, it must be emphasized that the oxidation step is of course only necessary, if *carbon-free* composites are to be prepared.

The obtained metal particles are highly dispersed, not agglomerated, and homogeneously distributed *throughout* the SiO_2 matrix. Their diameters are very small and uniform, even in the materials with high metal loadings. The main advantage of the method is that very narrow particle size distributions are obtained. Bimetallic particles can be prepared by the same approach [5-7].

The average metal particle size and the size distribution depend on several factors, such as the kind of metal, the reaction conditions, the nature of the organic anchoring group, and for some metals also on the metal loading [4,5]. We have recently shown that they are also influenced by the metal / $(RO)_3$Si-X-A ratio. The narrowest particle size *distribution* for Cu/Ni particles was obtained, if just the amount of $(EtO)_3Si(CH_2)_3NHCH_2CH_2NH_2$ needed for complexation of all metal ions was added to the starting solution [7].

The oxidation conditions are very critical with regard to the average metal particle size and the size distribution. In this step the organic groups are thermally removed, and metal oxide particles are formed. Therefore, the oxidation conditions should allow the complete removal of all organic parts, but should not be harsher than necessary to avoid an excessive growth of the metal particles. The present paper contains a systematic study on the influence of the oxidation temperature and the oxidation period on the metal particle size, the size distribution, and the properties of the SiO_2 matrix. To focus on these reaction parameters, all investigations were performed on one system with a constant chemical composition.

PREPARATION, CHARACTERIZATION AND THERMAL DEGRADATION OF THE METAL COMPLEX-CONTAINING GELS

The preparation of the nanocomposite Pt · 62 SiO_2 is outlined in Scheme 1 and was performed as previously described for other metal/SiO_2 nanocomposites [4]. We selected a Pt-containing composite, because no reduction step is required. PtO is not stable at the temperatures employed for the second step (see below) and decomposes by formation of elemental Pt. This simplifies the preparation procedure and eliminates the parameters of the reduction step.

When $Pt(acac)_2$ is reacted with 2 equivalents of $H_2NCH_2CH_2NH(CH_2)_3Si(OEt)_3$ in ethanolic solution (100 ml ethanol per 1 mmol Pt salt), the ethylenediamine moiety of the silane coordinates to the Pt^{2+} ion. Although the resulting complex was not identified in this case, we know from other studies [7, 8] that the same type of complexes is formed with $H_2NCH_2CH_2NH(CH_2)_3Si(OEt)_3$ as with ethylene diamine (en), i.e. substitution of ethylene diamine by a silylalkyl group does not severely influence its coordination behaviour.

To the solution of the in situ formed metal complex 60 molar equivalents of $Si(OEt)_4$ (TEOS) and then 1845 equivalents of water as aqueous 0.2 n ammonia were added (corresponding to a 7.5-fold excess of water relative to the hydrolysis of all Si-OR groups). The resulting homogeneous mixture was heated to 70°C for 72 h in a closed vessel. Then the solvent was removed at 60°C/12 Torr. The obtained solid, chemically homogeneous xerogel was dried at 70°C and 10^{-3} Torr to remove adsorbed water and alcohol. At this point the gel **1** has the idealized composition $(acac)_2Pt[H_2NCH_2CH_2NH(CH_2)_3SiO_{3/2}]_2 \cdot 60\ SiO_2$, in agreement with elemental analyses.

$$Pt(\mathit{acac})_2 + 2\ \mathit{H_2NCH_2CH_2NH(CH_2)_3}Si(O\mathit{Et})_3 + 60\ Si(O\mathit{Et})_4$$

$+NH_3/H_2O$ | $-$ ROH

$$(\mathit{acac})_2Pt[\mathit{H_2NCH_2CH_2NH(CH_2)_3}SiO_{3/2}]_2 \cdot 60\ SiO_2\ \ (\mathbf{1})$$

air / T↑

$$Pt \cdot 62\ SiO_2$$

Scheme 1: Preparation of the $Pt \cdot 62\ SiO_2$ nanocomposite (acac = acetylacetonate). The organic groups are printed in italics.

The nature of the Pt compound(s) in **1** was probed by X-Ray Photoelectron Spectroscopy (XPS). By comparison of the intensity of the XPS signals with those expected for a statistical distribution of all involved elements an enrichment of Pt and C on the surface of the sample was observed. The deconvoluted Pt signal is shown in Figure 1.

It consists of two doublets (Pt $4f_{5/2}$ and Pt $4f_{7/2}$) with maxima for Pt $4f_{7/2}$ at 73.7 eV (58%) and 72.6 eV (42%). The latter value is comparable to that of $Pt(en)_2Cl_2$ (72.6 eV [9]) and confirms that at least part of the sample contains the Pt^{2+} ions in the anticipated coordination environment. The signal at 73.7 eV compares to that of $Pt(OAc)_2$ (73.8 eV) [10]. Together with the Pt/C-enrichment at the surface this indicates that the other part of the Pt^{2+} ions is in an oxygen-rich environment. This could either be due to unreacted $Pt(acac)_2$ or to reaction of $(acac)_2Pt[H_2NCH_2CH_2NH(CH_2)_3SiO_{3/2}]_2$ with air during sample preparation. Although the sample was only slightly ground, the latter suspicion is corroborated by the observation that in another sample of the same material, intensively ground for 10 min, there was no nitrogen signal. There were two doublets with maxima for Pt $4f_{7/2}$ at 71.1 eV (23%) and 71.8 eV (77%). This observation is a strong indication that the energy input during grinding already results in a destruction of the chelating ligand and formation of elemental Pt (71.07 eV), as in the preparation procedure of the composites by external heating (see below).

The starting point for the systematic investigation of the calcination parameters is the thermogravimetric analysis of the Pt complex-containing gel **1**, shown in Figure 2. After the initial weight loss of about 3%, typical for the loss of adsorbed water in gels, there is a steep decrease at 220°C (which is lowered to 185°C at a lower heating rate of 2.5°/min). DSC shows that this weight loss of 6.5% is associated with a strongly exothermic reaction. There is only one additional, less intense and broad exothermic DSC signal at about 300°C, where further degradation of the gel continues.

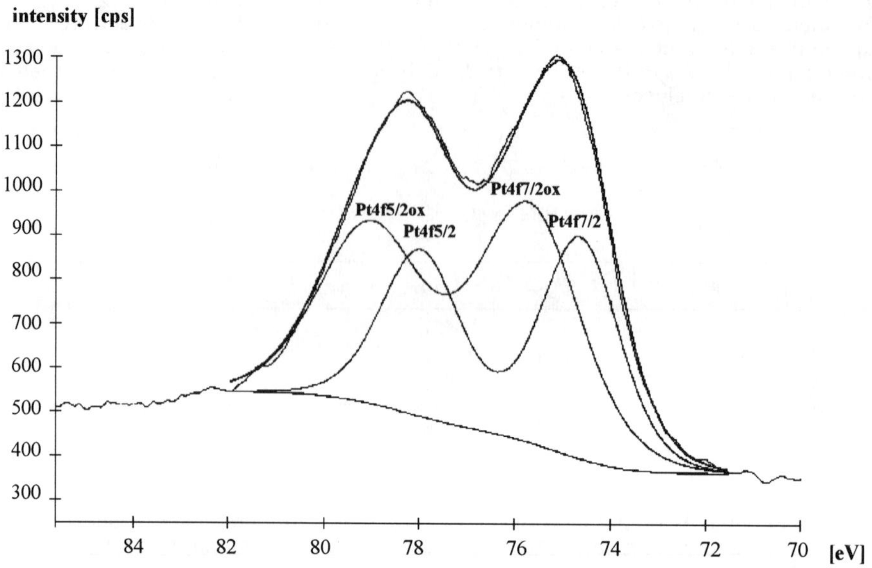

Figure 1. Deconvoluted Pt doublet in the XPS spectrum of the gel **1**.

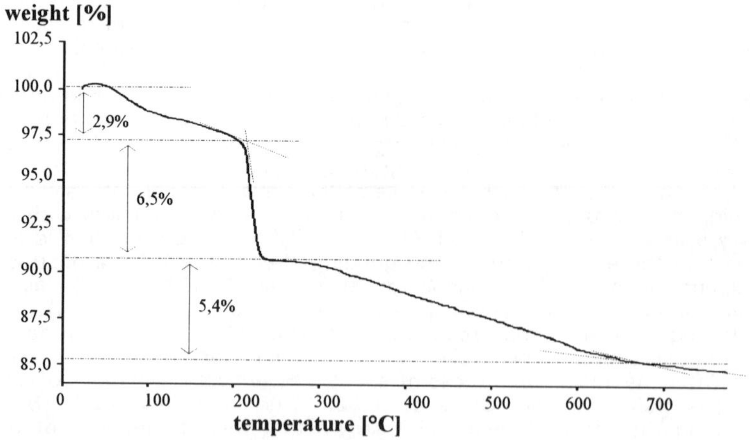

Figure 2. TGA of the Pt complex-containing gel **1** in air. Heating rate 10°/min.

To get an idea about the chemical process occuring at about 200°C, we prepared a gel from only $Pt(acac)_2$ and 62 equivalents of $Si(OEt)_4$. TGA/DSC of this gel shows a strong exothermic reaction at 186°C (heating rate of 2.5°/min) associated with the loss of all carbon, i.e. the complete degradation of the acac ions. Based on this observation, we assign the exothemic event at about 200°C in the Pt complex-containing gel **1** (Figure 2) also to the (total or partial) decomposition of the acac anions.

We have observed earlier that the metal particle size in metal/SiO_2 composites prepared by our method also depends on the complexing group used in the first step of the synthesis. The mean particle diameter in Ni · 15 SiO_2 decreased from 22.9 nm to 2.6 nm (same reaction conditions), if the acetylacetone derivative $[CH_3(O)C]_2CH(CH_2)_3Si(OR)_3$ was used for the complexation of the nickel ions during sol-gel processing instead of the ethylene diamine derivative $(EtO)_3Si(CH_2)_3NHCH_2CH_2NH_2$ [4]. An explanation for this observation may be the different local heat evolution during the thermal degradation of the complexing ligands.

The heat generated locally by decomposition of the acac groups in **1** already induces some thermal degradation of the $NH_2CH_2CH_2NH(CH_2)_3$ groups. This is concluded from the residual C, N, and H content determined by elemental analysis after heating samples of **1** to different temperatures and then holding them for 1h. Figure 3 shows that in the early stages (T < 300°C) degradation of carbon-containing moieties is faster than that of the nitrogen-containing ones, probably due to the decomposition of the acac groups. Nitrogen is completely removed between 300 and 400°C, while about 20% of the original carbon content is retained. Degradation of the $NH_2CH_2CH_2NH(CH_2)_3$ groups therefore seems to proceed from the nitrogen end of the chain. We cannot exclude, however, that non-volatile carbon-containing intermediates are formed first, which are only completely oxidized at higher temperatures.

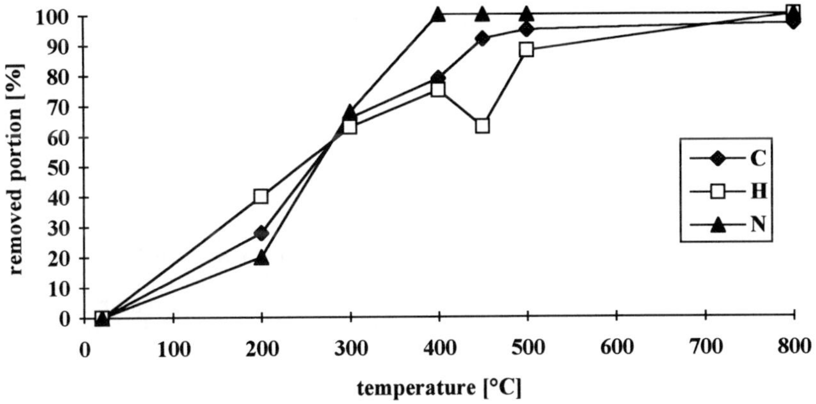

Figure 3. Percentage (wt%) of C, N, and H, removed by thermal degradation of **1** in air. The sample was heated to the corresponding temperature and then held for 1 h.

The oxidation *period* influences the carbon content (initial value before oxidation: 6.3%) only at relatively low temperatures. The residual carbon of 0.9 wt% after 1 h at 450°C in air was lowered to 0.36% when the sample was held at 450°C for additional 4 h. No additional carbon loss was achieved upon holding the sample at 450°C for 10 h altogether. The same trend was observed at 500°C (0.35% after 1 h, 0.15% after 5 and 10 h) and 550°C (0.24% after 1 h, 0.07% after 5 and 10 h). Oxidation at 750°C results in a residual carbon content of 0.07%, independent of the holding period. Carbon-free samples are only obtained at 950°C.

CHARACTERIZATION OF THE NANOCOMPOSITES

A homogeneous grey powder is obtained after thermal oxidation of **1**. Granulometric determination of the grain size showed a homogeneous distribution with a maximum of 0.11 mm. 10 wt% of the powder has a grain size below 0.02 mm or above 0.24 mm, respectively.

A sample of **1** was heated to 550°C for 1 h and then investigated by X-Ray Diffraction (XRD). The only detectable crystalline phase was elemental Pt, with reflections at $2\Theta = 39.8$, 46.3, and 67.5°. Platinum oxides, PtO_2 and PtO, are thermally unstable; above 507°C only elemental Pt is existent [11]. For the following discussion it can therefore be assumed that elemental Pt is the only metal-containing phase, if the oxidation step is carried out at $T > 500$°C. At lower temperatures the (partial) presence of the oxides can be assumed, but does not affect the general reasoning.

One of the objectives of this work was to investigate the influence of the oxidation temperature and the oxidation period on the average *metal particle diameter* and the *particle size distribution*. Samples of **1** were heated in air to temperatures between 450 and 950°C for 1, 5 or 10 h, as described above. The diameter of several hundred particles per sample was measured in Transition Electron Micrographs (TEM). The particle diameters were averaged in three different ways, assuming spherical particles (D_i = particle diameter, n_i = numbers of particles):

- Arithmetic mean: $D_N = \Sigma n_i D_i / \Sigma n_i$.
- Mean based on the particle surface: $D_A = \Sigma 4\pi n_i (D_i/2)^2 D_i \,/\, \Sigma 4\pi n_i (D_i/2)^2$.
- Mean based on the particle volume: $D_V = \Sigma {}^4/_3 \pi n_i (D_i/2)^3 D_i \,/\, \Sigma {}^4/_3 \pi n_i (D_i/2)^3$.

The closer the values of D_N, D_A, and D_V for a particular sample, the narrower is its particle size distribution. A few big metal particles affect D_A and D_V more than D_N, because their higher surface area and their larger volume gives them more weight in calculating D_A and D_V.

The results of the TEM investigations are compared in Figure 4 (logarithmic scale!). When the oxidation temperature does not exceed 550°C, the mean Pt particle diameter is between 3-5 nm, independent of the averaging method. The particle size distribution is therefore very narrow. A typical size distribution curve is shown in Figure 5. In this temperature range the mean particle size is hardly affected by the oxidation period. At oxidation temperatures of 750°C and 950°C the average particle size increases considerably. There are still small particles, but big particles are increasingly formed, resulting in an enormous broadening of the size distribution. This is due to ripening processes, as indicated by the strong influence of the oxidation period at these temperatures. Very big particles with diameters up to 1920 nm are found at 950°C.

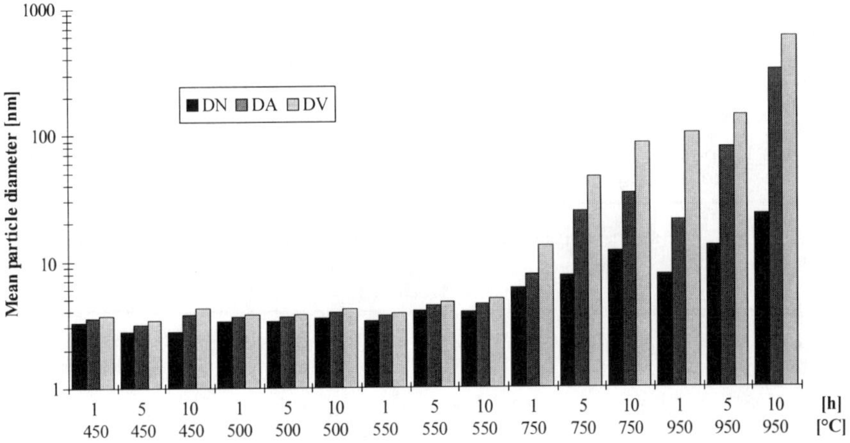

Figure 4. Mean Pt particle diameters depending on the oxidation temperature and oxidation period. D_N, D_A, and D_V are defined in the text.

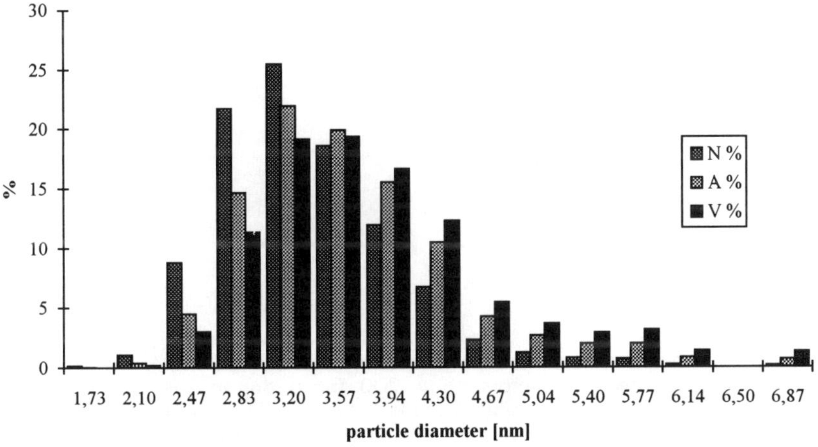

Figure 5. Pt particle size distribution of Pt · 62 SiO_2, oxidized at 550°C for 1 h. The three columns represent the percentage of the total number (N), the percentage of the total surface (A) and the percentage of the total volume (V) of the particles of a given size.

The conditions during the oxidation step also influence the *properties of the SiO_2 matrix*. When the large organic groups used for tethering of the metal ions and the organic conterions (acac) are oxidatively removed, new pores should be formed or existing pores should be enlarged. Since the oxidation process initially replaces Si-C bonds by Si-OH groups, these pores will be particularly prone to sintering processes.

The specific surface areas, specific pore volumes, and mean pore radii of the nanocomposites Pt · 62 SiO_2 depending on the conditions of the oxidation step are shown in Table I.

Table I. Specific surface areas, specific pore volumes, and mean pore radii depending on oxidation temperature and oxidation period

conditions	spec.surface area $[m^2/g]$	spec. pore volume $[m^3/g]$	mean pore radius [nm]
1 (before oxidation)	50	0.19	5.4
450°C / 1h	191	0.25	4.5
450°C / 5h	244	0.26	4.1
450°C / 10h	244	0.26	4.2
500°C / 1h	214	0.25	4.2
500°C / 5h	233	0.25	4.5
500°C / 10h	204	0.25	4.4
550°C / 1h	92	0.22	5.0
550°C / 5h	106	0.23	4.6
550°C / 10h	134	0.23	4.4
750°C / 1h	90	0.21	4.7
750°C / 5h	83	0.20	4.5
750°C / 10h	82	0.20	4.5
950°C / 1h	75	0.19	4.5
950°C / 5h	63	0.17	4.6
950°C / 10h	61	0.17	4.6

The specific surface area is increased from about 50 m^2/g in the metal complex-containing gel **1** to more than 200 m^2/g during the oxidation step. However, between 500°C and 550°C considerable sintering occurs. The surface area drops to about half its value in this intervall. The further decrease at higher temperatures is relatively small. While the period of heat treatment is without any influence at high temperatures, some increase of the surface area is observed between 450 and 550°C, when the period is extended.

The specific pore volume follows a similar trend. Starting from 0.19 m^3/g in the gel **1**, the maximum is obtained after heating to 450°C for 10 h. At higher temperature the pore volumes drop below their original value.

The influence of the oxidation temperature on the pore radii distribution is shown in Figure 6. Between 450°C and 500°C, and between 550°C and 750°C there is nearly no change, and therefore the corresponding distribution curves are not shown in Figure 6. The volume of pores with a radius larger than 25 nm is below 10%, and therefore this range is disregarded in the following discussion.

The maximum of the pore radii distribution is about 3 nm in the unoxidized gel **1**, and the percentage of pore radii below that value is rather small. There is only little change in the pore radii distribution above 10 nm when the gel is oxidized at temperatures between 450 and 550°C. The main difference between the unoxidized gel and that after heat treatment at 450 or 500°C is the creation of small pores. The volume of pores between 2 - 6 nm increases by about 40%. Particularly striking is the appearance of a great portion of pores with radii around 2 nm. The origin of these nanopores appears to be the destruction of the $(CH_2)_3NHCH_2CH_2NH_2$ chains, having a length of about 1 nm. The creation of nanoporosity therefore can be traced back to the removal of the organic moieties. The pores with radii below 2 nm appear to be particularly labile, because they largely disappear on heating the sample to 550°C. The portion of pores with radii of 3 - 4 nm also decreases in this temperature range. Heating to 950°C further reduces the pore volume. The pores below 10 nm approach the value in the unoxidized gel **1**. However, at high temperatures mainly the pores above 12 nm are affected. Their volume is reduced by 60%.

CONCLUSIONS

A molecular dispersion of metal ions in silica gel is achieved by their coordination with $(RO)_3Si\text{-}X\text{-}A$ and sol-gel processing of the resulting metal complexes $L_nM[A\text{-}X\text{-}Si(OR)_3]_n$. On heating the metal complex-containing gels in air, three processes have to be considered:

- removal of the organic groups by oxidation,
- formation of small metal oxide particles, or small metal particles, if the metal oxide is thermally unstable, and
- rearrangement processes of the matrix.

The conditions of the heating step are very critical, because they partially influence the three processes in a different manner.

In the present paper we have demonstrated these effects for the nanocomposite Pt · 62 SiO_2, prepared by using two equivalents of the ethylene diamine derivative $(EtO)_3Si(CH_2)_3NHCH_2CH_2NH_2$ for complexation of the Pt ions. It was shown that heating the gels to 500°C for 1 h is a good compromise. At this temperature the organic groups are mostly removed, and only 0.35 - 0.15 wt% carbon (depending on the period of heating) is retained. Nanopores are created by the removal of the organic groups, and a maximum surface area is obtained. The mean Pt particle diameter is very small (3 - 5 nm), and the size distribution is very narrow. Carbon-free composites are only obtained at 950°C. However, above 550°C the Pt particles grow and the particle size distribution gets broader. The temperature range of 500-550°C is most critical for the morphology of the silica matrix: the previously created nanopores disappear again, and the specific surface area is reduced.

In conclusion, the optimum values with regard to the Pt particles and the matrix morphology are obtained by heating to 500°C, although a small amount of carbon is retained at this temperature. It depends on the use of the composites whether this carbon content can be tolerated. In the present study we have neglected the influence of the kind and amount of the organic groups (complexing groups and counter-ions), and the influence of the metal. As a matter of fact, each system has to be individually optimized.

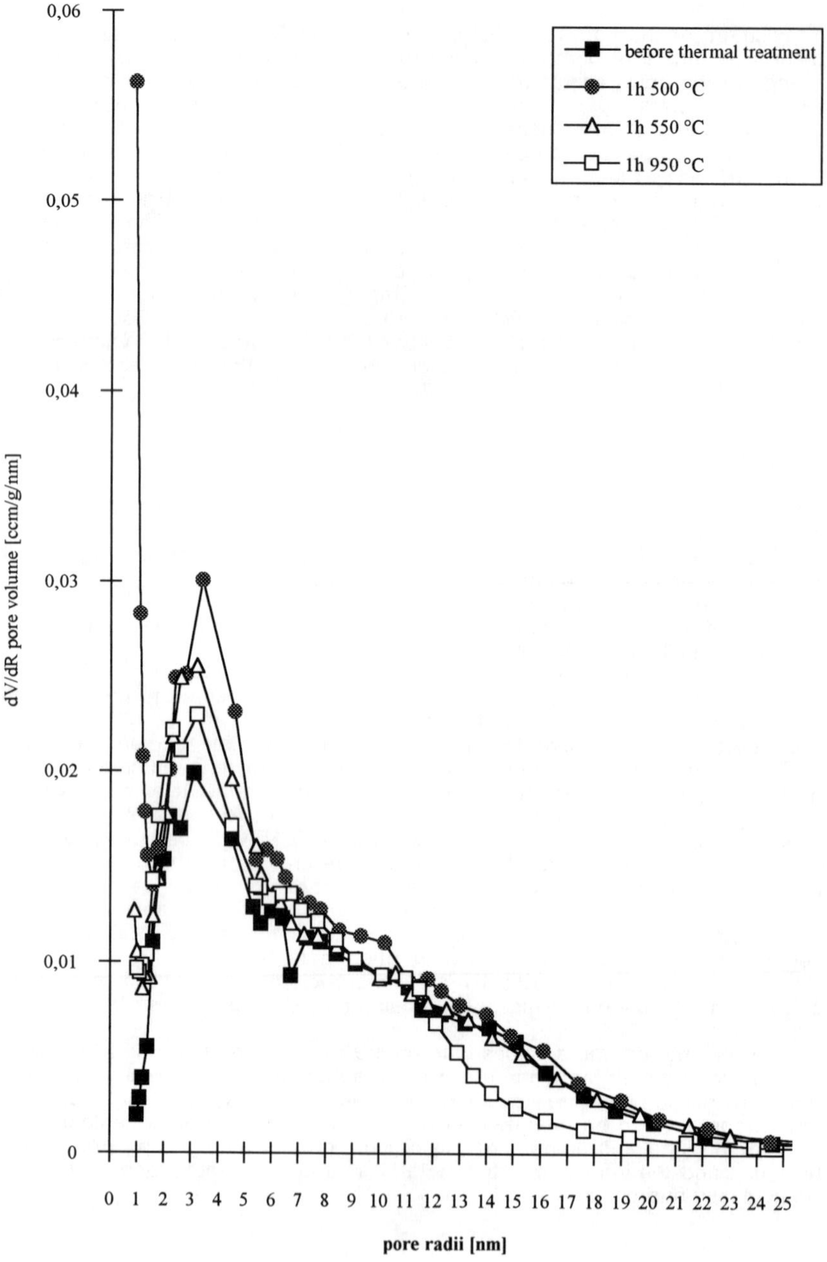

Figure 6. Pore radii distribution of Pt · 62 SiO_2 depending on the oxidation temperature.

REFERENCES

1. R. A. Roy, R. Roy, Mat. Res. Bull., **19**, 169 (1984).

2. U. Schubert, F. Schwertfeger, C. Görsmann in G.-M.Chow, K.Gonsalves (Eds.), Molecularly Designed Nanostructured Materials, ACS Symp.Ser. in press.

3. U. Schubert, N. Hüsing, A. Lorenz, Chem. Mater., **7**, 2010 (1995).

4. B. Breitscheidel, J. Zieder, U. Schubert, Chem. Mater., **3**, 559 (1991). U. Schubert, B. Breitscheidel, H. Buhler, C. Egger, W. Urbaniak, Mat. Res. Soc. Symp. Proc., **271**, 621 (1992).

5. U. Schubert, C. Görsmann, S. Tewinkel, A. Kaiser, T. Heinrich, Mat. Res. Soc. Symp. Proc., **351**, 141 (1994).

6. W. Mörke, R. Lamber, U. Schubert, B. Breitscheidel, Chem. Mater., **6**, 1659 (1994). W. Mörke, T. Bieruta, J. Jarsetz, C. Görsmann, U. Schubert, Colloid Surf. in press.

7. A. Kaiser, C. Görsmann, U. Schubert, J. Sol-Gel Sci. Technol., in press.

8. C. Görsmann, Ph.D. Thesis, University of Würzburg, 1996.

9. M. Yamashita, N. Matsumoto, S. Kida, Inorg. Chim. Acta, **31**, 381 (1978).

10. R. I. Rudyi, N.V. Cherkashina, G. Y. Mazo, J. Salin, I. I. Moiseev, Izv. Akad. Nauk. SSSR Ser. Khim., **4**, 754 (1980).

11. G. V. Samsonov, The Oxide Handbook, Plenum Press, New York, 1973, p.164.

TEM-CHARACTERIZATION OF METALLIC NANOPARTICLES EMBEDDED IN SOL-GEL PRODUCED GLASS-LIKE LAYERS

U. WERNER, M. SCHMITT, H. SCHMIDT
Institut für Neue Materialien, Im Stadtwald 43, D-66123 Saarbrücken, Germany

ABSTRACT

Size and shape of metallic nanoparticles, which are embedded in sol-gel produced glass-like layers, are able to influence the optical properties of coatings on glass. Since the shape of the particles is determined by its crystalline structure, and the structure is modified significantly during the particle growth, the development of both was observed in dependence on the stage of growth applying conventional and high resolution electron microscopy. These investigations carried out on Au and Ag particles have revealed different types of colloidal growth, a real structure often formed by dislocations and twins as well as a relatively high particle mobility.

INTRODUCTION

The embedding of metallic nanoparticles in glass-like layers, which are produced by sol-gel processing, allows the modification of the optical properties of these layers such as the optical absorption and the nonlinear optical properties. This modification is determined by the concentration, size, shape and the internal structure of the particles [1]. In order to improve the knowledge with respect to an optimum embedding, the electron-microscopical investigation aims especially at the characterization of vicinity, size, shape, internal structure and behavior of the nanoparticles, and the variation of these parameters during the particle growth. Comprehensive information is obtained by transmission electron microscopy (TEM) techniques starting from the global characterization of the layer using ordinary diffraction contrast up to the investigation of the atomic structure level of the nanoparticles using the interference contrast of high-resolution electron microscopy (HREM) at an acceleration voltage of 200 kV.

EXPERIMENT AND PREPARATION

The synthesis of nanocrystalline Au and Ag particles [2], which are embedded in glass-like coatings, starts with an SiO_2 sol to which the corresponding metal is added in ionic form ($H[AuCl_4]$, $AgNO_3$). By using functionalized aminosilanes the metal ions are enabled by their complexation to go into solution and, simultaneously, their reduction is suppressed. Glass slides are covered with this organically modified SiO_2 sol by applying the dip-coating method. The metal ions are reduced by the electrons released during the decomposition and combustion of the organic sol components in the subsequent heating procedure (densification temperature: T_d = 100-600°C). By this process the colloidal nucleation is initiated. An uncontrolled growth is prevented by the functionalized silanes which cover the colloid surfaces and thus avoid by their entropic interaction forces a premature coagulation and coalescence of the colloids. Additionally, the heating causes that the solvent of the sol evaporates, which leads to the development of an organic-inorganic (ormocer) or glass-like matrix (described in detail in [3,4]).

The embedding of the particles in an electrically insulating amorphous layer gives rise to unfavorable imaging conditions such as inelastic scattering of electrons (loss of coherence) and electrical charging of the layer (resulting in spontaneous specimen motions as well as astigmatism). These influences are minimized by extensively reducing the matrix material in two ways: firstly, the simple scratching off of layer splinters from the coatings on glass and, secondly, the ion beam etching of the covered glass slides. While the first case offers the advantage of easy structure imaging of nanoparticles positioned in the wedge-shaped edge region of the splinters, the second case offers the possibility to get relatively large electron transparent areas which allow survey images in diffraction contrast.

Mat. Res. Soc. Symp. Proc. Vol. 435 © 1996 Materials Research Society

RESULTS AND DISCUSSION

The particle shell

The colloidal growth is decisively governed by the mutual repulsion of the ligands enveloping the nanoparticles. Hence, for the assessment of the growth a characterization of the ligand shells is eminently important. However, the amorphous ligand shells embedded in the ormocer matrix (where they are not yet decomposed by the heating procedure) show no directly observable electron microscopical contrast. The application of ion beam etching opens a way for their indirect observation because of the strength-sensitive material erosion by the ion beam: Since the ligand molecules have a catalytic effect on the hydrolysis and condensation, the densification of the matrix in the range of the ligand shells becomes higher than in the remaining region. By this process the ion beam can uncover the higher densified particle shells without dismantling them.

Fig. 1a shows an Ag nanoparticle with a stronger densified ormocer shell (thickness: 9.0 nm) in an ion beam thinned ormocer matrix ($T_d = 200°C$) that was imaged with the incident electron beam parallel to the layer surface. The measured shell thickness within an extremely thin layer area - where the shells of relatively small particles also possess sufficient contrast - is depicted in the diagram of Fig. 1b. This diagram shows no systematic variation of the shell thickness in dependence on the particle diameter; the measurement yields a mean shell thickness of 7.5 nm. This behavior reveals the causal action of a shell forming process, such as the attachment of ligands to the particles and their catalytic effect during the densification of the SiO_2 network, which is independent of size and shape of the particles.

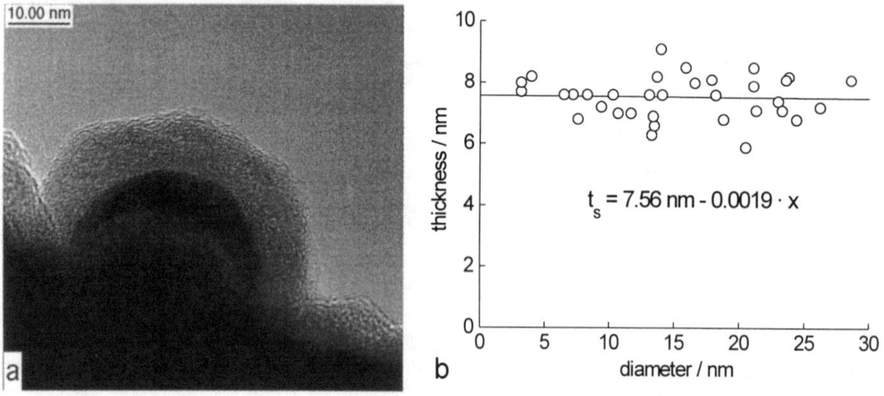

Fig. 1: a) Imaging of an enveloped Ag nanoparticle originally embedded in an ormocer matrix (Ag / aminosilane = 1 / 1, $T_d = 200°C$) after [4]. The stronger densified ormocer shell of the particle was uncovered due to the strength-sensitive material erosion of an ion beam. b) Thickness of the shells t_s depending on the particle diameter x. The fitted straight line ($t_s = 7.56$ nm - $0.0019 \cdot x$) yields no variation of thickness as function of the diameter.

Size distribution of the embedded particles

Using self-supporting organically modified SiO_2 layers (ormocer layers) with embedded Au colloids (Au / aminosilane = 1 / 4, $T_d = 120°C$) investigations of size distributions of metallic nanoparticles were carried out. By means of ion beam etching it was possible to thin sufficiently large areas with negligible variation in thickness that allowed the imaging of a statistically representative number of particles. The size distributions for different times after the layer densification are shown in Figs. 2a and b: The fraction of larger particles with partly pronounced crystalline habit is clearly greater in Fig. 2b than in 2a. The reason for this aging process recorded here must be mainly attributed to the ormocer layer with its moderate densification. Similar processes are hardly taking place with the same high intensity in fully densified glass-like layers. However,

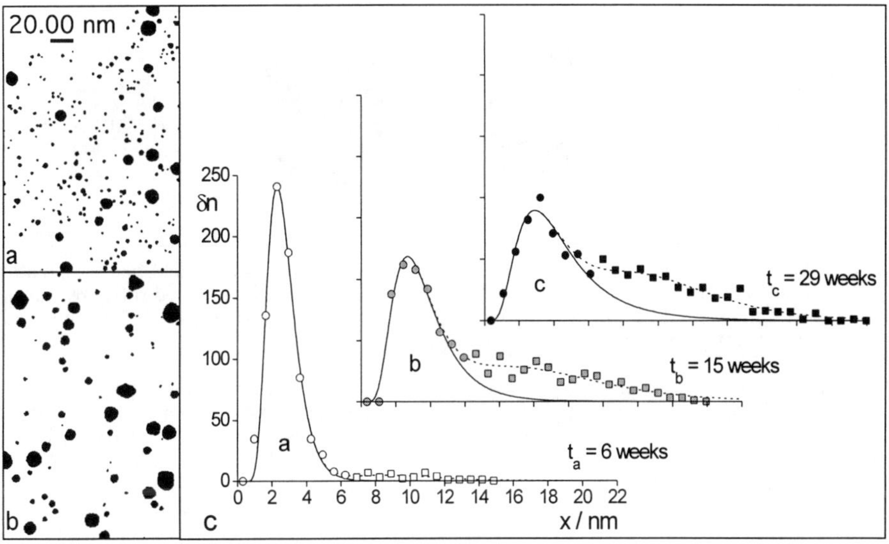

Fig. 2: Evolution of the size distribution of Au colloids in an ormocer layer (Au / aminosilane = 1 / 4, T_d = 120°C) determined 6 (image and graph a), 15 (image and graph b) and 29 weeks (graph c) after the densification (storage at room temperature). Fig. 2c: Particle number counted per class $\delta n = dn / dx \cdot \delta x$ (δx: width of classes) depending on the diameter x of the Au particles (because of different microscopical magnifications all graphs become absolutely comparable by multiplying δn of graph 'a' by 3.6). Fitting of log-normal distributions (equation (1), solid lines) to the frequencies in the low-diameter region.

the investigation of the aging process permits interesting conclusions on the crystal growth in sol-gel produced matrices.

The binary images of Figs. 2a and b were derived from the originals by image processing procedures such as shading correction, pixel editing, binarization and opening. The obtained binary images permit a computer-aided evaluation regarding the size distributions given in the graphs of Fig. 2c. In the range of small particle diameters the diameter frequencies can be fitted by a log-normal distribution with positive skewness

$$\frac{dn}{dx} = \frac{n_0}{\sqrt{2\pi} \cdot \sigma x} \cdot \exp\left[-\frac{(\ln x - \ln \xi)^2}{2\sigma^2} \right] \qquad (1)$$

with the parameters according to Table I.

Table I: Parameters of the log-normal distribution (1): x stands for the particle diameter, N_0 for the total particle number, n_0 for the particle number covered by the log-normal distribution and ξ as well as σ for the distribution parameters. The expectation value EX, standard deviation $\sqrt{D^2X}$ and skewness γ of the distribution are also given.

	t / weeks	N_0	n_0	ξ / nm	σ	EX /nm	$\sqrt{D^2X}$ /nm	γ
t_a	6	793	728	2.57	0.326	2.71	0.91	1.04
t_b	15	887	600	3.36	0.473	3.76	1.88	1.62
t_c	29	894	576	3.95	0.546	4.58	2.70	1.86

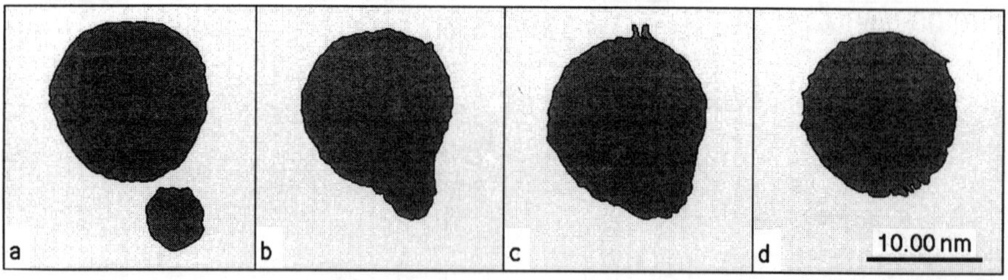

Fig. 3: Sequence of a liquid-like coalescence of two Au nanoparticles in a glass-like SiO_2 matrix (Au / aminosilane = 1 / 4, T_d = 500°C). The particle fusion was initiated solely by the energy transferred by the imaging electron beam at a TEM dose rate < 50000 e^-/nm^2 per image. Image processing: contour tracing.

The development of a positively-skewed log-normal distribution points to a growth process that proceeds on the basis of a liquid-like particle coalescence [5]. The probability of such a process can be realized if one considers that the particles covered by the log-normal distribution have diameters clearly below 6 nm. Gold clusters of this size show a liquid-like behavior in vacuum above room temperature because of their lowered melting points [6]. Provided that this behavior, or at least an extremely easy modifiability of the internal structure applies to the embedded clusters, too - and the observed behavior indicates this (see Fig. 3) - the measured distributions can be interpreted as follows: During the densification of the matrix at a temperature of 120°C the growth of the majority of the initially formed gold crystal nuclei takes place via liquid-like coalescence. A pronounced log-normal distribution results as shown in the graph 'a' in Fig. 2c. This growth process probably continues also after the densification, however, in a much reduced way. This is indicated by the statistical mean and standard deviation of the log-normal distribution increasing with time (cf. graphs 2a-c). Particles with diameters > 6 nm distinctly deviate from the log-normal distribution. They are essentially formed a long time after densification (T_d = 120°C) and show a partly pronounced crystal habit. An estimation of the volume fractions of the different groups of particles reveals that the growth of the larger ones cannot be attributed to the numerical shrinkage of the smaller ones (eventually by Ostwald ripening) because the whole volume of the larger particles exceeds that of the smaller ones to a large extent. However, since estimations have shown [7] that a great deal of the ionic Au (80%) was not reduced during the densification, it can be assumed that these ions were reduced during the aging of the matrix. Therefore, a growth by diffusion and absorption of reduced metal ions can be considered as the most obvious interpretation of the creation of the larger colloids.

Coalescence and internal structure of nanoparticles

Occasionally, particles embedded in glass-like matrices can be detected which are formed by coalescence, which however, neither reach an ideal equilibrium shape nor an irregular one. These particles show a needle-like shape with a preferred orientation as represented in Fig. 4. Such particles are technologically interesting due to their relatively high aspect ratio that causes nonlinear optical effects. For utilization of such particles a deeper understanding of their coalescence behavior and the internal structure of particles in glass-like layers is necessary.

Owing to the easy modifiability of the lattice structure of nanoparticles the energy transferred by a strongly focused intense electron beam can be sufficient to give rise to coalescence of particles. A structure image of a nanoparticle generated in this way by the fusion of two particles is shown in Fig. 5. In order to get an undisturbed insight into the crystalline structure the primary colloids were grown in solution controlled by ligands, and subsequently deposited onto a carbon film. Furthermore, for a clear representation of the crystalline structure a graphical computer-aided evaluation of the image via image processing was performed. This processing - represented in Fig. 6 - comprises a local contrast maximization in the image of the atomic rows, followed by image binarization and determination of the gravity centers of the rows of the atoms as well as, finally, a schematic drawing of the atomic rows and of (111) and (002) lattice planes.

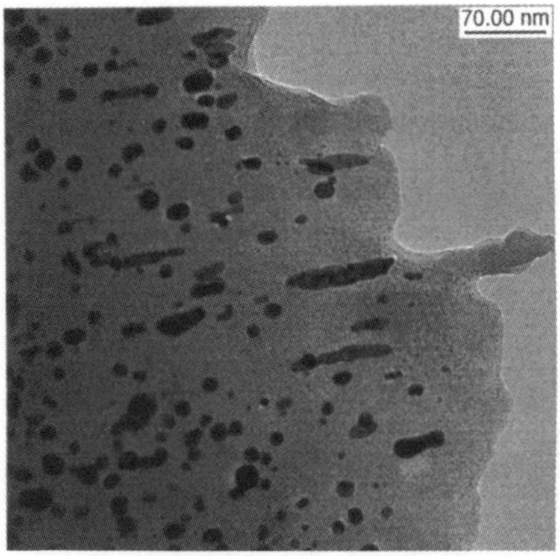

Fig. 4: Partially strongly oriented elongated Au particles (Au / aminosilane = 1 / 4, T_d = 500°C) formed by coalescence in an area of a highly densified glass-like layer.

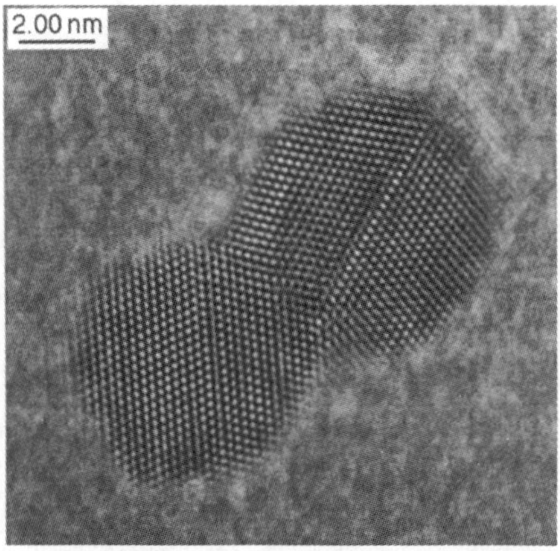

Fig. 5: Many beam imaging of a dumb-bell-shaped particle formed by intense electron irradiation induced coalescence of two Au colloids (Au / aminosilane = 1 / 10, grown in solution). Image processing: sum of the original image and its masked Fourier filtering.

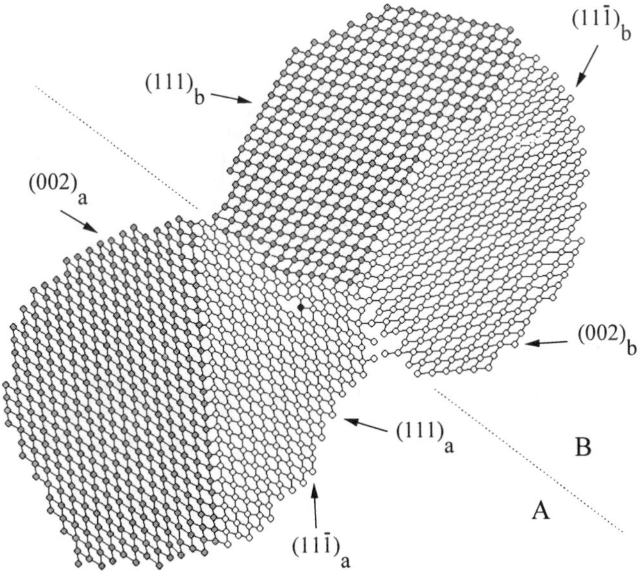

Fig. 6:

Graphical evaluation of Fig. 5 by the determination of the gravity centers of the atomic rows. The different grains are accentuated by different shading. The edge dislocation which compensates the misorientation between $(111)_a$ and $(111)_b$ lattice planes is marked by its black colored core.

The colloid shown in Fig. 6 was formed by two twinned primary particles. Initially their twin boundaries possessed an orientation difference of 25.5° and their centers of gravity were at a distance of 7.3 nm. During the course of the coalescence the particles have approached each other by 0.5 nm, and by this also rotated by 7.5°. As a result a fusion region was formed by a distorted twin boundary: E.g., the marked edge dislocation (dislocation core: black) compensates

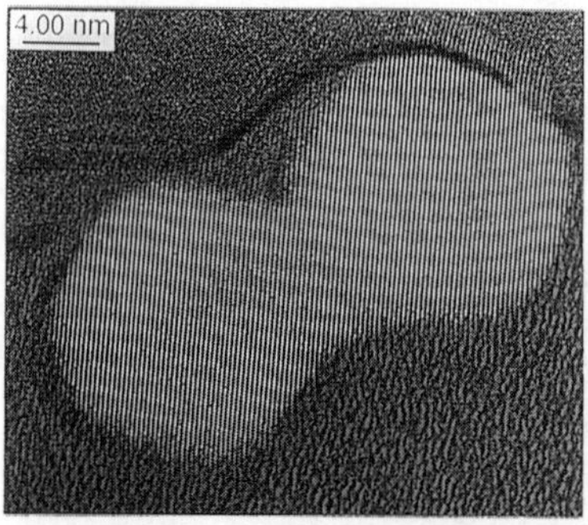

Fig. 7:

Dumb-bell-shaped particle with ideal lattice structure in the region of fusion formed by coalescence of two Au nanoparticles during densification of a glass-like coating (Au / aminosilane = 1 / 10, T_d = 500°C). Image processing: sum of the original image and its masked Fourier filtering.

for the remaining misorientation between the $(111)_a$ and $(111)_b$ lattice planes of the primary particles A and B, respectively. The resulting coalesced particle has minimized its lattice energy essentially by rotating the primary particles and forming a twin boundary (as it has been observed in the case of particles produced by inert gas evaporation [8]) because a twin only insignificantly deviates from the ideal lattice structure.

The coalescence of two colloids during the densification of a glass-like coating is verified by Fig. 7 (unfortunately, due to electrical charging the image is affected by astigmatism). The resulting shape of the particle is also dumb-bell-like and the imaged (111) lattice fringes exhibit no distortions in the contact region. This ideal fusion of the lattice structure of both primary particles suggests that an adaptation of the orientation of the particles by rotation must have preceded the coalescence within the matrix, too; otherwise, an ideal fusion would be very improbable.

CONCLUSIONS

The microscopic investigations of metallic colloids embedded in glass-like coatings have shown a stronger densification of glass-like particle shells caused by the growth-controlling ligand envelopes of the colloids. The observed positively asymmetric log-normal size distribution of the nanoparticles have revealed that in the densification process the crystalline particles mainly grow by means of liquid-like coalescence. On the basis of diffusion and condensation of metal atoms another growth process occurred as a result of aging of the investigated ormocer layer. Furthermore, the examinations have exhibited a relatively high mobility of the nanoparticles outside, but also inside the glass-like layer. This mobility comprises translational and rotational motions of the particles. It is assumed that the ability of the particles to carry out these motions combined with special conditions for these motions forced by the matrix structure during the densification could be the key to the understanding of the observed oriented coalescence of colloids forming highly elongated particles.

REFERENCES

1. U. Kreibig in Contributions of cluster physics to material science, edited by J. Davenas and P.M. Rabette (Dordrecht: Nijhoff 1986), pp. 373-423.
2. M. Mennig, M. Schmitt, U. Becker, G. Jung and H. Schmidt, Sol-Gel Optics III, SPIE **2288**, 131-139 (1994).
3. M. Mennig, U. Becker, M. Schmitt and H. Schmidt, Advances in Science and Technology **11**, 39-46 (1995).
4. Cl. Fink-Straube, PhD thesis, Saarbrücken, 1994.
5. C.G. Granqvist and R.A. Buhrman, J. Appl. Phys. **47**, 2200-2219 (1976).
6. Ph. Buffat and J.-P. Borel, Phys. Rev. A **13,** 2287-2298 (1976).
7. B. Kutsch, O. Lyon, M. Schmitt, M. Mennig and H. Schmidt, to be published 1996.
8. P. Gao and H. Gleiter, Acta metall. **35,** 1571-1575 (1987).

NEW SYSTEMS RELATED TO CdS NANOPARTICLES IN SOL-GEL MATRICES

T. GACOIN*, L. MALIER*, G. COUNIO*, S. ESNOUF*, J.P. BOILOT*,
L. AUDINET**, C. RICOLLEAU**, M. GANDAIS**
*Laboratoire PMC, CNRS URA 1254, Ecole Polytechnique, 91128 Palaiseau Cedex, France.
tga@pmcsun1.polytechnique.fr
**Laboratoire de Minéralogie-Cristallographie, Universités Paris VI et Paris VII,
CNRS URA 009, 4 Place Jussieu, 75252 Paris Cedex 05, France.

ABSTRACT

A process for the synthesis of CdS nanoparticles embedded in hybrid organic/inorganic silica matrices has been developed, based both on colloid and sol-gel chemistry. The large possibilities offered by these techniques allow the study of the influence of various parameters on the luminescence properties of the nanoparticles. Two directions were investigated: first, we characterized the surface of the nanoparticles complexed with an organic thiol using NMR spectroscopy. Then, we synthesized new systems to study the influence of chemical modifications, such as the composition of the aggregates (CdS:Mn), the surface (CdS/ZnS core/shell nanostructures) and the interface between the aggregate and the gel matrix.

INTRODUCTION

Since the beginning of the 80's, semiconductor nanocrystals have been the subject of intensive research. The great interest of the scientific community for this subject can be explained in three different ways : first, the theoretical description of the quantum size effect was an interesting challenge for many physicists. Then, the synthesis of small semiconductor particles was already possible for a few compounds, allowing the confrontation of theory with experiments. And finally, semiconductor nanoaggregates were supposed to behave like two level systems whose saturation could provide large optical nonlinearities.

For the past few years, we have seen an important evolution of the subject because many experimental results have shown that i) the synthesis of semiconductor nanocrystals, having a narrow size distribution, is well controlled only in a few number of systems (essentially II-VI compounds) and ii) a semiconductor nanocrystal cannot be described as an excised fragment of the bulk semiconductor, and the surface plays a role of major importance in the recombination of optically excited carriers.

Now, new developments of the subject are clearly dependent on a better control of the synthesis of the nanoparticles and a better characterization of the chemical structure of their surface. In a previous paper [1], we have described an original process for the synthesis of hybrid gel matrices doped with semiconductor nanocrystals (CdS, ZnS, PbS). This process takes advantage of the large possibilities offered both by colloid and sol-gel chemistry. This allows a good control of many parameters of the synthesis as well as proper conditions for the characterization of the nanoparticles.

We present here the latest developments of our work concerning the study of the relationship between the structure of the surface of the nanoparticles and their luminescence properties. A first part will give an account of the process we used for the synthesis of silica matrices homogeneously doped with CdS nanoparticles. Then, we will present results concerning the ^{19}F and ^{113}Cd NMR characterization of the chemical structure of the CdS nanoparticles complexed with 4 fluorophenylthiol, and finally, the synthesis and the characterization of two new systems: Mn^{2+} doped CdS and CdS/ZnS nanostructures.

CdS/SILICA NANOCOMPOSITES

Most of the works which have been published concerning the synthesis of sol-gel matrices doped with II-VI semiconductor nanoparticles use either the direct precipitation of the aggregates in a sol containing functionalized alkoxides or precipitation within a porous silica network [2].

We have developed a process in which the synthesis of the nanoparticles is completely separated from the sol-gel media [1]. The CdS nanoparticles are first precipitated within the inverted micelles of a microemulsion. This technique allows the production of nanoparticles with a variable size ranging between 1.2 and 10 nm of diameter with a sharp distribution. Then, a thiol is grafted on the surface of

Mat. Res. Soc. Symp. Proc. Vol. 435 © 1996 Materials Research Society

the particles allowing i) their separation from the inverted micelle solution ii) their redispertion in an appropriate solvent with a high concentration and iii) their homogeneous incorporation in a solution of hydrolyzed alkoxide, the further polymerization of which will form the final matrix.

Experimental

The typical procedure for the precipitation of CdS in inverted micelles consists in adding H_2S in excess to a solution of cadmium salt dissolved in the water droplets of a AOT/water/heptane microemulsion (AOT stands for the surfactant molecule bi-2-ethylhexylsulfosuccinate sodium salt). In all these experiments, the H_2O and AOT concentrations are kept constant, respectively equal to 2.5 mol.l^{-1} and 0.5 mol.l^{-1}. Under these conditions, the average size of the nanoparticles can be continuously modified between 1 and 10 nm, depending on the initial concentration of cadmium salt dissolved in the water droplets (0.01 to 0.2 mol.l^{-1}). Colloids thus obtained can be easily characterized by their optical absorption (figure 1). This allows a simple determination of the average size of the nanoparticles using size/gap correlations published by other authors [3].

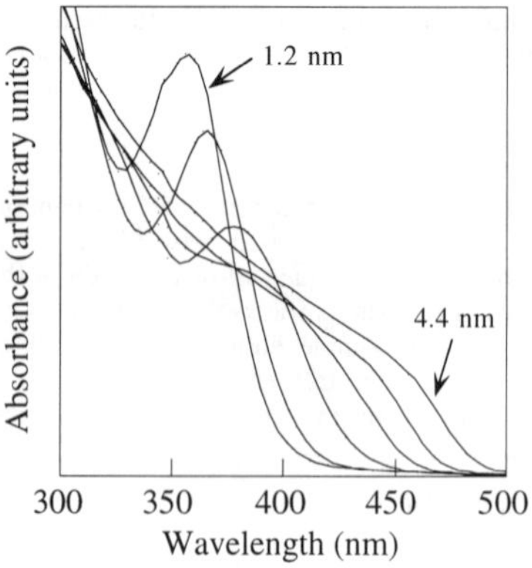

Figure 1 : Absorption spectra of CdS colloids obtained with initial Cd^{2+} concentrations ranging from 0.01 to 0.2 mol.l^{-1}

The separation of the aggregates from the inverted micelle medium is achieved by the grafting of a molecule at their surface in the presence of an excess of Cd ions. This reaction suppresses any stabilizing interaction between the aggregates and the sulfonate groups of the AOT, and makes the surface become hydrophobic. The nanoparticles are thus no more stable in the water droplets and precipitate. Centrifugation of the solution and washing with heptane finally give a powder of pure capped particles.

The chemical nature of the grafted molecule and the grafting efficiency determine the ability of dispersing the particles in an appropriate solvant. Phenylthiol [4] and pyridine [1] make it possible to obtain colloids with a high concentration only in pyridine. We found out that the use of 4-fluorophenylthiol allowed the dispersion of the aggregates in various solvents such as ethanol, THF and acetone with a very high concentration (up to 10% in volume).

The incorporation of the aggregates in a sol-gel media is achieved simply by addition of the colloid in a solution of hydrolyzed silicon precursors. Many silicon alkoxides may be used, but we observed that the use of alkoxide of the $R\text{-}Si(OEt)_3$ type with R= methyl (MTEOS) , vinyl (VTEOS) gives materials with an improved optical and mechanical quality. Furthermore, photostability and luminescence of the particles are enhanced.

Hydrolysis of the alkoxide precursor is achieved in alcohol, using pH 2.5 water. The molar alkoxide/water/ethanol composition in the initial solution is generally taken in the ratio 1/3/3 respectively. After 12 hours of hydrolysis, this silica sol is mixed with the colloidal solution of CdS nanoaggregates in acetone. The resulting solution may be either deposited as a thin film on a glass substrate by spin coating or poured in a beaker for drying at 60°C to produce monolithes. The final volume fraction of CdS in the silica matrix can be made to vary up to 10%.

NMR characterization of the surface complexation of the nanoparticules by 4-fluorophenylthiol

The complexation of the surface of the aggregates by an organic molecule plays a major role in the process we use for the fabrication of CdS/silica nanocomposite. Indeed, the chemical nature of the molecule and the grafting efficiency determine the ability to disperse the aggregates in a given solvent. Another point is that the presence of the molecule at the surface of the semiconductor certainly affects its luminescent properties. A better understanding of all these effects requires a good characterization of the chemical structure of the surface.

In this work, ^{19}F and ^{113}Cd NMR were used to study the grafting of 4-fluorophenylthiol at the surface of the CdS particles. Typical spectra are shown in figure 2 for aggregates with a diameter equal to 3.4 nm dispersed in acetone.

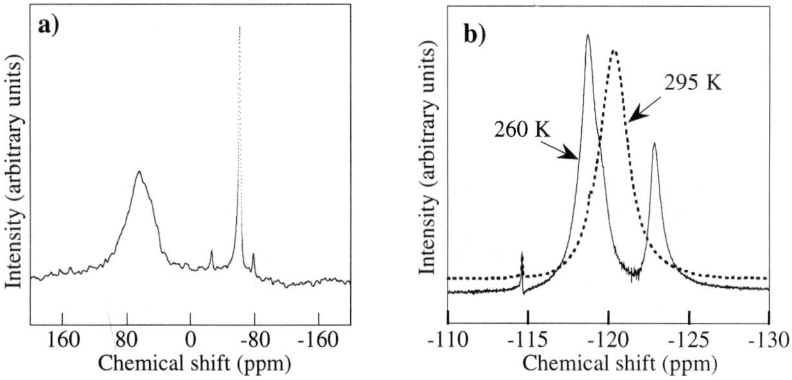

Figure 2 : ^{113}Cd (a) and ^{19}F (b) NMR spectra of a CdS colloid in acetone (chemical shifts are taken relative to $Cd(CH_3)_2$ and $CFCl_3$ respectively)

The ^{113}Cd spectrum shows a broad dissymetric band centered at 60 ppm, a sharp highly intense line located at -62 ppm and other unexplained weakly intense lines. Three kinds of Cd environments may be considered: Cd^i is a cadmium surrounded by 4 sulfide atoms (located inside the aggregate), Cd^o is complexed with 4 thiolate ligands (located outside the aggregate), and Cd^s is both complexed by sulfides and thiolates ligands (it is then located at the surface of the aggregate). Besides, the thiolate ligands can either be bridging ($\mu S\phi F$) or terminal ($^tS\phi F$). Dance [5] has studied in details the structure and the NMR spectra of well defined small $Cd_xS_yS\phi_z$ clusters ({x,y,z}={4,0,10};{8,1,16};{10,4,16} and {17,4,28}). His results suggest that the broad band in our spectra mainly consists in contributions from both Cd^iS_4 and the surface cadmium complex $Cd^sS_3S\phi F_1$. The two other Cd^s environments are not detected in our samples. The sharp line located at -62 ppm is attributed to $Cd^o\mu S\phi F_3{}^tS\phi F_1$. At ambient temperature, this latter complex is in rotation around a Cd-$\mu S\phi F$ bond, so that the $\mu S\phi F$ and the $^tS\phi F$ are in exchange. This is confirmed by the ^{19}F spectrum, which only shows a single band peaking at -120 ppm at room temperature, whereas two bands are observed at 220 K, peaking at -118 and -123 ppm and attributed to $\mu S\phi F$ and $^tS\phi F$ respectively.

A detailed quantitative study of NMR spectra is now in progress, in order to determine the grafting efficiency, which can be related to the number of Cd^s ions for a given size of the nanoparticles. As this number cannot be directly measured, it seems that additional information are first needed concerning the structure of the outer Cd complex.

Luminescence

Figure 3 shows the luminescence spectra of the same CdS nanoaggregates (about 2.5 nm of diameter) dispersed in a TEOS and a MTEOS sol-gel matrix respectively. In the TEOS sample, the luminescence spectra consist essentially in a broad band centered around 600 nm. This band is often observed in CdS colloid and is attributed to surface

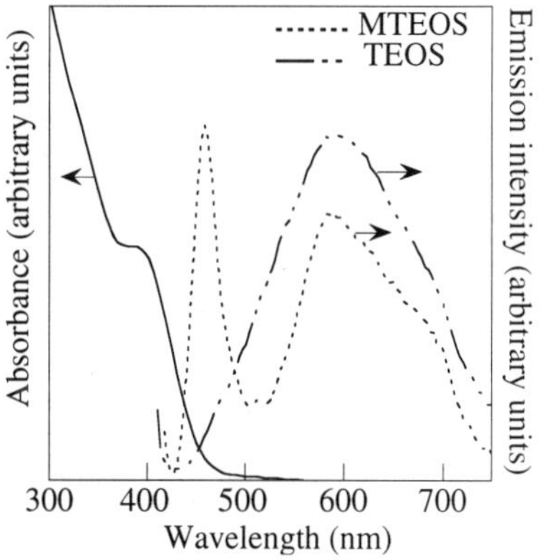

Figure 3: Luminescence of the CdS particles in MTEOS and TEOS sol-gel matrices

recombination from trapped carriers. In the MTEOS spectra, the same band is observed, but in addition we find a sharp peak near the absorption threshold. This band is attributed to band to band recombination of the excited carriers. Another difference between the two samples is the quantum yield which is about ten times higher in MTEOS than in TEOS.

The important difference between the luminescence of the CdS nanoaggregates in the two samples is not yet understood, but it might be explained in term of interaction between the surface of the aggregates and the surface of the pore in which they are localized. It seems that the presence of hydrophobic methyl group limits the number of deep traps as well as non radiative recombination centers.

CdS:Mn / SILICA NANOCOMPOSITES

II-VI semiconductors doped with Mn^{2+} are well known phosphors, the luminescence of which is attributed to an energy transfer between the semiconductor excited state and the 3d levels of the Mn^{2+} ion. The light emission corresponds to the $4T_1 \longrightarrow 6A_1$ transition with an energy peaking around 2.13 eV. The synthesis of CdS:Mn as small nanoaggregates will enable us to compare their luminescence with those of pure CdS nanocrystals, since the energy transfer could prevent the trapping of the excited carriers to the surface.

Experimental

The process we used for the synthesis of CdS particules doped with Mn^{2+} consisted in a coprecipitation of Cd and Mn salts in conditions similar to those previously described in the case of pure CdS. Nevertheless, the much higher solubility of MnS compared to CdS made the use of H_2S as the sulfide precursor inefficient for the coprecipitation. Na_2S was then found more suitable, and the reaction was achieved by mixing two solutions of inverted micelles, one containing both Cd^{2+} and Mn^{2+}, and the other containing Na_2S. In a more usual experiment, the molar concentrations of $Cd(NO_3)_2$, $Mn(NO_3)_2$, and Na_2S dissolved in the water droplets are equal to 0.2, 0.12 and 0.4 mol.l^{-1} respectively. After obtaining the colloid, the incorporation a sol-gel matrix is achieved following the same process as for pure CdS.

Characterization of the Mn ions in solid solution in CdS

The presence of manganese in solid solution was characterized using EPR experiments. This technique allowed to distinguish the different Mn^{2+} ions because of the influence of their structural and chemical environment. It became also possible to have a quantitative estimation of the number of Mn^{2+} per aggregates corresponding to each contribution.

The EPR spectra taken at 293 K and 4 K enabled us to consider three main contributions [6]: (A) a broad line contribution, from Mn^{2+} in strong exchange interaction, like in MnS, (B) a sharp line contribution, from Mn^{2+} ions in tetrahedral environment like in diluted bulk CdS:Mn and (C) a broad line contribution, from Mn ions in tetrahedral environment, attributed to Mn^{2+} in the aggregates, but close to the surface.

The intensity of contribution A is more than 90% of the total signal. This attests that the simple experimental conditions we use are not very efficient to obtain solid solutions of CdMnS. Nevertheless, the intensity of contributions B+C corresponds to one to four Mn^{2+} per aggregate. This seems to be sufficient to influence drastically the optical properties of the aggregates, probably because of their small size (2 to 3 nm).

Luminescence

The luminescence spectra of two samples of xerogels doped with CdS:Mn and ZnS:Mn (made following the same process as the one used for cadmium) are presented in figure 4. It is clear that the radiative recombination of the excited carriers at the surface of the aggregates has almost completely disappeared. The presence of Mn within the aggregates then modifies the mechanism of the recombination. What has not been made clear yet is whether the energy transfer occurs or not after the trapping of the carriers in surface defects. This will be discussed in a next paper.

At room temperature, the photoluminescence efficiency was found to be about 10 %, with a radiative lifetime of about 1.7 ms [7]. Those values are in agreement with the results reported for bulk CdS:Mn. It seems then that quantum confinment does not significantly modify the luminescence properties of CdS:Mn. Nevertheless, those results are at variance with those reported by Bhargava [8] who indicated a lifetime 5 orders of magnitude faster than the bulk value in the case of ZnS:Mn nanocrystals.

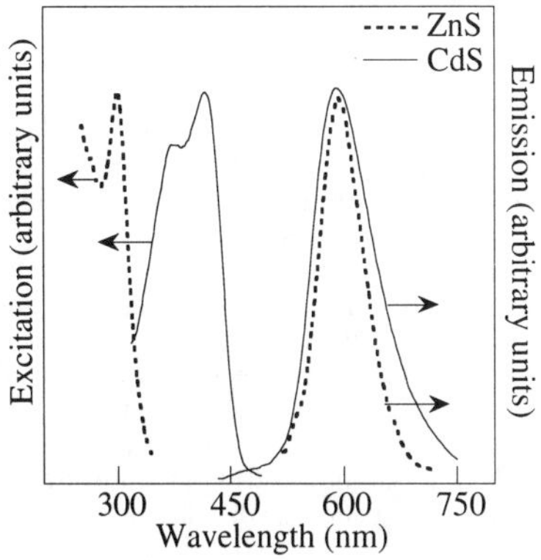

figure 4: Luminescence and excitation spectra of CdS:Mn and ZnS:Mn doped xerogels

CdS / ZnS CORE/SHELL NANOSTRUCTURES

The synthesis of CdS/ZnS core/shell nanoaggregates represents an interesting challenge based on the idea that i) the presence of a semiconductor with a higher gap at the surface of the aggregate could confine the carriers (electron and hole) inside the nanocrystal of CdS, and ii) an epitaxial growth of ZnS at the surface of CdS could eliminate all traps responsible for non radiative or surface recombination.

Experimental

In a first step, a CdS colloid was synthesized in inverted micelles following the process previously described. After the obtention of the colloid, the excess of H_2S was removed and a solution of zinc salt in inverted micelle was added with the same volume as the CdS colloid. Then, an excess of H_2S was slowly added (within a few minutes) to precipitate the ZnS. The incorporation of the core/shell nanoparticles in hybrid silica matrices was achieved following the same process as the one used to obtain pure CdS. The sizes of the core CdS and the shell ZnS depended on the initial concentration of Cd and Zn salt initially dissolved in the water pools. In a more classical experiment, these two concentrations were taken equal to 0.1 and 0.4 $mol.l^{-1}$ respectively.

structural characterization

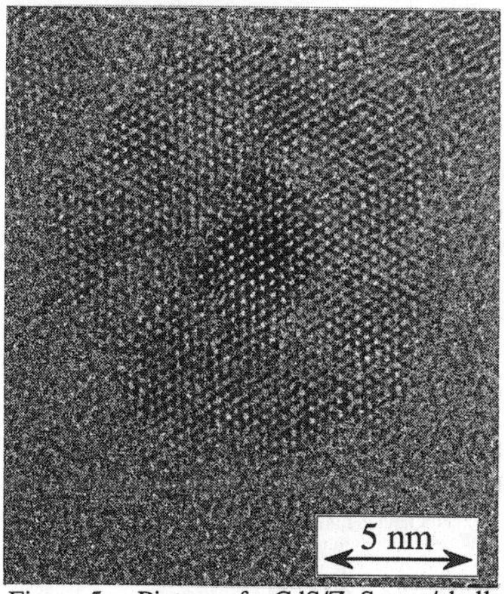

Figure 5a : Picture of a CdS/ZnS core/shell nanoparticle

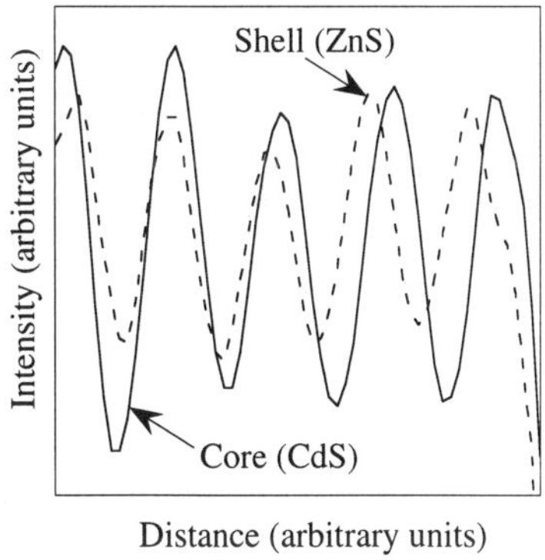

Figure 5b : Intensity profile of the {111} planes of the CdS/ZnS heterostructure

Structural characterization was performed both on the core and the core/shell structure by means of high resolution transmission electron microscopy. Observations were made through a carbon film on which a drop of a solution of the aggregates in acetone had been deposited. The initial CdS particles have an average diameter of about 8.6 nm. Both cubic blende-type and hexagonal wurtzite-type structures are observed as already reported in a previous study [9]. The CdS/ZnS nanoparticles have an average diameter of about 11.1 nm. The core/shell structure is indicated by the fact that small variations of the objective focus make a contrast appears between the core and the shell. Stronger evidence for an epitaxial growth of ZnS on CdS comes from the measurement of the difference of lattice parameter between the inner and the outer part of the particles (figure 5a,b). The result is about 7%, which is consistent with the lattice parameters of bulk CdS and ZnS. In most of the aggregates, misfit dislocations are observed in the neighbourhood of the core/shell interface. In such cases, the thickness of the epitaxial layer has overcome the limit value allowing the accommodation of the core/shell lattice misfit by elastic strains. Concerning the structure of the aggregates the two cristalline structures (cubic and hexagonal) are observed just as for the CdS cores. The ZnS shell grows with the same structure as the CdS core, and this leads to the formation of a ZnS phase with the wurtzite-type structure which is unusual for ZnS colloids, when the CdS core is in the wurtzite-type structure.

CONCLUSION

We have presented the synthesis and the characterization of new systems related to CdS nanocrystals trapped in sol-gel silica matrices. Our purpose was to characterize the nanoparticles and to study the influence of some chemical modifications (composition, environment) on luminescence properties. The main results we have shown here are that:

i) the grafting of 4-fluorophenylthiol at the surface of CdS nanoparticles is achieved trough the formation of a complex involving Cd ions which do not belong to the aggregate itself.

ii) The presence of hydrophobic methyl group within the pore structure of the xerogel in which the aggregates are embedded reduces the number of surface traps and of non radiative luminescence centers.

iii) The presence of Mn ions in solid solution in the CdS nanoparticles drastically affects their luminescence as the only signal observed comes from $4T_1 \longrightarrow 6A_1$ transition of the Mn d-electrons.

iv) We have succeeded in the synthesis of CdS/ZnS core/shell nanostructure. The luminescence properties of these structures are under investigation.

A lot of work has still to be performed to correlate the information we have collected concerning the structure of all these systems and their optical properties, but we believe that the large flexibility of the process based on colloid and sol-gel chemistry is a unique opportunity to understand the complex problem of the luminescence of small semiconductor nanoparticles.

REFERENCES

1. T. Gacoin, C. Train, F. Chaput, J.P. Boilot, P. Aubert, M. Gandais, Y. Wang and A. Lecomte in Sol-Gel Optics II, edited by the SPIE (SPIE Proc. 1758, San-Diego, CA, 1992) pp.565.
2. H. Schmidt, Sol-Gel Optics, edited by L. Klein (Kluwer Academic Publishers 1994), **20**, 451; M. Nogami, ibid., **15**, 329.
3. Y. Wang, N. Herron, Phys. Rev. B **42** (11), 7253 (1990).
4. M.L. Steigerwald, A.P. Alivisatos, J.M. Gibson, T.D. Harris, R. Kortan, A.J. Muller, A.M. Thayer, T.M. Duncan, D.C. Douglas, L.E. Brus, J. Am. Chem. Soc. **110**, 3046 (1988).
5. G.S. Lee, K.J. Fisher, A.M. Vassallo, J.V. Hanna, I.G. Dance, Inorg. Chem. **32**, 66 (1993).
6. G. Counio, S. Esnouf, J.P. Boilot and T. Gacoin, to be published.
7. M.A. Chamarro, V. Voliotis, R. Grousson, P. Lavallard, T. Gacoin, G. Counio, J.P. Boilot, R. Cases, accepted in J. Cryst. Growth (1996).
8. R.N. Bhargava, D. Gallager, X. Hong, A. Nurmikko, Phys. Rev. Lett. 72 (1994) 416.
9. M. Gandais, L. Audinet, C. Ricolleau, T. Gacoin, J.P. Boilot, accepted in J. Cryst. Growth (1996)

SOL-GEL PROCESS OF FLUORIDE AND FLUOROBROMIDE MATERIALS

O. Poncelet, J. Guilment, G. Paz-Pujalt
Kodak-Pathé, European Research Laboratory, Chalon-sur-Saône, France

ABSTRACT

The alkoxides $M(OR_X)_n$ wherein M is an alkaline-earth or rare-earth and OR_X is a fluoroalkoxo or bromoalkoxo group exhibit the properties to form pure fluoride materials by hydrolysis at room temperature. Using this property, we synthesized pure barium and europium fluoride and pure alkaline-earth bromide. The hydrolysis of an heteroleptic species, $Ba(OCH_2CH_2Br)[OCH(CF_3)_2]$ allowed pure and crystalline BaFBr materials to be obtained at room temperature.

INTRODUCTION

The most common molecular precursors for sol-gel technology are alkoxides, which are known for a large part of the periodic table [1, 2]. Mainly developed for the synthesis of oxide material, the sol-gel technology has also been described for the obtention of oxinitrides and sulfides. Recently the sol-gel synthesis of magnesium fluoride by the action of HF on magnesium ethoxide has been reported [3].
The purpose of this paper is to describe the possibility to obtain crystalline alkaline-earth or lanthanide fluoride materials at room temperature by hydrolysis of fluoroalkoxides: $M(OR_F)_n$ wherein M is an alkaline-earth or a lanthanide, OR_F is a fluoroalkoxo group.
With a similar approach, we synthesized crystalline alkaline-earth bromides by the hydrolysis at room temperature of bromoalkoxides: $M(OR_{Br})_2$ wherein M is an alkaline-earth, OR_{Br} is a bromoalkoxo group. Finally, the hydrolysis of $M(OR_{Br})(OR_F)$ species lead to pure BaFBr.

EXPERIMENTAL

All reactions were carried out under prepurified argon at room temperature by using Schlenk techniques. Pure chlorides, bromides and metals are from Aldrich and Strem. All solvents were appropriately dried, distilled prior to use and stored under argon [4]. The alkaline-earth and or lanthanide alkoxides were synthesized as described in literature [1,5]. The Raman Spectra of the alkoxide species and the reference materials were recorded at a resolution of 4 cm^{-1} on a FT-IR Bruker IFS66 spectrometer (FRA106 accessory).The XRD powder pattern of each fluoride sample was obtained with a Siemens D5000 diffractometer using filtered CuKα. Hydrolyses of fluoroalkoxides and bromoalkoxides were made in dry ethanol. We used four times the required stoichiometry of water versus metals in each case.

RESULTS AND DISCUSSION

Alkaline-earths are electropositive metals, so the more convenient route to synthesize their alkoxides is the action of an alcohol on the metal (equ. 1). In the case of bromoalcohols, the reaction needs very often to be catalyzed by addition of NH_3, Et_2NH or $(Me_3Si)_2NH$.

$$M + 2\,R_XOH \longrightarrow M(OR_F)_2 + H_2 \quad (1)$$

$(R_XOH = (CF_3)_2CHOH$,CF_3CH_2OH, $BrCH_2CH_2OH$, $BrCH_2CH(CH_3)OH)$

Mat. Res. Soc. Symp. Proc. Vol. 435 © 1996 Materials Research Society

The rare-earth fluoroalkoxides are obtained by alcoholysis of $Ln(OCH_2CH_2OCH_3)_3$.

Synthesis of BaF_2 and EuF_3

The reaction of hexafluoroisopropanol on barium leads to $Ba(OCH(CF_3)_2)_2$ characterized by FT-Raman (figure 1), the Ba-F interactions are characterized by a strong band at 739 cm^{-1}. The hydrolysis of $Ba(OCH(CF_3)_2)_2$ catalyzed by $HClO_4$ (the pH of the water was in the range 1 to 3) in dry ethanol quantitatively lead to BaF_2 characterized by XRD (figure 2).

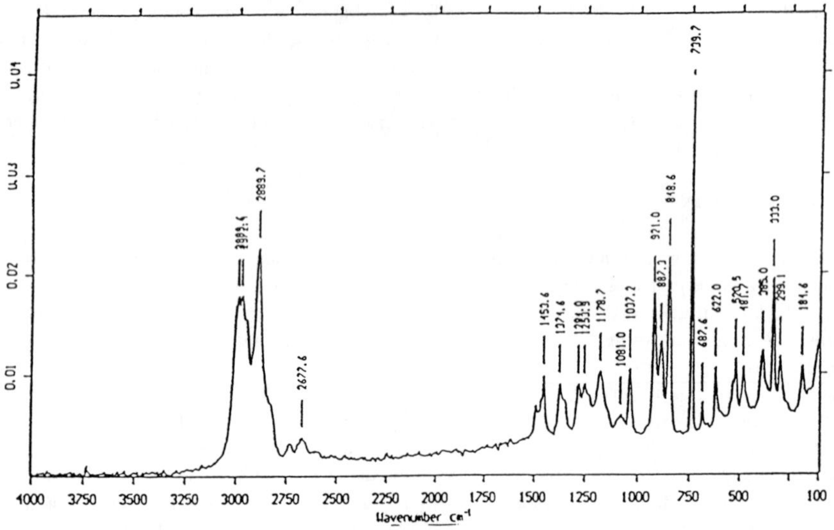

Figure 1 FT-Raman spectrum of $Ba[OCH(CF_3)_2]_2$

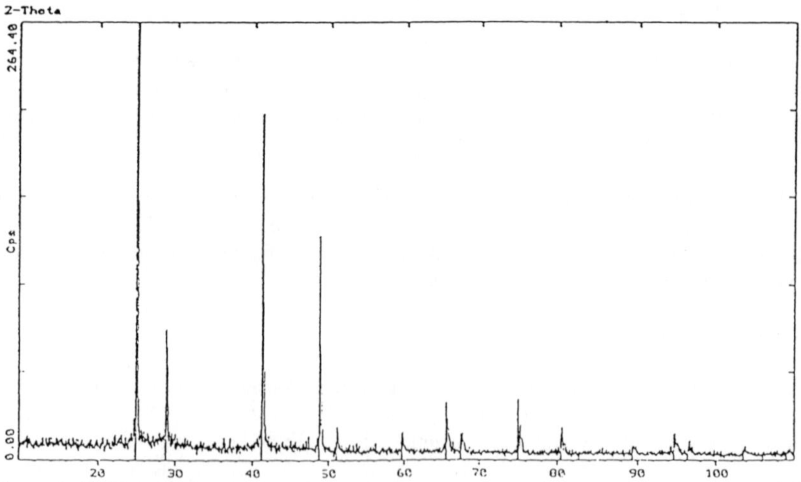

Figure 2 XRD spectrum of BaF_2 powders obtained by hydrolysis of $Ba[OCH(CF_3)_2]_2$ at room temperature

$Eu[OCH(CF_3)_2]]_3$ exhibit strong Eu-F interactions (FT-Raman band at 743.2 cm^{-1}). Its hydrolysis catalyzed by $HClO_4$ quantitatively lead at room temperature to EuF_3 characterized by XRD (see figure 3).

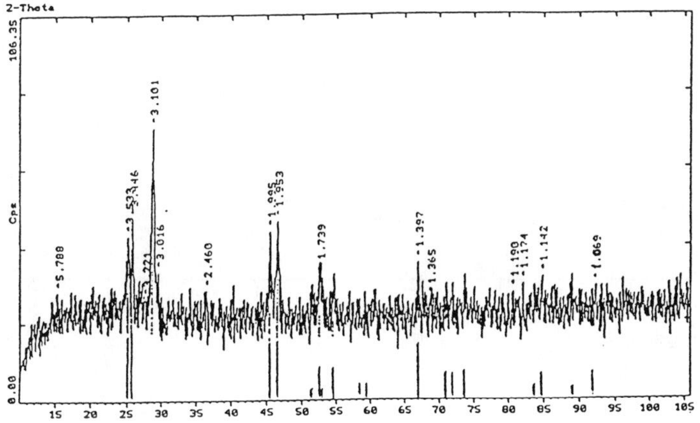

Figure 3 XRD spectrum of EuF_3 powders obtained by hydrolysis of $Eu[OCH(CF_3)_2]_3$

How to explain the possibility to obtain fluorides instead of oxides by hydrolysis of fluoroalkoxides ?

Homoleptic and heteroleptic fluoroalkoxides have been successfully used for the obtention of fluoride materials by MOCVD [6,7,8]. But it is the first time to our knowledge that such fluoride material has been obtained by sol-gel process without adding of HF to catalyze the reaction. Recent X-Ray data recorded on monocrystals show that short M--F contacts appear to be a common feature for fluorinated alkoxide or β-diketonate derivatives, especially for large elements such as alkaline-earth metals and lanthanides [9, 10, 11]. For example, in the case of $BaY_2[OCH(CF_3)_2](thd)_4$ [12], the coordination polyhedron of the central barium atom formed by four alkoxide type oxygen atomes is supplemented by interactions with eight fluorine atoms (Ba--F: 2.9-3.16 Å). The sum of the van der Waals radii is 3.57 Å. These interactions Ba-F really act as secondary bonds leading to a final degree of coordination of twelve.

These observations made on monocrystals (solid state) show that these molecular precursors could be formally considered both as oxide or fluoride precursors strictly depending on their hydrolysis conditions.

Our results confirm that it is possible to obtain fluoride materials by hydrolysis of precursors in alcohols proving that even in solution with this kind of molecules, the metal centers keep in their coordination spheres some M-F contacts allowing fluoride materials to be obtained during their hydrolysis. This strong "template" effect associated with thermodynamic driving forces probably explains the fact that fluoride can be formed instead of oxide. Their hydrolysis have to be catalyzed by an acid.

Syntheses of $BaBr_2$ and $SrBr_2$

We studied the products of the hydrolysis of alkaline-earth bromoalkoxides. These species were obtained by action of the alcohol on metals (equ. 1) or by alcoholysis of common alkoxide species such like an ethoxide species.

The hydrolysis of $Ba(OR_{Br})_2$ leads to the formation of a solvated gel of $BaBr_2$. FT-Raman spectra show the two bands respectively at 501.1 and 440.5 cm^{-1}corresponding to Br-Ba interaction. Similar bands, however more intense, can be found in the $BaBr_2{\cdot}H_2O$ (Aldrich) used as reference (see figure 4).

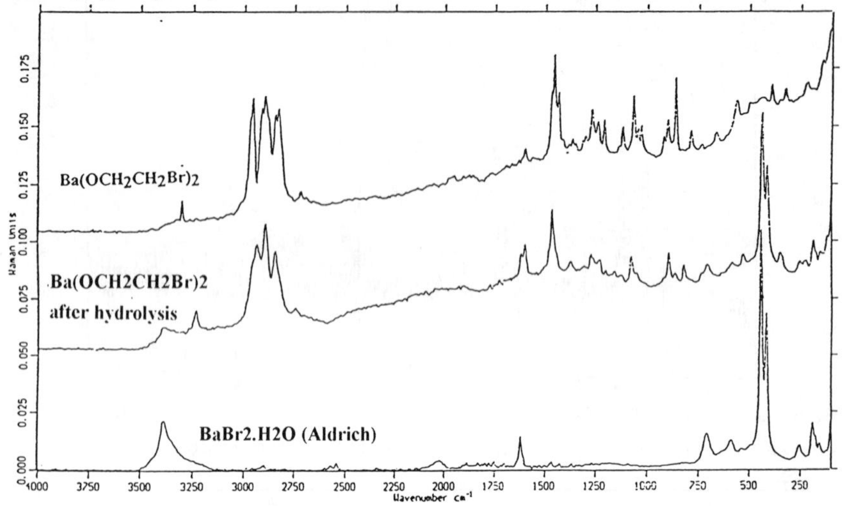

Figure 4 FT-Raman of the hydrolysis of $Ba(OCH_2CH_2Br)_2$

The powders obtained are characterized by XRD (see figure 5). Contrary to the fluoroalkoxides species, in the case of bromoalkoxides we did not use acidic catalysts.

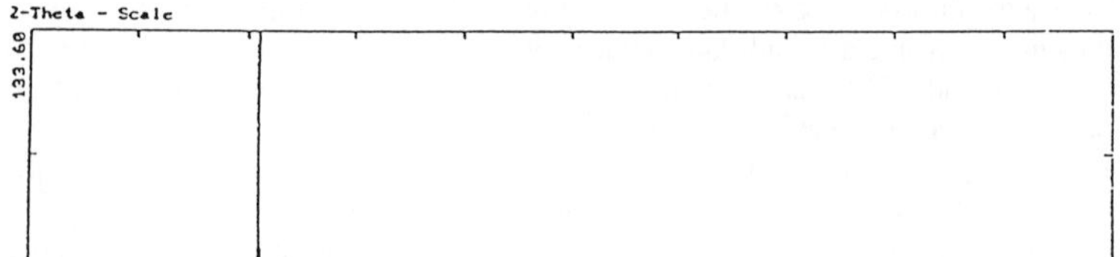

Figure 5 XRD spectrum of $BaBr_2$ powders obtained by hydrolysis of $Ba(OCH_2CH_2Br)_2$

The FT-Raman spectra of strontium bromoalkoxides species exhibit the two characteristic bands (at 490.4 and 459.2 cm^{-1}) proving specific Sr-Br interactions in the complexes The hydrolysis of "$Sr(OR_{Br})_2$" at room temperature leads to crystalline $SrBr_2$.

Syntheses of BaFBr materials

Using a similar approach, we synthesized pure and crystalline bromofluoride by hydrolysis of heteroleptic species of alkaline-earth : $Ba(OCH_2CH_2Br)[OCH(CF_3)_2]$. This species was obtained by metathesis reaction (equ.2).

$$Ba(OCH_2CH_2Br)_2 + Ba[OCH(CF_3)_2]_2 \longrightarrow 2\ Ba(OCH_2CH_2Br)[OCH(CF_3)_2] \quad (2)$$

The FT-Raman spectrum of $Ba(OR_{Br})(OR_F)$ clearly shows the strong Ba-F interaction but not the Ba-Br interactions which appeared on dried $Ba(OCH_2CH_2Br)_2$ species (see figure 6).

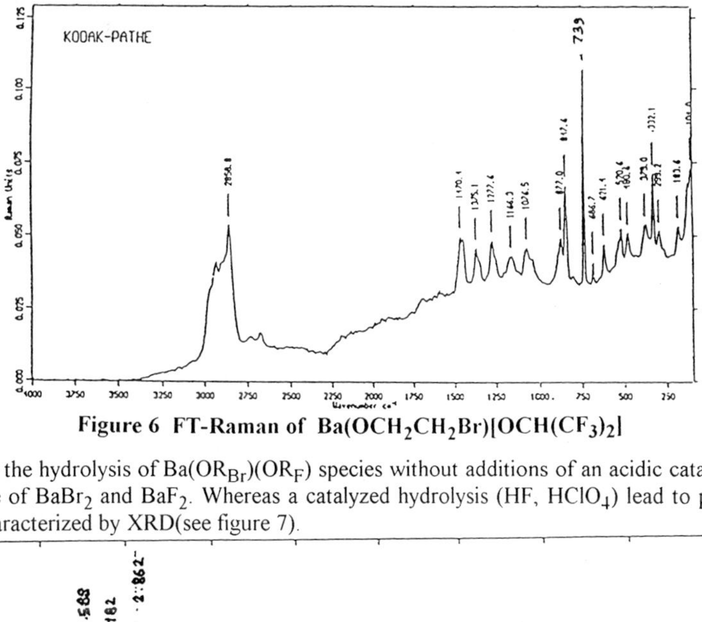

Figure 6 FT-Raman of $Ba(OCH_2CH_2Br)[OCH(CF_3)_2]$

However the hydrolysis of $Ba(OR_{Br})(OR_F)$ species without additions of an acidic catalyst lead to a mixture of $BaBr_2$ and BaF_2. Whereas a catalyzed hydrolysis (HF, $HClO_4$) lead to pure BaFBr phase characterized by XRD(see figure 7).

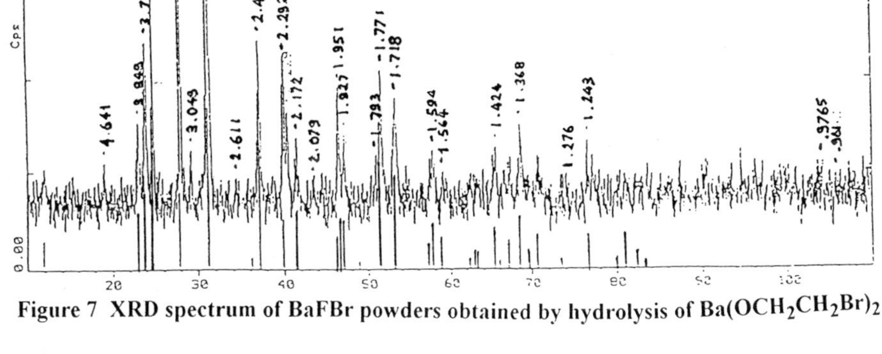

Figure 7 XRD spectrum of BaFBr powders obtained by hydrolysis of $Ba(OCH_2CH_2Br)_2$

The formation of M-X (X=Cl, Br) bond involves the cleavage of C-X bond, we observed that in the case of bromoalcohol the mechanism was easier than in the case of fluoroalcohol, this observation is corroborated by thermodynamical data: C-F = 485 kj/mole then C-Br = 285 kj/mole explaining the necessity to catalyze the reaction in the case of the fluoroalcohols.

CONCLUSION

We showed that it is possible to synthesize fluoride materials at room temperature by sol-gel process. The most attractive advantage of the sol-gel process is the possibility to directly obtain lanthanide or alkaline-earth fluoride materials under various shapes (thin films, fibres, bulks). A perfect control of the stoichiometry of the final fluoride materials is attainable using a mixture of homometallic fluoroalkoxides as starting materials. .So it would be possible to use the sol-gel process to manufacture fluoride materials on heat sensitive organic substrates or HF sensitive inorganic substrates.

Fluoro, bromo and fluorobromide alkoxides appeared to be attractive as precursors for the sol-gel process of non-oxide materials. These results confirmed too the interest to use tailor made molecular precursors for specific applications.

REFERENCES

[1] D.C. Bradley, R.C. Mehrotra and D.P. Gaur, Metal Alkoxides, Academic Press, London , 1978.

[2] L.G. Hubert-Pfalzgraf, New J. Chem. **11**, p.663, (1987).

[3] A.A. Rywak, J. M. Burlitch, Chem. Mater. **8**, pp 60-67, (1996).

[4] D.D. Perrin, W.L.F. Armarego, Purification of Laboratory Chemicals, Pergamon Press, London 1988.

[5] O. Poncelet, L.G. Hubert-Pfalzgraf, J.C. Daran, R. Astier, J.Chem.Soc., Chem. Com., p.1846, (1989).

[6] J.A. Samuels,.E.B. Lobkovsky, W.E. Streib, K. Folting, J.C. Huffman., J.W. Zwanziger; K.G. Caulton, J. Am. Chem. Soc. **115**, p.5093, (1993).

[7] J. A. Samuels, Wen-C. Chiang, Chung-Yu Yu, E. Apen, D.C. Smith, D.V. Baxter, K.G. Caulton, Chem.Mater.,**6**, p.1684 (1995).

[8] J.A. Samuels, K. Folting, J.C. Huffman, K.G. Caulton, Chem. Mater., 7, p .929, (1995),.

[9] A.P. Purdy, C.F. George, Inorg. Chem., **30**, p.1970, (1991).

[10] H. Vincent, F. Labrize, L.G. Hubert-Pfalzgraf, Polyhedron, p. 3323, (1994).

[11] D.C Bradley, H. Chudzynska, M.E. Hammond, M.B. Hursthouse, M. Motavelli, W. Ruowen , Polyhedron, **11**, p. 375, (1992).

[12] F. Labrize, L.G. Hubert-Pfalzgraf, J.C. Daran, S. Halut, J.Chem.Soc., Chem.Com., ,p.1556, (1993).

INVESTIGATION OF SECOND HARMONIC GENERATION IN GLUTAMIC ACID-METAL COMPLEXES

THOMAS M. COOPER, STEVEN M. CLINE, DAVID E. ZELMON, RAMA VUPPULADHADIUM, SAMHITA DAS GUPTA, AND UMA B. RAMABADRAN
Materials Directorate, Wright Laboratory, Wright-Patterson Air Force Base, OH 45433

ABSTRACT

To design new second order nonlinear crystals, we have characterized a series of dipeptide complexes and copper glutamate. We tested 16 materials using powder second harmonic generation. The best of these materials was copper glutamate. Results of initial nonlinear optical characterization of the copper glutamate powder determined by the Kurtz powder test are presented.

INTRODUCTION

Second order nonlinear optical materials have numerous applications as second-harmonic-generators, optical parametric oscillators and electro-optic modulators. Classes of materials investigated include inorganic crystals, organic crystals and polymeric thin films. In our laboratory we have been investigating the optical properties of biologically-derived materials including corona-poled polypeptide thin films [1] and thin films prepared by an electrostatic self-assembly technique [2]. Amino acids and short peptides have been investigated as harmonic-generating crystals [3-5].

Semiorganic crystals are also being extensively studied for frequency conversion and electro-optic applications[6-8]. Compounds such as L-arginine phosphate and zinc tris-thiourea sulphate are examples of such materials that are formed by coupling a polarizable organic molecule to an inorganic host to form a salt or a coordination complex. Because of the great flexibility in the choice of both components, it is possible to obtain a desirable combination of nonlinear optical, mechanical and growth properties. In addition, the cost of these crystals should be low because the starting materials are inexpensive and the crystals can be grown from aqueous solutions. In particular, Velsko[5] describes a search strategy for new harmonic-generating crystals. He demonstrates the probability of finding a successful harmonic generating crystal will be larger by searching chiral materials *vs.* non-chiral materials. We performed qualitative second harmonic measurements on a series of crystalline dipeptides. Certain peptides showed intense second harmonic generation. We are currently investigating metal salts of glutamic acid as candidate second harmonic generating crystals[9,10], including copper [9] and zinc [10] glutamate dihydrate: $CuC_5H_7NO_4(H_2O)_2$(Figure 1). These are chiral semiorganic materials that crystallize in the orthorhombic space group $P2_12_12_1$[9]. Copper glutamate has unusual coordination of nitrogen and oxygen atoms around the copper, forming a distorted octahedron[9], possibly leading to enhanced hyperpolarizability. We have found growth of millimeter-sized rhomboidal crystals to be straightforward, so their investigation as potential nonlinear crystals was undertaken. In this paper we demonstrate that copper glutamate has nonlinear optical activity when in the powder form. A quantitative Kurtz powder test demonstrates potential phase matchability. We also describe preliminary investigations of copper glutamate crystal growth.

Mat. Res. Soc. Symp. Proc. Vol. 435

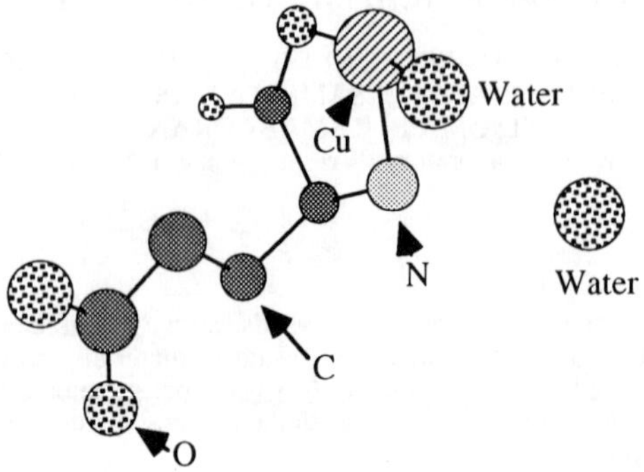

Figure 1 Copper-Glutamate Monomer plus two waters of hydration[9]

EXPERIMENT

Dipeptide powders were obtained from Sigma Chemical Corporation. Out of the 2000+ peptides available, we selected peptides which were described as crystalline for further investigation.

Copper Glutamate Powder Synthesis

Synthesis of Cu-glutamate salt was adapted from literature procedures[11,12]. A solution of Cu(II) acetate monohydrate(40 g) in deionized water(1L) was added dropwise(1/2 hr) to a stirred solution of L-glutamic acid(29 g) in deionized water(750 mL) kept at 80 ^{o}C. The copper glutamate began to form a cloudy blue precipitate after 2/3 of the Cu(II) acetate had been added. The mixture was kept at room temperature for two days until precipitation was complete. The product was isolated by suction filtration, washed with deionized water, ethanol and ether. Finally the product was dried at 50 ^{o}C, 20 torr for 48 hours. The yield was 45g(95%). Elemental analysis theoretical%: Cu 25.87, C 24.43, H 4.89, N 5.66; actual%: Cu 25.25, C 24.57, H 4.5, N 5.57.

Copper Glutamate Crystal Synthesis

Method 1: Copper glutamate crystals were grown by first solubilizing the $CuGlu(H_2O)_2$ powder in deionized water by heating the powder/water to boiling and adding 6M HCl to pH<2. The concentration of the solution was approximately 78 mg/ml(0.318M). The solution was cooled to room temperature, filtered(0.2 mM filter) and the pH was increased to 2.63 by adding 6M NaOH. The solution sat at room temperature for three days before prism-like crystals appeared.

Method 2: Rhomboidal copper glutamate crystals were formed by solubilizing the $CuGlu(H_2O)_2$ powder as above, cooling to room temperature and filtering. The powder concentration was approximately 20 mg/ml at pH 2.54. Three parts of the copper glutamate solution was mixed with 2 parts 2M NaCl/0.1 M Na acetate solution. The solution evaporated under dessication. Rhomboidal crystals formed several days later. A UV/Vis spectrum of the copper glutamate complex in the mother liquor is shown in Figure 2.

Second Harmonic Measurements

A mode-locked Nd:YAG laser with an average power of 1kW was used. The green fluorescent emission from the Krypton arc lamp of the laser was removed by passing the fundamental beam through a filter. The intensity of the fundamental laser beam was controlled by using in combination a half wave plate and a polarizer. The beam was chopped to avoid sample damage due to laser irradiation.

Copper glutamate powders were sieved into sizes ranging from 25-30 microns to 300-355 microns and packed in 1mm thick optical cells. Each sample was then placed in an integrating sphere and irradiated with laser light. The integrated second harmonic signal as a function of particle size was measured using a photo multiplier tube attached to a port on the integrating sphere. A filter was placed in front of the PMT to remove the fundamental frequency. A lock-in amplifier was used to measure the average second harmonic signal from each sample.

RESULTS

Powder second harmonic generation[13,14] is a commonly used technique for screening materials for their second order nonlinear properties. We examined 16 materials for the existence of second harmonic generation and the results of that screening process is shown in Table I. The dipeptides showing strong SHG had polarizable atoms(met-lys formate salt, phe-leu amide hydrobromide *vs* hydrochloride, gly-leu amide hydrochloride) the presence of aromatic side chains(L-carnosine, gly-tyr acetate) or absorption bands in resonance at 1.064 or 0.532μ(copper glutamate). Those dipeptides showing moderate SHG all had aromatic amino acids(lys-phe hydrochloride, lys-phe, gly-his hydrochloride and gly-tyr amide hydrochloride). With the exception of phe-leu amide hydrochloride, the peptides showing weak SHG had no aromatic residues. Finally the peptides showing no SHG were non-chiral(gly-gly ethyl ester and gly-gly amide hydrochloride).

The absorption maximum of copper glutamate in solution(796 nm) suggested enhancement of hyperpolarizability upon excitation in the near to mid-IR. We measured the copper glutamate second harmonic signal as a function particle size(Figure 3). The second harmonic intensity rose sharply as the particle size increased toward the coherence length and then remained constant as the particle size increased further. Kurtz's theory predicts that copper glutamate is phase matchable for frequency doubling at 1.06 microns.

Demonstration of phase matchability in copper glutamate powder prompted us to investigate crystal growth. We have been applying protein crystal growth methods[15-17] to the growth of copper glutamate crystals. Crystal growth needs to be in three dimensions so that a 5mm x 5mm frequency doubling cube can be cut from the crystal. We investigated the hanging drop crystal growth method where 15 μL 10 mg/ml Cu-glutamate solution was mixed with 15μL 20% poly(ethylene glycol). The copper glutamate formed an amorphous glass under these conditions. Crystals could be grown in 2.0M NaCl, pH 2.4 under slow evaporation.

CONCLUSION

We have synthesized copper glutamate and characterized it by x-ray diffraction and powder second harmonic generation. The materials demonstrates substantial second harmonic generation and appears to be phase matchable at 1.06 microns. In our lab we are preparing other metal-glutamate complexes. Synthesis, crystal growth, phase matching and other experiments will described in a later publication.

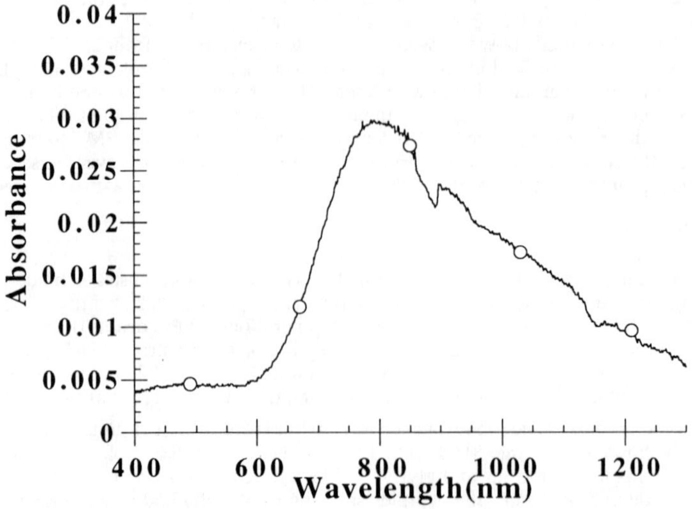

Figure 2. Absorption spectrum of Cu-glutamate complex in water, pH 1.55.

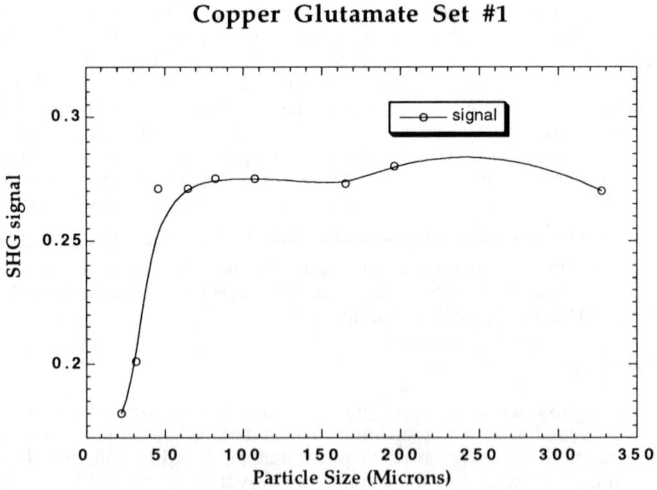

Figure 3 Second harmonic generation signal as a function of particle size

Table I-Qualitative Second Harmonic Intensity Observations

Material	SHG signal
Met-Lys formate salt	Strong
Phe-Leu amide hydrobromide	Strong
L-Carnosine	Strong
Gly-Leu amide hydrochloride	Strong
Gly-Tyr acetate	Strong
Copper Glutamate	Strong
Lys-Phe hydrochloride	Moderate
Lys-Phe	Moderate
Gly-His hydrochloride	Moderate
Gly-Tyr Amide hydrochloride	Moderate
Lys-Val dihydrobromide	Very Weak
Lys-Val hydrochloride	Very Weak
Lys-Leu monohydrobromide	Very Weak
Phe-Leu amide hydrochloride	Very Weak
Gly-Gly ethyl ester	None
Gly-Gly amide hydrochloride	None

REFERENCES

1. Z. Tokarski, L. Natarajan, B. Epling, T. Cooper, K. Hussong, T. Grinstead and W.W. Adams, Chem. Mater., **6**, pp. 2063-2069(1994).
2. T. Cooper, A. Campbell and R.L. Crane, Langmuir **11**, pp. 2713-2718(1995).
3. S. Monaco, L. Davis, S. Velsko, F. Wang and D. Eimerl, J. Cryst. Growth **85**, pp. 252-255(1987).
4. S. Tokutake, Y. Imanishi and M. Sisido, Mol. Cryst. Liq. Cryst. **170**, pp. 245-257(1989).
5. S. Velsko in Materials for Nonlinear Optics: Chemical Perspectives, ACS Symposium Series 455, edited by S. Marder, J. Sohn and G. Stucky, American Chemical Society, Washington D.C., (1991), pp. 343-359.
6. P. R. Newman, L. F. Warren, P. Cunningham, T. Y. Chang, D. E. Cooper, G. L. Burdge, P. Polak-Dingels, and C. K. Lowe-Ma, Mater. Res. Soc. Symp. Proc., **173**, 557 (1990)
7. H. O. Marcy, L. F. Warren, M. S. Webb, C. A. Ebbers, S. P. Velsko, G. C. Kennedy, and G. C. Catella, Appl. Opt., **31**, 5051 (1992)
8. U. B. Ramabadran, R. Vuppuladhadium, D. L. Small, D. E. Zelmon, and G. C. Kennedy, Appl. Opt., **35**, 903 (1996)
9. C. Gramaccioli and R. Marsh., Acta Cryst. **21**, pp. 594-600(1966).
10. C. Gramaccioli, Acta Cryst. **21**, pp. 600-605(1966).
11. R. Ledger and F. Stewart, Aust. J. Chem. **18**, pp. 1477-94(1965).
12. W. van Heeswijk, M. Eenink and J. Feden, Angew. Chem. **82**, pp. 744-747(1982).
13. S. K. Kurtz and T. T. Perry, J. Appl. Phys., 39, 3798-3813 (1968).
14. S. K. Kurtz and J. P. Dougherty, in Systematic Materials Analysis, J. H. Richardson and R. V. Peterson, eds., v4 (Academic Press, New York 1978) pp. 269-342.
15. A. McPherson, J. Cryst. Growth **110**, pp. 1-10(1991).
16. A. McPherson, Eur. J. Biochem **189**, pp. 1-23(1990).
17. P. Weber, Adv. Protein Chem. **41**, pp. 1-36(1991).

SECOND HARMONIC GENERATION FROM MULTILAYERS OF ORIENTED METAL BISPHOSPHONATES

Grace Ann Neff, Timothy M. Mahon, Travis A. Abshere and Catherine J. Page, Department of Chemistry, University of Oregon, Eugene, OR 97403.

ABSTRACT Second order nonlinear optical properties (NLO) require the presence of a polarizable moiety situated in an anharmonic potential. Our approach to incorporating such properties into self-assembled multilayers involves use of asymmetric α,ω bisphosphonates which meet this requirement by virtue of their chemical structure and binding properties. We have developed and optimized protection and deprotection schemes to allow for oriented layering of these molecules. Characterization by optical ellipsometry and grazing angle X-ray diffraction provides insight on average layer thicknesses and bulk film densities. Second harmonic generation (SHG) intensity from the bulk film is measured to verify NLO activity.

INTRODUCTION

The use of self-assembly to produce multilayer thin films has many advantages. These include simplicity of required equipment, ease of preparation, and stability of resulting products. Even more important, the layer-by-layer process imparts control over film thickness, composition and physical properties. For these reasons, the method has attracted great interest in materials research in recent years, with most of the work focusing on either the silanol-based[1-4] or metal bisphosphonate[5-11] chemistries. Our group has worked mainly with the latter system,[11-15] but is also developing new coordination chemistry-based assembly systems and is currently working to prepare hybrid superstructures[16] consisting of alternating cobalt-diisocyanide[17, 18] and metal bisphosphonate layers.

The Hf- bisphosphonate system we use is analogous to the prototype Zr-bisphosphonate system developed by Mallouk and co-workers.[5] In general, the metal bisphosphonate system is ideal for producing multilayers with tailored physical properties, because it offers great versatility in terms of functional groups that can be accommodated by the 24 Å^2 cross-sectional area allowed by the metal-phosphonate geometry.[19,20] This versatility is exploited in the work described herein. Our focus here is on the use of asymmetric α,ω bisphosphonates which bind in an oriented manner to produce multilayers which exhibit second-order nonlinear optical properties.

Second-order NLO properties require the presence of polarizable electrons situated in an anharmonic potential. We meet this requirement by incorporating π systems, such as phenyl or stilbyl rings, between electron donors and acceptors within our bisphosphonate molecules. These chromophores must also be oriented in the same direction within the multilayer structure, in order to produce directional polarizability. We accomplish this by functionalizing one end of the bisphosphonate with a phosphonic acid group, while leaving the other end "protected" in phosphonate ester form. This ensures oriented binding of the molecule to Hf on the surface through the acid end. A typical molecule, aryl-(4-diethylphosphonate)-10-decylphosphonic acid ether (**I**), is shown below in Figure 1.

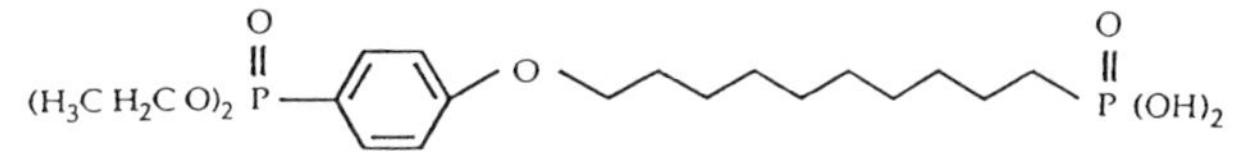

FIGURE 1: (**I**) aryl-(4-diethylphosphonate)- 10-decylphosphonic acid ether

Mat. Res. Soc. Symp. Proc. Vol. 435

Monolayers constructed using molecules such as (**I**) must be deprotected (i.e. hydrolyzed) in order to build multilayers. We have found that the common hydrolysis method of refluxing in 6M HCl[21] severely damages the film and substrate. We have modified this method by reducing the HCl concentration to 0.5 M and the heating temperature to 40-50°C to efficiently hydrolyze substrate-bound esters.[15] In the work described herein, we apply this modified hydrolysis to produce multilayers from Hf and (**I**), as shown in Figure 2. To characterize these multi-layers, we utilize optical ellipsometry, grazing-angle X-ray diffraction and second harmonic generation. Preliminary results from each method, as well as conclusions and future plans, are discussed.

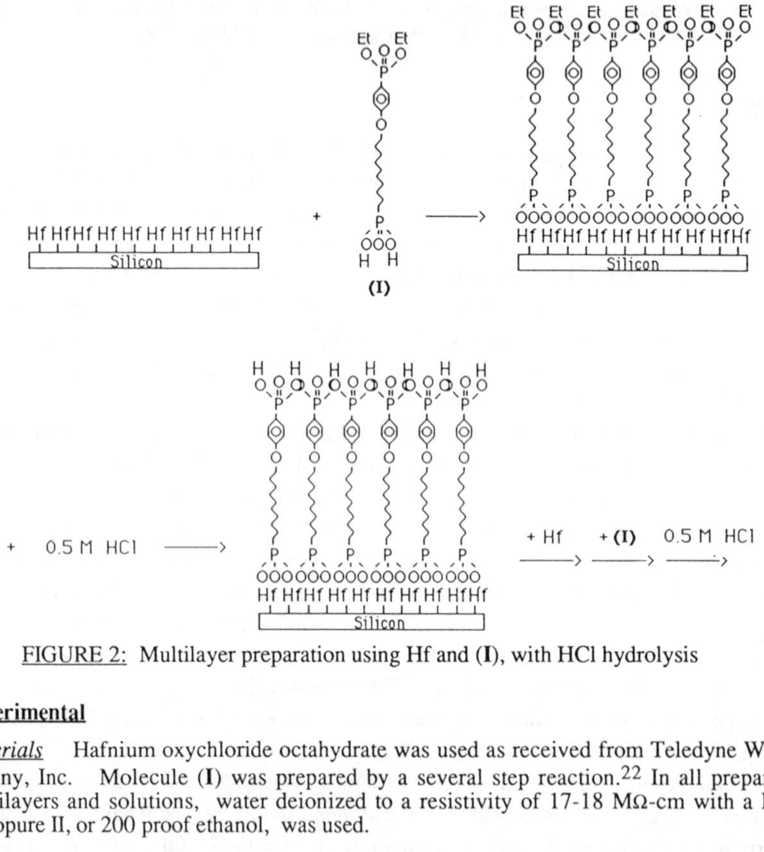

FIGURE 2: Multilayer preparation using Hf and (**I**), with HCl hydrolysis

Experimental

Materials Hafnium oxychloride octahydrate was used as received from Teledyne Wah Chang Albany, Inc. Molecule (**I**) was prepared by a several step reaction.[22] In all preparations of multilayers and solutions, water deionized to a resistivity of 17-18 MΩ-cm with a Barnstead Nanopure II, or 200 proof ethanol, was used.

Substrates, surface functionalization, and multilayer growth The silicon substrates used were Si (100) wafers from Silicon Quest. Wafers were degreased and then functionalized with hafnium as previously described.[13] The multilayer samples were prepared by immersing a Hf-functionalized wafer in a 1.25 mM solution of molecule (**I**) in 1:1 ethanol:water overnight, then hydrolyzed in .5M HCl at 45° C overnight, and finally immersed in 5 mM $HfOCl_2$ for 4-5 hours. Samples were rinsed in between immersion steps with the appropriate solvent.

Instrumentation and measurements Optical ellipsometry (at 632.8 nm) and grazing angle X-ray diffraction measurements were performed as previously described.[11] Second harmonic generation (SHG) measurements were performed using the 800 nm output of a regeneratively amplified Ti:Sapphire laser operating at 1 kHz. The laser light was collimated to a beam diameter of 3 mm. Approximately 80% of this was directed on the multilayer sample at a 45° angle of incidence. The SH signal at 400 nm was collected in reflection by a monochromator, PMT and gated electronics, for a period of five minutes to ensure the signal was stable. The remaining 20% of the laser light was passed through a 200 μm thick Type I phase-matched KDP crystal to provide a reference SH signal. The signal from the multlilayers was normalized to this reference signal from KDP.

RESULTS and DISCUSSION

Figure 3 below shows total thickness obtained from ellipsometry measurements (using an index of refraction of n=1.5 for the films) vs. number of layers for two samples of Hf and molecule (**I**).

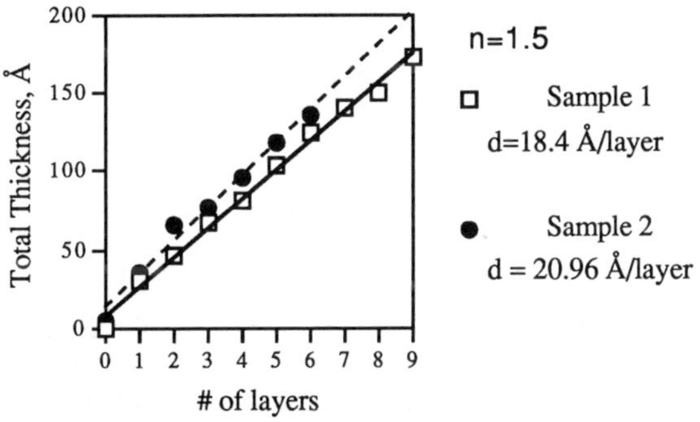

FIGURE 3: Total thickness by ellipsometry vs. # of layers for Samples 1 and 2, with n=1.5.

For such layers, with the alkyl chains oriented at 31° from surface normal (as would occur for a P-C bond normal to the metal-phosphonate plane,[11]) we expect an average layer thickness of ~21Å. The average thickness obtained for Sample 1 is just short of this (while that for Sample 2 is very close). There are several possible explanations for this. One possibility is that the molecules in Sample 1 are tilting at angles of greater than 30° from the surface normal (i.e. ~39°), resulting in layers that are thinner than expected. Such a tilt angle is not dramatically different from the accepted 30° or even uncommon compared to other samples made in this lab.[11] Another possibility is that this sample is less dense than sample 2 and thus has a smaller index of refraction than 1.5 Alternatively, the sample 2 film could be "patchy." These latter possibilities highlight the major limitation of ellipsometry for determining film thickness: in order to use this technique, one must know the index of refraction of the material. The index of refraction is not known *a priori*, and we have found that it can vary from one multilayer sample to the next of the same composition.[11] Thus, it is vital to have an alternative method for determining layer thickness that is independent of the index of refraction. Grazing angle X-ray diffraction[11] provides an alternative and more reliable method, if the film uniformity is such that interference between X-rays reflected off the front and the back of the film is observed.

Figure 4 shows X-ray diffraction patterns from 7 layers on both samples 1 and 2. Total film thickness *d* can be calculated from observed interference patterns as discussed in detail in reference 11. A plot of total film thickness obtained from X-ray data vs. # of layers gives a slope equal to the average film thickness which can be compared to ellipsometry data. By matching the slopes of ellipsometry data to that of the X-ray data, the index of refraction for a particular film can be determined.

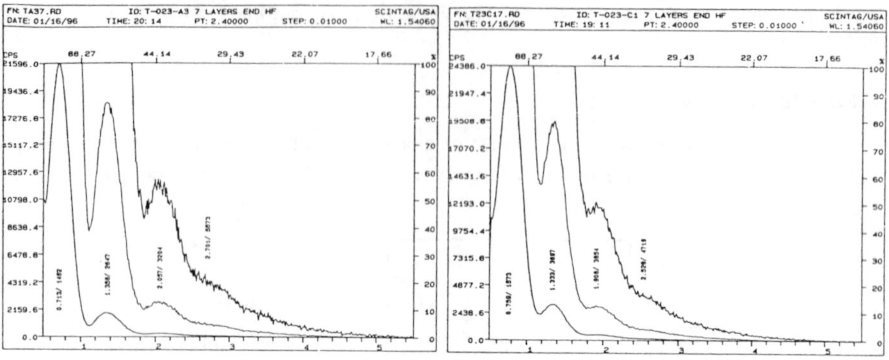

Figure 4: Grazing Angle X-ray Diffraction Patterns

Figure 5 shows such a comparison between ellipsometry and X-ray data for both samples. This shows that average layer thicknesses obtained from X-ray data are slightly larger than those from ellipsometry data using n=1.5 (see Figure 3). In fact, the actual index of refraction for Sample 1 is determined to be n=1.36. The index of refraction determined for sample 2 (1.475) is closer to accepted values for similar materials (usually 1.50).[11] The smaller indices in both cases indicate these materials are less dense or more "patchy" than simple Hf-decylbisphosphonate multilayers and the corresponding bulk materials.[11]

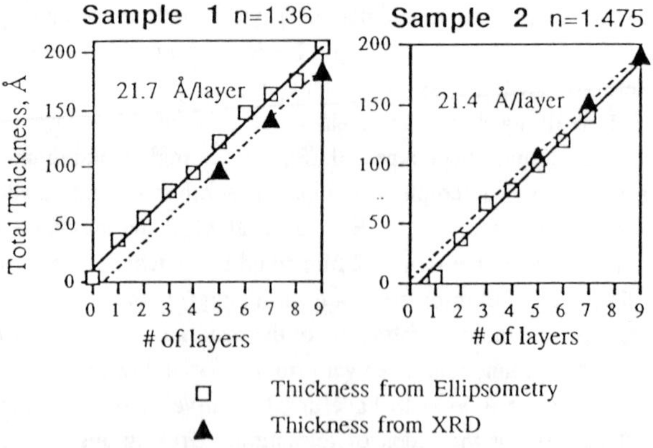

Figure 5: Ellipsometry data matched to X-ray data, Samples 1 & 2

Ellipsometry and X-ray data both indicate our method of assembling these multilayer structures produces uniform regular layers. To determine whether or not these multilayers exhibit NLO properties, SHG is utilized as an analytical tool. The susceptibility tensor responsible for second-order NLO effects from the bulk structure, χ^2, is proportional to the number of molecules, N, in that structure, as shown by the relationship:

$$\chi^2 = N < \beta >$$

where $< \beta >$ is the orientational average of the molecular hyperpolarizability. From this expression, one would expect that as the number of layers containing NLO chromophores increases, the SH signal should also increase. Figure 6 shows normalized SH signal vs. # of layers for both samples. The expected relationship is not observed for either sample. The behavior of Sample 2 is the easier of the two to rationalize: SH signal from the layer structure could be interfering destructively with SH signal from the Si substrate, and as the signal from the multilayers increases with increasing # of layers, the resultant signal observed continually decreases. Behavior of Sample 1 is more difficult to explain, but may be a consequence of a less-dense, more orientationally disordered film which also gives a SH response that destructively interferes with substrate signal.

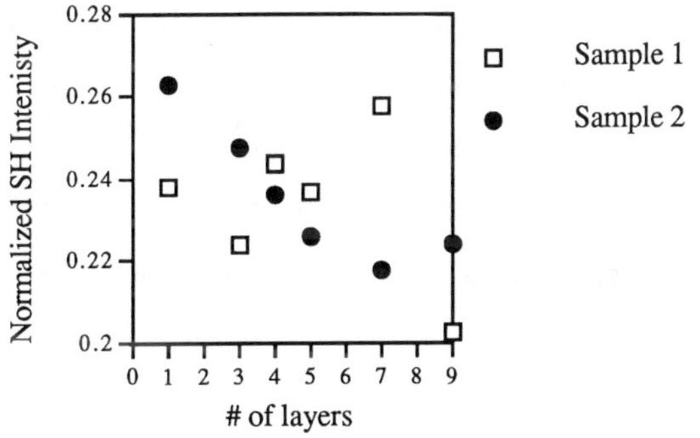

Figure 6: Average SH Intensity, normalized to KDP, vs. # of layers, for Samples 1 & 2.

In addition, because the chromophore in molecule (I) is relatively weak, the SH signal originating from the multilayers could be weak compared to that from the substrate. Work is currently underway with a molecule similar to (I) that contains a better chromophore with a stilbyl moiety rather than a phenyl moiety, to see if we can observe the expected SH increase with the number of chromophore layers. We also plan to make similar SHG measurements from samples of both chromophores grown on glass slides, to eliminate the competing SH signal from the silicon substrate. Sum frequency generation (SFG), also a second-order NLO process, can be utilized to measure the IR spectra of the multilayers in certain frequency ranges. We plan to make use of this technique, as well as perform SH polarization studies, to get information about the orientation of the chromophore molecules within the multilayers.

Conclusions

The preliminary results reported in this paper suggest that our method of constructing metal bisphosphonate multilayers via a three step self-assembly process produces regular, uniform layers. Both ellipsometry and X-ray diffraction results show incremental growth of reasonable thickness. Observation of the regular "fringes" in the grazing angle diffraction patterns suggests relatively uniform multilayers. The observance of a decrease in the SH signal from the multilayers suggests that the layers are producing a NLO signal that is interfering with that produced at the substrate surface. Further study of films on glass and with better chromophores, as well as studies to determine orientation of the molecules within the multilayers, are planned in order to further characterize the NLO activity of multilayers produced in this manner.

Acknowledgments

We thank Derek E. Gragson, Bayrn McCarty and Prof. Geraldine L. Richmond for use of the Ti:Sapphire laser and help with the SHG measurements, and the University of Oregon Department of Chemistry for graduate student support (GAN).

References

1. L. Netzer, J. Sagiv, J. Am. Chem. Soc., **105**, 674 (1983).
2. R. Maoz, L. Netzer, J. Gun, J. Sagiv, J. Chem. Phys., **85**, 1059 (1988).
3. N. Tillman, A. Ulman, T.L. Penner, Langmuir, **5**, 101 (1989).
4. D. Li, M.A. Ratner, T.J. Marks, C. Zhang, J. Yang, G. J. Wong, J. Am. Chem. Soc., **112**, 7389 (1990).
5. H. Lee, L.J. Kepley, H.-G. Hong, T.E. Mallouk, J. Am. Chem. Soc., **110**, 618 (1988).
6. H. Lee, L.J. Kepley, H.-G. Hong, S. Akhter, T.E. Mallouk, J. Phys. Chem., **92**, 2597 (1988).
7. H.E. Katz, G. Scheller, T.M. Putvinski, M.L. Schilling, W.L. Wilson, C.E.D. Chidsey, Science, **254**, 1485 (1991).
8. H.C. Yang, K. Aoki, H. Hong, D.D. Sackett, M.F. Rendt, S. Yau, C.M. Bell, and T.E. Mallouk, J. Am. Chem. Soc., **115**, 11855 (1993).
9. L.A. Vermuelen, J.L. Snover, L.S. Sapochak, M.E. Thompson, J. Am. Chem. Soc., **115**, 11767 (1993).
10. H.E. Katz, Chem. Mater., **6**, 2227, (1994).
11. A.C. Zeppenfeld, S.L. Fiddler, W.K. Ham, B.J. Klopfenstein, C.J.Page, J. Am. Chem. Soc., **116**, 9158 (1994).
12. A.C. Zeppenfeld, C. J. Page, Mater. Res. Soc. Symp. Proc. **351**, 77 (1994), Molecularly Designed Ultrafine/Nanostructured Materials, K.E. Gonsalves, G.-M. Chow, T.D. Xiao, R.C. Cammarata, Editors, Materials Research Society, Pittsburgh, PA.
13. G.A. Neff, A.C. Zeppenfeld, B.J. Klopfenstein, C. J. Page, Mater. Res. Soc. Symp. Proc., **351**, 269 (1994), Molecularly Designed Ultrafine/Nanostructured Materials, K.E. Gonsalves, G.-M. Chow, T.D. Xiao, R.C. Cammarata, Editors, Materials Research Society, Pittsburgh, PA .
14. J.T. O'Brien, A.C. Zeppenfeld, G.L. Richmond, C.J. Page, Langmuir, **10**, 4657 (1994).
15. G.A. Neff, C.J. Page, E. Meintjes, T. Tsuda, W.-C. Pilgrim, N. Roberts, W.W. Warren, Jr., Langmuir, **12**, 238 (1996).
16. M.A. Ansell, E.B. Cogan, G.A. Neff, C.J. Page, submitted to Supramolecular Science.
17. M.A. Ansell, A.C. Zeppenfeld, W.K. Ham, C.J. Page, Mater. Res. Soc. Symp. Proc., **351**, 171 (1994), Materials Research Society, Pittsburgh, PA.
18. M.A. Ansell, A.C. Zeppenfeld, K.Yoshimoto, E.B. Cogan, C.J. Page, Chem. Mater., **8**, 591 (1996).
19. M.B. Dines, P.M. DiGiacomo, Inorg. Chem., **20**, 92 (1981).
20. G. Cao, H.-G. Hong, T.E. Mallouk, Acc. Chem. Res., **25**, 420 (1992).
21. R.E. Engel, Synthesis of Carbon-Phosphorus Bonds, CRC Press, Boca Raton, FL, p.21 (1988).
22. Unpublished results.

FIBER OPTIC SENSING OF CYANIDES IN SOLUTIONS

S.S. PARK *, J.D. MACKENZIE, C.Y. LI **, P.GUERREIRO, N. PEYGHAMBARIAN
*MSE Department, University of California, Los Angeles, CA 90095, sspark@seas.ucla.edu
**Optical Sciences Center, University of Arizona, Tucson, AZ 85721

ABSTRACT

A novel sol-gel technique was used to immobilize malachite green ions (MG^+) in stable, optically transparent, porous silica gel films. A simple and sensitive method was developed for the detection of cyanides in solutions using spectrophotometry to measure changes caused by cyanide ions (CN^-) in the absorption spectra of the green-colored silica gel films. After reaction with cyanide ions, the absorption spectra of the films changed with a typical decrease in absorbance at 620 nm. On the basis of the absorption spectra of the films, a portable and easy to use fiber optic cyanide film sensor was fabricated. Decolorization undergone by the green-colored gel films, as they were exposed to cyanide ions, was detected through a fiber. Preliminary results indicate concentrations on the order of a few ppm are detected using the fiber optic sensor.

INTRODUCTION

Although cyanide compounds are strongly poisonous to biological systems, large amounts of cyanide compounds have been used in the fields of mines, factories and plants, which produce precious metals or integrated circuits. A simple and sensitive method for the determination of cyanide compounds is therefore required in these fields. Many methods have been developed for the determination of cyanide compounds in water, waste water, and plating solutions. Conventional methods generally rely on sophisticated sampling techniques and expensive laboratory-based analytical instruments such as GC and HPLC. Samples must be collected, stored and transported from the field into the laboratory before measurements can be made. These problems can be overcome using the fiber optic chemical sensors that enable real-time in-situ detection of pollutants without sample collection and at relatively low cost [1].

Non-leachable inorganic or organic molecule doped sol-gel glasses have been attracting increasing interest since they can be used as chemical sensors [2-3]. However, most of monolithic sol-gel glasses tend to fracture during the gelation process, upon immersion in aqueous solution or upon subsequent drying due to the formation of large pressure gradients between the wetted and dried pores. Thus, many efforts toward the development of film-type sol-gel microsensors are required. Sol-gel technique has other advantages such as excellent homogeneity, easy control of film thickness, and the ability to coat large and complex shapes such as fiber tip [4]. It is, also, well known that malachite green ions (MG^+) react with cyanide ions (CN^-) to form colorless malachite green leucocyanide (MGCN) in a stoichiometric manner [5].

In this paper, a detailed description of the fabrication of malachite green-doped silica gel films and fiber optic cyanide sensor, leaching procedure of malachite green from the films, the dependence of the absorbance of the films on cyanide ion concentration, and the sensing properties of the fiber optic sensor will be given.

EXPERIMENT

Malachite green-doped silica gel films were prepared by the following process. A mixture of tetraethoxysilane (TEOS) from Aldrich Chemical Co. Inc., malachite green oxalate from Fisher Chemical, and ethanol was stirred for 30 minutes in air at room temperature and mixed with $HC\ell$ in water for 30 minutes in air at room temperature. The molar ratio of TEOS : H_2O : EtOH : $HC\ell$ was $1 : 4 : 5 : 3\times10^{-3}$. Assuming that all tetraethoxysilane would hydrolyze and

Mat. Res. Soc. Symp. Proc. Vol. 435 © 1996 Materials Research Society

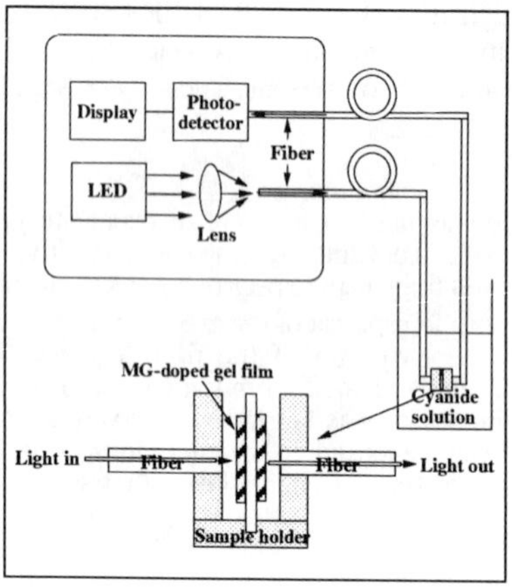

Figure 1. Block diagram of a fiber optic cyanide sensor.

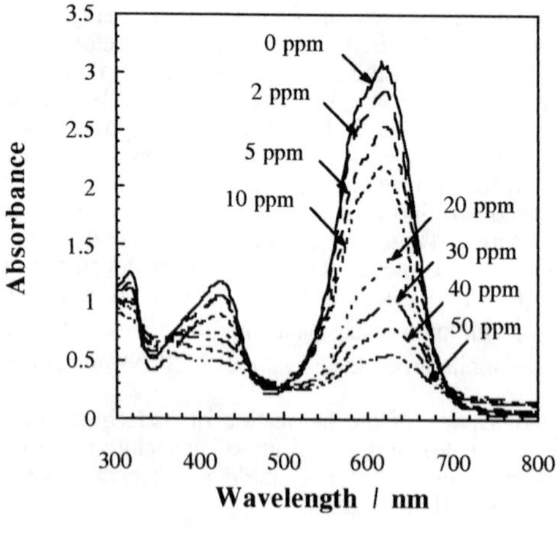

Figure 2. The absorbance spectra of malachite green-doped gel films after a 24 hour exposure to 50 mℓ cyanide solutions of varying cyanide ion concentration at room temperature.

polycondense to form silica, the amount of silica was calculated. On the basis of the calculated amount of silica, the doping amount of malachite green oxalate (15 wt.%) was calculated. The mixed solution was aged for 20 hours in air at room temperature. Slide glass substrates were ultrasonically cleaned, first in acetone, and then in ethanol. The pre-cleaned substrates were dipped into the aged solution using the withdrawal speed of 11 cm/min in air. The films were coated onto both sides of the substrates. The gel films were dried at 100 ^{0}C for one hour in air. The heating and cooling rate were 0.5 ^{0}C/min. The dimension of the films was 1" x 2".

The silica gel films were immersed into 50 mℓ cyanide solution and allowed to be equilibrated for 24 hours. Various cyanide ion concentrations were made by dissolving potassium cyanide (KCN) from Aldrich Chemical Co. Inc. in deionized water. Fresh sample solutions were prepared before measurement. After taking out the films from the sample solution, the optical absorbance spectra of them were measured in the wavelength range of 300-800 nm using a Diode Array Spectrophotometer (Hewlett Packard, HP8452A).

A portable fiber optic chemical sensor (FOCS) for the detection of cyanides in solutions was constructed as shown in Fig. 1. The complete FOCS system consists of: (i) a light emitting diode (LED) for light source, (ii) film sensor holder, (iii) a photodiode detector, (iv) signal collection and display electronics. The optical polymer fiber with the film sensor holder was immersed into 500 ml cyanide solution. The light from a LED was focused through a lens into the end of a fiber and the light guided by the fiber passed through the films. The transmitted light signal was passed via another fiber to a photodiode detector.

RESULTS AND DISCUSSION

The absorption spectra of the malachite green (MG) doped-silica gel films after a 24 hour exposure to cyanide solutions of varying cyanide ion concentration at room temperature is shown in Fig. 2. The thickness of the films was about 320 nm. The highest absorbance peak was found at the wavelength of 620 nm, which was corresponding to green color. Clearly, the absorbance decreased with increasing cyanide ion concentration. As stated earlier, the formation of malachite green leucocyanide resulted in the decolorization of the malachite green-doped gel films.

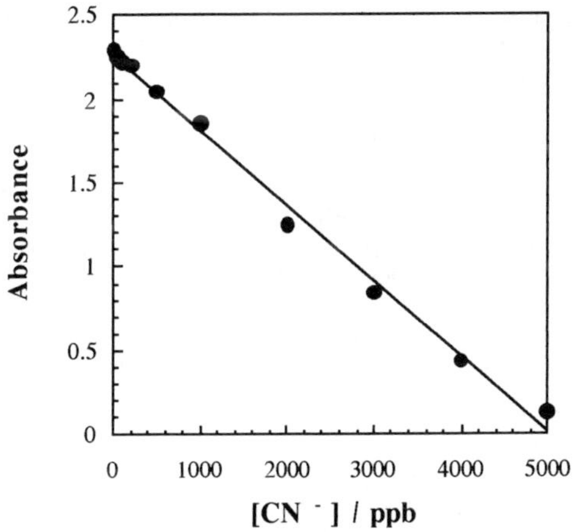

Figure 3. The change in absorbance at 615 nm for 10 g of malachite green oxalate solution (3×10^{-4} mole/ℓ) after a 20 minute exposure to 200 g of cyanide solutions of varying cyanide ion concentration at 90 ^{0}C.

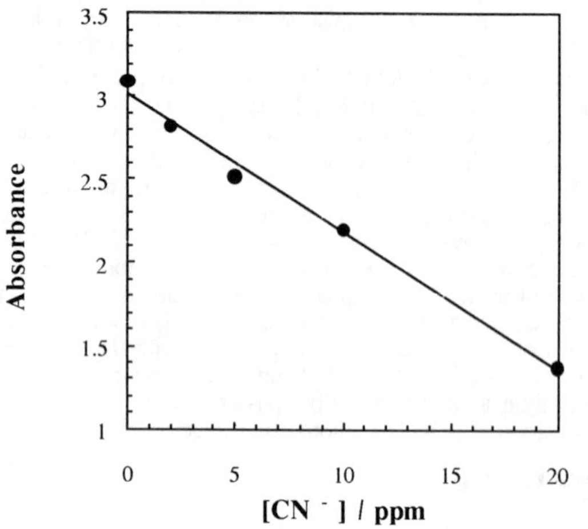

Figure 4. The change in absorbance at 620 nm for malachite green doped-gel films after a 24 hour exposure to 50 mℓ cyanide solutions of varying cyanide ion concentration at room temperature.

On the other hand, the highest absorbance peak of malachite green solution was found at the wavelength of 615 nm and only short equilibration time of 20 minutes was required to obtain enough absorbance change at 90 ^{0}C. Figure 3 shows the change in absorbance at 615 nm with cyanide ion concentration for malachite green solution. 10 g of malachite green oxalate solution ($3x10^{-4}$ mole/ℓ) was reacted with 200 g of cyanide solution. The absorbance decreased linearly with increasing cyanide ion concentration in the range of 50 ppb to 5 ppm. Note that the low ppb range is comparable with the level achievable by GC and HPLC.

Figure 4 shows the change in absorbance at 620 nm for the malachite green-doped (15 wt.%) silica gel films after a 24 hour exposure to cyanide solutions of varying cyanide ion concentration at room temperature. The absorbance decreased linearly with increasing cyanide ion concentration from 1 to 20 ppm. The films were allowed to remain in cyanide solution for 24 hours at room temperature before they were removed for measurement. The entrapped malachite green ions can react with cyanide ions in liquid only through open pores in the films. Thus, after much malachite green leucocyanide molecules form first, the molecules may block up the diffusion road of other unreacted cyanide ions and the formation rate of malachite green leucocyanide will be decreased. This seemed to result in the long equilibration time of 24 hours.

Doped reagents tend to leach from sol-gel glasses when immersed in aqueous solution. In some cases, a total leaching of the reagent from the glasses can take place upon prolonged leaching. To quantify the extent of leaching, the films were immersed in 50 mℓ deionized water that was renewed every day. In our case, approximately 40 % of the malachite green leached out of the films during 5 days. Since leaching of the malachite green was observed, the malachite green must not be bounded to the silica gel matrix. Thus, porosity may play an important role in terms of sensitivity. For example, if the dye molecules are trapped in closed pores, they will be unable to react with cyanide ions. For regions of open pores, if the pores are too small, then malachite green leucocyanide may not be able to form because it can not physically fit into the pore. More detailed investigations of the microstructural effects on the formation of malachite green leucocyanide in the silica gel films is currently being performed.

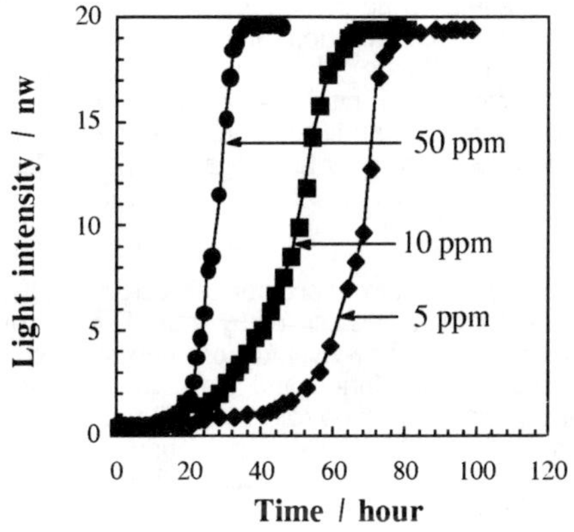

Figure 5. The light intensity change of a fiber optic cyanide sensor at room temperature with time against varying cyanide concentration.

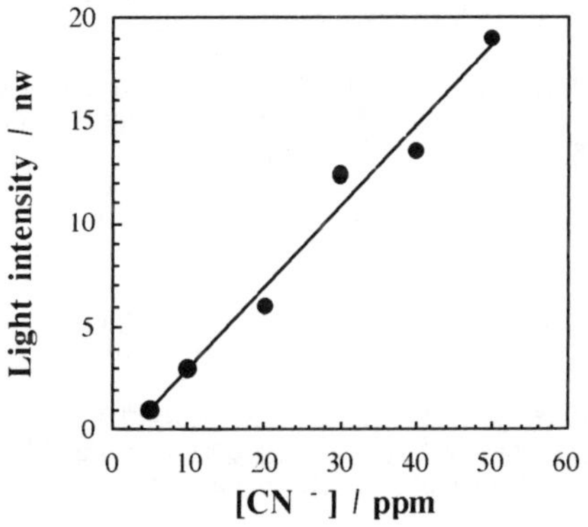

Figure 6. The light intensity change of a fiber optic cyanide sensor at room temperature after a 32 hour exposure to cyanide solutions of varying cyanide ion concentration.

Typical light intensity change of the fiber optic cyanide sensor with time against varying cyanide concentration is shown in Fig. 5. The measurement was performed at room temperature. As shown in the figure, light intensity increased slowly in the first stage and then, suddenly increased sharply. This behavior seemed to be contributed to the characteristic formation process of malachite green leucocyanide in the sol-gel derived films. The time required for the complete reaction of malachite green ions with cyanide ions in the films increased with decreasing cyanide ion concentration. This result seems to be reasonable because the formation rate of malachite green leucocyanide in the gel films will decrease with decreasing cyanide concentration. However, the formation reaction of malachite green leucocyanide was not reversible under our experimental conditions. More detailed investigations of the reversibility of the reaction is currently being performed. Figure 6 shows the light intensity change of the fiber optic sensor after a 32 hour exposure to cyanide solutions of varying cyanide ion concentration. The sensor yields a linear calibration graph over the range 5-50 ppm at least.

CONCLUSIONS

A simple and sensitive method was developed for the detection of cyanides in solutions using spectrophotometry to measure changes caused by cyanide ions in the absorption spectra of malachite green-doped sol- gel films. Low cyanide ion concentrations of 50 ppb and 1 ppm were measured using malachite green solution and malachite green-doped silica gel films, respectively. The basis for a portable fiber optic cyanide film sensor has been demonstrated using a laboratory test system. Cyanide ion concentration about a few ppm could be measured using the fiber optic sensor. However, present method may be useful for the detection of trace concentrations of cyanides in industrial wastewaters.

ACKNOWLEDGMENTS

This work was funded by UC-TSR&TP (University of California Toxic Substances Research & Teaching Program).

REFERENCES

[1] U.S. Environmental Protection Agency, In-Situ Monitoring at Superfund Sites with Fibre Optics: 1. Rationale, (U.S. E.P.A., Nevada, 1987), p. 1.

[2] L. M. Ellerby, C. R. Nishida, F. Nishida, S. A. Yamanaka, B. Dunn, J. S. Valentine, J. I. Zink, Encapsulation of Proteins in Transparent Porous Silicate Glasses Prepared by the Sol-Gel Method, Science, **255**, p. 1113 (1992).

[3] O. Lev, M.Tsionsky, L. Rabinovich, V. Glezer, S. Sampath, I. Pankratov, J. Gun, Organically Modified Sol-Gel Sensors, Analytical Chemistry, **67**, p. 22A (1995).

[4] L. C. Klein, Sol-Gel Technology for Thin Films, Fibres, Preforms, Electronics, and Speciality Shapes, (Noyes Publications, New Jersey, 1988), p. 49.

[5] E.O. Holmes, The Effects of the Properties of Solvents of Various Dielectric Constants and Structures on the Photoionization of the Leucocarbinols and Leucocyanides of Malachite Green, Crystal Violet, and Sunset Orange and Related Phenomena, J. Phys. Chem., **70**, p. 1037 (1966).

AUTHOR INDEX

Abshere, Travis A., 661
Aegerter, M., 363
Akbarian, F., 187, 565
Alam, T.M., 421
Alexander, T.P., 559
Apple, T., 513
Armelao, L., 387
Arnold, F.E., 93
Assink, Roger A., 33, 345, 421, 469
Atik, M., 363
Audinet, L., 643

Babonneau, Florence, 119, 437
Baertlein, C.D., 559
Banse, F., 43
Baranwal, R., 437
Barber, D.B., 575
Bates, S.E., 469
Bayle, J.P., 475
Beaucage, Greg, 301, 469
Becker, C., 237
Beecroft, L.L., 575
Bell, I.S., 73
Bescher, Eric Pascal, 449, 605
Bhawalkar, J., 535
Black, Eric P., 301, 469
Blum, Y.D., 431
Boilot, J.P., 583, 643
Bolf, Alan G., 3
Bonagamba, T.J., 363
Bonhomme, C., 437
Boulton, J.M., 559
Brasselet, S., 395
Bray, K.L., 443, 617
Brennan, A.B., 155
Brinker, C. Jeffrey, 271, 357
Bruinsma, P.J., 131
Brun, A., 583
Bukowski, T.J., 559
Burzynski, Ryszard, 595
Buss, Richard J., 301
Butler, B.D., 85

Campero, A., 527
Campostrini, R., 381
Canva, M., 583
Cao, G.Z., 271
Cardenas, A., 501
Carpenter, Joseph P., 33
Casstevens, Martin K., 595
Chaput, F., 583
Chaudhuri, S. Ray, 307, 351
Chen, J.P., 93
Chen, Xiaohe, 55
Chen, Y.L., 131
Choe, E., 25
Chu, L., 215, 221
Chung, D.D.L., 249

Clem, P.G., 521
Cline, Steven M., 655
Coltrain, Bradley K., 507
Cooper, Thomas M., 655
Costa, V.C., 443
Counio, G., 643
Coveney, P.V., 73
Cuney, S., 143

Dahmouche, K., 363
D'Andrea, G., 381
Dang, T.D., 93
Daniels, M.W., 215, 221
Darracq, B., 583
Dave, B.C., 187, 565, 611
Davis, Mark E., 263
Delattre, L., 271
Dessolle, V., 475
Dreier, U., 455
Dromzee, Y., 487
Duan, Z., 307, 351
Dunn, B.S., 187, 565, 611

Engel, Christine, 327
Esnouf, S., 643
Eychenne-Baron, C., 43

Francis, L.F., 215, 221
Fu, Xuli, 249

Gacoin, T., 583, 643
Gandais, M., 643
Garcia, L., 449
Gaw, Kevin, 165
Gellermann, C., 67
Gerard, J.F., 143
Gerhard, V., 455
Ghosal, Saswati, 595
Gonsalves, Kenneth E., 55
Görsmann, Claus, 625
Greaves, John, 33
Grey, Clare P., 179
Gualandris, Virginie, 119
Guerreiro, P., 667
Gui, Linlin, 173
Guilment, J., 375, 649
Guizard, C., 199, 283, 501
Gupta, Samhita Das, 655

Haddad, T.S., 25
Haluschka, Christoph, 327
Hanley, H.J.M., 85
Harrison, C.M., 427
He, G.S., 535
Heckenbenner, P., 283
Hirashima, Hiroshi, 409
Hoebbel, D., 461
Hong, Edward, 449, 605

Hoshino, Y., 307, 351
Hovnanian, N., 283, 501
Huang, Q., 229
Hubert-Pfalzgraf, L.G., 137
Hüsing, Nicola, 339
Hussain, M., 243, 369

Imai, Hiroaki, 409
Imai, Yoshio, 165
Interrante, L.V., 513
Iwamoto, T., 403

Jeon, N.L., 521
Jikei, Mitsutoshi, 165
Johnson, S.M., 431
Jones, W., 73
Jordens, K., 207
Judeinstein, P., 363, 475

Kakimoto, Masa-aki, 165
Kale, Rahul P., 271
Kanazawa, C., 431
Katayama, Shingo, 321, 481
Katsoulis, Dimitris E., 589
Kentgens, A.P.M., 415
Keryk, John R., 589
Kiefer, Wolfgang, 339
Kim, A.Y., 131
Kooli, F., 73
Kramer, S.J., 229, 295
Krug, H., 13, 237, 553
Kubota, Yashihiro, 263
Kumar, N.D., 535
Kumudinie, C., 93

Lafontaine, E., 475
Laine, R.M., 437
Lal, M., 535
Lan, Tie, 79
Landry, Christine J.T., 507
Lebeau, B., 395
Leidner, R.T., 575
Lerner, M.M., 495
Leroux, F., 611
Levy, Y., 583
Leyrer, Jürgen, 625
Li, C.Y., 667
Lichtenhan, Joseph D., 3, 25
Lindquist, D.A., 427
Liu, J., 131
Liu, Q., 513
Lopez, Gabriel P., 271
Lorenz, Anne, 333
Lox, Egbert, 625
Loy, Douglas A., 33, 277, 301, 421
Lu, Yunfeng, 271
Luo, Shengcheng, 173

Machowski, Walter J., 255
MacKenzie, J.D., 229, 295, 403, 449
605, 667
Macosko, Christopher W., 113

Mahon, Timothy M., 661
Malier, L., 643
Maquet, J., 437, 527
Mark, J.E., 93
Maurina, S., 381
McCarthy, K.C., 559
McClain, Mark D., 277
McCormick, Alon V., 113
Mello, N.C., 363
Michalczyk, Michael J., 105
Miller, T.M., 155
Moreno, J., 315
Morris, R.D., 427
Motakef, S., 559
Müller, P., 553
Muzny, C.D., 85
Myers, Sharon A., 33

Nakahira, A., 243, 369
Nazar, L.F., 611
Neff, Grace Ann, 661
Niihara, K., 243, 369
Nishijima, S., 243, 369
Noel, Charles J., 3
Nuzzo, R.G., 521

Ober, C.K., 575
Oh, Junrok, 409
Ohtsuki, C., 403
Okada, T., 243, 369
Oliveira, P.W., 13, 553
Oriakhi, C.O., 495
Osaka, A., 403

Page, Catherine J., 661
Pajot, N., 137
Panepucci, H., 363
Papiernik, R., 137
Park, S.S., 667
Parraud, S., 137
Pascault, J.P., 143
Paul, Partha P., 255
Pauthe, Monique, 119
Payne, D.A., 521
Paz-Pujalt, G., 649
Peeters, M.P.J., 415
Peiffer, D.G., 85
Peyghambarian, N., 667
Pinnavaia, Thomas J., 79
Pollock, C.R., 575
Poncelet, O., 375, 649
Popall, M., 547
Porter, J.R., 431
Prabakar, Sheshasayana, 277, 345, 469
Prasad, P.N., 535
Premachandra, J., 93

Ramabadran, Uma B., 655
Raman, N.K., 357
Rankin, Stephen E., 113
Reinert, T., 461
Ribot, F.O., 43, 387

Ricolleau, C., 643
Riedel, Ralf, 327
Riegel, Bernhard, 339
Riehl, D., 583
Robert, F., 487
Roscher, C., 547
Rubio-Alonso, F., 229, 295
Ruland, G., 535
Ruth, Patrick N., 3

Sanchez, C., 43, 387, 395, 487, 527
Sarkar, A., 307, 351
Schaefer, Dale W., 301, 469
Schmidt, H.K., 13, 237, 455, 461, 553, 637
Schmitt, M., 637
Schrotter, J.C., 199, 283
Schubert, Ulrich, 333, 339, 625
Sharp, Kenneth G., 105, 179
Shea, Kenneth J., 33, 301
Smaihi, M., 199, 283, 501
Small, James H., 33
Snijkers-Hendrickx, I.J.M., 415
Sorarù, G.D., 381
Soto, A.M., 527
Steunou, N., 487
Stone, B.T., 443, 617
Storch, W., 67
Suzuki, Hironori, 165

Tang, Youqi, 173
Tavizon, G., 315
Teegarden, David M., 507
Teowee, G., 559
Timmons, Scott F., 255
Truchet, S., 375
Tsukuma, Koji, 409
Tsuru, K., 403
Tsvetkov, F., 85

Uhlmann, D.R., 559
Ulibarri, Tamara A., 301, 469

Vicente, L., 315
Vigier, G., 143
Viveros, T., 315
Vuppuladhadium, Rama, 655

Wallace, S., 357
Wen, J., 207
Werner, U., 637
Wettling, D., 375
Wilkes, G.L., 207
Williams, B., 427
Wilson, D.M., 431
Wolter, H., 67
Wong, H.P., 611
Wu, H.J., 431

Xu, Yu-Huan, 605

Yamada, Noriko, 321, 481
Yan, Y., 307, 351
Yoshida, M., 535
Yoshinaga, Ikuko , 321,481

Zelmon, David E., 655
Zhang, C., 437
Zhang, Jun, 173
Zhang, Yue, 595
Zhao, W., 93
Zheng, Z., 513
Zink, J.I., 187, 565
Zyss, J., 395

SUBJECT INDEX

abrasion
  resistance, 255
  resistant coatings, 207
absorber materials, 455
acetylacetone, 387
ac impedance spectroscopy, 363
acrylamide, 85
adsorption, 221, 455
aerogels, 295, 333, 611
AEROMOSILS, 295
AFM image of hybrid film, 307
agglomeration, 55
aging, 179
α-$K_8SiW_{11}O_{39}$, 589
alkoxide, 369, 409, 649
  derivatives, 137
aluminum(-)
  $Al(OBu^s)_2E$, 237, 415
  alkyl, 427
aminofunctional silanes, 13
amphoteric surfactant, 131
anisotropy, 475
apatite, 403
aqueous solutions, 409
atomic force microscopy, 155
AZO dye grafted, 395

BaFBr materials, 649
$BaTiO_3$, 605
BET surface areas, 387
bioactivity, 403
biologically-derived materials, 655
biosystem, 565
β-keto ligands, 461
bonding agents, 93

CdS, 643
CdS:Mn, 643
cement, 249
Ceramers, 321
CERIUS^2 v1.6.2, 73
chemical homogeneity, 119
chemometric tools, 375
chromophores, 605
clay/polymer composite, 85
coatings(-), 215, 221, 395
  dip, 187, 307
  spin, 307
co-condensation, 119
colloidal silica, 215
compatibilizers, 507
composites, 155, 431, 575
condensation, 387, 513
conductivity, 611
controlled crosslinking, 307
copolymerization, 113
copper glutamate, 655
core-shell nanostructures, 643
Cr-forsterite, 575
cross(-)
  condensation, 345
  polarization, 119, 437
cryogenic temperature, 243
Cryptosporidium parvum, 449
curing, 431
cyanides, 667
cyclic disilsesquioxanes, 33

dental restorative material, 255
dielectric(-)
  constants, 481
  properties, 559
  relaxation, 559
  spectroscopy, 559
Diels-Alder, 277
diffuse reflectance, 215
diffusion, 553
dipeptide complexes, 655
diphenylsiloxanediol, 461
direct-deposition-process, 409
disiloxanediamine, 165
dispersion, 249
displacement, 495
drying, 357
DTA, 351
dynamic light scattering, 85
dynamic mechanical measurements, 321

ε_r, 559
EDAX measurements, 93
electrical
  properties, 327
  resistivity, 249
electrochemical properties, 611
electrooptic, 595
electrostatic interaction, 55
ellipsometry, 661
emulsion, 137
energy minimization, 73
epoxy, 79, 369
Er^{3+}-doped silica gels, 617
ethanol solutions, 409
ethyl cellulose, 481
Eu^{3+}-doped silica gels, 443
exfoliated, 79

facilitated
  diffusion dialysis, 283
  transport solid membrane, 501
fiber optic sensing, 667
films, 575
fluorescence(-)
  line narrowing, 443
  probe, 187
  properties, 617
fluoride materials, 649

fracture toughness, 243
FTIR spectra, 55, 321, 481

gas
  chromatography, 381
  mixture separation, 283
gelation, 143, 339
glass(-)
  like layers, 637
  transition, 179
3-glycidoxypropyltrimethoxysilane, 215, 221, 415
grazing angle X-ray diffraction, 661
GRIN-material, 13, 553

hard ORMOSILS, 229
hectorite, 85
heterometal bonds, 461
heteropolysiloxanes, 501
hexagonal mesophase, 131
high pressure, 165
high temperature polymers, 93
holography, 553
hybrid(-) 13, 25, 105, 369, 527, 617
  gels, 381
  materials, 243, 351
  membrane synthesis, 283
  processing, 327
  sol-gel, 395
hydrated silica, 403
hydrocarbon-bridged, 277
hydrolysis, 113, 345, 387, 421, 513
hydrolytic stability, 461
hydrophobicity, 339

imine, 427
immuno-suppressed patients, 449
impedance spectroscopy, 327
INEPT, 421
infrared spectra, 93
inorganic-organic (-)
  composite materials, 455
  co-polymers, 67, 527, 547
  hybrids, 321
  network, 513
in-situ precipitated silica, 93
intercalated, 79
interfacial bonding, 85
interpenetrating network, 155
intramolecular cyclization, 33
isocyanate, 143

kinetics, 113, 345

latex(-)
  modified, 249
  particles, 249
layered double hydroxides, 73
LDHs, 73
lifetime studies, 437

Li^+ ion conductivity, 363
$LiNbO_3$, 605
linear polymer, 25

magnetic susceptibility, 481
malachite green, 667
mass spectrometry, 381
mechanical properties, 155, 229, 255
mechanism(-)
  deagglomeration, 55
  stabilization, 55
membrane(s), 283
  top layer, 307
mesoporous
  silicates, 263
  zirconia, 131
metal alkoxides, 481, 507
metallic nanoparticles, 637
methyltrimethoxysilane, 421
microcontact printing, 521
microindentation, 55
microporous
  silica, 271
  thin film, 307
microstructure, 199, 357
modified
  silane monomers, 583
  silica, 199
molecular
  dynamics simulations, 73
  hybrids, 165
  probes, 187
Monte Carlo sorption simulation, 73

nanobuilding blocks, 43
nanocomposites, 55, 79, 237, 369, 495 535, 553, 625
nanoparticles, 575
nano-size, 25
nanosized, 55
nanostructured, 55
network, 105
NIR spectroscopy, 375
NMR spectroscopy (-), 461, 643
  ^{27}Al MAS, 415
  ^{13}C solid state, 437
  ^{2}H, 179, 475
  ^{7}Li, 363
  ^{7}Li NMR line widths
  ^{7}Li NMR T_1 relaxation time, 363
  ^{17}O, 119
  ^{31}P MAS, 501
  ^{29}Si, 199, 215, 345, 351, 421, 437
  ^{29}Si MAS, 415, 469, 501
  ^{119}Sn, 43
  solid state
non-aqueous, 55
non-hydrolytic synthesis, 487
norbornenylene, 277
nucleation, 403

oligo(meth)acrylcarboxy silanes, 67
oocyst concentration, 449
optical, 575
    waveguides, 547
organic(-)
    modified, 295, 357
    structure-directing molecules, 263
    template approach, 271
organically-modified Eu^{3+}-doped gels, 443
organic-inorganic(-)
    composites, 507
    hybrid materials, 173, 255, 559
    networks, 207
organofunctional
    alkoxides, 333, 625
    alkoxysilanes, 339
organotin, 43
ORMOCERs, 55, 547
Ormolyte, 363
ORMOSILS, 229, 295, 351, 403, 595
oxo clusters, 43
oxycarbide glasses, 381

pathogenic parasite, 449
perfluoroarylalkoxysilanes, 547
phase separation, 143
phenyltriethoxysilane, 415
phosphanyl-substituted metal alkoxides, 333
photochemical, 565
photochromic, 565
photorefractivity, 583, 595
PMMA, 173, 179
poly(cyclic disilsesquioxanes), 33
polyhedral oligomeric silsesquioxanes, 3
polymer(-), 85
    mobility, 179
    poly(amic acid), 55, 199
    polyamine, 221
    polybenzobisthiazoles, 93
    polybenzoxazoles, 93
    poly(diethoxysilylenemethylene), 513
    polydimethylsiloxane, 229, 321, 469, 559, 589
    poly(ethylene glycol), 43
    poly(ethylene oxide), 495
    polyimide, 55
    polyimide-silica, 165, 199
    polymerizable ligands, 137
    polymer-silica hybrid materials, 93
    poly(methyl methacrylate)-silicate IPNs, 179
    polypyrrole, 611
    polysilsesquioxane, 277
    polystyrene-poly(vinylphenol) copolymers, 507
polyoxometalates, 589
POM, 589
pore
    size, 271
    volume, 271
porosity, 301
porous
    hybrid coating, 307
    spherical materials, 137
positron annihilation method, 243, 369
POSS
    monomers, 3
    reagents, 3
power limiting, 535
preceramic polymers, 431
precursors, 547
pseudo-equilibrium, 113
pyrolysis, 381, 513

Raman spectroscopy, 339
random network, 229
resin transfer molding, 431
reverse micelles, 535
rotational mobility, 187
rubbery
    ORMOSILS, 229
    properties, 475

SAXS, 143
scanning electron microscopy, 93
Schiff base, 427
second(-)
    harmonic generation, 583
second order
    nonlinear optical materials, 655
    nonlinear optical properties 661
    nonlinearities, 395
self-assembly(-), 263
    monolayers, 521
    multilayers, 661
semiorganic crystals, 655
sensor, 449
shrinkage, 357
silane
    coupling agents, 215
    prepolymers, 143
silica(-), 221, 295
    organic hybrid films, 409
    /polymer materials, 475
silicon
    carbonitrides, 327
    oxycarbides, 513
silicone, 589
siloxane(-), 113
    silica, 119
silsesquioxane, 25, 301, 437
SiO_2-PEG, 363
SiO_2-TiO_2, 559
small angle
    neutron scattering, 301
    X-ray scattering, 155, 301
smectite clays, 79
sol-gel(-)
    chemistry, 105
    materials, 583
    matrices, 643
    method, 315

polymerization, 33, 301
precursors, 67, 521
process, 173, 187, 199, 207, 255, 363, 375, 427, 501, 535, 553, 565
processing, 67, 155, 547, 625
reactions, 165
SiO_2 films, 187
synthesis, 271
technique, 595, 667
solid state laser, 575
solvatothermal, 487
solvent(-), 387
absorption of solvents from gas phase, 455
stripping, 179
stability, 351
star gels, 105
steric interaction, 55
storage modulus, 321
styrene, 25
substitution effects, 113
sulfonate ligands, 333
supercritical drying, 611
surface
charge, 13
modification, 13
synthetic preparations, 263

Ta_2O_5, 521
TEM, 637
temporal XRD, 495
tensile
modulus, 93
strength, 93
terephthalate, 73
ternary Si-C-N ceramics, 327
tetraethoxysilane, 469
tetraethylammonium, 495
tetramethoxysilane, 199
thermal
conductivity, 55
stability, 93, 229, 415
thermochromic effect, 173
thermogravimetric analyses, 93, 351, 381
thermomechanical properties, 237
thermoplastic hybrid materials, 3, 237
tin(IV), 387
titanium(-), 137
alkoxides, 333
oxo-alkoxides, 487
oxo-alkoxo clusters, 487
titania, 173
toughness, 105
transition metal perovskites, 315
transmission electron microscopy, 637
transparency, 93
triethoxysilyl 2,4-dinitrophenyl-aminosilane, 605
triethyl aluminum, 427
tunable solid state dye, 535
two beam coupling, 583

unoccupied space, 243
upconversion lasing, 535
UV resistant, 207

vanadium(-)
oxide, 611
oxo, 527
viscoelastic measurements, 143
viscosity, 431

water absorption, 93

xerogels, 301

zeolites, 263
zirconia, 131
zirconium alkoxides, 333